# ENGINEERING MATHEMATICS

**Companion volume**

K. A. Stroud *Further Engineering Mathematics*

# Engineering Mathematics

## Programmes and Problems

**K. A. Stroud**

*formerly Principal Lecturer in*
*Mathematics*
*Lanchester Polytechnic*
*Coventry*

**Third Edition**

First edition 1970
Reprinted (with corrections) 1972, 1973, 1974, 1975, 1977, 1978, 1979 (twice), 1980, 1981 (twice)
Second edition 1982
Reprinted 1983, 1984, 1985 (twice), 1986
Third edition 1987
Reprinted 1987, 1988, 1990 (twice), 1991 (twice)

Published by
MACMILLAN EDUCATION LTD
Houndmills, Basingstoke, Hampshire RG21 2XS
and London
Companies and representatives
throughout the world

Printed in Hong Kong

British Library Cataloguing in Publication Data
Stroud, K. A.
Engineering mathematics: programmes and problems. — 3rd ed.
1. Engineering mathematics
I. Title
510'2462 TA330
ISBN 0–333–44886–3
ISBN 0–333–44887–1 Pbk

# CONTENTS

# PREFACE TO THE FIRST EDITION

The purpose of this book is to provide a complete year's course in mathematics for those studying in the engineering, technical and scientific fields. The material has been specially written for courses leading to

(i) Part I of B.Sc. Engineering Degrees,

(ii) Higher National Diploma and Higher National Certificate in technological subjects, and for other courses of a comparable level. While formal proofs are included where necessary to promote understanding, the emphasis throughout is on providing the student with sound mathematical skills and with a working knowledge and appreciation of the basic concepts involved. The programmed structure ensures that the book is highly suited for general class use and for individual self-study, and also provides a ready means for remedial work or subsequent revision.

The book is the outcome of some eight years' work undertaken in the development of programmed learning techniques in the Department of Mathematics at the Lanchester College of Technology, Coventry. For the past four years, the whole of the mathematics of the first year of various Engineering Degree courses has been presented in programmed form, in conjunction with seminar and tutorial periods. The results obtained have proved to be highly satisfactory, and further extension and development of these learning techniques are being pursued.

Each programme has been extensively validated before being produced in its final form and has consistently reached a success level above 80/80, i.e. at least 80% of the students have obtained at least 80% of the possible marks in carefully structured criterion tests. In a research programme, carried out against control groups receiving the normal lectures, students working from programmes have attained significantly higher mean scores than those in the control groups and the spread of marks has been considerably reduced. The general pattern has also been reflected in the results of the sessional examinations.

The advantages of working at one's own rate, the intensity of the student involvement, and the immediate assessment of responses, are well known to those already acquainted with programmed learning activities. Programmed learning in the first year of a student's course at a college or university provides the additional advantage of bridging the gap between the rather highly organised aspect of school life and the freer environment and greater personal responsibility for his own progress which face every student on entry to the realms of higher education.

Acknowledgement and thanks are due to all those who have assisted in any way in the development of the work, including those who have been actively engaged in validation processes. I especially wish to record my sincere thanks for the continued encouragement and support which I received from my present Head of Department at the College,

Mr. J. E. Sellars, M.Sc., A.F.R.Ae.S., F.I.M.A., and also from Mr. R. Wooldridge, M.C., B.Sc., F.I.M.A., formerly Head of Department, now Principal of Derby College of Technology. Acknowledgement is also made of the many sources, too numerous to list, from which the selected examples quoted in the programmes have been gleaned over the years. Their inclusion contributes in no small way to the success of the work.

K. A. Stroud

# PREFACE TO THE SECOND EDITION

The continued success of *Engineering Mathematics* since its first publication has been reflected in the numbers of courses for which it has been adopted as the official class text and also in the correspondence from numerous individuals who have welcomed the self-instructional aspects of the work.

Over the years, however, syllabuses of existing courses have undergone some modification and new courses have been established. As a result, suggestions have been received from time to time requesting the inclusion of further programme topics in addition to those already provided as core material for relevant undergraduate and comparable courses. Unlimited expansion of the book to accommodate all the topics requested is hardly feasible on account of the physical size of the book and the commercial aspects of production. However, in the light of these representations and as a result of further research undertaken by the author and the publishers, it has now been found possible to provide a new edition of *Engineering Mathematics* incorporating three of the topics for which there is clearly a wide demand.

The additional programmes cover the following topics:

(a) *Matrices*: Definitions; types of matrices; operations; transpose; inverse; solution of linear equations; eigenvalues and eigenvectors.

(b) *Curves and curve fitting*: Standard curves; asymptotes; systematic curve sketching; curve recognition; curve fitting; method of least squares.

(c) *Statistics*: Discrete and continuous data; grouped data; frequency and relative frequency; histograms; central tendency – mean, mode and median; coding; frequency polygons and frequency curves; dispersion – range, variance and standard deviation; normal distribution and standardised normal curve.

The three new programmes follow the structure of the previous material and each is provided with numerous worked examples and exercises. As before, each programme concludes with a short Test Exercise for self-assessment and a set of Further Problems provides valuable extra practice. A complete set of answers is available at the end of the book.

Advantage has also been taken during the revision of the book to amend a small number of minor points in other parts of the text and it is anticipated that, in its new up-dated form, the book will have an even greater appeal and continue to provide a worthwhile service.

K. A. Stroud
1982

# PREFACE TO THE THIRD EDITION

Following the publication of the enlarged second edition of *Engineering Mathematics*, which included a programme on the introduction to Statistics, requests were again received for an associated programme on Probability. This has now been incorporated as Programme XXVIII of the current third edition of the book.

The additional programme follows the established pattern and structure of the previous sections of the text, including the customary worked examples through which the student is guided with progressive responsibility and concluding with the Text Exercise and a set of Further Problems for essential practice. Answers to all problems are provided. The opportunity has also been taken to make one or two minor modifications to the remainder of the text.

*Engineering Mathematics*, originally devised as a first year mathematics course for engineering and science degree undergraduates and students of comparable courses, is widely sought both for general class use and for individual study. A companion volume and sequel, *Further Engineering Mathematics*, dealing with core material of a typical second/third year course, is also now available through the normal channels. The two texts together provide a comprehensive and integrated course of study and have been well received as such.

My thanks are due, once again, to the publishers for their ready cooperation and helpful advice in the preparation of the material for publication.

K. A. S.
1987

# HINTS ON USING THE BOOK

This book contains twenty-eight lessons, each of which has been written in such a way as to make learning more effective and more interesting. It is almost like having a personal tutor, for you proceed at your own rate of learning and any difficulties you may have are cleared before you have the chance to practise incorrect ideas or techniques.

You will find that each programme is divided into sections called frames, each of which normally occupies half a page. When you start a programme, begin at frame 1. Read each frame carefully and carry out any instructions or exercise which you are asked to do. In almost every frame, you are required to make a response of some kind, testing your understanding of the information in the frame, and you can immediately compare your answer with the correct answer given in the next frame. To obtain the greatest benefit, you are strongly advised to cover up the following frame until you have made your response. When a series of dots occurs, you are expected to supply the missing word, phrase, or number. At every stage, you will be guided along the right path. There is no need to hurry: read the frames carefully and follow the directions exactly. In this way, you must learn.

At the end of each programme, you will find a short Test Exercise. This is set directly on what you have learned in the lesson: the questions are straightforward and contain no tricks. To provide you with the necessary practice, a set of Further Problems is also included: do as many of these problems as you can. Remember that in mathematics, as in many other situations, practice makes perfect – or more nearly so.

Even if you feel you have done some of the topics before, work steadily through each programme: it will serve as useful revision and fill in any gaps in your knowledge that you may have.

# USEFUL BACKGROUND INFORMATION

## I. Algebraic Identities

$(a+b)^2 = a^2 + 2ab + b^2$ $\qquad$ $(a+b)^3 = a^3 + 3a^2b + 3ab^2 + b^3$

$(a-b)^2 = a^2 - 2ab + b^2$ $\qquad$ $(a-b)^3 = a^3 - 3a^2b + 3ab^2 - b^3$

$$(a+b)^4 = a^4 + 4a^3b + 6a^2b^2 + 4ab^3 + b^4$$
$$(a-b)^4 = a^4 - 4a^3b + 6a^2b^2 - 4ab^3 + b^4$$

$a^2 - b^2 = (a-b)(a+b).$ $\qquad$ $a^3 - b^3 = (a-b)(a^2 + ab + b^2)$

$a^3 + b^3 = (a+b)(a^2 - ab + b^2)$

## II. Trigonometrical Identities

(1) $\sin^2\theta + \cos^2\theta = 1;\quad \sec^2\theta = 1 + \tan^2\theta;\quad \text{cosec}^2\theta = 1 + \cot^2\theta$

(2)
$$\sin(A+B) = \sin A \cos B + \cos A \sin B$$
$$\sin(A-B) = \sin A \cos B - \cos A \sin B$$
$$\cos(A+B) = \cos A \cos B - \sin A \sin B$$
$$\cos(A-B) = \cos A \cos B + \sin A \sin B$$
$$\tan(A+B) = \frac{\tan A + \tan B}{1 - \tan A \tan B}$$
$$\tan(A-B) = \frac{\tan A - \tan B}{1 + \tan A \tan B}$$

(3) Let $A = B = \theta$. $\quad \therefore$
$$\sin 2\theta = 2\sin\theta\cos\theta$$
$$\cos 2\theta = \cos^2\theta - \sin^2\theta = 1 - 2\sin^2\theta = 2\cos^2\theta - 1$$
$$\tan 2\theta = \frac{2\tan\theta}{1 - \tan^2\theta}$$

(4) Let $\theta = \frac{\varnothing}{2}$ $\quad \therefore \sin \varnothing = 2 \sin \frac{\varnothing}{2} \cos \frac{\varnothing}{2}$

$$\cos \varnothing = \cos^2 \frac{\varnothing}{2} - \sin^2 \frac{\varnothing}{2}$$

$$= 1 - 2 \sin^2 \frac{\varnothing}{2}$$

$$= 2 \cos^2 \frac{\varnothing}{2} - 1$$

$$\tan \varnothing = \frac{2 \tan \frac{\varnothing}{2}}{1 - \tan^2 \frac{\varnothing}{2}}$$

(5) $\sin C + \sin D = 2 \sin \frac{C + D}{2} \cos \frac{C - D}{2}$

$\sin C - \sin D = 2 \cos \frac{C + D}{2} \sin \frac{C - D}{2}$

$\cos C + \cos D = 2 \cos \frac{C + D}{2} \cos \frac{C - D}{2}$

$\cos D - \cos C = 2 \sin \frac{C + D}{2} \sin \frac{C - D}{2}$

(6) $2 \sin A \cos B = \sin (A + B) + \sin (A - B)$
$2 \cos A \sin B = \sin (A + B) - \sin (A - B)$
$2 \cos A \cos B = \cos (A + B) + \cos (A - B)$
$2 \sin A \sin B = \cos (A - B) - \cos (A + B)$

(7) Negative angles: $\sin (-\theta) = -\sin \theta$
$\cos (-\theta) = \cos \theta$
$\tan (-\theta) = -\tan \theta$

(8) Angles having the same trig. ratios:

(i) Same sine: $\theta$ and $(180° - \theta)$
(ii) Same cosine: $\theta$ and $(360° - \theta)$, i.e. $(-\theta)$
(iii) Same tangent: $\theta$ and $(180° + \theta)$

(9) $a \sin\theta + b\cos\theta = A\sin(\theta + \alpha)$
$a \sin\theta - b\cos\theta = A\sin(\theta - \alpha)$
$a \cos\theta + b\sin\theta = A\cos(\theta - \alpha)$
$a \cos\theta - b\sin\theta = A\cos(\theta + \alpha)$

$$\text{where: } \begin{cases} A = \sqrt{(a^2 + b^2)} \\ \alpha = \tan^{-1}\dfrac{b}{a} \qquad (0° < \alpha < 90°) \end{cases}$$

## III. Standard Curves

(1) ***Straight line:***

Slope, $m = \dfrac{dy}{dx} = \dfrac{y_2 - y_1}{x_2 - x_1}$

Angle between two lines, $\tan\theta = \dfrac{m_2 - m_1}{1 + m_1 m_2}$

For parallel lines, $m_2 = m_1$

For perpendicular lines, $m_1 m_2 = -1$

Equation of a straight line (slope $= m$)

(i) Intercept $c$ on real $y$-axis: $y = mx + c$
(ii) Passing through $(x_1, y_1)$: $y - y_1 = m(x - x_1)$
(iii) Joining $(x_1, y_1)$ and $(x_2, y_2)$: $\dfrac{y - y_1}{y_2 - y_1} = \dfrac{x - x_1}{x_2 - x_1}$

(2) ***Circle:***

Centre at origin, radius $r$: $x^2 + y^2 = r^2$
Centre $(h, k)$, radius $r$: $(x - h)^2 + (y - k)^2 = r^2$
General equation: $x^2 + y^2 + 2gx + 2fy + c = 0$
with centre $(-g, -f)$; radius $= \sqrt{(g^2 + f^2 - c)}$
Parametric equations: $x = r\cos\theta$, $y = r\sin\theta$

(3) ***Parabola:***

Vertex at origin, focus $(a, 0)$: $y^2 = 4ax$
Parametric equations: $x = at^2$, $y = 2at$

(4) *Ellipse:*

Centre at origin, foci $(\pm\sqrt{[a^2 - b^2]}, 0)$: $\dfrac{x^2}{a^2} + \dfrac{y^2}{b^2} = 1$

where $a$ = semi major axis, $b$ = semi minor axis

Parametric equations: $x = a \cos\theta,\ y = b \sin\theta$

(5) *Hyperbola:*

Centre at origin, foci $(\pm\sqrt{[a^2 + b^2]}, 0)$: $\dfrac{x^2}{a^2} - \dfrac{y^2}{b^2} = 1$

Parametric equations: $x = a \sec\theta,\ y = b \tan\theta$

Rectangular hyperbola:

Centre at origin, vertex $\pm\left(\dfrac{a}{\sqrt{2}}, \dfrac{a}{\sqrt{2}}\right)$: $xy = \dfrac{a^2}{2} = c^2$ where $c = \dfrac{a}{\sqrt{2}}$

i.e. $xy = c^2$

Parametric equations: $x = ct,\ y = c/t$

## IV. Laws of Mathematics

(1) *Associative laws* – for addition and multiplication

$$a + (b + c) = (a + b) + c$$
$$a(bc) = (ab)c$$

(2) *Commutative laws* – for addition and multiplication

$$a + b = b + a$$
$$ab = ba$$

(3) *Distributive laws* – for multiplication and division

$$a(b + c) = ab + ac$$
$$\frac{b + c}{a} = \frac{b}{a} + \frac{c}{a} \text{ (provided } a \neq 0)$$

# Programme 1

## COMPLEX NUMBERS

### PART 1

## 1

### Introduction: the symbol j

The solution of a quadratic equation $ax^2 + bx + c = 0$ can, of course, be obtained by the formula, $x = \dfrac{-b \pm \sqrt{(b^2 - 4ac)}}{2a}$

For example, if $2x^2 + 9x + 7 = 0$, then we have

$$x = \frac{-9 \pm \sqrt{(81 - 56)}}{4} = \frac{-9 \pm \sqrt{25}}{4} = \frac{-9 \pm 5}{4}$$

$$\therefore \quad x = -\frac{4}{4} \text{ or } -\frac{14}{4}$$

$$\therefore \quad \underline{x = -1 \text{ or } -3{\cdot}5}$$

That was straight-forward enough, but if we solve the equation $5x^2 - 6x + 5 = 0$ in the same way, we get

$$x = \frac{6 \pm \sqrt{(36 - 100)}}{10} = \frac{6 \pm \sqrt{(-64)}}{10}$$

and the next stage is now to determine the square root of (−64).

Is it (i) 8, (ii) −8, (iii) neither?

## 2

**neither**

It is, of course, neither, since +8 and −8 are the square roots of 64 and not of (−64). In fact, $\sqrt{(-64)}$ cannot be represented by an ordinary number, for there is no real number whose square is a negative quantity.

However, $-64 = -1 \times 64$ and therefore we can write

$$\sqrt{(-64)} = \sqrt{(-1 \times 64)} = \sqrt{(-1)}\sqrt{64} = 8\sqrt{(-1)}$$

$$\text{i.e. } \sqrt{(-64)} = 8\sqrt{(-1)}$$

Of course, we are still faced with $\sqrt{(-1)}$, which cannot be evaluated as a real number, for the same reason as before, but, if we write the letter j to stand for $\sqrt{(-1)}$, then $\sqrt{(-64)} = \sqrt{(-1)}\,.\,8 = \mathrm{j}8$.

So although we cannot evaluate $\sqrt{(-1)}$, we can denote it by j and this makes our working a lot neater.

$$\sqrt{(-64)} = \sqrt{(-1)}\sqrt{64} = \mathrm{j}8$$

Similarly,

$$\sqrt{(-36)} = \sqrt{(-1)}\sqrt{36} = \mathrm{j}6$$

$$\sqrt{(-7)} = \sqrt{(-1)}\sqrt{7} = \mathrm{j}2{\cdot}646$$

So $\sqrt{(-25)}$ can be written ..............................

**3**

$$\boxed{j5}$$

We now have a way of finishing off the quadratic equation we started in frame 1.

$$5x^2 - 6x + 5 = 0 \quad \therefore\ x = \frac{6 \pm \sqrt{(36 - 100)}}{10} = \frac{6 \pm \sqrt{(-64)}}{10}$$

$$\therefore\ x = \frac{6 \pm j8}{10} \qquad \therefore\ x = 0{\cdot}6 \pm j0{\cdot}8$$

$$\therefore\ \underline{x = 0{\cdot}6 + j0{\cdot}8} \text{ or } \underline{x = 0{\cdot}6 - j0{\cdot}8}$$

We will talk about results like these later.

*For now, on to frame 4.*

**4**

**Powers of j**

Since j stands for $\sqrt{(-1)}$, let us consider some powers of j.

| | |
|---|---|
| $j = \sqrt{(-1)}$ | $j = \sqrt{(-1)}$ |
| $j^2 = -1$ | $j^2 = -1$ |
| $j^3 = (j^2)j = -1.j = -j$ | $j^3 = -j$ |
| $j^4 = (j^2)^2 = (-1)^2 = 1$ | $j^4 = 1$ |

Note especially the last result: $j^4 = 1$. Every time a factor $j^4$ occurs, it can be replaced by the factor 1, so that the power of j is reduced to one of the four results above.

e.g. $j^9 = (j^4)^2 j = (1)^2 j = 1.j = j$

$j^{20} = (j^4)^5 = (1)^5 = 1$

$j^{30} = (j^4)^7 j^2 = (1)^7(-1) = 1(-1) = -1$

and $j^{15} = (j^4)^3 j^3 = 1(-j) = -j$

So, in the same way, $j^5$ = ........................

**5**

| j |
|---|

since $j^5 = (j^4)j = 1.j = j$

Every one is done in the same way.

$$j^6 = (j^4)j^2 = 1(j^2) = 1(-1) = -1$$
$$j^7 = (j^4)j^3 = 1(-j) = -j$$
$$j^8 = (j^4)^2 = (1)^2 = 1$$

So (i) $j^{42} =$ ......................

(ii) $j^{12} =$ ......................

(iii) $j^{11} =$ ......................

and (iv) If $x^2 - 6x + 34 = 0$, $x =$ ......................

**6**

| (i) $-1$, (ii) $1$, (iii) $-j$, (iv) $x = 3 \pm j5$ |
|---|

The working in (iv) is as follows:

$$x^2 - 6x + 34 = 0 \quad \therefore x = \frac{6 \pm \sqrt{(36 - 136)}}{2} = \frac{6 \pm \sqrt{(-100)}}{2}$$

$$\therefore x = \frac{6 \pm j10}{2} = 3 \pm j5$$

$$\text{i.e. } \underline{x = 3 + j5} \text{ or } \underline{x = 3 - j5}$$

So remember, to simplify powers of j, we take out the highest power of $j^4$ that we can, and the result must then simplify to one of the four results: j, $-1$, $-j$, 1.

*Turn on now to frame 7.*

7

**Complex numbers**

The result $x = 3 + j5$ that we obtained, consists of two separate terms, 3 and j5. These terms cannot be combined any further, since the second is not a real number (due to its having the factor j).

In such an expression as $x = 3 + j5$,

3 is called the *real part* of $x$

5 is called the *imaginary part* of $x$

and the two together form what is called a *complex number*.

So, a Complex number = (Real part) + j(Imaginary part)

In the complex number $2 + j7$, the real part = ......................

and the imaginary part = .....................

8

| real part = 2; imaginary part = 7 (NOT j7!) |
|---|

Complex numbers have many applications in engineering. To use them, we must know how to carry out the usual arithmetical operations.

1. ***Addition and Subtraction of Complex Numbers.*** This is easy, as one or two examples will show.

*Example 1* $(4 + j5) + (3 - j2)$. Although the real and imaginary parts cannot be combined, we can remove the brackets and total up terms of the same kind.

$$(4 + j5) + (3 - j2) = 4 + j5 + 3 - j2 = (4 + 3) + j(5 - 2)$$
$$= \underline{7 + j3}$$

*Example 2*

$$(4 + j7) - (2 - j5) = 4 + j7 - 2 + j5 = (4 - 2) + j(7 + 5)$$
$$= \underline{2 + j12}$$

So, in general, $(a + jb) + (c + jd) = (a + c) + j(b + d)$

Now you do this one:

$$(5 + j7) + (3 - j4) - (6 - j3) = \text{......................}$$

**9**

$$2 + j6$$

since $(5 + j7) + (3 - j4) - (6 - j3)$

$$= 5 + j7 + 3 - j4 - 6 + j3$$

$$= (5 + 3 - 6) + j(7 - 4 + 3)$$

$$= \underline{2 + j6}$$

Now you do these in just the same way:

(i) $(6 + j5) - (4 - j3) + (2 - j7) =$ ......................

and (ii) $(3 + j5) - (5 - j4) - (-2 - j3) =$ ......................

**10**

(i) $4 + j$ (ii) $j12$

Here is the working:

(i) $(6 + j5) - (4 - j3) + (2 - j7)$

$$= 6 + j5 - 4 + j3 + 2 - j7$$

$$= (6 - 4 + 2) + j(5 + 3 - 7)$$

$$= \underline{4 + j}$$

(ii) $(3 + j5) - (5 - j4) - (-2 - j3)$

$$= 3 + j5 - 5 + j4 + 2 + j3$$

$$= (3 - 5 + 2) + j(5 + 4 + 3)$$

(Take care with signs!)

$$= 0 + j12 = \underline{j12}$$

This is very easy then, so long as you remember that the real and the imaginary parts must be treated quite separately – just like $x$'s and $y$'s in an algebraic expression.

*On to frame 11.*

11

2. *Multiplication of Complex Numbers*

*Example:* $(3 + j4)(2 + j5)$

These are multiplied together in just the same way as you would determine the product $(3x + 4y)(2x + 5y)$.

Form the product terms of (i) the two left-hand terms
(ii) the two inner terms
(iii) the two outer terms
(iv) the two right-hand terms

$(3 + j4)(2 + j5)$

1 4 2 3

$= 6 + j8 + j15 + j^2 20$

$= 6 + j23 - 20$ (since $j^2 = -1$)

$= -14 + j23$

Likewise, $(4 - j5)(3 + j2)$ ......................

12

$22 - j7$

for: $(4 - j5)(3 + j2) = 12 - j15 + j8 - j^2 10$

$= 12 - j7 + 10 \quad (j^2 = -1)$

$= 22 - j7$

If the expression contains more than two factors, we multiply the factors together in stages:

$(3 + j4)(2 - j5)(1 - j2)$

$= (6 + j8 - j15 - j^2 20)(1 - j2)$

$= (6 - j7 + 20)(1 - j2)$

$= (26 - j7)(1 - j2)$

$=$ ................................

*Finish it off.*

## 13

$$12 - j59$$

for:
$$(26 - j7)(1 - j2)$$
$$= 26 - j7 - j52 + j^2 14$$
$$= 26 - j59 - 14 = \underline{12 - j59}$$

Note that when we are dealing with complex numbers, the result of our calculations is also, in general, a complex number.

Now you do this one on your own.

$$(5 + j8)(5 - j8) = \ldots\ldots\ldots\ldots\ldots\ldots$$

## 14

$$89$$

Here it is:

$$(5 + j8)(5 - j8) = 25 + j40 - j40 - j^2 64$$
$$= 25 + 64$$
$$= \underline{89}$$

In spite of what we said above, here we have a result containing no j term. The result is therefore entirely real.

This is rather an exceptional case. Look at the two complex numbers we have just multiplied together. Can you find anything special about them? If so, what is it?

*When you have decided, turn on to the next frame.*

**15**

They are identical except for the middle sign in the brackets,
i.e. $(5 + j8)$ and $(5 - j8)$

A pair of complex numbers like these are called *conjugate* complex numbers and *the product of two conjugate complex numbers is always entirely real.*

Look at it this way –

$$(a + b)(a - b) = a^2 - b^2 \quad \text{Difference of two squares}$$

Similarly
$$(5 + j8)(5 - j8) = 5^2 - (j8)^2 = 5^2 - j^2 8^2$$
$$= 5^2 + 8^2 \quad (j^2 = -1)$$
$$= 25 + 64 = \underline{89}$$

Without actually working it out, will the product of $(7 - j6)$ and $(4 + j3)$ be
(i) a real number
(ii) an imaginary number
(iii) a complex number

**16**

a complex number

since $(7 - j6)(4 + j3)$ is a product of two complex numbers which are *not* conjugate complex numbers or multiples of conjugates.

**Remember**: Conjugate complex numbers are identical except for the signs in the middle of the brackets.

$(4 + j5)$ and $(4 - j5)$ *are* conjugate complex numbers
$(a + jb)$ and $(a - jb)$ *are* conjugate complex numbers
but $(6 + j2)$ and $(2 + j6)$ *are not* conjugate complex numbers
$(5 - j3)$ and $(-5 + j3)$ *are not* conjugate complex numbers

So what must we multiply $(3 - j2)$ by, to produce a result that is entirely real?

**17**

(3 + j2) or a multiple of it

because the conjugate of (3 − j2) is identical to it, except for the middle sign, i.e. (3 + j2), and we know that the product of two *conjugate* complex numbers is always real.

Here are some examples:

*Example 1* $(3 - j2)(3 + j2) = 3^2 - (j2)^2 = 9 - j^2 4$
$= 9 + 4 = \underline{13}$

*Example 2* $(2 + j7)(2 - j7) = 2^2 - (j7)^2 = 4 - j^2 49$
$= 4 + 49 = \underline{53}$

. . . and so on.

Complex numbers of the form $(a + jb)$ and $(a - jb)$ are called ...................... complex numbers.

**18**

conjugate

Now you should have no trouble with these–

(a) Write down the following products

(i) $(4 - j3)(4 + j3)$
(ii) $(4 + j7)(4 - j7)$
(iii) $(a + jb)(a - jb)$
(iv) $(x - jy)(x + jy)$

(b) Multiply (3 − j5) by a suitable factor to give a product that is entirely real.

*When you have finished, move on to frame 19.*

**19**

Here are the results in detail.

(a) (i) $(4 - j3)(4 + j3) = 4^2 - j^2 3^2 = 16 + 9 = \boxed{25}$

(ii) $(4 + j7)(4 - j7) = 4^2 - j^2 7^2 = 16 + 49 = \boxed{65}$

(iii) $(a + jb)(a - jb) = a^2 - j^2 b^2 = \boxed{a^2 + b^2}$

(iv) $(x - jy)(x + jy) = x^2 - j^2 y^2 = \boxed{x^2 + y^2}$

(b) To obtain a real product, we can multiply $(3 - j5)$ by its conjugate, i.e. $(3 + j5)$, giving

$$(3 - j5)(3 + j5) = 3^2 - j^2 5^2 = 9 + 25 = \boxed{34}$$

*Now move on to the next frame for a short revision exercise.*

**20**

**Revision exercise.**

1. Simplify (i) $j^{12}$ (ii) $j^{10}$ (iii) $j^{23}$
2. Simplify:

(i) $(5 - j9) - (2 - j6) + (3 - j4)$

(ii) $(6 - j3)(2 + j5)(6 - j2)$

(iii) $(4 - j3)^2$

(iv) $(5 - j4)(5 + j4)$

3. Multiply $(4 - j3)$ by an appropriate factor to give a product that is entirely real. What is the result?

*When you have completed the exercise, turn on to frame 21.*

**21** Here are the results. Check yours.

1. (i) $j^{12} = (j^4)^3 = 1^3 = \boxed{1}$

   (ii) $j^{10} = (j^4)^2 j^2 = 1^2(-1) = \boxed{-1}$

   (iii) $j^{23} = (j^4)^5 j^3 = j^3 = \boxed{-j}$

2. (i) $(5 - j9) - (2 - j6) + (3 - j4)$

   $= 5 - j9 - 2 + j6 + 3 - j4$

   $= (5 - 2 + 3) + j(6 - 9 - 4) = \boxed{6 - j7}$

   (ii) $(6 - j3)(2 + j5)(6 - j2)$

   $= (12 - j6 + j30 - j^2 15)(6 - j2)$

   $= (27 + j24)(6 - j2)$

   $= 162 + j144 - j54 + 48 = \boxed{210 + j90}$

   (iii) $(4 - j3)^2 = 16 - j24 - 9$

   $= \boxed{7 - j24}$

   (iv) $(5 - j4)(5 + j4)$

   $= 25 - j^2 16 = 25 + 16 = \boxed{41}$

3. Suitable factor is the conjugate of the given complex number.

$$(4 - j3)(4 + j3) = 16 + 9 = \boxed{25}$$

All correct? Right. *Now turn on to the next frame to continue the programme.*

**22**

Now let us deal with division.

Division of a complex number by a real number is easy enough.

$$\frac{5-j4}{3}=\frac{5}{3}-j\frac{4}{3}=1{\cdot}67-j1{\cdot}33$$

But how do we manage with $\dfrac{7-j4}{4+j3}$?

If we could, somehow, convert the denominator into a real number, we could divide out as in the example above. So our problem is really, how can we convert $(4+j3)$ into a completely real denominator – and this is where our last piece of work comes in.

We know that we can convert $(4+j3)$ into a completely real number by multiplying it by its c ...................

**23**

Conjugate i.e. the same complex number but with the opposite sign in the middle, in the case $(4-j3)$

□□□□□□□□□□□□□□□□□□□□□□□□□□□□□□□□□□□□□□

But if we multiply the denominator by $(4-j3)$, we must also multiply the numerator by the same factor.

$$\frac{7-j4}{4+j3}=\frac{(7-j4)(4-j3)}{(4+j3)(4-j3)}=\frac{28-j37-12}{16+9}=\frac{16-j37}{25}$$

$$\frac{16}{25}-j\frac{37}{25}=\underline{0{\cdot}64-j1{\cdot}48}$$

and the job is done.

To divide one complex number by another, therefore, we multiply numerator and denominator by the conjugate of the denominator. This will convert the denominator into a real number and the final step can then be completed.

Thus, to simplify $\dfrac{4-j5}{1+j2}$, we shall multiply top and bottom by ................

## 24

the conjugate of the denominator, i.e. $(1 - j2)$

□□□□□□□□□□□□□□□□□□□□□□□□□□□□□□□□□□□□□□□□

If we do that, we get:

$$\frac{4-j5}{1+j2} = \frac{(4-j5)(1-j2)}{(1+j2)(1-j2)} = \frac{4-j13-10}{1+4}$$

$$= \frac{-6-j13}{5} = \frac{-6}{5} - j\frac{13}{5}$$

$$= \underline{-1{\cdot}2 - j2{\cdot}6}$$

Now here is one for you to do:

Simplify $\quad \dfrac{3+j2}{1-j3}$

*When you have done it, move on to the next frame.*

## 25

*Result*

$$\boxed{-0{\cdot}3 + j1{\cdot}1}$$

$$\frac{3+j2}{1-j3} = \frac{(3+j2)(1+j3)}{(1-j3)(1+j3)} = \frac{3+j11-6}{1+9}$$

$$= \frac{-3+j11}{10} = \underline{-0{\cdot}3 + j1{\cdot}1}$$

□□□□□□□□□□□□□□□□□□□□□□□□□□□□□□□□□□□□□□□□

Now do these in the same way:

(i) $\dfrac{4-j5}{2-j}$ $\qquad$ (ii) $\dfrac{3+j5}{5-j3}$

(iii) $\dfrac{(2+j3)(1-j2)}{3+j4}$

*When you have worked these, turn on to frame 26 to check your results.*

**26**

***Results:*** Here are the solutions in detail.

(i)
$$\frac{4-j5}{2-j}=\frac{(4-j5)(2+j)}{(2-j)(2+j)}=\frac{8-j6+5}{4+1}$$
$$=\frac{13-j6}{5}=\boxed{2{\cdot}6-j1{\cdot}2}$$

(ii)
$$\frac{3+j5}{5-j3}=\frac{(3+j5)(5+j3)}{(5-j3)(5+j3)}=\frac{15+j34-15}{25+9}$$
$$=\frac{j34}{34}=\boxed{j}$$

(iii)
$$\frac{(2+j3)(1-j2)}{(3+j4)}=\frac{2-j+6}{3+j4}=\frac{8-j}{3+j4}$$
$$=\frac{(8-j)(3-j4)}{(3+j4)(3-j4)}$$
$$=\frac{24-j35-4}{9+16}=\frac{20-j35}{25}$$
$$\boxed{=0{\cdot}8-j1{\cdot}4}$$

And now you know how to apply the four rules to complex numbers.

**27**

**Equal Complex Numbers**

Now let us see what we can find out about two complex numbers which we are told are equal.

Let the numbers be

$$a+jb \text{ and } c+jd$$

Then we have

$$a+jb=c+jd$$

Re-arranging terms, we get

$$a-c=j(d-b)$$

In this last statement, the quantity on the left-hand side is entirely real, while that on the right-hand side is entirely imaginary, i.e. a real quantity equals an imaginary quantity! This seems contradictory and in general it just cannot be true. But there is *one* special case for which the statement can be true. That is when ................

**28**

each side is zero

$$a - c = j(d - b)$$

can be true only if

$$a - c = 0, \quad \text{i.e. } a = c$$

$$\text{and if} \quad d - b = 0, \quad \text{i.e. } b = d$$

So we get this important result:

If two complex numbers are equal

(i) the two real parts are equal

(ii) the two imaginary parts are equal

For example, if $x + jy = 5 + j4$, then we know $x = 5$ and $y = 4$
and if $a + jb = 6 - j3$, then $a =$ .................. and $b =$ ..................

**29**

$a = 6$ and $b = -3$

Be careful to include the sign!

□□□□□□□□□□□□□□□□□□□□□□□□□□□□□□□□□□□□□□□

Now what about this one?

If $(a + b) + j(a - b) = 7 + j2$, find the values of $a$ and $b$.

Well now, following our rule about two equal complex numbers, what can we say about $(a + b)$ and $(a - b)$?

**30**

$\boxed{a + b = 7}$ and $\boxed{a - b = 2}$

since the two real parts are equal and the two imaginary parts are equal.

□□□□□□□□□□□□□□□□□□□□□□□□□□□□□□□□□□□□□□

This gives you two simultaneous equations, from which you can determine the values of $a$ and $b$.

So what are they?

**31**

$\boxed{a = 4{\cdot}5; \quad b = 2{\cdot}5}$

For

$$\left.\begin{array}{l} a + b = 7 \\ a - b = 2 \end{array}\right\} \begin{array}{ll} 2a = 9 & \therefore\ a = 4{\cdot}5 \\ 2b = 5 & \therefore\ b = 2{\cdot}5 \end{array}$$

□□□□□□□□□□□□□□□□□□□□□□□□□□□□□□□□□□□□□□

We see then that an equation involving complex numbers leads to a pair of simultaneous equations by putting

(i) the two real parts equal
(ii) the two imaginary parts equal

This is quite an important point to remember.

**32**

## Graphical Representation of a Complex Number

Although we cannot evaluate a complex number as a real number, we can represent it diagrammatically, as we shall now see.

In the usual system of plotting numbers, the number 3 could be represented by a line from the origin to the point 3 on the scale. Likewise, a line to represent (−3) would be drawn from the origin to the point (−3). These two lines are equal in length but are drawn in opposite directions. Therefore, we put an arrow head on each to distinguish between them.

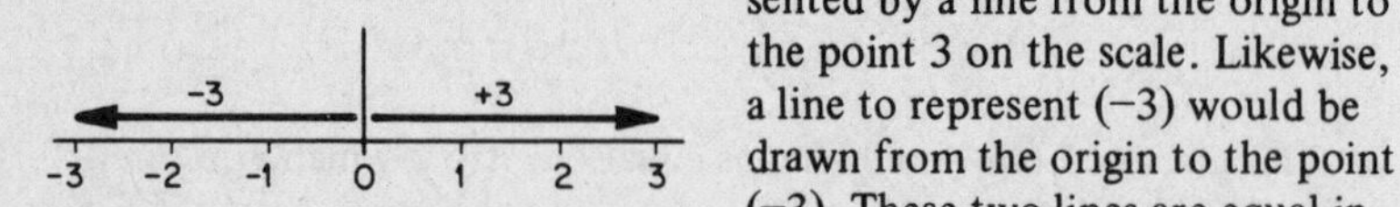


A line which represents a magnitude (by its length) and direction (by the arrow head) is called a **vector**. We shall be using this word quite a lot.

Any vector therefore must include both magnitude (or size) and ........................

**33**

direction

□□□□□□□□□□□□□□□□□□□□□□□□□□□□□□□□□□□□□

If we multiply (+3) by the factor (−1), we get (−3), i.e. the factor (−1) has the effect of turning the vector through 180°

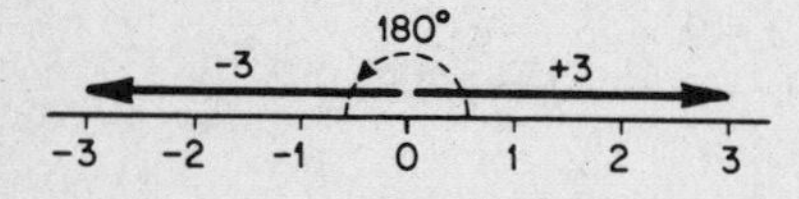


Multiplying by (−1) is equivalent to multiplying by $j^2$, i.e. by the factor j twice. Therefore multiplying by a single factor j will have half the effect and rotate the vector through only ...................... °

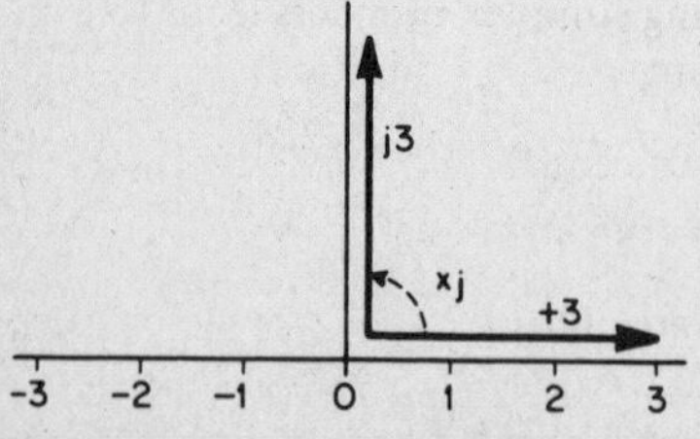

**34**

$90^\circ$

□□□□□□□□□□□□□□□□□□□□□□□□□□□□□□□□□□□□□□

The factor j always turns a vector through $90^\circ$ in the positive direction of measuring angles, i.e. anticlockwise.

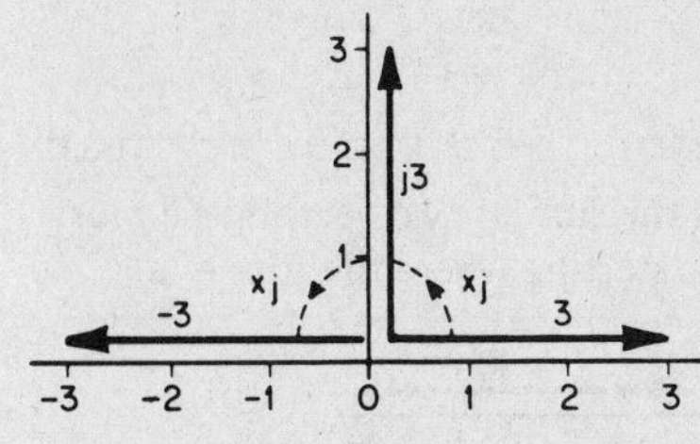


If we now multiply j3 by a further factor j, we get $j^2 3$, i.e. $(-3)$ and the diagram agrees with this result.

If we multiply $(-3)$ by a further factor j, sketch the new position of the vector on a similar diagram.

**35**

*Result:*

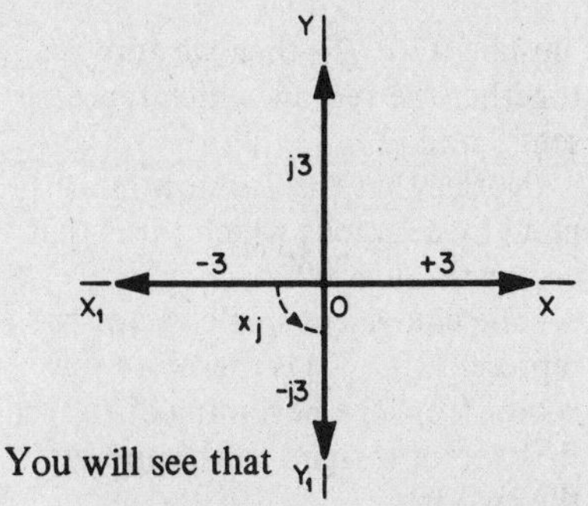


Let us denote the two reference lines by $XX_1$ and $YY_1$ as usual.

You will see that

(i) The scale on the X-axis represents real numbers.
$XX_1$ is therefore called the *real axis.*

(ii) The scale on the Y-axis represents imaginary numbers.
$YY_1$ is therefore called the *imaginary axis.*

On a similar diagram, sketch vectors to represent

(i) 5, (ii) $-4$, (iii) j2, (iv) $-$j

**36** *Results:*

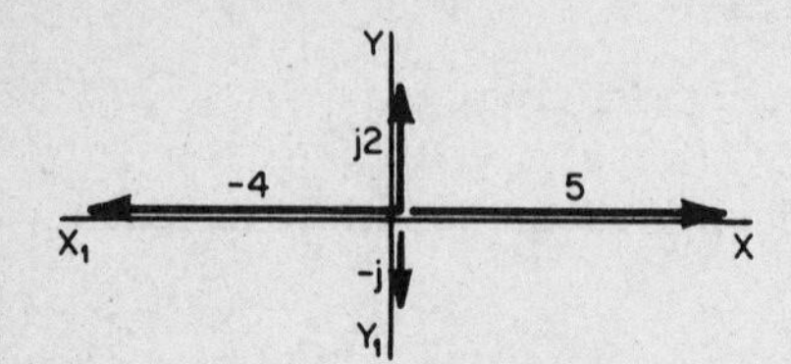


Check that each of your vectors carries an arrow head to show direction.

□□□□□□□□□□□□□□□□□□□□□□□□□□□□□□□□□□□□□□

If we now wish to represent 3 + 2 as the sum of two vectors, we must draw them as a chain, the second vector starting where the first one finishes.

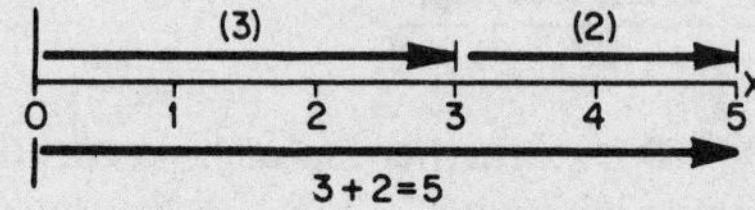


The two vectors, 3 and 2, are together equivalent to a single vector drawn from the origin to the end of the final vector (giving naturally that 3 + 2 = 5).

*Continue*

**37** If we wish to represent the complex number $(3 + j2)$, then we add together the vectors which represent 3 and j2.

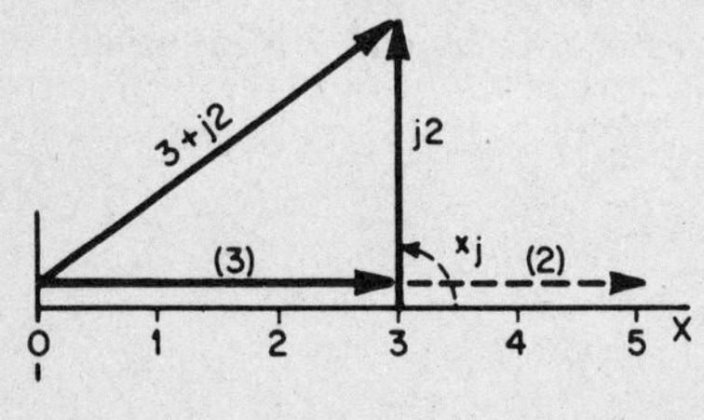


Notice that the 2 is now multiplied by a factor j which turns that vector through 90°.

The equivalent single vector to represent $(3 + j2)$ is therefore the vector from the beginning of the first vector (origin) to the end of the last one.

This graphical representation constitutes an *Argand diagram.*

Draw an Argand diagram to represent the vectors

(i) $z_1 = 2 + j3$ (ii) $z_2 = -3 + j2$

(iii) $z_3 = 4 - j3$ (iv) $z_4 = -4 - j5$

*Label each one clearly.*

**38**

Here they are. Check yours.

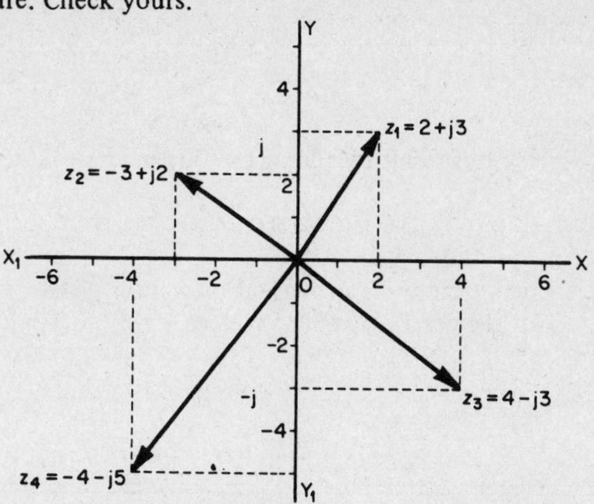


Note once again that the end of each vector is plotted very much like plotting $x$ and $y$ co-ordinates.

The real part corresponds to the $x$-value.

The imaginary part corresponds to the $y$-value.

*Move on to frame 39.*

**39**

## Graphical Addition of Complex Numbers

Let us find the sum of $z_1 = 5 + j2$ and $z_2 = 2 + j3$ by Argand diagram. If we are adding vectors, they must be drawn as a chain. We therefore draw at the end of $z_1$, a vector AP representing $z_2$ in magnitude and direction, i.e. AP = OB and is parallel to it. Therefore OAPB is a parallelogram. Thus the sum of $z_1$ and $z_2$ is given by the vector joining the starting point to the end of the last vector, i.e. OP.

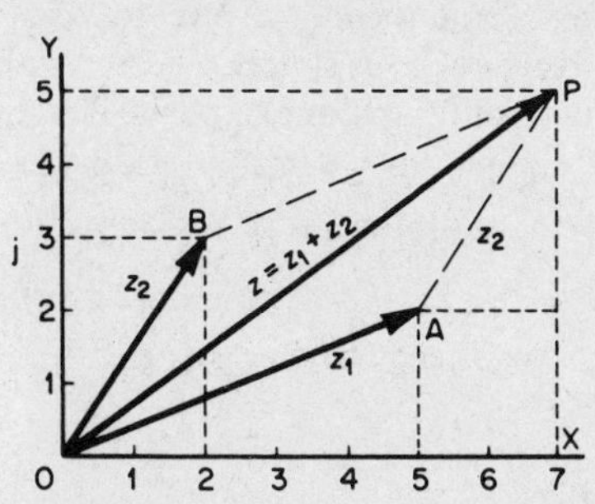


The complex numbers $z_1$ and $z_2$ can thus be added together by drawing the diagonal of the parallelogram formed by $z_1$ and $z_2$.

If OP represents the complex number $a + jb$, what are the values of $a$ and $b$ in this case?

**40**

$a = 5 + 2 = 7$ $b = 2 + 3 = 5$

$$\therefore\ \text{OP} = z = 7 + \text{j}5$$

You can check this result by adding (5 + j2) and (2 + j3) algebraically.

□□□□□□□□□□□□□□□□□□□□□□□□□□□□□□□□□□□□□□

So the sum of two vectors on an Argand diagram is given by the ..................... of the parallelogram of vectors.

**41**

diagonal

□□□□□□□□□□□□□□□□□□□□□□□□□□□□□□□□□□□□□□

How do we do subtraction by similar means? We do this rather craftily without learning any new methods. The trick is simply this:

$$z_1 - z_2 = z_1 + (-z_2)$$

That is, we draw the vector representing $z_1$ and the *negative* vector of $z_2$ and add them as before. The negative vector of $z_2$ is simply a vector with the same magnitude (or length) as $z_2$ but pointing in the opposite direction.

e.g. If $z_1 = 5 + \text{j}2$ and $z_2 = 2 + \text{j}3$

vector $\text{OA} = z_1 = 5 + \text{j}2$

$\text{OP} = -z_2 = -(2 + \text{j}3)$

Then $\text{OQ} = z_1 + (-z_2)$

$= z_1 - z_2$

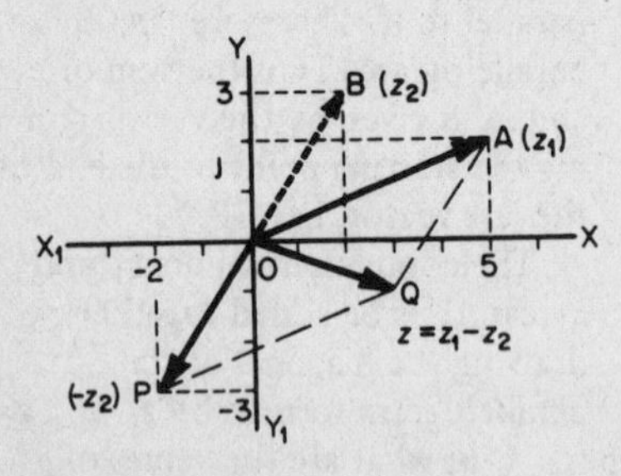


Determine on an Argand diagram (4 + j2) + (−2 + j3) − (−1 + j6)

42

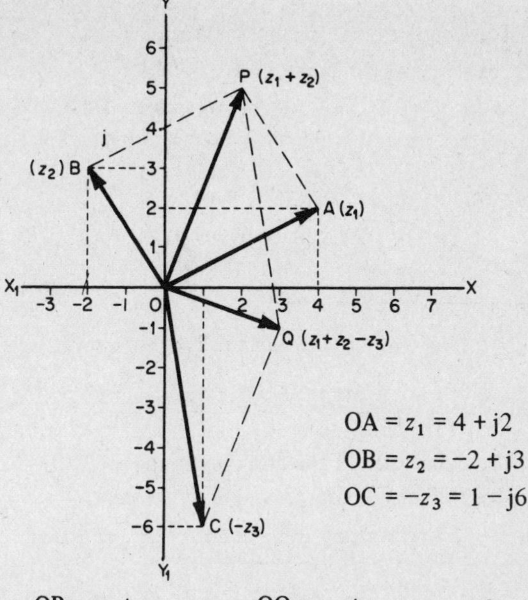


$OA = z_1 = 4 + j2$

$OB = z_2 = -2 + j3$

$OC = -z_3 = 1 - j6$

Then $\quad OP = z_1 + z_2 \qquad OQ = z_1 + z_2 - z_3 = \underline{3 - j}$

43

**Polar Form of a Complex Number**

It is convenient sometimes to express a complex number $a + jb$ in a different form. On an Argand diagram, let OP be a vector $a + jb$. Let $r$ = length of the vector and $\theta$ the angle made with OX.

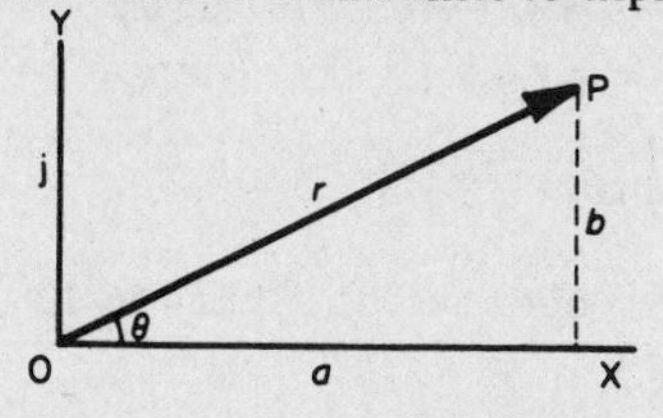


Then $\quad r^2 = a^2 + b^2 \qquad r = \sqrt{(a^2 + b^2)}$

and $\quad \tan\theta = \dfrac{b}{a} \qquad \theta = \tan^{-1}\dfrac{b}{a}$

Also $\quad a = r\cos\theta$ and $b = r\sin\theta$

Since $z = a + jb$, this can be written

$z = r\cos\theta + jr\sin\theta \qquad$ i.e. $z = r(\cos\theta + j\sin\theta)$

This is called the *polar form* of the complex number $a + jb$, where

$$r = \sqrt{(a^2 + b^2)} \text{ and } \theta = \tan^{-1}\frac{b}{a}$$

Let us take a numerical example.

**44** *Example:* To express $z = 4 + j3$ in polar form.

First draw a sketch diagram (that always helps)

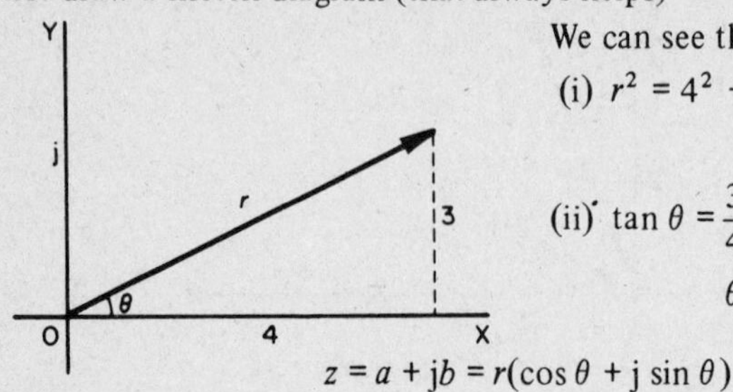


We can see that –

(i) $r^2 = 4^2 + 3^2 = 16 + 9 = 25$

$r = 5$

(ii) $\tan\theta = \dfrac{3}{4} = 0{\cdot}75$

$\theta = 36°52'$

$$z = a + jb = r(\cos\theta + j\sin\theta)$$

So in this case $\underline{z = 5(\cos 36°52' + j\sin 36°52')}$

Now here is one for you to do–

Find the polar form of the complex number $(2 + j3)$

*When you have finished it, consult the next frame.*

---

**45**

$$\boxed{z = 3{\cdot}606\,(\cos 56°19' + j\sin 56°19')}$$

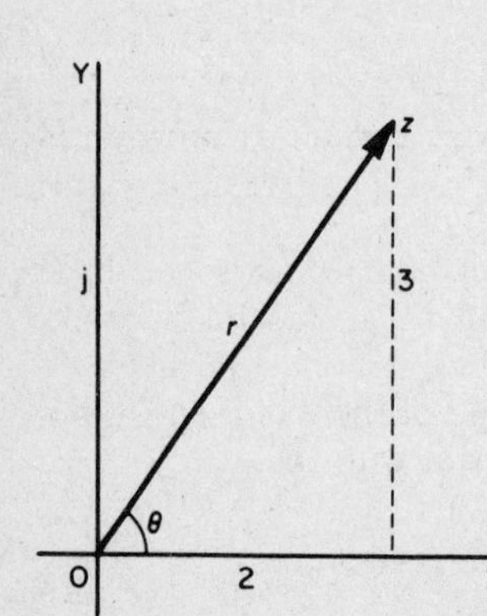


Here is the working

$z = 2 + j3 = r(\cos\theta + j\sin\theta)$

$r^2 = 4 + 9 = 13 \qquad r = 3{\cdot}606$

$\tan\theta = \dfrac{3}{2} = 1{\cdot}5 \qquad \theta = 56°19'$

$\underline{z = 3{\cdot}606\,(\cos 56°19' + j\sin 56°19')}$

□□□□□□□□□□□□□□□□□□□□□□□□□□□□□□□□□□□□□□

We have special names for the values of $r$ and $\theta$.

$$z = a + jb = r(\cos\theta + j\sin\theta)$$

(i) $r$ is called the *modulus* of the complex number $z$ and is often abbreviated to 'mod z' or indicated by $|z|$.

Thus if $z = 2 + j5$, then $|z| = \sqrt{(2^2 + 5^2)} = \sqrt{(4 + 25)} = \sqrt{29}$

(ii) $\theta$ is called the *argument* of the complex number and can be abbreviated to 'arg $z$'.

So if $z = 2 + j5$, then arg $z$ = ......................

**46**

$\arg z = 68°12'$

$$z = 2 + j5.\ \text{Then } \arg z = \theta = \tan^{-1}\frac{5}{2} = 68°12'$$

□ □ □ □ □ □ □ □ □ □ □ □ □ □ □ □ □ □ □ □ □ □ □ □ □ □ □ □ □ □ □ □ □ □ □ □ □ □ □ □

Warning. In finding $\theta$, there are of course two angles between 0° and 360°, the tangent of which has the value $\frac{b}{a}$. We must be careful to use the angle in the correct quadrant. *Always* draw a sketch of the vector to ensure you have the right one.

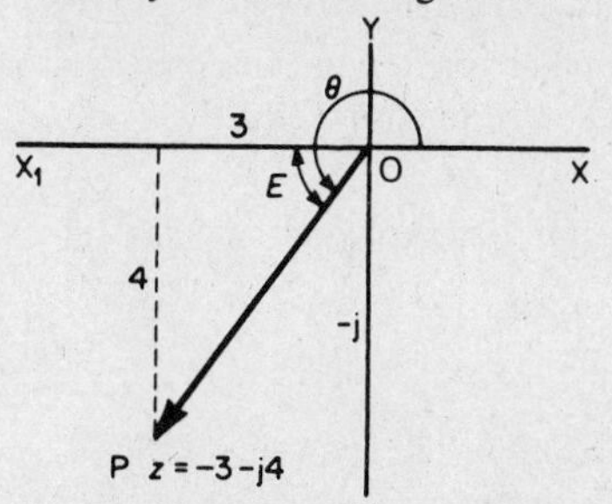


e.g. Find $\arg z$ when $z = -3 - j4$.

$\theta$ is measured from OX to OP. We first find $E$ the equivalent acute angle from the triangle shown.

$\tan E = \frac{4}{3} = 1{\cdot}333 \quad \therefore E = 53°7'$

Then in this case,

$\theta = 180° + E = 233°7' \quad \underline{\arg z = 233°7'}$

Now you find $\arg(-5 + j2)$

*Move on when finished.*

**47**

$\arg z = 158°12'$

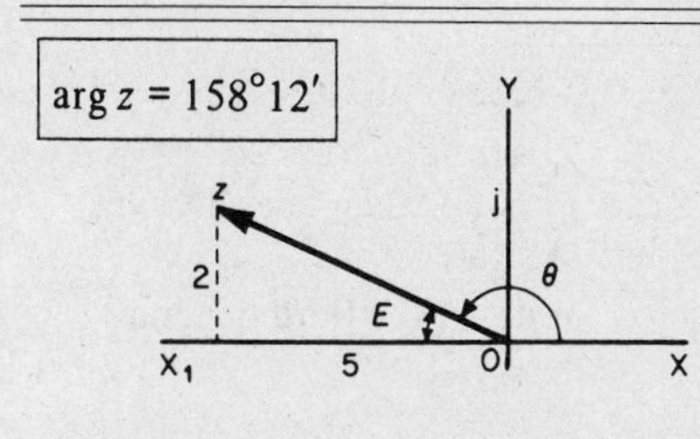


$$z = -5 + j2$$

$\tan E = \frac{2}{5} = 0{\cdot}4 \quad \therefore E = 21°48'$

In this particular case, $\theta = 180° - E$

$\therefore \theta = 158°12'$

□ □ □ □ □ □ □ □ □ □ □ □ □ □ □ □ □ □ □ □ □ □ □ □ □ □ □ □ □ □ □ □ □ □ □ □ □ □ □ □

Complex numbers in polar form are always of the same shape and differ only in the actual values of $r$ and $\theta$. We often use the shorthand version $r\underline{|\theta}$ to denote the polar form.

e.g. If $z = -5 + j2$, $r = \sqrt{(25 + 4)} = \sqrt{29} = 5{\cdot}385$ and from above $\theta = 158°12'$

$\therefore$ The full polar form is $z = 5{\cdot}385\,(\cos 158°12' + j \sin 158°12')$ and this can be shortened to $z = 5{\cdot}385\ \underline{|158°12'}$

Express in shortened form, the polar form of $(4 - j3)$

Do not forget to draw a sketch diagram first.

**48**

$$z = 5\ \underline{|323°8'}$$

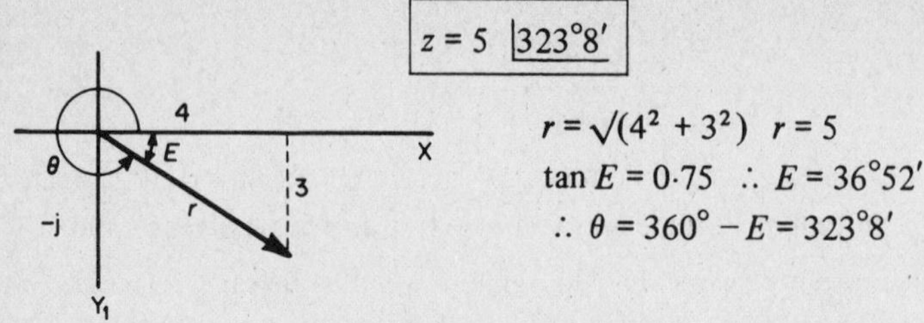


$r = \sqrt{(4^2 + 3^2)}$ $r = 5$

$\tan E = 0{\cdot}75$ $\therefore E = 36°52'$

$\therefore \theta = 360° - E = 323°8'$

$$\therefore z = 5(\cos 323°8' + \mathrm{j} \sin 323°8') = 5\ \underline{|323°8'}$$

□□□□□□□□□□□□□□□□□□□□□□□□□□□□□□□□□□□□□□

Of course, given a complex number in polar form, you can convert it into the basic form $a + \mathrm{j}b$ simply by evaluating the cosine and the sine and multiplying by the value of $r$.

$$\text{e.g. } z = 5(\cos 35° + \mathrm{j} \sin 35°) = 5(0{\cdot}8192 + \mathrm{j}0{\cdot}5736)$$

$$\underline{z = 4{\cdot}0960 + \mathrm{j}3{\cdot}8680}$$

Now you do this one—

Express in the form $a + \mathrm{j}b$, $4(\cos 65° + \mathrm{j} \sin 65°)$

**49**

$$z = 1{\cdot}6904 + \mathrm{j}3{\cdot}6252$$

for $z = 4(\cos 65° + \mathrm{j} \sin 65°) = 4(0{\cdot}4226 + \mathrm{j}0{\cdot}9063) = 1{\cdot}6904 + \mathrm{j}3{\cdot}6252$

□□□□□□□□□□□□□□□□□□□□□□□□□□□□□□□□□□□□□□

If the argument is greater than 90°, care must be taken in evaluating the cosine and sine to include the appropriate signs.

e.g. If $z = 2(\cos 210° + \mathrm{j} \sin 210°)$ the vector lies in the third quadrant.

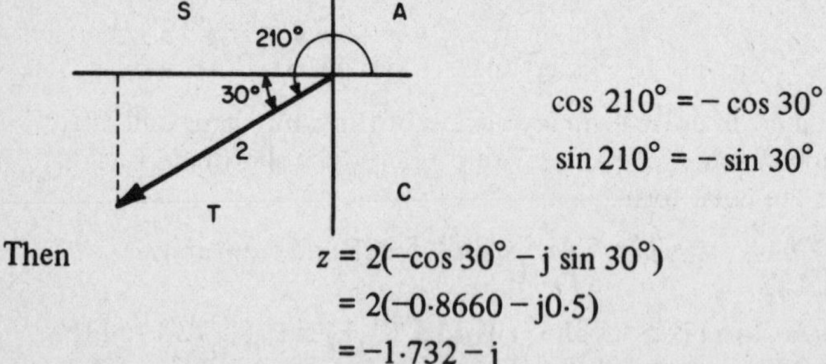


$\cos 210° = -\cos 30°$

$\sin 210° = -\sin 30°$

Then

$$\begin{aligned} z &= 2(-\cos 30° - \mathrm{j} \sin 30°) \\ &= 2(-0{\cdot}8660 - \mathrm{j}0{\cdot}5) \\ &= \underline{-1{\cdot}732 - \mathrm{j}} \end{aligned}$$

Here you are. What about this one?

Express $z = 5(\cos 140° + \mathrm{j} \sin 140°)$ in the form $a + \mathrm{j}b$

What do you make it?

**50**

$z = -3{\cdot}8300 + j3{\cdot}2140$

Here are the details –

$\cos 140° = -\cos 40°$

$\sin 140° = \sin 40°$

$$z = 5(\cos 140° + j \sin 140°) = 5(-\cos 40° + j \sin 40°)$$
$$= 5(-0{\cdot}7660 + j0{\cdot}6428)$$
$$= -3{\cdot}8300 + j3{\cdot}2140$$

□ □ □ □ □ □ □ □ □ □ □ □ □ □ □ □ □ □ □ □ □ □ □ □ □ □ □ □ □ □ □ □ □ □ □ □ □ □

Fine. Now by way of revision, work out the following.

(i) Express $-5 + j4$ in polar form

(ii) Express $3 \underline{|300°}$ in the form $a + jb$

*When you have finished both of them, check your results with those on frame 51.*

**51**

*Results*

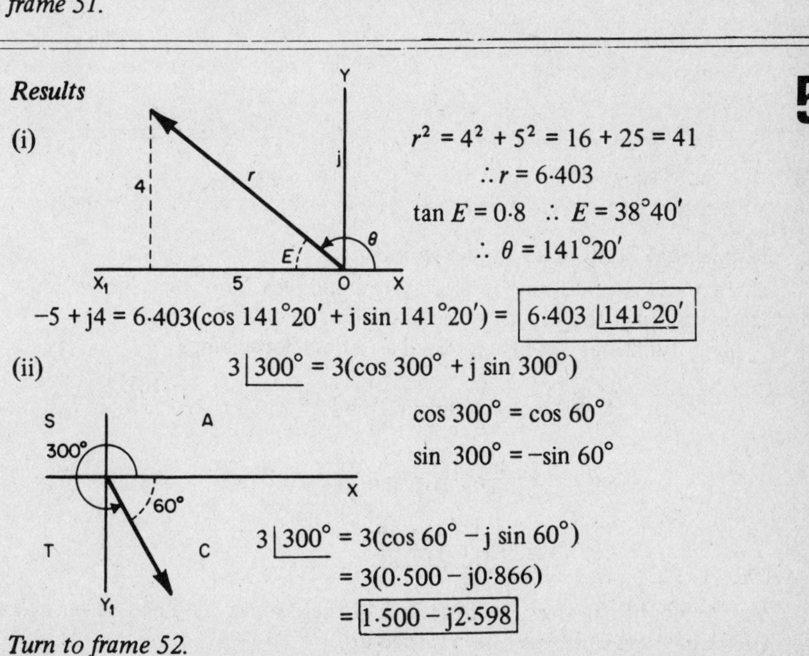


(i)

$r^2 = 4^2 + 5^2 = 16 + 25 = 41$

$\therefore r = 6{\cdot}403$

$\tan E = 0{\cdot}8 \quad \therefore E = 38°40'$

$\therefore \theta = 141°20'$

$$-5 + j4 = 6{\cdot}403(\cos 141°20' + j \sin 141°20') = \boxed{6{\cdot}403 \underline{|141°20'}}$$

(ii)

$$3\underline{|300°} = 3(\cos 300° + j \sin 300°)$$

$\cos 300° = \cos 60°$

$\sin 300° = -\sin 60°$

$$3\underline{|300°} = 3(\cos 60° - j \sin 60°)$$
$$= 3(0{\cdot}500 - j0{\cdot}866)$$
$$= \boxed{1{\cdot}500 - j2{\cdot}598}$$

*Turn to frame 52.*

**52**

We see then that there are two ways of expressing a complex number:

(i) in standard form: $z = a + \text{j}b$

(ii) in polar form: $z = r(\cos\theta + \text{j}\sin\theta)$

where $r = \sqrt{(a^2 + b^2)}$

and $\theta = \tan^{-1}\dfrac{b}{a}$

If we remember the simple diagram, we can easily convert from one system to the other.

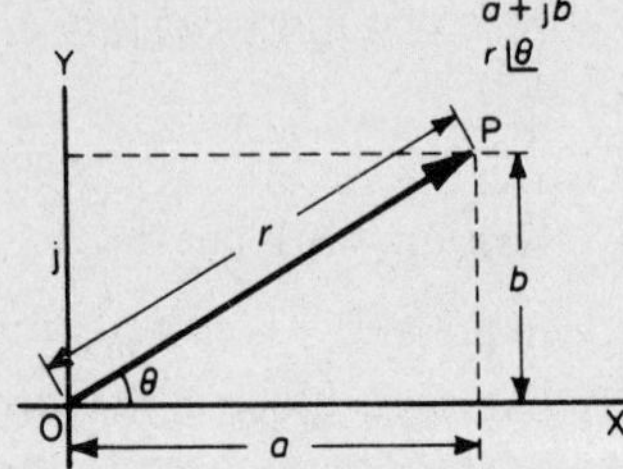


*So on now to frame 53.*

**53**

**Exponential Form of a complex number.**

There is still another way of expressing a complex number which we must deal with, for it too has its uses. We shall arrive at it this way:

Many functions can be expressed as series. For example,

$$e^x = 1 + x + \frac{x^2}{2!} + \frac{x^3}{3!} + \frac{x^4}{4!} + \frac{x^5}{5!} + \dots$$

$$\sin x = x - \frac{x^3}{3!} + \frac{x^5}{5!} - \frac{x^7}{7!} + \frac{x^9}{9!} + \dots$$

$$\cos x = 1 - \frac{x^2}{2!} + \frac{x^4}{4!} - \frac{x^6}{6!} + \dots$$

You no doubt have hazy recollections of these series You had better make a note of them since they have turned up.

**54**

If we now take the series for $e^x$ and write $j\theta$ in place of $x$, we get

$$e^{j\theta} = 1 + j\theta + \frac{(j\theta)^2}{2!} + \frac{(j\theta)^3}{3!} + \frac{(j\theta)^4}{4!} + \dots$$

$$= 1 + j\theta + \frac{j^2\theta^2}{2!} + \frac{j^3\theta^3}{3!} + \frac{j^4\theta^4}{4!} \dots$$

$$= 1 + j\theta - \frac{\theta^2}{2!} - \frac{j\theta^3}{3!} + \frac{\theta^4}{4!} + \dots$$

$$= (1 - \frac{\theta^2}{2!} + \frac{\theta^4}{4!} - \dots)$$

$$+ j(\theta - \frac{\theta^3}{3!} + \frac{\theta^5}{5!} - \dots)$$

$$= \cos\theta + j\sin\theta$$

Therefore, $r(\cos\theta + j\sin\theta)$ can now be written as $re^{j\theta}$. This is called the *exponential form* of the complex number. It can be obtained from the polar form quite easily since the $r$ value is the same and the angle $\theta$ is the same in both. It is important to note, however, that in the exponential form, the angle must be in *radians*.

*Move on to the next frame.*

**55**

The three ways of expressing a complex number are therefore

(i) $z = a + jb$

(ii) $z = r(\cos\theta + j\sin\theta)$ .. .. Polar form

(iii) $z = r.e^{j\theta}$ .. .. .. .. Exponential form

Remember that the exponential form is obtained from the polar form.

(i) the $r$ value is the same in each case.

(ii) the angle is also the same in each case, but in the exponential form the angle must be in radians.

So, knowing that, change the polar form $5(\cos 60^\circ + j\sin 60^\circ)$ into the exponential form.

*Then turn to frame 56.*

**56**

Exponential form $5\,e^{j\frac{\pi}{3}}$

for we have $5(\cos 60° + j \sin 60°)$ $r = 5$

$\theta = 60° = \frac{\pi}{3}$ radians

$\therefore$ Exponential form is $5\,e^{j\frac{\pi}{3}}$

□□□□□□□□□□□□□□□□□□□□□□□□□□□□□□□□□□□□□□

And now a word about negative angles

We know $e^{j\theta} = \cos\theta + j\sin\theta$

If we replace $\theta$ by $-\theta$ in this result, we get

$$e^{-j\theta} = \cos(-\theta) + j\sin(-\theta)$$
$$= \cos\theta - j\sin\theta$$

So we have

$$\left.\begin{aligned} e^{j\theta} &= \cos\theta + j\sin\theta \\ e^{-j\theta} &= \cos\theta - j\sin\theta \end{aligned}\right\}$$

Make a note of these.

**57**

There is one operation that we have been unable to carry out with complex numbers before this. That is to find the logarithm of a complex number. The exponential form now makes this possible, since the exponential form consists only of products and powers.

For, if we have

$$z = r\,e^{j\theta}$$

Then we can say

$$\underline{\ln z = \ln r + j\theta}$$

e.g. If

$$z = 6{\cdot}42\,e^{j1{\cdot}57}$$

then

$$\ln z = \ln 6{\cdot}42 + j1{\cdot}57$$
$$\underline{= 1{\cdot}8594 + j1{\cdot}57}$$

and the result is once again a complex number.

And if $z = 3{\cdot}8\,e^{-j0{\cdot}236}$, then $\ln z =$ ......................

**58**

$$\ln z = \ln 3{\cdot}8 - j0{\cdot}236 = \boxed{1{\cdot}3350 - j0{\cdot}236}$$

□□□□□□□□□□□□□□□□□□□□□□□□□□□□□□□□□□□□□□

Finally, here is an example of a rather different kind. Once you have seen it done, you will be able to deal with others of this kind. Here it is.

Express $e^{1-j\pi/4}$ in the form $a + jb$

Well now, we can write

$$e^{1-j\pi/4} \text{ as } e^{1}e^{-j\pi/4}$$

$$= e(\cos \pi/4 - j \sin \pi/4)$$

$$= e\left\{\frac{1}{\sqrt{2}} - j\frac{1}{\sqrt{2}}\right\}$$

$$= \underline{\frac{e}{\sqrt{2}}(1 - j)}$$

**59**

This brings us to the end of this programme, except for the test exercise. Before you do that, read down the Revision Sheet that follows in the next frame and revise any points on which you are not completely sure.

Then turn on and work through the test exercise: you will find the questions quite straightforward and easy.

*But first, turn to frame 60.*

## 60 Revision Summary

1. *Powers of j*

$$j = \sqrt{(-1)}, \quad j^2 = -1, \quad j^3 = -j, \quad j^4 = 1.$$

A factor j turns a vector through 90° in the positive direction.

2. *Complex numbers*

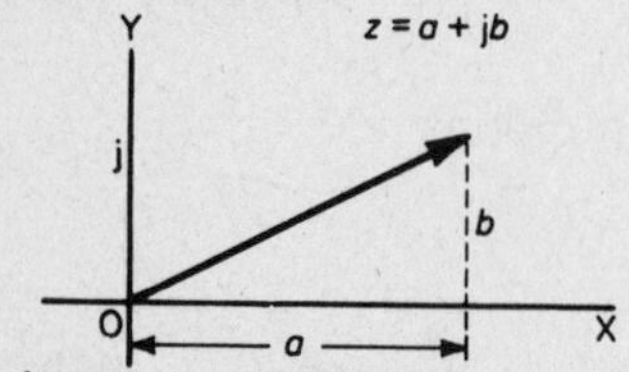


$a$ = real part
$b$ = imaginary part

3. *Conjugate complex numbers* $(a + jb)$ and $(a - jb)$

The product of two conjugate complex numbers is always real.

$$(a + jb)(a - jb) = a^2 + b^2$$

4. *Equal complex numbers*

If $a + jb = c + jd$, then $a = c$ and $b = d$.

5. *Polar form of a complex number*

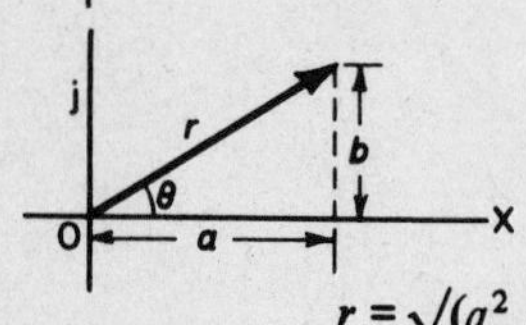


$$\begin{aligned} z &= a + jb \\ &= r(\cos\theta + j\sin\theta) \\ &= r\underline{|\theta} \end{aligned}$$

$$r = \sqrt{(a^2 + b^2)}; \quad \theta = \tan^{-1}\left\{\frac{b}{a}\right\}$$

also $\quad a = r\cos\theta; \quad b = r\sin\theta$

$r$ = the modulus of $z$, written 'mod $z$' or $|z|$

$\theta$ = the argument of $z$, written 'arg $z$'

6. *Exponential form of a complex number*

$$\left.\begin{aligned} z &= r(\cos\theta + j\sin\theta) = re^{j\theta} \\ \text{and} \quad & r(\cos\theta - j\sin\theta) = re^{-j\theta} \end{aligned}\right\} \theta \text{ in radians}$$

7. *Logarithm of a complex number*

$$z = re^{j\theta} \quad \therefore \ln z = \ln r + j\theta$$

or if $\quad z = re^{-j\theta} \quad \therefore \ln z = \ln r - j\theta$

**61**

**Test Exercise – I**

1. Simplify (i) $j^3$, (ii) $j^5$, (iii) $j^{12}$, (iv) $j^{14}$.

2. Express in the form $a + jb$

   (i) $(4 - j7)(2 + j3)$ (ii) $(-1 + j)^2$

   (iii) $(5 + j2)(4 - j5)(2 + j3)$ (iv) $\dfrac{4 + j3}{2 - j}$

3. Express in polar form

   (i) $3 + j5$ (ii) $-6 + j3$ (iii) $-4 - j5$

4. Express in the form $a + jb$

   (i) $5(\cos 225° + j \sin 225°)$ (ii) $4 \underline{|330°}$

5. Find the values of $x$ and $y$ that satisfy the equation

   $(x + y) + j(x - y) = 14{\cdot}8 + j6{\cdot}2$

6. Express in exponential form

   (i) $z_1 = 10 \underline{|37°15'}$ and (ii) $z_2 = 10 \underline{|322°45'}$

   Hence find $\ln z_1$ and $\ln z_2$.

7. Express $z = e^{1+j\pi/2}$ in the form $a + jb$.

*Now you are ready to start Part 2 of the work on complex numbers.*

**Further Problems – I**

1. Simplify (i) $(5 + j4)(3 + j7)(2 - j3)$

   (ii) $\dfrac{(2 - j3)(3 + j2)}{(4 - j3)}$ (iii) $\dfrac{\cos 3x + j \sin 3x}{\cos x + j \sin x}$

2. Express $\dfrac{2 + j3}{j(4 - j5)} + \dfrac{2}{j}$ in the form $a + jb$.

3. If $z = \dfrac{1}{2 + j3} + \dfrac{1}{1 - j2}$, express $z$ in the form $a + jb$.

4. If $z = \dfrac{2 + j}{1 - j}$, find the real and imaginary parts of the complex number $z + \dfrac{1}{z}$.

5. Simplify $(2 + j5)^2 + \dfrac{5(7 + j2)}{3 - j4} - j(4 - j6)$, expressing the result in the form $a + jb$.

6. If $z_1 = 2 + j$, $z_2 = -2 + j4$ and $\dfrac{1}{z_3} = \dfrac{1}{z_1} + \dfrac{1}{z_2}$, evaluate $z_3$ in the form $a + jb$. If $z_1, z_2, z_3$ are represented on an Argand diagram by the points P, Q, R, respectively, prove that R is the foot of the perpendicular from the origin on to the line PQ.

7. Points A, B, C, D, on an Argand diagram, represent the complex numbers $9 + j$, $4 + j13$, $-8 + j8$, $-3 - j4$ respectively. Prove that ABCD is a square.

8. If $(2 + j3)(3 - j4) = x + jy$, evaluate $x$ and $y$.

9. If $(a + b) + j(a - b) = (2 + j5)^2 + j(2 - j3)$, find the values of $a$ and $b$.

10. If $x$ and $y$ are real, solve the equation

$$\frac{jx}{1 + jy} = \frac{3x + j4}{x + 3y}$$

11. If $z = \dfrac{a + jb}{c + jd}$, where $a, b, c, d$, are real quantities, show that (i) if $z$ is

real then $\frac{a}{b} = \frac{c}{d}$, and (ii) if $z$ is entirely imaginary then $\frac{a}{b} = -\frac{d}{c}$.

12. Given that $(a + b) + \mathrm{j}(a - b) = (1 + \mathrm{j})^2 + \mathrm{j}(2 + \mathrm{j})$, obtain the values of $a$ and $b$.

13. Express $(-1 + \mathrm{j})$ in the form $r\,\mathrm{e}^{\mathrm{j}\theta}$, where $r$ is positive and $-\pi < \theta < \pi$.

14. Find the modulus of $z = (2 - \mathrm{j})(5 + \mathrm{j}12)/(1 + \mathrm{j}2)^3$.

15. If $x$ is real, show that $(2 + \mathrm{j})\,\mathrm{e}^{(1+\mathrm{j}3)x} + (2 - \mathrm{j})\,\mathrm{e}^{(1-\mathrm{j}3)x}$ is also real.

16. Given that $z_1 = R_1 + R + \mathrm{j}\omega L$; $z_2 = R_2$; $z_3 = \dfrac{1}{\mathrm{j}\omega C_3}$; and $z_4 = R_4 + \dfrac{1}{\mathrm{j}\omega C_4}$; and also that $z_1 z_3 = z_2 z_4$, express $R$ and $L$ in terms of the real constants $R_1, R_2, R_4, C_3$ and $C_4$.

17. If $z = x + \mathrm{j}y$, where $x$ and $y$ are real, and if the real part of $(z + 1)/(z + \mathrm{j})$ is equal to 1, show that the point $z$ lies on a straight line in the Argand diagram.

18. When $z_1 = 2 + \mathrm{j}3$, $z_2 = 3 - \mathrm{j}4$, $z_3 = -5 + \mathrm{j}12$, then $z = z_1 + \dfrac{z_2 z_3}{z_2 + z_3}$.
If $E = Iz$, find $E$ when $I = 5 + \mathrm{j}6$.

19. If $\dfrac{R_1 + \mathrm{j}\omega L}{R_3} = \dfrac{R_2}{R_4 - \mathrm{j}\dfrac{1}{\omega C}}$, where $R_1, R_2, R_3, R_4, \omega, L$ and $C$ are real,

show that

$$L = \frac{CR_2R_3}{\omega^2 C^2 R_4^2 + 1}$$

20. If $z$ and $\bar{z}$ are conjugate complex numbers, find two complex numbers, $z = z_1$ and $z = z_2$, that satisfy the equation

$$3z\bar{z} + 2(z - \bar{z}) = 39 + \mathrm{j}12$$

On an Argand diagram, these two numbers are represented by the points P and Q. If R represents the number j1, show that the angle PRQ is a right angle.

# Programme 2

## COMPLEX NUMBERS

### PART 2

**1**

## Introduction

In Part 1 of this programme on Complex Numbers, we discovered how to manipulate them in adding, subtracting, multiplying and dividing. We also finished Part 1 by seeing that a complex number $a + jb$ can also be expressed in Polar Form, which is always of the form $r(\cos\theta + j\sin\theta)$.

You will remember that values of $r$ and $\theta$ can easily be found from the diagram of the given vector.

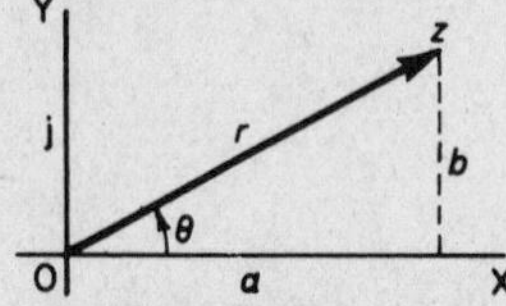


$$r^2 = a^2 + b^2 \quad \therefore\ r = \sqrt{(a^2 + b^2)}$$

$$\text{and } \tan\theta = \frac{b}{a} \quad \therefore\ \theta = \tan^{-1}\frac{b}{a}$$

To be sure that you have taken the correct value of $\theta$, always DRAW A SKETCH DIAGRAM to see which quadrant the vector is in.

Remember that $\theta$ is always measured from ....................

**2**

OX   i.e. the positive axis OX.

□□□□□□□□□□□□□□□□□□□□□□□□□□□□□□□□□□□□□□

Right. Just by way of revision and as a warming up exercise, do the following:

Express $z = 12 - j5$ in polar form.

Do not forget the sketch diagram. It ensures that you get the correct value for $\theta$.

*When you have finished, and not before, turn on to frame 3 to check your result.*

**3**

*Result:*

$$13(\cos 337°23' + j \sin 337°23')$$

Here it is, worked out in full.

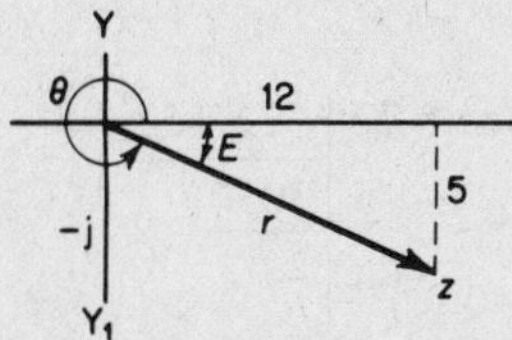


$r^2 = 12^2 + 5^2 = 144 + 25 = 169$

$\therefore r = 13$

$\tan E = \frac{5}{12} = 0{\cdot}4167 \quad \therefore E = 22°37'$

In this case, $\theta = 360° - E = 360° - 22°37' \quad \therefore \theta = 337°23'$

$$z = r(\cos\theta + j\sin\theta) = 13(\cos 337°23' + j \sin 337°23')$$

□□□□□□□□□□□□□□□□□□□□□□□□□□□□□□□□□□□□□□□□□

Did you get that right? Here is one more, done in just the same way.

Express $-5 - j4$ in polar form.

Diagram first of all! Then you cannot go wrong.

*When you have the result, on to frame 4.*

---

**4**

*Result:*

$$z = 6{\cdot}403(\cos 218°40' + j \sin 218°40')$$

Here is the working: check yours.

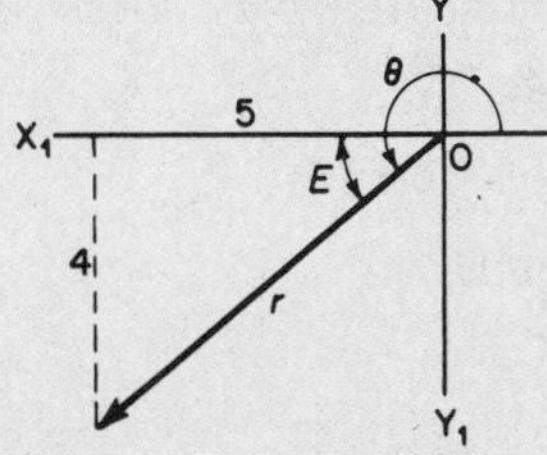


$r^2 = 5^2 + 4^2 = 25 + 16 = 41$

$\therefore r = \sqrt{41} = 6{\cdot}403$

$\tan E = \frac{4}{5} = 0{\cdot}8 \quad \therefore E = 38°40'$

In this case, $\theta = 180° + E = 218°40'$

So $z = -5 - j4 = 6{\cdot}403(\cos 218°40' + j \sin 218°40')$

□□□□□□□□□□□□□□□□□□□□□□□□□□□□□□□□□□□□□□□□□

Since every complex number in polar form is of the same shape, i.e. $r(\cos\theta + j\sin\theta)$ and differs from another complex number simply by the values of $r$ and $\theta$, we have a shorthand method of quoting the result in polar form. Do you remember what it is? The shorthand way of writing the result above, i.e. $6{\cdot}403(\cos 218°40' + j \sin 218°40')$ is ......................

**5**

$6{\cdot}403\,\underline{|218°40'}$

□□□□□□□□□□□□□□□□□□□□□□□□□□□□□□□□□□□□□□

Correct. Likewise:

$5{\cdot}72(\cos 322°15' + \text{j}\sin 322°15')$ is written $5{\cdot}72\,\underline{|322°15'}$

$5(\cos 105° + \text{j}\sin 105°)$ " " $5\,\underline{|105°}$

$3{\cdot}4(\cos\frac{\pi}{6} + \text{j}\sin\frac{\pi}{6})$ " " $3{\cdot}4\,\underline{|\frac{\pi}{6}}$

They are all complex numbers in polar form. They are all the same shape and differ one from another simply by the values of .......... and .......... .

**6**

$r$ and $\theta$

□□□□□□□□□□□□□□□□□□□□□□□□□□□□□□□□□□□□□□

Now let us consider the following example.

Express $z = 4 - \text{j}3$ in polar form.

First the diagram.

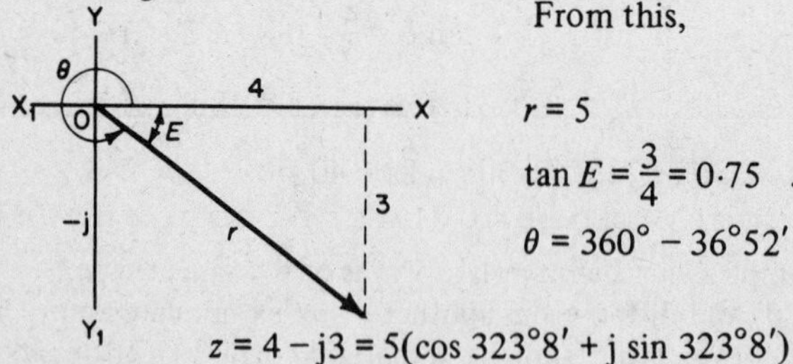


From this,

$r = 5$

$\tan E = \frac{3}{4} = 0{\cdot}75 \quad \therefore\ E = 36°52'$

$\theta = 360° - 36°52' = 323°8'$

$$z = 4 - \text{j}3 = 5(\cos 323°8' + \text{j}\sin 323°8')$$

or in shortened form, $z =$ ....................

**7**

$z = 5 \;\underline{|323^\circ 8'}$

□□□□□□□□□□□□□□□□□□□□□□□□□□□□□□□□□□□□□□

In this last example, we have

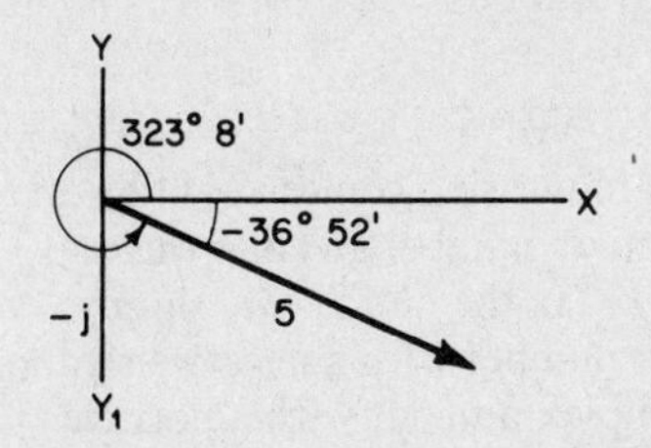


$$z = 5(\cos 323^\circ 8' + \text{j} \sin 323^\circ 8')$$

But the direction of the vector, measured from OX, could be given as $-36^\circ 52'$, the minus sign showing that we are measuring the angle in the opposite sense from the usual positive direction.

We could write $z = 5(\cos [-36^\circ 52'] + \text{j} \sin [-36^\circ 52'])$. But you already know that $\cos[-\theta] = \cos\theta$ and $\sin[-\theta] = -\sin\theta$.

$$\underline{z = 5(\cos 36^\circ 52' - \text{j} \sin 36^\circ 52')}$$

i.e. very much like the polar form but with a minus sign in the middle. This comes about whenever we use negative angles.

In the same way, $z = 4(\cos 250^\circ + \text{j} \sin 250^\circ) = 4(\cos [-110^\circ] + \text{j} \sin[-110^\circ])$
$= 4(\text{..........} \;\; \text{..........})$

**8**

$z = 4(\cos 110^\circ - \text{j} \sin 110^\circ)$

since $\cos(-110^\circ) = \cos 110^\circ$
and $\sin(-110^\circ) = -\sin 110^\circ$

□□□□□□□□□□□□□□□□□□□□□□□□□□□□□□□□□□□□□□

It is sometimes convenient to use this form when the value of $\theta$ is greater than 180°, i.e. in the 3rd and 4th quadrants.

*Ex. 1*

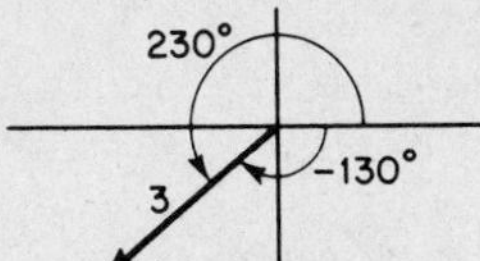


$$z = 3(\cos 230^\circ + \text{j} \sin 230^\circ)$$
$$= 3(\cos 130^\circ - \text{j} \sin 130^\circ).$$

Similarly, *Ex. 2* $\quad z = 3(\cos 300^\circ + \text{j} \sin 300^\circ) = 3(\cos 60^\circ - \text{j} \sin 60^\circ)$

*Ex. 3* $\quad z = 4(\cos 290^\circ + \text{j} \sin 290^\circ) = 4(\cos 70^\circ - \text{j} \sin 70^\circ)$

*Ex. 4* $\quad z = 2(\cos 215^\circ + \text{j} \sin 215^\circ) = 2(\cos 145^\circ - \text{j} \sin 145^\circ)$

and *Ex. 5* $\quad z = 6(\cos 310^\circ + \text{j} \sin 310^\circ) = \text{........................}$

**9**

$$z = 6(\cos 50^\circ - j \sin 50^\circ)$$

since $\cos 310^\circ = \cos 50^\circ$

and $\sin 310^\circ = -\sin 50^\circ$

□□□□□□□□□□□□□□□□□□□□□□□□□□□□□□□□□□□□□□□□

One moment ago, we agreed that the minus sign comes about by the use of negative angles. To convert a complex number given in this way back into proper polar form, i.e. with a '+' in the middle, we simply work back the way we came. A complex number with a negative sign in the middle is equivalent to the same complex number with a positive sign, but with the angles made negative.

e.g. $z = 4(\cos 30^\circ - j \sin 30^\circ)$

$= 4(\cos [-30^\circ] + j \sin [-30^\circ])$

$= 4(\cos 330^\circ + j \sin 330^\circ)$ and we are back in the proper polar form.

You do this one. Convert $z = 5(\cos 40^\circ - j \sin 40^\circ)$ into proper polar form.

*Then on to frame 10.*

**10**

$$z = 5(\cos 320^\circ + j \sin 320^\circ)$$

since $z = 5(\cos 40^\circ - j \sin 40^\circ) = 5(\cos [-40^\circ] + j \sin [-40^\circ])$

$= 5(\cos 320^\circ + j \sin 320^\circ)$

□□□□□□□□□□□□□□□□□□□□□□□□□□□□□□□□□□□□□□□□

Here is another for you to do.

Express $z = 4(\cos 100^\circ - j \sin 100^\circ)$ in proper polar form.

Do not forget, it all depends on the use of negative angles.

**11**

$$z = 4(\cos 260^\circ + \mathrm{j} \sin 260^\circ)$$

for $\quad z = 4(\cos 100^\circ - \mathrm{j} \sin 100^\circ) = 4(\cos [-100^\circ] + \mathrm{j} \sin [-100^\circ])$
$$= 4(\cos 260^\circ + \mathrm{j} \sin 260^\circ)$$

□□□□□□□□□□□□□□□□□□□□□□□□□□□□□□□□□□□□□□□□□□

We ought to see how this modified polar form affects our shorthand notation.

Remember, $5(\cos 60^\circ + \mathrm{j} \sin 60^\circ)$ is written $5\,\underline{|\,60^\circ}$

How then shall we write $5(\cos 60^\circ - \mathrm{j} \sin 60^\circ)$?

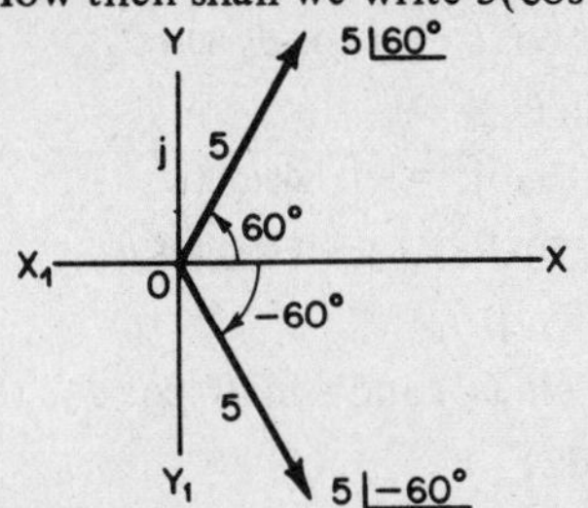


We know that this really stands for $5(\cos [-60^\circ] + \mathrm{j} \sin [-60^\circ])$ so we could write $5\,\underline{|-60^\circ}$. But instead of using the negative angle we use a different symbol i.e. $5\,\underline{|-60^\circ}$ becomes $5\,\overline{|\,60^\circ}$

Similarly, $3(\cos 45^\circ - \mathrm{j} \sin 45^\circ) = 3\,\underline{|-45^\circ} =$ ....................

**12**

$$3\,\overline{|\,45^\circ}$$

□□□□□□□□□□□□□□□□□□□□□□□□□□□□□□□□□□□□□□□□□□

This is easy to remember,

for the sign $\underline{|\;\;}$ resembles the first quadrant and indicates measuring angles ↶ i.e. in the positive direction,

while the sign $\overline{|\;\;}$ resembles the fourth quadrant and indicates measuring angles ↷ i.e. in the negative direction.

e.g. $(\cos 15^\circ + \mathrm{j} \sin 15^\circ)$ is written $\underline{|\,15^\circ}$

but $(\cos 15^\circ - \mathrm{j} \sin 15^\circ)$, which is really $(\cos [-15^\circ] + \mathrm{j} \sin [-15^\circ])$ is written $\overline{|\,15^\circ}$

So how do we write (i) $(\cos 120^\circ + \mathrm{j} \sin 120^\circ)$
and (ii) $(\cos 135^\circ - \mathrm{j} \sin 135^\circ)$
in the shorthand way?

**13**

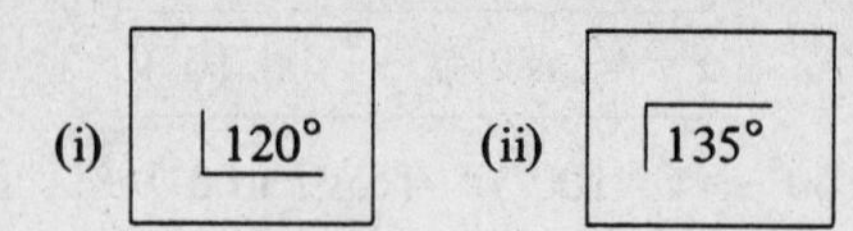


The polar form at first sight seems to be a complicated way of representing a complex number. However it is very useful as we shall see. Suppose we multiply together two complex numbers in this form.

Let $z_1 = r_1(\cos\theta_1 + \text{j}\sin\theta_1)$ and $z_2 = r_2(\cos\theta_2 + \text{j}\sin\theta_2)$

Then $z_1 z_2 = r_1(\cos\theta_1 + \text{j}\sin\theta_1)\, r_2(\cos\theta_2 + \text{j}\sin\theta_2)$

$$= r_1 r_2(\cos\theta_1\cos\theta_2 + \text{j}\sin\theta_1\cos\theta_2 + \text{j}\cos\theta_1\sin\theta_2 + \text{j}^2\sin\theta_1\sin\theta_2)$$

Re-arranging the terms and remembering that $\text{j}^2 = -1$, we get

$$z_1 z_2 = r_1 r_2\,[(\cos\theta_1\cos\theta_2 - \sin\theta_1\sin\theta_2) + \text{j}(\sin\theta_1\cos\theta_2 + \cos\theta_1\sin\theta_2)]$$

Now the brackets $(\cos\theta_1\cos\theta_2 - \sin\theta_1\sin\theta_2)$ and $(\sin\theta_1\cos\theta_2 + \cos\theta_1\sin\theta_2)$ ought to ring a bell. What are they?

**14**

$$\cos\theta_1\cos\theta_2 - \sin\theta_1\sin\theta_2 = \cos(\theta_1 + \theta_2)$$
$$\sin\theta_1\cos\theta_2 + \cos\theta_1\sin\theta_2 = \sin(\theta_1 + \theta_2)$$

In that case, $z_1 z_2 = r_1 r_2\,[\cos(\theta_1 + \theta_2) + \text{j}\sin(\theta_1 + \theta_2)]$

Note this important result. We have just shown that

$$r_1(\cos\theta_1 + \text{j}\sin\theta_1).\, r_2(\cos\theta_2 + \text{j}\sin\theta_2) = \underline{r_1 r_2\,[\cos(\theta_1 + \theta_2) + \text{j}\sin(\theta_1 + \theta_2)]}$$

i.e. To multiply together two complex numbers in polar form,

(i) multiply the $r$'s together, (ii) add the angles, $\theta$, together.

It is just as easy as that!

e.g. $2(\cos 30° + \text{j}\sin 30°) \times 3(\cos 40° + \text{j}\sin 40°)$

$$= 2 \times 3(\cos[30° + 40°] + \text{j}\sin[30° + 40°])$$
$$= \underline{6(\cos 70° + \text{j}\sin 70°)}$$

So if we multiply together $5(\cos 50° + \text{j}\sin 50°)$ and $2(\cos 65° + \text{j}\sin 65°)$ we get ................................ .

**15**

$$10(\cos 115° + j \sin 115°)$$

Remember, multiply the $r$'s; add the $\theta$'s.
Here you are then; all done the same way:

(i) $2(\cos 120° + j \sin 120°) \times 4(\cos 20° + j \sin 20°)$
$= 8(\cos 140° + j \sin 140°)$

(ii) $a(\cos\theta + j \sin\theta) \times b(\cos\phi + j \sin\phi)$
$= ab(\cos[\theta + \phi] + j \sin[\theta + \phi])$

(iii) $6(\cos 210° + j \sin 210°) \times 3(\cos 80° + j \sin 80°)$
$= 18(\cos 290° + j \sin 290°)$

(iv) $5(\cos 50° + j \sin 50°) \times 3(\cos[-20°] + j \sin[-20°])$
$= 15(\cos 30° + j \sin 30°)$

Have you got it? No matter what the angles are, all we do is

(i) multiply the moduli, (ii) add the arguments.

So therefore, $4(\cos 35° + j \sin 35°) \times 3(\cos 20° + j \sin 20°)$
$= \ldots\ldots\ldots\ldots\ldots\ldots\ldots$

**16**

$$12(\cos 55° + j \sin 55°)$$

Now let us see if we can discover a similar set of rules for Division.

We already know that to simplify $\dfrac{5 + j6}{3 + j4}$ we first obtain a denominator that is entirely real by multiplying top and bottom by ..................... .

**17**

the conjugate of the denominator i.e. $3 - j4$

□□□□□□□□□□□□□□□□□□□□□□□□□□□□□□□□□□□□□□

Right. Then let us do the same thing with

$$\frac{r_1(\cos\theta_1 + j\sin\theta_1)}{r_2(\cos\theta_2 + j\sin\theta_2)}$$

$$\frac{r_1(\cos\theta_1 + j\sin\theta_1)}{r_2(\cos\theta_2 + j\sin\theta_2)} = \frac{r_1(\cos\theta_1 + j\sin\theta_1)(\cos\theta_2 - j\sin\theta_2)}{r_2(\cos\theta_2 + j\sin\theta_2)(\cos\theta_2 - j\sin\theta_2)}$$

$$= \frac{r_1}{r_2}\frac{(\cos\theta_1\cos\theta_2 + j\sin\theta_1\cos\theta_2 - j\cos\theta_1\sin\theta_2 + \sin\theta_1\sin\theta_2)}{(\cos^2\theta_2 + \sin^2\theta_2)}$$

$$= \frac{r_1}{r_2}\frac{[(\cos\theta_1\cos\theta_2 + \sin\theta_1\sin\theta_2) + j(\sin\theta_1\cos\theta_2 - \cos\theta_1\sin\theta_2)]}{1}$$

$$= \frac{r_1}{r_2}[\cos(\theta_1 - \theta_2) + j\sin(\theta_1 - \theta_2)]$$

So, for division, the rule is ..........................

**18**

divide the $r$'s and subtract the angle

□□□□□□□□□□□□□□□□□□□□□□□□□□□□□□□□□□□□□□

That is correct.

e.g. $$\frac{6(\cos 72^\circ + j\sin 72^\circ)}{2(\cos 41^\circ + j\sin 41^\circ)} = 3(\cos 31^\circ + j\sin 31^\circ)$$

So we now have two important rules

If $z_1 = r_1(\cos\theta_1 + j\sin\theta_1)$ and $z_2 = r_2(\cos\theta_2 + j\sin\theta_2)$

then (i) $z_1z_2 = r_1r_2[\cos(\theta_1 + \theta_2) + j\sin(\theta_1 + \theta_2)]$

and (ii) $\frac{z_1}{z_2} = \frac{r_1}{r_2}[\cos(\theta_1 - \theta_2) + j\sin(\theta_1 - \theta_2)]$

The results are still, of course, in proper polar form.

Now here is one for you to think about.

If $z_1 = 8(\cos 65^\circ + j\sin 65^\circ)$ and $z_2 = 4(\cos 23^\circ + j\sin 23^\circ)$

then (i) $z_1z_2 =$ .................... and (ii) $\frac{z_1}{z_2} =$ ......................

**19**

$$z_1 z_2 = 32(\cos 88^\circ + j \sin 88^\circ)$$

$$\frac{z_1}{z_2} = 2(\cos 42^\circ + j \sin 42^\circ)$$

□□□□□□□□□□□□□□□□□□□□□□□□□□□□□□□□□□□□□□□□□□

Of course, we can combine the rules in a single example.

e.g.

$$\frac{5(\cos 60^\circ + j \sin 60^\circ) \times 4(\cos 30^\circ + j \sin 30^\circ)}{2(\cos 50^\circ + j \sin 50^\circ)}$$

$$= \frac{20(\cos 90^\circ + j \sin 90^\circ)}{2(\cos 50^\circ + j \sin 50^\circ)}$$

$$= \underline{10(\cos 40^\circ + j \sin 40^\circ)}$$

What does the following product become?

$4(\cos 20^\circ + j \sin 20^\circ) \times 3(\cos 30^\circ + j \sin 30^\circ) \times 2(\cos 40^\circ + j \sin 40^\circ)$

**20**

***Result:***

$$24(\cos 90^\circ + j \sin 90^\circ)$$

i.e. $(4 \times 3 \times 2)\,[\cos(20^\circ + 30^\circ + 40^\circ) + j \sin(20^\circ + 30^\circ + 40^\circ)]$

$$= 24(\cos 90^\circ + j \sin 90^\circ)$$

□□□□□□□□□□□□□□□□□□□□□□□□□□□□□□□□□□□□□□□□□□

Now what about a few revision examples on the work we have done so far?

*Turn to the next frame.*

**21** **Revision Exercise**

Work all these questions and then turn on to frame 22 and check your results.

1. Express in polar form, $z = -4 + j2$.
2. Express in true polar form, $z = 5(\cos 55^\circ - j \sin 55^\circ)$
3. Simplify the following, giving the results in polar form

   (i) $3(\cos 143^\circ + j \sin 143^\circ) \times 4(\cos 57^\circ + j \sin 57^\circ)$

   (ii) $\dfrac{10(\cos 126^\circ + j \sin 126^\circ)}{2(\cos 72^\circ + j \sin 72^\circ)}$

4. Express in the form $a + jb$,

   (i) $2(\cos 30^\circ + j \sin 30^\circ)$

   (ii) $5(\cos 57^\circ - j \sin 57^\circ)$

*Solutions are on frame 22. Turn on and see how you have fared.*

**Solutions**

1.

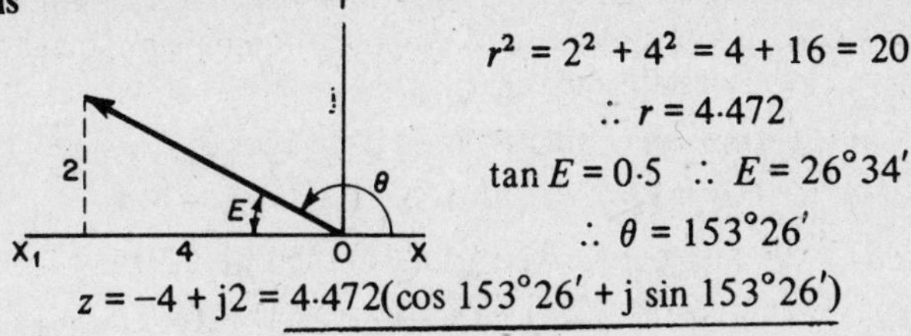


$r^2 = 2^2 + 4^2 = 4 + 16 = 20$

$\therefore r = 4{\cdot}472$

$\tan E = 0{\cdot}5 \quad \therefore E = 26°34'$

$\therefore \theta = 153°26'$

$z = -4 + j2 = \underline{4{\cdot}472(\cos 153°26' + j \sin 153°26')}$

2. $z = 5(\cos 55° - j \sin 55°) = 5[\cos(-55°) + j \sin(-55°)]$
$= \underline{5(\cos 305° + j \sin 305°)}$

3. (i) $3(\cos 143° + j \sin 143°) \times 4(\cos 57° + j \sin 57°)$
$= 3 \times 4[\cos(143° + 57°) + j \sin(143° + 57°)]$
$= \underline{12(\cos 200° + j \sin 200°)}$

(ii) $\dfrac{10(\cos 126° + j \sin 126°)}{2(\cos 72° + j \sin 72°)}$
$= \dfrac{10}{2}[\cos(126° - 72°) + j \sin(126° - 72°)]$
$= \underline{5(\cos 54° + j \sin 54°)}$

4. (i) $2(\cos 30° + j \sin 30°)$
$= 2(0{\cdot}866 + j0{\cdot}5) = \underline{1{\cdot}732 + j}$

(ii) $5(\cos 57° - j \sin 57°)$
$= 5(0{\cdot}5446 - j0{\cdot}8387)$
$= \underline{2{\cdot}7230 - j4{\cdot}1935}$

*Now continue the programme on frame 23.*

## 23

Now we are ready to go on to a very important section which follows from our work on multiplication of complex numbers in polar form.

We have already established that –

if $\quad z_1 = r_1(\cos\theta_1 + \text{j}\sin\theta_1)$ and $z_2 = r_2(\cos\theta_2 + \text{j}\sin\theta_2)$

then $\quad z_1 z_2 = r_1 r_2[\cos(\theta_1 + \theta_2) + \text{j}\sin(\theta_1 + \theta_2)]$

So if $\quad z_3 = r_3(\cos\theta_3 + \text{j}\sin\theta_3)$ then we have

$$z_1 z_2 z_3 = r_1 r_2[\cos(\theta_1 + \theta_2) + \text{j}\sin(\theta_1 + \theta_2)]\, r_3(\cos\theta_3 + \text{j}\sin\theta_3)$$
$$= \text{........................}$$

## 24

$$z_1 z_2 z_3 = r_1 r_2 r_3[\cos(\theta_1 + \theta_2 + \theta_3) + \text{j}\sin(\theta_1 + \theta_2 + \theta_3)]$$

for in multiplication, we multiply the moduli and add the arguments.

□□□□□□□□□□□□□□□□□□□□□□□□□□□□□□□□□□□□□□

Now suppose that $z_1$, $z_2$, $z_3$ are all alike and that each is equal to $z = r(\cos\theta + \text{j}\sin\theta)$. Then the result above becomes

$$z_1 z_2 z_3 = z^3 = r.r.r[\cos(\theta + \theta + \theta) + \text{j}\sin(\theta + \theta + \theta)]$$
$$= r^3(\cos 3\theta + \text{j}\sin 3\theta).$$

or
$$z^3 = [r(\cos\theta + \text{j}\sin\theta)]^3 = r^3(\cos\theta + \text{j}\sin\theta)^3$$
$$= r^3(\cos 3\theta + \text{j}\sin 3\theta).$$

That is: If we wish to cube a complex number in polar form, we just cube the modulus ($r$ value) and multiply the argument ($\theta$) by 3.

Similarly, to square a complex number in polar form, we square the modulus ($r$ value) and multiply the argument ($\theta$) by ......................

**25**

| 2 | i.e. $[r(\cos\theta + \mathrm{j}\sin\theta)]^2 = r^2(\cos 2\theta + \mathrm{j}\sin 2\theta)$ |
|---|---|

□□□□□□□□□□□□□□□□□□□□□□□□□□□□□□□□□□□□□□

Let us take another look at these results.

$$[r(\cos\theta + \mathrm{j}\sin\theta)]^2 = r^2(\cos 2\theta + \mathrm{j}\sin 2\theta)$$
$$[r(\cos\theta + \mathrm{j}\sin\theta)]^3 = r^3(\cos 3\theta + \mathrm{j}\sin 3\theta)$$

Similarly,

$$[r(\cos\theta + \mathrm{j}\sin\theta)]^4 = r^4(\cos 4\theta + \mathrm{j}\sin 4\theta)$$
$$[r(\cos\theta + \mathrm{j}\sin\theta)]^5 = r^5(\cos 5\theta + \mathrm{j}\sin 5\theta)$$

and so on.

In general, then, we can say

$$[r(\cos\theta + \mathrm{j}\sin\theta)]^n = \ldots\ldots\ldots\ldots\ldots\ldots$$

**26**

$$[r(\cos\theta + \mathrm{j}\sin\theta)]^n = \boxed{r^n(\cos n\theta + \mathrm{j}\sin n\theta)}$$

□□□□□□□□□□□□□□□□□□□□□□□□□□□□□□□□□□□□□□

This general result is very important and is called *DeMoivre's Theorem.* It says that to raise a complex number in polar form to any power $n$, we raise the $r$ to the power $n$ and multiply the angle by $n$.

e.g. $[4(\cos 50° + \mathrm{j}\sin 50°]^2 = 4^2[\cos(2 \times 50°) + \mathrm{j}\sin(2 \times 50°)]$
$= 16(\cos 100° + \mathrm{j}\sin 100°)$

and $[3(\cos 110° + \mathrm{j}\sin 110°)]^3 = 27(\cos 330° + \mathrm{j}\sin 330°)$

and in the same way,

$[2(\cos 37° + \mathrm{j}\sin 37°)]^4 = \ldots\ldots\ldots\ldots\ldots\ldots\ldots\ldots$

**27**

$$16(\cos 148^\circ + j \sin 148^\circ)$$

□□□□□□□□□□□□□□□□□□□□□□□□□□□□□□□□□□□□□□

This is where the polar form really comes into its own! For DeMoivre's theorem also applies when we are raising the complex number to a fractional power, i.e. when we are finding the roots of a complex number.

e.g. To find the square root of $z = 4(\cos 70^\circ + j \sin 70^\circ)$.

We have $\sqrt{z} = z^{\frac{1}{2}} = [4(\cos 70^\circ + j \sin 70^\circ)]^{\frac{1}{2}}$ i.e. $n = \frac{1}{2}$

$$= 4^{\frac{1}{2}}(\cos \frac{70^\circ}{2} + j \sin \frac{70^\circ}{2})$$

$$= \underline{2(\cos 35^\circ + j \sin 35^\circ)}$$

It works every time, no matter whether the power is positive, negative, whole number or fraction. In fact, DeMoivre's theorem is so important, let us write it down again. Here goes –

If $z = r(\cos \theta + j \sin \theta)$, then $z^n =$ ......................

**28**

$z = r(\cos \theta + j \sin \theta)$, then $\boxed{z^n = r^n(\cos n\theta + j \sin n\theta)}$

for any value of $n$.

□□□□□□□□□□□□□□□□□□□□□□□□□□□□□□□□□□□□□□

Look again at finding a root of a complex number. Let us find the cube root of $z = 8(\cos 120^\circ + j \sin 120^\circ)$.

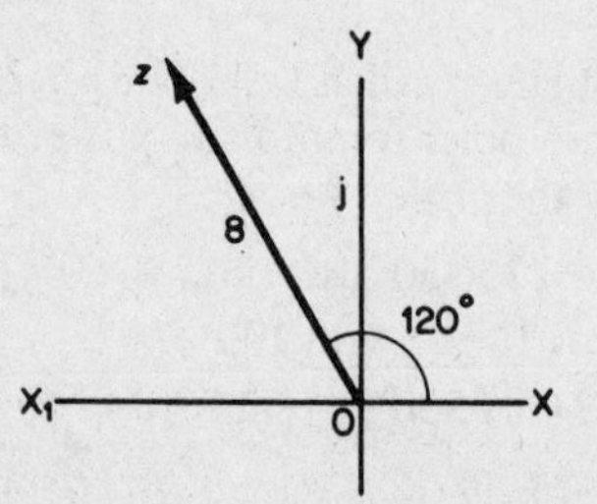


Here is the given complex number shown on an Argand diagram.

$z = 8 \underline{|120^\circ}$

Of course, we could say that $\theta$ was '1 revolution + 120°': the vector would still be in the same position, or, for that matter, (2 revs. + 120°), (3 revs. + 120°), etc.

i.e. $z = 8 \underline{|120^\circ}$, or $8 \underline{|480^\circ}$, or $8 \underline{|840^\circ}$, or $8 \underline{|1200^\circ}$, etc. and if we now apply DeMoivre's theorem to each of these, we get

$z^{\frac{1}{3}} = 8^{\frac{1}{3}} \underline{\left|\frac{120^\circ}{3}\right.}$ or $8^{\frac{1}{3}} \underline{\left|\frac{480^\circ}{3}\right.}$ or ............ or ............ etc.

**29**

$$z^{\frac{1}{3}} = 8^{\frac{1}{3}} \angle \frac{120^\circ}{3} \text{ or } 8^{\frac{1}{3}} \angle \frac{480^\circ}{3} \text{ or } 8^{\frac{1}{3}} \angle \frac{840^\circ}{3} \text{ or } 8^{\frac{1}{3}} \angle \frac{1200^\circ}{3}$$

If we simplify these, we get

$z^{\frac{1}{3}} = 2 \angle 40^\circ$ or $2 \angle 160^\circ$ or $2 \angle 280^\circ$ or $2 \angle 400^\circ$ etc.

If we put each of these on an Argand diagram, as follows,

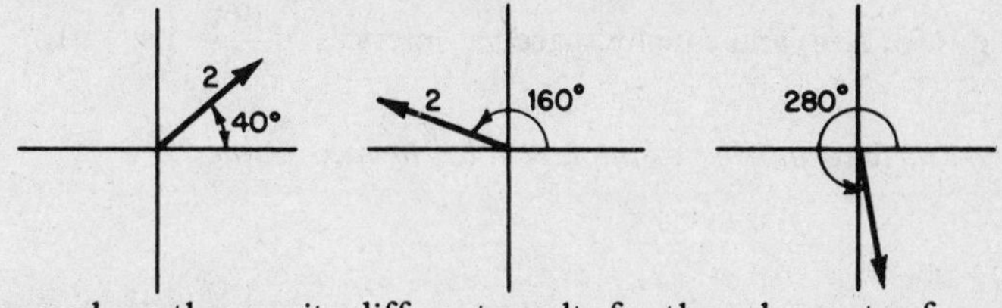


we see we have three quite different results for the cube roots of $z$ and also that the fourth diagram is a repetition of the first. Any subsequent calculations merely repeat these three positions.

*Make a sketch of the first three vectors on a single Argand diagram.*

**30**

Here they are: The cube roots of $z = 8(\cos 120^\circ + j \sin 120^\circ)$.

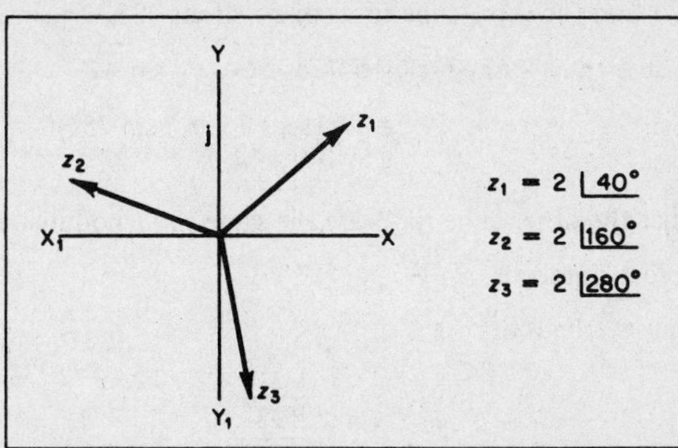


We see, therefore, that there are 3 cube roots of a complex number. Also, if you consider the angles, you see that the 3 roots are equally spaced round the diagram, any two adjacent vectors being separated by .................... degrees.

**31**

120°

That is right. Therefore all we need to do in practice is to find the first of the roots and simply add 120° on to get the next – and so on.

Notice that the three cube roots of a complex number are equal in modulus (or size) and equally spaced at intervals of $\frac{360^\circ}{3}$ i.e. 120°.

*Now let us take another example. On to the next frame.*

**32**

*Example.* To find the three cube roots of $z = 5(\cos 225^\circ + \text{j} \sin 225^\circ)$

The first root is given by $z_1 = z^{\frac{1}{3}} = 5^{\frac{1}{3}}(\cos \frac{225^\circ}{3} + \text{j} \sin \frac{225^\circ}{3})$

$$= 1{\cdot}71(\cos 75^\circ + \text{j} \sin 75^\circ)$$

$$z_1 = 1{\cdot}71 \underline{|75^\circ}$$

We know that the other cube roots are the same size (modulus), i.e. 1·71, and separated at intervals of $\frac{360^\circ}{3}$, i.e. 120°.

So the three cube roots are:

$$z_1 = 1{\cdot}71 \underline{|75^\circ}$$
$$z_2 = 1{\cdot}71 \underline{|195^\circ}$$
$$z_3 = 1{\cdot}71 \underline{|315^\circ}$$

It helps to see them on an Argand diagram, so sketch them on a combined diagram.

**33**

Here they are:

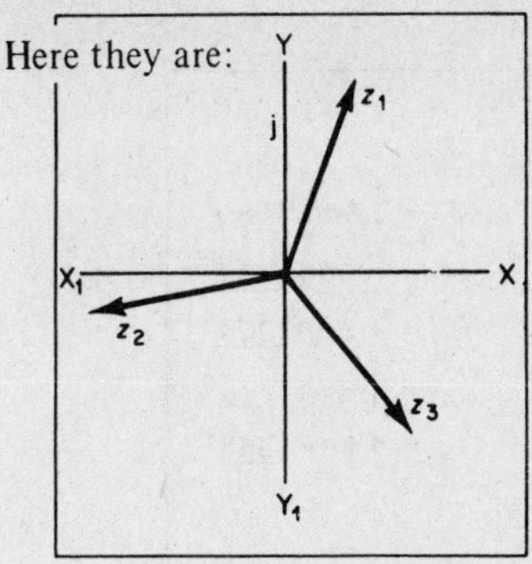


We find any roots of a complex number in the same way.

(i) Apply DeMoivre's theorem to find the first of the $n$ roots.

(ii) The other roots will then be distributed round the diagram at regular intervals of $\frac{360°}{n}$

A complex number, therefore, has

2 square roots, separated by $\frac{360°}{2}$ i.e. 180°

3 cube roots, " " $\frac{360°}{3}$ i.e. 120°

4 fourth roots, " " $\frac{360°}{4}$ i.e. 90°

5 fifth roots, " " .................... etc.

**34**

There would be 5 fifth roots separated by $\boxed{\frac{360°}{5} \text{ i.e. } 72°}$

□□□□□□□□□□□□□□□□□□□□□□□□□□□□□□□□□□□□□□□□□□

And now: To find the 5 fifth roots of $12\,\underline{|300°}$

$z = 12\,\underline{|300°} \quad \therefore \; z_1 = 12^{\frac{1}{5}}\,\underline{\Big|\frac{300°}{5}} = 12^{\frac{1}{5}}\,\underline{|60°}$

We now have to find the value of $12^{\frac{1}{5}}$. Do it by logs.

$$\left[\begin{array}{l} \text{Let } A = 12^{\frac{1}{5}}\text{. Then } \log A = \frac{1}{5}\log 12 = \frac{1}{5}(1{\cdot}0792) = 0{\cdot}2158 \\ \quad \text{Taking antilogs,} \quad A = 1{\cdot}644 \end{array}\right]$$

The first of the 5 fifth roots is therefore, $z_1 = 1{\cdot}644\,\underline{|60°}$

The others will be of the same magnitude, i.e. 1·644, and equally separated at intervals of $\frac{360°}{5}$ i.e. 72°

So the required 5 fifth roots of $12\,\underline{|300°}$ are

$$z_1 = 1{\cdot}644\,\underline{|60°}, \quad z_2 = 1{\cdot}644\,\underline{|132°}, \quad z_3 = 1{\cdot}644\,\underline{|204°}$$
$$z_4 = 1{\cdot}644\,\underline{|276°}, \quad z_5 = 1{\cdot}644\,\underline{|348°}$$

Sketch them on an Argand diagram, as before.

**35**

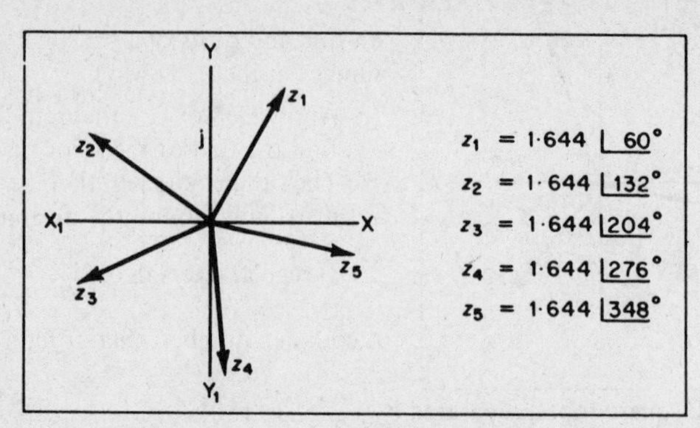


*Principal root.* Although there are 5 fifth roots of a complex number, we are sometimes asked to find the *principal* root. This is always the root whose vector is nearest to the positive OX axis.

In some cases, it may be the first root. In others, it may be the last root. The only test is to see which root is nearest to the positive OX axis.

In the example above, the *principal* root is therefore ..........................

**36**

Principal root $z_5 = 1{\cdot}644 \underline{|348^\circ}$

□□□□□□□□□□□□□□□□□□□□□□□□□□□□□□□□□□□□□□□□□

Good. Now here is another example worked in detail. Follow it.

We have to find the 4 fourth roots of $z = 7(\cos 80^\circ + \text{j} \sin 80^\circ)$

The first root, $z_1 = 7^{\frac{1}{4}} \underline{\left| \frac{80^\circ}{4} \right.} = 7^{\frac{1}{4}} \underline{|20^\circ}$

[ Now find $7^{\frac{1}{4}}$ by logs. Let $A = 7^{\frac{1}{4}}$
Then $\log A = \frac{1}{4} \log 7 = \frac{1}{4}(0{\cdot}8451) = 0{\cdot}2113$ and $A = 1{\cdot}627$ ]

$$z_1 = 1{\cdot}627 \underline{|20^\circ}$$

The other roots will be separated by intervals of $\frac{360^\circ}{4} = 90^\circ$

Therefore the four fourth roots are –

$$z_1 = 1{\cdot}627 \underline{|20^\circ} \qquad z_2 = 1{\cdot}627 \underline{|110^\circ}$$
$$z_3 = 1{\cdot}627 \underline{|200^\circ} \qquad z_4 = 1{\cdot}627 \underline{|290^\circ}$$

And once again, draw an Argand diagram to illustrate these roots.

37

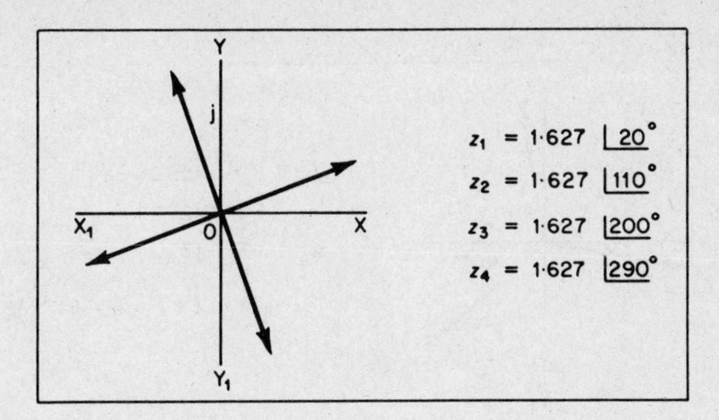


□□□□□□□□□□□□□□□□□□□□□□□□□□□□□□□□□□□□□□

And in this example, the principal fourth root is ......................

38

Principal root: $z_1 = 1{\cdot}627 \, \underline{|20^\circ}$

since it is the root nearest to the positive OX axis.

□□□□□□□□□□□□□□□□□□□□□□□□□□□□□□□□□□□□□□

Now you can do one entirely on your own. Here it is.

Find the three cube roots of $6(\cos 240^\circ + j \sin 240^\circ)$. Represent them on an Argand diagram and indicate which is the principal cube root.

*When you have finished it, turn on to frame 39 and check your results.*

**39** *Result:*

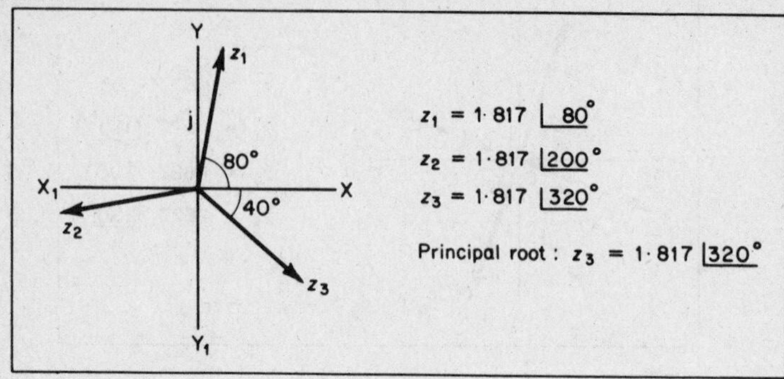


□□□□□□□□□□□□□□□□□□□□□□□□□□□□□□□□□□□□□□

Here is the working.

$$z = 6\,\underline{|240°} \qquad z_1 = 6^{\frac{1}{3}}\,\underline{\left|\frac{240°}{3}\right.} = 1{\cdot}817\,\underline{|80°}$$

$$\text{Interval between roots} = \frac{360°}{3} = 120°$$

Therefore the roots are:

$$z_1 = 1{\cdot}817\,\underline{|80°} \qquad z_2 = 1{\cdot}817\,\underline{|200°} \qquad z_3 = 1{\cdot}817\,\underline{|320°}$$

The principal root is the root nearest to the positive OX axis. In this case, then, the principal root is $z_3 = 1{\cdot}817\,\underline{|320°}$

*On to the next frame.*

---

**40** *Expansion of* $\sin n\theta$ *and* $\cos n\theta$, where $n$ is a positive integer.

By DeMoivre's theorem, we know that

$$\cos n\theta + \mathrm{j}\sin n\theta = (\cos\theta + \mathrm{j}\sin\theta)^n$$

The method is simply to expand the right-hand side as a binomial series, after which we can equate real and imaginary parts.

An example will soon show you how it is done:

*Ex. 1.* To find expansions for $\cos 3\theta$ and $\sin 3\theta$.

We have

$$\cos 3\theta + \mathrm{j}\sin 3\theta = (\cos\theta + \mathrm{j}\sin\theta)^3$$
$$= (\mathrm{c} + \mathrm{js})^3 \qquad \text{where } \mathrm{c} \equiv \cos\theta,\ \mathrm{s} \equiv \sin\theta$$

Now expand this by the binomial series – like $(a + b)^3$ so that

$$\cos 3\theta + \mathrm{j}\sin 3\theta = \ldots\ldots\ldots\ldots\ldots\ldots$$

**41**

$$c^3 + j3c^2 s - 3cs^2 - js^3$$

for: $\cos 3\theta + j \sin 3\theta = c^3 + 3c^2(js) + 3c(js)^2 + (js)^3$

$= c^3 + j3c^2 s - 3cs^2 - js^3$ since $j^2 = -1$

$= (c^3 - 3cs^2) + j(3c^2 s - s^3)$ $\quad j^3 = -j$

Now, equating real parts and imaginary parts, we get

$\cos 3\theta =$ ..................................

and $\sin 3\theta =$ ..................................

**42**

$$\cos 3\theta = \cos^3\theta - 3\cos\theta\sin^2\theta$$
$$\sin 3\theta = 3\cos^2\theta\sin\theta - \sin^3\theta$$

If we wish, we can replace $\sin^2\theta$ by $(1 - \cos^2\theta)$
and $\cos^2\theta$ by $(1 - \sin^2\theta)$
so that we could write the results above as

$\cos 3\theta =$ ........................ (all in terms of $\cos\theta$)

$\sin 3\theta =$ ........................ (all in terms of $\sin\theta$)

**43**

$$\cos 3\theta = 4\cos^3\theta - 3\cos\theta$$
$$\sin 3\theta = 3\sin\theta - 4\sin^3\theta$$

since $\cos 3\theta = \cos^3\theta - 3\cos\theta\,(1 - \cos^2\theta)$

$= \cos^3\theta - 3\cos\theta + 3\cos^3\theta$

$= \underline{4\cos^3\theta - 3\cos\theta}$

and $\sin 3\theta = 3(1 - \sin^2\theta)\sin\theta - \sin^3\theta$

$= 3\sin\theta - 3\sin^3\theta - \sin^3\theta$

$= \underline{3\sin\theta - 4\sin^3\theta}$

While these results are useful, it is really the method that counts.
So now do this one in just the same way:

*Ex. 2.* Obtain an expansion for $\cos 4\theta$ in terms of $\cos\theta$.

*When you have finished, check your result with the next frame.*

**44**

$$\boxed{\cos 4\theta = 8\cos^4\theta - 8\cos^2\theta + 1}$$

Working: $\cos 4\theta + j\sin 4\theta = (\cos\theta + j\sin\theta)^4$

$$= (c + js)^4$$
$$= c^4 + 4c^3(js) + 6c^2(js)^2 + 4c(js)^3 + (js)^4$$
$$= c^4 + j4c^3 s - 6c^2 s^2 - j4cs^3 + s^4$$
$$= (c^4 - 6c^2 s^2 + s^4) + j(4c^3 s - 4cs^3)$$

Equating real parts:

$$\cos 4\theta = c^4 - 6c^2 s^2 + s^4$$
$$= c^4 - 6c^2(1 - c^2) + (1 - c^2)^2$$
$$= c^4 - 6c^2 + 6c^4 + 1 - 2c^2 + c^4$$
$$= 8c^4 - 8c^2 + 1$$
$$= \underline{8\cos^4\theta - 8\cos^2\theta + 1}$$

Now for a different problem. *On to the next frame.*

**45**

*Expansions for $\cos^n\theta$ and $\sin^n\theta$ in terms of sines and cosines of miltiples of $\theta$.*

Let $z = \cos\theta + j\sin\theta$

then $\dfrac{1}{z} = z^{-1} = \cos\theta - j\sin\theta$

$$\therefore\ z + \frac{1}{z} = 2\cos\theta \text{ and } z - \frac{1}{z} = j\,2\sin\theta$$

Also, by DeMoivre's theorem,

$$z^n = \cos n\theta + j\sin n\theta$$

and $\dfrac{1}{z^n} = z^{-n} = \cos n\theta - j\sin n\theta$

$$\therefore\ z^n + \frac{1}{z^n} = 2\cos n\theta \text{ and } z^n - \frac{1}{z^n} = j\,2\sin n\theta$$

Let us collect these four results together: $z = \cos\theta + j\sin\theta$

| | |
|---|---|
| $z + \frac{1}{z} = 2\cos\theta$ | $z - \frac{1}{z} = j\,2\sin\theta$ |
| $z^n + \frac{1}{z^n} = 2\cos n\theta$ | $z^n - \frac{1}{z^n} = j\,2\sin n\theta$ |

*Make a note of these results in your record book. Then turn on and we will see how we use them.*

**46**

*Ex. 1.* To expand $\cos^3\theta$

From our results, $$z + \frac{1}{z} = 2\cos\theta$$

$$\therefore\ (2\cos\theta)^3 = (z + \frac{1}{z})^3$$
$$= z^3 + 3z^2(\frac{1}{z}) + 3z(\frac{1}{z^2}) + \frac{1}{z^3}$$
$$= z^3 + 3z + 3\frac{1}{z} + \frac{1}{z^3}$$

Now here is the trick: we re-write this, collecting the terms up in pairs from the two extreme ends, thus –

$$(2\cos\theta)^3 = (z^3 + \frac{1}{z^3}) + 3(z + \frac{1}{z})$$

And, from the four results that we noted,

$$z + \frac{1}{z} = \ldots\ldots\ldots\ldots$$

and $$z^3 + \frac{1}{z^3} = \ldots\ldots\ldots\ldots$$

---

**47**

$$\boxed{z + \frac{1}{z} = 2\cos\theta;\ z^3 + \frac{1}{z^3} = 2\cos 3\theta}$$

$$\therefore\ (2\cos\theta)^3 = 2\cos 3\theta + 3.2\cos\theta$$
$$8\cos^3\theta = 2\cos 3\theta + 6\cos\theta$$
$$4\cos^3\theta = \cos 3\theta + 3\cos\theta$$
$$\underline{\cos^3\theta = \frac{1}{4}(\cos 3\theta + 3\cos\theta)}$$

Now one for you:

*Ex. 2.* Find an expansion for $\sin^4\theta$

Work in the same way, but, this time, remember that

$$z - \frac{1}{z} = \mathrm{j}\,2\sin\theta \text{ and } z^n - \frac{1}{z^n} = \mathrm{j}\,2\sin n\theta$$

*When you have obtained a result, check it with the next frame.*

**48**

$$\sin^4\theta = \frac{1}{8}\,[\cos 4\theta - 4\cos 2\theta + 3]$$

for, we have:

$$z - \frac{1}{z} = \text{j}\,2\sin\theta;\; z^n - \frac{1}{z^n} = \text{j}\,2\sin n\theta$$

$$\therefore\ (\text{j}\,2\sin\theta)^4 = \left(z - \frac{1}{z}\right)^4$$

$$= z^4 - 4z^3\left(\frac{1}{z}\right) + 6z^2\left(\frac{1}{z^2}\right) - 4z\left(\frac{1}{z^3}\right) + \frac{1}{z^4}$$

$$= \left(z^4 + \frac{1}{z^4}\right) - 4\left(z^2 + \frac{1}{z^2}\right) + 6$$

Now $\quad z^n + \dfrac{1}{z^n} = 2\cos n\theta$

$$\therefore\ 16\sin^4\theta = 2\cos 4\theta - 4.2\cos 2\theta + 6$$

$$\therefore\ \sin^4\theta = \frac{1}{8}[\cos 4\theta - 4\cos 2\theta + 3]$$

They are all done the same way: once you know the trick, the rest is easy.

*Now let us move on to something new.*

**49**

**Loci Problems**

We are sometimes required to find the locus of a point which moves in the Argand diagram according to some stated condition. Before we work through one or two examples of this kind, let us just revise a couple of useful points.

You will remember that when we were representing a complex number in polar form, i.e., $z = a + \text{j}b = r(\cos\theta + \text{j}\sin\theta)$, we said that
(i) $r$ is called the modulus of $z$ and is written 'mod $z$' or $|z|$ and
(ii) $\theta$ " " " argument of $z$ " " " 'arg $z$'

Also, $r = \sqrt{(a^2 + b^2)}$ and $\theta = \tan^{-1}\left\{\dfrac{b}{a}\right\}$

so that $|z| = \sqrt{(a^2 + b^2)}$ and $\arg z = \tan^{-1}\left\{\dfrac{b}{a}\right\}$

Similarly, if $z = x + \text{j}y$, then $|z| =$ ......................
and $\arg z =$ ....................

**50**

If $z = x + jy$, $\boxed{|z| = \sqrt{(x^2 + y^2)} \text{ and } \arg z = \tan^{-1}\left\{\frac{y}{x}\right\}}$

Keep those in mind and we are now ready to tackle some examples.

*Ex. 1.* If $z = x + jy$, find the locus defined as $|z| = 5$.

Now we know that in this case, $|z| = \sqrt{(x^2 + y^2)}$

The locus is defined as $\sqrt{(x^2 + y^2)} = 5$

$$\therefore x^2 + y^2 = 25$$

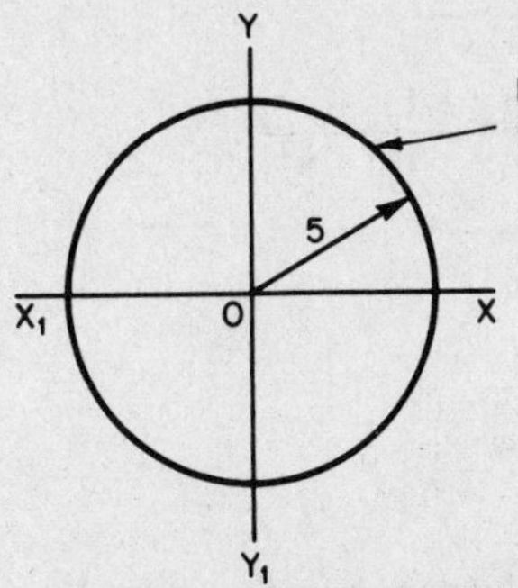


This is a circle, with centre at the origin and with radius 5.

*That was easy enough. Turn on for Example 2.*

**51**

*Ex. 2.* If $z = x + jy$, find the locus defined as $\arg z = \frac{\pi}{4}$

In this case, $\arg z = \tan^{-1}\left\{\frac{y}{x}\right\} \therefore \tan^{-1}\left\{\frac{y}{x}\right\} = \frac{\pi}{4}$

$$\therefore \frac{y}{x} = \tan\frac{\pi}{4} = \tan 45° = 1 \quad \therefore \frac{y}{x} = 1 \quad \therefore y = x$$

So the locus $\arg z = \frac{\pi}{4}$ is therefore the straight line $y = x$ for $y > 0$

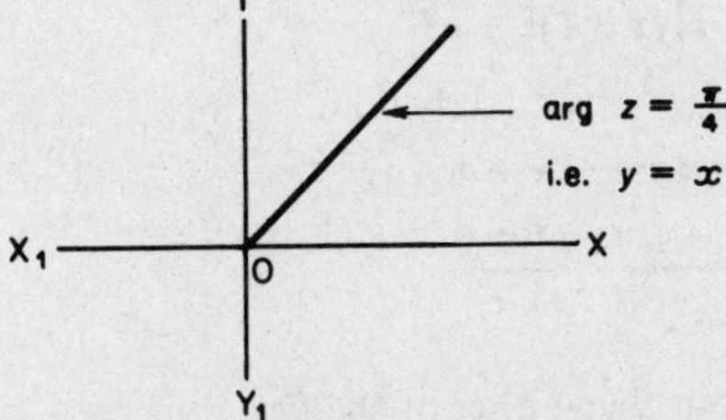


All locus problems at this stage are fundamentally of one of these kinds. Of course, the given condition may look a trifle more involved, but the approach is always the same.

*Let us look at a more complicated one. Next frame.*

**52** *Ex. 3.* If $z = x + jy$, find the equation of the locus $\left|\frac{z+1}{z-1}\right| = 2$.

Since $z = x + jy$,

$$z + 1 = x + jy + 1 = (x + 1) + jy \quad = r_1 \angle\theta_1 \quad = z_1$$
$$z - 1 = x + jy - 1 = (x - 1) + jy \quad = r_2 \angle\theta_2 \quad = z_2$$

$$\therefore \frac{z+1}{z-1} = \frac{r_1 \angle\theta_1}{r_2 \angle\theta_2} = \frac{r_1}{r_2} \angle\theta_1 - \theta_2$$

$$\therefore \left|\frac{z+1}{z-1}\right| = \frac{r_1}{r_2} = \frac{|z_1|}{|z_2|} = \frac{\sqrt{[(x+1)^2 + y^2]}}{\sqrt{[(x-1)^2 + y^2]}}$$

$$\therefore \frac{\sqrt{[(x+1)^2 + y^2]}}{\sqrt{[(x-1)^2 + y^2]}} = 2$$

$$\therefore \frac{(x+1)^2 + y^2}{(x-1)^2 + y^2} = 4$$

All that now remains is to multiply across by the denominator and tidy up the result. So finish it off in its simplest form.

**53**

We had $$\frac{(x+1)^2 + y^2}{(x-1)^2 + y^2} = 4$$

So therefore
$$(x+1)^2 + y^2 = 4\{(x-1)^2 + y^2\}$$
$$x^2 + 2x + 1 + y^2 = 4(x^2 - 2x + 1 + y^2)$$
$$= 4x^2 - 8x + 4 + 4y^2$$
$$\therefore \underline{3x^2 - 10x + 3 + 3y^2 = 0}$$

This is the equation of the given locus.

Although this takes longer to write out than either of the first two examples, the basic principle is the same. The given condition must be a function of either the modulus or the argument.

*Move on now to frame 54 for Example 4.*

**54**

*Ex. 4.* If $z = x + jy$, find the equation of the locus $\arg(z^2) = -\frac{\pi}{4}$.

$$z = x + jy = r\lfloor\theta \quad \therefore \arg z = \theta = \tan^{-1}\left\{\frac{y}{x}\right\}$$

$$\therefore \tan\theta = \frac{y}{x}$$

$\therefore$ By DeMoivre's theorem, $z^2 = r^2 \lfloor 2\theta$

$$\therefore \arg(z^2) = 2\theta = -\frac{\pi}{4}$$

$$\therefore \tan 2\theta = \tan(-\frac{\pi}{4}) = -1$$

$$\therefore \frac{2\tan\theta}{1-\tan^2\theta} = -1$$

$$\therefore 2\tan\theta = \tan^2\theta - 1$$

But $\quad \tan\theta = \frac{y}{x} \qquad \therefore \frac{2y}{x} = \frac{y^2}{x^2} - 1$

$$2xy = y^2 - x^2 \quad \therefore \underline{y^2 = x^2 + 2xy}$$

In that example, the given condition was a function of the argument.

Here is one for you to do:

If $z = x + jy$, find the equation of the locus $\arg(z + 1) = \frac{\pi}{3}$

*Do it carefully; then check with the next frame.*

---

**55**

Here is the solution set out in detail.

If $z = x + jy$, find the locus $\arg(z + 1) = \frac{\pi}{3}$.

$$z = x + jy \quad \therefore z + 1 = x + jy + 1 = (x + 1) + jy$$

$$\arg(z + 1) = \tan^{-1}\left\{\frac{y}{x+1}\right\} = \frac{\pi}{3}$$

$$\therefore \frac{y}{x+1} = \tan\frac{\pi}{3} = \sqrt{3}$$

$$\underline{y = \sqrt{3}(x + 1) \text{ for } y > 0}$$

And that is all there is to that.

Now do this one. You will have no trouble with it.

If $z = x + jy$, find the equation of the locus $|z - 1| = 5$

*When you have finished it, turn on to frame 56.*

**56** Here it is: $z = x + \text{j}y$; given locus $|z - 1| = 5$

$$z - 1 = x + \text{j}y - 1 = (x - 1) + \text{j}y$$

$$\therefore |z - 1| = \sqrt{[(x - 1)^2 + y^2]} = 5$$

$$\therefore (x - 1)^2 + y^2 = 25$$

$$\therefore x^2 - 2x + 1 + y^2 = 25$$

$$\therefore \underline{x^2 - 2x + y^2 = 24}$$

Every one is very much the same.

This brings us to the end of this programme, except for the final test exercise. Before you work through it, read down the Revision Sheet (frame 57), just to refresh your memory of what we have covered in this programme.

*So on now to frame 57.*

**57**

**Revision Sheet**

1. *Polar form of a complex number*

$$z = a + \mathrm{j}b = r(\cos\theta + \mathrm{j}\sin\theta) = r\,\underline{|\theta}$$

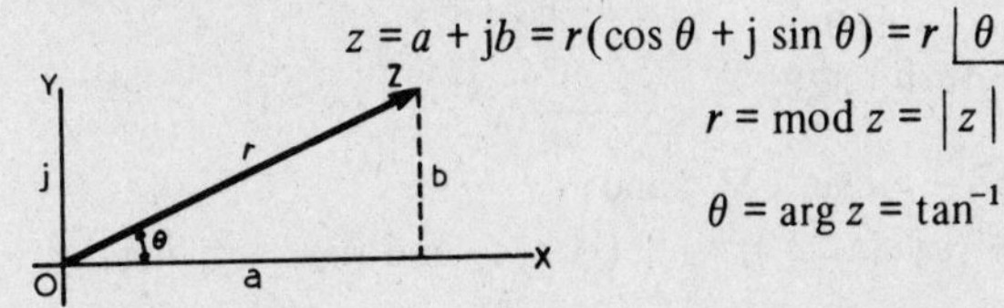


$$r = \text{mod}\ z = |z| = \sqrt{a^2 + b^2}$$

$$\theta = \arg z = \tan^{-1}\left\{\frac{b}{a}\right\}$$

2. *Negative angles*

$$z = r(\cos[-\theta] + \mathrm{j}\sin[-\theta])$$

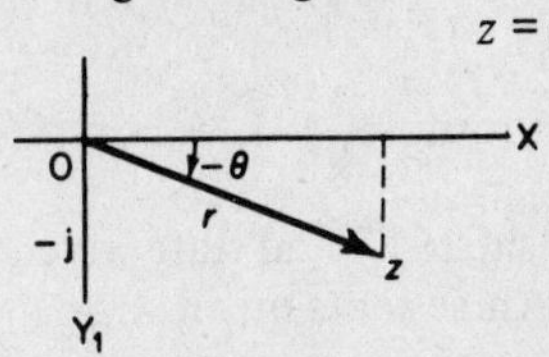


$$\cos[-\theta] = \cos\theta$$

$$\sin[-\theta] = -\sin\theta$$

$$\therefore\ z = r(\cos\theta - \mathrm{j}\sin\theta) = r\,\overline{|\theta}$$

3. *Multiplication and division in polar form*

If $\quad z_1 = r_1\,\underline{|\theta_1}; \quad z_2 = r_2\,\underline{|\theta_2}$

then $\quad z_1 z_2 = r_1 r_2\,\underline{|\theta_1 + \theta_2}$

$$\frac{z_1}{z_2} = \frac{r_1}{r_2}\,\underline{|\theta_1 - \theta_2}$$

4. *DeMoivre's theorem*

If $\quad z = r(\cos\theta + \mathrm{j}\sin\theta)$, then $z^n = r^n(\cos n\theta + \mathrm{j}\sin n\theta)$

5. *Exponential form of a complex number*

$z = a + \mathrm{j}b$ . . . . . . . . . . . . . standard form

$= r(\cos\theta + \mathrm{j}\sin\theta)$ . . . . . polar form

$= r\,\mathrm{e}^{\mathrm{j}\theta}$ [$\theta$ in radians] . . . . exponential form

Also

$$\mathrm{e}^{\mathrm{j}\theta} = \cos\theta + \mathrm{j}\sin\theta$$

$$\mathrm{e}^{-\mathrm{j}\theta} = \cos\theta - \mathrm{j}\sin\theta$$

6. *Logarithm of a complex number*

$$z = r\,\mathrm{e}^{\mathrm{j}\theta} \quad \therefore\ \ln z = \ln r + \mathrm{j}\theta$$

7. *Loci problems*

If $\quad z = x + \mathrm{j}y, \quad |z| = \sqrt{(x^2 + y^2)}$

$$\arg z = \tan^{-1}\left\{\frac{y}{x}\right\}$$

*That's it! Now you are ready for the Test Exercise on Frame 58.*

## 58 Test Exercise–II

1. Express in polar form, $z = -5 - j3$.

2. Express in the form $a + jb$, (i) $2\,\underline{|156^\circ}$, (ii) $5\,\overline{|37^\circ}$.

3. If $z_1 = 12(\cos 125^\circ + j \sin 125^\circ)$ and
   $z_2 = 3(\cos 72^\circ + j \sin 72^\circ)$, find (i) $z_1 z_2$ and (ii) $\dfrac{z_1}{z_2}$ giving the results in polar form.

4. If $z = 2(\cos 25^\circ + j \sin 25^\circ)$, find $z^3$ in polar form.

5. Find the three cube roots of $8(\cos 264^\circ + j \sin 264^\circ)$ and state which of them is the principal cube root. Show all three roots on an Argand diagram.

6. Expand $\sin 4\theta$ in powers of $\sin\theta$ and $\cos\theta$.

7. Express $\cos^4\theta$ in terms of cosines of multiples of $\theta$.

8. If $z = x + jy$, find the equations of the two loci defined by

   (i) $|z - 4| = 3$ (ii) $\arg(z + 2) = \dfrac{\pi}{6}$

**Further Problems–II**

1. If $z = x + jy$, where $x$ and $y$ are real, find the values of $x$ and $y$ when

$$\frac{3z}{1-j} + \frac{3z}{j} = \frac{4}{3-j}$$

2. In the Argand diagram, the origin is the centre of an equilateral triangle and one vertex of the triangle is the point $3 + j\sqrt{3}$. Find the complex numbers representing the other vertices.

3. Express $2 + j3$ and $1 - j2$ in polar form and apply DeMoivre's theorem to evaluate $\dfrac{(2 + j3)^4}{1 - j2}$. Express the result in the form $a + jb$ and in exponential form.

4. Find the fifth roots of $-3 + j3$ in polar form and in exponential form.

5. Express $5 + j12$ in polar form and hence evaluate the principal value of $\sqrt[3]{(5 + j12)}$, giving the results in the form $a + jb$ and in form $r\,e^{j\theta}$.

6. Determine the fourth roots of $-16$, giving the results in the form $a + jb$.

7. Find the fifth roots of $-1$, giving the results in polar form. Express the principal root in the form $r\,e^{j\theta}$.

8. Determine the roots of the equation $x^3 + 64 = 0$ in the form $a + jb$, where $a$ and $b$ are real.

9. Determine the three cube roots of $\dfrac{2-j}{2+j}$ giving the results in modulus/argument form. Express the principal root in the form $a + jb$.

10. Show that the equation $z^3 = 1$ has one real root and two other roots which are not real, and that, if one of the non-real roots is denoted by $\omega$, the other is then $\omega^2$. Mark on the Argand diagram the points which represent the three roots and show that they are the vertices of an equilateral triangle.

11. Determine the fifth roots of $(2 - j5)$, giving the results in modulus/argument form. Express the principal root in the form $a + jb$ and in the form $r\,e^{j\theta}$.

12. Solve the equation $z^2 + 2(1 + j)z + 2 = 0$, giving each result in the form $a + jb$, with $a$ and $b$ correct to 2 places of decimals.

13. Express $e^{1 - j\pi/2}$ in the form $a + jb$.

14. Obtain the expansion of $\sin 7\theta$ in powers of $\sin\theta$.

15. Express $\sin^6 x$ as a series of terms which are cosines of angles that are multiples of $x$.

16. If $z = x + jy$, where $x$ and $y$ are real, show that the locus $\left|\dfrac{z-2}{z+2}\right| = 2$ is a circle and determine its centre and radius.

17. If $z = x + jy$, show that the locus $\arg\left\{\dfrac{z-1}{z-j}\right\} = \dfrac{\pi}{6}$ is a circle. Find its centre and radius.

18. If $z = x + jy$, determine the Cartesian equation of the locus of the point $z$ which moves in the Argand diagram so that

$$|z + j2|^2 + |z - j2|^2 = 40$$

19. If $z = x + jy$, determine the equations of the two loci:

    (i) $\left|\dfrac{z+2}{z}\right| = 3$ (ii) $\arg\left\{\dfrac{z+2}{z}\right\} = \dfrac{\pi}{4}$

20. If $z = x + jy$, determine the equations of the loci in the Argand diagram, defined by

    (i) $\left|\dfrac{z+2}{z-1}\right| = 2$, and (ii) $\arg\left\{\dfrac{z-1}{z+2}\right\} = \dfrac{\pi}{2}$

21. Prove that

    (i) if $|z_1 + z_2| = |z_1 - z_2|$, the difference of the arguments of $z_1$ and $z_2$ is $\dfrac{\pi}{2}$.

(ii) if $\arg\left\{\frac{z_1 + z_2}{z_1 - z_2}\right\} = \frac{\pi}{2}$, then $|z_1| = |z_2|$

22. If $z = x + \mathrm{j}y$, determine the loci in the Argand diagram, defined by

(i) $|z + \mathrm{j}2|^2 - |z - \mathrm{j}2|^2 = 24$

(ii) $|z + \mathrm{j}k|^2 + |z - \mathrm{j}k|^2 = 10k^2 \quad (k > 0)$

# Programme 3

## HYPERBOLIC FUNCTIONS

**Introduction**

When you were first introduced to trigonometry, it is almost certain that you defined the trig. ratios – sine, cosine and tangent – as ratios between the sides of a right-angled triangle. You were then able, with the help of trig. tables, to apply these new ideas from the start to solve simple right-angled triangle problems . . . . . and away you went.

You could, however, have started in quite a different way. If a circle of unit radius is drawn and various constructions made from an external point, the lengths of the lines so formed can be defined as the sine, cosine and tangent of one of the angles in the figure. In fact, trig. functions are sometimes referred to as 'circular functions'.

This would be a geometrical approach and would lead in due course to all the results we already know in trigonometry. But, in fact, you did *not* start that way, for it is more convenient to talk about right-angled triangles and simple practical applications.

Now if the same set of constructions is made with a hyperbola instead of a circle, the lengths of the lines now formed can similarly be called the hyperbolic sine, hyperbolic cosine and hyperbolic tangent of a particular angle in the figure, and, as we might expect, all these hyperbolic functions behave very much as trig. functions (or circular functions) do.

This parallel quality is an interesting fact and important, as you will see later for we shall certainly refer to it again. But, having made the point, we can say this: that just as the trig. ratios were *not* in practice defined geometrically from the circle, so the hyperbolic functions are *not* in practice defined geometrically from the hyperbola. In fact, the definitions we are going to use have apparently no connection with the hyperbola at all.

So now the scene is set. *Turn on to Frame 1 and start the programme.*

**1**

You may remember that of the many functions that can be expressed as a series of powers of $x$, a common one is $e^x$.

$$e^x = 1 + x + \frac{x^2}{2!} + \frac{x^3}{3!} + \frac{x^4}{4!} + \ldots$$

If we replace $x$ by $-x$, we get

$$e^{-x} = 1 - x + \frac{x^2}{2!} - \frac{x^3}{3!} + \frac{x^4}{4!} - \ldots$$

and these two functions $e^x$ and $e^{-x}$ are the foundations of the definitions we are going to use.

(i) If we take the value of $e^x$, subtract $e^{-x}$, and divide by 2, we form what is defined as the hyperbolic sine of $x$.

$$\frac{e^x - e^{-x}}{2} = \textit{hyperbolic sine of } x$$

This is a lot to write every time we wish to refer to it, so we shorten it to *sinh x*, the *h* indicating its connection with the hyperbola. We pronounce it 'shine $x$'.

$$\frac{e^x - e^{-x}}{2} = \sinh x$$

So, in the same way, $\dfrac{e^y - e^{-y}}{2}$ would be written as .......................

---

**2**

sinh $y$

In much the same way, we have two other definitions:

(ii) $\dfrac{e^x + e^{-x}}{2}$ = *hyperbolic cosine* of $x$

$= \cosh x$ [pronounced 'cosh $x$']

(iii) $\dfrac{e^x - e^{-x}}{e^x + e^{-x}}$ = *hyperbolic tangent* of $x$

$= \tanh x$ [pronounced 'than $x$']

We must start off by learning these definitions, for all the subsequent developments depend on them.

So now then; what was the definition of sinh $x$?

$\sinh x =$ ......................

**3**

$$\sinh x = \frac{e^x - e^{-x}}{2}$$

Here they are together so that you can compare them.

$$\sinh x = \frac{e^x - e^{-x}}{2}$$

$$\cosh x = \frac{e^x + e^{-x}}{2}$$

$$\tanh x = \frac{e^x - e^{-x}}{e^x + e^{-x}}$$

Make a copy of these in your record book for future reference when necessary.

**4**

$$\sinh x = \frac{e^x - e^{-x}}{2}\,;\ \cosh x = \frac{e^x + e^{-x}}{2};\ \tanh x = \frac{e^x - e^{-x}}{e^x + e^{-x}}$$

We started the programme by referring to $e^x$ and $e^{-x}$ as series of powers of $x$. It should not be difficult therefore to find series at least for $\sinh x$ and for $\cosh x$. Let us try.

(i) *Series for sinh x*

$$e^x = 1 + x + \frac{x^2}{2!} + \frac{x^3}{3!} + \frac{x^4}{4!} + \ \ldots$$

$$e^{-x} = 1 - x + \frac{x^2}{2!} - \frac{x^3}{3!} + \frac{x^4}{4!} - \ \ldots$$

If we subtract, we get

$$e^x - e^{-x} = 2x + \frac{2x^3}{3!} + \frac{2x^5}{5!} \ldots$$

Divide by 2

$$\frac{e^x - e^{-x}}{2} = \underline{\sinh x = x + \frac{x^3}{3!} + \frac{x^5}{5!} +} \ \ldots$$

(ii) If we add the series for $e^x$ and $e^{-x}$, we get a similar result. What is it?

*When you have decided, turn on to Frame 5.*

**5**

$$\cosh x = 1 + \frac{x^2}{2!} + \frac{x^4}{4!} + \frac{x^6}{6!} + \ldots$$

□□□□□□□□□□□□□□□□□□□□□□□□□□□□□□□□□□□□□□□□□□

For we have:

$$e^x = 1 + x + \frac{x^2}{2!} + \frac{x^3}{3!} + \frac{x^4}{4!} + \ldots$$

$$e^{-x} = 1 - x + \frac{x^2}{2!} - \frac{x^3}{3!} + \frac{x^4}{4!} - \ldots$$

$$\therefore\ e^x + e^{-x} = 2 + \frac{2x^2}{2!} + \frac{2x^4}{4!} + \ldots$$

$$\therefore\ \frac{e^x + e^{-x}}{2} = \underline{\cosh x = 1 + \frac{x^2}{2!} + \frac{x^4}{4!} + \ldots}$$

*Move on to Frame 6.*

**6**

So we have:

$$\sinh x = x + \frac{x^3}{3!} + \frac{x^5}{5!} + \frac{x^7}{7!} + \ldots$$

$$\cosh x = 1 + \frac{x^2}{2!} + \frac{x^4}{4!} + \frac{x^6}{6!} + \ldots$$

*Note:* All terms positive: $\sinh x$ has all the odd powers,
$\cosh x$ has all the even powers.

We cannot easily get a series for $\tanh x$ by this process, so we will leave that one to some other time.

Make a note of these two series in your record book. Then, cover up what you have done so far and see if you can write down the definitions of:

(i) $\sinh x =$ ...................... (ii) $\cosh x =$ ......................

(iii) $\tanh x =$ ...................... No looking!

**7**

$$\sinh x = \frac{e^x - e^{-x}}{2}; \quad \cosh x = \frac{e^x + e^{-x}}{2}; \quad \tanh x = \frac{e^x - e^{-x}}{e^x + e^{-x}}$$

All correct? Right.

□□□□□□□□□□□□□□□□□□□□□□□□□□□□□□□□□□□□□□□

**Graphs of Hyperbolic Functions**

We shall get to know quite a lot about these hyperbolic functions if we sketch the graphs of these functions. Since they depend on the values of $e^x$ and $e^{-x}$, we had better just refresh our memories of what these graphs look like.

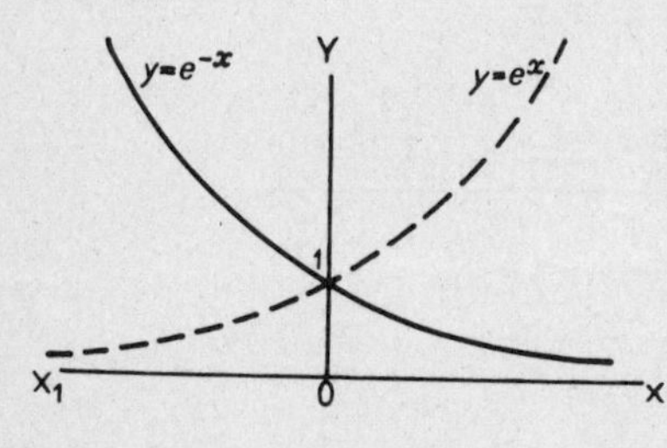


$y = e^x$ and $y = e^{-x}$ cross the $y$-axis at the point $y = 1$ ($e^\circ = 1$). Each graph then approaches the $x$-axis as an asymptote, getting nearer and nearer to it as it goes away to infinity in each direction, without actually crossing it.

So, for what range of values of $x$ are $e^x$ and $e^{-x}$ positive?

**8**

$e^x$ and $e^{-x}$ are positive for all values of $x$

Correct, since the graphs are always above the $x$-axis.

□□□□□□□□□□□□□□□□□□□□□□□□□□□□□□□□□□□□□□□

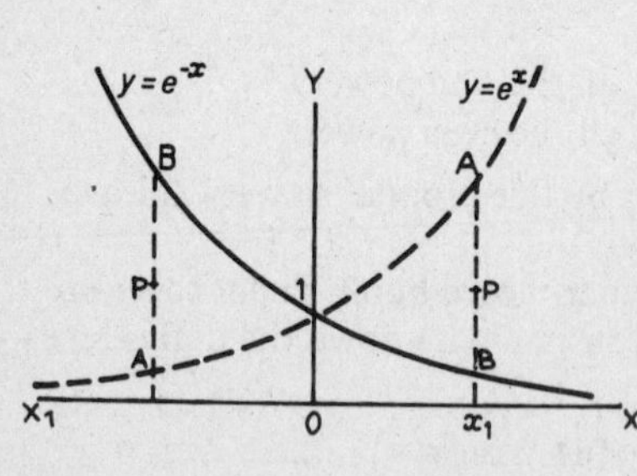


At any value of $x$, e.g. $x = x_1$,

$\cosh x = \frac{e^x + e^{-x}}{2}$, i.e. the value of $\cosh x$ is the average of the values of $e^x$ and $e^{-x}$ at that value of $x$. This is given by P, the mid point of AB.

If we can imagine a number of ordinates (or verticals) like AB and we plot their mid-points, we shall obtain the graph of $y = \cosh x$.

Can you sketch in what the graph will look like?

**9**

Here it is:

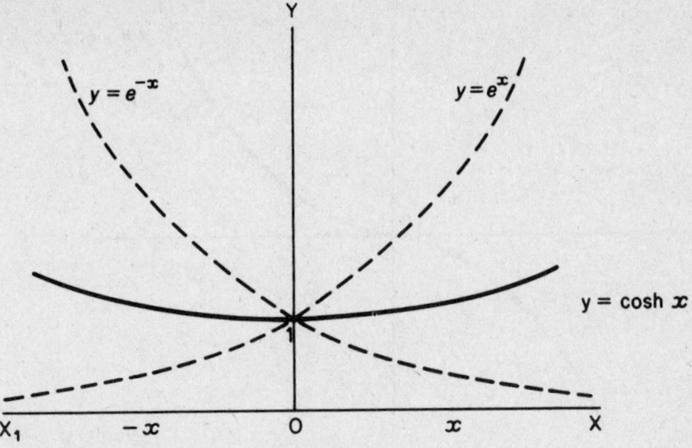


We see from the graph of $y = \cosh x$ that:

(i) $\cosh 0 = 1$

(ii) the value of $\cosh x$ is never less than 1

(iii) the curve is symmetrical about the $y$-axis, i.e.

$$\cosh(-x) = \cosh x$$

(iv) for any given value of $\cosh x$, there are two values of $x$, equally spaced about the origin, i.e. $x = \pm a$.

Now let us see about the graph of $y = \sinh x$ in the same sort of way.

**10**

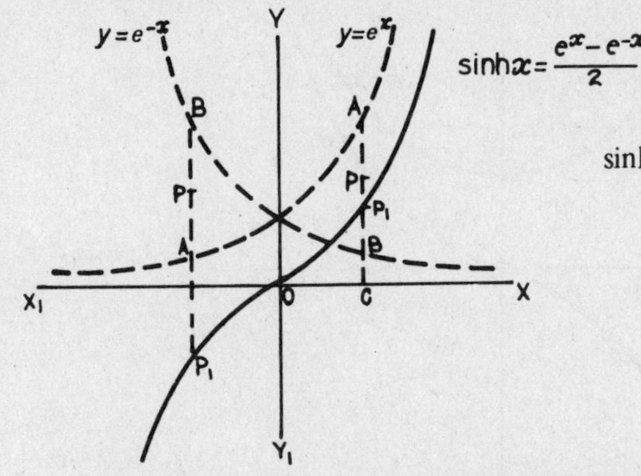


$$\sinh x = \frac{e^x - e^{-x}}{2}$$

$$\sinh x = \frac{e^x - e^{-x}}{2}$$

On the diagram,

$CA = e^x$

$CB = e^{-x}$

$BA = e^x - e^{-x}$

$$BP = \frac{e^x - e^{-x}}{2}$$

The corresponding point on the graph of $y = \sinh x$ is thus obtained by standing the ordinate BP on the $x$-axis at C, i.e. $P_1$.

Note that on the left of the origin, BP is negative and is therefore placed below the $x$-axis.

So what can we say about $y = \sinh x$?

**11**

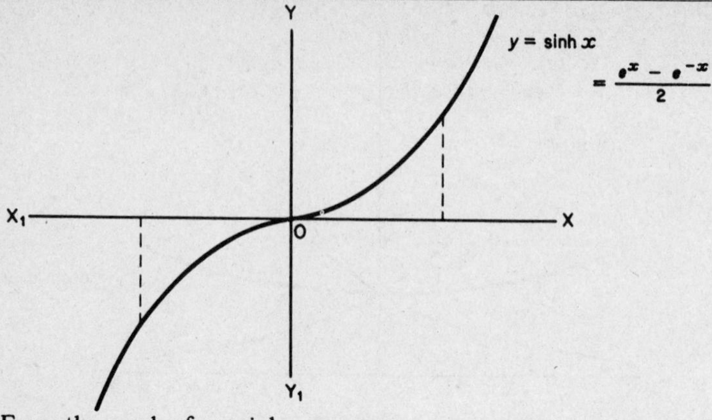


From the graph of $y = \sinh x$, we see

(i) $\sinh 0 = 0$

(ii) $\sinh x$ can have all values from $-\infty$ to $+\infty$

(iii) the curve is symmetrical about the origin, i.e.

$$\sinh(-x) = -\sinh x$$

(iv) for a given value of $\sinh x$, there is only one real value of $x$.

If we draw $y = \sinh x$ and $y = \cosh x$ on the same graph, what do we get?

**12**

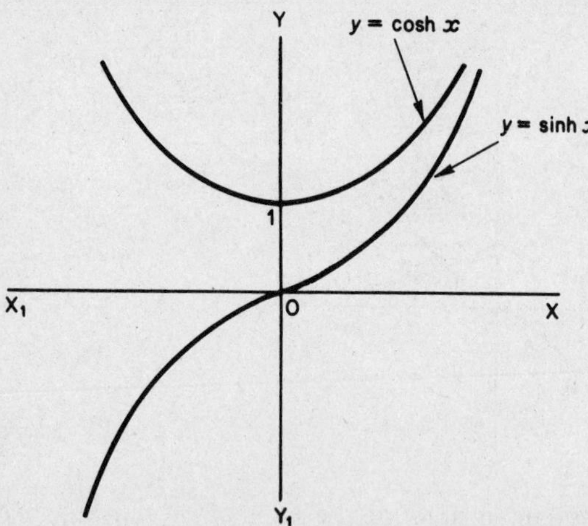


Note that $y = \sinh x$ is always outside $y = \cosh x$, but gets nearer to it as $x$ increases

$$\text{i.e. as } x \to \infty, \ \sinh x \to \cosh x$$

And now let us consider the graph of $y = \tanh x$. *Turn on.*

**13**

It is not easy to build $y = \tanh x$ directly from the graphs of $y = e^x$ and $y = e^{-x}$. If, however, we take values of $e^x$ and $e^{-x}$ and then calculate

$y = \dfrac{e^x - e^{-x}}{e^x + e^{-x}}$ and plot points, we get a graph as shown.

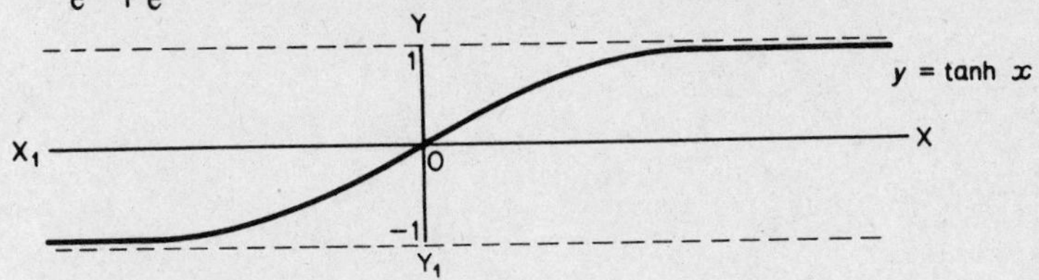


We see (i) $\tanh 0 = 0$

(ii) $\tanh x$ always lies between $y = -1$ and $y = 1$

(iii) $\tanh(-x) = -\tanh x$

(iv) as $x \to \infty$, $\tanh x \to 1$

as $x \to -\infty$, $\tanh x \to -1$.

Finally, let us now sketch all three graphs on one diagram so that we can compare them and distinguish between them.

Here they are:

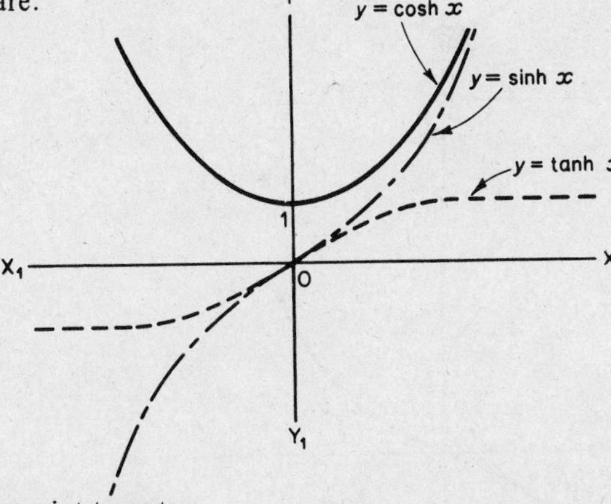


One further point to note:

At the origin, $y = \sinh x$ and $y = \tanh x$ have the same slope. The two graphs therefore slide into each other and out again. They do not cross each other at three distinct points (as some people think).

It is worth while to remember this combined diagram: sketch it in your record book for reference.

**15** **Revision Exercise**

Fill in the following–

(i) $\dfrac{e^x + e^{-x}}{2} = \ldots\ldots\ldots\ldots\ldots\ldots$

(ii) $\dfrac{e^x - e^{-x}}{e^x + e^{-x}} = \ldots\ldots\ldots\ldots\ldots\ldots$

(iii) $\dfrac{e^x - e^{-x}}{2} = \ldots\ldots\ldots\ldots\ldots\ldots$

(iv)

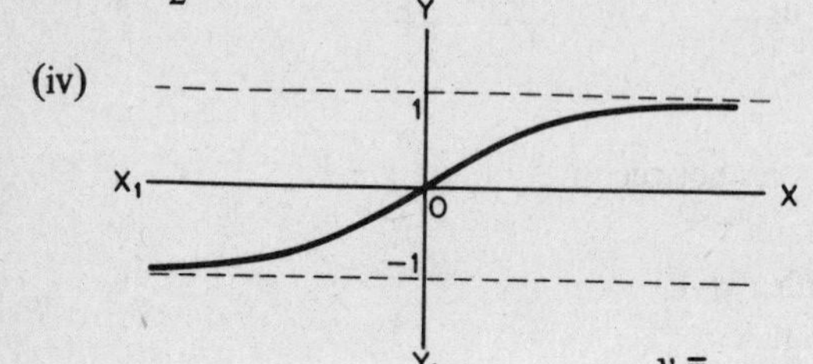


$y = \ldots\ldots\ldots\ldots\ldots\ldots$

(v)

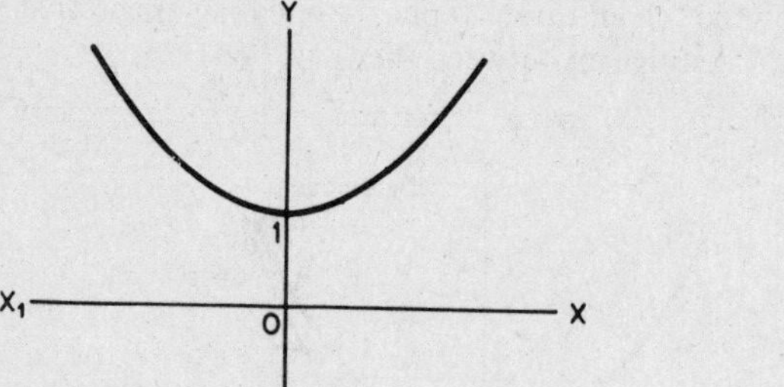


$y = \ldots\ldots\ldots\ldots\ldots\ldots$

(vi)

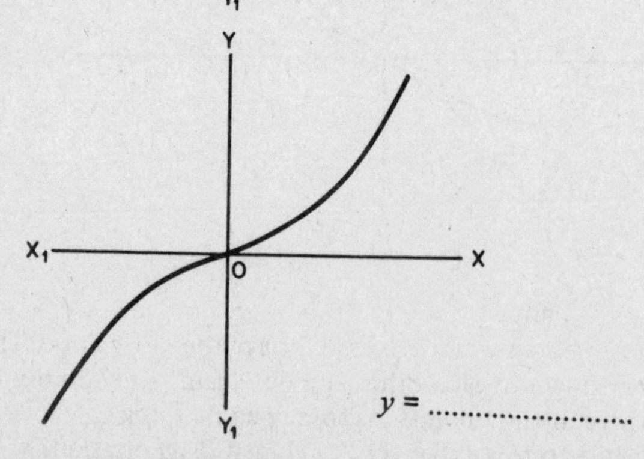


$y = \ldots\ldots\ldots\ldots\ldots\ldots$

*Results on the next frame. Check your answers carefully.*

**16**

***Results:*** Here they are: check yours.

(i) $\dfrac{e^x + e^{-x}}{2} = \cosh x$

(ii) $\dfrac{e^x - e^{-x}}{e^x + e^{-x}} = \tanh x$

(iii) $\dfrac{e^x - e^{-x}}{2} = \sinh x$

(iv)

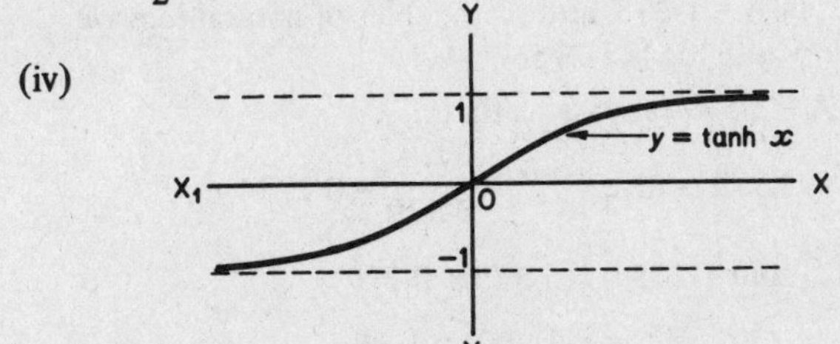


(v)

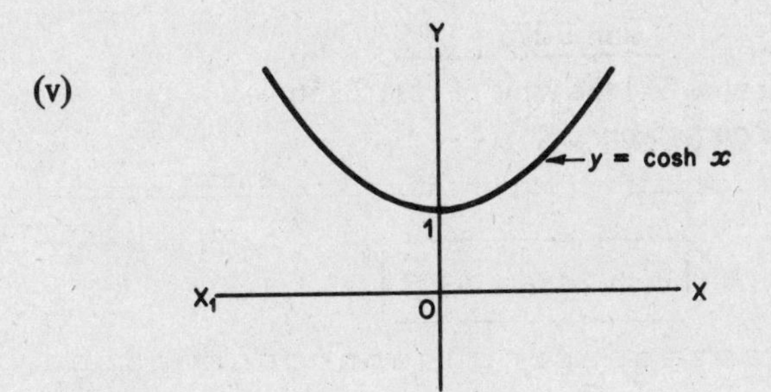


(vi)

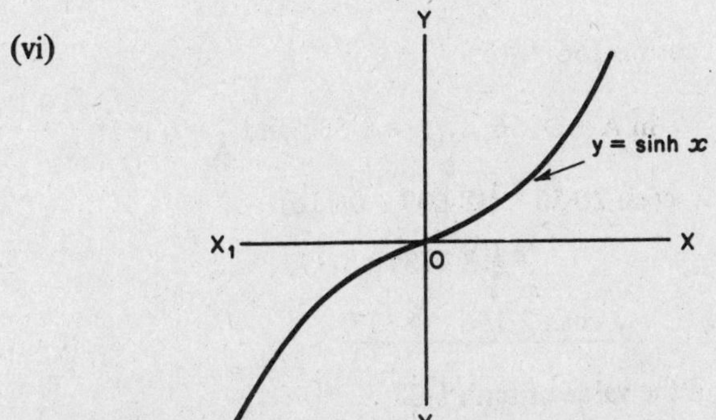


*Now we can continue with the next piece of work.*

**17**

## Evaluation of Hyperbolic Functions

The values of sinh $x$, cosh $x$ and tanh $x$ for some values of $x$ are given in the tables. But for other values of $x$ it is necessary to calculate the value of the hyperbolic functions. One or two examples will soon show how this is done.

*Example 1.* To evaluate sinh 1·275

Now $\sinh x = \frac{1}{2}(e^{x} - e^{-x}) \quad \therefore \sinh 1{\cdot}275 = \frac{1}{2}(e^{1{\cdot}275} - e^{-1{\cdot}275})$. We now have to evaluate $e^{1{\cdot}275}$. Note that when we have done that, $e^{-1{\cdot}275}$ is merely its reciprocal and can be found from tables. Here goes then:

Let $A = e^{1{\cdot}275} \quad \therefore \ln A = 1{\cdot}275$ and from tables of natural logs we now find the number whose log is 1·275.

This is 3·579 $\therefore$ A = 3·579 (as easy as that!)

So $$e^{1{\cdot}275} = 3{\cdot}579 \text{ and } e^{-1{\cdot}275} \doteq \frac{1}{3{\cdot}579} = 0{\cdot}2794$$

$$\therefore \sinh 1{\cdot}275 = \tfrac{1}{2}(3{\cdot}579 - 0{\cdot}279)$$

$$= \tfrac{1}{2}(3{\cdot}300) = 1{\cdot}65$$

$$\therefore \underline{\sinh 1{\cdot}275 = 1{\cdot}65}$$

In the same way, you now find the value of cosh 2·156.

*When finished, move on to frame 18.*

---

**18**

| cosh 2·156 = 4·377 |
|---|

□□□□□□□□□□□□□□□□□□□□□□□□□□□□□□□□□□□□□□□□□□□□□□□□□□□□□□□□□□□□□□□□□□□□□

Here is the working:

*Example 2.* $$\cosh 2{\cdot}156 = \tfrac{1}{2}(e^{2{\cdot}156} + e^{-2{\cdot}156})$$

Let $A = e^{2{\cdot}156} \quad \therefore \ln A = 2{\cdot}156 \quad \therefore A = 8{\cdot}637$ and $\frac{1}{A} = 0{\cdot}1158$

$$\therefore \cosh 2{\cdot}156 = \tfrac{1}{2}(8{\cdot}637 + 0{\cdot}116)$$

$$= \tfrac{1}{2}(8{\cdot}753) = 4{\cdot}377$$

$$\therefore \underline{\cosh 2{\cdot}156 = 4{\cdot}377}$$

Right, one more. Find the value of tanh 1·27.

*When you have finished, move on to frame 19.*

**19**

$$\boxed{\tanh 1{\cdot}27 = 0{\cdot}8539}$$

□□□□□□□□□□□□□□□□□□□□□□□□□□□□□□□□□□□□□□□□□□□□□□

Working:

*Example 3.* $\tanh 1{\cdot}27 = \dfrac{e^{1\cdot27} - e^{-1\cdot27}}{e^{1\cdot27} + e^{-1\cdot27}}$

Let $A = e^{1\cdot27}$ $\therefore \ln A = 1{\cdot}27$ $\therefore A = 3{\cdot}561$ and $\dfrac{1}{A} = 0{\cdot}2808$

$$\therefore \tanh 1{\cdot}27 = \frac{3{\cdot}561 - 0{\cdot}281}{3{\cdot}561 + 0{\cdot}281} = \frac{3{\cdot}280}{3{\cdot}842}$$

$$\underline{\tanh 1{\cdot}27 = 0{\cdot}8539}$$

0·5159
0·5845
$\bar{1}{\cdot}9314$

So, evaluating sinh, cosh and tanh is easy enough and depends mainly on being able to evaluate $e^k$, where $k$ is a given number – and that is most easily done by using natural logs as we have seen.

*And now let us look at the reverse process. So on to frame 20.*

**20**

**Inverse Hyperbolic Functions**

*Example 1.* To find $\sinh^{-1} 1{\cdot}475$, i.e. to find the value of $x$ such that $\sinh x = 1{\cdot}475$.

Here it is: $\sinh x = 1{\cdot}475$ $\therefore \frac{1}{2}(e^x - e^{-x}) = 1{\cdot}475$

$$\therefore e^x - \frac{1}{e^x} = 2{\cdot}950$$

Multiplying both sides by $e^x$: $(e^x)^2 - 1 = 2{\cdot}95(e^x)$

$$(e^x)^2 - 2{\cdot}95(e^x) - 1 = 0$$

This is a quadratic equation and can be solved as usual, giving

$$e^x = \frac{2{\cdot}95 \pm \sqrt{(2{\cdot}95^2 + 4)}}{2} = \frac{2{\cdot}95 \pm \sqrt{(8{\cdot}703 + 4)}}{2}$$

$$= \frac{2{\cdot}95 \pm \sqrt{12{\cdot}703}}{2} = \frac{2{\cdot}95 \pm 3{\cdot}564}{2}$$

$$= \frac{6{\cdot}514}{2} \text{ or } -\frac{0{\cdot}614}{2} = 3{\cdot}257 \text{ or } -0{\cdot}307$$

But $e^x$ is always positive for real values of $x$. Therefore the only real solution is given by $e^x = 3{\cdot}257$.

$$\therefore x = \ln 3{\cdot}257 = 1{\cdot}1809$$

$$\therefore \underline{x = 1{\cdot}1809}$$

*Exercise 2.*

Now you find $\cosh^{-1} 2{\cdot}364$ in the same way.

**21**

$$\boxed{\cosh^{-1} 2{\cdot}364 = \pm 1{\cdot}507}$$

□□□□□□□□□□□□□□□□□□□□□□□□□□□□□□□□□□□□□□

For: To evaluate $\cosh^{-1} 2{\cdot}364$, let $x = \cosh^{-1} 2{\cdot}364$

$$\therefore \cosh x = 2{\cdot}364 \quad \therefore \frac{e^x + e^{-x}}{2} = 2{\cdot}364 \quad \therefore e^x + \frac{1}{e^x} = 4{\cdot}728$$

$$(e^x)^2 - 4{\cdot}728(e^x) + 1 = 0$$

$$e^x = \frac{4{\cdot}728 \pm \sqrt{(22{\cdot}36 - 4)}}{2} \qquad \sqrt{18{\cdot}36} = 4{\cdot}285$$

$$= \tfrac{1}{2}(4{\cdot}728 \pm 4{\cdot}285) = \tfrac{1}{2}(9{\cdot}013) \text{ or } \tfrac{1}{2}(0{\cdot}443)$$

$$e^x = 4{\cdot}5065 \text{ or } 0{\cdot}2215$$

$$\therefore x = \ln 4{\cdot}5065 \text{ or } \ln 0{\cdot}2215$$

$$= 1{\cdot}5056 \text{ or } \bar{2}{\cdot}4926 \text{ i.e. } -1{\cdot}5074$$

$$\underline{x = \pm 1{\cdot}507}$$

Before we do the next one, do you remember the exponential definition of $\tanh x$? Well, what is it?

**22**

$$\boxed{\tanh x = \frac{e^x - e^{-x}}{e^x + e^{-x}}}$$

□□□□□□□□□□□□□□□□□□□□□□□□□□□□□□□□□□□□□□

That being so, we can now evaluate $\tanh^{-1} 0{\cdot}623$.

Let $\qquad x = \tanh^{-1} 0{\cdot}623 \quad \therefore \tanh x = 0{\cdot}623$

$$\therefore \frac{e^x - e^{-x}}{e^x + e^{-x}} = 0{\cdot}623$$

$$\therefore e^x - e^{-x} = 0{\cdot}623(e^x + e^{-x})$$

$$\therefore (1 - 0{\cdot}623)\, e^x = (1 + 0{\cdot}623)\, e^{-x}$$

$$0{\cdot}377\, e^x = 1{\cdot}623\, e^{-x}$$

$$= \frac{1{\cdot}623}{e^x}$$

$$\therefore (e^x)^2 = \frac{1{\cdot}623}{0{\cdot}377}$$

$$\therefore e^x = 2{\cdot}075$$

$$\begin{array}{r} 0{\cdot}2103 \\ \bar{1}{\cdot}5763 \\ 2)\,\overline{0{\cdot}6340} \\ \underline{0{\cdot}3170} \end{array}$$

$$\therefore x = \ln 2{\cdot}075 = 0{\cdot}7299$$

$$\therefore \underline{\tanh^{-1} 0{\cdot}623 = 0{\cdot}730}$$

Now one for you to do on your own. Evaluate $\sinh^{-1} 0{\cdot}5$.

**23**

$$\sinh^{-1} 0{\cdot}5 = 0{\cdot}4810$$

□□□□□□□□□□□□□□□□□□□□□□□□□□□□□□□□□□□□□□

Check your working.

Let $\quad x = \sinh^{-1} 0{\cdot}5 \quad \therefore \sinh x = 0{\cdot}5$

$$\therefore \frac{e^x - e^{-x}}{2} = 0{\cdot}5 \quad \therefore e^x - \frac{1}{e^x} = 1$$

$$\therefore (e^x)^2 - 1 = e^x$$

$$(e^x)^2 - (e^x) - 1 = 0$$

$$e^x = \frac{1 \pm \sqrt{(1+4)}}{2} = \frac{1 \pm \sqrt{5}}{2}$$

$$= \frac{3{\cdot}2361}{2} \text{ or } \frac{-1{\cdot}2361}{2}$$

$$= 1{\cdot}6181 \text{ or } -0{\cdot}6181$$

$$\therefore x = \ln 1{\cdot}6181 = 0{\cdot}4810$$

$$\underline{\sinh^{-1} 0{\cdot}5 = 0{\cdot}4810}$$

$e^x = -0{\cdot}6181$ gives no real value of $x$.

And just one more! Evaluate $\tanh^{-1} 0{\cdot}75$.

**24**

$$\tanh^{-1} 0{\cdot}75 = 0{\cdot}9731$$

□□□□□□□□□□□□□□□□□□□□□□□□□□□□□□□□□□□□□□

Let $\quad x = \tanh^{-1} 0{\cdot}75 \quad \therefore \tanh x = 0{\cdot}75$

$$\therefore \frac{e^x - e^{-x}}{e^x + e^{-x}} = 0{\cdot}75$$

$$e^x - e^{-x} = 0{\cdot}75(e^x + e^{-x})$$

$$(1 - 0{\cdot}75)\,e^x = (1 + 0{\cdot}75)\,e^{-x}$$

$$0{\cdot}25\,e^x = 1{\cdot}75\,e^{-x}$$

$$(e^x)^2 = \frac{1{\cdot}75}{0{\cdot}25} = 7$$

$$e^x = \pm\sqrt{7} = \pm 2{\cdot}6458$$

But remember that $e^x$ cannot be negative for real values of $x$.

Therefore $e^x = 2{\cdot}6458$ is the only real solution.

$$\therefore x = \ln 2{\cdot}6458 = 0{\cdot}9731$$

$$\underline{\tanh^{-1} 0{\cdot}75 = 0{\cdot}9731}$$

25

**Log. Form of the Inverse Hyperbolic Functions**

Let us do the same thing in a general way.

To find $\tanh^{-1} x$ in log. form.

As usual, we start off with: Let $y = \tanh^{-1} x \quad \therefore x = \tanh y$

$$\therefore \frac{e^y - e^{-y}}{e^y + e^{-y}} = x \quad \therefore e^y - e^{-y} = x(e^y + e^{-y})$$

$$e^y(1 - x) = e^{-y}(1 + x) = \frac{1}{e^y}(1 + x)$$

$$e^{2y} = \frac{1 + x}{1 - x}$$

$$\therefore 2y = \ln\left\{\frac{1 + x}{1 - x}\right\}$$

$$\therefore y = \tanh^{-1} x = \tfrac{1}{2} \ln\left\{\frac{1 + x}{1 - x}\right\}$$

So that $\tanh^{-1} 0{\cdot}5 = \tfrac{1}{2} \ln\left\{\frac{1{\cdot}5}{0{\cdot}5}\right\}$

$$= \tfrac{1}{2} \ln 3 = \tfrac{1}{2}(1{\cdot}0986) = 0{\cdot}5493$$

And similarly, $\tanh^{-1}(-0{\cdot}6) = \ldots\ldots\ldots\ldots\ldots\ldots$

26

$$\boxed{\tanh^{-1}(-0{\cdot}6) = -0{\cdot}6932}$$

□□□□□□□□□□□□□□□□□□□□□□□□□□□□□□□□□□□□□□

For, $\tanh^{-1} x = \tfrac{1}{2} \ln\left\{\frac{1 + x}{1 - x}\right\}$

$$\therefore \tanh^{-1}(-0{\cdot}6) = \tfrac{1}{2} \ln\left\{\frac{1 - 0{\cdot}6}{1 + 0{\cdot}6}\right\} = \tfrac{1}{2} \ln\left\{\frac{0{\cdot}4}{1{\cdot}6}\right\}$$

$$= \tfrac{1}{2} \ln 0{\cdot}25$$

$$= \tfrac{1}{2}(\bar{2}{\cdot}6137)$$

$$= \tfrac{1}{2}(-1{\cdot}3863)$$

$$= -0{\cdot}6932$$

| | |
|---|---|
| 2·5 | 0·9163 |
| 10 | 2·3026 |
| | $\bar{2}{\cdot}6137$ |

Now, in the same way, find an expression for $\sinh^{-1} x$.

Start off by saying: Let $y = \sinh^{-1} x \quad \therefore x = \sinh y$

$$\therefore \frac{e^y - e^{-y}}{2} = x \quad \therefore e^y - e^{-y} = 2x \quad \therefore e^y - \frac{1}{e^y} = 2x$$

$$(e^y)^2 - 2x(e^y) - 1 = 0$$

*Now finish it off.*

**27**

***Result:*** $\boxed{\sinh^{-1}x = \ln\{x + \sqrt{(x^2+1)}\}}$

□□□□□□□□□□□□□□□□□□□□□□□□□□□□□□□□□□□□□□□□□□□□□□□

For $$(e^y)^2 - 2x(e^y) - 1 = 0$$

$$e^y = \frac{2x \pm \sqrt{(4x^2+4)}}{2} = \frac{2x \pm 2\sqrt{(x^2+1)}}{2}$$

$$= x \pm \sqrt{(x^2+1)}$$

$$e^y = x + \sqrt{(x^2+1)} \text{ or } e^y = x - \sqrt{(x^2+1)}$$

At first sight, there appear to be two results, but notice this:

In the second result, $\sqrt{(x^2+1)} > x$

$$\therefore\ e^y = x - (\text{something} > x) \quad \text{i.e. negative.}$$

Therefore we can discard the second result as far as we are concerned since powers of e are always positive. (Remember the graph of $e^x$.)

The only real solution then is given by $e^y = x + \sqrt{(x^2+1)}$

$$\underline{y = \sinh^{-1}x = \ln\{x + \sqrt{(x^2+1)}\}}$$

**28**

Finally, let us find the general expression for $\cosh^{-1}x$.

Let $$y = \cosh^{-1}x \quad \therefore\ x = \cosh y = \frac{e^y + e^{-y}}{2}$$

$$\therefore\ e^y + \frac{1}{e^y} = 2x \quad \therefore\ (e^y)^2 - 2x(e^y) + 1 = 0$$

$$\therefore\ e^y = \frac{2x \pm \sqrt{(4x^2-4)}}{2} = x \pm \sqrt{(x^2-1)}$$

$$\therefore\ e^y = x + \sqrt{(x^2-1)} \text{ and } e^y = x - \sqrt{(x^2-1)}$$

Both these results are positive, since $\sqrt{(x^2-1)} < x$.

However, $$\frac{1}{x + \sqrt{(x^2-1)}} = \frac{1}{x + \sqrt{(x^2-1)}} \cdot \frac{x - \sqrt{(x^2-1)}}{x - \sqrt{(x^2-1)}}$$

$$= \frac{x - \sqrt{(x^2-1)}}{x^2 - (x^2-1)} = x - \sqrt{(x^2-1)}$$

So our results can be written

$$e^y = x + \sqrt{(x^2-1)} \text{ and } e^y = \frac{1}{x + \sqrt{(x^2-1)}}$$

$$e^y = x + \sqrt{(x^2-1)} \text{ or } \{x + \sqrt{(x^2-1)}\}^{-1}$$

$$\therefore\ y = \ln\{x + \sqrt{(x^2-1)}\} \text{ or } -\ln\{x + \sqrt{(x^2-1)}\}$$

$$\therefore\ \underline{\cosh^{-1}x = \pm\ln\{x + \sqrt{(x^2-1)}\}}$$

Notice that the plus and minus signs give two results which are symmetrical about the $y$-axis (agreeing with the graph of $y = \cosh x$).

**29**

Here are the three general results collected together.

$$\sinh^{-1} x = \ln\left\{x + \sqrt{(x^2 + 1)}\right\}$$

$$\cosh^{-1} x = \pm\ln\left\{x + \sqrt{(x^2 - 1)}\right\}$$

$$\tanh^{-1} x = \tfrac{1}{2}\ln\left\{\frac{1 + x}{1 - x}\right\}$$

Add these to your list in your record book. They will be useful. Compare the first two carefully, for they are very nearly alike. Note also that

(i) $\sinh^{-1} x$ has only one value.

(ii) $\cosh^{-1} x$ has two values.

*So what comes next? We shall see in frame 30.*

**30**

**Hyperbolic Identities**

There is no need to recoil in horror. You will see before long that we have an easy way of doing these. First of all, let us consider one or two relationships based on the basic definitions.

(1) The first set are really definitions themselves. Like the trig. ratios, we have reciprocal hyperbolic functions:

(i) $\coth x$ (i.e. hyperbolic cotangent) $= \dfrac{1}{\tanh x}$

(ii) $\operatorname{sech} x$ (i.e. hyperbolic secant) $= \dfrac{1}{\cosh x}$

(iii) $\operatorname{cosech} x$ (i.e. hyperbolic cosecant) $= \dfrac{1}{\sinh x}$

These, by the way, are pronounced (i) coth, (ii) sheck and (iii) co-sheck respectively.

These remind us, once again, how like trig. functions these hyperbolic functions are.

*Make a list of these three definitions: then turn on to frame 31.*

**31**

(2) Let us consider $\dfrac{\sinh x}{\cosh x} = \dfrac{e^x - e^{-x}}{2} \div \dfrac{e^x + e^{-x}}{2}$

$$= \frac{e^x - e^{-x}}{e^x + e^{-x}} = \tanh x$$

$$\therefore \tanh x = \frac{\sinh x}{\cosh x} \qquad \left\{ \text{Very much like } \tan\theta = \frac{\sin\theta}{\cos\theta} \right\}$$

(3) $\operatorname{Cosh} x = \frac{1}{2}(e^x + e^{-x})$; $\sinh x = \frac{1}{2}(e^x - e^{-x})$

Add these results: $\cosh x + \sinh x = e^x$

Subtract: $\cosh x - \sinh x = e^{-x}$

Multiply these two expressions together:

$$(\cosh x + \sinh x)(\cosh x - \sinh x) = e^x.e^{-x}$$

$$\therefore \underline{\cosh^2 x - \sinh^2 x = 1}$$

{In trig., we have $\cos^2\theta + \sin^2\theta = 1$, so there is a difference in sign here.}

*On to frame 32.*

**32**

(4) We just established that $\cosh^2 x - \sinh^2 x = 1$.

Divide by $\cosh^2 x$: $\quad 1 - \dfrac{\sinh^2 x}{\cosh^2 x} = \dfrac{1}{\cosh^2 x}$

$$\therefore 1 - \tanh^2 x = \operatorname{sech}^2 x$$

$$\therefore \underline{\operatorname{sech}^2 x = 1 - \tanh^2 x}$$

{Something like $\sec^2\theta = 1 + \tan^2\theta$, isn't it?}

(5) If we start again with $\cosh^2 x - \sinh^2 x = 1$ and divide this time by $\sinh^2 x$, we get

$$\frac{\cosh^2 x}{\sinh^2 x} - 1 = \frac{1}{\sinh^2 x}$$

$$\therefore \coth^2 x - 1 = \operatorname{cosech}^2 x$$

$$\therefore \underline{\operatorname{cosech}^2 x = \coth^2 x - 1}$$

{In trig., we have $\operatorname{cosec}^2\theta = 1 + \cot^2\theta$, so there is a sign difference here too.}

*Turn on to frame 33.*

**33** (6) We have already used the fact that

$$\cosh x + \sinh x = e^{x} \quad \text{and} \quad \cosh x - \sinh x = e^{-x}$$

If we square each of these statements, we obtain

(i) ..........................................................................

(ii) ..........................................................................

**34**

$$\boxed{\begin{aligned}\cosh^2 x + 2\sinh x \cosh x + \sinh^2 x = e^{2x}\\ \cosh^2 x - 2\sinh x \cosh x + \sinh^2 x = e^{-2x}\end{aligned}}$$

So if we subtract as they stand, we get

$$4\sinh x \cosh x = e^{2x} - e^{-2x}$$

$$\therefore\ 2\sinh x \cosh x = \frac{e^{2x} - e^{-2x}}{2} = \sinh 2x$$

$$\therefore\ \underline{\sinh 2x = 2\sinh x \cosh x}$$

If however we add the two lines together, we get ..........................

**35**

$$2(\cosh^2 x + \sinh^2 x) = e^{2x} + e^{-2x}$$

$$\therefore\ \cosh^2 x + \sinh^2 x = \frac{e^{2x} + e^{-2x}}{2} = \cosh 2x$$

$$\therefore\ \underline{\cosh 2x = \cosh^2 x + \sinh^2 x}$$

We already know that $\quad \cosh^2 x - \sinh^2 x = 1$

$$\therefore\ \cosh^2 x \qquad = 1 + \sinh^2 x$$

Substituting this in our last result, we have

$$\cosh 2x = 1 + \sinh^2 x + \sinh^2 x$$

$$\therefore\ \underline{\cosh 2x = 1 + 2\sinh^2 x}$$

Or we could say $\quad \cosh^2 x - 1 = \sinh^2 x$

$$\therefore\ \cosh 2x = \cosh^2 x + (\cosh^2 x - 1)$$

$$\therefore\ \underline{\cosh 2x = 2\cosh^2 x - 1}$$

Now we will collect all these hyperbolic identities together and compare them with the corresponding trig. identities.

*These are all listed in the next frame, so turn on.*

| | *Trig. Identities* | *Hyperbolic Identities* |
|---|---|---|
| (1) | $\cot x = 1/\tan x$ | $\coth x = 1/\tanh x$ |
| | $\sec x = 1/\cos x$ | $\text{sech}\, x = 1/\cosh x$ |
| | $\text{cosec}\, x = 1/\sin x$ | $\text{cosech}\, x = 1/\sinh x$ |
| (2) | $\cos^2 x + \sin^2 x = 1$ | $\cosh^2 x - \sinh^2 x = 1$ |
| | $\sec^2 x = 1 + \tan^2 x$ | $\text{sech}^2 x = 1 - \tanh^2 x$ |
| | $\text{cosec}^2 x = 1 + \cot^2 x$ | $\text{cosech}^2 x = \coth^2 x - 1$ |
| (3) | $\sin 2x = 2 \sin x \cos x$ | $\sinh 2x = 2 \sinh x \cosh x$ |
| | $\cos 2x = \cos^2 x - \sin^2 x$ | $\cosh 2x = \cosh^2 x + \sinh^2 x$ |
| | $= 1 - 2 \sin^2 x$ | $= 1 + 2 \sinh^2 x$ |
| | $= 2 \cos^2 x - 1$ | $= 2 \cosh^2 x - 1$ |

If we look at these results, we find that some of the hyperbolic identities follow exactly the trig. identities: others have a difference in sign. This change of sign occurs whenever $\sin^2 x$ in the trig. results is being converted into $\sinh^2 x$ to form the corresponding hyperbolic identities. This sign change also occurs when $\sin^2 x$ is involved without actually being written as such. For example, $\tan^2 x$ involves $\sin^2 x$ since $\tan^2 x$ could be written as $\dfrac{\sin^2 x}{\cos^2 x}$. The change of sign therefore occurs

with $\tan^2 x$ when it is being converted into $\tanh^2 x$
$\cot^2 x$ " " " " " " $\coth^2 x$
$\text{cosec}^2 x$ " " " " " " $\text{cosech}^2 x$

The sign change also occurs when we have a product of two sinh terms, e.g. the trig. identity $\cos(A + B) = \cos A \cos B - \sin A \sin B$ gives the hyperbolic identity $\cosh(A + B) = \cosh A \cosh B + \sinh A \sinh B$.

Apart from this one change, the hyperbolic identities can be written down from the trig. identities which you already know.

For example:

$$\tan 2x = \frac{2 \tan x}{1 - \tan^2 x} \quad \text{becomes} \quad \tanh 2x = \frac{2 \tanh x}{1 + \tanh^2 x}$$

So providing you know your trig. identities, you can apply the rule to form the corresponding hyperbolic identities.

**37**

**Relationship between Trigonometric and Hyperbolic Functions**

From our previous work on complex numbers, we know that:

$$e^{j\theta} = \cos\theta + j\sin\theta$$

and $$e^{-j\theta} = \cos\theta - j\sin\theta$$

Adding these two results together, we have

$$e^{j\theta} + e^{-j\theta} = \dots\dots\dots\dots$$

**38**

$$\boxed{2\cos\theta}$$

So that, $$\cos\theta = \frac{e^{j\theta} + e^{-j\theta}}{2}$$

which is of the form $\dfrac{e^x + e^{-x}}{2}$, with $x$ replaced by $(j\theta)$

$$\therefore \cos\theta = \dots\dots\dots\dots$$

**39**

$$\boxed{\cosh j\theta}$$

Here, then, is our first relationship.

$$\underline{\cos\theta = \cosh j\theta}$$

*Make a note of that for the moment: then on to frame 40.*

**40**

If we return to our two original statements

$$e^{j\theta} = \cos\theta + j\sin\theta$$

$$e^{-j\theta} = \cos\theta - j\sin\theta$$

and this time subtract, we get a similar kind of result

$$e^{j\theta} - e^{-j\theta} = \dots\dots\dots\dots$$

**41**

$$\boxed{2j\sin\theta}$$

So that, $$j\sin\theta = \frac{e^{j\theta} - e^{-j\theta}}{2}$$

$$= \dots\dots\dots\dots$$

**42**

$$\boxed{\sinh \mathrm{j}\theta}$$

So, $\underline{\sinh \mathrm{j}\theta = \mathrm{j} \sin \theta}$

*Make a note of that also.*

**43**

So far, we have two important results:

$$\text{(i)} \quad \cosh \mathrm{j}\theta = \cos \theta$$
$$\text{(ii)} \quad \sinh \mathrm{j}\theta = \mathrm{j} \sin \theta$$

Now if we substitute $\theta = \mathrm{j}x$ in the first of these results, we have

$$\cos \mathrm{j}x = \cosh(\mathrm{j}^2 x)$$
$$= \cosh(-x)$$
$$\therefore \quad \cos \mathrm{j}x = \cosh x \qquad [\text{since } \cosh(-x) = \cosh x]$$

Writing this in reverse order, gives

$\underline{\cosh x = \cos \mathrm{j}x}$ Another result to note.

Now do exactly the same with the second result above, i.e. put $\theta = \mathrm{j}x$ in the relationship $\mathrm{j} \sin \theta = \sinh \mathrm{j}\theta$ and simplify the result. What do you get?

**44**

$$\boxed{\mathrm{j} \sinh x = \sin \mathrm{j}x}$$

For we have:

$$\mathrm{j} \sin \theta = \sinh \mathrm{j}\theta$$
$$\mathrm{j} \sin \mathrm{j}x = \sinh(\mathrm{j}^2 x)$$
$$= \sinh(-x)$$
$$= -\sinh x \qquad [\text{since } \sinh(-x) = -\sinh x]$$

Finally, divide both sides by j, and we have

$\underline{\sin \mathrm{j}x = \mathrm{j} \sinh x}$

*Now on to the next frame.*

**45**

Now let us collect together the results we have established. They are so nearly alike, that we must distinguish between them.

| $\sin jx = j\sinh x$ | $\sinh jx = j\sin x$ |
|---|---|
| $\cos jx = \cosh x$ | $\cosh jx = \cos x$ |

and, by division, we can also obtain

| $\tan jx = j\tanh x$ | $\tanh jx = j\tan x$ |
|---|---|

Copy the complete table into your record book for future use.

**46**

Here is one application of these results:

*Example 1.* Find an expansion for $\sin(x + jy)$.

Now we know that

$$\sin(A + B) = \sin A\cos B + \cos A\sin B$$

$$\therefore \sin(x + jy) = \sin x\cos jy + \cos x\sin jy$$

so using the results we have listed, we can replace

$\cos jy$ by ......................

and $\sin jy$ by ......................

**47**

| $\cos jy = \cosh y$ | $\sin jy = j\sinh y$ |
|---|---|

So that

$$\sin(x + jy) = \sin x\cos jy + \cos x\sin jy$$

becomes

$$\underline{\sin(x + jy) = \sin x\cosh y + j\cos x\sinh y}$$

*Note:* $\sin(x + jy)$ is a function of the angle $(x + jy)$, which is, of course, a complex quantity. In this case, $(x + jy)$ is referred to as a *Complex Variable* and you will most likely deal with this topic at a later stage of your course.

Meanwhile, here is just one example for you to work through.

Find an expansion for $\cos(x - jy)$.

*Then check with frame 48.*

**48**

$$\boxed{\cos(x - jy) = \cos x \cosh y + j \sin x \sinh y}$$

Here is the working:

$$\cos(A - B) = \cos A \cos B + \sin A \sin B$$

$$\therefore \cos(x - jy) = \cos x \cos jy + \sin x \sin jy$$

$$\text{But } \cos jy = \cosh y$$

$$\text{and } \sin jy = j \sinh y$$

$$\therefore \underline{\cos(x - jy) = \cos x \cosh y + j \sin x \sinh y}$$

**49**

All that now remains is the test exercise, but before working through it, look through your notes, or revise any parts of the programme on which you are not perfectly clear.

*Then, when you are ready, turn on to the next frame.*

## 50 Test Exercise – III

1. If $L = 2C \sinh \dfrac{H}{2C}$, find L when H = 63 and C = 50.

2. If $v^2 = 1{\cdot}8\ L \tanh \dfrac{6{\cdot}3d}{L}$, find $v$ when $d = 40$ and L = 315.

3. On the same axes, draw sketch graphs of (i) $y = \sinh x$, (ii) $y = \cosh x$, (iii) $y = \tanh x$.

4. Simplify $\dfrac{1 + \sinh 2A + \cosh 2A}{1 - \sinh 2A - \cosh 2A}$

5. Calculate from first principles, the value of

   (i) $\sinh^{-1} 1{\cdot}532$  (ii) $\cosh^{-1} 1{\cdot}25$

6. If $\tanh x = \dfrac{1}{3}$, find $e^{2x}$ and hence evaluate $x$.

7. The curve assumed by a heavy chain or cable is

$$y = C \cosh \frac{x}{C}$$

   If C = 50, calculate (i) the value of $y$ when $x = 109$,
   (ii) the value of $x$ when $y = 75$.

8. Obtain the expansion of $\sin(x - jy)$ in terms of the trigonometric and hyperbolic functions of $x$ and $y$.

**Further Problems – III**

1. Prove that $\cosh 2x = 1 + 2\sinh^2 x$.

2. Express $\cosh 2x$ and $\sinh 2x$ in exponential form and hence solve, for real values of $x$, the equation
$$2\cosh 2x - \sinh 2x = 2$$

3. If $\sinh x = \tan y$, show that $x = \ln(\sec y + \tan y)$.

4. If $a = c\cosh x$ and $b = c\sinh x$, prove that
$$(a + b)^2 e^{-2x} = a^2 - b^2$$

5. Evaluate (i) $\tanh^{-1} 0{\cdot}75$, (ii) $\cosh^{-1} 2$.

6. Prove that $\tanh^{-1}\left\{\dfrac{x^2 - 1}{x^2 + 1}\right\} = \ln x$.

7. Express (i) $\cosh \dfrac{1 + j}{2}$ and (ii) $\sinh \dfrac{1 + j}{2}$ in the form $a + jb$, giving $a$ and $b$ to 4 significant figures.

8. Prove that (i) $\sinh(x + y) = \sinh x \cosh y + \cosh x \sinh y$
(ii) $\cosh(x + y) = \cosh x \cosh y + \sinh x \sinh y$

Hence prove that
$$\tanh(x + y) = \frac{\tanh x + \tanh y}{1 + \tanh x \tanh y}$$

9. Show that the co-ordinates of any point on the hyperbola $\dfrac{x^2}{a^2} - \dfrac{y^2}{b^2} = 1$ can be represented in the form $x = a\cosh u$, $y = b\sinh u$.

10. Solve for real values of $x$
$$3\cosh 2x = 3 + \sinh 2x$$

11. Prove that $\dfrac{1 + \tanh x}{1 - \tanh x} = e^{2x}$

12. It $t = \tanh \dfrac{x}{2}$, prove that $\sinh x = \dfrac{2t}{1 - t^2}$ and $\cosh x = \dfrac{1 + t^2}{1 - t^2}$. Hence solve the equation
$$7\sinh x + 20\cosh x = 24$$

13. If $x = \ln \tan\left\{\frac{\pi}{4} + \frac{\theta}{2}\right\}$, find $e^x$ and $e^{-x}$, and hence show that $\sinh x = \tan\theta$.

14. Given that $\sinh^{-1}x = \ln\left\{x + \sqrt{(x^2+1)}\right\}$, determine $\sinh^{-1}(2+j)$ in the form $a + jb$.

15. If $\tan\left\{\frac{x}{2}\right\} = \tan A \tanh B$, prove that

$$\tan x = \frac{\sin 2A \sinh 2B}{1 + \cos 2A \cosh 2B}.$$

16. Prove that $\sinh 3\theta = 3\sinh\theta + 4\sinh^3\theta$.

17. If $x + jy = \tan^{-1}(e^{a+jb})$, show that $\tan 2x = \frac{-\cos b}{\sinh a}$, and that $\tanh 2y = \frac{\sin b}{\cosh a}$.

18. If $\lambda = \frac{at}{2}\cdot\left\{\frac{\sinh at + \sin at}{\cosh at - \cos at}\right\}$, calculate $\lambda$ when $a = 0{\cdot}215$ and $t = 5$.

19. Prove that $\tanh^{-1}\left\{\frac{x^2 - a^2}{x^2 + a^2}\right\} = \ln\frac{x}{a}$.

20. Given that $\sinh^{-1}x = \ln\{x + \sqrt{(x^2+1)}\}$, show that, for small values of $x$,

$$\sinh^{-1}x \triangleq x - \frac{x^3}{6} + \frac{3x^5}{40}.$$

# Programme 4

## DETERMINANTS

**1**

## Determinants

You are quite familiar with the method of solving a pair of simultaneous equations by elimination.

e.g. To solve

$$2x + 3y + 2 = 0 \quad \ldots \text{(i)}$$
$$3x + 4y + 6 = 0 \quad \ldots \text{(ii)}$$

we could first find the value of $x$ by eliminating $y$. To do this, of course, we should multiply (i) by 4 and (ii) by 3 to make the coefficient of $y$ the same in each equation.

So

$$8x + 12y + 8 = 0$$
$$9x + 12y + 18 = 0$$

Then, by subtraction, we get $x + 10 = 0$, i.e. $x = -10$. By substituting back in either equation, we then obtain $y = 6$.

So finally, $\underline{x = -10, \; y = 6}$

That was trivial. You have done similar ones many times before. In just the same way, if

$$a_1x + b_1y + d_1 = 0 \quad \ldots \text{(i)}$$
$$a_2x + b_2y + d_2 = 0 \quad \ldots \text{(ii)}$$

then to eliminate $y$ we make the coefficients of $y$ in the two equations identical by multiplying (i) by .................... and (ii) by ....................

**2**

(i) by $b_2$ and (ii) by $b_1$

Correct, of course. So the equations

$$a_1x + b_1y + d_1 = 0$$
$$a_2x + b_2y + d_2 = 0$$

become

$$a_1b_2x + b_1b_2y + b_2d_1 = 0$$
$$a_2b_1x + b_1b_2y + b_1d_2 = 0$$

Subtracting, we get

$$(a_1b_2 - a_2b_1)x + b_2d_1 - b_1d_2 = 0$$

so that

$$(a_1b_2 - a_2b_1)x = b_1d_2 - b_2d_1$$

Then $x =$ .......................

**3**

$$x = \frac{b_1 d_2 - b_2 d_1}{a_1 b_2 - a_2 b_1}$$

In practice, this result can give a finite value for $x$ only if the denominator is not zero. That is, the equations

$$a_1 x + b_1 y + d_1 = 0$$
$$a_2 x + b_2 y + d_2 = 0$$

give a finite value for $x$ provided that $(a_1 b_2 - a_2 b_1) \neq 0$.

Consider these equations:

$$3x + 2y - 5 = 0$$
$$4x + 3y - 7 = 0$$

In this case, $a_1 = 3,\ b_1 = 2,\ a_2 = 4,\ b_2 = 3$

$$a_1 b_2 - a_2 b_1 = 3.3 - 4.2$$
$$= 9 - 8 = 1$$

This is not zero, so there $\left\{\begin{matrix}\text{will}\\ \text{will not}\end{matrix}\right\}$ be a finite value of $x$.

**4**

will

The expression $a_1 b_2 - a_2 b_1$ is therefore an important one in the solution of simultaneous equations. We have a shorthand notation for this.

$$a_1 b_2 - a_2 b_1 = \begin{vmatrix} a_1 & b_1 \\ a_2 & b_2 \end{vmatrix}$$

For $\begin{vmatrix} a_1 & b_1 \\ a_2 & b_2 \end{vmatrix}$ to represent $a_1 b_2 - a_2 b_1$ then we must multiply the terms diagonally to form the product terms in the expansion: we multiply $\begin{vmatrix} a_1 & \\ & b_2 \end{vmatrix}$ and then subtract the product $\begin{vmatrix} & b_1 \\ a_2 & \end{vmatrix}$ i.e. + ↘ and − ↗

e.g. $\begin{vmatrix} 3 & 7 \\ 5 & 2 \end{vmatrix} = \begin{vmatrix} 3 & \\ & 2 \end{vmatrix} - \begin{vmatrix} & 7 \\ 5 & \end{vmatrix} = 3.2 - 5.7 = 6 - 35 = -29$

So $\begin{vmatrix} 6 & 5 \\ 1 & 2 \end{vmatrix} = \begin{vmatrix} 6 & \\ & 2 \end{vmatrix} - \begin{vmatrix} & 5 \\ 1 & \end{vmatrix} =$ ........................

**5**

$$\begin{vmatrix} 6 & 5 \\ 1 & 2 \end{vmatrix} = 12 - 5 = \boxed{7}$$

□□□□□□□□□□□□□□□□□□□□□□□□□□□□□□□□□□□□□□

$\begin{vmatrix} a_1 & b_1 \\ a_2 & b_2 \end{vmatrix}$ is called a *determinant* of the second order (since it has two rows and two columns) and represents $a_1b_2 - a_2b_1$. You can easily remember this as $+ \searrow - \nearrow$.

Just for practice, evaluate the following determinants:

$$\text{(i)} \begin{vmatrix} 4 & 2 \\ 5 & 3 \end{vmatrix}, \quad \text{(ii)} \begin{vmatrix} 7 & 4 \\ 6 & 3 \end{vmatrix}, \quad \text{(iii)} \begin{vmatrix} 2 & 1 \\ 4 & -3 \end{vmatrix}$$

*Finish all three: then turn on to frame 6.*

**6**

$$\text{(i)} \begin{vmatrix} 4 & 2 \\ 5 & 3 \end{vmatrix} = 4.3 - 5.2 = 12 - 10 = \boxed{2}$$

$$\text{(ii)} \begin{vmatrix} 7 & 4 \\ 6 & 3 \end{vmatrix} = 7.3 - 6.4 = 21 - 24 = \boxed{-3}$$

$$\text{(iii)} \begin{vmatrix} 2 & 1 \\ 4 & -3 \end{vmatrix} = 2(-3) - 4.1 = -6 - 4 = \boxed{-10}$$

□□□□□□□□□□□□□□□□□□□□□□□□□□□□□□□□□□□□□□

Now, in solving the equations $\begin{cases} a_1x + b_1y + d_1 = 0 \\ a_2x + b_2y + d_2 = 0 \end{cases}$

we found that $x = \dfrac{b_1d_2 - b_2d_1}{a_1b_2 - a_2b_1}$ and the numerator and the denominator can each be written as a determinant.

$b_1d_2 - b_2d_1 = \ldots\ldots\ldots\ldots$ ; $\quad a_1b_2 - a_2b_1 = \ldots\ldots\ldots\ldots$

**7**

$$\begin{vmatrix} b_1 & d_1 \\ b_2 & d_2 \end{vmatrix} ; \begin{vmatrix} a_1 & b_1 \\ a_2 & b_2 \end{vmatrix}$$

If we eliminate $x$ from the original equations and find an expression for $y$, we obtain

$$y = -\left\{\frac{a_1d_2 - a_2d_1}{a_1b_2 - a_2b_1}\right\}$$

So, for any pair of simultaneous equations

$$a_1x + b_1y + d_1 = 0$$

$$a_2x + b_2y + d_2 = 0$$

we have $$x = \frac{b_1d_2 - b_2d_1}{a_1b_2 - a_2b_1} \quad \text{and} \quad y = -\frac{a_1d_2 - a_2d_1}{a_1b_2 - a_2b_1}$$

Each of these numerators and denominators can be expressed as a determinant.

So, $x =$ ........................ and $y =$ ........................

**8**

$$x = \frac{\begin{vmatrix} b_1 & d_1 \\ b_2 & d_2 \end{vmatrix}}{\begin{vmatrix} a_1 & b_1 \\ a_2 & b_2 \end{vmatrix}} \quad \text{and} \quad y = -\frac{\begin{vmatrix} a_1 & d_1 \\ a_2 & d_2 \end{vmatrix}}{\begin{vmatrix} a_1 & b_1 \\ a_2 & b_2 \end{vmatrix}}$$

$$\therefore \frac{x}{\begin{vmatrix} b_1 & d_1 \\ b_2 & d_2 \end{vmatrix}} = \frac{1}{\begin{vmatrix} a_1 & b_1 \\ a_2 & b_2 \end{vmatrix}} \quad \text{and} \quad \frac{y}{\begin{vmatrix} a_1 & d_1 \\ a_2 & d_2 \end{vmatrix}} = \frac{-1}{\begin{vmatrix} a_1 & b_1 \\ a_2 & b_2 \end{vmatrix}}$$

We can combine these results, thus:

$$\frac{x}{\begin{vmatrix} b_1 & d_1 \\ b_2 & d_2 \end{vmatrix}} = \frac{-y}{\begin{vmatrix} a_1 & d_1 \\ a_2 & d_2 \end{vmatrix}} = \frac{1}{\begin{vmatrix} a_1 & b_1 \\ a_2 & b_2 \end{vmatrix}}$$

*Make a note of these results and then turn on to the next frame.*

**9** So if

$$\begin{cases} a_1x + b_1y + d_1 = 0 \\ a_2x + b_2y + d_2 = 0 \end{cases}$$

Then

$$\frac{x}{\begin{vmatrix} b_1 & d_1 \\ b_2 & d_2 \end{vmatrix}} = \frac{-y}{\begin{vmatrix} a_1 & d_1 \\ a_2 & d_2 \end{vmatrix}} = \frac{1}{\begin{vmatrix} a_1 & b_1 \\ a_2 & b_2 \end{vmatrix}}$$

Each variable is divided by a determinant. Let us see how we can get them from the original equations.

(i) Consider $\dfrac{x}{\begin{vmatrix} b_1 & d_1 \\ b_2 & d_2 \end{vmatrix}}$. Let us denote the determinant in the denominator by $\Delta_1$, i.e. $\Delta_1 = \begin{vmatrix} b_1 & d_1 \\ b_2 & d_2 \end{vmatrix}$.

To form $\Delta_1$ from the given equations, omit the $x$-terms and write down the coefficients and constant terms in the order in which they stand.

$$\begin{cases} a_1x + b_1y + d_1 = 0 \\ a_2x + b_2y + d_2 = 0 \end{cases} \text{ gives } \begin{vmatrix} b_1 & d_1 \\ b_2 & d_2 \end{vmatrix}$$

(ii) Similarly for $\dfrac{-y}{\begin{vmatrix} a_1 & d_1 \\ a_2 & d_2 \end{vmatrix}}$, let $\Delta_2 = \begin{vmatrix} a_1 & d_1 \\ a_2 & d_2 \end{vmatrix}$

To form $\Delta_2$ from the given equations, omit the $y$-terms and write down the coefficients and constant terms in the order in which they stand.

$$\begin{cases} a_1x + b_1y + d_1 = 0 \\ a_2x + b_2y + d_2 = 0 \end{cases} \text{ gives } \Delta_2 = \begin{vmatrix} a_1 & d_1 \\ a_2 & d_2 \end{vmatrix}$$

(iii) For the expression $\dfrac{1}{\begin{vmatrix} a_1 & b_1 \\ a_2 & b_2 \end{vmatrix}}$, denote the determinant by $\Delta_0$.

To form $\Delta_0$ from the given equations, omit the constant terms and write down the coefficients in the order in which they stand

$$\begin{cases} a_1x + b_1y + d_1 = 0 \\ a_2x + b_2y + d_2 = 0 \end{cases} \text{ gives } \begin{vmatrix} a_1 & b_1 \\ a_2 & b_2 \end{vmatrix}$$

Note finally that

$$\frac{x}{\Delta_1} = -\frac{y}{\Delta_2} = \frac{1}{\Delta_0}$$

*Now let us do some examples, so on to frame 10.*

**10**

*Example 1.* To solve the equations $\begin{cases} 5x + 2y + 19 = 0 \\ 3x + 4y + 17 = 0 \end{cases}$

The key to the method is

$$\frac{x}{\Delta_1} = \frac{-y}{\Delta_2} = \frac{1}{\Delta_0}$$

To find $\Delta_0$, omit the constant terms

$$\therefore \Delta_0 = \begin{vmatrix} 5 & 2 \\ 3 & 4 \end{vmatrix} = 5.4 - 3.2 = 20 - 6 = 14$$

$$\therefore \Delta_0 = 14 \quad \ldots \text{ (i)}$$

Now, to find $\Delta_1$, omit the $x$-terms.

$$\therefore \Delta_1 = \ldots\ldots\ldots\ldots\ldots\ldots\ldots$$

**11**

$$\boxed{\Delta_1 = -42}$$

for $$\Delta_1 = \begin{vmatrix} 2 & 19 \\ 4 & 17 \end{vmatrix} = 34 - 76 = -42 \quad \ldots \text{ (ii)}$$

Similarly, to find $\Delta_2$, omit the $y$-terms

$$\Delta_2 = \begin{vmatrix} 5 & 19 \\ 3 & 17 \end{vmatrix} = 85 - 57 = 28 \quad \ldots \text{ (iii)}$$

Substituting the values of $\Delta_1$, $\Delta_2$, $\Delta_0$ in the key, we get

$$\frac{x}{-42} = \frac{-y}{28} = \frac{1}{14}$$

from which $x = \ldots\ldots\ldots\ldots\ldots\ldots$ and $y = \ldots\ldots\ldots\ldots\ldots\ldots$

**12**

$$\boxed{x = \frac{-42}{14} = \underline{\underline{-3}};\ -y = \frac{28}{14},\ y = \underline{\underline{-2}}}$$

Now for another example.

*Example 2.* Solve by determinants $\begin{cases} 2x + 3y - 14 = 0 \\ 3x - 2y + 5 = 0 \end{cases}$

First of all, write down the key:

$$\frac{x}{\Delta_1} = \frac{-y}{\Delta_2} = \frac{1}{\Delta_0}$$

(Note that the terms are alternately positive and negative.)

Then $$\Delta_0 = \begin{vmatrix} 2 & 3 \\ 3 & -2 \end{vmatrix} = -4 - 9 = -13 \quad \ldots \text{ (i)}$$

Now you find $\Delta_1$ and $\Delta_2$ in the same way.

**13**

$$\boxed{\Delta_1 = -13; \quad \Delta_2 = 52}$$

For we have
$$\begin{cases} 2x + 3y - 14 = 0 \\ 3x - 2y + \phantom{1}5 = 0 \end{cases}$$

$$\therefore \Delta_1 = \begin{vmatrix} 3 & -14 \\ -2 & 5 \end{vmatrix} = \begin{vmatrix} 3 & \\ & 5 \end{vmatrix} - \begin{vmatrix} & -14 \\ -2 & \end{vmatrix}$$

$$= 15 - 28 = -13. \quad \therefore \Delta_1 = -13$$

$$\Delta_2 = \begin{vmatrix} 2 & -14 \\ 3 & 5 \end{vmatrix} = \begin{vmatrix} 2 & \\ & 5 \end{vmatrix} - \begin{vmatrix} & -14 \\ 3 & \end{vmatrix}$$

$$= 10 - (-42) = 52 \quad \therefore \Delta_2 = 52$$

So that
$$\frac{x}{\Delta_1} = \frac{-y}{\Delta_2} = \frac{1}{\Delta_0}$$

and $\Delta_1 = -13; \; \Delta_2 = 52; \; \Delta_0 = -13$

$$\therefore \quad x = \frac{\Delta_1}{\Delta_0} = \frac{-13}{-13} = 1 \quad \therefore \underline{x = 1}$$

$$-y = \frac{\Delta_2}{\Delta_0} = \frac{52}{-13} = -4 \quad \therefore \underline{y = 4}$$

Do not forget the key
$$\frac{x}{\Delta_1} = \frac{-y}{\Delta_2} = \frac{1}{\Delta_0}$$

with alternate plus and minus signs.

*Make a note of this in your record book.*

---

**14** Here is another one: do it on your own.

*Example 3.* Solve by determinants

$$\begin{cases} 4x - 3y + 20 = 0 \\ 3x + 2y - \phantom{2}2 = 0 \end{cases}$$

First of all, write down the key.

Then off you go: find $\Delta_0$, $\Delta_1$ and $\Delta_2$ and hence determine the values of $x$ and $y$.

*When you have finished, turn on to frame 15.*

**15**

$$x = -2; \quad y = 4$$

Here is the working in detail:

$$\begin{cases} 4x - 3y + 20 = 0 \\ 3x + 2y - \;\; 2 = 0 \end{cases} \qquad \frac{x}{\Delta_1} = \frac{-y}{\Delta_2} = \frac{1}{\Delta_0}$$

$$\Delta_0 = \begin{vmatrix} 4 & -3 \\ 3 & 2 \end{vmatrix} = 8 - (-9) = 8 + 9 = 17$$

$$\Delta_1 = \begin{vmatrix} -3 & 20 \\ 2 & -2 \end{vmatrix} = 6 - 40 = -34$$

$$\Delta_2 = \begin{vmatrix} 4 & 20 \\ 3 & -2 \end{vmatrix} = -8 - 60 = -68$$

$$x = \frac{\Delta_1}{\Delta_0} = \frac{-34}{17} = -2 \qquad \therefore \underline{x = -2}$$

$$-y = \frac{\Delta_2}{\Delta_0} = \frac{-68}{17} = -4 \qquad \therefore \underline{y = 4}$$

□□□□□□□□□□□□□□□□□□□□□□□□□□□□□□□□□□□□□□

Now, by way of revision, complete the following:

(i) $\begin{vmatrix} 5 & 6 \\ 7 & 4 \end{vmatrix} =$ ..........................

(ii) $\begin{vmatrix} 5 & -2 \\ -3 & -4 \end{vmatrix} =$ ..........................

(iii) $\begin{vmatrix} a & d \\ b & c \end{vmatrix} =$ ..........................

(iv) $\begin{vmatrix} p & q \\ r & s \end{vmatrix} =$ ..........................

---

**16**

Here are the results. You must have got them correct.

(i) $20 - 42 = -22$
(ii) $-20 - 6 = -26$
(iii) $ac - bd$
(iv) $ps - rq$

*For the next section of the work, turn on to frame 17.*

**17**

**Determinants of the third order**

A determinant of the third order will contain 3 rows and 3 columns, thus:

$$\begin{vmatrix} a_1 & b_1 & c_1 \\ a_2 & b_2 & c_2 \\ a_3 & b_3 & c_3 \end{vmatrix}$$

Each element in the determinant is associated with its MINOR, which is found by omitting the row and column containing the element concerned.

e.g. the minor of $a_1$ is $\begin{vmatrix} b_2 & c_2 \\ b_3 & c_3 \end{vmatrix}$ obtained $\begin{matrix} a_1 & b_1 & c_1 \\ a_2 & b_2 & c_2 \\ a_3 & b_3 & c_3 \end{matrix}$

the minor of $b_1$ is $\begin{vmatrix} a_2 & c_2 \\ a_3 & c_3 \end{vmatrix}$ obtained $\begin{matrix} a_1 & b_1 & c_1 \\ a_2 & b_2 & c_2 \\ a_3 & b_3 & c_3 \end{matrix}$

the minor of $c_1$ is $\begin{vmatrix} a_2 & b_2 \\ a_3 & b_3 \end{vmatrix}$ obtained $\begin{matrix} a_1 & b_1 & c_1 \\ a_2 & b_2 & c_2 \\ a_3 & b_3 & c_3 \end{matrix}$

So, in the same way, the minor of $a_2$ is ................

**18**

Minor of $a_2$ is $\begin{vmatrix} b_1 & c_1 \\ b_3 & c_3 \end{vmatrix}$

since, to find the minor of $a_2$, we simply ignore the row and column containing $a_2$, i.e.

$$\begin{matrix} a_1 & b_1 & c_1 \\ a_2 & b_2 & c_2 \\ a_3 & b_3 & c_3 \end{matrix}$$

Similarly, the minor of $b_3$ is ................

**19**

Minor of $b_3$ = $\begin{vmatrix} a_1 & c_1 \\ a_2 & c_2 \end{vmatrix}$

i.e. omit the row and column containing $b_3$.

$$\begin{matrix} a_1 & b_1 & c_1 \\ a_2 & b_2 & c_2 \\ a_3 & b_3 & c_3 \end{matrix}$$

*Now on to frame 20.*

20

**Evaluation of a third order determinant**

To expand a determinant of the third order, we can write down each element along the top row, multiply it by its minor and give the terms a plus or minus sign alternately.

$$\begin{vmatrix} a_1 & b_1 & c_1 \\ a_2 & b_2 & c_2 \\ a_3 & b_3 & c_3 \end{vmatrix} = a_1\begin{vmatrix} b_2 & c_2 \\ b_3 & c_3 \end{vmatrix} - b_1\begin{vmatrix} a_2 & c_2 \\ a_3 & c_3 \end{vmatrix} + c_1\begin{vmatrix} a_2 & b_2 \\ a_3 & b_3 \end{vmatrix}$$

Then, of course, we already know how to expand a determinant of the second order by multiplying diagonally, $+ \searrow \; - \nearrow$

*Example 1.*

$$\begin{vmatrix} 1 & 3 & 2 \\ 4 & 5 & 7 \\ 2 & 4 & 8 \end{vmatrix} = 1\begin{vmatrix} 5 & 7 \\ 4 & 8 \end{vmatrix} - 3\begin{vmatrix} 4 & 7 \\ 2 & 8 \end{vmatrix} + 2\begin{vmatrix} 4 & 5 \\ 2 & 4 \end{vmatrix}$$

$$= 1(5.8 - 4.7) - 3(4.8 - 2.7) + 2(4.4 - 2.5)$$
$$= 1(40 - 28) - 3(32 - 14) + 2(16 - 10)$$
$$= 1(12) - 3(18) + 2(6)$$
$$= 12 - 54 + 12 = \underline{-30}$$

21

Here is another.

*Example 2.*

$$\begin{vmatrix} 3 & 2 & 5 \\ 4 & 6 & 7 \\ 2 & 9 & 2 \end{vmatrix} = 3\begin{vmatrix} 6 & 7 \\ 9 & 2 \end{vmatrix} - 2\begin{vmatrix} 4 & 7 \\ 2 & 2 \end{vmatrix} + 5\begin{vmatrix} 4 & 6 \\ 2 & 9 \end{vmatrix}$$

$$= 3(12 - 63) - 2(8 - 14) + 5(36 - 12)$$
$$= 3(-51) - 2(-6) + 5(24)$$
$$= -153 + 12 + 120 = \underline{-21}$$

Now here is one for you to do.

*Example 3.* Evaluate

$$\begin{vmatrix} 2 & 7 & 5 \\ 4 & 6 & 3 \\ 8 & 9 & 1 \end{vmatrix}$$

Expand along the top row, multiply each element by its minor, and assign alternate + and − signs to the products.

*When you are ready, move on to frame 22.*

**22** *Result*

38

For

$$\begin{vmatrix} 2 & 7 & 5 \\ 4 & 6 & 3 \\ 8 & 9 & 1 \end{vmatrix} = 2\begin{vmatrix} 6 & 3 \\ 9 & 1 \end{vmatrix} - 7\begin{vmatrix} 4 & 3 \\ 8 & 1 \end{vmatrix} + 5\begin{vmatrix} 4 & 6 \\ 8 & 9 \end{vmatrix}$$

$$= 2(6-27) - 7(4-24) + 5(36-48)$$
$$= 2(-21) - 7(-20) + 5(-12)$$
$$= -42 + 140 - 60 = \underline{38}$$

We obtained the result above by expanding along the top row of the given determinant. If we expand down the first column in the same way, still assigning alternate + and − signs to the products, we get

$$\begin{vmatrix} 2 & 7 & 5 \\ 4 & 6 & 3 \\ 8 & 9 & 1 \end{vmatrix} = 2\begin{vmatrix} 6 & 3 \\ 9 & 1 \end{vmatrix} - 4\begin{vmatrix} 7 & 5 \\ 9 & 1 \end{vmatrix} + 8\begin{vmatrix} 7 & 5 \\ 6 & 3 \end{vmatrix}$$

$$= 2(6-27) - 4(7-45) + 8(21-30)$$
$$= 2(-21) - 4(-38) + 8(-9)$$
$$= -42 + 152 - 72 = \underline{38}$$

which is the same result as that which we obtained before.

**23** We can, if we wish, expand along any row or column in the same way, multiplying each element by its minor, so long as we assign to each product the appropriate + or − sign. The appropriate 'place signs' are given by

$$\begin{matrix} + & - & + & - & + & \dots & \dots \\ - & + & - & + & - & \dots & \dots \\ + & - & + & - & + & \dots & \dots \\ - & + & - & + & - & \dots & \dots \end{matrix}$$

etc., etc.

The key element (in the top left-hand corner) is always +. The others are then alternately + or −, as you proceed along any row or down any column.

So in the determinant

$$\begin{vmatrix} 1 & 3 & 7 \\ 5 & 6 & 9 \\ 4 & 2 & 8 \end{vmatrix}$$

the "place sign" of the element 9 is ......................

**24**

$$\boxed{-}$$

since in a third order determinant, the 'place signs' are

$$\begin{vmatrix} + & - & + \\ - & + & - \\ + & - & + \end{vmatrix}$$

Remember that the top left-hand element always has a + place sign. The others follow from it.

Now consider this one

$$\begin{vmatrix} 3 & 7 & 2 \\ 6 & 8 & 4 \\ 1 & 9 & 5 \end{vmatrix}$$

If we expand down the middle column, we get

$$\begin{vmatrix} 3 & 7 & 2 \\ 6 & 8 & 4 \\ 1 & 9 & 5 \end{vmatrix} = -7\begin{vmatrix} 6 & 4 \\ 1 & 5 \end{vmatrix} + 8\begin{vmatrix} 3 & 2 \\ 1 & 5 \end{vmatrix} - 9\begin{vmatrix} 3 & 2 \\ 6 & 4 \end{vmatrix}$$

$= \ldots\ldots\ldots\ldots\ldots\ldots\ldots\ldots$

*Finish it off. Then move on.*

**25**

*Result*

$$\boxed{-78}$$

for

$$-7\begin{vmatrix} 6 & 4 \\ 1 & 5 \end{vmatrix} + 8\begin{vmatrix} 3 & 2 \\ 1 & 5 \end{vmatrix} - 9\begin{vmatrix} 3 & 2 \\ 6 & 4 \end{vmatrix}$$

$$= -7(30 - 4) + 8(15 - 2) - 9(12 - 12)$$
$$= -7(26) + 8(13) - 9(0)$$
$$= -182 + 104 = \underline{-78}$$

So now you do this one:

Evaluate $\begin{vmatrix} 2 & 3 & 4 \\ 6 & 1 & 3 \\ 5 & 7 & 2 \end{vmatrix}$ by expanding along the *bottom* row.

*When you have done it,* *turn to frame 26.*

**26**

Answer $\boxed{119}$

We have $\begin{vmatrix} 2 & 3 & 4 \\ 6 & 1 & 3 \\ 5 & 7 & 2 \end{vmatrix}$ and remember $\begin{vmatrix} + & - & + \\ - & + & - \\ + & - & + \end{vmatrix}$

$$= 5\begin{vmatrix} 3 & 4 \\ 1 & 3 \end{vmatrix} - 7\begin{vmatrix} 2 & 4 \\ 6 & 3 \end{vmatrix} + 2\begin{vmatrix} 2 & 3 \\ 6 & 1 \end{vmatrix}$$

$$= 5(9-4) - 7(6-24) + 2(2-18)$$

$$= 5(5) - 7(-18) + 2(-16)$$

$$= 25 + 126 - 32 = \underline{119}$$

One more:

Evaluate $\begin{vmatrix} 1 & 2 & 8 \\ 7 & 3 & 1 \\ 4 & 6 & 9 \end{vmatrix}$ by expanding along the *middle* row.

**27**

*Result* $\boxed{143}$

For $$\begin{vmatrix} 1 & 2 & 8 \\ 7 & 3 & 1 \\ 4 & 6 & 9 \end{vmatrix} = -7\begin{vmatrix} 2 & 8 \\ 6 & 9 \end{vmatrix} + 3\begin{vmatrix} 1 & 8 \\ 4 & 9 \end{vmatrix} - 1\begin{vmatrix} 1 & 2 \\ 4 & 6 \end{vmatrix}$$

$$= -7(18-48) + 3(9-32) - 1(6-8)$$

$$= -7(-30) + 3(-23) - 1(-2)$$

$$= 210 - 69 + 2 = \underline{143}$$

□□□□□□□□□□□□□□□□□□□□□□□□□□□□□□□□□□□□□□

We have seen how we can use second order determinants to solve simultaneous equations in 2 unknowns.

We can now extend the method to solve simultaneous equations in 3 unknowns.

*So turn on to frame 28.*

28

## Simultaneous equations in three unknowns

Consider the equations

$$\begin{cases} a_1x + b_1y + c_1z + d_1 = 0 \\ a_2x + b_2y + c_2z + d_2 = 0 \\ a_3x + b_3y + c_3z + d_3 = 0 \end{cases}$$

If we find $x, y$ and $z$ by the elimination method, we obtain results that can be expressed in determinant form thus:

$$\frac{x}{\begin{vmatrix} b_1 & c_1 & d_1 \\ b_2 & c_2 & d_2 \\ b_3 & c_3 & d_3 \end{vmatrix}} = \frac{-y}{\begin{vmatrix} a_1 & c_1 & d_1 \\ a_2 & c_2 & d_2 \\ a_3 & c_3 & d_3 \end{vmatrix}} = \frac{z}{\begin{vmatrix} a_1 & b_1 & d_1 \\ a_2 & b_2 & d_2 \\ a_3 & b_3 & d_3 \end{vmatrix}} = \frac{-1}{\begin{vmatrix} a_1 & b_1 & c_1 \\ a_2 & b_2 & c_2 \\ a_3 & b_3 & c_3 \end{vmatrix}}$$

We can remember this more easily in this form:–

$$\frac{x}{\Delta_1} = \frac{-y}{\Delta_2} = \frac{z}{\Delta_3} = \frac{-1}{\Delta_0}$$

where $\Delta_1$ = the det. of the coefficients omitting the $x$-terms
$\Delta_2$ = " " " " " " " $y$-terms
$\Delta_3$ = " " " " " " " $z$-terms
$\Delta_0$ = " " " " " " " constant terms.

Notice that the signs are alternately plus and minus.

Let us work through a numerical example.

*Example 1.* Find the value of $x$ from the equations

$$\begin{cases} 2x + 3y - z - 4 = 0 \\ 3x + y + 2z - 13 = 0 \\ x + 2y - 5z + 11 = 0 \end{cases}$$

First the key: $$\frac{x}{\Delta_1} = \frac{-y}{\Delta_2} = \frac{z}{\Delta_3} = \frac{-1}{\Delta_0}$$

To find the value of $x$, we use $\dfrac{x}{\Delta_1} = \dfrac{-1}{\Delta_0}$, i.e. we must find $\Delta_1$ and $\Delta_0$.

(i) to find $\Delta_0$, omit the constant terms.

$$\therefore \Delta_0 = \begin{vmatrix} 2 & 3 & -1 \\ 3 & 1 & 2 \\ 1 & 2 & -5 \end{vmatrix} = 2\begin{vmatrix} 1 & 2 \\ 2 & -5 \end{vmatrix} - 3\begin{vmatrix} 3 & 2 \\ 1 & -5 \end{vmatrix} - 1\begin{vmatrix} 3 & 1 \\ 1 & 2 \end{vmatrix}$$

$$= -18 + 51 - 5 = 28$$

(ii) Now you find $\Delta_1$, in the same way.

**29**

$$\boxed{\Delta_1 = -56}$$

for
$$\Delta_1 = \begin{vmatrix} 3 & -1 & -4 \\ 1 & 2 & -13 \\ 2 & -5 & 11 \end{vmatrix} \begin{aligned} &= 3(22-65) + 1(11+26) - 4(-5-4) \\ &= 3(-43) + 1(37) - 4(-9) \\ &= -129 + 37 + 36 \\ &= -129 + 73 = -56 \end{aligned}$$

But
$$\frac{x}{\Delta_1} = \frac{-1}{\Delta_0} \qquad \therefore \frac{x}{-56} = \frac{-1}{28}$$

$$\therefore x = \frac{56}{28} = 2 \qquad \therefore \underline{x = 2}$$

Note that by this method we can evaluate any one of the variables, without necessarily finding the others. Let us do another example.

*Example 2.* Find $y$, given that

$$\begin{cases} 2x + \ \ y - 5z + 11 = 0 \\ \ \ x - \ \ y + \ \ z - \ \ 6 = 0 \\ 4x + 2y - 3z + \ \ 8 = 0 \end{cases}$$

First, the key, which is ........................

**30**

$$\boxed{\frac{x}{\Delta_1} = \frac{-y}{\Delta_2} = \frac{z}{\Delta_3} = \frac{-1}{\Delta_0}}$$

To find $y$, we use
$$\frac{-y}{\Delta_2} = \frac{-1}{\Delta_0}$$

Therefore, we must find $\Delta_2$ and $\Delta_0$.

The equations are
$$\begin{cases} 2x + \ \ y - 5z + 11 = 0 \\ \ \ x - \ \ y + \ \ z - \ \ 6 = 0 \\ 4x + 2y - 3z + \ \ 8 = 0 \end{cases}$$

To find $\Delta_2$, omit the $y$-terms.

$$\therefore \Delta_2 = \begin{vmatrix} 2 & -5 & 11 \\ 1 & 1 & -6 \\ 4 & -3 & 8 \end{vmatrix} = 2\begin{vmatrix} 1 & -6 \\ -3 & 8 \end{vmatrix} + 5\begin{vmatrix} 1 & -6 \\ 4 & 8 \end{vmatrix} + 11\begin{vmatrix} 1 & 1 \\ 4 & -3 \end{vmatrix}$$

$$= 2(8-18) + 5(8+24) + 11(-3-4)$$
$$= -20 + 160 - 77 = 63$$

To find $\Delta_0$, omit the constant terms

$$\therefore \Delta_0 = \text{......................}$$

**31**

$$\boxed{\Delta_0 = -21}$$

for

$$\Delta_0 = \begin{vmatrix} 2 & 1 & -5 \\ 1 & -1 & 1 \\ 4 & 2 & -3 \end{vmatrix} = 2\begin{vmatrix} -1 & 1 \\ 2 & -3 \end{vmatrix} - 1\begin{vmatrix} 1 & 1 \\ 4 & -3 \end{vmatrix} - 5\begin{vmatrix} 1 & -1 \\ 4 & 2 \end{vmatrix}$$

$$= 2(3-2) - 1(-3-4) - 5(2+4)$$

$$= 2 + 7 - 30 = -21$$

So we have $\dfrac{-y}{\Delta_2} = \dfrac{-1}{\Delta_0} \quad \therefore\ y = \dfrac{\Delta_2}{\Delta_0} = \dfrac{63}{-21}$

$$\therefore\ \underline{y = -3}$$

The important things to remember are

(i) The key: $\dfrac{x}{\Delta_1} = \dfrac{-y}{\Delta_2} = \dfrac{z}{\Delta_3} = \dfrac{-1}{\Delta_0}$

with alternate + and − signs.

(ii) To find $\Delta_1$, which is associated with $x$ in this case, omit the $x$-terms and form a determinant with the remaining coefficients and constant terms. Similarly for $\Delta_2, \Delta_3, \Delta_0$.

*Next frame.*

**32**

Here is a short revision exercise on the work so far.

**Revision Exercise**

Find the following by the use of determinants.

1. $\left\{\begin{aligned} x + 2y - 3z - 3 &= 0 \\ 2x - y - z - 11 &= 0 \\ 3x + 2y + z + 5 &= 0 \end{aligned}\right\}$ Find $y$.

2. $\left\{\begin{aligned} 3x - 4y + 2z + 8 &= 0 \\ x + 5y - 3z + 2 &= 0 \\ 5x + 3y - z + 6 &= 0 \end{aligned}\right\}$ Find $x$ and $z$.

3. $\left\{\begin{aligned} 2x - 2y - z - 3 &= 0 \\ 4x + 5y - 2z + 3 &= 0 \\ 3x + 4y - 3z + 7 &= 0 \end{aligned}\right\}$ Find $x, y$ and $z$.

*When you have finished them all, check your answers with those given in the next frame.*

## 33

Here are the answers:

1. $y = -4$
2. $x = -2;\ z = 5$
3. $x = 2;\ y = -1;\ z = 3$

If you have them *all* correct, turn straight on to *frame 52*.

If you have not got them all correct, it is well worth spending a few minutes seeing where you may have gone astray, for one of the main applications of determinants is in the solution of simultaneous equations.

*If you made any slips, move to frame 34.*

## 34

The answer to question No. 1 in the revision test was $\boxed{y = -4}$

Did you get that one right? If so, move on straight away to *frame 41*.
If you did not manage to get it right, let us work through it in detail.

The equations were

$$\begin{cases} x + 2y - 3z - 3 = 0 \\ 2x - y - z - 11 = 0 \\ 3x + 2y + z + 5 = 0 \end{cases}$$

Copy them down on your paper so that we can refer to them as we go along.

The first thing, always, is to write down the key to the solutions. In this case:

$$\frac{x}{\Delta_1} = \ldots = \ldots = \ldots$$

To fill in the missing terms, take each variable in turn, divide it by the associated determinant, and include the appropriate sign.

So what do we get?

*On to frame 35.*

**35**

$$\boxed{\frac{x}{\Delta_1} = \frac{-y}{\Delta_2} = \frac{z}{\Delta_3} = \frac{-1}{\Delta_0}}$$

The signs go alternately + and −.

In this question, we have to find $y$, so we use the second and last terms in the key.

$$\text{i.e. } \frac{-y}{\Delta_2} = \frac{-1}{\Delta_0} \qquad \therefore\ y = \frac{\Delta_2}{\Delta_0}.$$

So we have to find $\Delta_2$ and $\Delta_0$.

To find $\Delta_2$, we ..............................................................................

**36**

form a determinant of the coefficients omitting those of the $y$-terms.

So

$$\Delta_2 = \begin{vmatrix} 1 & -3 & -3 \\ 2 & -1 & -11 \\ 3 & 1 & 5 \end{vmatrix}$$

Expanding along the top row, this gives

$$\Delta_2 = 1\begin{vmatrix} -1 & -11 \\ 1 & 5 \end{vmatrix} - (-3)\begin{vmatrix} 2 & -11 \\ 3 & 5 \end{vmatrix} + (-3)\begin{vmatrix} 2 & -1 \\ 3 & 1 \end{vmatrix}$$

We now evaluate each of these second order determinants by the usual process of multiplying diagonally, remembering the sign convention that

$+ \searrow$ and $- \nearrow$

So we get $\Delta_2$ = ..................................

**37**

$$\boxed{\Delta_2 = 120}$$

for
$$\Delta_2 = 1(-5+11) + 3(10+33) - 3(2+3)$$
$$= 6 + 3(43) - 3(5)$$
$$= 6 + 129 - 15 = 135 - 15 = 120$$
$$\therefore \underline{\Delta_2 = 120}$$

We also have to find $\Delta_0$, i.e. the determinant of the coefficients omitting the constant terms.

So
$$\Delta_0 = \begin{vmatrix} \ldots & \ldots & \ldots \\ \ldots & \ldots & \ldots \\ \ldots & \ldots & \ldots \end{vmatrix}$$

**38**

$$\boxed{\Delta_0 = \begin{vmatrix} 1 & 2 & -3 \\ 2 & -1 & -1 \\ 3 & 2 & 1 \end{vmatrix}}$$

If we expand this along the top row, we get

$$\Delta_0 = \ldots\ldots\ldots\ldots\ldots$$

**39**

$$\Delta_0 = 1\begin{vmatrix} -1 & -1 \\ 2 & 1 \end{vmatrix} - 2\begin{vmatrix} 2 & -1 \\ 3 & 1 \end{vmatrix} - 3\begin{vmatrix} 2 & -1 \\ 3 & 2 \end{vmatrix}$$

Now, evaluating the second order determinants in the usual way gives that

$$\Delta_0 = \ldots\ldots\ldots\ldots\ldots$$

**40**

$$\boxed{\Delta_0 = -30}$$

for
$$\Delta_0 = 1(-1+2) - 2(2+3) - 3(4+3)$$
$$= 1(1) - 2(5) - 3(7)$$
$$= 1 - 10 - 21 = -30$$
$$\text{So} \quad \underline{\Delta_0 = -30.}$$

So we have
$$y = \frac{\Delta_2}{\Delta_0} = \frac{120}{-30} = -4$$
$$\therefore \underline{y = -4}$$

Every one is done in the same way.

Did you get No. 2 of the revision questions correct?

If so, turn straight on to *frame 51*.

If not, have another go at it, now that we have worked through No. 1 in detail.

*When you have finished, move to frame 41.*

**41**

The answers to No. 2 in the revision exercise were $\boxed{\begin{array}{l} x = -2 \\ z = 5 \end{array}}$

Did you get those correct? If so, turn on right away to *frame 51*. If not, follow through the working. Here it is:

*No. 2* The equations were

$$\begin{cases} 3x - 4y + 2z + 8 = 0 \\ x + 5y - 3z + 2 = 0 \\ 5x + 3y - z + 6 = 0 \end{cases}$$

Copy them down on to your paper.

The key to the solutions is:

$$\frac{x}{\Delta_1} = \ldots = \ldots = \ldots$$

*Fill in the missing terms and then turn on to frame 42.*

**42**

$$\boxed{\frac{x}{\Delta_1} = \frac{-y}{\Delta_2} = \frac{z}{\Delta_3} = \frac{-1}{\Delta_0}}$$

We have to find $x$ and $z$. $\therefore$ We shall use

$$\frac{x}{\Delta_1} = -\frac{1}{\Delta_0} \quad \text{i.e.} \quad x = -\frac{\Delta_1}{\Delta_0}$$

and $$\frac{z}{\Delta_3} = \frac{-1}{\Delta_0} \quad \text{i.e.} \quad z = -\frac{\Delta_3}{\Delta_0}$$

So we must find $\Delta_1$, $\Delta_3$ and $\Delta_0$.

(i) To find $\Delta_1$, form the determinant of coefficients omitting those of the $x$-terms.

$$\therefore \Delta_1 = \begin{vmatrix} \dots & \dots & \dots \\ \dots & \dots & \dots \\ \dots & \dots & \dots \end{vmatrix}$$

**43**

$$\boxed{\Delta_1 = \begin{vmatrix} -4 & 2 & 8 \\ 5 & -3 & 2 \\ 3 & -1 & 6 \end{vmatrix}}$$

Now expand along the top row.

$$\Delta_1 = -4\begin{vmatrix} -3 & 2 \\ -1 & 6 \end{vmatrix} - 2\begin{vmatrix} 5 & 2 \\ 3 & 6 \end{vmatrix} + 8\begin{vmatrix} 5 & -3 \\ 3 & -1 \end{vmatrix}$$

$$= \dots\dots\dots\dots\dots\dots\dots\dots\dots\dots$$

*Finish it off: then on to frame 44.*

**44**

$$\boxed{\Delta_1 = 48}$$

for

$$\begin{aligned}\Delta_1 &= -4(-18+2) - 2(30-6) + 8(-5+9)\\ &= -4(-16) - 2(24) + 8(4)\\ &= 64 - 48 + 32 = 96 - 48 = 48\end{aligned}$$

$$\therefore \underline{\Delta_1 = 48}$$

(ii) To find $\Delta_3$, form the determinant of coefficients omitting the $z$-terms.

$$\therefore \Delta_3 = \begin{vmatrix} \dots & \dots & \dots \\ \dots & \dots & \dots \\ \dots & \dots & \dots \end{vmatrix}$$

**45**

$$\boxed{\Delta_3 = \begin{vmatrix} 3 & -4 & 8 \\ 1 & 5 & 2 \\ 5 & 3 & 6 \end{vmatrix}}$$

Expanding this along the top row gives

$$\Delta_3 = \dots\dots\dots\dots\dots\dots\dots\dots$$

**46**

$$\boxed{\Delta_3 = 3\begin{vmatrix} 5 & 2 \\ 3 & 6 \end{vmatrix} + 4\begin{vmatrix} 1 & 2 \\ 5 & 6 \end{vmatrix} + 8\begin{vmatrix} 1 & 5 \\ 5 & 3 \end{vmatrix}}$$

Now evaluate the second order determinants and finish it off. So that

$$\Delta_3 = \dots\dots\dots\dots\dots\dots\dots\dots$$

*On to frame 47.*

## 47

$$\boxed{\Delta_3 = -120}$$

since
$$\begin{aligned}\Delta_3 &= 3(30-6)+4(6-10)+8(3-25)\\ &= 3(24)+4(-4)+8(-22)\\ &= 72-16-176.\\ &= 72-192=-120\end{aligned}$$

$$\therefore \Delta_3 = -120$$

(iii) Now we want to find $\Delta_0$.

$$\Delta_0 = \begin{vmatrix} \dots & \dots & \dots \\ \dots & \dots & \dots \\ \dots & \dots & \dots \end{vmatrix}$$

## 48

$$\boxed{\Delta_0 = \begin{vmatrix} 3 & -4 & 2 \\ 1 & 5 & -3 \\ 5 & 3 & -1 \end{vmatrix}}$$

Now expand this along the top row as we have done before. Then evaluate the second order determinants which will appear and so find the value of $\Delta_0$.

Work it right through: so that

$$\Delta_0 = \dots\dots\dots\dots\dots\dots$$

**49**

$$\boxed{\Delta_0 = 24}$$

for

$$\Delta_0 = 3\begin{vmatrix} 5 & -3 \\ 3 & -1 \end{vmatrix} + 4\begin{vmatrix} 1 & -3 \\ 5 & -1 \end{vmatrix} + 2\begin{vmatrix} 1 & 5 \\ 5 & 3 \end{vmatrix}$$

$$= 3(-5+9) + 4(-1+15) + 2(3-25)$$

$$= 3(4) + 4(14) + 2(-22)$$

$$= 12 + 56 - 44$$

$$= 68 - 44 = 24$$

$$\therefore \underline{\Delta_0 = 24}$$

So we have: $\Delta_1 = 48, \quad \Delta_3 = -120, \quad \Delta_0 = 24$

Also we know that

$$x = -\frac{\Delta_1}{\Delta_0} \text{ and } z = -\frac{\Delta_3}{\Delta_0}$$

So that $x =$ .......................... and $z =$ ..........................

**50**

$$x = -\frac{48}{24} = -2 \quad \boxed{x = -2}$$

$$z = -\frac{(-120)}{24} = 5 \quad \boxed{z = 5}$$

Well, there you are. The method is the same every time – but take care not to make a slip with the signs.

Now what about question No. 3 in the revision exercise. Did you get that right? If so, move on straight away to *frame 52*.

If not, have another go at it. Here are the equations again: copy them down and then find $x, y$ and $z$.

$$2x - 2y - z - 3 = 0$$

$$4x + 5y - 2z + 3 = 0$$

$$3x + 4y - 3z + 7 = 0$$

*When you have finished this one, turn on to the next frame and check your results.*

**51** *Answers to No. 3*

$$\boxed{x = 2, \quad y = -1, \quad z = 3}$$

Here are the main steps, so that you can check your own working.

$$\frac{x}{\Delta_1} = \frac{-y}{\Delta_2} = \frac{z}{\Delta_3} = \frac{-1}{\Delta_0}$$

$$\Delta_1 = \begin{vmatrix} -2 & -1 & -3 \\ 5 & -2 & 3 \\ 4 & -3 & 7 \end{vmatrix} = 54$$

$$\Delta_2 = \begin{vmatrix} 2 & -1 & -3 \\ 4 & -2 & 3 \\ 3 & -3 & 7 \end{vmatrix} = 27$$

$$\Delta_3 = \begin{vmatrix} 2 & -2 & -3 \\ 4 & 5 & 3 \\ 3 & 4 & 7 \end{vmatrix} = 81$$

$$\Delta_0 = \begin{vmatrix} 2 & -2 & -1 \\ 4 & 5 & -2 \\ 3 & 4 & -3 \end{vmatrix} = -27$$

$$\frac{x}{\Delta_1} = -\frac{1}{\Delta_0} \qquad \therefore\ x = -\frac{\Delta_1}{\Delta_0} = -\frac{54}{-27} = 2$$

$$\underline{x = 2}$$

$$\frac{-y}{\Delta_2} = -\frac{1}{\Delta_0} \qquad \therefore\ y = \frac{\Delta_2}{\Delta_0} = \frac{27}{-27} = -1$$

$$\underline{y = -1}$$

$$\frac{z}{\Delta_3} = -\frac{1}{\Delta_0} \qquad \therefore\ z = -\frac{\Delta_3}{\Delta_0} = \frac{81}{-27} = -3$$

$$\underline{z = -3}$$

All correct now?

*On to frame 52 then for the next section of the work.*

52

**Consistency of a set of equations**

Let us consider the following three equations in two unknowns.

$$3x - y - 4 = 0 \qquad \text{(i)}$$
$$2x + 3y - 8 = 0 \qquad \text{(ii)}$$
$$x - 2y + 3 = 0 \qquad \text{(iii)}$$

If we solve equations (ii) and (iii) in the usual way, we find that $x = 1$ and $y = 2$.

If we now substitute these values in the left-hand side of (i), we obtain $3x - y - 4 = 3 - 2 - 4 = -3$ (and not 0 as the equation states).

The solutions of (ii) and (iii) do not satisfy (i) and the three given equations do not have a common solution. They are thus not *consistent*. There are no values of $x$ and $y$ which satisfy all three equations.

If equations are consistent, they have a ................ ..................

53

common solution

Let us now consider the three equations

$$3x + y - 5 = 0 \qquad \text{(i)}$$
$$2x + 3y - 8 = 0 \qquad \text{(ii)}$$
$$x - 2y + 3 = 0 \qquad \text{(iii)}$$

The solutions of (ii) and (iii) are, as before, $x = 1$ and $y = 2$. Substituting these in (i) gives

$$3x + y - 5 = 3 + 2 - 5 = 0$$

i.e. all three equations have the common solution $x = 1, y = 2$ and the equations are said to be c ....................

**54**

consistent

Now we will take the general case

$$a_1x + b_1y + d_1 = 0 \qquad \text{(i)}$$
$$a_2x + b_2y + d_2 = 0 \qquad \text{(ii)}$$
$$a_3x + b_3y + d_3 = 0 \qquad \text{(iii)}$$

If we solve equations (ii) and (iii),

i.e. $$\begin{cases} a_2x + b_2y + d_2 = 0 \\ a_3x + b_3y + d_3 = 0 \end{cases}$$

we get $$\frac{x}{\Delta_1} = \frac{-y}{\Delta_2} = \frac{1}{\Delta_0}$$

where $$\Delta_1 = \begin{vmatrix} b_2 & d_2 \\ b_3 & d_3 \end{vmatrix}, \quad \Delta_2 = \begin{vmatrix} a_2 & d_2 \\ a_3 & d_3 \end{vmatrix}, \quad \Delta_0 = \begin{vmatrix} a_2 & b_2 \\ a_3 & b_3 \end{vmatrix}$$

so that $$x = \frac{\Delta_1}{\Delta_0} \text{ and } y = -\frac{\Delta_2}{\Delta_0}$$

If these results also satisfy equation (i), then

$$a_1 \cdot \frac{\Delta_1}{\Delta_0} + b_1 \cdot \frac{-\Delta_2}{\Delta_0} + d_1 = 0$$

i.e. $$a_1 \cdot \Delta_1 - b_1 \cdot \Delta_2 + d_1 \cdot \Delta_0 = 0$$

i.e. $$a_1 \begin{vmatrix} b_2 & d_2 \\ b_3 & d_3 \end{vmatrix} - b_1 \begin{vmatrix} a_2 & d_2 \\ a_3 & d_3 \end{vmatrix} + d_1 \begin{vmatrix} a_2 & b_2 \\ a_3 & b_3 \end{vmatrix} = 0$$

i.e. $$\begin{vmatrix} a_1 & b_1 & d_1 \\ a_2 & b_2 & d_2 \\ a_3 & b_3 & d_3 \end{vmatrix} = 0$$

which is therefore the condition that the three given equations are *consistent.*

So three simultaneous equations in two unknowns are consistent if the determinant of coefficients is ........................

zero

**55**

*Example 1.* Test for consistency
$$\begin{cases} 2x + y - 5 = 0 \\ x + 4y + 1 = 0 \\ 3x - y - 10 = 0 \end{cases}$$

For the equations to be consistent $\begin{vmatrix} 2 & 1 & -5 \\ 1 & 4 & 1 \\ 3 & -1 & -10 \end{vmatrix}$ must be zero.

$$\begin{vmatrix} 2 & 1 & -5 \\ 1 & 4 & 1 \\ 3 & -1 & -10 \end{vmatrix} = 2\begin{vmatrix} 4 & 1 \\ -1 & -10 \end{vmatrix} - 1\begin{vmatrix} 1 & 1 \\ 3 & -10 \end{vmatrix} - 5\begin{vmatrix} 1 & 4 \\ 3 & -1 \end{vmatrix}$$

$$= 2(-40 + 1) - 1(-10 - 3) - 5(-1 - 12)$$
$$= 2(-39) - (-13) - 5(-13)$$
$$= -78 + 13 + 65 = -78 + 78 = 0$$

The given equations therefore.................... consistent.
(are/are not)

---

are

**56**

*Example 2.* Find the value of $k$ for which the equations are consistent.

$$\begin{cases} 3x + y + 2 = 0 \\ 4x + 2y - k = 0 \\ 2x - y + 3k = 0 \end{cases}$$

For consistency, $\begin{vmatrix} 3 & 1 & 2 \\ 4 & 2 & -k \\ 2 & -1 & 3k \end{vmatrix} = 0$

$$\therefore 3\begin{vmatrix} 2 & -k \\ -1 & 3k \end{vmatrix} - 1\begin{vmatrix} 4 & -k \\ 2 & 3k \end{vmatrix} + 2\begin{vmatrix} 4 & 2 \\ 2 & -1 \end{vmatrix} = 0$$

$$3(6k - k) - 1(12k + 2k) + 2(-4 - 4) = 0$$

$$\therefore 15k - 14k - 16 = 0 \quad \therefore k - 16 = 0 \quad \therefore \underline{k = 16}$$

Now one for you, done in just the same way.

*Example 3.* Given
$$\begin{cases} x + (k+1)y + 1 = 0 \\ 2kx + 5y - 3 = 0 \\ 3x + 7y + 1 = 0 \end{cases}$$

Find the values of $k$ for which the equations are consistent.

## 57

$$\boxed{k = 2 \text{ or } -\frac{1}{2}}$$

The condition for consistency is that

$$\begin{vmatrix} 1 & k+1 & 1 \\ 2k & 5 & -3 \\ 3 & 7 & 1 \end{vmatrix} = 0$$

$$\therefore 1\begin{vmatrix} 5 & -3 \\ 7 & 1 \end{vmatrix} - (k+1)\begin{vmatrix} 2k & -3 \\ 3 & 1 \end{vmatrix} + 1\begin{vmatrix} 2k & 5 \\ 3 & 7 \end{vmatrix} = 0$$

$$(5 + 21) - (k + 1)(2k + 9) + (14k - 15) = 0$$

$$26 - 2k^2 - 11k - 9 + 14k - 15 = 0$$

$$-2k^2 + 3k + 2 = 0$$

$$\therefore 2k^2 - 3k - 2 = 0 \quad \therefore (2k + 1)(k - 2) = 0$$

$$\therefore \underline{k = 2 \text{ or } k = -\frac{1}{2}}$$

Finally, one more for you to do.

*Example 4.*

Find the values of $k$ for consistency when $\begin{cases} x + y - k = 0 \\ kx - 3y + 11 = 0 \\ 2x + 4y - 8 = 0 \end{cases}$

## 58

$$\boxed{k = 1 \text{ or } -\frac{1}{2}}$$

For

$$\begin{vmatrix} 1 & 1 & -k \\ k & -3 & 11 \\ 2 & 4 & -8 \end{vmatrix} = 0$$

$$1\begin{vmatrix} -3 & 11 \\ 4 & -8 \end{vmatrix} - 1\begin{vmatrix} k & 11 \\ 2 & -8 \end{vmatrix} - k\begin{vmatrix} k & -3 \\ 2 & 4 \end{vmatrix} = 0$$

$$\therefore (24 - 44) - (-8k - 22) - k(4k + 6) = 0$$

$$\therefore -20 + 8k + 22 - 4k^2 - 6k = 0$$

$$-4k^2 + 2k + 2 = 0$$

$$\therefore 2k^2 - k - 1 = 0 \quad \therefore (2k + 1)(k - 1) = 0$$

$$\therefore \underline{k = 1 \text{ or } k = -\frac{1}{2}}$$

**59**

**Properties of determinants**

Expanding a determinant in which the elements are large numbers can be a very tedious affair. It is possible, however, by knowing something of the properties of determinants, to simplify the working. So here are some of the main properties. Make a note of them in your record book for future reference.

1. *The value of a determinant remains unchanged if rows are changed to columns and columns to rows.*

$$\begin{vmatrix} a_1 & a_2 \\ b_1 & b_2 \end{vmatrix} = \begin{vmatrix} a_1 & b_1 \\ a_2 & b_2 \end{vmatrix}$$

2. *If two rows (or two columns) are interchanged, the sign of the determinant is changed.*

$$\begin{vmatrix} a_2 & b_2 \\ a_1 & b_1 \end{vmatrix} = -\begin{vmatrix} a_1 & b_1 \\ a_2 & b_2 \end{vmatrix}$$

3. *If two rows (or two columns) are identical, the value of the determinant is zero.*

$$\begin{vmatrix} a_1 & a_1 \\ a_2 & a_2 \end{vmatrix} = 0$$

4. *If the elements of any one row (or column) are all multiplied by a common factor, the determinant is multiplied by that factor.*

$$\begin{vmatrix} ka_1 & kb_1 \\ a_2 & b_2 \end{vmatrix} = k\begin{vmatrix} a_1 & b_1 \\ a_2 & b_2 \end{vmatrix}$$

5. *If the elements of any one row (or column) are increased (or decreased) by equal multiples of the corresponding elements of any other row (or column), the value of the determinant is unchanged.*

$$\begin{vmatrix} a_1 + kb_1 & b_1 \\ a_2 + kb_2 & b_2 \end{vmatrix} = \begin{vmatrix} a_1 & b_1 \\ a_2 & b_2 \end{vmatrix}$$

□ □ □ □ □ □ □ □ □ □ □ □ □ □ □ □ □ □ □ □ □ □ □ □ □ □ □ □ □ □ □ □ □ □ □ □ □ □ □ □

*NOTE:* The properties stated above are general and apply not only to second order determinants, but to determinants of any order.

*Turn on now to the next frame for one or two examples.*

**60** *Example 1.* Evaluate $\begin{vmatrix} 427 & 429 \\ 369 & 371 \end{vmatrix}$

Of course, we could evaluate this by the usual method

$$(427)(371)-(369)(429)$$

which is rather deadly! On the other hand, we could apply our knowledge of the properties of determinants, thus:

$$\begin{vmatrix} 427 & 429 \\ 369 & 371 \end{vmatrix} = \begin{vmatrix} 427 & 429-427 \\ 369 & 371-369 \end{vmatrix} \quad \text{(Rule 5)}$$

$$= \begin{vmatrix} 427 & 2 \\ 369 & 2 \end{vmatrix}$$

$$= \begin{vmatrix} 58 & 0 \\ 369 & 2 \end{vmatrix} \quad \text{(Rule 5)}$$

$$= (58)(2)-(0) = \underline{116}$$

Naturally, the more zero elements we can arrange, the better.

*For another example, move to frame 61.*

---

**61** *Example 2.* Evaluate

$$\begin{vmatrix} 1 & 2 & 2 \\ 4 & 3 & 5 \\ 4 & 2 & 7 \end{vmatrix}$$ column 2 minus column 3 will give us one zero

$$= \begin{vmatrix} 1 & 0 & 2 \\ 4 & -2 & 5 \\ 4 & -5 & 7 \end{vmatrix}$$ column 3 minus twice (column 1) will give another zero

$$= \begin{vmatrix} 1 & 0 & 0 \\ 4 & -2 & -3 \\ 4 & -5 & -1 \end{vmatrix}$$ Now expand along the top row

$$= \begin{vmatrix} -2 & -3 \\ -5 & -1 \end{vmatrix}$$ We could take a factor (−1) from the top row and another factor (−1) from the bottom row.

$$= (-1)(-1)\begin{vmatrix} 2 & 3 \\ 5 & 1 \end{vmatrix}$$

$$= 1(2-15) = \underline{-13}$$

*Next frame.*

**62**

*Example 3.* Evaluate $\begin{vmatrix} 4 & 2 & 2 \\ 2 & 4 & 2 \\ 2 & 2 & 4 \end{vmatrix}$

You do that one, but by way of practice, apply as many of the listed properties as possible. It is quite fun.

*When you have finished it, turn on to frame 63.*

**63**

The answer is $\boxed{32}$, but what we are more interested in is the method of applying the properties, so follow it through. This is one way of doing it; not the only way by any means.

$$\begin{vmatrix} 4 & 2 & 2 \\ 2 & 4 & 2 \\ 2 & 2 & 4 \end{vmatrix}$$

We can take out a factor 2 from each row, giving a factor $2^3$, i.e. 8 outside the determinant.

$$= 8\begin{vmatrix} 2 & 1 & 1 \\ 1 & 2 & 1 \\ 1 & 1 & 2 \end{vmatrix}$$

column 2 minus column 3 will give one zero in the top row.

$$= 8\begin{vmatrix} 2 & 0 & 1 \\ 1 & 1 & 1 \\ 1 & -1 & 2 \end{vmatrix}$$

column 1 minus twice (column 3) will give another zero in the same row.

$$= 8\begin{vmatrix} 0 & 0 & 1 \\ -1 & 1 & 1 \\ -3 & -1 & 2 \end{vmatrix}$$

Expanding along the top row will now reduce this to a second order determinant.

$$= 8\begin{vmatrix} -1 & 1 \\ -3 & -1 \end{vmatrix}$$

Now row 2 + row 1

$$= 8\begin{vmatrix} -1 & 1 \\ -4 & 0 \end{vmatrix}$$

$$= -8\begin{vmatrix} 1 & 1 \\ 4 & 0 \end{vmatrix} = -8(-4) = \underline{32}$$

**64**

Here is another type of problem.

*Example 4.* Solve the equation $\begin{vmatrix} x & 5 & 3 \\ 5 & x+1 & 1 \\ -3 & -4 & x-2 \end{vmatrix} = 0$

In this type of question, we try to establish common factors wherever possible. For example, if we add row 2 and row 3 to row 1, we get

$$\begin{vmatrix} (x+2) & (x+2) & (x+2) \\ 5 & x+1 & 1 \\ -3 & -4 & x-2 \end{vmatrix} = 0$$

Taking out the common factor $(x + 2)$ gives

$$(x+2)\begin{vmatrix} 1 & 1 & 1 \\ 5 & x+1 & 1 \\ -3 & -4 & x-2 \end{vmatrix} = 0$$

Now if we take column 1 from column 2 and also from column 3, what do we get?

*When you have done it, move on to the next frame.*

**65**

We now have $(x+2)\begin{vmatrix} 1 & 0 & 0 \\ 5 & x-4 & -4 \\ -3 & -1 & x+1 \end{vmatrix} = 0$

Expanding along the top row, reduces this to a second order determinant.

$$(x+2)\begin{vmatrix} x-4 & -4 \\ -1 & x+1 \end{vmatrix} = 0$$

If we now multiply out the determinant, we get

$$(x+2)\,[(x-4)(x+1)-4] = 0$$

$$\therefore\ (x+2)(x^2-3x-8) = 0$$

$$\therefore\ x+2=0 \quad \text{or} \quad x^2-3x-8=0$$

which finally gives $\underline{x=-2 \quad \text{or} \quad x=\dfrac{3\pm\sqrt{41}}{2}}$

Finally, here is one for you to do on your own.

*Example 5.* Solve the equation

$$\begin{vmatrix} 5 & x & 3 \\ x+2 & 2 & 1 \\ -3 & 2 & x \end{vmatrix} = 0$$

*Check your working with that given in the next frame.*

**66**

*Result:*

$$x = -4 \quad \text{or} \quad 1 \pm \sqrt{6}$$

Here is one way of doing the problem:

$$\begin{vmatrix} 5 & x & 3 \\ x+2 & 2 & 1 \\ -3 & 2 & x \end{vmatrix} = 0$$

Adding row 2 and row 3 to row 1, gives

$$\begin{vmatrix} x+4 & x+4 & x+4 \\ x+2 & 2 & 1 \\ -3 & 2 & x \end{vmatrix} = 0$$

Take out the common factor $(x + 4)$

$$(x+4)\begin{vmatrix} 1 & 1 & 1 \\ x+2 & 2 & 1 \\ -3 & 2 & x \end{vmatrix} = 0$$

Take column 3 from column 1 and from column 2

$$(x+4)\begin{vmatrix} 0 & 0 & 1 \\ x+1 & 1 & 1 \\ -x-3 & 2-x & x \end{vmatrix} = 0$$

This now reduces to second order

$$(x+4)\begin{vmatrix} x+1 & 1 \\ -x-3 & 2-x \end{vmatrix} = 0$$

Subtract column 2 from column 1

$$(x+4)\begin{vmatrix} x & 1 \\ -5 & 2-x \end{vmatrix} = 0$$

We now finish it off

$\therefore\ (x + 4)(2x - x^2 + 5) = 0$

$\therefore\ x + 4 = 0 \quad \text{or} \quad x^2 - 2x - 5 = 0$

which gives $\underline{x = -4 \quad \text{or} \quad x = 1 \pm \sqrt{6}}$

□□□□□□□□□□□□□□□□□□□□□□□□□□□□□□□□□□□□□□

You have now reached the end of this programme on determinants except for the Test Exercise which follows in frame 67. Before you work through it, brush up any parts of the work about which you are at all uncertain. If you have worked steadily through the programme, you should have no difficulty with the exercise.

## 67 Test Exercise – IV

Answer *all* the questions. Take your time and work carefully. There is no extra credit for speed.

Off you go then. They are all quite straightforward.

1. Evaluate (a) $\begin{vmatrix} 1 & 1 & 2 \\ 2 & 1 & 1 \\ 1 & 2 & 1 \end{vmatrix}$ (b) $\begin{vmatrix} 1 & 2 & 3 \\ 3 & 1 & 2 \\ 2 & 3 & 1 \end{vmatrix}$

2. By determinants, find the value of $x$, given

$$\begin{cases} 2x + 3y - z - 13 = 0 \\ x - 2y + 2z + 3 = 0 \\ 3x + y + z - 10 = 0 \end{cases}$$

3. Use determinants to solve completely

$$\begin{cases} x - 3y + 4z - 5 = 0 \\ 2x + y + z - 3 = 0 \\ 4x + 3y + 5z - 1 = 0 \end{cases}$$

4. Find the values of $k$ for which the following equations are consistent

$$\begin{cases} 3x + 5y + k = 0 \\ 2x + y - 5 = 0 \\ (k + 1)x + 2y - 10 = 0 \end{cases}$$

5. Solve the equation $\begin{vmatrix} x+1 & -5 & -6 \\ -1 & x & 2 \\ -3 & 2 & x+1 \end{vmatrix} = 0$

*Now you can continue with the next programme.*

**Further Problems – IV**

1. Evaluate (i) $\begin{vmatrix} 3 & 5 & 7 \\ 11 & 9 & 13 \\ 15 & 17 & 19 \end{vmatrix}$ (ii) $\begin{vmatrix} 1 & 428 & 861 \\ 2 & 535 & 984 \\ 3 & 642 & 1107 \end{vmatrix}$

2. Evaluate (i) $\begin{vmatrix} 25 & 3 & 35 \\ 16 & 10 & -18 \\ 34 & 6 & 38 \end{vmatrix}$ (ii) $\begin{vmatrix} 155 & 226 & 81 \\ 77 & 112 & 39 \\ 74 & 111 & 37 \end{vmatrix}$

3. Solve by determinants

$$\begin{aligned} 4x - 5y + 7z &= -14 \\ 9x + 2y + 3z &= 47 \\ x - y - 5z &= 11 \end{aligned}$$

4. Use determinants to solve the equations

$$\begin{aligned} 4x - 3y + 2z &= -7 \\ 6x + 2y - 3z &= 33 \\ 2x - 4y - z &= -3 \end{aligned}$$

5. Solve by determinants

$$\begin{aligned} 3x + 2y - 2z &= 16 \\ 4x + 3y + 3z &= 2 \\ 2x - y + z &= -1 \end{aligned}$$

6. Find the values of $\lambda$ for which the following equations are consistent

$$\begin{aligned} 5x + (\lambda + 1)y - 5 &= 0 \\ (\lambda - 1)x + 7y + 5 &= 0 \\ 3x + 5y + 1 &= 0 \end{aligned}$$

7. Determine the values of $k$ for which the following equations have solutions

$$\begin{aligned} 4x - (k - 2)y - 5 &= 0 \\ 2x + y - 10 &= 0 \\ (k + 1)x - 4y - 9 &= 0 \end{aligned}$$

8. (a) Find the values of $k$ which satisfy the equation

$$\begin{vmatrix} k & 1 & 0 \\ 1 & k & 1 \\ 0 & 1 & k \end{vmatrix} = 0$$

(b) Factorise

$$\begin{vmatrix} 1 & 1 & 1 \\ a & b & c \\ a^3 & b^3 & c^3 \end{vmatrix}$$

9. Solve the equation

$$\begin{vmatrix} x & 2 & 3 \\ 2 & x+3 & 6 \\ 3 & 4 & x+6 \end{vmatrix} = 0$$

10. Find the values of $x$ that satisfy the equation

$$\begin{vmatrix} x & 3+x & 2+x \\ 3 & -3 & -1 \\ 2 & -2 & -2 \end{vmatrix} = 0$$

11. Express

$$\begin{vmatrix} 1 & 1 & 1 \\ a^2 & b^2 & c^2 \\ (b+c)^2 & (c+a)^2 & (a+b)^2 \end{vmatrix}$$

as a product of linear factors.

12. A resistive network gives the following equations.

$$2(i_3 - i_2) + 5(i_3 - i_1) = 24$$
$$(i_2 - i_3) + 2i_2 + (i_2 - i_1) = 0$$
$$5(i_1 - i_3) + 2(i_1 - i_2) + i_1 = 6$$

Simplify the equations and use determinants to find the value of $i_2$ correct to two significant figures.

13. Show that $(a + b + c)$ is a factor of the determinant

$$\begin{vmatrix} b+c & a & a^3 \\ c+a & b & b^3 \\ a+b & c & c^3 \end{vmatrix}$$

and express the determinant as a product of five factors.

14. Find values of $k$ for which the following equations are consistent.

$$\begin{aligned} x + (1+k)y + 1 &= 0 \\ (2+k)x + 5y - 10 &= 0 \\ x + 7y + 9 &= 0 \end{aligned}$$

15. Express $\begin{vmatrix} 1+x^2 & yz & 1 \\ 1+y^2 & zx & 1 \\ 1+z^2 & xy & 1 \end{vmatrix}$ as a product of four linear factors.

16. Solve the equation $\begin{vmatrix} x+1 & x+2 & 3 \\ 2 & x+3 & x+1 \\ x+3 & 1 & x+2 \end{vmatrix} = 0$

17. If $x, y, z$, satisfy the equations

$$\begin{aligned} (\tfrac{1}{2}M_1 + M_2)x - M_2 y &= W \\ -M_2 x + 2M_2 y + (M_1 - M_2)z &= 0 \\ -M_2 y + (\tfrac{1}{2}M_1 + M_2)z &= 0 \end{aligned}$$

evaluate $x$ in terms of W, $M_1$ and $M_2$.

18. Three currents, $i_1, i_2, i_3$, in a network are related by the following equations.

$$\begin{aligned} 2i_1 + 3i_2 + 8i_3 &= 30 \\ 6i_1 - i_2 + 2i_3 &= 4 \\ 3i_1 - 12i_2 + 8i_3 &= 0 \end{aligned}$$

By the use of determinants, find the value of $i_1$ and hence solve completely the three equations.

19. If $k(x-a) + 2x - z = 0$
$k(y-a) + 2y - z = 0$
$k(z-a) - x - y + 2z = 0$

show that $x = \dfrac{ak(k+3)}{k^2+4k+2}$

20. Find the angles between $\theta = 0$ and $\theta = \pi$ that satisfy the equation

$$\begin{vmatrix} 1+\sin^2\theta & \cos^2\theta & 4\sin 2\theta \\ \sin^2\theta & 1+\cos^2\theta & 4\sin 2\theta \\ \sin^2\theta & \cos^2\theta & 1+4\sin 2\theta \end{vmatrix} = 0$$

# Programme 5

## MATRICES

**1**

## Matrices – definitions

A *matrix* is a set of real or complex numbers (or *elements*) arranged in rows and columns to form a rectangular array.
A matrix having $m$ rows and $n$ columns is called an $m \times n$ (i.e. '$m$ by $n$') matrix and is referred to as having *order* $m \times n$.
A matrix is indicated by writing the array within large square brackets

e.g. $\begin{bmatrix} 5 & 7 & 2 \\ 6 & 3 & 8 \end{bmatrix}$ is a $2 \times 3$ matrix, i.e. a '2 by 3' matrix, where

5, 7, 2, 6, 3, 8 are the elements of the matrix.

Note that, in describing the matrix, the number of rows is stated first and the number of columns second.

$\begin{bmatrix} 5 & 6 & 4 \\ 2 & -3 & 2 \\ 7 & 8 & 7 \\ 6 & 7 & 5 \end{bmatrix}$ is a matrix of order $4 \times 3$, i.e. 4 rows and 3 columns.

So the matrix $\begin{bmatrix} 6 & 4 \\ 0 & 1 \\ 2 & 3 \end{bmatrix}$ is of order .........

and the matrix $\begin{bmatrix} 2 & 5 & 3 & 4 \\ 6 & 7 & 4 & 9 \end{bmatrix}$ is of order .........

**2**

| $3 \times 2$; $\quad 2 \times 4$ |
|---|

A matrix is simply an array of numbers: there is no arithmetical connection between the elements and it therefore differs from a determinant in that the elements cannot be multiplied together in any way to find a numerical value of the matrix. A matrix has no numerical value. Also, in general, rows and columns cannot be interchanged as was the case with determinants.

*Line matrix*: A line matrix consists of 1 row only.

e.g. $[4 \quad 3 \quad 7 \quad 2]$ is a line matrix of order $1 \times 4$.

*Column matrix*: A column matrix consists of 1 column only.

e.g. $\begin{bmatrix} 6 \\ 3 \\ 8 \end{bmatrix}$ is a column matrix of order $3 \times 1$.

To conserve space in printing, a column matrix is sometimes written on one line but with 'curly' brackets, e.g. $\{6 \quad 3 \quad 8\}$ is the same column matrix of order $3 \times 1$.

*Move on to the next frame.*

**3**

So, from what we have already said:

(a) $\begin{bmatrix} 5 \\ 2 \end{bmatrix}$ is a ............ matrix of order .........

(b) $[4 \quad 0 \quad 7 \quad 3]$ is a .......... matrix of order ..........

(c) $\{2 \quad 6 \quad 9\}$ is a .......... matrix of order .......

**4**

| (a) column, 2 x 1; | (b) line, 1 x 4; | (c) column, 3 x 1 |
|---|---|---|

We use a simple row matrix in stating the $x$ and $y$ coordinates of a point relative to the $x$ and $y$ axes, though in this case it is customary to use round brackets. For example, if P is the point (3, 5) then the 3 is the $x$-coordinate and the 5 the $y$-coordinate. In matrices generally, however, no commas are used to separate the elements.

*Single element matrix*: A single number may be regarded as a 1 x 1 matrix, i.e. having 1 row and 1 column.

*Double suffix notation*: Each element in a matrix has its own particular 'address' or location which can be defined by a system of double suffixes, the first indicating the row, the second the column, thus:

$$\begin{bmatrix} a_{11} & a_{12} & a_{13} & a_{14} \\ a_{21} & a_{22} & a_{23} & a_{24} \\ a_{31} & a_{32} & a_{33} & a_{34} \end{bmatrix}$$

$\therefore a_{23}$ indicates the element in the second row and third column.

Therefore, in the matrix

$$\begin{bmatrix} 6 & -5 & 1 & -3 \\ 2 & -4 & 8 & 3 \\ 4 & -7 & -6 & 5 \\ -2 & 9 & 7 & -1 \end{bmatrix}$$

the location of (a) the element 3 can be stated as ........
(b) the element −1 can be stated as ........
(c) the element 9 can be stated as ........

**5**

(a) $a_{24}$; (b) $a_{44}$; (c) $a_{42}$

**Matrix notation**: Where there is no ambiguity, a whole matrix can be denoted by a single general element enclosed in square brackets, or by a single letter printed in bold type. This is a very neat shorthand and saves much space and writing. For example,

$$\begin{bmatrix} a_{11} & a_{12} & a_{13} & a_{14} \\ a_{21} & a_{22} & a_{23} & a_{24} \\ a_{31} & a_{32} & a_{33} & a_{34} \end{bmatrix}$$ can be denoted by $[a_{ij}]$ or $[a]$ or by $\mathbf{A}$.

Similarly, $$\begin{bmatrix} x_1 \\ x_2 \\ x_3 \end{bmatrix}$$ can be denoted by $[x_i]$ or $[x]$ or simply $\mathbf{x}$.

For an $(m \times n)$ matrix, we shall use a bold face capital letter, e.g. **A**.
For a line or column matrix, we shall use a lower-case bold letter, e.g. **x**.
(In *handwritten* work, we can indicate boldface type by a wavy line placed under the letter, e.g. $\utilde{A}$ or $\utilde{x}$.)
So, if **B** represents a 2 x 3 matrix, write out the elements $b$ in the matrix, using the double suffix notation. This gives ..............

**6**

$$\mathbf{B} = \begin{bmatrix} b_{11} & b_{12} & b_{13} \\ b_{21} & b_{22} & b_{23} \end{bmatrix}$$

**Equal matrices**: By definition, two matrices are said to be equal if corresponding elements throughout are equal. Therefore, the two matrices must also be of the same order.

So, if $$\begin{bmatrix} a_{11} & a_{12} & a_{13} \\ a_{21} & a_{22} & a_{23} \end{bmatrix} = \begin{bmatrix} 4 & 6 & 5 \\ 2 & 3 & 7 \end{bmatrix}$$

then $a_{11} = 4$; $a_{12} = 6$; $a_{13} = 5$; $a_{21} = 2$; etc.

Therefore, if $[a_{ij}] = [x_{ij}]$ then $a_{ij} = x_{ij}$ for all values of $i$ and $j$.

So, if $$\begin{bmatrix} a & b & c \\ d & e & f \\ g & h & k \end{bmatrix} = \begin{bmatrix} 5 & -7 & 3 \\ 1 & 2 & 6 \\ 0 & 4 & 8 \end{bmatrix}$$

then $d =$ ........; $b =$ ........; $a - k =$ ........

**7**

$$d = 1; \quad b = -7; \quad a - k = -3$$

**Addition and subtraction of matrices:** To be added or subtracted, two matrices must be of the *same order*. The sum or difference is then determined by adding or subtracting corresponding elements.

e.g. $$\begin{bmatrix} 4 & 2 & 3 \\ 5 & 7 & 6 \end{bmatrix} + \begin{bmatrix} 1 & 8 & 9 \\ 3 & 5 & 4 \end{bmatrix} = \begin{bmatrix} 4+1 & 2+8 & 3+9 \\ 5+3 & 7+5 & 6+4 \end{bmatrix}$$

$$= \begin{bmatrix} 5 & 10 & 12 \\ 8 & 12 & 10 \end{bmatrix}$$

and $$\begin{bmatrix} 6 & 5 & 12 \\ 9 & 4 & 8 \end{bmatrix} - \begin{bmatrix} 3 & 7 & 1 \\ 2 & 10 & -5 \end{bmatrix} = \begin{bmatrix} 6-3 & 5-7 & 12-1 \\ 9-2 & 4-10 & 8+5 \end{bmatrix}$$

$$= \begin{bmatrix} 3 & -2 & 11 \\ 7 & -6 & 13 \end{bmatrix}$$

So, (a) $$\begin{bmatrix} 6 & 5 & 4 & 1 \\ 2 & 3 & -7 & 8 \end{bmatrix} + \begin{bmatrix} 1 & 4 & 2 & 3 \\ 6 & -1 & 0 & 5 \end{bmatrix} = \text{..............}$$

(b) $$\begin{bmatrix} 8 & 3 & 6 \\ 5 & 2 & 7 \\ 1 & 0 & 4 \end{bmatrix} - \begin{bmatrix} 1 & 2 & 3 \\ 4 & 5 & 6 \\ 7 & 8 & 9 \end{bmatrix} = \text{..............}$$

**8**

(a) $$\begin{bmatrix} 7 & 9 & 6 & 4 \\ 8 & 2 & -7 & 13 \end{bmatrix};$$ (b) $$\begin{bmatrix} 7 & 1 & 3 \\ 1 & -3 & 1 \\ -6 & -8 & -5 \end{bmatrix}$$

**Multiplication of matrices:**

(a) *Scalar multiplication*: To multiply a matrix by a single number (i.e. a scalar), each individual element of the matrix is multiplied by that factor.

e.g. $$4 \times \begin{bmatrix} 3 & 2 & 5 \\ 6 & 1 & 7 \end{bmatrix} = \begin{bmatrix} 12 & 8 & 20 \\ 24 & 4 & 28 \end{bmatrix}$$

i.e., in general, $k[a_{ij}] = [ka_{ij}]$.

It also means that, in reverse, we can take a common factor out of every element – not just one row or one column as in determinants.

Therefore, $\begin{bmatrix} 10 & 25 & 45 \\ 35 & 15 & 50 \end{bmatrix}$ can be written ..............

**9**

$$5 \times \begin{bmatrix} 2 & 5 & 9 \\ 7 & 3 & 10 \end{bmatrix}$$

(b) *Multiplication of two matrices*: Two matrices can be multiplied together only when the number of columns in the first is equal to the number of rows in the second.

e.g. if $\mathbf{A} = [a_{ij}] = \begin{bmatrix} a_{11} & a_{12} & a_{13} \\ a_{21} & a_{22} & a_{23} \end{bmatrix}$ and $\mathbf{b} = [b_i] = \begin{bmatrix} b_1 \\ b_2 \\ b_3 \end{bmatrix}$

then $\mathbf{A} \cdot \mathbf{b} = \begin{bmatrix} a_{11} & a_{12} & a_{13} \\ a_{21} & a_{22} & a_{23} \end{bmatrix} \cdot \begin{bmatrix} b_1 \\ b_2 \\ b_3 \end{bmatrix}$

$$= \begin{bmatrix} a_{11}b_1 + a_{12}b_2 + a_{13}b_3 \\ a_{21}b_1 + a_{22}b_2 + a_{23}b_3 \end{bmatrix}$$

i.e. each element in the top row of **A** is multiplied by the corresponding element in the first column of **b** and the products added. Similarly, the second row of the product is found by multiplying each element in the second row of **A** by the corresponding element in the first column of **b**.

*Example 1*

$$\begin{bmatrix} 4 & 7 & 6 \\ 2 & 3 & 1 \end{bmatrix} \cdot \begin{bmatrix} 8 \\ 5 \\ 9 \end{bmatrix} = \begin{bmatrix} 4{\cdot}8 + 7{\cdot}5 + 6{\cdot}9 \\ 2{\cdot}8 + 3{\cdot}5 + 1{\cdot}9 \end{bmatrix} = \begin{bmatrix} 32 + 35 + 54 \\ 16 + 15 + 9 \end{bmatrix} = \begin{bmatrix} 121 \\ 40 \end{bmatrix}$$

Similarly $\begin{bmatrix} 2 & 3 & 5 & 1 \\ 4 & 6 & 0 & 7 \end{bmatrix} \cdot \begin{bmatrix} 3 \\ 4 \\ 2 \\ 9 \end{bmatrix} =$ ............

**10**

$$\begin{bmatrix} 6 + 12 + 10 + 9 \\ 12 + 24 + 0 + 63 \end{bmatrix} = \begin{bmatrix} 37 \\ 99 \end{bmatrix}$$

In just the same way, if $\mathbf{A} = \begin{bmatrix} 3 & 6 & 8 \\ 1 & 0 & 2 \end{bmatrix}$ and $\mathbf{b} = \begin{bmatrix} 7 \\ 4 \\ 5 \end{bmatrix}$ then $\mathbf{A} \cdot \mathbf{b} =$ ............

**11**

$$\begin{bmatrix} 85 \\ 17 \end{bmatrix}$$

The same process is carried out for each row and column.

*Example 2*

If $\mathbf{A} = [a_{ij}] = \begin{bmatrix} 1 & 5 \\ 2 & 7 \\ 3 & 4 \end{bmatrix}$ and $\mathbf{B} = [b_{ij}] = \begin{bmatrix} 8 & 4 & 3 & 1 \\ 2 & 5 & 8 & 6 \end{bmatrix}$

then $\mathbf{A} \cdot \mathbf{B} = \begin{bmatrix} 1 & 5 \\ 2 & 7 \\ 3 & 4 \end{bmatrix} \cdot \begin{bmatrix} 8 & 4 & 3 & 1 \\ 2 & 5 & 8 & 6 \end{bmatrix}$

$$= \begin{bmatrix} 1{\cdot}8 + 5{\cdot}2 & 1{\cdot}4 + 5{\cdot}5 & 1{\cdot}3 + 5{\cdot}8 & 1{\cdot}1 + 5{\cdot}6 \\ 2{\cdot}8 + 7{\cdot}2 & 2{\cdot}4 + 7{\cdot}5 & 2{\cdot}3 + 7{\cdot}8 & 2{\cdot}1 + 7{\cdot}6 \\ 3{\cdot}8 + 4{\cdot}2 & 3{\cdot}4 + 4{\cdot}5 & 3{\cdot}3 + 4{\cdot}8 & 3{\cdot}1 + 4{\cdot}6 \end{bmatrix}$$

$$= \begin{bmatrix} 8 + 10 & 4 + 25 & 3 + 40 & 1 + 30 \\ 16 + 14 & 8 + 35 & 6 + 56 & 2 + 42 \\ 24 + 8 & 12 + 20 & 9 + 32 & 3 + 24 \end{bmatrix}$$

$$= \begin{bmatrix} 18 & 29 & 43 & 31 \\ 30 & 43 & 62 & 44 \\ 32 & 32 & 41 & 27 \end{bmatrix}$$

Note that multiplying a $(3 \times 2)$ matrix and a $(2 \times 4)$ matrix gives a product matrix of order $(3 \times 4)$

i.e. order $(3 \times 2) \times$ order $(2 \times 4) \rightarrow$ order $(3 \times 4)$.

(same)

In general then, the product of a $(l \times m)$ matrix and an $(m \times n)$ matrix has order $(l \times n)$.

If $\mathbf{A} = \begin{bmatrix} 2 & 4 & 6 \\ 3 & 9 & 5 \end{bmatrix}$ and $\mathbf{B} = \begin{bmatrix} 7 & 1 \\ -2 & 9 \\ 4 & 3 \end{bmatrix}$

then $\mathbf{A} \cdot \mathbf{B} =$ ...............

**12**

$$\boxed{\begin{bmatrix} 30 & 56 \\ 23 & 99 \end{bmatrix}}$$

since $\mathbf{A} \cdot \mathbf{B} = \begin{bmatrix} 2 & 4 & 6 \\ 3 & 9 & 5 \end{bmatrix} \cdot \begin{bmatrix} 7 & 1 \\ -2 & 9 \\ 4 & 3 \end{bmatrix}$

$$= \begin{bmatrix} 14 - \ \ 8 + 24 & 2 + 36 + 18 \\ 21 - 18 + 20 & 3 + 81 + 15 \end{bmatrix} = \begin{bmatrix} 30 & 56 \\ 23 & 99 \end{bmatrix}$$

*Example 3*

It follows that a matrix can be squared only if it is itself a square matrix, i.e. the number of rows equals the number of columns.

If $\mathbf{A} = \begin{bmatrix} 4 & 7 \\ 5 & 2 \end{bmatrix}$

$$\mathbf{A}^2 = \begin{bmatrix} 4 & 7 \\ 5 & 2 \end{bmatrix} \cdot \begin{bmatrix} 4 & 7 \\ 5 & 2 \end{bmatrix}$$

$$= \begin{bmatrix} 16 + 35 & 28 + 14 \\ 20 + 10 & 35 + \ \ 4 \end{bmatrix} = \begin{bmatrix} 51 & 42 \\ 30 & 39 \end{bmatrix}$$

Remember that multiplication of matrices is defined only when ...............

**13**

| the number of columns in the first = the number of rows in the second |
|---|

That is correct. $\begin{bmatrix} 1 & 5 & 6 \\ 4 & 9 & 7 \end{bmatrix} \cdot \begin{bmatrix} 2 & 3 & 5 \\ 8 & 7 & 1 \end{bmatrix}$ has no meaning.

If **A** is an $(m \times n)$ matrix
and **B** is an $(n \times m)$ matrix } then products $\mathbf{A} \cdot \mathbf{B}$ and $\mathbf{B} \cdot \mathbf{A}$ are possible.

*Example*:

$$\text{If } \mathbf{A} = \begin{bmatrix} 1 & 2 & 3 \\ 4 & 5 & 6 \end{bmatrix} \text{ and } \mathbf{B} = \begin{bmatrix} 7 & 10 \\ 8 & 11 \\ 9 & 12 \end{bmatrix}$$

$$\text{then } \mathbf{A} \cdot \mathbf{B} = \begin{bmatrix} 1 & 2 & 3 \\ 4 & 5 & 6 \end{bmatrix} \cdot \begin{bmatrix} 7 & 10 \\ 8 & 11 \\ 9 & 12 \end{bmatrix}$$

$$= \begin{bmatrix} 7 + 16 + 27 & 10 + 22 + 36 \\ 28 + 40 + 54 & 40 + 55 + 72 \end{bmatrix} = \begin{bmatrix} 50 & 68 \\ 122 & 167 \end{bmatrix}$$

$$\text{and } \mathbf{B} \cdot \mathbf{A} = \begin{bmatrix} 7 & 10 \\ 8 & 11 \\ 9 & 12 \end{bmatrix} \cdot \begin{bmatrix} 1 & 2 & 3 \\ 4 & 5 & 6 \end{bmatrix}$$

$$= \begin{bmatrix} 7 + 40 & 14 + 50 & 21 + 60 \\ 8 + 44 & 16 + 55 & 24 + 66 \\ 9 + 48 & 18 + 60 & 27 + 72 \end{bmatrix} = \begin{bmatrix} 47 & 64 & 81 \\ 52 & 71 & 90 \\ 57 & 78 & 99 \end{bmatrix}$$

Note that, in matrix multiplication, $\mathbf{A} \cdot \mathbf{B} \neq \mathbf{B} \cdot \mathbf{A}$, i.e. multiplication is not commutative. The order of the factors is important!

In the product $\mathbf{A} \cdot \mathbf{B}$, **B** is *pre-multiplied* by **A**
and **A** is *post-multiplied* by **B**.

$$\text{So, if } \mathbf{A} = \begin{bmatrix} 5 & 2 \\ 7 & 4 \\ 3 & 1 \end{bmatrix} \text{ and } \mathbf{B} = \begin{bmatrix} 9 & 2 & 4 \\ -2 & 3 & 6 \end{bmatrix}$$

then $\mathbf{A} \cdot \mathbf{B} =$ ............... and $\mathbf{B} \cdot \mathbf{A} =$ ...............

**14**

$$\mathbf{A}\cdot\mathbf{B} = \begin{bmatrix} 41 & 16 & 32 \\ 55 & 26 & 52 \\ 25 & 9 & 18 \end{bmatrix}; \quad \mathbf{B}\cdot\mathbf{A} = \begin{bmatrix} 71 & 30 \\ 29 & 14 \end{bmatrix}$$

**Transpose of a matrix**: If the rows and columns of a matrix are interchanged, i.e. the first row becomes the first column,
the second row becomes the second column,
the third row becomes the third column, etc.,

the new matrix so formed is called the *transpose* of the original matrix. If **A** is the original matrix, its transpose is denoted by $\tilde{\mathbf{A}}$ or $\mathbf{A}^{\mathrm{T}}$. We shall use the latter.

$$\therefore \text{ If } \mathbf{A} = \begin{bmatrix} 4 & 6 \\ 7 & 9 \\ 2 & 5 \end{bmatrix}, \quad \text{then } \mathbf{A}^{\mathrm{T}} = \begin{bmatrix} 4 & 7 & 2 \\ 6 & 9 & 5 \end{bmatrix}$$

Therefore, given that

$$\mathbf{A} = \begin{bmatrix} 2 & 7 & 6 \\ 3 & 1 & 5 \end{bmatrix} \quad \text{and } \mathbf{B} = \begin{bmatrix} 4 & 0 \\ 3 & 7 \\ 1 & 5 \end{bmatrix}$$

then $\mathbf{A}\cdot\mathbf{B} =$ ............ and $(\mathbf{A}\cdot\mathbf{B})^{\mathrm{T}} =$ ...............

**15**

$$\mathbf{A} \cdot \mathbf{B} = \begin{bmatrix} 35 & 79 \\ 20 & 32 \end{bmatrix}; \quad (\mathbf{A} \cdot \mathbf{B})^{\mathrm{T}} = \begin{bmatrix} 35 & 20 \\ 79 & 32 \end{bmatrix}$$

**Special matrices:**

(a) *Square matrix* is a matrix of order $m \times m$.

e.g. $\begin{bmatrix} 1 & 2 & 5 \\ 6 & 8 & 9 \\ 1 & 7 & 4 \end{bmatrix}$ is a $3 \times 3$ matrix

A square matrix $[a_{ij}]$ is *symmetric* if $a_{ij} = a_{ji}$, e.g. $\begin{bmatrix} 1 & 2 & 5 \\ 2 & 8 & 9 \\ 5 & 9 & 4 \end{bmatrix}$

i.e. it is symmetrical about the leading diagonal.

Note that $\mathbf{A} = \mathbf{A}^{\mathrm{T}}$.

A square matrix $[a_{ij}]$ is *skew-symmetric* if $a_{ij} = -a_{ji}$ $\begin{bmatrix} 1 & 2 & 5 \\ -2 & 8 & 9 \\ -5 & -9 & 4 \end{bmatrix}$

In that case, $\mathbf{A} = -\mathbf{A}^{\mathrm{T}}$.

(b) *Diagonal matrix* is a square matrix with all elements zero except those on the leading diagonal, thus $\begin{bmatrix} 5 & 0 & 0 \\ 0 & 2 & 0 \\ 0 & 0 & 7 \end{bmatrix}$

(c) *Unit matrix* is a diagonal matrix in which the elements on the leading diagonal are all unity, i.e. $\begin{bmatrix} 1 & 0 & 0 \\ 0 & 1 & 0 \\ 0 & 0 & 1 \end{bmatrix}$

The unit matrix is denoted by **I**.

If $\mathbf{A} = \begin{bmatrix} 5 & 2 & 4 \\ 1 & 3 & 8 \\ 7 & 9 & 6 \end{bmatrix}$ and $\mathbf{I} = \begin{bmatrix} 1 & 0 & 0 \\ 0 & 1 & 0 \\ 0 & 0 & 1 \end{bmatrix}$ then $\mathbf{A} \cdot \mathbf{I} =$ ....................

**16**

$$\begin{bmatrix} 5 & 2 & 4 \\ 1 & 3 & 8 \\ 7 & 9 & 6 \end{bmatrix} \quad \text{i.e. } \mathbf{A} \cdot \mathbf{I} = \mathbf{A}$$

Similarly, if we form the product **I** · **A** we obtain

$$\mathbf{I} \cdot \mathbf{A} = \begin{bmatrix} 1 & 0 & 0 \\ 0 & 1 & 0 \\ 0 & 0 & 1 \end{bmatrix} \cdot \begin{bmatrix} 5 & 2 & 4 \\ 1 & 3 & 8 \\ 7 & 9 & 6 \end{bmatrix}$$

$$= \begin{bmatrix} 5+0+0 & 2+0+0 & 4+0+0 \\ 0+1+0 & 0+3+0 & 0+8+0 \\ 0+0+7 & 0+0+9 & 0+0+6 \end{bmatrix} = \begin{bmatrix} 5 & 2 & 4 \\ 1 & 3 & 8 \\ 7 & 9 & 6 \end{bmatrix} = \mathbf{A}.$$

$$\therefore \mathbf{A} \cdot \mathbf{I} = \mathbf{I} \cdot \mathbf{A} = \mathbf{A}$$

Therefore, the unit matrix **I**, behaves very much like the unit factor in ordinary algebra and arithmetic.

(d) *Null matrix*: A null matrix is one whose elements are all zero,

i.e. $\begin{bmatrix} 0 & 0 & 0 \\ 0 & 0 & 0 \\ 0 & 0 & 0 \end{bmatrix}$ and is denoted by **0** or simply 0.

If $\mathbf{A} \cdot \mathbf{B} = \mathbf{0}$, we cannot say that therefore $\mathbf{A} = \mathbf{0}$ or $\mathbf{B} = \mathbf{0}$

for if $\mathbf{A} = \begin{bmatrix} 2 & 1 & -3 \\ 6 & 3 & -9 \end{bmatrix}$ and $\mathbf{B} = \begin{bmatrix} 1 & 9 \\ 4 & -6 \\ 2 & 4 \end{bmatrix}$

then $\mathbf{A} \cdot \mathbf{B} = \begin{bmatrix} 2 & 1 & -3 \\ 6 & 3 & -9 \end{bmatrix} \cdot \begin{bmatrix} 1 & 9 \\ 4 & -6 \\ 2 & 4 \end{bmatrix}$

$$= \begin{bmatrix} 2+4-6 & 18-6-12 \\ 6+12-18 & 54-18-36 \end{bmatrix} = \begin{bmatrix} 0 & 0 \\ 0 & 0 \end{bmatrix}$$

That is, $\mathbf{A} \cdot \mathbf{B} = \mathbf{0}$, but clearly $\mathbf{A} \neq \mathbf{0}$ and $\mathbf{B} \neq \mathbf{0}$.

*Now a short revision exercise.* Do these without looking back.

1. If $\mathbf{A} = \begin{bmatrix} 4 & 6 & 5 & 7 \\ 3 & 1 & 9 & 4 \end{bmatrix}$ and $\mathbf{B} = \begin{bmatrix} 2 & 8 & 3 & -1 \\ 5 & 2 & -4 & 6 \end{bmatrix}$

   determine (a) $\mathbf{A} + \mathbf{B}$ and (b) $\mathbf{A} - \mathbf{B}$.

2. If $\mathbf{A} = \begin{bmatrix} 4 & 3 \\ 2 & 7 \\ 6 & 1 \end{bmatrix}$ and $\mathbf{B} = \begin{bmatrix} 5 & 9 & 2 \\ 4 & 0 & 8 \end{bmatrix}$

   determine (a) $5\mathbf{A}$; (b) $\mathbf{A} \cdot \mathbf{B}$; (c) $\mathbf{B} \cdot \mathbf{A}$.

3. If $\mathbf{A} = \begin{bmatrix} 2 & 6 \\ 5 & 7 \\ 4 & 1 \end{bmatrix}$ and $\mathbf{B} = \begin{bmatrix} 3 & 2 \\ 0 & 7 \\ 2 & 3 \end{bmatrix}$ then $\mathbf{A} \cdot \mathbf{B} =$ ............

4. Given that $\mathbf{A} = \begin{bmatrix} 4 & 2 & 6 \\ 1 & 8 & 7 \end{bmatrix}$ determine (a) $\mathbf{A}^{\mathrm{T}}$ and (b) $\mathbf{A} \cdot \mathbf{A}^{\mathrm{T}}$

*When you have completed them, check your results with the next frame.*

**17**

Here are the solutions. Check your results.

1. (a) $\mathbf{A}+\mathbf{B}=\begin{bmatrix}6 & 14 & 8 & 6\\ 8 & 3 & 5 & 10\end{bmatrix}$; (b) $\mathbf{A}-\mathbf{B}=\begin{bmatrix}2 & -2 & 2 & 8\\ -2 & -1 & 13 & -2\end{bmatrix}$

2. (a) $5\mathbf{A}=\begin{bmatrix}20 & 15\\ 10 & 35\\ 30 & 5\end{bmatrix}$; (b) $\mathbf{A}\cdot\mathbf{B}=\begin{bmatrix}32 & 36 & 32\\ 38 & 18 & 60\\ 34 & 54 & 20\end{bmatrix}$;

(c) $\mathbf{B}\cdot\mathbf{A}=\begin{bmatrix}50 & 80\\ 64 & 20\end{bmatrix}$.

3. $\mathbf{A}\cdot\mathbf{B}=\begin{bmatrix}2 & 6\\ 5 & 7\\ 4 & 1\end{bmatrix}\cdot\begin{bmatrix}3 & 2\\ 0 & 7\\ 2 & 3\end{bmatrix}$ is not possible since the number of columns in the first must be equal to the number of rows in the second.

4. $\mathbf{A}=\begin{bmatrix}4 & 2 & 6\\ 1 & 8 & 7\end{bmatrix}\quad\therefore\ \mathbf{A}^{\mathrm{T}}=\begin{bmatrix}4 & 1\\ 2 & 8\\ 6 & 7\end{bmatrix}$

$$\mathbf{A}\cdot\mathbf{A}^{\mathrm{T}}=\begin{bmatrix}4 & 2 & 6\\ 1 & 8 & 7\end{bmatrix}\cdot\begin{bmatrix}4 & 1\\ 2 & 8\\ 6 & 7\end{bmatrix}=\begin{bmatrix}16+\ 4+36 & 4+16+42\\ 4+16+42 & 1+64+49\end{bmatrix}$$

$$=\begin{bmatrix}56 & 62\\ 62 & 114\end{bmatrix}$$

*Now move on to the next frame.*

**18**

**Determinant of a square matrix:** The determinant of a square matrix is the determinant having the same elements as those of the matrix. For example,

the determinant of $\begin{bmatrix} 5 & 2 & 1 \\ 0 & 6 & 3 \\ 8 & 4 & 7 \end{bmatrix}$ is $\begin{vmatrix} 5 & 2 & 1 \\ 0 & 6 & 3 \\ 8 & 4 & 7 \end{vmatrix}$ and the value of this

determinant is $5(42 - 12) - 2(0 - 24) + 1(0 - 48)$

$$= 5(30) - 2(-24) + 1(-48) = 150 + 48 - 48 = \underline{150}$$

Note that the transpose of the matrix is $\begin{bmatrix} 5 & 0 & 8 \\ 2 & 6 & 4 \\ 1 & 3 & 7 \end{bmatrix}$ and the

determinant of the transpose is $\begin{vmatrix} 5 & 0 & 8 \\ 2 & 6 & 4 \\ 1 & 3 & 7 \end{vmatrix}$ the value of which is

$5(42 - 12) - 0(14 - 4) + 8(6 - 6) = 5(30) = \underline{150}.$

That is, the determinant of a square matrix has the same value as that of the determinant of the transposed matrix.
A matrix whose determinant is zero is called a *singular* matrix.

The determinant of the matrix $\begin{bmatrix} 3 & 2 & 5 \\ 4 & 7 & 9 \\ 1 & 8 & 6 \end{bmatrix}$ has the value ...........

and the determinant of the diagonal matrix $\begin{bmatrix} 2 & 0 & 0 \\ 0 & 5 & 0 \\ 0 & 0 & 4 \end{bmatrix}$ has the

value ..........

**19**

$$\begin{vmatrix} 3 & 2 & 5 \\ 4 & 7 & 9 \\ 1 & 8 & 6 \end{vmatrix} = 3(-30) - 2(15) + 5(25) = \underline{5}.$$

$$\begin{vmatrix} 2 & 0 & 0 \\ 0 & 5 & 0 \\ 0 & 0 & 4 \end{vmatrix} = 2(20) + 0 + 0 = \underline{40}.$$

**Cofactors** If $\mathbf{A} = [a_{ij}]$ is a square matrix, we can form a determinant of its elements

$$\begin{vmatrix} a_{11} & a_{12} & a_{13} & \dots & a_{1n} \\ a_{21} & a_{22} & a_{23} & \dots & a_{2n} \\ a_{31} & a_{32} & a_{33} & \dots & a_{3n} \\ \cdot & \cdot & \cdot & & \cdot \\ \cdot & \cdot & \cdot & & \cdot \\ \cdot & \cdot & \cdot & & \cdot \\ a_{n1} & a_{n2} & a_{n3} & \dots & a_{nn} \end{vmatrix}$$

Each element gives rise to a *cofactor*, which is simply the minor of the element in the determinant together with its 'place sign', which was described in detail in the previous programme.

For example, the determinant of the matrix $\mathbf{A} = \begin{bmatrix} 2 & 3 & 5 \\ 4 & 1 & 6 \\ 1 & 4 & 0 \end{bmatrix}$ is

$$\det \mathbf{A} = |\mathbf{A}| = \begin{vmatrix} 2 & 3 & 5 \\ 4 & 1 & 6 \\ 1 & 4 & 0 \end{vmatrix}$$ which has a value of 45.

The minor of the element 2 is $\begin{vmatrix} 1 & 6 \\ 4 & 0 \end{vmatrix} = 0 - 24 = -24.$

The place sign is +. Therefore the cofactor of the element 2 is +(– 24) i.e. $\underline{-24}$.

Similarly, the minor of the element 3 is $\begin{vmatrix} 4 & 6 \\ 1 & 0 \end{vmatrix} = 0 - 6 = -6.$

The place sign is –. Therefore the cofactor of the element 3 is $-(-6) = \underline{6}$.

In each case, the minor is found by striking out the line and column containing the element in question and forming a determinant of the remaining elements. The appropriate place signs are given by

$$\begin{vmatrix} + & - & + & - & \cdots \\ - & + & - & + & \\ + & - & + & & \\ \cdot & & & & \\ \cdot & & & & \\ \cdot & & & & \end{vmatrix}$$

alternately plus and minus from the top left-hand corner which carries a +.

Therefore, in the example above, the minor of the element 6 is $\begin{vmatrix} 2 & 3 \\ 1 & 4 \end{vmatrix}$ i.e. $8 - 3 = 5$. The place sign is –. Therefore the cofactor of the element 6 is – 5.

So, for the matrix $\begin{bmatrix} 7 & 1 & -2 \\ 6 & 5 & 4 \\ 3 & 8 & 9 \end{bmatrix}$, the cofactor of the element 3 is .............. and that of the element 4 is ..............

| Cofactor of 3 is $4 - (-10) = 14$ |
|---|
| Cofactor of 4 is $-(56 - 3) = -53$ |

**Adjoint of a square matrix:**

If we start afresh with $\mathbf{A} = \begin{bmatrix} 2 & 3 & 5 \\ 4 & 1 & 6 \\ 1 & 4 & 0 \end{bmatrix}$, its determinant

$$\det \mathbf{A} = |\mathbf{A}| = \begin{vmatrix} 2 & 3 & 5 \\ 4 & 1 & 6 \\ 1 & 4 & 0 \end{vmatrix}$$ from which we can form a new matrix **C**

of the cofactors.

$$\mathbf{C} = \begin{bmatrix} A_{11} & A_{12} & A_{13} \\ A_{21} & A_{22} & A_{23} \\ A_{31} & A_{32} & A_{33} \end{bmatrix}$$ where $A_{11}$ is the cofactor of $a_{11}$

$A_{ij}$ is the cofactor of $a_{ij}$ etc.

$$A_{11} = + \begin{vmatrix} 1 & 6 \\ 4 & 0 \end{vmatrix} = +(0 - 24) = -24.$$

$$A_{12} = - \begin{vmatrix} 4 & 6 \\ 1 & 0 \end{vmatrix} = -(0 - 6) = 6$$

$$A_{13} = + \begin{vmatrix} 4 & 1 \\ 1 & 4 \end{vmatrix} = +(16 - 1) = 15$$

$$A_{21} = - \begin{vmatrix} 3 & 5 \\ 4 & 0 \end{vmatrix} = -(0 - 20) = 20$$

$$A_{22} = + \begin{vmatrix} 2 & 5 \\ 1 & 0 \end{vmatrix} = +(0 - 5) = -5$$

$$A_{23} = - \begin{vmatrix} 2 & 3 \\ 1 & 4 \end{vmatrix} = -(8 - 3) = -5$$

$$A_{31} = + \begin{vmatrix} 3 & 5 \\ 1 & 6 \end{vmatrix} = +(18 - 5) = 13$$

$$A_{32} = - \begin{vmatrix} 2 & 5 \\ 4 & 6 \end{vmatrix} = -(12 - 20) = 8$$

$$A_{33} = + \begin{vmatrix} 2 & 3 \\ 4 & 11 \end{vmatrix} = +(2 - 12) = -10.$$

$\therefore$ The matrix of cofactors is $\mathbf{C} = \begin{bmatrix} -24 & 6 & 15 \\ 20 & -5 & -5 \\ 13 & 8 & -10 \end{bmatrix}$

and the transpose of $\mathbf{C}$, i.e. $\mathbf{C}^{\mathrm{T}} = \begin{bmatrix} -24 & 20 & 13 \\ 6 & -5 & 8 \\ 15 & -5 & -10 \end{bmatrix}$. This is called the *adjoint* of the original matrix $\mathbf{A}$ and is written adj $\mathbf{A}$.

Therefore, to find the adjoint of a square matrix $\mathbf{A}$

(a) we form the matrix $\mathbf{C}$ of cofactors,
(b) we write the transpose of $\mathbf{C}$, i.e. $\mathbf{C}^{\mathrm{T}}$.

Hence the adjoint of $\begin{bmatrix} 5 & 2 & 1 \\ 3 & 1 & 4 \\ 4 & 6 & 3 \end{bmatrix}$ is ...............

**21**

$$\text{adj } \mathbf{A} = \mathbf{C}^{\mathrm{T}} = \begin{bmatrix} -21 & 0 & 7 \\ 7 & 11 & -17 \\ 14 & -22 & -1 \end{bmatrix}$$

**Inverse of a square matrix**

The adjoint of a square matrix is important, since it enables us to form the inverse of the matrix. If each element of the adjoint of $\mathbf{A}$ is divided by the value of the determinant of $\mathbf{A}$, i.e. $|\mathbf{A}|$, (provided $|\mathbf{A}| \neq 0$), the resulting matrix is called the *inverse* of $\mathbf{A}$ and is denoted by $\mathbf{A}^{-1}$.

For the matrix which we used in the last frame, $\mathbf{A} = \begin{bmatrix} 2 & 3 & 5 \\ 4 & 1 & 6 \\ 1 & 4 & 0 \end{bmatrix}$,

$$\det \mathbf{A} = |\mathbf{A}| = \begin{vmatrix} 2 & 3 & 5 \\ 4 & 1 & 6 \\ 1 & 4 & 0 \end{vmatrix} = 2(0 - 24) - 3(0 - 6) + 5(16 - 1) = 45.$$

the matrix of cofactors $\mathbf{C} = \begin{bmatrix} -24 & 6 & 15 \\ 20 & -5 & -5 \\ 13 & 8 & -10 \end{bmatrix}$

and the adjoint of $\mathbf{A}$, i.e. $\mathbf{C}^{\mathrm{T}} = \begin{bmatrix} -24 & 20 & 13 \\ 6 & -5 & 8 \\ 15 & -5 & -10 \end{bmatrix}$

Then the inverse of $\mathbf{A}$ is given by

$$\mathbf{A}^{-1} = \begin{bmatrix} -\dfrac{24}{45} & \dfrac{20}{45} & \dfrac{13}{45} \\[2ex] \dfrac{6}{45} & -\dfrac{5}{45} & \dfrac{8}{45} \\[2ex] \dfrac{15}{45} & -\dfrac{5}{45} & -\dfrac{10}{45} \end{bmatrix} = \frac{1}{45} \begin{bmatrix} -24 & 20 & 13 \\ 6 & -5 & 8 \\ 15 & -5 & -10 \end{bmatrix}$$

*Therefore, to form the inverse of a square matrix* $\mathbf{A}$:

(a) Evaluate the determinant of $\mathbf{A}$, i.e. $|\mathbf{A}|$.
(b) Form a matrix $\mathbf{C}$ of the cofactors of the elements of $|\mathbf{A}|$.
(c) Write the transpose of $\mathbf{C}$, i.e. $\mathbf{C}^{\mathrm{T}}$ to obtain the adjoint of $\mathbf{A}$.
(d) Divide each element of $\mathbf{C}^{\mathrm{T}}$ by $|\mathbf{A}|$.
(e) The resulting matrix is the inverse $\mathbf{A}^{-1}$ of the original matrix $\mathbf{A}$.

Let us work through an example in detail.

To find the inverse of $\mathbf{A} = \begin{bmatrix} 1 & 2 & 3 \\ 4 & 1 & 5 \\ 6 & 0 & 2 \end{bmatrix}$,

(a) evaluate the determinant of $\mathbf{A}$, i.e. $|\mathbf{A}|$. $\quad |\mathbf{A}| = \ldots\ldots\ldots\ldots\ldots$

**22**

$$|\mathbf{A}| = 28$$

for $|\mathbf{A}| = \begin{vmatrix} 1 & 2 & 3 \\ 4 & 1 & 5 \\ 6 & 0 & 2 \end{vmatrix} = 1(2-0) - 2(8-30) + 3(0-6) = 28.$

(b) Now form the matrix of the cofactors. $\mathbf{C} =$ ...............

**23**

$$\mathbf{C} = \begin{bmatrix} 2 & 22 & -6 \\ -4 & -16 & 12 \\ 7 & 7 & -7 \end{bmatrix}$$

for $A_{11} = +(2-0) = 2$; $A_{12} = -(8-30) = 22$; $A_{13} = +(0-6) = -6$
$A_{21} = -(4-0) = -4$; $A_{22} = +(2-18) = -16$; $A_{23} = -(0-12) = 12$
$A_{31} = +(10-3) = 7$; $A_{32} = -(5-12) = 7$; $A_{33} = +(1-8) = -7$

(c) Next we have to write down the transpose of **C** to obtain the adjoint of **A**.

$$\text{adj } \mathbf{A} = \mathbf{C}^{\mathrm{T}} = \text{...............}$$

**24**

$$\text{adj } \mathbf{A} = \mathbf{C}^{\mathrm{T}} = \begin{bmatrix} 2 & -4 & 7 \\ 22 & -16 & 7 \\ -6 & 12 & -7 \end{bmatrix}$$

(d) Finally, we divide the elements of adj **A** by the value of $|\mathbf{A}|$, i.e. 28, to arrive at $\mathbf{A}^{-1}$, the inverse of **A**.

$$\therefore \mathbf{A}^{-1} = \text{...........}$$

**25**

$$\mathbf{A}^{-1} = \begin{bmatrix} \frac{2}{28} & -\frac{4}{28} & \frac{7}{28} \\ \frac{22}{28} & -\frac{16}{28} & \frac{7}{28} \\ -\frac{6}{28} & \frac{12}{28} & -\frac{7}{28} \end{bmatrix} = \frac{1}{28}\begin{bmatrix} 2 & -4 & 7 \\ 22 & -16 & 7 \\ -6 & 12 & -7 \end{bmatrix}$$

Every one is done in the same way. Work the next one right through on your own.

Determine the inverse of the matrix $\mathbf{A} = \begin{bmatrix} 2 & 7 & 4 \\ 3 & 1 & 6 \\ 5 & 0 & 8 \end{bmatrix}$

$$\mathbf{A}^{-1} = \ldots\ldots\ldots\ldots\ldots$$

**26**

$$\mathbf{A}^{-1} = \frac{1}{38}\begin{bmatrix} 8 & -56 & 38 \\ 6 & -4 & 0 \\ -5 & 35 & -19 \end{bmatrix}$$

Here are the details.

$$\det \mathbf{A} = |\,\mathbf{A}\,| = \begin{vmatrix} 2 & 7 & 4 \\ 3 & 1 & 6 \\ 5 & 0 & 8 \end{vmatrix} = 2(8) - 7(-6) + 4(-5) = 38.$$

cofactors:

$A_{11} = +(8 - 0) = 8$; $\quad A_{12} = -(24 - 30) = 6$; $\quad A_{13} = +(0 - 5) = -5$

$A_{21} = -(56 - 0) = -56$; $\quad A_{22} = +(16 - 20) = -4$; $\quad A_{23} = -(0 - 35) = 35$

$A_{31} = +(42 - 4) = 38$; $\quad A_{32} = -(12 - 12) = 0$; $\quad A_{33} = +(2 - 21) = -19$

$$\therefore\ \mathbf{C} = \begin{bmatrix} 8 & 6 & -5 \\ -56 & -4 & 35 \\ 38 & 0 & -19 \end{bmatrix} \qquad \therefore\ \mathbf{C}^{\mathrm{T}} = \begin{bmatrix} 8 & -56 & 38 \\ 6 & -4 & 0 \\ -5 & 35 & -19 \end{bmatrix}$$

$$\text{then} \quad \mathbf{A}^{-1} = \frac{1}{38}\begin{bmatrix} 8 & -56 & 38 \\ 6 & -4 & 0 \\ -5 & 35 & -19 \end{bmatrix}$$

Now let us find some uses for the inverse.

*Product of a square matrix and its inverse*

From a previous example, we have seen that when $\mathbf{A} = \begin{bmatrix} 1 & 2 & 3 \\ 4 & 1 & 5 \\ 6 & 0 & 2 \end{bmatrix}$

$$\mathbf{A}^{-1} = \frac{1}{28}\begin{bmatrix} 2 & -4 & 7 \\ 22 & -16 & 7 \\ -6 & 12 & -7 \end{bmatrix}$$

$$\text{Then } \mathbf{A}^{-1} \cdot \mathbf{A} = \frac{1}{28}\begin{bmatrix} 2 & -4 & 7 \\ 22 & -16 & 7 \\ -6 & 12 & -7 \end{bmatrix} \cdot \begin{bmatrix} 1 & 2 & 3 \\ 4 & 1 & 5 \\ 6 & 0 & 2 \end{bmatrix}$$

$$= \frac{1}{28}\begin{bmatrix} 2-16+42 & 4-4+0 & 6-20+14 \\ 22-64+42 & 44-16+0 & 66-80+14 \\ -6+48-42 & -12+12+0 & -18+60-14 \end{bmatrix}$$

$$= \frac{1}{28}\begin{bmatrix} 28 & 0 & 0 \\ 0 & 28 & 0 \\ 0 & 0 & 28 \end{bmatrix}$$

$$= \begin{bmatrix} 1 & 0 & 0 \\ 0 & 1 & 0 \\ 0 & 0 & 1 \end{bmatrix} = \mathrm{I} \qquad \therefore \underline{\mathbf{A}^{-1} \cdot \mathbf{A} = \mathbf{I}}$$

$$\text{Also } \mathbf{A} \cdot \mathbf{A}^{-1} = \begin{bmatrix} 1 & 2 & 3 \\ 4 & 1 & 5 \\ 6 & 0 & 2 \end{bmatrix} \times \frac{1}{28}\begin{bmatrix} 2 & -4 & 7 \\ 22 & -16 & 7 \\ -6 & 12 & -7 \end{bmatrix}$$

$$= \frac{1}{28}\begin{bmatrix} 1 & 2 & 3 \\ 4 & 1 & 5 \\ 6 & 0 & 2 \end{bmatrix} \cdot \begin{bmatrix} 2 & -4 & 7 \\ 22 & -16 & 7 \\ -6 & 12 & -7 \end{bmatrix}$$

= ......................... Finish it off.

$$\mathbf{A} \cdot \mathbf{A}^{-1} = \frac{1}{28}\begin{bmatrix} 28 & 0 & 0 \\ 0 & 28 & 0 \\ 0 & 0 & 28 \end{bmatrix} = \begin{bmatrix} 1 & 0 & 0 \\ 0 & 1 & 0 \\ 0 & 0 & 1 \end{bmatrix} = \mathbf{I}$$

$$\therefore \mathbf{A} \cdot \mathbf{A}^{-1} = \mathbf{A}^{-1} \cdot \mathbf{A} = \mathbf{I}.$$

That is, the product of a square matrix and its inverse, in whatever order the factors are written, is the unit matrix of the same matrix order.

**Solution of a set of linear equations**

Consider the set of linear equations

$$\begin{aligned} a_{11}x_1 + a_{12}x_2 + a_{13}x_3 + \ldots + a_{1n}x_n &= b_1 \\ a_{21}x_1 + a_{22}x_2 + a_{23}x_3 + \ldots + a_{2n}x_n &= b_2 \\ \vdots \\ a_{n1}x_1 + a_{n2}x_2 + a_{n3}x_3 + \ldots + a_{nn}x_n &= b_n \end{aligned}$$

From our knowledge of matrix multiplication, this can be written in matrix form.

$$\begin{bmatrix} a_{11} & a_{12} & a_{13} & \ldots & a_{1n} \\ a_{21} & a_{22} & a_{23} & \ldots & a_{2n} \\ \vdots & \vdots & \vdots & & \vdots \\ a_{n1} & a_{n2} & a_{n3} & \ldots & a_{nn} \end{bmatrix} \cdot \begin{bmatrix} x_1 \\ x_2 \\ \vdots \\ x_n \end{bmatrix} = \begin{bmatrix} b_1 \\ b_2 \\ \vdots \\ b_n \end{bmatrix}$$

i.e. $\mathbf{A} \cdot \mathbf{x} = \mathbf{b}$

where $\mathbf{A} = \begin{bmatrix} a_{11} & a_{12} & \ldots & a_{1n} \\ a_{21} & a_{22} & \ldots & a_{2n} \\ \vdots & \vdots & & \vdots \\ a_{n1} & a_{n2} & \ldots & a_{nn} \end{bmatrix}$; $\mathbf{x} = \begin{bmatrix} x_1 \\ x_2 \\ \vdots \\ x_n \end{bmatrix}$; and $\mathbf{b} = \begin{bmatrix} b_1 \\ b_2 \\ \vdots \\ b_n \end{bmatrix}$

If we multiply both sides of the matrix equation by the inverse of **A**, we have

$$\mathbf{A}^{-1} \cdot \mathbf{A} \cdot \mathbf{x} = \mathbf{A}^{-1} \cdot \mathbf{b}$$

But $\mathbf{A}^{-1} \cdot \mathbf{A} = \mathbf{I} \therefore \mathbf{I} \cdot \mathbf{x} = \mathbf{A}^{-1} \cdot \mathbf{b}$ i.e. $\mathbf{x} = \mathbf{A}^{-1} \cdot \mathbf{b}$

Therefore, if we form the inverse of the matrix of coefficients and pre-multiply matrix **b** by it, we shall determine the matrix of the solutions of **x**.

*Example* To solve the set of equations

$$x_1 + 2x_2 + x_3 = 4$$
$$3x_1 - 4x_2 - 2x_3 = 2$$
$$5x_1 + 3x_2 + 5x_3 = -1$$

First write the set of equations in matrix form, which gives

. . . . . . . . . . . . . . . . . .

**28**

$$\begin{bmatrix} 1 & 2 & 1 \\ 3 & -4 & -2 \\ 5 & 3 & 5 \end{bmatrix} \cdot \begin{bmatrix} x_1 \\ x_2 \\ x_3 \end{bmatrix} = \begin{bmatrix} 4 \\ 2 \\ -1 \end{bmatrix}$$

i.e. $\mathbf{A} \cdot \mathbf{x} = \mathbf{b} \quad \therefore \mathbf{x} = \mathbf{A}^{-1} \cdot \mathbf{b}$

So the next step is to find the inverse of **A** where **A** is the matrix of the coefficients of **x**. We have already seen how to determine the inverse of a matrix, so in this case $\mathbf{A}^{-1}$ = ...................

**29**

$$\mathbf{A}^{-1} = -\frac{1}{35}\begin{bmatrix} -14 & -7 & 0 \\ -25 & 0 & 5 \\ 29 & 7 & -10 \end{bmatrix}$$

For: $|\mathbf{A}| = \begin{vmatrix} 1 & 2 & 1 \\ 3 & -4 & -2 \\ 5 & 3 & 5 \end{vmatrix} = -14 - 50 + 29 = 29 - 64 \therefore |A| = -35$

Cofactors

$A_{11} = +(-20 + 6) = -14$; $A_{12} = -(15 + 10) = -25$; $A_{13} = +(9 + 20) = 29$

$A_{21} = -(10 - 3) = -7$; $A_{22} = +(5 - 5) = 0$; $A_{23} = -(3 - 10) = 7$

$A_{31} = +(-4 + 4) = 0$; $A_{32} = -(-2 - 3) = 5$; $A_{33} = +(-4 - 6) = -10$

$$\therefore \mathbf{C} = \begin{bmatrix} -14 & -25 & 29 \\ -7 & 0 & 7 \\ 0 & 5 & -10 \end{bmatrix} \quad \therefore \text{adj } \mathbf{A} = \mathbf{C}^{\mathrm{T}} = \begin{bmatrix} -14 & -7 & 0 \\ -25 & 0 & 5 \\ 29 & 7 & -10 \end{bmatrix}$$

Now $|\mathbf{A}| = -35 \quad \therefore \mathbf{A}^{-1} = \dfrac{\text{adj } A}{|A|} = -\dfrac{1}{35}\begin{bmatrix} -14 & -7 & 0 \\ -25 & 0 & 5 \\ 29 & 7 & -10 \end{bmatrix}$

$$\therefore \mathbf{x} = \mathbf{A}^{-1} \cdot \mathbf{b} = -\frac{1}{35}\begin{bmatrix} -14 & -7 & 0 \\ -25 & 0 & 5 \\ 29 & 7 & -10 \end{bmatrix} \cdot \begin{bmatrix} 4 \\ 2 \\ -1 \end{bmatrix}$$

= .................... Multiply it out.

**30**

$$\mathbf{x} = -\frac{1}{35}\begin{bmatrix} -70 \\ -105 \\ 140 \end{bmatrix} = \begin{bmatrix} 2 \\ 3 \\ -4 \end{bmatrix}$$

So finally $\mathbf{x} = \begin{bmatrix} x_1 \\ x_2 \\ x_3 \end{bmatrix} = \begin{bmatrix} 2 \\ 3 \\ -4 \end{bmatrix}$ $\therefore \underline{x_1 = 2;\ x_2 = 3;\ x_3 = -4.}$

Once you have found the inverse, the rest is simply $\mathbf{x} = \mathbf{A}^{-1} \cdot \mathbf{b}$.
Here is another example to solve in the same way.

If
$$2x_1 - x_2 + 3x_3 = 2$$
$$x_1 + 3x_2 - x_3 = 11$$
$$2x_1 - 2x_2 + 5x_3 = 3$$

then $x_1 =$ ...........; $x_2 =$ ...........; $x_3 =$ ...........

**31**

$$x_1 = -1; \quad x_2 = 5; \quad x_3 = 3$$

The essential intermediate results are as follows:

$$\begin{bmatrix} 2 & -1 & 3 \\ 1 & 3 & -1 \\ 2 & -2 & 5 \end{bmatrix} \cdot \begin{bmatrix} x_1 \\ x_2 \\ x_3 \end{bmatrix} = \begin{bmatrix} 2 \\ 11 \\ 3 \end{bmatrix} \quad \text{i.e. } \mathbf{A} \cdot \mathbf{x} = \mathbf{b} \ \therefore \mathbf{x} = \mathbf{A}^{-1} \cdot \mathbf{b}$$

$\det \mathbf{A} = |\mathbf{A}| = 9.$

$$\mathbf{C} = \begin{bmatrix} 13 & -7 & -8 \\ -1 & 4 & 2 \\ -8 & 5 & 7 \end{bmatrix} \quad \therefore \text{adj } \mathbf{A} = \mathbf{C}^{\mathrm{T}} = \begin{bmatrix} 13 & -1 & -8 \\ -7 & 4 & 5 \\ -8 & 2 & 7 \end{bmatrix}$$

$$\mathbf{A}^{-1} = \frac{\mathbf{C}^{\mathrm{T}}}{|\mathbf{A}|} = \frac{1}{9}\begin{bmatrix} 13 & -1 & -8 \\ -7 & 4 & 5 \\ -8 & 2 & 7 \end{bmatrix}$$

$$\mathbf{x} = \mathbf{A}^{-1} \cdot \mathbf{b} = \frac{1}{9}\begin{bmatrix} 13 & -1 & -8 \\ -7 & 4 & 5 \\ -8 & 2 & 7 \end{bmatrix} \cdot \begin{bmatrix} 2 \\ 11 \\ 3 \end{bmatrix} = \frac{1}{9}\begin{bmatrix} -9 \\ 45 \\ 27 \end{bmatrix} = \begin{bmatrix} -1 \\ 5 \\ 3 \end{bmatrix}$$

$$\therefore \mathbf{x} = \begin{bmatrix} x_1 \\ x_2 \\ x_3 \end{bmatrix} = \begin{bmatrix} -1 \\ 5 \\ 3 \end{bmatrix} \qquad \therefore \underline{x_1 = -1; \quad x_2 = 5; \quad x_3 = 3.}$$

**Gaussian elimination method** for solving a set of linear equations

$$\begin{bmatrix} a_{11} & a_{12} & a_{13} & \dots & a_{1n} \\ a_{21} & a_{22} & a_{23} & \dots & a_{2n} \\ \cdot & \cdot & \cdot & & \cdot \\ \cdot & \cdot & \cdot & & \cdot \\ \cdot & \cdot & \cdot & & \cdot \\ a_{n1} & a_{n2} & a_{n3} & \dots & a_{nn} \end{bmatrix} \cdot \begin{bmatrix} x_1 \\ x_2 \\ \cdot \\ \cdot \\ \cdot \\ x_n \end{bmatrix} = \begin{bmatrix} b_1 \\ b_2 \\ \cdot \\ \cdot \\ \cdot \\ b_n \end{bmatrix} \quad \text{i.e. } \mathbf{A} \cdot \mathbf{x} = \mathbf{b}$$

All the information for solving the set of equations is provided by the matrix of coefficients **A** and the column matrix **b**. If we write the elements of **b** within the matrix **A**, we obtain the *augmented matrix* **B** of the given set of equations.

$$\text{i.e. } \mathbf{B} = \left[\begin{array}{ccccc:c} a_{11} & a_{12} & a_{13} & \dots & a_{1n} & b_1 \\ a_{21} & a_{22} & a_{23} & \dots & a_{2n} & b_2 \\ \cdot & \cdot & \cdot & & \cdot & \cdot \\ \cdot & \cdot & \cdot & & \cdot & \cdot \\ \cdot & \cdot & \cdot & & \cdot & \cdot \\ a_{n1} & a_{n2} & a_{n3} & \dots & a_{nn} & b_n \end{array}\right]$$

(a) We then eliminate the elements other than $a_{11}$ from the first column by subtracting $a_{21}/a_{11}$ times the first row from the second row and $a_{31}/a_{11}$ times the first row from the third row, etc.

(b) This gives a new matrix of the form

$$\left[\begin{array}{ccccc:c} a_{11} & a_{12} & a_{13} & \dots & a_{1n} & b_1 \\ 0 & c_{22} & c_{23} & \dots & c_{2n} & d_2 \\ \cdot & \cdot & \cdot & & \cdot & \cdot \\ \cdot & \cdot & \cdot & & \cdot & \cdot \\ \cdot & \cdot & \cdot & & \cdot & \cdot \\ 0 & c_{n2} & c_{n3} & \dots & c_{nn} & d_n \end{array}\right]$$

The process is then repeated to eliminate $c_{i2}$ from the third and subsequent rows.

*A specific example will explain the method, so move on to the next frame.*

**32** *Example* To solve

$$x_1 + 2x_2 - 3x_3 = 3$$
$$2x_1 - x_2 - x_3 = 11$$
$$3x_1 + 2x_2 + x_3 = -5$$

This can be written

$$\begin{bmatrix} 1 & 2 & -3 \\ 2 & -1 & -1 \\ 3 & 2 & 1 \end{bmatrix} \cdot \begin{bmatrix} x_1 \\ x_2 \\ x_3 \end{bmatrix} = \begin{bmatrix} 3 \\ 11 \\ -5 \end{bmatrix}$$

The augmented matrix becomes $\left[\begin{array}{ccc:c} 1 & 2 & -3 & 3 \\ 2 & -1 & -1 & 11 \\ 3 & 2 & 1 & -5 \end{array}\right]$

Now subtract $\frac{2}{1}$ times the first row from the second row

and $\frac{3}{1}$ times the first row from the third row.

This gives $\left[\begin{array}{ccc:c} 1 & 2 & -3 & 3 \\ 0 & -5 & 5 & 5 \\ 0 & -4 & 10 & -14 \end{array}\right]$

Now subtract $\frac{-4}{-5}$, i.e. $\frac{4}{5}$, times the second row from the third row.

The matrix becomes $\left[\begin{array}{ccc:c} 1 & 2 & -3 & 3 \\ 0 & -5 & 5 & 5 \\ 0 & 0 & 6 & -18 \end{array}\right]$

Note that as a result of these steps, the matrix of coefficients of $x$ has been reduced to a triangular matrix.

Finally, we detach the right-hand column back to its original position

$$\begin{bmatrix} 1 & 2 & -3 \\ 0 & -5 & 5 \\ 0 & 0 & 6 \end{bmatrix} \cdot \begin{bmatrix} x_1 \\ x_2 \\ x_3 \end{bmatrix} = \begin{bmatrix} 3 \\ 5 \\ -18 \end{bmatrix}$$

Then, by 'back-substitution', starting from the bottom row we get

$$6x_3 = -18 \quad \therefore x_3 = -3 \qquad x_3 = -3$$

$$-5x_2 + 5x_3 = 5 \quad \therefore -5x_2 = 5 + 15 = 20 \quad \therefore x_2 = -4$$

$$x_1 + 2x_2 - 3x_3 = 3 \quad \therefore x_1 - 8 + 9 = 3 \qquad \therefore x_1 = 2$$

$$\therefore x_1 = 2; \quad x_2 = -4; \quad x_3 = -3$$

Note that when dealing with the augmented matrix, we may, if we wish,

(a) interchange two rows,
(b) multiply any row by a non-zero factor,
(c) add (or subtract) a constant multiple of any one row to (or from) another.

These operations are permissible since we are really dealing with the coefficients of both sides of the equations.

*Now for another example: turn on to the next frame.*

**33**

*Example* Solve the following set of equations

$$x_1 - 4x_2 - 2x_3 = 21$$

$$2x_1 + x_2 + 2x_3 = 3$$

$$3x_1 + 2x_2 - x_3 = -2$$

First write the equations in matrix form, which is ..................

**34**

$$\begin{bmatrix} 1 & -4 & -2 \\ 2 & 1 & 2 \\ 3 & 2 & -1 \end{bmatrix} \cdot \begin{bmatrix} x_1 \\ x_2 \\ x_3 \end{bmatrix} = \begin{bmatrix} 21 \\ 3 \\ -2 \end{bmatrix}$$

The augmented matrix is then ..................

**35**

$$\left[\begin{array}{ccc:c} 1 & -4 & -2 & 21 \\ 2 & 1 & 2 & 3 \\ 3 & 2 & -1 & -2 \end{array}\right]$$

We can now eliminate the $x_1$ coefficients from the second and third rows by ..........................................................................

and ..........................................................................

**36**

subtracting 2 times the first row from the second row
and 3 times the first row from the third row.

So the matrix now becomes

$$\left[\begin{array}{ccc:c} 1 & -4 & -2 & 21 \\ 0 & 9 & 6 & -39 \\ 0 & 14 & 5 & -65 \end{array}\right]$$

and the next stage is to subtract from the third row ....... times the second row.

**37**

$$\frac{14}{9}$$

If we do this, the matrix becomes

$$\left[\begin{array}{ccc:c} 1 & -4 & -2 & 21 \\ 0 & 9 & 6 & -39 \\ 0 & 0 & -4.33 & -4.33 \end{array}\right]$$

Re-forming the matrix equation

$$\begin{bmatrix} 1 & -4 & -2 \\ 0 & 9 & 6 \\ 0 & 0 & -4.33 \end{bmatrix} \cdot \begin{bmatrix} x_1 \\ x_2 \\ x_3 \end{bmatrix} = \begin{bmatrix} 21 \\ -39 \\ -4.33 \end{bmatrix}$$

Now, starting from the bottom row, you can finish it off.

$x_1 =$ ........; $x_2 =$ .......; $x_3 =$ ........

**38**

$$x_1 = 3; \quad x_2 = -5; \quad x_3 = 1$$

Now for something rather different.

**Eigenvalues and eigenvectors**

In many applications of matrices to technological problems involving coupled oscillations and vibrations, equations of the form

$$\mathbf{A} \cdot \mathbf{x} = \lambda \mathbf{x}$$

occur, where $\mathbf{A} = [a_{ij}]$ is a square matrix and $\lambda$ is a number (scalar). Clearly, $\mathbf{x} = \mathbf{0}$ is a solution for any value of $\lambda$ and is not normally useful. For *non-trivial solutions*, i.e. $\mathbf{x} \neq \mathbf{0}$, the values of $\lambda$ are called the *eigenvalues, characteristic values*, or *latent roots* of the matrix **A** and the corresponding solutions of the given equations $\mathbf{A} \cdot \mathbf{x} = \lambda \mathbf{x}$ are called the *eigenvectors* or characteristic vectors of **A**.

Expressed as a set of separate equations, we have

$$\begin{bmatrix} a_{11} & a_{12} & \dots & a_{1n} \\ a_{21} & a_{22} & \dots & a_{2n} \\ \cdot & \cdot & & \cdot \\ \cdot & \cdot & & \cdot \\ \cdot & \cdot & & \cdot \\ a_{n1} & a_{n2} & \dots & a_{nn} \end{bmatrix} \cdot \begin{bmatrix} x_1 \\ x_2 \\ \cdot \\ \cdot \\ \cdot \\ x_n \end{bmatrix} = \lambda \begin{bmatrix} x_1 \\ x_2 \\ \cdot \\ \cdot \\ \cdot \\ x_n \end{bmatrix}$$

i.e.

$$\begin{array}{l} a_{11}x_1 + a_{12}x_2 + \dots + a_{1n}x_n = \lambda x_1 \\ a_{21}x_1 + a_{22}x_2 + \dots + a_{2n}x_n = \lambda x_2 \\ \quad \cdot \qquad \cdot \qquad\qquad \cdot \qquad \cdot \\ \quad \cdot \qquad \cdot \qquad\qquad \cdot \qquad \cdot \\ \quad \cdot \qquad \cdot \qquad\qquad \cdot \qquad \cdot \\ a_{n1}x_1 + a_{n2}x_2 + \dots + a_{nn}x_n = \lambda x_n \end{array}$$

Bringing the right-hand-side terms to the left-hand side, this simplifies to

$$\begin{array}{l} (a_{11} - \lambda)x_1 + a_{12}x_2 + \dots + a_{1n}x_n = 0 \\ a_{21}x_1 + (a_{22} - \lambda)x_2 + \dots + a_{2n}x_n = 0 \\ \quad \cdot \qquad \cdot \qquad\qquad \cdot \qquad \cdot \\ \quad \cdot \qquad \cdot \qquad\qquad \cdot \qquad \cdot \\ \quad \cdot \qquad \cdot \qquad\qquad \cdot \qquad \cdot \\ a_{n1}x_1 + a_{n2}x_2 + \dots + (a_{nn} - \lambda)x_n = 0 \end{array}$$

i.e. $\begin{bmatrix} (a_{11}-\lambda) & a_{12} & \dots & a_{1n} \\ a_{21} & (a_{22}-\lambda) & \dots & a_{2n} \\ \cdot & \cdot & & \cdot \\ \cdot & \cdot & & \cdot \\ \cdot & \cdot & & \cdot \\ a_{n1} & a_{n2} & \dots & (a_{nn}-\lambda) \end{bmatrix} \cdot \begin{bmatrix} x_1 \\ x_2 \\ \cdot \\ \cdot \\ \cdot \\ x_n \end{bmatrix} = \begin{bmatrix} 0 \\ 0 \\ \cdot \\ \cdot \\ \cdot \\ 0 \end{bmatrix}$

$\mathbf{A} \cdot \mathbf{x} = \lambda\mathbf{x}$ becomes $\mathbf{A} \cdot \mathbf{x} - \lambda\mathbf{x} = \mathbf{0}$

and then $(\mathbf{A} - \lambda\mathbf{I})\mathbf{x} = \mathbf{0}$

Note that the unit matrix is introduced since we can subtract only a matrix from another matrix.

For this set of homogeneous linear equations (i.e. right-hand constants all zero) to have a non-trivial solution, $|\mathbf{A} - \lambda\mathbf{I}|$ must be zero.

$$|\mathbf{A} - \lambda\mathbf{I}| = \begin{vmatrix} (a_{11}-\lambda) & a_{12} & \dots & a_{1n} \\ a_{21} & (a_{22}-\lambda) & \dots & a_{2n} \\ \cdot & \cdot & & \cdot \\ \cdot & \cdot & & \cdot \\ \cdot & \cdot & & \cdot \\ a_{n1} & a_{n2} & \dots & (a_{nn}-\lambda) \end{vmatrix} = 0.$$

$|\mathbf{A} - \lambda\mathbf{I}|$ is called the *characteristic determinant* of $\mathbf{A}$ and $|\mathbf{A} - \lambda\mathbf{I}| = 0$ is the *characteristic equation*. On expanding the determinant, this gives a polynomial of degree $n$ and solution of the characteristic equation gives the values of $\lambda$, i.e. the eigenvalues of $\mathbf{A}$.

*Example 1* To find the eigenvalues of the matrix $\mathbf{A} = \begin{bmatrix} 4 & -1 \\ 2 & 1 \end{bmatrix}$.

$\mathbf{A} \cdot \mathbf{x} = \lambda\mathbf{x}$ i.e. $(\mathbf{A} - \lambda\mathbf{I})\mathbf{x} = \mathbf{0}$

Characteristic determinant: $|\mathbf{A} - \lambda\mathbf{I}| = \begin{vmatrix} (4-\lambda) & -1 \\ 2 & (1-\lambda) \end{vmatrix}$

Characteristic equation: $|\mathbf{A} - \lambda\mathbf{I}| = 0$

$\therefore (4-\lambda)(1-\lambda) + 2 = 0 \quad \therefore 4 - 5\lambda + \lambda^2 + 2 = 0$

$\therefore \lambda^2 - 5\lambda + 6 = 0 \quad \therefore (\lambda - 2)(\lambda - 3) = 0$

$\therefore \lambda = 2$ or $3 \quad \therefore \underline{\lambda_1 = 2; \quad \lambda_2 = 3.}$

*Example 2* To find the eigenvalues of the matrix $\mathbf{A} = \begin{bmatrix} 2 & 3 & -2 \\ 1 & 4 & -2 \\ 2 & 10 & -5 \end{bmatrix}$

The characteristic determinant is ...............

**39**

$$\begin{vmatrix} (2-\lambda) & 3 & -2 \\ 1 & (4-\lambda) & -2 \\ 2 & 10 & (-5-\lambda) \end{vmatrix}$$

Expanding this, we get

$$(2-\lambda)\{(4-\lambda)(-5-\lambda)+20\} - 3\{(-5-\lambda)+4\} - 2\{10-2(4-\lambda)\}$$
$$= (2-\lambda)\{-20+\lambda+\lambda^2+20\} + 3(1+\lambda) - 2(2+2\lambda)$$
$$= (2-\lambda)\{\lambda^2+\lambda\} + 3(1+\lambda) - 4(1+\lambda)$$
$$= (2-\lambda)\lambda(\lambda+1) - (1+\lambda) = (1+\lambda)(2\lambda - \lambda^2 - 1) = -(1+\lambda)(1-\lambda)^2$$

$\therefore$ Characteristic equation: $(1+\lambda)(1-\lambda)^2 = 0 \quad \therefore \lambda = -1, 1, 1$

$$\therefore \underline{\lambda_1 = -1; \quad \lambda_2 = 1; \quad \lambda_3 = 1.}$$

Now one for you to do.

Find the eigenvalues of the matrix $\mathbf{A} = \begin{bmatrix} 1 & -1 & 0 \\ 1 & 2 & 1 \\ -2 & 1 & -1 \end{bmatrix}$

Work through the steps in the same manner.

$\lambda =$ ............

**40**

$$\boxed{\lambda_1 = -1; \quad \lambda_2 = 1; \quad \lambda_3 = 2}$$

Here is the working:

Characteristic equation: $\begin{vmatrix} (1-\lambda) & -1 & 0 \\ 1 & (2-\lambda) & 1 \\ -2 & 1 & (-1-\lambda) \end{vmatrix} = 0$

$$\therefore \ (1-\lambda)\{(2-\lambda)(-1-\lambda)-1\}+1(-1-\lambda+2)+0=0$$

$$(1-\lambda)\{\lambda^2-\lambda-3\}+1-\lambda=0$$

$$\therefore \ 1-\lambda=0 \quad \text{or} \quad \lambda^2-\lambda-2=0$$

$$\therefore \ \lambda=1 \quad \text{or} \quad (\lambda+1)(\lambda-2)=0 \text{ i.e. } \lambda=-1 \text{ or } 2$$

$$\therefore \ \underline{\lambda_1=-1; \quad \lambda_2=1; \quad \lambda_3=2}$$

*Eigenvectors*: Each eigenvalue obtained has corresponding to it a solution of **x** called an *eigenvector*. In matrices, the term 'vector' indicates a line matrix or column matrix.

*Example 1* Consider the equation $\mathbf{A}\cdot\mathbf{x}=\lambda\mathbf{x}$ where $\mathbf{A}=\begin{bmatrix} 4 & 1 \\ 3 & 2 \end{bmatrix}$

The characteristic equation is $\begin{vmatrix} (4-\lambda) & 1 \\ 3 & (2-\lambda) \end{vmatrix} = 0.$

$$\therefore \ (4-\lambda)(2-\lambda)-3=0 \quad \therefore \ \lambda^2-6\lambda+5=0$$

$$\therefore \ (\lambda-1)(\lambda-5)=0 \quad \therefore \ \lambda=1 \text{ or } 5.$$

$$\underline{\lambda_1=1; \quad \lambda_2=5}$$

For $\lambda_1 = 1$, the equation $\mathbf{A}\cdot\mathbf{x}=\lambda\mathbf{x}$ becomes

$$\begin{bmatrix} 4 & 1 \\ 3 & 2 \end{bmatrix} \cdot \begin{bmatrix} x_1 \\ x_2 \end{bmatrix} = 1 \cdot \begin{bmatrix} x_1 \\ x_2 \end{bmatrix}$$

$$\left.\begin{aligned} 4x_1 + x_2 &= x_1 \\ 3x_1 + 2x_2 &= x_2 \end{aligned}\right\} \text{ Either of these gives } x_2 = -3x_1$$

This result merely tells us that whatever value $x_1$ has, the value of $x_2$ is $-3$ times it. Therefore, the eigenvector $\mathbf{x}_1 = \begin{bmatrix} k \\ -3k \end{bmatrix}$ is the general form of an infinite number of such eigenvectors. The simplest eigenvector is therefore $\mathbf{x}_1 = \begin{bmatrix} 1 \\ -3 \end{bmatrix}$.

For $\lambda_2 = 5$, a similar result can be obtained. Determine the eigenvector in the same way.

$$\mathbf{x}_2 = \text{............}$$

**41**

$\mathbf{x}_2 = \begin{bmatrix} k \\ k \end{bmatrix}$ is the general solution; $\mathbf{x}_2 = \begin{bmatrix} 1 \\ 1 \end{bmatrix}$ is a solution.

For, when $\lambda_2 = 5$, $\begin{bmatrix} 4 & 1 \\ 3 & 2 \end{bmatrix} \cdot \begin{bmatrix} x_1 \\ x_2 \end{bmatrix} = 5 \begin{bmatrix} x_1 \\ x_2 \end{bmatrix} = \begin{bmatrix} 5x_1 \\ 5x_2 \end{bmatrix}$

$$\therefore\ 4x_1 + x_2 = 5x_1 \quad \therefore\ x_1 = x_2 \quad \therefore\ \mathbf{x}_2 = \begin{bmatrix} 1 \\ 1 \end{bmatrix} \text{ is a solution.}$$

Therefore, $\mathbf{x}_1 = \begin{bmatrix} 1 \\ -3 \end{bmatrix}$ is an eigenvector corresponding to $\lambda_1 = 1$

and $\mathbf{x}_2 = \begin{bmatrix} 1 \\ 1 \end{bmatrix}$ is an eigenvector corresponding to $\lambda_2 = 5$.

*Example 2* Determine the eigenvalues and eigenvectors for the equation

$$\mathbf{A} \cdot \mathbf{x} = \lambda\mathbf{x} \quad \text{where} \quad \mathbf{A} = \begin{bmatrix} 2 & 0 & 1 \\ -1 & 4 & -1 \\ -1 & 2 & 0 \end{bmatrix}$$

The characteristic equation is $\begin{vmatrix} (2-\lambda) & 0 & 1 \\ -1 & (4-\lambda) & -1 \\ -1 & 2 & -\lambda \end{vmatrix} = 0$

$$\therefore\ (2-\lambda)\{-\lambda(4-\lambda)+2\} + 1\{-2+(4-\lambda)\} = 0$$

$$\therefore\ (2-\lambda)\{\lambda^2 - 4\lambda + 2\} + (2-\lambda) = 0$$

$$\therefore\ (2-\lambda)\{\lambda^2 - 4\lambda + 3\} = 0 \qquad \therefore\ \lambda = \text{..................}$$

## 42

$$\lambda = 1, 2, 3.$$

For $\lambda_1 = 1$

$$\begin{bmatrix} 1 & 0 & 1 \\ -1 & 3 & -1 \\ -1 & 2 & -1 \end{bmatrix} \cdot \begin{bmatrix} x_1 \\ x_2 \\ x_3 \end{bmatrix} = \begin{bmatrix} 0 \\ 0 \\ 0 \end{bmatrix}$$

$$x_1 + x_3 = 0 \quad \therefore x_3 = -x_1$$

$$-x_1 + 2x_2 - x_3 = 0 \quad \therefore -x_1 + 2x_2 + x_1 = 0 \quad \therefore x_2 = 0$$

$\therefore \mathbf{x}_1 = \begin{bmatrix} 1 \\ 0 \\ -1 \end{bmatrix}$ is an eigenvector corresponding to $\lambda_1 = 1$.

For $\lambda_2 = 2$

$$\begin{bmatrix} 0 & 0 & 1 \\ -1 & 2 & -1 \\ -1 & 2 & -2 \end{bmatrix} \cdot \begin{bmatrix} x_1 \\ x_2 \\ x_3 \end{bmatrix} = \begin{bmatrix} 0 \\ 0 \\ 0 \end{bmatrix}$$

Therefore, an eigenvector corresponding to $\lambda_2 = 2$ is

$\mathbf{x}_2 =$ ............

## 43

$$\mathbf{x}_2 = \begin{bmatrix} 2 \\ 1 \\ 0 \end{bmatrix}$$

since $x_3 = 0$ and $-x_1 + 2x_2 - x_3 = 0 \quad \therefore x_1 = 2x_2$.

For $\lambda_3 = 3$, we can find an eigenvector in the same way. This gives

$\mathbf{x}_3 =$ ............

**44**

$$\mathbf{x}_3 = \begin{bmatrix} 1 \\ 2 \\ 1 \end{bmatrix}$$

For, with $\lambda_3 = 3$ $\begin{bmatrix} -1 & 0 & 1 \\ -1 & 1 & -1 \\ -1 & 2 & -3 \end{bmatrix} \cdot \begin{bmatrix} x_1 \\ x_2 \\ x_3 \end{bmatrix} = \begin{bmatrix} 0 \\ 0 \\ 0 \end{bmatrix}$

$\therefore\ -x_1 + x_3 = 0 \quad \therefore\ x_3 = x_1$

$-x_1 + x_2 - x_3 = 0 \quad \therefore\ -2x_1 + x_2 = 0 \quad \therefore\ x_2 = 2x_1$

So, collecting our results together, we have

$\mathbf{x}_1 = \begin{bmatrix} 1 \\ 0 \\ -1 \end{bmatrix}$ is an eigenvector corresponding to the eigenvalue $\lambda_1 = 1$

$\mathbf{x}_2 = \begin{bmatrix} 2 \\ 1 \\ 0 \end{bmatrix}$ is an eigenvector corresponding to the eigenvalue $\lambda_2 = 2$

$\mathbf{x}_3 = \begin{bmatrix} 1 \\ 2 \\ 1 \end{bmatrix}$ is an eigenvector corresponding to the eigenvalue $\lambda_3 = 3$.

Here is one for you to do on your own. The method is the same as before.

If $\mathbf{A} \cdot \mathbf{x} = \lambda\mathbf{x}$ where $\mathbf{A} = \begin{bmatrix} 1 & -1 & 0 \\ 1 & 2 & 1 \\ -2 & 1 & -1 \end{bmatrix}$ and the eigenvalues are known to be $\lambda_1 = -1$, $\lambda_2 = 1$ and $\lambda_3 = 2$, determine corresponding eigenvectors.

$\mathbf{x}_1 = \ldots\ldots\ldots\ldots$ ; $\mathbf{x}_2 = \ldots\ldots\ldots\ldots$ ; $\mathbf{x}_3 = \ldots\ldots\ldots\ldots$

**45**

$$\mathbf{x}_1 = \begin{bmatrix} 1 \\ 2 \\ -7 \end{bmatrix}; \quad \mathbf{x}_2 = \begin{bmatrix} 1 \\ 0 \\ -1 \end{bmatrix}; \quad \mathbf{x}_3 = \begin{bmatrix} 1 \\ -1 \\ -1 \end{bmatrix}$$

Using $\begin{bmatrix} (1-\lambda) & -1 & 0 \\ 1 & (2-\lambda) & 1 \\ -2 & 1 & (-1-\lambda) \end{bmatrix} \cdot \begin{bmatrix} x_1 \\ x_2 \\ x_3 \end{bmatrix} = \begin{bmatrix} 0 \\ 0 \\ 0 \end{bmatrix}$

simple substitution of the values of $\lambda$ in turn and the knowledge of how to multiply matrices together give the results indicated.

As we have seen, a basic knowledge of matrices provides a neat and concise way of dealing with sets of linear equations. In practice, the numerical coefficients are not always simple numbers, neither is the number of equations in the set limited to three. In more extensive problems, recourse to computing facilities is a great help, but the underlying methods are still the same.

All that now remains is to check down the Revision Summary and then you can work through the Test Exercise. The problems are all straightforward and based on the work covered. You will have no trouble.

**Revision Summary**

1. *Matrix* – a rectangular array of numbers (elements).
2. *Order* – a matrix order of $(m \times n)$ denotes $m$ rows and $n$ columns.
3. *Line matrix* – one row only.
4. *Column matrix* – one column only.
5. *Double suffix notation* – $a_{34}$ denotes element in 3rd row and 4th column.
6. *Equal matrices* – corresponding elements equal.
7. *Addition and subtraction of matrices* – add or subtract corresponding elements. Therefore, for addition or subtraction, matrices must be of the same order.
8. *Multiplication of matrices*
   (a) *Scalar multiplier* – every element multiplied by the same constant, i.e. $k[a_{ij}] = [ka_{ij}]$.
   (b) *Matrix multiplier* – product $\mathbf{A} \cdot \mathbf{B}$ possible only if the number of columns in **A** equals the number of rows in **B**.

$$\begin{bmatrix} a & b & c \\ d & e & f \end{bmatrix} \cdot \begin{bmatrix} g & j \\ h & k \\ i & l \end{bmatrix} = \begin{bmatrix} ag + bh + ci & aj + bk + cl \\ dg + eh + fi & dj + ek + fl \end{bmatrix}.$$

9. *Square matrix* – of order $(m \times m)$

   (a) *Symmetric* if $a_{ij} = a_{ji}$, e.g. $\begin{bmatrix} 3 & 2 & 4 \\ 2 & 6 & 1 \\ 4 & 1 & 5 \end{bmatrix}$

   (b) *Skew symmetric* if $a_{ij} = -a_{ji}$, e.g. $\begin{bmatrix} 3 & 2 & 4 \\ -2 & 6 & 1 \\ -4 & -1 & 5 \end{bmatrix}$

10. *Diagonal matrix* – all elements zero except those on the leading diagonal.
11. *Unit matrix* – a diagonal matrix with elements on the leading diagonal all unity, i.e. $\begin{bmatrix} 1 & 0 & 0 \\ 0 & 1 & 0 \\ 0 & 0 & 1 \end{bmatrix}$ denoted by **I**.
12. *Null matrix* – all elements zero.

13. *Transpose of a matrix* – rows and columns interchanged. Transpose of $\mathbf{A} = \mathbf{A}^T$.
14. *Cofactors* – minors of the elements of $|\mathbf{A}|$ together with the respective 'place signs' of the elements.
15. *Adjoint of a square matrix* **A** – Form matrix **C** of the cofactors of the elements of $|\mathbf{A}|$ : then the adjoint of $\mathbf{A} = \mathbf{C}^T$, i.e. the transpose of **C**. $\therefore$ adj $\mathbf{A} = \mathbf{C}^T$.
16. *Inverse of a square matrix* **A**

$$\mathbf{A}^{-1} = \frac{\text{adj } \mathbf{A}}{|\mathbf{A}|} = \frac{\mathbf{C}^T}{|\mathbf{A}|}.$$

17. *Product of a square matrix and its inverse*

$$\mathbf{A} \cdot \mathbf{A}^{-1} = \mathbf{A}^{-1} \cdot \mathbf{A} = \mathbf{I}.$$

18. *Solution of a set of linear equations*

$$\mathbf{A} \cdot \mathbf{x} = \mathbf{b} \qquad \mathbf{x} = \mathbf{A}^{-1} \cdot \mathbf{b}.$$

19. *Gaussian elimination method* – reduce augmented matrix to triangular form: then use 'back substitution'.
20. *Eigenvalues* – values of $\lambda$ for which $\mathbf{A} \cdot \mathbf{x} = \lambda\mathbf{x}$.
21. *Eigenvectors* – solutions of **x** corresponding to particular values of $\lambda$.

*Now for the Test Exercise.*

## Test Exercise – V

1. If $\mathbf{A} = \begin{bmatrix} 2 & 4 & 6 & 3 \\ 1 & 7 & 0 & 4 \end{bmatrix}$ and $\mathbf{B} = \begin{bmatrix} 3 & 5 & 2 & 7 \\ 9 & 1 & 6 & 3 \end{bmatrix}$

   determine (a) $\mathbf{A} + \mathbf{B}$ and (b) $\mathbf{A} - \mathbf{B}$.

2. Given that $\mathbf{A} = \begin{bmatrix} 6 & 0 & 4 \\ 1 & 5 & -3 \end{bmatrix}$ and $\mathbf{B} = \begin{bmatrix} 2 & 9 \\ 8 & 0 \\ -4 & 7 \end{bmatrix}$

   determine (a) $3 \cdot \mathbf{A}$, (b) $\mathbf{A} \cdot \mathbf{B}$, (c) $\mathbf{B} \cdot \mathbf{A}$

3. If $\mathbf{A} = \begin{bmatrix} 2 & 3 & 5 \\ 1 & 7 & 4 \\ 8 & 0 & 6 \end{bmatrix}$, form the transpose $\mathbf{A}^{\mathrm{T}}$ and determine the

   matrix product $\mathbf{A}^{\mathrm{T}} \cdot \mathbf{I}$.

4. Show that the square matrix $\mathbf{A} = \begin{bmatrix} 3 & 2 & 4 \\ 1 & 5 & 3 \\ -1 & 8 & 2 \end{bmatrix}$ is a singular matrix.

5. If $\mathbf{A} = \begin{bmatrix} 1 & 4 & 3 \\ 6 & 2 & 5 \\ 1 & 7 & 0 \end{bmatrix}$, determine (a) $|\mathbf{A}|$ and (b) adj $\mathbf{A}$.

6. Find the inverse of the matrix $\mathbf{A} = \begin{bmatrix} 2 & 1 & 4 \\ 3 & 5 & 1 \\ 2 & 0 & 6 \end{bmatrix}$

7. Express the following set of linear equations in matrix form

$$\begin{aligned} 2x_1 + 4x_2 - 5x_3 &= -7 \\ x_1 - 3x_2 + x_3 &= 10 \\ 3x_1 + 5x_2 + 3x_3 &= 2 \end{aligned}$$

8. Solve the following set of linear equations by matrix method

$$\begin{aligned} x_1 + 3x_2 + 2x_3 &= 3 \\ 2x_1 - x_2 - 3x_3 &= -8 \\ 5x_1 + 2x_2 + x_3 &= 9 \end{aligned}$$

9. For the following set of simultaneous equations
   (a) form the augmented coefficient matrix
   (b) solve the equations by Gaussian elimination.

$$\begin{aligned} x_1 + 2x_2 + 3x_3 &= 5 \\ 3x_1 - x_2 + 2x_3 &= 8 \\ 4x_1 - 6x_2 - 4x_3 &= -2 \end{aligned}$$

10. If $\mathbf{A} \cdot \mathbf{x} = \lambda \mathbf{x}$, where $\mathbf{A} = \begin{bmatrix} 2 & 2 & -2 \\ 1 & 3 & 1 \\ 1 & 2 & 2 \end{bmatrix}$ determine the eigenvalues of the matrix $\mathbf{A}$ and an eigenvector corresponding to each eigenvalue.

**Further Problems – V**

1. If $\mathbf{A} = \begin{bmatrix} 7 & 2 \\ 3 & 1 \end{bmatrix}$ and $\mathbf{B} = \begin{bmatrix} 4 & 6 \\ 5 & 8 \end{bmatrix}$, determine

   (a) $\mathbf{A} + \mathbf{B}$, (b) $\mathbf{A} - \mathbf{B}$, (c) $\mathbf{A} \cdot \mathbf{B}$, (d) $\mathbf{B} \cdot \mathbf{A}$

2. If $\mathbf{A} = \begin{bmatrix} j & 0 \\ 0 & -j \end{bmatrix}$, $\mathbf{B} = \begin{bmatrix} 0 & 1 \\ -1 & 0 \end{bmatrix}$, $\mathbf{C} = \begin{bmatrix} 0 & j \\ j & 0 \end{bmatrix}$, $\mathbf{I} = \begin{bmatrix} 1 & 0 \\ 0 & 1 \end{bmatrix}$,

   where $j = \sqrt{-1}$, express (a) $\mathbf{A} \cdot \mathbf{B}$, (b) $\mathbf{B} \cdot \mathbf{C}$, (c) $\mathbf{C} \cdot \mathbf{A}$, and (d) $\mathbf{A}^2$,

in terms of other matrices.

3. If $\mathbf{A} = \begin{bmatrix} 1 & 0.5 \\ 0.5 & 0.1 \end{bmatrix}$ and $\mathbf{B} = \begin{bmatrix} 1 & 2 \\ 2 & 3 \end{bmatrix}$, determine

   (a) $\mathbf{B}^{-1}$, (b) $\mathbf{A} \cdot \mathbf{B}$, (c) $\mathbf{B}^{-1} \cdot \mathbf{A}$

4. Determine the value of $k$ for which the following set of homogeneous equations has non-trivial solutions

$$\begin{aligned} 4x_1 + 3x_2 - x_3 &= 0 \\ 7x_1 - x_2 - 3x_3 &= 0 \\ 3x_1 - 4x_2 + kx_3 &= 0 \end{aligned}$$

5. Express the following sets of simultaneous equations in matrix form

   (a)
$$\begin{aligned} 2x_1 - 3x_2 - x_3 &= 2 \\ x_1 + 4x_2 + 2x_3 &= 3 \\ x_1 - x_2 + x_3 &= 5 \end{aligned}$$

   (b)
$$\begin{aligned} x_1 - 2x_2 - x_3 + 3x_4 &= 10 \\ 2x_1 + 3x_2 \qquad + x_4 &= 8 \\ x_1 \qquad - 4x_3 - 2x_4 &= 3 \\ - x_2 + 3x_3 + x_4 &= -7 \end{aligned}$$

In *Questions 6 to 10* solve, where possible, the sets of equations by a matrix method

6. $$\begin{aligned} 2i_1 + i_2 + i_3 &= 8 \\ 5i_1 - 3i_2 + 2i_3 &= 3 \\ 7i_1 + i_2 + 3i_3 &= 20 \end{aligned}$$

7. $$\begin{aligned} 3x + 2y + 4z &= 3 \\ x + y + z &= 2 \\ 2x - y + 3z &= -3 \end{aligned}$$

8. $$\begin{aligned} 4i_1 - 5i_2 + 6i_3 &= 3 \\ 8i_1 - 7i_2 - 3i_3 &= 9 \\ 7i_1 - 8i_2 + 9i_3 &= 6 \end{aligned}$$

9. $$\begin{aligned} 3x + 2y + 5z &= 1 \\ x - y + z &= 4 \\ 6x + 4y + 10z &= 7 \end{aligned}$$

10. $$\begin{aligned} 3x_1 + 2x_2 - 2x_3 &= 16 \\ 4x_1 + 3x_2 + 3x_3 &= 2 \\ -2x_1 + x_2 - x_3 &= 1 \end{aligned}$$

In *Questions 11 to 13*, form the augmented matrix and solve the sets of equations by Gaussian elimination.

11. $$\begin{aligned} 5i_1 - i_2 + 2i_3 &= 3 \\ 2i_1 + 4i_2 + i_3 &= 8 \\ i_1 + 3i_2 - 3i_3 &= 2 \end{aligned}$$

12. $$\begin{aligned} i_1 + 2i_2 + 3i_3 &= -4 \\ 2i_1 + 6i_2 - 3i_3 &= 33 \\ 4i_1 - 2i_2 + i_3 &= 3 \end{aligned}$$

13. $$\begin{aligned} 7i - 4i_2 \qquad &= 12 \\ -4i_1 + 12i_2 - 6i_3 &= 0 \\ - 6i_2 + 14i_3 &= 0 \end{aligned}$$

14. In a star-connected circuit, currents $i_1, i_2, i_3$ flowing through impedances $Z_1, Z_2, Z_3$, are given by

$$\begin{aligned} i_1 + i_2 + i_3 &= 0 \\ Z_1 i_1 - Z_2 i_2 &= e_1 - e_2 \\ Z_2 i_2 - Z_3 i_3 &= e_2 - e_3. \end{aligned}$$

If $Z_1 = 10$; $Z_2 = 8$; $Z_3 = 3$; $e_1 - e_2 = 65$; $e_2 - e_3 = 120$; apply matrix methods to determine the values of $i_1, i_2, i_3$.

15. Currents $i_1, i_2, i_3$ in a network are related by the following equations

$$\begin{aligned} Z_1 i_1 + Z_3 i_3 &= V \\ Z_2 i_2 - Z_3 i_3 &= 0 \\ i_1 - i_2 - i_3 &= 0. \end{aligned}$$

Determine expressions for $i_1, i_2, i_3$, in terms of $Z_1, Z_2, Z_3$ and $V$.

*Questions 16 to 20* refer to the vector equation $\mathbf{A} \cdot \mathbf{x} = \lambda \mathbf{x}$. For the coefficient matrix $\mathbf{A}$ given in each case, determine the eigenvalues and an eigenvector corresponding to each eigenvalue.

16. $$\mathbf{A} = \begin{bmatrix} 2 & 1 & 1 \\ 1 & 3 & 2 \\ -1 & 1 & 2 \end{bmatrix}$$

17. $$\mathbf{A} = \begin{bmatrix} 1 & 2 & 2 \\ 1 & 3 & 1 \\ 2 & 2 & 1 \end{bmatrix}$$

18. $$\mathbf{A} = \begin{bmatrix} 2 & 0 & 1 \\ -1 & 4 & -1 \\ -1 & 2 & 0 \end{bmatrix}$$

19. $$\mathbf{A} = \begin{bmatrix} 1 & -4 & -2 \\ 0 & 3 & 1 \\ 1 & 2 & 4 \end{bmatrix}$$

20. $$\mathbf{A} = \begin{bmatrix} 3 & 0 & 3 \\ 0 & 3 & 3 \\ 2 & 3 & 1 \end{bmatrix}$$

# Programme 6

## VECTORS

**1**

**Introduction**: **scalar and vector quantities**

Physical quantities can be divided into two main groups, scalar quantities and vector quantities.

(a) A *scalar quantity* is one that is defined completely by a single number with appropriate units, e.g. length, area, volume, mass, time, etc. Once the units are stated, the quantity is denoted entirely by its size or *magnitude*.

(b) A *vector quantity* is defined completely when we know not only its magnitude (with units) but also the direction in which it operates, e.g. force, velocity, acceleration. A vector quantity necessarily involves *direction* as well as magnitude.

So, (i) a speed of 10 km/h is a scalar quantity, but
(ii) a velocity of '10 km/h due North' is a .................... quantity.

**2**

vector

A force F acting at a point P is a vector quantity, since to define it completely we must give

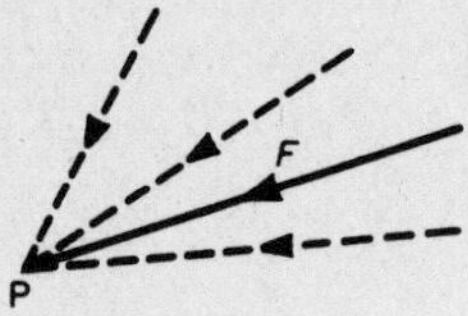

F
P

(i) its magnitude, and also
(ii) its ........................

**3**

direction

So that:

(i) A temperature of 100°C is a .................... quantity.
(ii) An acceleration of 9·8 m/s$^2$ vertically downwards is a .................... quantity.
(iii) The weight of a 7 kg mass is a .................... quantity.
(iv) The sum of £500 is a .................... quantity.
(v) A north-easterly wind of 20 knots is a .................... quantity.

**4**

| (i) scalar, (ii) vector, (iii) vector, (iv) scalar, (v) vector |
|---|

Since, in (ii), (iii) and (v), the complete description of the quantity includes not only its magnitude, but also its ....................

**5**

| direction |
|---|

**Vector representation**

A vector quantity can be represented graphically by a line, drawn so that:

(i) the *length* of the line denotes the magnitude of the quantity, according to some stated vector scale,

(ii) the *direction* of the line denotes the direction in which the vector quantity acts. The sense of the direction is indicated by an arrow head.

e.g. A horizontal force of 35 N acting to the right, would be indicated by a line ——→—— and if the chosen vector scale were 1 cm ≡ 10 N, the line would be .................... cm long.

**6**

| 3·5 |
|---|

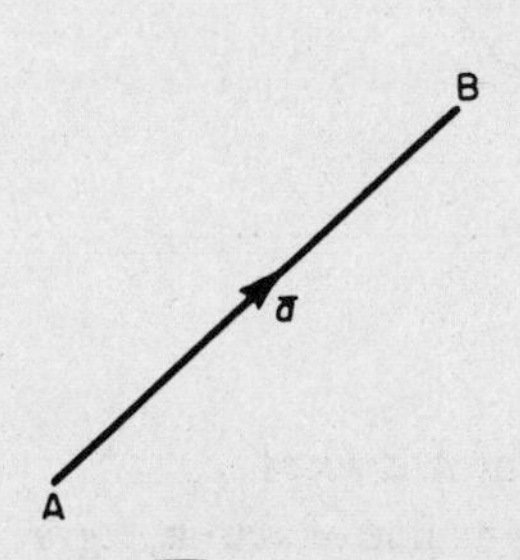


The vector quantity AB is referred to as

$$\overline{AB} \quad \text{or} \quad \bar{a}$$

The magnitude of the vector quantity is written $|\overline{AB}|$, or $|\bar{a}|$, or simply AB, or $a$ (i.e. without the bar over it).

Note that $\overline{BA}$ would represent a vector quantity of the same magnitude but with opposite sense.

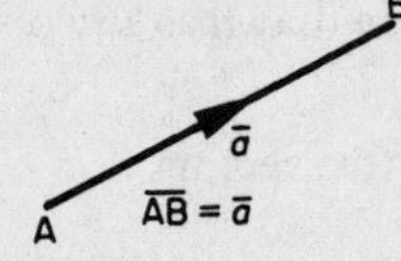


B

A

$\overline{BA} = -\overline{AB} = -\bar{a}$

*On to frame 7.*

**7** **Two equal vectors**

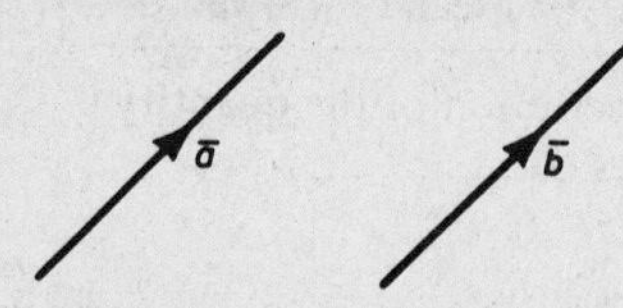


If two vectors, $\bar{a}$ and $\bar{b}$, are said to be equal, they have the same magnitude and the same direction.

If $\bar{a} = \bar{b}$, then (i) $a = b$ (magnitudes equal)

(ii) the direction of $\bar{a}$ = direction of $\bar{b}$, i.e. the two vectors are parallel and in the same sense.

Similarly, if two vectors $\bar{a}$ and $\bar{b}$ are such that $\bar{b} = -\bar{a}$, what can we say about (i) their magnitudes,

(ii) their directions?

**8**

| (i) Magnitudes are equal. |
| --- |
| (ii) The vectors are parallel but opposite in sense. |

i.e. if $\bar{b} = -\bar{a}$, then

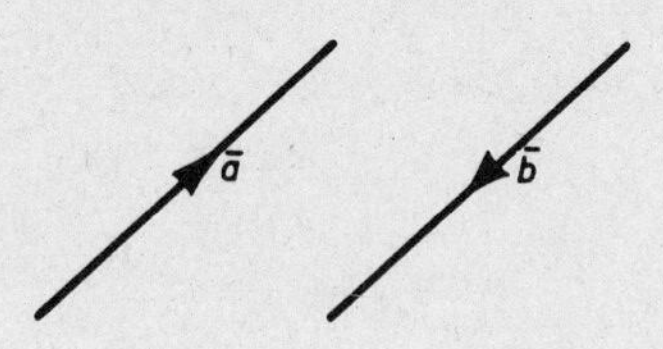


**9** **Types of vectors**

(i) A *position vector* $\overline{AB}$ occurs when the point A is fixed.

(ii) A *line vector* is such that it can slide along its line of action, e.g. a mechanical force acting on a body.

(iii) A *free vector* is not restricted in any way. It is completely defined by its magnitude and direction and can be drawn as any one of a set of equal-length parallel lines.

Most of the vectors we shall consider will be free vectors.

*So on now to frame 10.*

**10**

**Addition of vectors**

The sum of two vectors, $\overline{AB}$ and $\overline{BC}$, is defined as the single or equivalent or resultant vector $\overline{AC}$

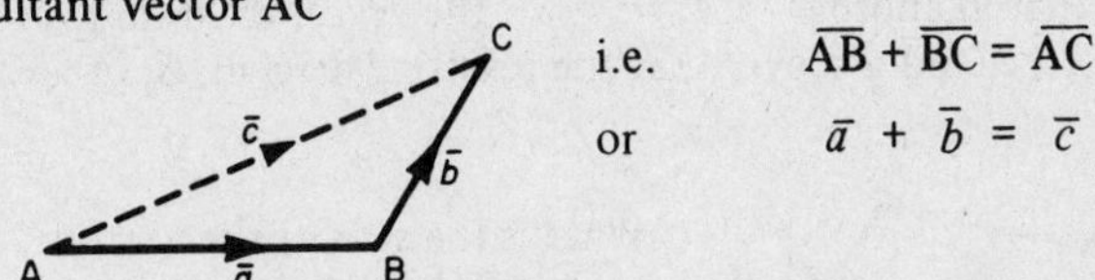


i.e. $\overline{AB} + \overline{BC} = \overline{AC}$

or $\bar{a} + \bar{b} = \bar{c}$

To find the sum of two vectors $\bar{a}$ and $\bar{b}$, then, we draw them as a chain, starting the second where the first ends: the sum $\bar{c}$ is then given by the single vector joining the start of the first to the end of the second.

e.g. If $\bar{p} \equiv$ a force of 40 N, acting in the direction due East
$\bar{q} \equiv$ a force of 30 N, " " " " due North

then the magnitude of the vector sum $r$ of these two forces will be ..........

**11**

$r = 50$ N

for

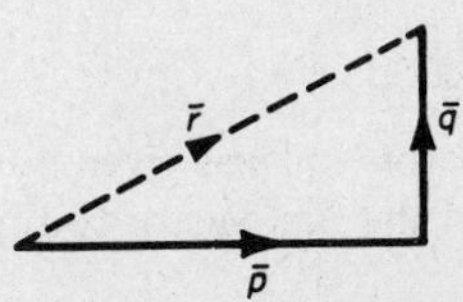


$$r^2 = p^2 + q^2$$
$$= 1600 + 900 = 2500$$
$$r = \sqrt{2500} = \underline{50 \text{ N}}$$

**The sum of a number of vectors** $\bar{a} + \bar{b} + \bar{c} + \bar{d} + \ldots..$

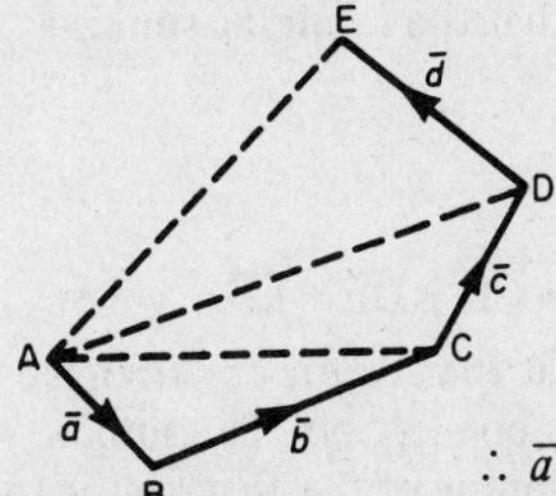


(i) Draw the vectors as a chain.
(ii) Then:

$$\bar{a} + \bar{b} = \overline{AC}$$
$$\overline{AC} + \bar{c} = \overline{AD}$$
$$\therefore \bar{a} + \bar{b} + \bar{c} = \overline{AD}$$
$$\overline{AD} + \bar{d} = \overline{AE}$$
$$\therefore \bar{a} + \bar{b} + \bar{c} + \bar{d} = \overline{AE}$$

i.e. the sum of all vectors, $\bar{a}, \bar{b}, \bar{c}, \bar{d}$, is given by the single vector joining the start of the first to the end of the last – in this case, $\overline{AE}$. This follows directly from our previous definition of the sum of two vectors.

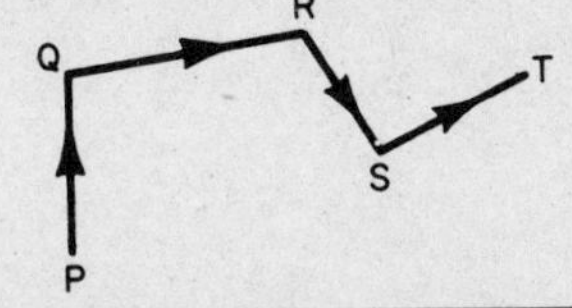


Similarly,

$\overline{PQ} + \overline{QR} + \overline{RS} + \overline{ST} =$ ......................

**12**

$\overline{PT}$

Now suppose that in another case, we draw the vector diagram to find the sum of $\bar{a}, \bar{b}, \bar{c}, \bar{d}, \bar{e}$, and discover that the resulting diagram is, in fact, a closed figure.

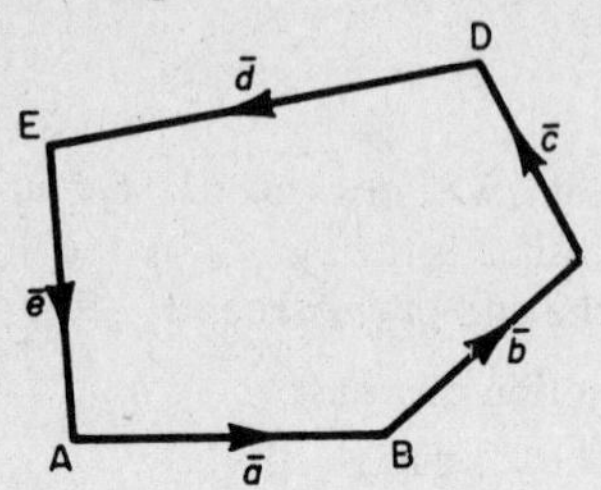


What is the sum of the vectors $\bar{a}, \bar{b}, \bar{c}, \bar{d}, \bar{e}$, in this case?

*Think carefully and when you have decided, move on to frame 13.*

**13**

Sum of the vectors = 0

For we said in the previous case, that the vector sum was given by the single equivalent vector joining the beginning of the first vector to the end of the last.

But, if the vector diagram is a *closed* figure, the end of the last vector coincides with the beginning of the first, so that the resultant sum is a vector with *no magnitude.*

*Now for one or two examples.*

*Example 1.* Find the vector sum $\overline{AB} + \overline{BC} + \overline{CD} + \overline{DE} + \overline{EF}$.

Without drawing a diagram, we can see that the vectors are arranged in a chain, each beginning where the previous one left off. The sum is therefore given by the vector joining the beginning of the first vector to the end of the last.

$$\therefore \text{ Sum} = \overline{AF}$$

In the same way,

$$\overline{AK} + \overline{KL} + \overline{LP} + \overline{PQ} = \ldots\ldots\ldots\ldots\ldots$$

**14**

$\boxed{\overline{AQ}}$

Right. Now what about this one?

Find the sum of $\overline{AB} - \overline{CB} + \overline{CD} - \overline{ED}$

We must beware of the negative vectors. Remember that $-\overline{CB} = \overline{BC}$, i.e. the same magnitude and direction but in the opposite sense.

Also $-\overline{ED} = \overline{DE}$

$$\therefore\ \overline{AB} - \overline{CB} + \overline{CD} - \overline{ED} = \overline{AB} + \overline{BC} + \overline{CD} + \overline{DE}$$
$$= \underline{\overline{AE}}.$$

Now you do this one:

Find the vector sum $\overline{AB} + \overline{BC} - \overline{DC} - \overline{AD}$

*When you have the result, move on to frame 15.*

**15**

$\boxed{0}$

For:

$$\overline{AB} + \overline{BC} - \overline{DC} - \overline{AD} = \overline{AB} + \overline{BC} + \overline{CD} + \overline{DA}$$

and the lettering indicates that the end of the last vector coincides with the beginning of the first. The vector diagram is thus a closed figure and therefore the sum of the vectors is 0.

Now here are some for you to do:

(i) $\overline{PQ} + \overline{QR} + \overline{RS} + \overline{ST} =$ ....................

(ii) $\overline{AC} + \overline{CL} - \overline{ML} =$ ....................

(iii) $\overline{GH} + \overline{HJ} + \overline{JK} + \overline{KL} + \overline{LG} =$ ....................

(iv) $\overline{AB} + \overline{BC} + \overline{CD} + \overline{DB} =$ ....................

*When you have finished all four, check with the results in the next frame.*

**16** Here are the results:

(i) $\overline{PQ} + \overline{QR} + \overline{RS} + \overline{ST} = \overline{PT}$

(ii) $\overline{AC} + \overline{CL} - \overline{ML} = \overline{AC} + \overline{CL} + \overline{LM} = \overline{AM}$

(iii) $\overline{GH} + \overline{HJ} + \overline{JK} + \overline{KL} + \overline{LG} = 0$

[Since the end of the last vector coincides with the beginning of the first.]

(iv) $\overline{AB} + \overline{BC} + \overline{CD} + \overline{DB} = \overline{AB}$

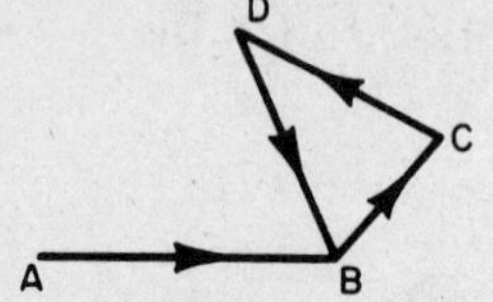


The last three vectors form a closed figure and therefore the sum of these three vectors is zero, leaving only $\overline{AB}$ to be considered.

*Now on to frame 17.*

**17**

**Components of a given vector**

Just as $\overline{AB} + \overline{BC} + \overline{CD} + \overline{DE}$ can be replaced by $\overline{AE}$, so any single vector $\overline{PT}$ can be replaced by any number of component vectors so long as they form a chain in the vector diagram, beginning at P and ending at T.

e.g.

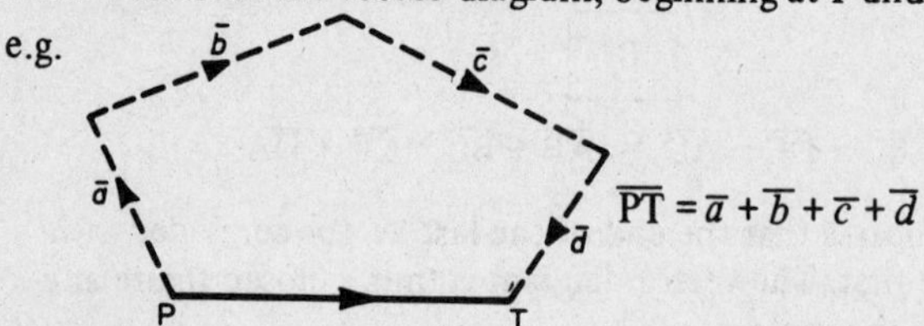


$\overline{PT} = \overline{a} + \overline{b} + \overline{c} + \overline{d}$

*Example 1.*

ABCD is a quadrilateral, with G and H the mid-points of DA and BC respectively. Show that $\overline{AB} + \overline{DC} = 2\,\overline{GH}$

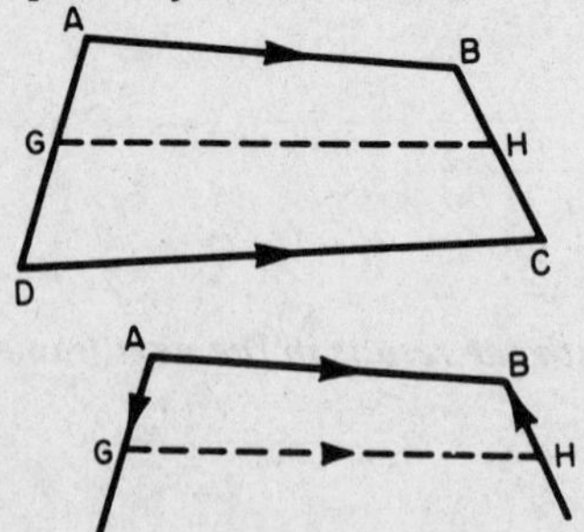


We can replace vector $\overline{AB}$ by any chain of vectors so long as they start at A and end at B

e.g. we could say

$$\overline{AB} = \overline{AG} + \overline{GH} + \overline{HB}$$

Similarly, we could say

$$\overline{DC} = \text{........................}$$

**18**

$$\boxed{\overline{DC} = \overline{DG} + \overline{GH} + \overline{HC}}$$

So we have

$$\overline{AB} = \overline{AG} + \overline{GH} + \overline{HB}$$

$$\overline{DC} = \overline{DG} + \overline{GH} + \overline{HC}$$

$$\begin{aligned} \therefore\ \overline{AB} + \overline{DC} &= \overline{AG} + \overline{GH} + \overline{HB} + \overline{DG} + \overline{GH} + \overline{HC} \\ &= 2\,\overline{GH} + (\overline{AG} + \overline{DG}) + (\overline{HB} + \overline{HC}) \end{aligned}$$

Now, G is the mid point of AD. Therefore, vectors $\overline{AG}$ and $\overline{DG}$ are equal in length but opposite in sense.

$$\therefore\ \overline{DG} = -\overline{AG}$$

Similarly $\quad \overline{HC} = -\overline{HB}$

$$\begin{aligned} \therefore\ \overline{AB} + \overline{DC} &= 2\,\overline{GH} + (\overline{AG} - \overline{AG}) + (\overline{HB} - \overline{HB}) \\ &= 2\,\overline{GH} \end{aligned}$$

*Next frame.*

**19**

*Example 2.*

Points L, M, N are mid points of the sides AB, BC, CA, of the triangle ABC. Show that

(i) $\overline{AB} + \overline{BC} + \overline{CA} = 0$

(ii) $2\,\overline{AB} + 3\,\overline{BC} + \overline{CA} = 2\,\overline{LC}$

(iii) $\overline{AM} + \overline{BN} + \overline{CL} = 0$.

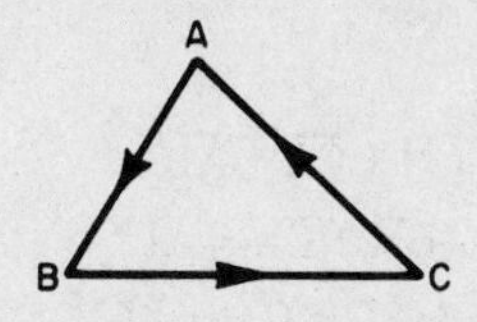


(i) We can dispose of the first part straight away without any trouble. We can see from the vector diagram that $\overline{AB} + \overline{BC} + \overline{CA} = 0$ since these three vectors form a .......... ............

**20**

closed figure

Now for part (ii).

To show that $2\overline{AB} + 3\overline{BC} + \overline{CA} = 2\overline{LC}$

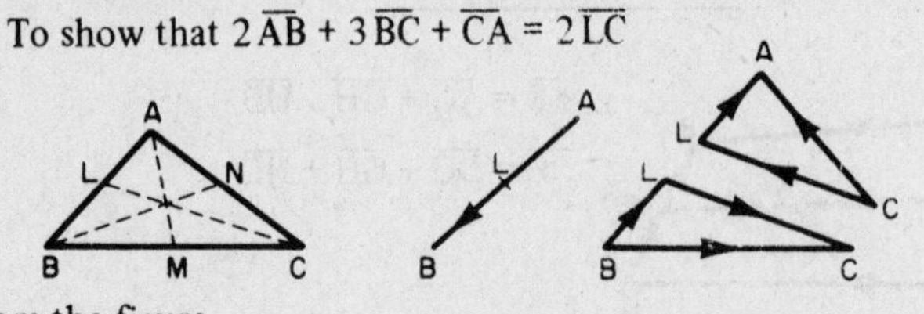


From the figure

$$\overline{AB} = 2\overline{AL}; \quad \overline{BC} = \overline{BL} + \overline{LC}; \quad \overline{CA} = \overline{CL} + \overline{LA}$$

$$\therefore\ 2\overline{AB} + 3\overline{BC} + \overline{CA} = 4\overline{AL} + 3\overline{BL} + 3\overline{LC} + \overline{CL} + \overline{LA}$$

$$\text{Now}\ \overline{BL} = -\overline{AL}; \quad \overline{CL} = -\overline{LC}; \quad \overline{LA} = -\overline{AL}$$

Substituting these in the previous line, gives

$$2\overline{AB} + 3\overline{BC} + \overline{CA} = \ldots\ldots\ldots\ldots$$

**21**

$2\overline{LC}$

For
$$\begin{aligned} 2\overline{AB} + 3\overline{BC} + \overline{CA} &= 4\overline{AL} + 3\overline{BL} + 3\overline{LC} + \overline{CL} + \overline{LA} \\ &= 4\overline{AL} - 3\overline{AL} + 3\overline{LC} - \overline{LC} - \overline{AL} \\ &= 4\overline{AL} - 4\overline{AL} + 3\overline{LC} - \overline{LC} \\ &= \underline{2\overline{LC}} \end{aligned}$$

Now part (iii)

To prove that $\overline{AM} + \overline{BN} + \overline{CL} = 0$

From the figure in frame 20, we can say

$$\overline{AM} = \overline{AB} + \overline{BM}$$
$$\overline{BN} = \overline{BC} + \overline{CN}$$

Similarly $\overline{CL} = \ldots\ldots\ldots\ldots$

**22**

$\overline{CL} = \overline{CA} + \overline{AL}$

So
$$\begin{aligned} \overline{AM} + \overline{BN} + \overline{CL} &= \overline{AB} + \overline{BM} + \overline{BC} + \overline{CN} + \overline{CA} + \overline{AL} \\ &= (\overline{AB} + \overline{BC} + \overline{CA}) + (\overline{BM} + \overline{CN} + \overline{AL}) \\ &= (\overline{AB} + \overline{BC} + \overline{CA}) + \tfrac{1}{2}(\overline{BC} + \overline{CA} + \overline{AB}) \\ &= \ldots\ldots\ldots\ldots\ldots\ldots \end{aligned}$$
Finish it off.

**23**

$$\overline{AM} + \overline{BN} + \overline{CL} = 0$$

Since $\overline{AM} + \overline{BN} + \overline{CL} = (\overline{AB} + \overline{BC} + \overline{CA}) + \frac{1}{2}(\overline{BC} + \overline{CA} + \overline{AB})$

Now $\overline{AB} + \overline{BC} + \overline{CA}$ is a closed figure $\therefore$ Vector sum = 0

and $\overline{BC} + \overline{CA} + \overline{AB}$ is a closed figure $\therefore$ Vector sum = 0

$$\therefore \underline{\overline{AM} + \overline{BN} + \overline{CL} = 0}$$

Here is another.

*Example 3.*

ABCD is a quadrilateral in which P and Q are the mid points of the diagonals AC and BD respectively.

Show that $\qquad \overline{AB} + \overline{AD} + \overline{CB} + \overline{CD} = 4\,\overline{PQ}$

First, just draw the figure: *then move on to frame 24.*

**24**

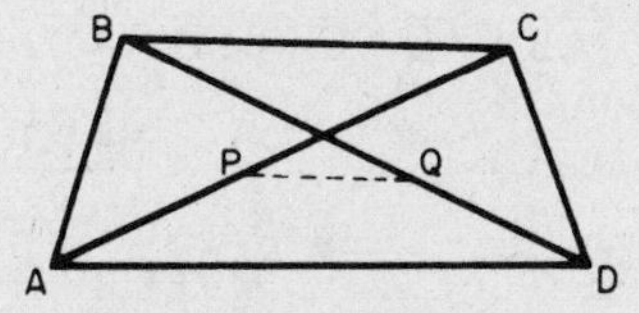


To prove that $\overline{AB} + \overline{AD} + \overline{CB} + \overline{CD} = 4\,\overline{PQ}$

Taking the vectors on the left-hand side, one at a time, we can write

$$\overline{AB} = \overline{AP} + \overline{PQ} + \overline{QB}$$
$$\overline{AD} = \overline{AP} + \overline{PQ} + \overline{QD}$$
$$\overline{CB} = \ldots\ldots\ldots\ldots\ldots$$
$$\overline{CD} = \ldots\ldots\ldots\ldots\ldots$$

**25**

$$\overline{CB} = \overline{CP} + \overline{PQ} + \overline{QB}\ ; \quad \overline{CD} = \overline{CP} + \overline{PQ} + \overline{QD}$$

Adding all four lines together, we have

$$\begin{aligned}\overline{AB} + \overline{AD} + \overline{CB} + \overline{CD} &= 4\,\overline{PQ} + 2\,\overline{AP} + 2\,\overline{CP} + 2\,\overline{QB} + 2\,\overline{QD}\\ &= 4\,\overline{PQ} + 2(\overline{AP} + \overline{CP}) + 2(\overline{QB} + \overline{QD})\end{aligned}$$

Now what can we say about $(\overline{AP} + \overline{CP})$?

## 26

$$\boxed{\overline{AP} + \overline{CP} = 0}$$

Since P is the mid point of AC $\therefore$ AP = PC

$$\therefore \overline{CP} = -\overline{PC} = -\overline{AP}$$

$$\therefore \overline{AP} + \overline{CP} = \overline{AP} - \overline{AP} = 0.$$

In the same way, $(\overline{QB} + \overline{QD}) =$ ......................

## 27

$$\boxed{\overline{QB} + \overline{QD} = 0}$$

Since Q is the mid point of BD $\therefore$ $\overline{QD} = -\overline{QB}$

$$\therefore \overline{QB} + \overline{QD} = \overline{QB} - \overline{QB} = 0$$

$$\therefore \overline{AB} + \overline{AD} + \overline{CB} + \overline{CD} = 4\,\overline{PQ} + 0 + 0$$

$$= \underline{4\,\overline{PQ}}$$

## 28

Here is one more.

*Example 4.*

Prove by vectors that the line joining the mid-points of two sides of a triangle is parallel to the third side and half its length.

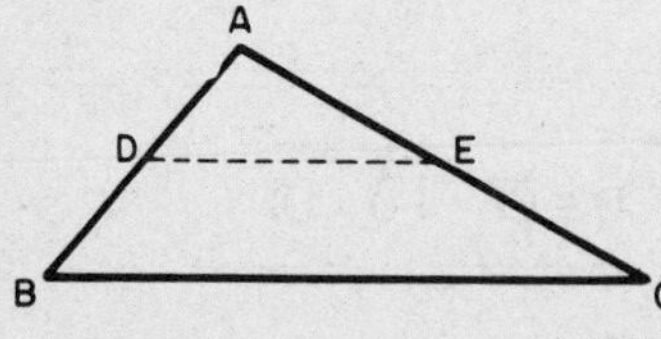


Let D and E be the mid-points of AB and AC respectively.

We have $\overline{DE} = \overline{DA} + \overline{AE}$

Now express $\overline{DA}$ and $\overline{AE}$ in terms of $\overline{BA}$ and $\overline{AC}$ respectively and see if you can get the required results.

*Then on to frame 29.*

**29**

Here is the working. Check through it.

$$\overline{DE} = \overline{DA} + \overline{AE}$$
$$= \tfrac{1}{2}\overline{BA} + \tfrac{1}{2}\overline{AC}$$
$$= \tfrac{1}{2}(\overline{BA} + \overline{AC})$$
$$\therefore \quad \underline{\overline{DE} = \tfrac{1}{2}\overline{BC}}$$

$\therefore$ $\overline{DE}$ is half the magnitude (length) of $\overline{BC}$ and acts in the same direction.

i.e. DE and BC are parallel.

*Now for the next section of the work: turn on to frame 30.*

**30**

**Components of a vector** in terms of **unit vectors**

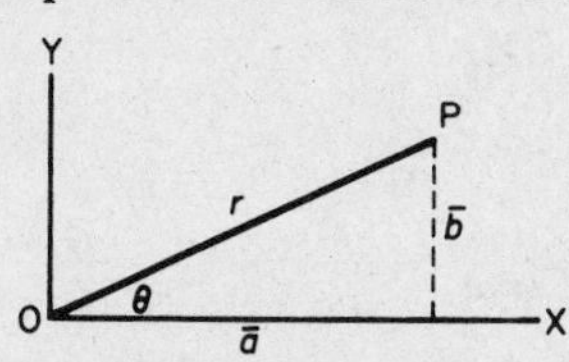


The vector $\overline{OP}$ is defined by its magnitude ($r$) and its direction ($\theta$). It could also be defined by its two components in the OX and OY directions.

i.e. $\overline{OP}$ is equivalent to a vector $\bar{a}$ in the OX direction + a vector $\bar{b}$ in the OY direction.

i.e. $\overline{OP} = \bar{a}$ (along OX) + $\bar{b}$ (along OY)

If we now define $\bar{i}$ to be a *unit vector* in the OX direction,

then $\quad \bar{a} = a\bar{i}$

Similarly, if we define $j$ to be a *unit vector* in the OY direction,

then $\quad \bar{b} = b\bar{j}$

So that the vector OP can be written as

$$\underline{\bar{r} = a\bar{i} + b\bar{j}}$$

where $\bar{i}$ and $\bar{j}$ are unit vectors in the OX and OY directions.

Having defined the unit vectors above, we shall in practice omit the bars over the $i$ and $j$, in the interest of clarity. But remember they are vectors.

**31** Let $\overline{z}_1 = 2i + 4j$ and $\overline{z}_2 = 5i + 2j$

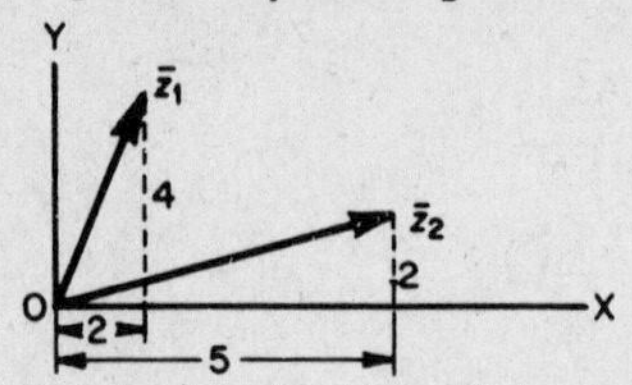


To find $\overline{z}_1 + \overline{z}_2$, draw the two vectors in a chain.

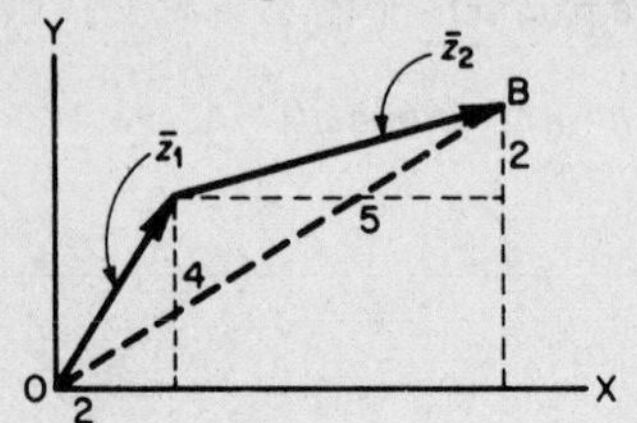


$$\overline{z}_1 + \overline{z}_2 = \overline{OB}$$
$$= (2 + 5)i + (4 + 2)j$$
$$= \underline{7i + 6j}$$

i.e. total up the vector components along OX,
and " " " " " OY

Of course, we can do this without a diagram:

If $\overline{z}_1 = 3i + 2j$ and $\overline{z}_2 = 4i + 3j$

$$\overline{z}_1 + \overline{z}_2 = 3i + 2j + 4i + 3j$$
$$= \underline{7i + 5j}$$

And in much the same way, $\overline{z}_2 - \overline{z}_1 =$ ....................

**32**

$$\boxed{\overline{z}_2 - \overline{z}_1 = 1i + 1j}$$

for
$$\overline{z}_2 - \overline{z}_1 = (4i + 3j) - (3i + 2j)$$
$$= 4i + 3j - 3i - 2j$$
$$= \underline{1i + 1j}$$

Similarly, if $\overline{z}_1 = 5i - 2j$; $\overline{z}_2 = 3i + 3j$; $\overline{z}_3 = 4i - 1j$,

then (i) $\overline{z}_1 + \overline{z}_2 + \overline{z}_3 =$ ....................

and (ii) $\overline{z}_1 - \overline{z}_2 - \overline{z}_3 =$ ....................

*When you have the results, turn on to frame 33.*

**33**

$$\boxed{\text{(i) } 12i \;\; ; \;\; \text{(ii) } -2i - 4j}$$

Here is the working:

$$\text{(i)}\quad \bar{z}_1 + \bar{z}_2 + \bar{z}_3 = 5i - 2j + 3i + 3j + 4i - 1j$$
$$= (5 + 3 + 4)i + (3 - 2 - 1)j$$
$$= \underline{12i}$$
$$\text{(ii)}\quad \bar{z}_1 - \bar{z}_2 - \bar{z}_3 = (5i - 2j) - (3i + 3j) - (4i - 1j)$$
$$= (5 - 3 - 4)i + (-2 - 3 + 1)j$$
$$= \underline{-2i - 4j}$$

Now this one.

If $\overline{\text{OA}} = 3i + 5j$ and $\overline{\text{OB}} = 5i - 2j$, find $\overline{\text{AB}}$.

As usual, a diagram will help. Here it is:

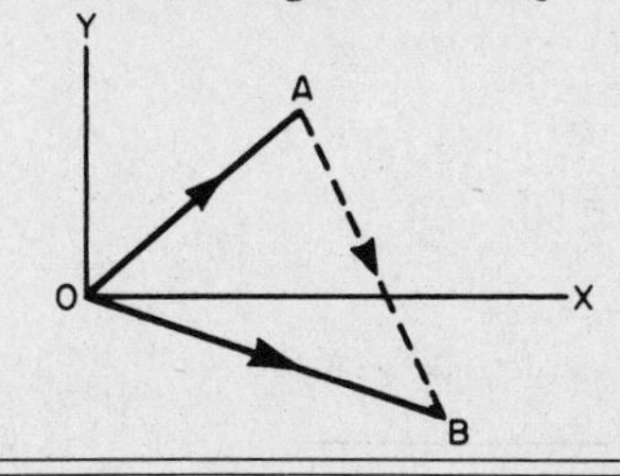


First of all, from the diagram, write down a relationship between the vectors. Then express them in terms of the unit vectors.

$\overline{\text{AB}}$ = ..................................

**34**

$$\boxed{\overline{\text{AB}} = 2i - 7j}$$

for we have $\quad \overline{\text{OA}} + \overline{\text{AB}} = \overline{\text{OB}}$ (from the diagram)

$$\therefore\; \overline{\text{AB}} = \overline{\text{OB}} - \overline{\text{OA}}$$
$$= (5i - 2j) - (3i + 5j) = \underline{2i - 7j}$$

*On to frame 35.*

**35**

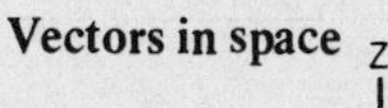

**Vectors in space**

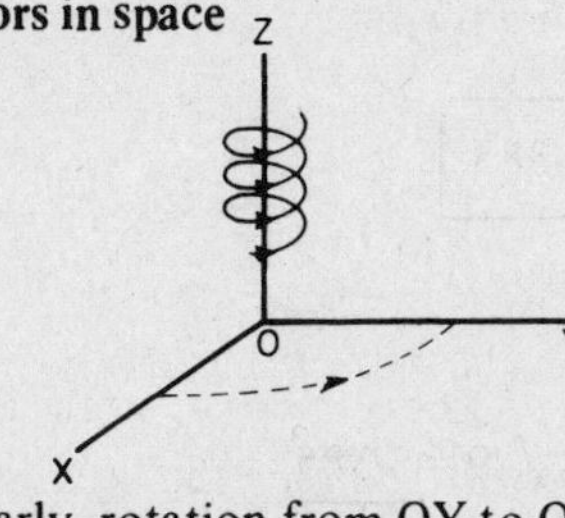


The axes of reference are defined by the 'right-hand' rule.

OX, OY, OZ form a right-handed set if rotation from OX to OY takes a right-handed corkscrew action along the positive direction of OZ.

Similarly, rotation from OY to OZ gives right-hand corkscrew action along the positive direction of ......................

**36**

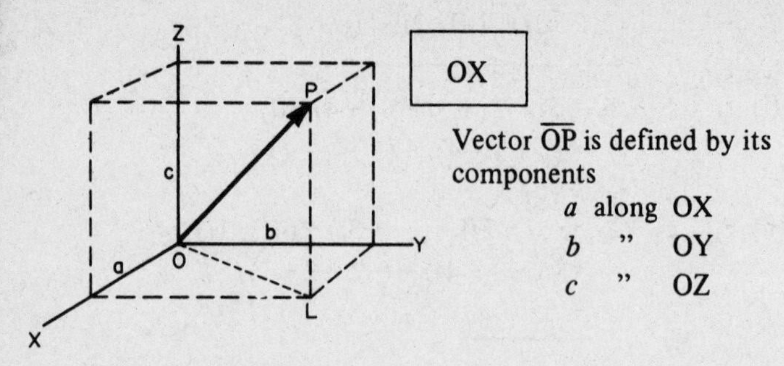


OX

Vector $\overline{OP}$ is defined by its components

$a$ along OX
$b$ " OY
$c$ " OZ

Let $i$ = unit vector in OX direction,
$j$ = " " " OY "
$k$ = " " " OZ "

Then $\overline{OP} = ai + bj + ck$

Also $OL^2 = a^2 + b^2$ and $OP^2 = OL^2 + c^2$

$OP^2 = a^2 + b^2 + c^2$

So, if $\underline{\bar{r} = ai + bj + ck}$, then $\underline{r = \sqrt{(a^2 + b^2 + c^2)}}$

This gives us an easy way of finding the magnitude of a vector expressed in terms of the unit vectors.

Now you can do this one:

If $\overline{PQ} = 4i + 3j + 2k$, then $|\overline{PQ}| =$ ........................

---

**37**

$$|\overline{PQ}| = \sqrt{29} = 5{\cdot}385$$

For, if $\overline{PQ} = 4i + 3j + 2k$

$$|\overline{PQ}| = \sqrt{(4^2 + 3^2 + 2^2)}$$
$$= \sqrt{(16 + 9 + 4)} = \sqrt{29} = \underline{5{\cdot}385}$$

*Now move on to frame 38.*

**38**

## Direction cosines

The direction of a vector in three dimensions is determined by the angles which the vector makes with the three axes of reference.

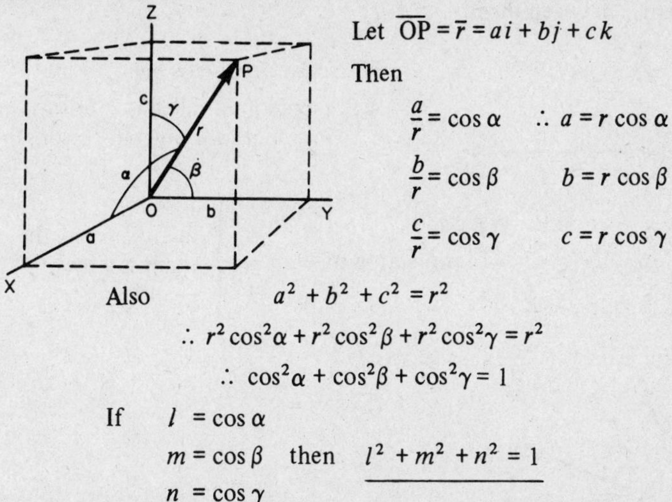


Let $\overline{\text{OP}} = \bar{r} = ai + bj + ck$

Then

$$\frac{a}{r} = \cos\alpha \qquad \therefore\ a = r\cos\alpha$$

$$\frac{b}{r} = \cos\beta \qquad b = r\cos\beta$$

$$\frac{c}{r} = \cos\gamma \qquad c = r\cos\gamma$$

Also $\quad a^2 + b^2 + c^2 = r^2$

$$\therefore\ r^2\cos^2\alpha + r^2\cos^2\beta + r^2\cos^2\gamma = r^2$$

$$\therefore\ \cos^2\alpha + \cos^2\beta + \cos^2\gamma = 1$$

If $\quad l = \cos\alpha$

$m = \cos\beta \quad$ then $\quad \underline{l^2 + m^2 + n^2 = 1}$

$n = \cos\gamma$

*Note:* $[l, m, n]$ written in square brackets are called the *direction cosines* of the vector $\overline{\text{OP}}$ and are the values of the cosines of the angles which the vector makes with the three axes of reference.

So for the vector $\bar{r} = ai + bj + ck$

$$l = \frac{a}{r};\quad m = \frac{b}{r};\quad n = \frac{c}{r} \text{ and, of course } r = \sqrt{(a^2 + b^2 + c^2)}$$

So, with that in mind, find the direction cosines $[l, m, n]$ of the vector

$$\bar{r} = 3i - 2j + 6k$$

*Then to frame 39.*

---

**39**

$$\bar{r} = 3i - 2j + 6k$$

$$\therefore\ a = 3, b = -2, c = 6 \qquad r = \sqrt{(9 + 4 + 36)}$$

$$\therefore\ r = \sqrt{49} = 7$$

$$\therefore\ \underline{l = \frac{3}{7};\quad m = -\frac{2}{7};\quad n = \frac{6}{7}}$$

Just as easy as that! *On to the next frame.*

**40**

**Scalar product of two vectors**

If $\bar{A}$ and $\bar{B}$ are two vectors, the *scalar product* of $\bar{A}$ and $\bar{B}$ is defined as $AB\cos\theta$, where A and B are the magnitudes of the vectors $\bar{A}$ and $\bar{B}$, and $\theta$ is the angle between them.

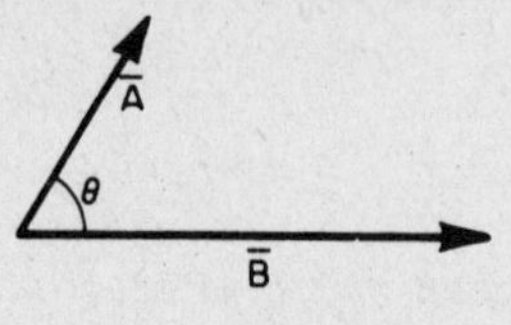


The scalar product is denoted by $\bar{A}.\bar{B}$ (sometimes called the 'dot product', for obvious reasons)

$$\therefore\ \bar{A}.\bar{B} = AB\cos\theta$$
$$= A \times \text{projection of B on A}$$
$$\text{or}\quad B \times \quad\text{"}\quad\text{"}\quad A\text{"}\ B$$

In either case, the result is a *scalar* quantity.

*For example*

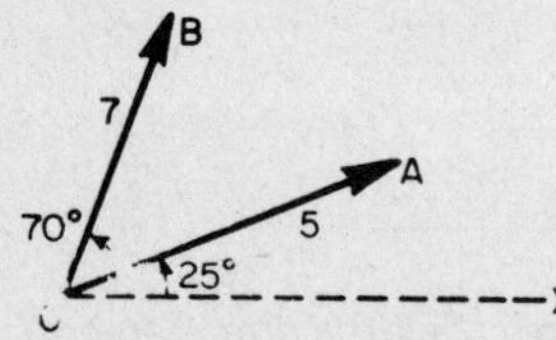


$\overline{OA}.\overline{OB}$ = ........................

**41**

$$\overline{OA}.\overline{OB} = \frac{35\sqrt{2}}{2}$$

For, we have:

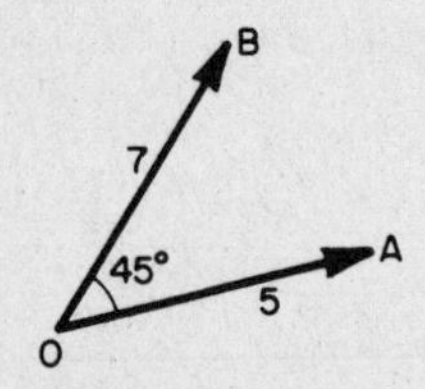


$$\overline{OA}.\overline{OB} = OA.OB.\cos\theta$$
$$= 5.7.\cos 45°$$
$$= 35.\frac{1}{\sqrt{2}} = \frac{35\sqrt{2}}{2}$$

Now what about this case:

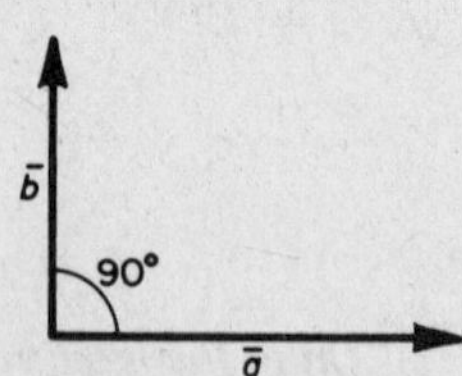


The scalar product of $\bar{a}$ and $\bar{b}$
$= \bar{a}.\bar{b} =$ ........................

**42**

$$\boxed{0}$$

since, in this case, $\bar{a}.\bar{b} = a.b.\cos 90° = a.b.0 = 0$

So, the scalar product of any two vectors at right-angles to each other is always zero.

And in this case now, with two vectors in the same direction, $\theta = 0°$,

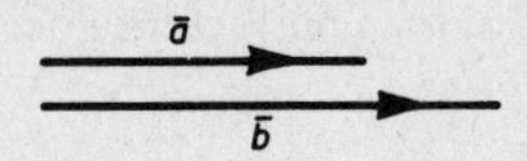


so $\bar{a}.\bar{b}$ = ........................

**43**

$$\boxed{a.b}$$

since $\bar{a}.\bar{b}. = a.b.\cos 0° = a.b.1 = a.b$

Now suppose our two vectors are expressed in terms of the unit vectors.

Let $\bar{A} = a_1 i + b_1 j + c_1 k$

and $\bar{B} = a_2 i + b_2 j + c_2 k$

Then
$$\begin{aligned}\bar{A}.\bar{B} &= (a_1 i + b_1 j + c_1 k).(a_2 i + b_2 j + c_2 k)\\ &= a_1 a_2 i.i + a_1 b_2 i.j + a_1 c_2 i.k + b_1 a_2 j.i + b_1 b_2 j.j\\ &\quad + b_1 c_2 j.k + c_1 a_2 k.i + c_1 b_2 k.j + c_1 c_2 k.k\end{aligned}$$

This will simplify very soon, so do not get worried.

For $i.i = 1.1.\cos 0° = 1$

$\therefore\ i.i = 1;\ \ j.j = 1;\ \ k.k = 1$ .............. (i)

Also $i.j = 1.1 \cos 90° = 0$

$i.j = 0;\ \ j.k = 0;\ \ k.i = 0$ .............. (ii)

So, using the results (i) and (ii), we can simplify the expression for $\bar{A}.\bar{B}$ above to give

$\bar{A}.\bar{B}$ = ........................

## 44

$$\overline{A}.\overline{B} = a_1a_2 + b_1b_2 + c_1c_2$$

since $\overline{A}.\overline{B} = a_1a_2 1 + a_1b_2 0 + a_1c_2 0 + b_1a_2 0 + b_1b_2 1 + b_1c_2 0 + c_1a_2 0 + c_1b_2 0 + c_1c_2 1$

$$\therefore \ \overline{A}.\overline{B} = a_1a_2 + b_1b_2 + c_1c_2$$

i.e. we just sum the products of coefficients of the unit vectors along corresponding axes.

e.g. If $\overline{A} = 2i + 3j + 5k$ and $\overline{B} = 4i + 1j + 6k$

then $\overline{A}.\overline{B} = 2.4 + 3.1 + 5.6$

$= 8 + 3 + 30 = 41 \quad \therefore \ \overline{A}.\overline{B} = 41$

One for you: If $\overline{P} = 3i - 2j + 1k$; $\overline{Q} = 2i + 3j - 4k$,

then $\overline{P}.\overline{Q} = \ldots\ldots\ldots\ldots\ldots\ldots\ldots$

## 45

$$-4$$

for $\overline{P}.\overline{Q} = 3.2 + (-2).3 + 1(-4)$

$= 6 - 6 - 4 \quad \therefore \ \text{P.Q} = -4$

Now we come to:

**Vector product of two vectors**

The *vector product* of $\overline{A}$ and $\overline{B}$ is written $\overline{A} \times \overline{B}$ (sometimes called the 'cross product') and is defined as a *vector* having the magnitude A B sin $\theta$, where $\theta$ is the angle between the two given vectors. The product vector acts in a direction perpendicular to $\overline{A}$ and $\overline{B}$ in such a sense that $\overline{A}$, $\overline{B}$, and $(\overline{A} \times \overline{B})$ form a right-handed set – in that order

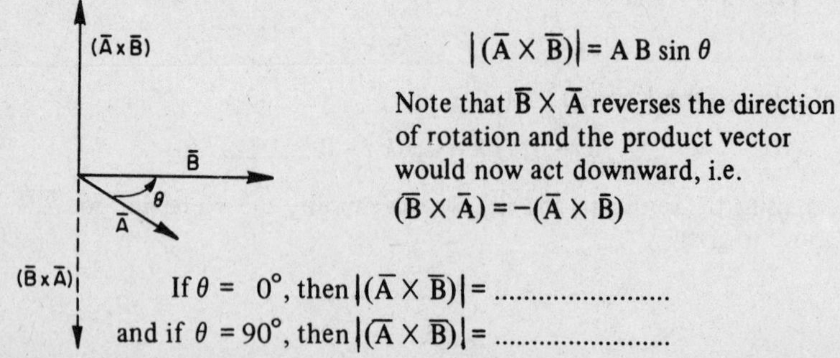


$$|(\overline{A} \times \overline{B})| = \text{A B} \sin\theta$$

Note that $\overline{B} \times \overline{A}$ reverses the direction of rotation and the product vector would now act downward, i.e.
$(\overline{B} \times \overline{A}) = -(\overline{A} \times \overline{B})$

If $\theta = 0°$, then $|(\overline{A} \times \overline{B})| = \ldots\ldots\ldots\ldots\ldots\ldots\ldots$

and if $\theta = 90°$, then $|(\overline{A} \times \overline{B})| = \ldots\ldots\ldots\ldots\ldots\ldots\ldots$

**46**

$$\theta = 0°, |(\overline{A} \times \overline{B})| = 0$$
$$\theta = 90°, |(\overline{A} \times \overline{B})| = A\,B$$

If $\overline{A}$ and $\overline{B}$ are given in terms of the unit vectors, then

$$\overline{A} = a_1 i + b_1 j + c_1 k \quad \text{and} \quad \overline{B} = a_2 i + b_2 j + c_2 k$$

Then $\quad \overline{A} \times \overline{B} = (a_1 i + b_1 j + c_1 k) \times (a_2 i + b_2 j + c_2 k)$

$$= a_1 a_2 i \times i + a_1 b_2 i \times j + a_1 c_2 i \times k + b_1 a_2 j \times i$$
$$+ b_1 b_2 j \times j + b_1 c_2 j \times k + c_1 a_2 k \times i + c_1 b_2 k \times j$$
$$+ c_1 c_2 k \times k$$

But $i \times i = 1.1.\sin 0° = 0$

$$\therefore i \times i = j \times j = k \times k = 0 \quad \ldots\ldots\ldots\ldots \text{(i)}$$

Also $i \times j = 1.1.\sin 90° = 1$ in direction OZ, i.e. $i \times j = k$

$$\left.\begin{aligned} \therefore i \times j &= k \\ j \times k &= i \\ k \times i &= j \end{aligned}\right\} \quad \ldots\ldots\ldots\ldots \text{(ii)}$$

And remember too that

$$\left.\begin{aligned} i \times j &= -(j \times i) \\ j \times k &= -(k \times j) \\ k \times i &= -(i \times k) \end{aligned}\right\} \quad \text{since the sense of rotation is reversed.}$$

Now with the results of (i) and (ii), and this last reminder, you can simplify the expression for $\overline{A} \times \overline{B}$.

Remove the zero terms and tidy up what is left.

*Then on to frame 47.*

**47**

$$\overline{A} \times \overline{B} = (b_1c_2 - b_2c_1)i + (a_2c_1 - a_1c_2)j + (a_1b_2 - a_2b_1)k$$

for $\overline{A} \times \overline{B} = a_1a_20 + a_1b_2k + a_1c_2(-j) + b_1a_2(-k) + b_1b_20$

$$+ b_1c_2i + c_1a_2j + c_1b_2(-i) + c_1c_20$$

$$= (b_1c_2 - b_2c_1)i + (a_2c_1 - a_1c_2)j + (a_1b_2 - a_2b_1)k$$

Now we could rearrange the middle term slightly and rewrite it thus:

$$\overline{A} \times \overline{B} = (b_1c_2 - b_2c_1)i - (a_1c_2 - a_2c_1)j + (a_1b_2 - a_2b_1)k$$

and you may recognize this pattern as the expansion of a determinant. So we now have that:

if $\overline{A} = a_1i + b_1j + c_1k$ and $\overline{B} = a_2i + b_2j + c_2k$

then
$$\overline{A} \times \overline{B} = \begin{vmatrix} i & j & k \\ a_1 & b_1 & c_1 \\ a_2 & b_2 & c_2 \end{vmatrix}$$

and that is the easiest way to write out the vector product of two vectors.

*Note:* (i) the top row consists of the unit vectors in order, $i, j, k$
(ii) the second row consists of the coefficients of $\overline{A}$
(iii) the third row consists of the coefficients of $\overline{B}$.

*Example.* If $\overline{P} = 2i + 4j + 3k$ and $\overline{Q} = 1i + 5j - 2k$ first write down the determinant that represents the vector product $\overline{P} \times \overline{Q}$.

**48**

$$\overline{P} \times \overline{Q} = \begin{vmatrix} i & j & k \\ 2 & 4 & 3 \\ 1 & 5 & -2 \end{vmatrix}$$

And now, expanding the determinant, we get

$\overline{P} \times \overline{Q} =$ ..........................

**49**

$$\overline{P} \times \overline{Q} = -23i + 7j + 6k$$

$$\overline{P} \times \overline{Q} = \begin{vmatrix} i & j & k \\ 2 & 4 & 3 \\ 1 & 5 & -2 \end{vmatrix}$$

$$= i\begin{vmatrix} 4 & 3 \\ 5 & -2 \end{vmatrix} - j\begin{vmatrix} 2 & 3 \\ 1 & -2 \end{vmatrix} + k\begin{vmatrix} 2 & 4 \\ 1 & 5 \end{vmatrix}$$

$$= i(-8 - 15) - j(-4 - 3) + k(10 - 4)$$

$$= -23i + 7j + \underline{6k}$$

So, by way of revision,

(i) *Scalar product* ('dot product')

$\overline{A}.\overline{B} = A\,B \cos\theta$ a scalar quantity.

(ii) *Vector Product* ('cross product')

$\overline{A} \times \overline{B}$ = vector of magnitude $A\,B \sin\theta$, acting in a direction to make $\overline{A}$, $\overline{B}$, $(\overline{A} \times \overline{B})$ a right-handed set.

Also
$$\overline{A} \times \overline{B} = \begin{vmatrix} i & j & k \\ a_1 & b_1 & c_1 \\ a_2 & b_2 & c_2 \end{vmatrix}$$

And here is one final example on this point.

*Example.* Find the vector product of $\overline{P}$ and $\overline{Q}$, where

$$\overline{P} = 3i - 4j + 2k \text{ and } \overline{Q} = 2i + 5j - 1k.$$

**50**

$$\overline{P} \times \overline{Q} = -6i + 7j + 23k$$

for
$$\overline{P} \times \overline{Q} = \begin{vmatrix} i & j & k \\ 3 & -4 & 2 \\ 2 & 5 & -1 \end{vmatrix}$$

$$= i\begin{vmatrix} -4 & 2 \\ 5 & -1 \end{vmatrix} - j\begin{vmatrix} 3 & 2 \\ 2 & -1 \end{vmatrix} + k\begin{vmatrix} 3 & -4 \\ 2 & 5 \end{vmatrix}$$

$$= i(4 - 10) - j(-3 - 4) + k(15 + 8)$$

$$= -6i + 7j + 23k$$

*On to frame 51.*

**51**

**Angle between two vectors**

Let $\overline{A}$ be one vector with direction cosines $[l, m, n]$
" $\overline{B}$ be the other vector with direction cosines $[l', m', n']$

We have to find the angle between these two vectors.

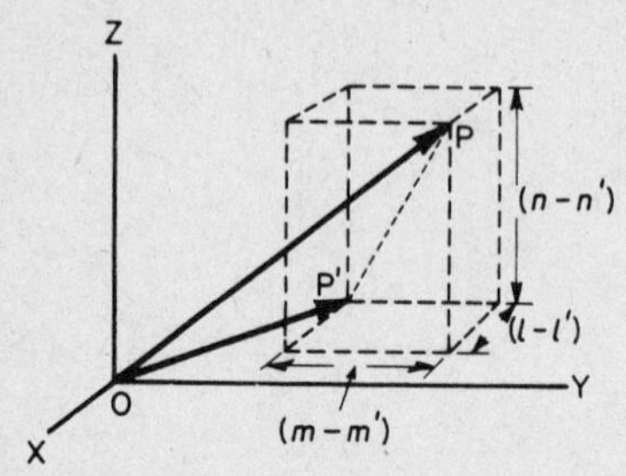


Let $\overline{OP}$ and $\overline{OP}'$ be *unit* vectors parallel to $\overline{A}$ and $\overline{B}$ respectively. Then P has co-ordinates $(l, m, n)$ and P' " " $(l', m', n')$

Then $(PP')^2 = (l - l')^2 + (m - m')^2 + (n - n')^2$
$= l^2 - 2.l.l' + l'^2 + m^2 - 2.m.m' + m'^2 + n^2 - 2nn' + n'^2$
$= (l^2 + m^2 + n^2) + (l'^2 + m'^2 + n'^2) - 2(ll' + mm' + nn')$

But $(l^2 + m^2 + n^2) = 1$ and $(l'^2 + m'^2 + n'^2) = 1$ as was proved earlier.

$$\therefore \ (PP')^2 = 2 - 2(ll' + mm' + nn') \ldots\ldots\ldots\ldots \text{(i)}$$

Also, by the cosine rule,

$$(PP')^2 = OP^2 + OP'^2 - 2.OP.OP'.\cos\theta$$
$$= 1 + 1 - 2.1.1.\cos\theta \quad \left\{ \begin{array}{l} \text{OP and OP' are} \\ \text{unit vectors} \end{array} \right\}$$
$$= 2 - 2\cos\theta \ldots\ldots\ldots \text{(ii)}$$

So, from (i) and (ii), we have:

$$(PP')^2 = 2 - 2(ll' + mm' + nn')$$
and $$(PP')^2 = 2 - 2\cos\theta$$
$$\therefore \ \cos\theta = \ldots\ldots\ldots\ldots$$

**52**

$$\boxed{\cos\theta = ll' + mm' + nn'}$$

i.e. just sum the products of the corresponding direction cosines of the two given vectors

So, if $[l, m, n] = [0.54, 0.83, -0.14]$
and $[l', m', n'] = [0.25, 0.60, 0.76]$
the angle between the vectors is $\theta = \ldots\ldots\ldots\ldots$

**53**

$$\boxed{\theta = 58°13'}$$

for, we have

$$\begin{aligned}\cos\theta &= ll' + mm' + nn' \\ &= (0.54)\ (0.25) + (0.83)\ (0.60) + (-0.14)(0.76) \\ &= 0.1350 + 0.4980 - 0.1064 \\ &= 0.6330 - 0.1064 = 0.5266\end{aligned}$$

$$58°13'$$

*NOTE:* For *parallel* vectors, $\theta = 0°$ $\therefore\ ll' + mm' + nn' = 1$
For *perpendicular* vectors, $\theta = 90°$, $\therefore\ ll' + mm' + nn' = 0$

*Now an example* for you to work:
Find the angle between the vectors

$$\overline{\text{P}} = 2i + 3j + 4k \text{ and } \overline{\text{Q}} = 4i - 3j + 2k$$

First of all, find the direction cosines of $\overline{\text{P}}$. You do that.

**54**

$$\boxed{l = \frac{2}{\sqrt{29}},\quad m = \frac{3}{\sqrt{29}},\quad n = \frac{4}{\sqrt{29}}}$$

for $\quad r = |\overline{\text{P}}| = \sqrt{(2^2 + 3^2 + 4^2)} = \sqrt{(4 + 9 + 16)} = \sqrt{29}$

$$\therefore\ l = \frac{a}{r} = \frac{2}{\sqrt{29}}$$

$$m = \frac{b}{r} = \frac{3}{\sqrt{29}}$$

$$n = \frac{c}{r} = \frac{4}{\sqrt{29}}$$

$$\therefore\ [l, m, n] = \left[\frac{2}{\sqrt{29}}, \frac{3}{\sqrt{29}}, \frac{4}{\sqrt{29}}\right]$$

Now find the direction cosines $[l', m', n']$ of $\overline{\text{Q}}$ in just the same way.

*When you have done that, turn on to the next frame.*

**55**

$$l' = \frac{4}{\sqrt{29}}, \quad m' = \frac{-3}{\sqrt{29}}, \quad n' = \frac{2}{\sqrt{29}}$$

since $\quad r' = |\overline{Q}| = \sqrt{(4^2 + 3^2 + 2^2)} = \sqrt{(16 + 9 + 4)} = \sqrt{29}$

$$\therefore \; [l', m', n'] = \left[\frac{4}{\sqrt{29}}, \frac{-3}{\sqrt{29}}, \frac{2}{\sqrt{29}}\right]$$

We already know that, for $\overline{P}$,

$$[l, m, n] = \left[\frac{2}{\sqrt{29}}, \frac{3}{\sqrt{29}}, \frac{4}{\sqrt{29}}\right]$$

So, using $\cos\theta = ll' + mm' + nn'$, you can finish it off and find the angle $\theta$. Off you go.

**56**

$$\theta = 76°2'$$

for $\quad \cos\theta = \frac{2}{\sqrt{29}} \cdot \frac{4}{\sqrt{29}} + \frac{3}{\sqrt{29}} \cdot \frac{(-3)}{\sqrt{29}} + \frac{4}{\sqrt{29}} \cdot \frac{2}{\sqrt{29}}$

$$= \frac{8}{29} - \frac{9}{29} + \frac{8}{29}$$

$$= \frac{7}{29} = 0{\cdot}2414 \qquad \therefore \; \underline{\theta = 76°2'}$$

*Now on to frame 57.*

**57**

**Direction ratios**

If $\overline{OP} = ai + bj + ck$, we know that

$$|\overline{OP}| = r = \sqrt{a^2 + b^2 + c^2}$$

and that the direction cosines of $\overline{OP}$ are given by

$$l = \frac{a}{r}, \quad m = \frac{b}{r}, \quad n = \frac{c}{r}$$

We can see that the components, $a, b, c$, are proportional to the direction cosines, $l, m, n$, respectively and they are sometimes referred to as the *direction ratios* of the vector $\overline{OP}$.

*Note* that the direction ratios can be converted into the direction cosines by dividing each of them by $r$ (the magnitude of the vector).

*Now turn on to frame 58.*

**58**

Here is a short summary of the work we have covered. Read through it.

**Summary**

1. A *scalar* quantity has magnitude only ; a *vector* quantity has both magnitude and direction.

2. The axes of reference, OX, OY, OZ, are chosen so that they form a right-handed set. The symbols $i, j, k$ denote *unit vectors* in the directions OX, OY, OZ, respectively.

   If $\overline{\mathrm{OP}} = ai + bj + ck$, then $\left|\overline{\mathrm{OP}}\right| = r = \sqrt{(a^2 + b^2 + c^2)}$

3. The *direction cosines* $[l, m, n]$ are the cosines of the angles between the vector and the axes OX, OY, OZ respectively.

   For any vector $l = \dfrac{a}{r}, \quad m = \dfrac{b}{r}, \quad n = \dfrac{c}{r}$

   and $l^2 + m^2 + n^2 = 1$

4. ***Scalar product*** ('dot product')

   $\overline{\mathrm{A}}.\overline{\mathrm{B}} = \mathrm{A\,B} \cos\theta$ where $\theta$ is angle between $\overline{\mathrm{A}}$ and $\overline{\mathrm{B}}$.

   If $\overline{\mathrm{A}} = a_1 i + b_1 j + c_1 k$ and $\overline{\mathrm{B}} = a_2 i + b_2 j + c_2 k$

   then $\overline{\mathrm{A}}.\overline{\mathrm{B}} = a_1 a_2 + b_1 b_2 + c_1 c_2$

5. ***Vector product*** ('cross product')

   $\overline{\mathrm{A}} \times \overline{\mathrm{B}} = (\mathrm{A\,B} \sin\theta)$ in direction perpendicular to $\overline{\mathrm{A}}$ and $\overline{\mathrm{B}}$, so that $\overline{\mathrm{A}}, \overline{\mathrm{B}}, (\overline{\mathrm{A}} \times \overline{\mathrm{B}})$ form a right-handed set.

   Also $$\overline{\mathrm{A}} \times \overline{\mathrm{B}} = \begin{vmatrix} i & j & k \\ a_1 & b_1 & c_1 \\ a_2 & b_2 & c_2 \end{vmatrix}$$

6. ***Angle between two vectors***

   $$\cos\theta = ll' + mm' + nn'$$

   For perpendicular vectors, $ll' + mm' + nn' = 0$.

All that now remains is the Test Exercise. Check through any points that may need brushing up and then turn on to the next frame.

**59**

Now you are ready for the Test Exercise below. Work through all the questions. Take your time over the exercise: the problems are all straightforward so avoid careless slips. Diagrams often help where appropriate. So off you go.

**Test Exercise – VI**

1. If $\overline{OA} = 4i + 3j$, $\overline{OB} = 6i - 2j$, $\overline{OC} = 2i - j$, find $\overline{AB}$, $\overline{BC}$ and $\overline{CA}$, and deduce the lengths of the sides of the triangle ABC.

2. If $\overline{A} = 2i + 2j - k$ and $\overline{B} = 3i - 6j + 2k$, find (i) $\overline{A}.\overline{B}$ and (ii) $\overline{A} \times \overline{B}$.

3. Find the direction cosines of the vector joining the two points $(4, 2, 2)$ and $(7, 6, 14)$.

4. If $\overline{A} = 5i + 4j + 2k$, $\overline{B} = 4i - 5j + 3k$, and $\overline{C} = 2i - j - 2k$, where $i, j, k$, are the unit vectors, determine

   (i) the value of $\overline{A}.\overline{B}$ and the angle between the vectors $\overline{A}$ and $\overline{B}$.

   (ii) the magnitude and the direction cosines of the product vector $(\overline{A} \times \overline{B})$ and also the angle which this product vector makes with the vector $\overline{C}$.

**Further Problems – VI**

1. The centroid of the triangle OAB is denoted by G. If O is the origin and $\overline{OA} = 4i + 3j$, $\overline{OB} = 6i - j$, find $\overline{OG}$ in terms of the unit vectors, $i$ and $j$.

2. Find the direction cosines of the vectors whose direction ratios are (3, 4, 5) and (1, 2, −3). Hence find the acute angle between the two vectors.

3. Find the modulus and the direction cosines of the vectors $3i + 7j - 4k$, $i - 5j - 8k$, and $6i - 2j + 12k$. Find also the modulus and the direction cosines of their sum.

4. If $\overline{A} = 2i + 4j - 3k$, and $\overline{B} = i + 3j + 2k$, determine the scalar and vector products, and the angle between the two given vectors.

5. If $\overline{OA} = 2i + 3j - k$, $\overline{OB} = i - 2j + 3k$, determine
   (i) the value of $\overline{OA}.\overline{OB}$
   (ii) the product $\overline{OA} \times \overline{OB}$ in terms of the unit vectors
   (iii) the cosine of the angle between $\overline{OA}$ and $\overline{OB}$

6. Find the cosine of the angle between the vectors $2i + 3j - k$ and $3i - 5j + 2k$.

7. Find the scalar product $(\overline{A}.\overline{B})$ and the vector product $(\overline{A} \times \overline{B})$, when
   (i) $\overline{A} = i + 2j - k$, $\overline{B} = 2i + 3j + k$
   (ii) $\overline{A} = 2i + 3j + 4k$, $\overline{B} = 5i - 2j + k$

8. Find the unit vector perpendicular to each of the vectors $2i - j + k$ and $3i + 4j - k$, where $i, j, k$ are the mutually perpendicular unit vectors. Calculate the sine of the angle between the two vectors.

9. If A is the point (1, −1, 2) and B is (−1, 2, 2) and C is the point (4, 3, 0), find the direction cosines of $\overline{BA}$ and $\overline{BC}$, and hence show that the angle ABC = 69°14′.

10. If $\overline{A} = 3i - j + 2k$, $\overline{B} \doteq i + 3j - 2k$, determine the magnitude and direction cosines of the product vector $(\overline{A} \times \overline{B})$ and show that it is perpendicular to a vector $\overline{C} = 9i + 2j + 2k$.

11. $\overline{A}, \overline{B}, \overline{C}$ are vectors defined by $\overline{A} = 8i + 2j - 3k$, $\overline{B} = 3i - 6j + 4k$, and $\overline{C} = 2i - 2j - k$, where $i, j, k$ are mutually perpendicular unit vectors.
    (i) Calculate $\overline{A}.\overline{B}$ and show that $\overline{A}$ and $\overline{B}$ are perpendicular to each other
    (ii) Find the magnitude and the direction cosines of the product vector $(\overline{A} \times \overline{B})$

12. If the position vectors of P and Q are $i + 3j - 7k$ and $5i - 2j + 4k$ respectively, find $\overline{PQ}$ and determine its direction cosines.

13. If position vectors, $\overline{OA}$, $\overline{OB}$, $\overline{OC}$, are defined by $\overline{OA} = 2i - j + 3k$, $\overline{OB} = 3i + 2j - 4k$, $\overline{OC} = -i + 3j - 2k$, determine
    (i) the vector $\overline{AB}$
    (ii) the vector $\overline{BC}$
    (iii) the vector product $\overline{AB} \times \overline{BC}$
    (iv) the unit vector perpendicular to the plane ABC

# Programme 7

## DIFFERENTIATION

# 1

**Standard Differential Coefficients**

Here is a revision list of the standard differential coefficients which you have no doubt used many times before. Copy out the list into your notebook and memorize those with which you are less familiar – possibly Nos. 4, 6, 10, 11, 12. Here they are:

| | $y = f(x)$ | $\dfrac{dy}{dx}$ |
|---|---|---|
| 1. | $x^n$ | $nx^{n-1}$ |
| 2. | $e^x$ | $e^x$ |
| 3. | $e^{kx}$ | $ke^{kx}$ |
| 4. | $a^x$ | $a^x . \ln a$ |
| 5. | $\ln x$ | $\dfrac{1}{x}$ |
| 6. | $\log_a x$ | $\dfrac{1}{x . \ln a}$ |
| 7. | $\sin x$ | $\cos x$ |
| 8. | $\cos x$ | $-\sin x$ |
| 9. | $\tan x$ | $\sec^2 x$ |
| 10. | $\cot x$ | $-\operatorname{cosec}^2 x$ |
| 11. | $\sec x$ | $\sec x . \tan x$ |
| 12. | $\operatorname{cosec} x$ | $-\operatorname{cosec} x . \cot x$ |
| 13. | $\sinh x$ | $\cosh x$ |
| 14. | $\cosh x$ | $\sinh x$ |

*The last two are proved on frame 2, so turn on.*

**2**

The differential coefficients of sinh $x$ and cosh $x$ are easily obtained by remembering the exponential definitions, and also that

$$\frac{d}{dx}\{e^x\} = e^x \quad \text{and} \quad \frac{d}{dx}\{e^{-x}\} = -e^{-x}$$

(i) $y = \sinh x \qquad y = \dfrac{e^x - e^{-x}}{2}$

$$\therefore \frac{dy}{dx} = \frac{e^x - (-e^{-x})}{2} = \frac{e^x + e^{-x}}{2} = \cosh x$$

$$\therefore \frac{d}{dx}(\sinh x) = \cosh x$$

(ii) $y = \cosh x \qquad y = \dfrac{e^x + e^{-x}}{2}$

$$\therefore \frac{dy}{dx} = \frac{e^x + (-e^{-x})}{2} = \frac{e^x - e^{-x}}{2} = \sinh x$$

$$\therefore \frac{d}{dx}(\cosh x) = \sinh x$$

Note that there is no minus sign involved as there is when differentiating the trig. function cos $x$.

We will find the differential coefficient of tanh $x$ later on.

*Move on to frame 3.*

**3**

Let us see if you really do know those basic differential coefficients. First of all cover up the list you have copied and then write down the differential coefficients of the following. All very easy.

1. $x^5$
2. $\sin x$
3. $e^{3x}$
4. $\ln x$
5. $\tan x$
6. $2^x$
7. $\sec x$
8. $\cosh x$
9. $\log_{10} x$
10. $e^x$
11. $\cos x$
12. $\sinh x$
13. $\operatorname{cosec} x$
14. $a^3$
15. $\cot x$
16. $a^x$
17. $x^{-4}$
18. $\log_a x$
19. $\sqrt{x}$
20. $e^{x/2}$

*When you have finished them all, turn on to the next frame to check your results.*

**4**

Here are the results. Check yours carefully and make a special note of any where you may have slipped up.

1. $5x^4$
2. $\cos x$
3. $3e^{3x}$
4. $1/x$
5. $\sec^2 x$
6. $2^x \ln 2$
7. $\sec x.\tan x$
8. $\sinh x$
9. $1/(x \ln 10)$
10. $e^x$
11. $-\sin x$
12. $\cosh x$
13. $-\operatorname{cosec} x.\cot x$
14. $0$
15. $-\operatorname{cosec}^2 x$
16. $a^x \ln a$
17. $-4x^{-5}$
18. $1/(x \ln a)$
19. $\frac{1}{2}x^{-\frac{1}{2}} = 1/(2\sqrt{x})$
20. $\frac{1}{2}e^{x/2}$

If by chance you have not got them all correct, it is well worth while returning to frame 1, or to the list you copied, and brushing up where necessary. These are the tools for all that follows.

*When you are sure you know the basic results, move on.*

**5**

**Functions of a function**

Sin $x$ is a function of $x$ since the value of sin $x$ depends on the value of the angle $x$. Similarly, $\sin(2x + 5)$ is a function of the angle $(2x + 5)$ since the value of the sine depends on the value of this angle.

i.e. $\sin(2x + 5)$ is a function of $(2x + 5)$

But $(2x + 5)$ is itself a function of $x$, since its value depends on $x$.

i.e. $(2x + 5)$ is a function of $x$

If we combine these two statements, we have

$\sin(2x + 5)$ is a function of $(2x + 5)$
" " " " " a function of $x$

Sin$(2x + 5)$ is therefore a function of a function of $x$ and such expressions are referred to generally as *functions of a function.*

So $e^{\sin y}$ is a function of a function of .......................

**6**

| $y$ | since $e^{\sin y}$ depends on the value of the index $\sin y$ and $\sin y$ depends on $y$. Therefore $e^{\sin y}$ is a function of a function of $y$. |
|---|---|

□□□□□□□□□□□□□□□□□□□□□□□□□□□□□□□□□□□□□□□

We very often need to find the differential coefficients of such functions of a function. We could do them from first principles:

*Example 1.* Differentiate with respect to $x$, $y = \cos(5x - 4)$.

Let $u = (5x - 4)$ $\therefore$ $y = \cos u$ $\therefore$ $\dfrac{dy}{du} = -\sin u = -\sin(5x - 4)$. But this gives us $\dfrac{dy}{du}$, not $\dfrac{dy}{dx}$. To convert our result into the required coefficient we use $\dfrac{dy}{dx} = \dfrac{dy}{du} \cdot \dfrac{du}{dx}$, i.e. we multiply $\dfrac{dy}{du}$ (which we have) by $\dfrac{du}{dx}$ to obtain $\dfrac{dy}{dx}$ (which we want); $\dfrac{du}{dx}$ is found from the substitution $u = (5x-4)$,

i.e. $\dfrac{du}{dx} = 5$.

$$\therefore \quad \frac{d}{dx}\{\cos(5x - 4)\} = -\sin(5x - 4) \times 5 = \underline{-5\sin(5x - 4)}$$

So you now find from first principles the differential coefficient of $y = e^{\sin x}$. (As before, put $u = \sin x$.)

**7**

$$\boxed{\frac{d}{dx}\{e^{\sin x}\} = \cos x \, . \, e^{\sin x}}$$

For: $y = e^{\sin x}$. Put $u = \sin x$ $\therefore$ $y = e^u$ $\therefore$ $\dfrac{dy}{du} = e^u$

$$\text{But } \frac{dy}{dx} = \frac{dy}{du} \cdot \frac{du}{dx} \text{ and } \frac{du}{dx} = \cos x$$

$$\therefore \quad \underline{\frac{d}{dx}\{e^{\sin x}\} = e^{\sin x} \, . \cos x}$$

This is quite general.

If $y = f(u)$ and $u = F(x)$, then $\dfrac{dy}{dx} = \dfrac{dy}{du} \cdot \dfrac{du}{dx}$, i.e. if $y = \ln F$, where $F$ is a function of $x$, then

$$\frac{dy}{dx} = \frac{dy}{dF} \cdot \frac{dF}{dx} = \frac{1}{F} \cdot \frac{dF}{dx}$$

So, if $y = \ln \sin x$ $\qquad \dfrac{dy}{dx} = \dfrac{1}{\sin x} \cdot \cos x = \cot x$

It is of utmost important not to forget this factor $\dfrac{dF}{dx}$, so beware!

**8**

Just two more examples:

(i) $y = \tan(5x - 4)$ Basic standard form is $y = \tan x$, $\frac{dy}{dx} = \sec^2 x$

In this case $(5x - 4)$ replaces the single $x$

$$\therefore \frac{dy}{dx} = \sec^2(5x - 4) \times \text{the diff. of the function } (5x - 4)$$
$$= \sec^2(5x - 4) \times 5 = \underline{5 \sec^2(5x - 4)}$$

(ii) $y = (4x - 3)^5$ Basic standard form is $y = x^5$, $\frac{dy}{dx} = 5x^4$

Here, $(4x - 3)$ replaces the single $x$

$$\therefore \frac{dy}{dx} = 5(4x - 3)^4 \times \text{the diff. of the function } (4x - 3)$$
$$= 5(4x - 3)^4 \times 4 = \underline{20(4x - 3)^4}$$

So, what about this one?

If $y = \cos(7x + 2)$, then $\frac{dy}{dx} = \text{.........................}$

**9**

$$y = \cos(7x + 2) \quad \boxed{\frac{dy}{dx} = -7 \sin(7x + 2)}$$

Right, now you differentiate these:

1. $y = (4x - 5)^6$
2. $y = e^{3-x}$
3. $y = \sin 2x$
4. $y = \cos(x^2)$
5. $y = \ln(3 - 4 \cos x)$

*The results are on frame 10. Check to see that yours are correct.*

**10**

*Results:*

1. $y = (4x - 5)^6 \qquad \dfrac{dy}{dx} = 6(4x - 5)^5.4 = 24(4x - 5)^5$
2. $y = e^{3-x} \qquad \dfrac{dy}{dx} = e^{3-x}(-1) = -e^{3-x}$
3. $y = \sin 2x \qquad \dfrac{dy}{dx} = \cos 2x.2 = 2\cos 2x$
4. $y = \cos(x^2) \qquad \dfrac{dy}{dx} = -\sin(x^2).2x = -2x\sin(x^2)$
5. $y = \ln(3 - 4\cos x) \qquad \dfrac{dy}{dx} = \dfrac{1}{3 - 4\cos x}.(4\sin x) = \dfrac{4\sin x}{3 - 4\cos x}$

□□□□□□□□□□□□□□□□□□□□□□□□□□□□□□□□□□□□□□

Now do these:

6. $y = e^{\sin 2x}$
7. $y = \sin^2 x$
8. $y = \ln \cos 3x$
9. $y = \cos^3(3x)$
10. $y = \log_{10}(2x - 1)$

Take your time to do them.

*When you are satisfied with your results, check them against the results in frame 11.*

**11**

*Results:*

6. $y = e^{\sin 2x} \qquad \dfrac{dy}{dx} = e^{\sin 2x}.2\cos 2x = 2\cos 2x.e^{\sin 2x}$
7. $y = \sin^2 x \qquad \dfrac{dy}{dx} = 2\sin x \cos x = \sin 2x$
8. $y = \ln \cos 3x \qquad \dfrac{dy}{dx} = \dfrac{1}{\cos 3x}(-3\sin 3x) = -3\tan 3x$
9. $y = \cos^3(3x) \qquad \dfrac{dy}{dx} = 3\cos^2(3x).(-3\sin 3x) = -9\sin 3x\cos^2 3x$
10. $y = \log_{10}(2x - 1) \qquad \dfrac{dy}{dx} = \dfrac{1}{(2x - 1)\ln 10}.2 = \dfrac{2}{(2x - 1)\ln 10}$

*All correct? Now on with the programme. Next frame please.*

**12**

Of course, we may need to differentiate functions which are products or quotients of two of the functions.

1. *Products*

If $y = uv$, where $u$ and $v$ are functions of $x$, then you already know that

$$\frac{dy}{dx} = u\frac{dv}{dx} + v\frac{du}{dx}$$

e.g. If $y = x^3 . \sin 3x$

then $$\frac{dy}{dx} = x^3 . 3 \cos 3x + 3x^2 \sin 3x$$
$$= 3x^2(x \cos 3x + \sin 3x)$$

Every one is done the same way. To differentiate a product

(i) put down the first, differentiate the second; plus
(ii) put down the second, differentiate the first.

So what is the differential coefficient of $e^{2x} \ln 5x$?

**13**

$$\boxed{\frac{dy}{dx} = e^{2x}(\frac{1}{x} + 2 \ln 5x)}$$

for $y = e^{2x} \ln 5x$, i.e. $u = e^{2x}$, $v = \ln 5x$

$$\frac{dy}{dx} = e^{2x}\frac{1}{5x} \cdot 5 + 2e^{2x} \ln 5x$$
$$= e^{2x}(\frac{1}{x} + 2 \ln 5x)$$

Now here is a short set for you to do. Find $\frac{dy}{dx}$ when

1. $y = x^2 \tan x$
2. $y = e^{5x}(3x + 1)$
3. $y = x \cos 2x$
4. $y = x^3 \sin 5x$
5. $y = x^2 \ln \sinh x$

*When you have completed all five move on to frame 14.*

**14**

*Results:*

1. $y = x^2 \tan x \quad \therefore \dfrac{dy}{dx} = x^2 \sec^2 x + 2x \tan x$
$$= x(x \sec^2 x + 2 \tan x)$$

2. $y = e^{5x}(3x + 1) \quad \therefore \dfrac{dy}{dx} = e^{5x}.3 + 5e^{5x}(3x + 1)$
$$= e^{5x}(3 + 15x + 5) = e^{5x}(8 + 15x)$$

3. $y = x \cos 2x \quad \therefore \dfrac{dy}{dx} = x(-2 \sin 2x) + 1.\cos 2x$
$$= \cos 2x - 2x \sin 2x$$

4. $y = x^3 \sin 5x \quad \therefore \dfrac{dy}{dx} = x^3 5 \cos 5x + 3x^2 \sin 5x$
$$= x^2(5x \cos 5x + 3 \sin 5x)$$

5. $y = x^2 \ln \sinh x \quad \therefore \dfrac{dy}{dx} = x^2 \dfrac{1}{\sinh x} \cosh x + 2x \ln \sinh x$
$$= x(x \coth x + 2 \ln \sinh x)$$

So much for the product. What about the quotient?

*Next frame.*

**15**

2. *Quotients*

In the case of the quotient, if $u$ and $v$ are functions of $x$, and $y = \dfrac{u}{v}$

then
$$\frac{dy}{dx} = \frac{v\dfrac{du}{dx} - u\dfrac{dv}{dx}}{v^2}$$

*Example 1.* If $y = \dfrac{\sin 3x}{x + 1}$, $\dfrac{dy}{dx} = \dfrac{(x + 1)\, 3 \cos 3x - \sin 3x.1}{(x + 1)^2}$

*Example 2.* If $y = \dfrac{\ln x}{e^{2x}}$, $\dfrac{dy}{dx} = \dfrac{e^{2x}\dfrac{1}{x} - \ln x.\, 2e^{2x}}{e^{4x}}$
$$= \frac{e^{2x}\left(\dfrac{1}{x} - 2 \ln x\right)}{e^{4x}}$$
$$= \frac{\dfrac{1}{x} - 2 \ln x}{e^{2x}}$$

If you can differentiate the separate functions, the rest is easy.

You do this one. If $y = \dfrac{\cos 2x}{x^2}$, $\dfrac{dy}{dx} =$ ........................

**16**

$$\frac{d}{dx}\left\{\frac{\cos 2x}{x^2}\right\} = \frac{-2(x \sin 2x + \cos 2x)}{x^3}$$

for

$$\frac{d}{dx}\left\{\frac{\cos 2x}{x^2}\right\} = \frac{x^2(-2 \sin 2x) - \cos 2x.2x}{x^4}$$

$$= \frac{-2x(x \sin 2x + \cos 2x)}{x^4}$$

$$= \frac{-2(x \sin 2x + \cos 2x)}{x^3}$$

So: For $y = u\,v$,
$$\frac{dy}{dx} = u\frac{dv}{dx} + v\frac{du}{dx} \quad \text{................ (i)}$$

for $y = \frac{u}{v}$
$$\frac{dy}{dx} = \frac{v\frac{du}{dx} - u\frac{dv}{dx}}{v^2} \quad \text{................ (ii)}$$

Be sure that you know these.

You can prove the differential coefficient of $\tan x$ by the quotient method, for if $y = \tan x$, $y = \frac{\sin x}{\cos x}$

Then by the quotient rule, $\frac{dy}{dx} =$ ...................... (Work it through in detail)

**17**

$$y = \tan x \quad \boxed{\frac{dy}{dx} = \sec^2 x}$$

for
$$y = \frac{\sin x}{\cos x} \quad \therefore \frac{dy}{dx} = \frac{\cos x.\cos x + \sin x.\sin x}{\cos^2 x}$$

$$= \frac{1}{\cos^2 x} = \sec^2 x$$

In the same way we can obtain the diff. coefft. of $\tanh x$

$$y = \tanh x = \frac{\sinh x}{\cosh x} \quad \therefore \frac{dy}{dx} = \frac{\cosh x.\cosh x - \sinh x.\sinh x}{\cosh^2 x}$$

$$= \frac{\cosh^2 x - \sinh^2 x}{\cosh^2 x}$$

$$= \frac{1}{\cosh^2 x} = \operatorname{sech}^2 x$$

$$\therefore \frac{d}{dx}(\tanh x) = \operatorname{sech}^2 x$$

Add this last result to your list of differential coefficients in your note-book.

So what is the diff. coefft. of $\tanh(5x + 2)$?

**18**

$$\frac{d}{dx}\left\{\tanh(5x+2)\right\} = \boxed{5\,\text{sech}^2(5x+2)}$$

for we have: If $\frac{d}{dx}\left\{\tanh x\right\} = \text{sech}^2 x$

then $\frac{d}{dx}\left\{\tanh(5x+2)\right\} = \text{sech}^2(5x+2) \times \text{diff. of } (5x+2)$

$$= \text{sech}^2(5x+2) \times 5$$

$$= \underline{5\,\text{sech}^2(5x+2)}$$

*Fine. Now move on to frame 19 for the next part of the programme.*

**19**

**Logarithmic differentiation**

The rules for differentiating a product or a quotient that we have revised are used when there are just two-factor functions, i.e. $uv$ or $\frac{u}{v}$. When there are more than two functions in any arrangement top or bottom, the diff. coefft. is best found by what is known as 'logarithmic differentiation'.

It all depends on the basic fact that $\frac{d}{dx}\left\{\ln x\right\} = \frac{1}{x}$ and that if $x$ is replaced by a function $F$ then $\frac{d}{dx}\left\{\ln F\right\} = \frac{1}{F}\cdot\frac{dF}{dx}$. Bearing that in mind, let us consider the case where $y = \frac{uv}{w}$, where $u$, $v$ and $w$ – and also $y$ – are functions of $x$.

First take logs to the base e.

$$\ln y = \ln u + \ln v - \ln w$$

Now differentiate each side with respect to $x$, remembering that $u$, $v$, $w$ and $y$ are all functions of $x$. What do we get?

**20**

$$\frac{1}{y}\cdot\frac{dy}{dx}=\frac{1}{u}\cdot\frac{du}{dx}+\frac{1}{v}\cdot\frac{dv}{dx}-\frac{1}{w}\cdot\frac{dw}{dx}$$

So to get $\frac{dy}{dx}$ by itself, we merely have to multiply across by $y$. Note that when we do this, we put the grand function that $y$ represents.

$$\frac{dy}{dx}=\frac{u\,v}{w}\left\{\frac{1}{u}\cdot\frac{du}{dx}+\frac{1}{v}\cdot\frac{dv}{dx}-\frac{1}{w}\cdot\frac{dw}{dx}\right\}$$

This is not a formula to memorize, but a *method* of working, since the actual terms on the right-hand side will depend on the functions you start with.

Let us do an example to make it quite clear.

$$\text{If } y=\frac{x^2\sin x}{\cos 2x},\ \text{find } \frac{dy}{dx}$$

The first step in the process is ............................

**21**

To take logs of both sides

$$y=\frac{x^2\sin x}{\cos 2x}\quad\therefore\ \ln y=\ln(x^2)+\ln(\sin x)-\ln(\cos 2x)$$

Now diff. both sides w.r.t. $x$, remembering that $\frac{d}{dx}(\ln F)=\frac{1}{F}\cdot\frac{dF}{dx}$

$$\frac{1}{y}\cdot\frac{dy}{dx}=\frac{1}{x^2}\cdot 2x+\frac{1}{\sin x}\cdot\cos x-\frac{1}{\cos 2x}\cdot(-2\sin 2x)$$

$$=\frac{2}{x}+\cot x+2\tan 2x$$

$$\therefore\ \frac{dy}{dx}=\frac{x^2\sin x}{\cos 2x}\left\{\frac{2}{x}+\cot x+2\tan 2x\right\}$$

This is a pretty complicated result, but the original function was also somewhat involved!

You do this one on your own:

If $y=x^4\,e^{3x}\tan x$, then $\frac{dy}{dx}=$ ..........................

**22**

$$\frac{dy}{dx} = x^4 e^{3x} \tan x\left\{\frac{4}{x} + 3 + \frac{\sec^2 x}{\tan x}\right\}$$

Here is the working. Follow it through.

$$y = x^4 e^{3x} \tan x \quad \therefore \ln y = \ln(x^4) + \ln(e^{3x}) + \ln(\tan x)$$

$$\frac{1}{y}\cdot\frac{dy}{dx} = \frac{1}{x^4}\cdot 4x^3 + \frac{1}{e^{3x}}\cdot 3e^{3x} + \frac{1}{\tan x}\cdot\sec^2 x$$

$$= \frac{4}{x} + 3 + \frac{\sec^2 x}{\tan x}$$

$$\therefore \frac{dy}{dx} = x^4 e^{3x} \tan x\left\{\frac{4}{x} + 3 + \frac{\sec^2 x}{\tan x}\right\}$$

There it is.

Always use the log. diff. method where there are more than two functions involved in a product or quotient (or both).

Here is just one more for you to do. Find $\frac{dy}{dx}$, given that

$$y = \frac{e^{4x}}{x^3 \cosh 2x}$$

**23**

$$\frac{dy}{dx} = \frac{e^{4x}}{x^3 \cosh 2x}\left\{4 - \frac{3}{x} - 2\tanh 2x\right\}$$

Working. Check yours.

$$y = \frac{e^{4x}}{x^3 \cosh 2x} \quad \therefore \ln y = \ln(e^{4x}) - \ln(x^3) - \ln(\cosh 2x)$$

$$\therefore \frac{1}{y}\frac{dy}{dx} = \frac{1}{e^{4x}}\cdot 4e^{4x} - \frac{1}{x^3}\cdot 3x^2 - \frac{1}{\cosh 2x}\cdot 2\sinh 2x$$

$$= 4 - \frac{3}{x} - 2\tanh 2x$$

$$\therefore \frac{dy}{dx} = \frac{e^{4x}}{x^3 \cosh 2x}\left\{4 - \frac{3}{x} - 2\tanh 2x\right\}$$

Well now, before continuing with the rest of the programme, here is a revision exercise for you to deal with.

*Turn on for details.*

**24** **Revision Exercise** on the work so far.

Differentiate with respect to $x$:

1. (i) $\ln 4x$ (ii) $\ln(\sin 3x)$

2. $e^{3x} \sin 4x$

3. $\dfrac{\sin 2x}{2x+5}$

4. $\dfrac{(3x+1)\cos 2x}{e^{2x}}$

5. $x^5 \sin 2x \cos 4x$

When you have finished them all (and not before) turn on to frame 25 to check your results.

**25**

*Solutions*

1. (i) $y = \ln 4x \quad \therefore \frac{dy}{dx} = \frac{1}{4x} \cdot 4 = \underline{\frac{1}{x}}$

(ii) $y = \ln \sin 3x \quad \therefore \frac{dy}{dx} = \frac{1}{\sin 3x} \cdot 3 \cos 3x$

$$= \underline{3 \cot 3x}$$

2. $y = e^{3x} \sin 4x \quad \therefore \frac{dy}{dx} = e^{3x}\, 4 \cos 4x + 3e^{3x} \sin 4x$

$$= \underline{e^{3x}(4 \cos 4x + 3 \sin 4x)}$$

3. $y = \frac{\sin 2x}{2x + 5} \quad \therefore \frac{dy}{dx} = \underline{\frac{(2x + 5)\, 2 \cos 2x - 2 \sin 2x}{(2x + 5)^2}}$

4. $y = \frac{(3x + 1) \cos 2x}{e^{2x}}$

$$\therefore \ln y = \ln(3x + 1) + \ln(\cos 2x) - \ln(e^{2x})$$

$$\therefore \frac{1}{y}\frac{dy}{dx} = \frac{1}{3x + 1} \cdot 3 + \frac{1}{\cos 2x} \cdot (-2 \sin 2x) - \frac{1}{e^{2x}} \cdot 2e^{2x}$$

$$= \frac{3}{3x + 1} - 2 \tan 2x - 2$$

$$\underline{\frac{dy}{dx} = \frac{(3x + 1) \cos 2x}{e^{2x}} \left\{ \frac{3}{3x + 1} - 2 \tan 2x - 2 \right\}}$$

5. $y = x^5 \sin 2x \cos 4x$

$$\therefore \ln y = \ln(x^5) + \ln(\sin 2x) + \ln(\cos 4x)$$

$$\therefore \frac{1}{y}\frac{dy}{dx} = \frac{1}{x^5} \cdot 5x^4 + \frac{2 \cos 2x}{\sin 2x} + \frac{1}{\cos 4x}(-4 \sin 4x)$$

$$= \frac{5}{x} + 2 \cot 2x - 4 \tan 4x$$

$$\underline{\frac{dy}{dx} = x^5 \sin 2x \cos 4x \left\{ \frac{5}{x} + 2 \cot 2x - 4 \tan 4x \right\}}$$

*So far so good. Now on to the next part of the programme on frame 26.*

**26**

## Implicit functions

If $y = x^2 - 4x + 2$, $y$ is completely defined in terms of $x$ and $y$ is called an *explicit function* of $x$.

When the relationship between $x$ and $y$ is more involved, it may not be possible (or desirable) to separate $y$ completely on the left-hand side, e.g. $xy + \sin y = 2$. In such a case as this, $y$ is called an *implicit function* of $x$, because a relationship of the form $y = f(x)$ is implied in the given equation.

It may still be necessary to determine the differential coefficients of $y$ with respect to $x$ and in fact this is not at all difficult. All we have to remember is that $y$ is a function of $x$, even if it is difficult to see what it is. In fact, this is really an extension of our 'function of a function' routine.

$x^2 + y^2 = 25$, as it stands, is an example of an .................... function.

---

**27**

$x^2 + y^2 = 25$ is an example of an **implicit** function.

□□□□□□□□□□□□□□□□□□□□□□□□□□□□□□□□□□□□□□

Once again, all we have to remember is that $y$ is a function of $x$. So, if $x^2 + y^2 = 25$, let us find $\dfrac{dy}{dx}$.

If we differentiate as it stands with respect to $x$, we get

$$2x + 2y\frac{dy}{dx} = 0$$

Note that we differentiate $y^2$ as a function squared, giving 'twice times the function, times the diff. coefft. of the function'. The rest is easy.

$$2x + 2y\frac{dy}{dx} = 0$$

$$\therefore\; y\frac{dy}{dx} = -x \qquad \therefore\; \underline{\frac{dy}{dx} = -\frac{x}{y}}$$

As you will have noticed, with an implicit function the differential coefficient may contain (and usually does) both $x$ and ....................

$\boxed{y}$

28

Let us look at one or two examples.

*Example 1.* If $x^2 + y^2 - 2x - 6y + 5 = 0$, find $\frac{dy}{dx}$ and $\frac{d^2y}{dx^2}$ at $x = 3, y = 2$.

Differentiate as it stands with respect to $x$.

$$2x + 2y\frac{dy}{dx} - 2 - 6\frac{dy}{dx} = 0$$

$$\therefore\ (2y - 6)\frac{dy}{dx} = 2 - 2x$$

$$\therefore\ \frac{dy}{dx} = \frac{2 - 2x}{2y - 6} = \frac{1 - x}{y - 3}$$

$$\therefore \text{ at } (3, 2) \qquad \frac{dy}{dx} = \frac{1 - 3}{2 - 3} = \frac{-2}{-1} = 2$$

Then $$\frac{d^2y}{dx^2} = \frac{d}{dx}\left\{\frac{1 - x}{y - 3}\right\} = \frac{(y - 3)(-1) - (1 - x)\frac{dy}{dx}}{(y - 3)^2}$$

$$= \frac{(3 - y) - (1 - x)\frac{dy}{dx}}{(y - 3)^2}$$

at (3,2) $$\frac{d^2y}{dx^2} = \frac{(3 - 2) - (1 - 3)\,2}{(2 - 3)^2} = \frac{1 - (-4)}{1} = 5$$

$$\therefore \text{ At } (3,2) \quad \frac{dy}{dx} = 2, \quad \frac{d^2y}{dx^2} = 5$$

Now this one. If $x^2 + 2xy + 3y^2 = 4$, find $\frac{dy}{dx}$.

Away you go, but beware of the product term. When you come to $2xy$ treat this as $(2x)(y)$.

29

$$x^2 + 2xy + 3y^2 = 4$$

$$2x + 2x\frac{dy}{dx} + 2y + 6y\frac{dy}{dx} = 0$$

$$\therefore\ (2x + 6y)\frac{dy}{dx} = -(2x + 2y)$$

$$\therefore\ \frac{dy}{dx} = -\frac{(2x + 2y)}{(2x + 6y)} = -\frac{(x + y)}{(x + 3y)}$$

And now, just one more:

If $x^3 + y^3 + 3xy^2 = 8$, find $\frac{dy}{dx}$

*Turn to frame 30 for the solution.*

**30** Solution in detail:

$$x^3 + y^3 + 3xy^2 = 8$$

$$3x^2 + 3y^2\frac{dy}{dx} + 3x \,.\, 2y\frac{dy}{dx} + 3y^2 = 0$$

$$\therefore\ (y^2 + 2xy)\frac{dy}{dx} = -(x^2 + y^2)$$

$$\therefore\ \frac{dy}{dx} = -\frac{(x^2 + y^2)}{(y^2 + 2xy)}$$

That is really all there is to it. All examples are tackled the same way. The key to it is simply that '$y$ is a function of $x$' and then apply the 'function of a function' routine.

*Now on to the last section of this particular programme, which starts on frame 31.*

**31**

**Parametric equations**

In some cases, it is more convenient to represent a function by expressing $x$ and $y$ separately in terms of a third independent variable, e.g. $y = \cos 2t$, $x = \sin t$. In this case, any value we give to $t$ will produce a pair of values for $x$ and $y$, which could if necessary be plotted and provide one point of the curve of $y = f(x)$.

The third variable, e.g. $t$, is called a *parameter*, and the two expressions for $x$ and $y$ *parametric equations*. We may still need to find the differential coefficients of the function with respect to $x$, so how do we go about it?

Let us take the case already quoted above. The parametric equations of a function are given as $y = \cos 2t$, $x = \sin t$. We are required to find expressions for $\dfrac{dy}{dx}$ and $\dfrac{d^2y}{dx^2}$

*Turn to the next frame to see how we go about it.*

**32**

$$y = \cos 2t, \quad x = \sin t. \text{ Find } \frac{dy}{dx} \text{ and } \frac{d^2y}{dx^2}$$

From $y = \cos 2t$, we can get $\dfrac{dy}{dt} = -2 \sin 2t$

From $x = \sin t$, we can get $\dfrac{dx}{dt} = \cos t$

We now use the fact that $\dfrac{dy}{dx} = \dfrac{dy}{dt} \cdot \dfrac{dt}{dx}$

so that
$$\frac{dy}{dx} = -2 \sin 2t \cdot \frac{1}{\cos t}$$
$$= -4 \sin t \cos t \cdot \frac{1}{\cos t}$$
$$\therefore \quad \frac{dy}{dx} = -4 \sin t$$

That was easy enough. Now how do we find the second diff. coefft.? We *cannot* get it by finding $\dfrac{d^2y}{dt^2}$ and $\dfrac{d^2x}{dt^2}$ from the parametric equations and joining them together as we did for the first diff. coefft. That method could only give us something called $\dfrac{d^2y}{d^2x}$ which has no meaning and is certainly not what we want. So what do we do?

*On to the next frame and all will be revealed!*

**33**

To find the second differential coefficient, we must go back to the very meaning of $\dfrac{d^2y}{dx^2}$

$$\text{i.e. } \frac{d^2y}{dx^2} = \frac{d}{dx}\left(\frac{dy}{dx}\right) = \frac{d}{dx}\left(-4 \sin t\right)$$

But we cannot differentiate a function of $t$ directly with respect to $x$.

Therefore we say $\dfrac{d}{dx}\left(-4 \sin t\right) = \dfrac{d}{dt}\left(-4 \sin t\right) \cdot \dfrac{dt}{dx}$.

$$\therefore \quad \frac{d^2y}{dx^2} = -4 \cos t \cdot \frac{1}{\cos t} = -4$$
$$\therefore \quad \frac{d^2y}{dx^2} = -4$$

Let us work through another one. What about this?
The parametric equations of a function are given as

$$y = 3 \sin \theta - \sin^3\theta, \quad x = \cos^3\theta$$

Find $\dfrac{dy}{dx}$ and $\dfrac{d^2y}{dx^2}$

*Turn on to frame 34.*

**34**

$$y = 3\sin\theta - \sin^3\theta \quad \therefore \frac{dy}{d\theta} = 3\cos\theta - 3\sin^2\theta\cos\theta$$

$$x = \cos^3\theta \quad \ldots\ldots \quad \therefore \frac{dx}{d\theta} = 3\cos^2\theta(-\sin\theta)$$

$$= -3\cos^2\theta\sin\theta$$

$$\frac{dy}{dx} = \frac{dy}{d\theta}\cdot\frac{d\theta}{dx} = 3\cos\theta\,(1-\sin^2\theta)\cdot\frac{1}{-3\cos^2\theta\sin\theta}$$

$$= \frac{3\cos^3\theta}{-3\cos^2\theta\sin\theta} \quad \therefore \underline{\frac{dy}{dx} = -\cot\theta}$$

Also

$$\frac{d^2y}{dx^2} = \frac{d}{dx}\left(-\cot\theta\right) = \frac{d}{d\theta}\left(-\cot\theta\right)\frac{d\theta}{dx}$$

$$= -(-\operatorname{cosec}^2\theta)\frac{1}{-3\cos^2\theta\sin\theta}$$

$$\therefore \underline{\frac{d^2y}{dx^2} = \frac{-1}{3\cos^2\theta\sin^3\theta}}$$

Now here is one for you to do in just the same way.

$$\text{If } x = \frac{2-3t}{1+t}, \quad y = \frac{3+2t}{1+t}, \quad \text{find } \frac{dy}{dx}$$

*When you have done it, move on to frame 35.*

---

**35**

$$\boxed{\frac{dy}{dx} = \frac{1}{5}}$$

For

$$x = \frac{2-3t}{1+t} \quad \therefore \frac{dx}{dt} = \frac{(1+t)(-3)-(2-3t)}{(1+t)^2}$$

$$y = \frac{3+2t}{1+t} \quad \therefore \frac{dy}{dt} = \frac{(1+t)(2)-(3+2t)}{(1+t)^2}$$

$$\frac{dx}{dt} = \frac{-3-3t-2+3t}{(1+t)^2} = \frac{-5}{(1+t)^2}$$

$$\frac{dy}{dt} = \frac{2+2t-3-2t}{(1+t)^2} = \frac{-1}{(1+t)^2}$$

$$\frac{dy}{dx} = \frac{dy}{dt}\cdot\frac{dt}{dx} = \frac{-1}{(1+t)^2}\cdot\frac{(1+t)^2}{-5} = \frac{1}{5} \quad \therefore \underline{\frac{dy}{dx} = \frac{1}{5}}$$

And now here is one more for you to do to finish up this part of the work. It is done in just the same way as the others.

$$\text{If } x = a(\cos\theta + \theta\sin\theta) \text{ and } y = a(\sin\theta - \theta\cos\theta)$$

$$\text{find } \frac{dy}{dx} \text{ and } \frac{d^2y}{dx^2}$$

**36**

Here it is, set out like the previous examples.

$$x = a(\cos\theta + \theta\sin\theta)$$

$$\therefore \frac{dx}{d\theta} = a(-\sin\theta + \theta\cos\theta + \sin\theta) = a\,\theta\cos\theta$$

$$y = a(\sin\theta - \theta\cos\theta)$$

$$\therefore \frac{dy}{d\theta} = a(\cos\theta + \theta\sin\theta - \cos\theta) = a\,\theta\sin\theta$$

$$\frac{dy}{dx} = \frac{dy}{d\theta}\cdot\frac{d\theta}{dx} = a\,\theta\sin\theta\,.\frac{1}{a\,\theta\cos\theta} = \tan\theta$$

$$\underline{\frac{dy}{dx} = \tan\theta}$$

$$\frac{d^2y}{dx^2} = \frac{d}{dx}(\tan\theta) = \frac{d}{d\theta}(\tan\theta)\,.\frac{d\theta}{dx}$$

$$= \sec^2\theta\,.\frac{1}{a\,\theta\cos\theta}$$

$$\therefore \underline{\frac{d^2y}{dx^2} = \frac{1}{a\,\theta\cos^3\theta}}$$

You have now reached the end of this programme on differentiation, much of which has been useful revision of what you have done before. This brings you to the final Test Exercise so turn on to it and work through it carefully.

*Next frame please.*

**37**

## Test Exercise – VII

Do all the questions. Write out the solutions carefully. They are all quite straightforward.

1. Differentiate the following with respect to $x$:

   (i) $\tan 2x$ (ii) $(5x+3)^6$ (iii) $\cosh^2 x$

   (iv) $\log_{10}(x^2-3x-1)$ (v) $\ln\cos 3x$ (vi) $\sin^3 4x$

   (vii) $e^{2x}\sin 3x$ (viii) $\dfrac{x^4}{(x+1)^2}$ (ix) $\dfrac{e^{4x}\sin x}{x\cos 2x}$

2. If $x^2+y^2-2x+2y=23$, find $\dfrac{dy}{dx}$ and $\dfrac{d^2y}{dx^2}$ at the point where $x=-2$, $y=3$.

3. Find an expression for $\dfrac{dy}{dx}$ when

$$x^3+y^3+4xy^2=5$$

4. If $x=3(1-\cos\theta)$ and $y=3(\theta-\sin\theta)$ find $\dfrac{dy}{dx}$ and $\dfrac{d^2y}{dx^2}$ in their simplest forms.

## Further Problems – VII

1. Differentiate with respect to $x$:

   (i) $\ln\left\{\dfrac{\cos x + \sin x}{\cos x - \sin x}\right\}$ (ii) $\ln(\sec x + \tan x)$ (iii) $\sin^4 x \cos^3 x$

2. Find $\dfrac{dy}{dx}$ when (i) $y = \dfrac{x \sin x}{1 + \cos x}$ (ii) $y = \ln\left\{\dfrac{1 - x^2}{1 + x^2}\right\}$

3. If $y$ is a function of $x$, and $x = \dfrac{e^t}{e^t + 1}$

   show that $\dfrac{dy}{dt} = x(1 - x)\dfrac{dy}{dx}$

4. Find $\dfrac{dy}{dx}$ when $x^3 + y^3 - 3xy^2 = 8$.

5. Differentiate: (i) $y = e^{\sin^2 5x}$ (ii) $y = \ln\left\{\dfrac{\cosh x - 1}{\cosh x + 1}\right\}$

   (iii) $y = \ln\left\{e^x\left(\dfrac{x - 2}{x + 2}\right)^{3/4}\right\}$

6. Differentiate: (i) $y = x^2 \cos^2 x$ (ii) $y = \ln\left\{x^2\sqrt{(1 - x^2)}\right\}$

   (iii) $y = \dfrac{e^{2x} \ln x}{(x - 1)^3}$

7. If $(x - y)^3 = A(x + y)$, prove that $(2x + y)\dfrac{dy}{dx} = x + 2y$.

8. If $x^2 - xy + y^2 = 7$, find $\dfrac{dy}{dx}$ and $\dfrac{d^2y}{dx^2}$ at $x = 3, y = 2$.

9. If $x^2 + 2xy + 3y^2 = 1$, prove that $(x + 3y)^3\dfrac{d^2y}{dx^2} + 2 = 0$.

10. If $x = \ln \tan\dfrac{\theta}{2}$ and $y = \tan\theta - \theta$, prove that

    $$\frac{d^2y}{dx^2} = \tan^2\theta \sin\theta\ (\cos\theta + 2\sec\theta)$$

11. If $y = 3\,e^{2x}\cos(2x - 3)$, verify that $\dfrac{d^2y}{dx^2} - 4\dfrac{dy}{dx} + 8y = 0$.

12. The parametric equations of a curve are $x = \cos 2\theta$, $y = 1 + \sin 2\theta$. Find $\dfrac{dy}{dx}$ and $\dfrac{d^2y}{dx^2}$ at $\theta = \pi/6$. Find also the equation of the curve as a relationship between $x$ and $y$.

13. If $y = \left\{x + \sqrt{(1 + x^2)}\right\}^{3/2}$, show that

$$4(1 + x^2)\frac{d^2y}{dx^2} + 4x\frac{dy}{dx} - 9y = 0$$

14. Find $\dfrac{dy}{dx}$ and $\dfrac{d^2y}{dx^2}$ if $x = a\cos^3\theta$, $y = a\sin^3\theta$.

15. If $x = 3\cos\theta - \cos^3\theta$, $y = 3\sin\theta - \sin^3\theta$, express $\dfrac{dy}{dx}$ and $\dfrac{d^2y}{dx^2}$ in terms of $\theta$.

16. Show that $y = e^{-2mx}\sin 4mx$ is a solution of the equation

$$\frac{d^2y}{dx^2} + 4m\frac{dy}{dx} + 20m^2y = 0$$

17. If $y = \sec x$, prove that $y\dfrac{d^2y}{dx^2} = \left(\dfrac{dy}{dx}\right)^2 + y^4$

18. Prove that $x = A\,e^{-kt}\sin pt$, satisfies the equation

$$\frac{d^2x}{dt^2} + 2k\frac{dx}{dt} + (p^2 + k^2)x = 0$$

19. If $y = e^{-kt}(A\cosh qt + B\sinh qt)$ where A, B, $q$ and $k$ are constants, show that

$$\frac{d^2y}{dt^2} + 2k\frac{dy}{dt} + (k^2 - q^2)y = 0$$

20. If $\sinh y = \dfrac{4\sinh x - 3}{4 + 3\sinh x}$, show that $\dfrac{dy}{dx} = \dfrac{5}{4 + 3\sinh x}$

# Programme 8

## DIFFERENTIATION APPLICATIONS

### PART 1

**1**

**Equation of a straight line**

The basic equation of a straight line is $y = mx + c$,

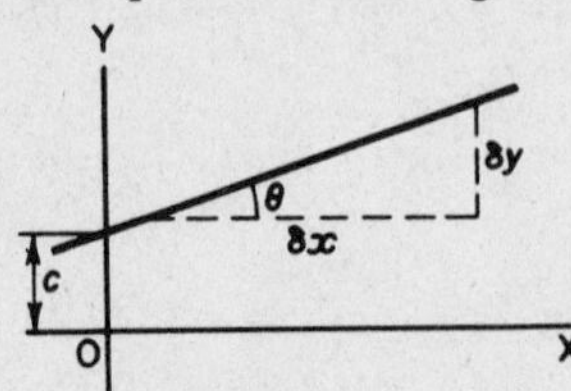


where $m$ = slope $= \dfrac{\delta y}{\delta x} = \dfrac{dy}{dx}$

$c$ = intercept on real $y$-axis

Note that if the scales of $x$ and $y$ are identical, $\dfrac{dy}{dx} = \tan\theta$

e.g. To find the equation of the straight line passing through P(3,2) and Q(−2,1), we could argue thus:

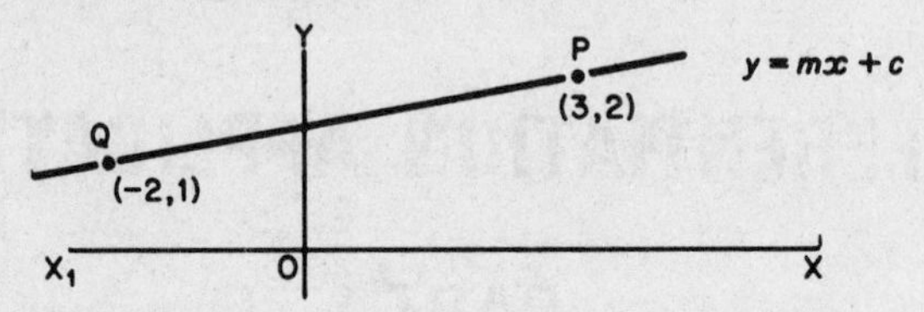


Line passes through P, i.e. when $x = 3, y = 2 \quad \therefore \; 2 = m3 + c$

Line passes through Q, i.e. when $x = -2, y = 1 \quad \therefore \; 1 = m(-2) + c.$

So we obtain a pair of simultaneous equations from which the values of $m$ and $c$ can be found. Therefore the equation is ........................

**2**

We find $m = 1/5$ and $c = 7/5$. Therefore the equation of the line is

$$y = \frac{x}{5} + \frac{7}{5}, \text{ i.e. } \boxed{5y = x + 7}$$

□□□□□□□□□□□□□□□□□□□□□□□□□□□□□□□□□□□□□

Sometimes we are given the slope, $m$, of a straight line passing through a given point $(x_1, y_1)$ and we are required to find its equation. In that case, it is more convenient to use the form

$$y - y_1 = m(x - x_1)$$

For example, the equation of the line passing through the point (5,3) with slope 2 is simply ...................... which simplifies to ......................

*Turn on to the next frame.*

**3**

$$\boxed{y - 3 = 2(x - 5)}$$

i.e. $y - 3 = 2x - 10 \quad \therefore \quad \boxed{y = 2x - 7}$

□□□□□□□□□□□□□□□□□□□□□□□□□□□□□□□□□□□□□□

Similarly, the equation of the line through the point (−2,−1) and having a slope $\frac{1}{2}$ is

$$y - (-1) = \frac{1}{2}\{x - (-2)\}$$
$$\therefore \; y + 1 = \frac{1}{2}(x + 2)$$
$$2y + 2 = x + 2$$
$$\therefore \; \underline{y = \frac{x}{2}}$$

So, in the same way, the line passing through (2,−3) and having slope (−2) is ......................

**4**

$$\boxed{y = 1 - 2x}$$

For $$y - (-3) = -2(x - 2)$$
$$\therefore \; y + 3 = -2x + 4 \quad \therefore \; y = 1 - 2x$$

□□□□□□□□□□□□□□□□□□□□□□□□□□□□□□□□□□□□□□

Right. So in general terms, the equation of the line passing through the point $(x_1, y_1)$ with slope $m$ is ...........................

*Turn on to frame 5.*

**5**

$$\boxed{y - y_1 = m(x - x_1)}$$ It is well worth remembering.

□□□□□□□□□□□□□□□□□□□□□□□□□□□□□□□□□□□□□□

So for one last time:

If a point P has co-ordinates (4,3) and the slope $m$ of a straight line through P is 2, then the equation of the line is thus

$$y - 3 = 2(x - 4)$$
$$= 2x - 8$$
$$\therefore \quad \underline{y = 2x - 5}$$

The equation of the line through P, perpendicular to the line we have just considered, will have a slope $m_1$, such that $m\,m_1 = -1$

i.e. $m_1 = -\dfrac{1}{m}$. And since $m = 2$, then $m_1 = -\dfrac{1}{2}$. This line passes through (4,3) and its equation is therefore

$$y - 3 = -\frac{1}{2}(x - 4)$$
$$= -x/2 + 2$$
$$y = -\frac{x}{2} + 5 \quad \underline{2y = 10 - x}$$

**6**

If $m$ and $m_1$ represent the slopes of two lines perpendicular to each other, then $m\,m_1 = -1$ or $m_1 = -\dfrac{1}{m}$

Consider the two straight lines

$$2y = 4x - 5 \quad \text{and} \quad 6y = 2 - 3x$$

If we convert each of these to the form $y = m\,x + c$, we get

(i) $y = 2x - \dfrac{5}{2}$ and (ii) $y = -\dfrac{1}{2}x + \dfrac{1}{3}$

So in (i) the slope $m = 2$ and in (ii) the slope $m_1 = -\dfrac{1}{2}$

We notice that, in this case, $m_1 = -\dfrac{1}{m}$ or that $mm_1 = -1$

Therefore we know that the two given lines are at right-angles to each other.

Which of these represents a pair of lines perpendicular to each other:

(i) $y = 3x - 5$ and $3y = x + 2$.
(ii) $2y = x - 5$ and $y = 6 - x$.
(iii) $y - 3x - 2 = 0$ and $3y + x + 9 = 0$.
(iv) $5y - x = 4$ and $2y + 10x + 3 = 0$.

**7**

*Result:*

| (iii) and (iv) |
|---|

□□□□□□□□□□□□□□□□□□□□□□□□□□□□□□□□□□□□□□□□□□□□□□□□

For if we convert each to the form $y = mx + c$, we get

(i) $y = 3x - 5$ and $y = \frac{x}{3} + \frac{2}{3}$

$m = 3\,;\, m_1 = \frac{1}{3} \quad \therefore\; m\,m_1 \neq -1$ Not perpendicular.

(ii) $y = \frac{x}{2} - \frac{5}{2}$ and $y = -x + 6$

$m = \frac{1}{2}\,;\, m_1 = -1 \quad \therefore\; m\,m_1 \neq -1$ Not perpendicular.

(iii) $y = 3x + 2$ and $y = -\frac{x}{3} - 3$

$m = 3\,;\, m_1 = -\frac{1}{3} \quad \therefore\; m\,m_1 = -1$ Perpendicular.

(iv) $y = \frac{x}{5} + \frac{4}{5}$ and $y = -5x - \frac{3}{2}$

$m = \frac{1}{5}\,;\, m_1 = -5 \quad \therefore\; m\,m_1 = -1$ Perpendicular

Do you agree with these?

---

**8**

Remember that if $y = m\,x + c$ and $y = m_1 x + c_1$ are perpendicular to each other, then

$$m\,m_1 = -1, \text{ i.e. } m_1 = -\frac{1}{m}$$

Here is one further example:

A line AB passes through the point P(3, −2) with slope $-\frac{1}{2}$. Find its equation and also the equation of the line CD through P perpendicular to AB.

*When you have finished, check your results with those on frame 9.*

**9**

Equation of AB:
$$y-(-2) = -\frac{1}{2}(x-3)$$
$$\therefore \quad y+2 = -\frac{x}{2}+\frac{3}{2}$$
$$\therefore \quad y = -\frac{x}{2}-\frac{1}{2}$$
$$\therefore \quad \underline{2y+x+1=0}$$

Equation of CD:
$$\text{slope } m_1 = -\frac{1}{m} = -\frac{1}{-\frac{1}{2}} = 2$$
$$y-(-2) = 2(x-3)$$
$$y+2 = 2x-6$$
$$\underline{y = 2x-8}$$

So we have:

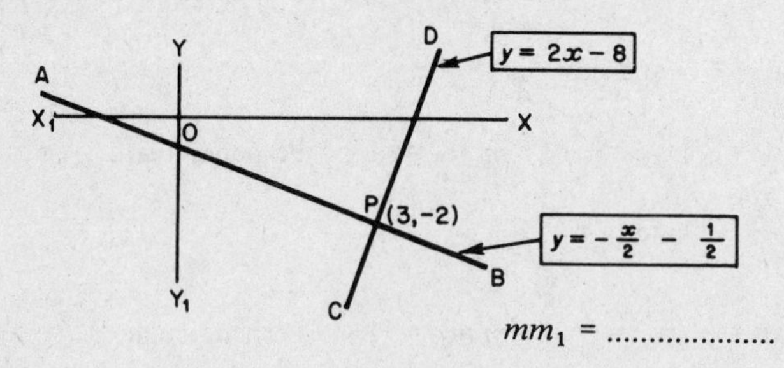


$mm_1 = \ldots\ldots\ldots\ldots$

**10**

$$\boxed{m\,m_1 = -1}$$

□□□□□□□□□□□□□□□□□□□□□□□□□□□□□□□□□□□□□□

And now, just one more to do on your own.

The point P(3, 4) is a point on the line $y = 5x - 11$.
Find the equation of the line through P which is perpendicular to the given line.

*That should not take long. When you have finished it, turn on to the next frame.*

**11**

$$\boxed{5y + x = 23}$$

For: slope of the given line, $y = 5x - 11$ is 5.

slope of required line $= -\frac{1}{5}$

The line passes through P, i.e. when $x = 3, y = 4$.

$$y - 4 = -\frac{1}{5}(x - 3)$$

$$5y - 20 = -x + 3 \quad \therefore \quad \underline{5y + x = 23}$$

□□□□□□□□□□□□□□□□□□□□□□□□□□□□□□□□□□□□□□

**Tangents and normals** to a curve at a given point.

The slope of a curve, $y = f(x)$, at a point P on the curve is given by the slope of the tangent at P. It is also given by the value of $\frac{dy}{dx}$ at the point P, which we can calculate, knowing the equation of the curve. Thus we can calculate the slope of the tangent to the curve at any point P.

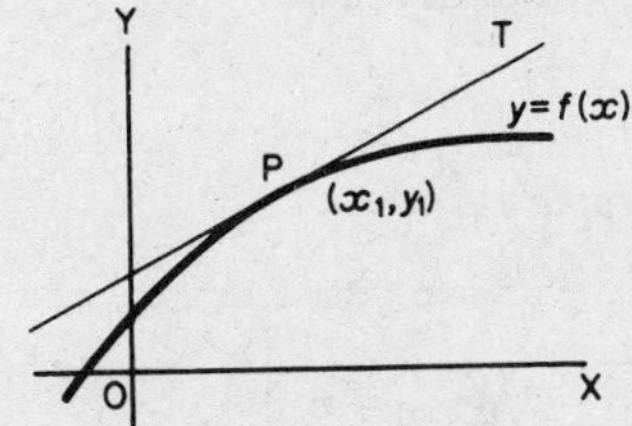


What else do we know about the tangent which will help us to determine its equation?

---

**12**

We know that the tangent passes through P, i.e. when $x = x_1, y = y_1$.

□□□□□□□□□□□□□□□□□□□□□□□□□□□□□□□□□□□□□□

Correct. This is sufficient information for us to find the equation of the tangent. Let us do an example.

e.g. Find the equation of the tangent to the curve $y = 2x^3 + 3x^2 - 2x - 3$ at the point P, $x = 1, y = 0$.

$$\frac{dy}{dx} = 6x^2 + 6x - 2$$

Slope of tangent $= \left\{\frac{dy}{dx}\right\}_{x=1} = 6 + 6 - 2 = 10$, i.e. $m = 10$

Passes through P, i.e. $x = 1, y = 0$.

$$y - y_1 = m(x - x_1) \text{ gives } y - 0 = 10(x - 1)$$

Therefore the tangent is $\underline{y = 10x - 10}$

We could also, if required, find the equation of the normal at P which is defined as the line through P perpendicular to the tangent at P. We know, for example, that the slope of the normal is ........................

**13**

$$\text{Slope of normal} = \frac{-1}{\text{Slope of tangent}} = -\frac{1}{10}$$

□□□□□□□□□□□□□□□□□□□□□□□□□□□□□□□□□□□□□□

The normal also passes through P, i.e. when $x = 1, y = 0$.

$\therefore$ Equation of normal is: $y - 0 = -\frac{1}{10}(x - 1)$

$$10y = -x + 1 \quad \underline{10y + x = 1}$$

That was very easy. Do this one just to get your hand in:

Find the equations of the tangent and normal to the curve $y = x^3 - 2x^2 + 3x - 1$ at the point (2, 5).

Off you go. Do it in just the same way.

*When you have got the results, move on to frame 14.*

**14**

Tangent: $y = 7x - 9$ | Normal: $7y + x = 37$

Here are the details:

$$y = x^3 - 2x^2 + 3x - 1$$

$$\therefore \frac{dy}{dx} = 3x^2 - 4x + 3 \quad \therefore \text{ At P}(2, 5), \frac{dy}{dx} = 12 - 8 + 3 = 7$$

Tangent passes through (2, 5), i.e. $x = 2, y = 5$

$$y - 5 = 7(x - 2) \quad \textit{Tangent} \text{ is } \underline{y = 7x - 9}$$

For normal, slope $= \dfrac{-1}{\text{slope of tangent}} = -\dfrac{1}{7}$

Normal passes through P (2, 5)

$$\therefore y - 5 = -\frac{1}{7}(x - 2)$$

$$7y - 35 = -x + 2$$

$$\textit{Normal} \text{ is } \underline{7y + x = 37}$$

You will perhaps remember doing all this long ago.

*Anyway, on to frame 15.*

**15**

The equation of the curve may, of course, be presented as an implicit function or as a pair of parametric equations. But this will not worry you for you already know how to differentiate functions in these two forms. Let us have an example or two.

Find the equations of the tangent and normal to the curve $x^2 + y^2 + 3xy - 11 = 0$ at the point $x = 1, y = 2$.

First of all we must find $\frac{dy}{dx}$ at $(1, 2)$. So differentiate right away.

$$2x + 2y\frac{dy}{dx} + 3x\frac{dy}{dx} + 3y = 0$$

$$(2y + 3x)\frac{dy}{dx} = -(2x + 3y)$$

$$\frac{dy}{dx} = -\frac{2x + 3y}{2y + 3x}$$

Therefore, at $x = 1, y = 2$,

$$\frac{dy}{dx} = \ldots\ldots\ldots\ldots$$

---

**16**

$$\frac{dy}{dx} = -\frac{2 + 6}{4 + 3} = -\frac{8}{7} \qquad \boxed{\frac{dy}{dx} = -\frac{8}{7}}$$

Now we proceed as for the previous cases.

Tangent passes through $(1, 2)$ $\therefore\ y - 2 = -\frac{8}{7}(x - 1)$

$$7y - 14 = -8x + 8$$

$$\therefore \text{ Tangent is } \underline{7y + 8x = 22}$$

Now to find the equation of the normal.

$$\text{Slope} = \frac{-1}{\text{Slope of tangent}} = \frac{7}{8}$$

Normal passes through $(1, 2)$ $\therefore\ y - 2 = \frac{7}{8}(x - 1)$

$$8y - 16 = 7x - 7$$

$$\therefore \text{ Normal is } \underline{8y = 7x + 9}$$

That's that!

Now try this one:

Find the equations of the tangent and normal to the curve $x^3 + x^2y + y^3 - 7 = 0$ at the point $x = 2, y = 3$.

**17** *Results:*

Tangent: $31y + 24x = 141$ | Normal: $24y = 31x + 10$

Here is the working:.

$$x^3 + x^2y + y^3 - 7 = 0$$

$$3x^2 + x^2\frac{dy}{dx} + 2xy + 3y^2\frac{dy}{dx} = 0$$

$$(x^2 + 3y^2)\frac{dy}{dx} = -(3x^2 + 2xy) \quad \therefore \frac{dy}{dx} = -\frac{3x^2 + 2xy}{x^2 + 3y^2}$$

$\therefore$ At (2,3) $$\frac{dy}{dx} = -\frac{12 + 12}{4 + 27} = -\frac{24}{31}$$

(i) Tangent passes through (2,3) $\therefore y - 3 = -\frac{24}{31}(x - 2)$

$$31y - 93 = -24x + 48 \qquad \therefore \underline{31y + 24x = 141}$$

(ii) Normal: slope $= \frac{31}{24}$. Passes through (2,3) $\therefore y - 3 = \frac{31}{24}(x - 2)$

$$24y - 72 = 31x - 62 \quad \therefore \underline{24y = 31x + 10}$$

*Now on to the next frame for another example.*

**18** Now what about this one?

The parametric equations of a curve are $x = \frac{3t}{1 + t}$, $y = \frac{t^2}{1 + t}$

Find the equations of the tangent and normal at the point for which $t = 2$.

First find the value of $\frac{dy}{dx}$ when $t = 2$.

$$x = \frac{3t}{1 + t} \quad \therefore \frac{dx}{dt} = \frac{(1 + t)3 - 3t}{(1 + t)^2} = \frac{3 + 3t - 3t}{(1 + t)^2} = \frac{3}{(1 + t)^2}$$

$$y = \frac{t^2}{1 + t} \quad \therefore \frac{dy}{dt} = \frac{(1 + t)2t - t^2}{(1 + t)^2} = \frac{2t + 2t^2 - t^2}{(1 + t)^2} = \frac{2t + t^2}{(1 + t)^2}$$

$$\frac{dy}{dx} = \frac{dy}{dt}\cdot\frac{dt}{dx} = \frac{2t + t^2}{(1 + t)^2}\cdot\frac{(1 + t)^2}{3} = \frac{2t + t^2}{3} \quad \therefore \text{At } t = 2, \frac{dy}{dx} = \frac{8}{3}$$

To get the equation of the tangent, we must know the $x$ and $y$ values of a point through which it passes. At P–

$$x = \frac{3t}{1 + t} = \frac{6}{1 + 2} = \frac{6}{3} = 2, \quad y = \frac{t^2}{1 + t} = \frac{4}{3}$$

*Continued on frame 19.*

**19**

So the tangent has a slope of $\frac{8}{3}$ and passes through $(2, \frac{4}{3})$

$\therefore$ Its equation is $y - \frac{4}{3} = \frac{8}{3}(x - 2)$

$$3y - 4 = 8x - 16 \quad \therefore \quad \underline{3y = 8x - 12} \text{ (Tangent)}$$

For the normal, slope $= \dfrac{-1}{\text{slope of tangent}} = -\dfrac{3}{8}$

Also passes through $(2, \frac{4}{3})$ $\quad \therefore \; y - \frac{4}{3} = -\frac{3}{8}(x - 2)$

$$24y - 32 = -9x + 18 \quad \therefore \; \underline{24y + 9x = 50} \text{ (Normal)}$$

Now you do this one. When you are satisfied with your result, check it with the results on frame 20. Here it is:

If $y = \cos 2t$ and $x = \sin t$, find the equations of the tangent and normal to the curve at $t = \frac{\pi}{6}$.

---

**20**

*Results:*

| Tangent: $2y + 4x = 3$ | Normal: $4y = 2x + 1$ |
|---|---|

Working:

$$y = \cos 2t \quad \therefore \; \frac{dy}{dt} = -2 \sin 2t = -4 \sin t \cos t$$

$$x = \sin t \quad \therefore \; \frac{dx}{dt} = \cos t$$

$$\frac{dy}{dx} = \frac{dy}{dt} \cdot \frac{dt}{dx} = \frac{-4 \sin t \cos t}{\cos t} = -4 \sin t$$

At $\quad t = \frac{\pi}{6}$, $\quad \dfrac{dy}{dx} = -4 \sin \dfrac{\pi}{6} = -4(\frac{1}{2}) = -2$

$$\therefore \text{ slope of tangent} = -2$$

Passes through $\quad x = \sin \frac{\pi}{6} = 0{\cdot}5; \quad y = \cos \frac{\pi}{3} = 0{\cdot}5$

$$\therefore \text{ Tangent is } y - \tfrac{1}{2} = -2(x - \tfrac{1}{2}) \quad \therefore \; 2y - 1 = -4x + 2$$

$$\therefore \; \underline{2y + 4x = 3} \text{ (Tangent)}$$

Slope of normal $= \frac{1}{2}$. Line passes through $(0{\cdot}5, 0{\cdot}5)$

Equation is $\quad y - \frac{1}{2} = \frac{1}{2}(x - \frac{1}{2})$

$$\therefore \; 4y - 2 = 2x - 1$$

$$\therefore \; \underline{4y = 2x \pm 1} \text{ (Normal)}$$

**21** Before we leave this part of the programme, let us revise the fact that we can easily find the angle between two intersecting curves.

Since the slope of a curve at $(x_1, y_1)$ is given by the value of $\frac{dy}{dx}$ at that point, and $\frac{dy}{dx} = \tan\theta$, where $\theta$ is the angle of slope, then we can use these facts to determine the angle between the curves at their point of intersection. One example will be sufficient.

*e.g.* Find the angle between $y^2 = 8x$ and $x^2 + y^2 = 16$ at their point of intersection for which $y$ is positive.

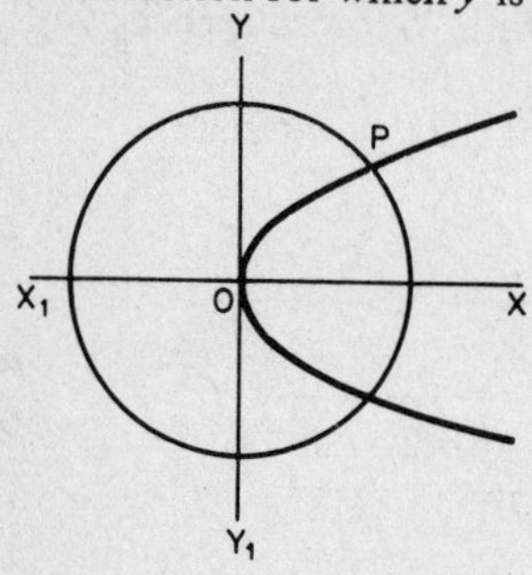


First find the point of intersection.
i.e. solve $y^2 = 8x$ and

$$x^2 + y^2 = 16$$

We have $x^2 + 8x = 16 \therefore x^2 + 8x - 16 = 0$

$$x = \frac{-8 \pm \sqrt{(64+64)}}{2} = \frac{-8 \pm \sqrt{128}}{2}$$

$$= \frac{-8 \pm 11{\cdot}314}{2} = \frac{3{\cdot}314}{2} \text{ or } \frac{-19{\cdot}314}{2}$$

$x = 1{\cdot}657$ or $[-9{\cdot}655]$ Not a real point of intersection.

When $x = 1{\cdot}657$, $y^2 = 8(1{\cdot}657) = 13{\cdot}256$, $y = 3{\cdot}641$

Co-ordinates of P are $x = 1{\cdot}657$, $y = 3{\cdot}641$

Now we have to find $\frac{dy}{dx}$ for each of the two curves. Do that.

---

**22** (i) $y^2 = 8x \therefore 2y\frac{dy}{dx} = 8 \therefore \frac{dy}{dx} = \frac{4}{y} = \frac{4}{3{\cdot}641} = \frac{1}{0{\cdot}910} = 1{\cdot}099$

$\tan\theta_1 = 1{\cdot}099 \therefore \theta_1 = 47°42'$

(ii) Similarly for $x^2 + y^2 = 16$

$$2x + 2y\frac{dy}{dx} = 0 \therefore \frac{dy}{dx} = -\frac{x}{y} = -\frac{1{\cdot}657}{3{\cdot}641} = -0{\cdot}4551$$

$\tan\theta_2 = -0{\cdot}4551 \therefore \theta_2 = -24°28'$

Finally,

$$\theta = \theta_1 - \theta_2 = 47°42' - (-24°28')$$
$$= 47°42' + 24°28'$$
$$= 72°10'$$

**23**

That just about covers all there is to know about finding tangents and normals to a curve. We now look at another application of differentiation.

**Curvature**

The value of $\frac{dy}{dx}$ at any point on a curve denotes the slope (or direction) of the curve at that point. Curvature is concerned with how quickly the curve is changing direction in the neighbourhood of that point.

Let us see in the next few frames what it is all about.

**24**

Let us first consider the change in direction of a curve $y = f(x)$ between the points P and Q as shown. The direction of a curve is measured by the slope of the tangent.

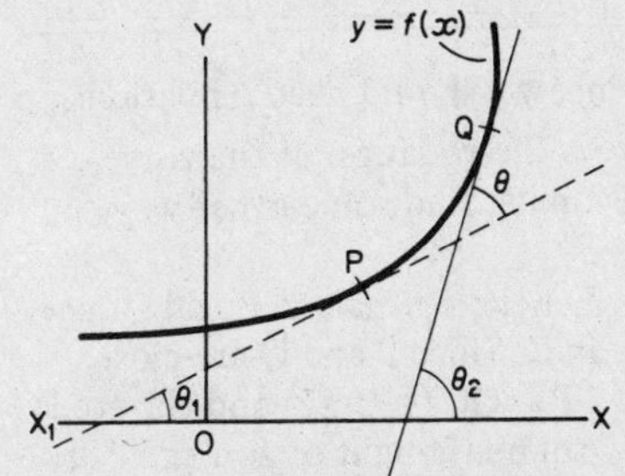


Slope at P $= \tan \theta_1 = \left\{\frac{dy}{dx}\right\}_P$

Slope at Q $= \tan \theta_2 = \left\{\frac{dy}{dx}\right\}_Q$

These can be calculated, knowing the equation of the curve.

From the values of $\tan \theta_1$ and $\tan \theta_2$, the angles $\theta_1$ and $\theta_2$ can be found from tables. Then from the diagram, $\theta = \theta_2 - \theta_1$.

If we are concerned with how fast the curve is bending, we must consider not only the change in direction from P to Q, but also the length of ...................... which provides this change in direction.

**25**

the arc PQ

i.e. we must know the change of direction, but also how far along the curve we must go to obtain this change in direction.

Now let us consider the two points, P and Q, near to each other, so that PQ is a small arc ($= \delta s$). The change in direction will not be great, so that if $\theta$ is the slope at P, then the angle of slope at Q can be put as $\theta + \delta\theta$.

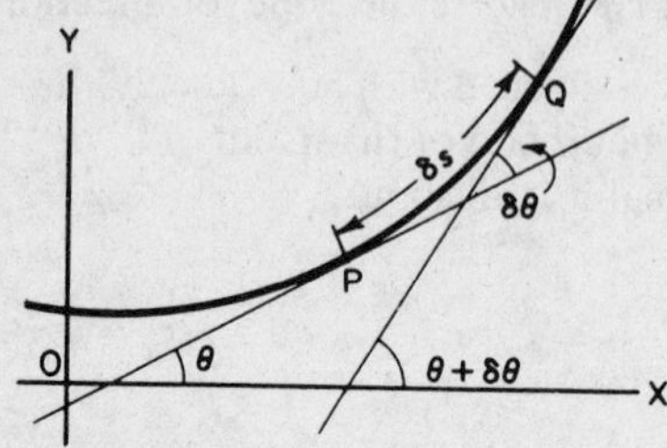


The change in direction from P to Q is therefore $\delta\theta$.
The length of arc from P to Q is $\delta s$.
The average rate of change of direction with arc from P to Q is

$$\frac{\text{the change in direction from P to Q}}{\text{the length of arc from P to Q}} = \frac{\delta\theta}{\delta s}$$

This could be called the average curvature from P to Q. If Q now moves down towards P, i.e. $\delta s \to 0$, we finally get $\frac{d\theta}{ds}$, which is the *curvature* at P. It tells us how quickly the curve is bending in the immediate neighbourhood of P.

**26**

In practice, it is difficult to find $\frac{d\theta}{ds}$ since we should need a relationship between $\theta$ and $s$, and usually all we have is the equation of the curve, $y = f(x)$ and the co-ordinates of P. So we must find some other way round it.

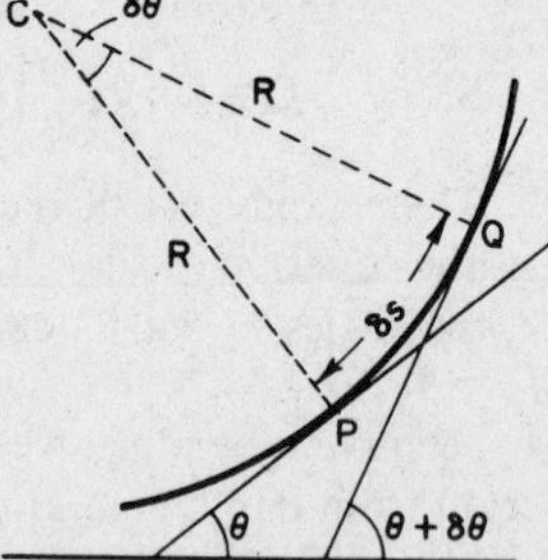


Let the normals at P and Q meet in C. Since P and Q are close, CP $\triangleq$ QC (=R say) and the arc PQ can be thought of as a small arc of a circle of radius R. Note that PCQ = $\delta\theta$ (for if the tangent turns through $\delta\theta$, the radius at right angles to it will also turn through the same angle).

You remember that the arc of a circle of radius $r$ which subtends an angle $\theta$ radians at the centre is given by arc = $r\theta$. So, in the diagram above,
arc PQ = $\delta s$ = ......................

**27**

$$\boxed{\text{arc } PQ = \delta s = R\delta\theta}$$

$$\delta s = R\delta\theta \quad \therefore \frac{\delta\theta}{\delta s} = \frac{1}{R}$$

If $\delta s \to 0$, this becomes $\frac{d\theta}{ds} = \frac{1}{R}$ which is the curvature at P.

That is, we can state the curvature at a point, in terms of the radius R of the circle we have considered. This is called the *radius of curvature*, and the point C the *centre of curvature*.

So we have now found that we can obtain the curvature $\frac{d\theta}{ds}$ if we have some way of finding the radius of curvature R.

If R is large, is the curvature large or small?

If you think 'large', move on to frame 28.
If you think 'small' turn on to frame 29.

**28**

Your answer was : 'If R is large, the curvature is large.'

□□□□□□□□□□□□□□□□□□□□□□□□□□□□□□□□□□□□□□□

This is not so. For the curvature $= \frac{d\theta}{ds}$ and we have just shown that $\frac{d\theta}{ds} = \frac{1}{R}$. R is the denominator, so that a large value for R gives a small value for the fraction $\frac{1}{R}$ and hence a small value for the curvature.

You can see it this way. If you walk round a circle with a large radius R, then the curve is relatively a gentle one, i.e. small value of curvature, but if R is small, the curve is more abrupt.

So once again, if R is large, the curvature is .....................

**29** If R is large, the curvature is small

Correct, since the curvature $\frac{d\theta}{ds} = \frac{1}{R}$

□□□□□□□□□□□□□□□□□□□□□□□□□□□□□□□□□□□□□□□□□□□□□□□□

In practice, we often indicate the curvature in terms of the radius of curvature R, since this is something we can appreciate.

Let us consider our two points P and Q again. Since $\delta s$ is very small, there is little difference between the arc PQ and the chord PQ, or between the direction of the chord and that of the tangent.

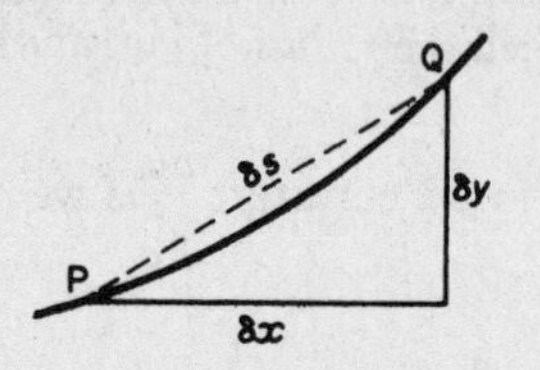


So, when $\delta s \to 0$, $\frac{dy}{dx} = \tan\theta$

$\frac{dx}{ds} = \cos\theta$

$\frac{dy}{dx} = \tan\theta$. Differentiate with respect to $s$.

Then
$$\frac{d}{ds}\left\{\frac{dy}{dx}\right\} = \frac{d}{ds}\left\{\tan\theta\right\}$$

$$\frac{d}{dx}\left\{\frac{dy}{dx}\right\}.\frac{dx}{ds} = \frac{d}{d\theta}\left\{\tan\theta\right\}.\frac{d\theta}{ds}$$

$$\therefore \frac{d^2y}{dx^2}\cos\theta = \sec^2\theta\frac{d\theta}{ds}$$

$$\sec^3\theta\frac{d\theta}{ds} = \frac{d^2y}{dx^2}$$

Now $\sec^3\theta = (\sec^2\theta)^{3/2} = (1 + \tan^2\theta)^{3/2} = \left\{1 + \left(\frac{dy}{dx}\right)^2\right\}^{3/2}$

$$\therefore \frac{d\theta}{ds} = \frac{1}{R} = \frac{\frac{d^2y}{dx^2}}{\left\{1 + \left(\frac{dy}{dx}\right)^2\right\}^{3/2}} \qquad \therefore R = \frac{\left\{1 + \left(\frac{dy}{dx}\right)^2\right\}^{3/2}}{\frac{d^2y}{dx^2}}$$

Now we have got somewhere. For knowing the equation $y = f(x)$ of the curve, we can calculate the first and second differential coefficients at the point P and substitute these values in the formula for R.

This is an important result. Copy it down and learn it. You may never be asked to prove it, but you will certainly be expected to know it and to apply it.

*So now for one or two examples.* *Turn on to frame 30.*

**30**

*Example 1.* Find the radius of curvature for the hyperbola $xy = 4$ at the point $x = 2, y = 2$.

$$R = \frac{\left\{1 + \left(\frac{dy}{dx}\right)^2\right\}^{3/2}}{\frac{d^2y}{dx^2}}$$

So all we need to find are $\frac{dy}{dx}$ and $\frac{d^2y}{dx^2}$ at $(2,2)$

$$xy = 4 \quad \therefore\ y = \frac{4}{x} = 4x^{-1} \quad \therefore\ \frac{dy}{dx} = -4x^{-2} = \frac{-4}{x^2}$$

and
$$\frac{d^2y}{dx^2} = 8x^{-3} = \frac{8}{x^3}$$

At $(2,2)$
$$\frac{dy}{dx} = -\frac{4}{4} = -1; \quad \frac{d^2y}{dx^2} = \frac{8}{8} = 1$$

$$\therefore\ R = \frac{\{1 + (-1)^2\}^{3/2}}{1} = \frac{\{1+1\}^{3/2}}{1} = (2)^{3/2} = 2\sqrt{2}$$

$$\therefore\ R = 2\sqrt{2} = \underline{2{\cdot}828 \text{ units.}}$$

*There we are. Another example on frame 31.*

**31**

*Example 2.* If $y = x + 3x^2 - x^3$, find R at $x = 0$.

$$R = \frac{\left\{1 + \left(\frac{dy}{dx}\right)^2\right\}^{3/2}}{\frac{d^2y}{dx^2}}$$

$$\frac{dy}{dx} = 1 + 6x - 3x^2 \quad \therefore\ \text{At } x = 0, \frac{dy}{dx} = 1 \qquad \therefore\ \left(\frac{dy}{dx}\right)^2 = 1$$

$$\frac{d^2y}{dx^2} = 6 - 6x \qquad \therefore\ \text{At } x = 0, \frac{d^2y}{dx^2} = 6$$

$$R = \frac{\{1+1\}^{3/2}}{6} = \frac{2^{3/2}}{6} = \frac{2\sqrt{2}}{6} = \frac{\sqrt{2}}{3}$$

$$\therefore\ \underline{R = 0{\cdot}471 \text{ units}}$$

Now you do this one:

Find the radius of curvature of the curve $y^2 = \frac{x^3}{4}$ at the point $\left(1, \frac{1}{2}\right)$

*When you have finished, check with the solution on frame 32.*

# 32

R = 5·21 units

Here is the solution in full.

$$y^2 = \frac{x^3}{4} \quad \therefore \; 2y\frac{dy}{dx} = \frac{3x^2}{4} \quad \therefore \; \frac{dy}{dx} = \frac{3x^2}{8y}$$

$$\therefore \text{ At } \left(1, \frac{1}{2}\right), \frac{dy}{dx} = \frac{3}{4} \quad \therefore \; \left(\frac{dy}{dx}\right)^2 = \frac{9}{16}$$

$$\frac{dy}{dx} = \frac{3x^2}{8y} \quad \therefore \; \frac{d^2y}{dx^2} = \frac{8y\,(6x) - 3x^2\,8\frac{dy}{dx}}{64\,y^2}$$

$$\therefore \text{ At } \left(1, \frac{1}{2}\right), \frac{d^2y}{dx^2} = \frac{24 - 24.\frac{3}{4}}{16} = \frac{24 - 18}{16} = \frac{3}{8}$$

$$R = \frac{\left\{1 + \left(\frac{dy}{dx}\right)^2\right\}^{3/2}}{\frac{d^2y}{dx^2}} = \frac{\left\{1 + \frac{9}{16}\right\}^{3/2}}{\frac{3}{8}} = \frac{\left\{\frac{25}{16}\right\}^{3/2}}{\frac{3}{8}} = \frac{8}{3}\cdot\frac{125}{64} = \frac{125}{24} = 5\frac{5}{24}$$

$\therefore$ R = 5·21 units

# 33

Of course, the equation of the curve could be an implicit function, as in the last example, or a pair of parametric equations.

*e.g.* If $x = \theta - \sin\theta$ and $y = 1 - \cos\theta$, find R when $\theta = 60° = \frac{\pi}{3}$

$$x = \theta - \sin\theta \quad \therefore \frac{dx}{d\theta} = 1 - \cos\theta$$

$$y = 1 - \cos\theta \quad \therefore \frac{dy}{d\theta} = \sin\theta$$

$$\frac{dy}{dx} = \frac{dy}{d\theta}\cdot\frac{d\theta}{dx}$$

$$\therefore \frac{dy}{dx} = \sin\theta.\frac{1}{1 - \cos\theta} = \frac{\sin\theta}{1 - \cos\theta}$$

$$\text{At } \theta = 60°, \sin\theta = \frac{\sqrt{3}}{2}, \cos\theta = \frac{1}{2}; \frac{dy}{dx} = \frac{\sqrt{3}}{1}$$

$$\frac{d^2y}{dx^2} = \frac{d}{dx}\left\{\frac{\sin\theta}{1 - \cos\theta}\right\} = \frac{d}{d\theta}\left\{\frac{\sin\theta}{1 - \cos\theta}\right\}\cdot\frac{d\theta}{dx}$$

$$= \frac{(1 - \cos\theta)\cos\theta - \sin\theta.\sin\theta}{(1 - \cos\theta)^2}\cdot\frac{1}{1 - \cos\theta}$$

$$= \frac{\cos\theta - \cos^2\theta - \sin^2\theta}{(1 - \cos\theta)^3} = \frac{\cos\theta - 1}{(1 - \cos\theta)^3} = \frac{-1}{(1 - \cos\theta)^2}$$

$$\therefore \text{ At } \theta = 60°, \frac{d^2y}{dx^2} = \frac{-1}{(1 - \frac{1}{2})^2} = \frac{-1}{\frac{1}{4}} = -4$$

$$\therefore R = \frac{\{1 + 3\}^{3/2}}{-4} = \frac{2^3}{-4} = \frac{8}{-4} = -2 \qquad \therefore \; R = -2 \text{ units}$$

**34**

You notice in this last example that the value of R is negative. This merely indicates which way the curve is bending. Since R is a physical length, then for all practical purposes, R is taken as 2 units long.

If the value of R is to be used in further calculations however, it is usually necessary to maintain the negative sign. You will see an example of this later.

Here is one for you to do in just the same way as before:

Find the radius of curvature of the curve $x = 2\cos^3\theta, y = 2\sin^3\theta$, at the point for which $\theta = \frac{\pi}{4} = 45°$.

*Work through it and then go to frame 35 to check your work.*

**35**

*Result:*

$$\boxed{R = 3 \text{ units}}$$

For $\quad x = 2\cos^3\theta \quad \therefore \frac{dx}{d\theta} = 6\cos^2\theta(-\sin\theta) = -6\sin\theta\cos^2\theta$

$$y = 2\sin^3\theta \quad \therefore \frac{dy}{d\theta} = 6\sin^2\theta\cos\theta$$

$$\frac{dy}{dx} = \frac{dy}{d\theta}\cdot\frac{d\theta}{dx} = \frac{6\sin^2\theta\cos\theta}{-6\sin\theta\cos^2\theta} = -\frac{\sin\theta}{\cos\theta} = -\tan\theta$$

$$\text{At } \theta = 45°, \frac{dy}{dx} = -1 \quad \therefore \left(\frac{dy}{dx}\right)^2 = 1$$

Also $\quad \frac{d^2y}{dx^2} = \frac{d}{dx}\left\{-\tan\theta\right\} = \frac{d}{d\theta}\left\{-\tan\theta\right\}\frac{d\theta}{dx} = \frac{-\sec^2\theta}{-6\sin\theta\cos^2\theta}$

$$= \frac{1}{6\sin\theta\cos^4\theta}$$

$$\therefore \text{ At } \theta = 45°, \frac{d^2y}{dx^2} = \frac{1}{6(\frac{1}{\sqrt{2}})(\frac{1}{4})} = \frac{4\sqrt{2}}{6} = \frac{2\sqrt{2}}{3}$$

$$R = \frac{\left\{1 + \left(\frac{dy}{dx}\right)^2\right\}^{3/2}}{\frac{d^2y}{dx^2}} = \frac{\left\{1+1\right\}^{3/2}}{\frac{2\sqrt{2}}{3}} = \frac{3}{2\sqrt{2}}\, 2^{3/2}$$

$$= \frac{3.2\sqrt{2}}{2\sqrt{2}} = 3$$

$$\therefore \underline{R = 3 \text{ units}}$$

**36**

**Centre of curvature.** To get a complete picture, we need to know also the position of the centre of the circle of curvature for the point $P(x_1, y_1)$.

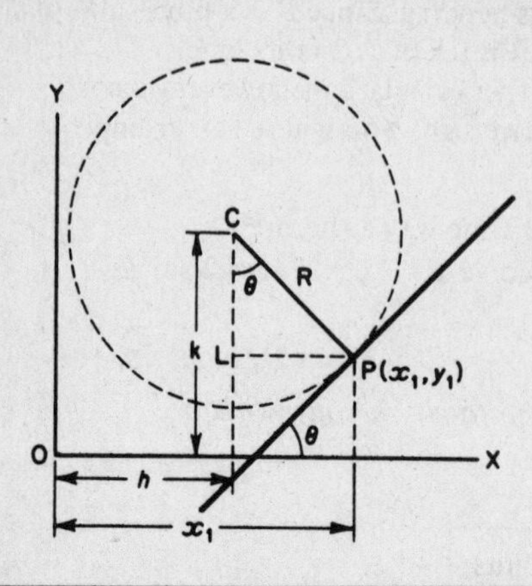


If the centre C is the point $(h, k)$, we can see from the diagram that:

$h = x_1 - LP = x_1 - R \sin\theta$

$k = y_1 + LC = y_1 + R \cos\theta$

That is, $\left\{ \begin{array}{l} h = x_1 - R \sin\theta \\ k = y_1 + R \cos\theta \end{array} \right\}$

where $x_1$ and $y_1$ are the co-ordinates of P, R is the radius of curvature at P, $\theta$ is the angle of slope at P, i.e. $\tan\theta = \left\{ \dfrac{dy}{dx} \right\}_P$

**37**

*Example.* Find the radius of curvature and the co-ordinates of the centre of curvature of the curve $y = \dfrac{11 - 4x}{3 - x}$ at the point $(2, 3)$.

$$\frac{dy}{dx} = \frac{(3-x)(-4) - (11-4x)(-1)}{(3-x)^2} = \frac{-12 + 4x + 11 - 4x}{(3-x)^2} = \frac{-1}{(3-x)^2}$$

$$\therefore \text{ At } x = 2, \quad \frac{dy}{dx} = \frac{-1}{1} = -1 \quad \therefore \left(\frac{dy}{dx}\right)^2 = 1$$

$$\frac{d^2y}{dx^2} = \frac{d}{dx}\{-(3-x)^{-2}\} = 2(3-x)^{-3}(-1) = \frac{-2}{(3-x)^3}$$

$$\therefore \text{ At } x = 2, \quad \frac{d^2y}{dx^2} = \frac{-2}{1} = -2$$

$$R = \frac{\left\{1 + \left(\frac{dy}{dx}\right)^2\right\}^{3/2}}{\frac{d^2y}{dx^2}} = \frac{\{1 + 1\}^{3/2}}{-2} = \frac{2\sqrt{2}}{-2} = -\sqrt{2}$$

$$\underline{R = -\sqrt{2}}$$

Now before we find the centre of curvature $(h, k)$ we must find the angle of slope $\theta$ from the fact that $\tan\theta = \dfrac{dy}{dx}$ at P.

i.e. $\tan\theta = -1 \quad \therefore \theta = -45°$ ($\theta$ measured between $\pm 90°$)

$\therefore \sin\theta =$ .............. and $\cos\theta =$ ..............

**38**

$\theta = -45°$ | $\sin\theta = -\frac{1}{\sqrt{2}}$ | $\cos\theta = \frac{1}{\sqrt{2}}$

□□□□□□□□□□□□□□□□□□□□□□□□□□□□□□□□□□□□□□□□□□

So we have: $x_1 = 2, y_1 = 3$

$$R = -\sqrt{2}$$

$$\sin\theta = -\frac{1}{\sqrt{2}}, \cos\theta = \frac{1}{\sqrt{2}}$$

$$\therefore h = x_1 - R\sin\theta = 2 - (-\sqrt{2})\left(-\frac{1}{\sqrt{2}}\right) = 2 - 1 = 1, h = 1$$

$$k = y_1 + R\cos\theta = 3 + (-\sqrt{2})\left(\frac{1}{\sqrt{2}}\right) = 3 - 1 = 2, k = 2$$

$\therefore$ centre of curvature C is the point (1,2)

*NOTE:* If, by chance, the calculated value of R is negative, the minus sign must be included when we substitute for R in the expressions for $h$ and $k$.

*Next frame for a final example.*

**39**

*Example.* Find the radius of curvature and the centre of curvature for the curve $y = \sin^2\theta, x = 2\cos\theta$, at the point for which $\theta = \frac{\pi}{3}$.

Before we rush off and deal with this one, let us heed an important *WARNING.* You will remember that the centre of curvature $(h, k)$ is given by

$$\left.\begin{aligned} h &= x_1 - R\sin\theta \\ k &= y_1 + R\cos\theta \end{aligned}\right\} \text{ and in these expressions}$$

$\theta$ is the angle of slope of the curve at the point being considered

$$\text{i.e. } \tan\theta = \left\{\frac{dy}{dx}\right\}_P$$

Now, in the problem stated above, $\theta$ is a parameter and *not* the angle of slope at any particular point. In fact, if we proceed with our usual notation, we shall be using $\theta$ to stand for two completely different things – and that can be troublesome, to say the least.

So the safest thing to do is this. *Where you have to find the centre of curvature of a curve given in parametric equations involving $\theta$, change the symbol of the parameter to something other than $\theta$.* Then you will be safe. The trouble occurs only when we find C, not when we are finding R only.

**40**

So, in this case, we will re-write the problem thus:

Find the radius of curvature and the centre of curvature for the curve $y = \sin^2 t, x = 2\cos t$, at the point for which $t = \frac{\pi}{3}$

Start off by finding the radius of curvature only. Then check your result so far with the solution given in the next frame before setting out to find the centre of curvature.

**41**

| R = −2·795, i.e. 2·795 units |
|---|

Here is the working.

$$y = \sin^2 t \quad \therefore \frac{dy}{dt} = 2\sin t\cos t$$

$$x = 2\cos t \therefore \frac{dx}{dt} = -2\sin t$$

$$\frac{dy}{dx} = \frac{dy}{dt}\cdot\frac{dt}{dx} = \frac{2\sin t\cos t}{-2\sin t} = -\cos t$$

$$\text{At } t = 60°, \frac{dy}{dx} = -\cos 60° = -\frac{1}{2} \quad \therefore \frac{dy}{dx} = -\frac{1}{2}$$

$$\text{Also } \frac{d^2y}{dx^2} = \frac{d}{dx}\left\{-\cos t\right\} = \frac{d}{dt}\left\{-\cos t\right\}.\frac{dt}{dx} = \frac{\sin t}{-2\sin t} = -\frac{1}{2}$$

$$\therefore \frac{d^2y}{dx^2} = -\frac{1}{2}$$

$$R = \frac{\left\{1 + \left(\frac{dy}{dx}\right)^2\right\}^{3/2}}{\frac{d^2y}{dx^2}} = \frac{\left\{1 + \frac{1}{4}\right\}^{3/2}}{-\frac{1}{2}} = -2\left\{\frac{5}{4}\right\}^{3/2}$$

$$= \frac{-10\sqrt{5}}{8} = \frac{-5\sqrt{5}}{4} = \frac{-5}{4}(2{\cdot}2361)$$

$$= \frac{-11{\cdot}1805}{4} = -2{\cdot}7951$$

$$\underline{R = -2{\cdot}795}$$

*All correct so far? Move on to the next frame, then.*

**42**

Now to find the centre of curvature $(h, k)$

$$h = x_1 - R\sin\theta$$
$$k = y_1 + R\cos\theta$$

where $\tan\theta = \dfrac{dy}{dx} = -\dfrac{1}{2}$ $\therefore$ $\theta = -26°34'$ ($\theta$ between $\pm$ 90°)

$\therefore$ $\sin(-26°34') = -0{\cdot}4472$; $\cos(-26°34') = 0{\cdot}8944$

Also $x_1 = 2\cos 60° = 2.\dfrac{1}{2} = 1$

$$y_1 = \sin^2 60° = \left\{\frac{\sqrt{3}}{2}\right\}^2 = \frac{3}{4}$$

and you have already proved that $R = -2{\cdot}795$.

What then are the co-ordinates of the centre of curvature?

*Calculate them and when you have finished, move on to the next frame.*

**43**

*Results:* $\boxed{h = -0{\cdot}25;\ k = -1{\cdot}75}$

For: $h = 1 - (-2{\cdot}795)(-0{\cdot}4472)$ | $0{\cdot}4464$, $\bar{1}{\cdot}6505$, $0{\cdot}0969$

$= 1 - 1{\cdot}250$

$\therefore$ $\underline{h = -0{\cdot}25}$

and $k = 0{\cdot}75 + (-2{\cdot}795)(0{\cdot}8944)$ | $0{\cdot}4464$, $\bar{1}{\cdot}9515$, $0{\cdot}3979$

$= 0{\cdot}75 - 2{\cdot}50$

$\therefore$ $\underline{k = -1{\cdot}75}$

Therefore, the centre of curvature is the point $\underline{(-0{\cdot}25, -1{\cdot}75)}$

This brings us to the end of this particular programme. If you have followed it carefully and carried out the exercises set, you must know quite a lot about the topics we have covered. So turn on now and work the Test Exercise. It is all very straightforward.

**44** Test Exercise – VIII

*Answer all questions*

1. Find the angle between the curves $x^2 + y^2 = 4$ and $5x^2 + y^2 = 5$ at their point of intersection for which $x$ and $y$ are positive.

2. Find the equations of the tangent and normal to the curve $y^2 = 11 - \dfrac{10}{4-x}$ at the point (6, 4).

3. The parametric equations of a function are $x = 2\cos^3\theta, y = 2\sin^3\theta$. Find the equation of the normal at the point for which $\theta = \dfrac{\pi}{4} = 45°$.

4. If $x = 1 + \sin 2\theta, y = 1 + \cos\theta + \cos 2\theta$, find the equation of the tangent at $\theta = 60°$.

5. Find the radius of curvature and the co-ordinates of the centre of curvature at the point $x = 4$ on the curve whose equation is $y = x^2 + 5\ln x - 24$.

6. Given that $x = 1 + \sin\theta, y = \sin\theta - \frac{1}{2}\cos 2\theta$, show that $\dfrac{d^2y}{dx^2} = 2$. Find the radius of curvature and the centre of curvature for the point on this curve where $\theta = 30°$.

*Now you are ready for the next programme.*

## Further Problems – VIII

1. Find the equation of the normal to the curve $y = \dfrac{2x}{x^2 + 1}$ at the point (3, 0·6) and the equation of the tangent at the origin.

2. Find the equations of the tangent and normal to the curve $4x^3 + 4xy + y^2 = 4$ at (0, 2), and find the co-ordinates of a further point of intersection of the tangent and the curve.

3. Obtain the equations of the tangent and normal to the ellipse $\dfrac{x^2}{169} + \dfrac{y^2}{25} = 1$ at the point $(13 \cos \theta, 5 \sin \theta)$. If the tangent and normal meet the $x$-axis at the points T and N respectively, show that ON.OT is constant, O being the origin of co-ordinates.

4. If $x^2y + xy^2 - x^3 - y^3 + 16 = 0$, find $\dfrac{dy}{dx}$ in its simplest form. Hence find the equation of the normal to the curve at the point (1,3).

5. Find the radius of curvature of the catenary $y = c \cosh\left(\dfrac{x}{c}\right)$ at the point $(x_1, y_1)$.

6. If $2x^2 + y^2 - 6y - 9x = 0$, determine the equation of the normal to the curve at the point (1,7).

7. Show that the equation of the tangent to the curve $x = 2a \cos^3 t$, $y = a \sin^3 t$, at any point $P(0 \leqslant t \leqslant \frac{\pi}{2})$ is
$$x \sin t + 2y \cos t - 2a \sin t \cos t = 0$$
If the tangent at P cuts the $y$-axis at Q, determine the area of the triangle POQ.

8. Find the equation of the normal at the point $x = a \cos \theta, y = b \sin \theta$, of the ellipse $\dfrac{x^2}{a^2} + \dfrac{y^2}{b^2} = 1$. The normal at P on the ellipse meets the major axis of the ellipse at N. Show that the locus of the mid-point of PN is an ellipse and state the lengths of its principal axes.

9. For the point where the curve $y = \frac{x - x^2}{1 + x^2}$ passes through the origin, determine:

   (i) the equations of the tangent and normal to the curve,
   (ii) the radius of curvature,
   (iii) the co-ordinates of the centre of curvature.

10. In each of the following cases, find the radius of curvature and the co-ordinates of the centre of curvature for the point stated.

    (i) $\frac{x^2}{25} + \frac{y^2}{16} = 1$ at $(0, 4)$
    (ii) $y^2 = 4x - x^2 - 3$ at $x = 2{\cdot}5$
    (iii) $y = 2 \tan \theta, x = 3 \sec \theta$ at $\theta = 45°$

11. Find the radius of curvature at the point $(1, 1)$ on the curve $x^3 - 2xy + y^3 = 0$.

12. If $3ay^2 = x(x - a)^2$ with $a > 0$, prove that the radius of curvature at the point $(3a, 2a)$ is $\frac{50a}{3}$.

13. If $x = 2\theta - \sin 2\theta$ and $y = 1 - \cos 2\theta$, show that $\frac{dy}{dx} = \cot \theta$ and that $\frac{d^2y}{dx^2} = \frac{-1}{4 \sin^4 \theta}$. If $\rho$ is the radius of curvature at any point on the curve, show that $\rho^2 = 8y$.

14. Find the radius of curvature of the curve $2x^2 + y^2 - 6y - 9x = 0$ at the point $(1, 7)$.

15. Prove that the centre of curvature $(h, k)$ at the point $P(at^2, 2at)$ on the parabola $y^2 = 4ax$ has co-ordinates $h = 2a + 3at^2$, $k = -2at^3$.

16. If $\rho$ is the radius of curvature at any point P on the parabola $x^2 = 4ay$, S is the point $(0, a)$, show that $\rho = 2\sqrt{[(SP)^3/SO]}$, where O is the origin of co-ordinates.

17. The parametric equations of a curve are $x = \cos t + t \sin t$, $y = \sin t - t \cos t$. Determine an expression for the radius of curvature $(\rho)$ and for the co-ordinates $(h, k)$ of the centre of curvature in terms of $t$.

18. Find the radius of curvature and the co-ordinates of the centre of curvature of the curve $y = 3 \ln x$, at the point where it meets the $x$-axis.

19. Show that the numerical value of the radius of curvature at the point $(x_1, y_1)$ on the parabola $y^2 = 4ax$ is $\dfrac{2(a + x_1)^{3/2}}{a^{1/2}}$. If C is the centre of curvature at the origin O and S is the point $(a, O)$, show that OC = 2(OS).

20. The equation of a curve is $4y^2 = x^2(2 - x^2)$.
    (i) Determine the equations of the tangents at the origin.
    (ii) Show that the angle between these tangents is $\tan^{-1}(2\sqrt{2})$.
    (iii) Find the radius of curvature at the point $(1, 1/2)$.

# Programme 9

## DIFFERENTIATION APPLICATIONS

### PART 2

## 1

**Inverse trigonometrical functions**

You already know that the symbol $\sin^{-1} x$ (sometimes referred to as 'arcsine $x$') indicates 'the angle whose sine is the value $x$'.

e.g. $\sin^{-1} 0{\cdot}5$ = the angle whose sine is the value 0·5
$= 30°$

There are, of course, many angles whose sine is 0·5, e.g. 30°, 150°, 390°, 510°, 750°, 870°, .. .. etc., so would it not be true to write that $\sin^{-1} 0{\cdot}5$ was any one (or all) of these possible angles?

The answer is *no*, for the simple reason that we have been rather lax in our definition of $\sin^{-1} x$ above. We should have said that $\sin^{-1} x$ indicates the *principal* value of the angle whose sine is the value $x$; to see what we mean by that, move on to frame 2.

## 2

The *principal* value of $\sin^{-1} 0{\cdot}5$ is the numerically smallest angle (measured between 0° and 180°, or 0° and −180°) whose sine is 0·5. Note that in this context, we quote the angle as being measured from 0° to 180°, or from 0° to −180°.

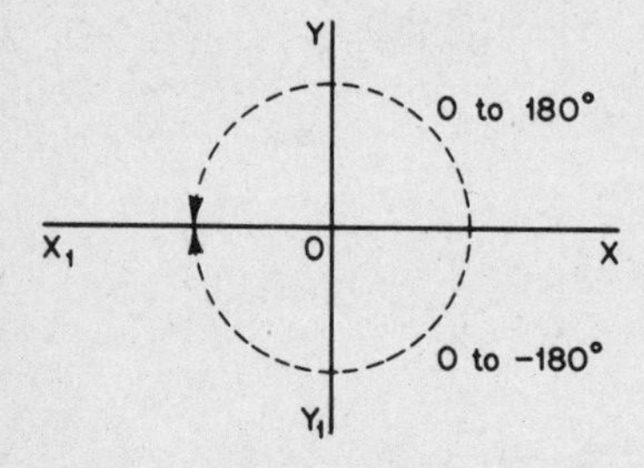


In this range, there are two angles whose sine is 0·5, i.e. 30° and 150°. The *principal* value of the angle is the one nearer to the positive OX direction, i.e. 30°.

$$\sin^{-1} 0{\cdot}5 = 30°$$

and no other angle!

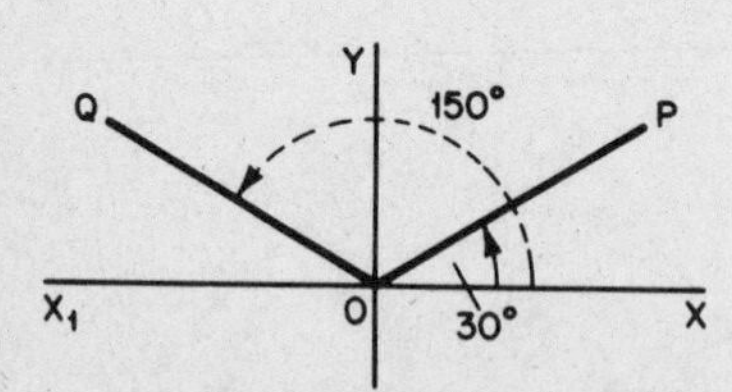


Similarly, if $\sin\theta = 0{\cdot}7071$, what is the principal value of the angle $\theta$?

*When you have decided, turn on.*

**3**

Principal value of $\theta = 45°$

for: $\sin\theta = 0{\cdot}7071$ $\therefore$ In the range 0° to 180°, or 0° to −180°, the possible angles are 45° and 135°.

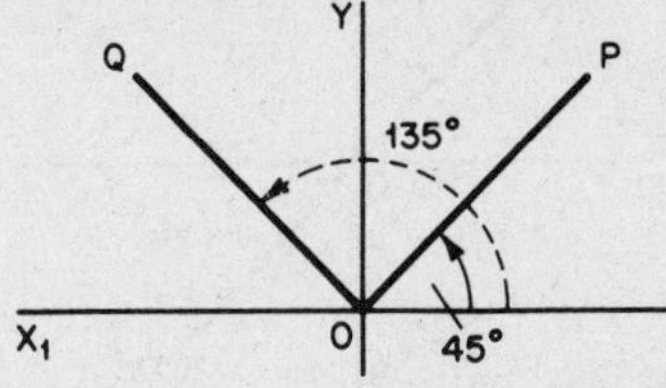


The principal value of the angle is the one nearer to the positive OX axis, i.e. 45°.

$\sin^{-1} 0{\cdot}7071 = 45°$

□□□□□□□□□□□□□□□□□□□□□□□□□□□□□□□□□□□□□□□□□□□

In the same way, we can find the value of $\tan^{-1}\sqrt{3}$.
If $\tan\theta = \sqrt{3} = 1{\cdot}7321$, then $\theta = 60°$ or 240°. Quoted in the range 0° to 180° or 0° to −180°, these angles are $\theta = 60°$ or −120°.

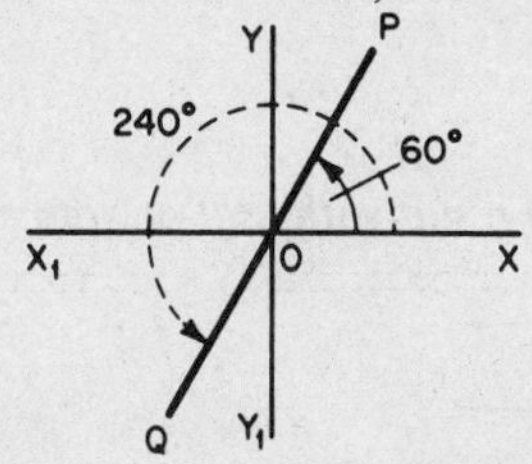


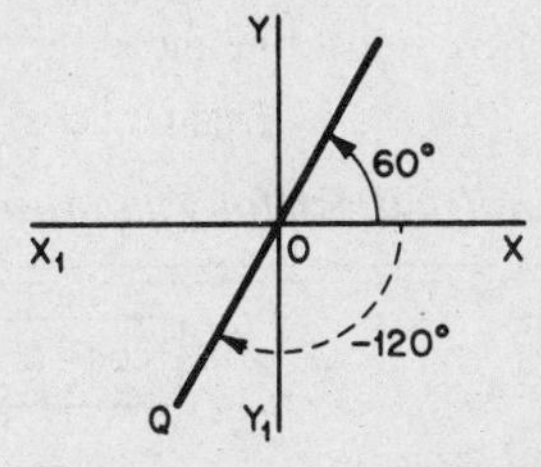


The principal value of the angle is the one nearer to the positive OX direction, i.e. in this case, $\tan^{-1}\sqrt{3}$ = .......................

**4**

$\tan^{-1}\sqrt{3} = 60°$

Now let us consider the value of $\cos^{-1} 0{\cdot}8192$.

From the cosine tables, we find one angle whose cosine is 0·8192 to be 35°. The other is therefore 360° − 35°, i.e. 325° (or −35°).

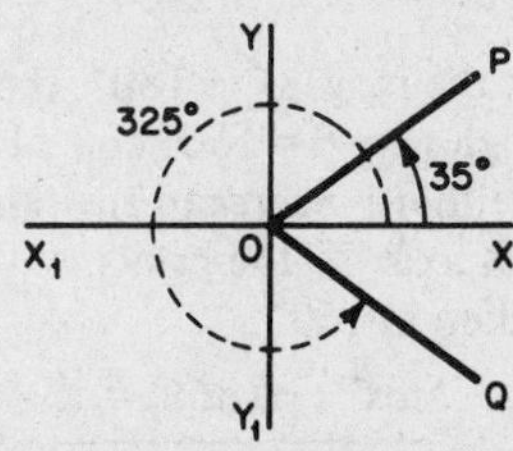


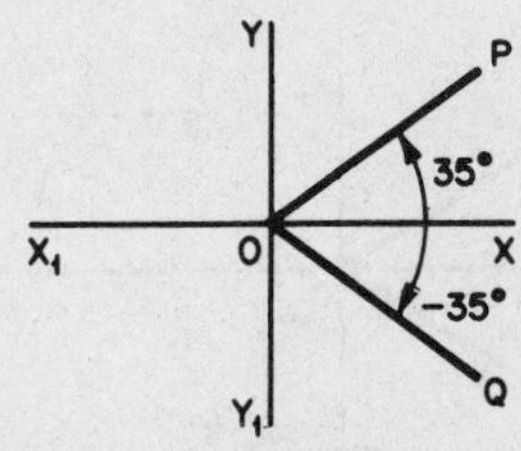


Of course, neither is *nearer* to OX: they are symmetrically placed. In such a situation as this, it is the accepted convention that the positive angle is taken as the principal value, i.e. 35°, $\therefore$ $\cos^{-1} 0{\cdot}8192 = 35°$

So, on your own, find $\tan^{-1}(-1)$. *Then on to frame 5.*

## 5

$$\boxed{\tan^{-1}(-1) = -45^\circ}$$

For, if $\tan\theta = -1$, $\theta = 135^\circ$ or $315^\circ$

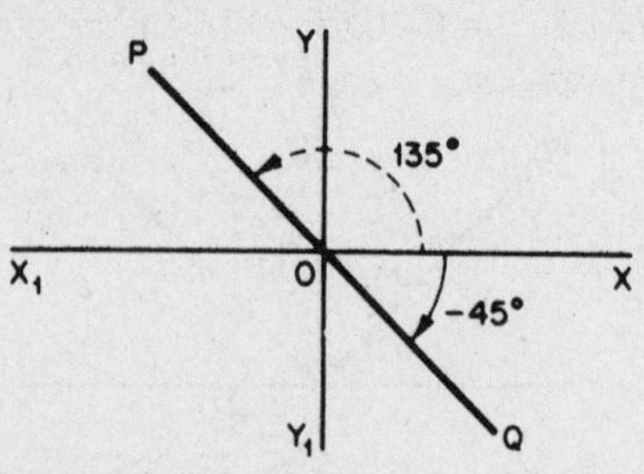


In the range $0^\circ$ to $\pm 180^\circ$, these angles are $135^\circ$ and $-45^\circ$.
The one nearer to the OX axis is $-45^\circ$. $\therefore$ Principal value $= -45^\circ$.

$$\underline{\tan^{-1}(-1) = -45^\circ}$$

Now here is just one more:

Evaluate $\cos^{-1}(-0{\cdot}866)$

*Work through it carefully and then check your result with that on frame 6.*

## 6

$$\boxed{\cos^{-1}(-0{\cdot}866) = 150^\circ}$$

For we have:

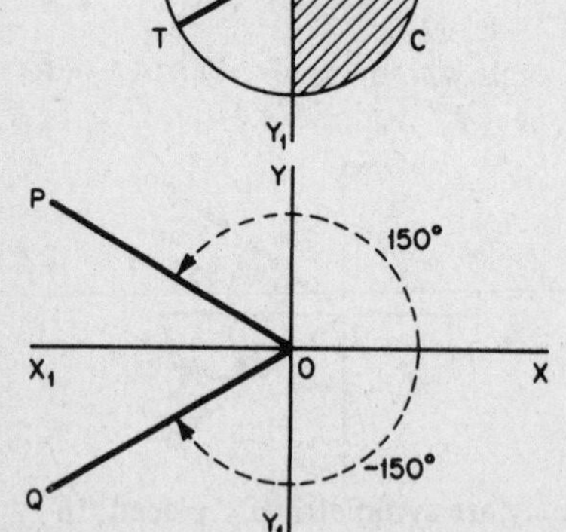


$\cos E = 0{\cdot}866 \quad \therefore E = 30^\circ$

$\therefore \theta \;= 150^\circ$ or $210^\circ$

In the range $0^\circ \pm 180^\circ$, these angles are $\theta = 150^\circ$ and $-150^\circ$
Neither is *nearer* to the positive OX axis. So the principal value is taken as $150^\circ$.

$$\underline{\cos^{-1}(-0{\cdot}866) = 150^\circ}$$

So to sum up, the inverse trig. functions, $\sin^{-1}x$, $\cos^{-1}x$, $\tan^{-1}x$ indicate the *p*.................... *v*.................... of the angles having the value of the trig. ratio stated.

**7**

principal value

**Differentiation of inverse trig. functions**

$\mathrm{Sin}^{-1}x$, $\cos^{-1}x$, $\tan^{-1}x$ depend, of course, on the values assigned to $x$. They are therefore functions of $x$ and we may well be required to find their differential coefficients. So let us deal with them in turn.

(i) Let $y = \sin^{-1}x$. We have to find $\dfrac{dy}{dx}$

First of all, write this inverse statement as a direct statement.

$$y = \sin^{-1}x \quad \therefore\ x = \sin y$$

Now we can differentiate this with respect to $y$ and obtain $\dfrac{dx}{dy}$

$$\frac{dx}{dy} = \cos y \quad \therefore\ \frac{dy}{dx} = \text{...........}$$

**8**

$$\frac{dy}{dx} = \frac{1}{\cos y}$$

Now we express $\cos y$ in terms of $x$, thus:

We know that $\cos^2 y + \sin^2 y = 1$

$$\therefore\ \cos^2 y = 1 - \sin^2 y = 1 - x^2 \text{ (since } x = \sin y)$$

$$\therefore\ \cos y = \sqrt{(1 - x^2)}$$

$$\therefore\ \frac{dy}{dx} = \frac{1}{\sqrt{(1 - x^2)}}$$

$$\underline{\frac{d}{dx}\left\{\sin^{-1}x\right\} = \frac{1}{\sqrt{(1 - x^2)}}}$$

Now you can determine $\dfrac{d}{dx}\left\{\cos^{-1}x\right\}$ in exactly the same way.

*Go through the same steps and finally check your result with that on frame 9.*

**9**

$$\frac{d}{dx}\left\{\cos^{-1}x\right\} = \frac{-1}{\sqrt{(1-x^2)}}$$

Here is the working:

Let $y = \cos^{-1}x \quad \therefore \ x = \cos y$

$$\therefore \ \frac{dx}{dy} = -\sin y \quad \therefore \ \frac{dy}{dx} = \frac{-1}{\sin y}$$

$$\cos^2 y + \sin^2 y = 1 \quad \therefore \ \sin^2 y = 1 - \cos^2 y = 1 - x^2$$

$$\sin y = \sqrt{(1-x^2)}$$

$$\therefore \ \frac{dy}{dx} = \frac{-1}{\sqrt{(1-x^2)}} \quad \therefore \ \frac{d}{dx}\left\{\cos^{-1}x\right\} = \frac{-1}{\sqrt{(1-x^2)}}$$

So we have two very similar results

(i) $\frac{d}{dx}\left\{\sin^{-1}x\right\} = \frac{1}{\sqrt{(1-x^2)}}$

(ii) $\frac{d}{dx}\left\{\cos^{-1}x\right\} = \frac{-1}{\sqrt{(1-x^2)}}$

} Different only in sign.

Now you find the differential coefficient of $\tan^{-1}x$. The working is slightly different, but the general method the same. See what you get and then move to frame 10 where the detailed working is set out.

**10**

$$\frac{d}{dx}\left\{\tan^{-1}x\right\} = \frac{1}{1+x^2}$$

Working: Let $y = \tan^{-1}x \quad \therefore \ x = \tan y.$

$$\frac{dx}{dy} = \sec^2 y = 1 + \tan^2 y = 1 + x^2$$

$$\frac{dx}{dy} = 1 + x^2 \quad \therefore \ \frac{dy}{dx} = \frac{1}{1+x^2}$$

$$\frac{d}{dx}\left\{\tan^{-1}x\right\} = \frac{1}{1+x^2}$$

Let us collect these three results together. Here they are:-

$$\frac{d}{dx}\left\{\sin^{-1}x\right\} = \frac{1}{\sqrt{(1-x^2)}} \quad \text{.................. (i)}$$

$$\frac{d}{dx}\left\{\cos^{-1}x\right\} = \frac{-1}{\sqrt{(1-x^2)}} \quad \text{.................. (ii)}$$

$$\frac{d}{dx}\left\{\tan^{-1}x\right\} = \frac{1}{1+x^2} \quad \text{.................. (iii)}$$

Copy these results into your record book. You will need to remember them.

*On to the next frame.*

**11**

Of course, these differential coefficients can occur in all the usual combinations, e.g. products, quotients, etc.

*Example 1.* Find $\frac{dy}{dx}$, given that $y = (1 - x^2)\sin^{-1}x$

Here we have a product

$$\therefore \frac{dy}{dx} = (1 - x^2)\frac{1}{\sqrt{(1-x^2)}} + \sin^{-1}x.(-2x)$$

$$= \sqrt{(1-x^2)} - 2x.\sin^{-1}x$$

*Example 2.* If $y = \tan^{-1}(2x - 1)$, find $\frac{dy}{dx}$

This time, it is a function of a function.

$$\frac{dy}{dx} = \frac{1}{1 + (2x - 1)^2}.\ 2 = \frac{2}{1 + 4x^2 - 4x + 1}$$

$$= \frac{2}{2 + 4x^2 - 4x} = \frac{1}{2x^2 - 2x + 1}$$

and so on.

**12**

Here you are. Here is a short exercise. Do them all: then check your results with those on the next frame.

**Revision Exercise**

Differentiate with respect to $x$:

1. $y = \sin^{-1} 5x$
2. $y = \cos^{-1} 3x$
3. $y = \tan^{-1} 2x$
4. $y = \sin^{-1}(x^2)$
5. $y = x^2.\sin^{-1}\left(\frac{x}{2}\right)$

*When you have finished them all, move on to frame 13.*

**13** *Results:*

1. $y = \sin^{-1} 5x \quad \therefore \frac{dy}{dx} = \frac{1}{\sqrt{\{1-(5x)^2\}}} . 5 = \frac{5}{\sqrt{\{1-25x^2\}}}$

2. $y = \cos^{-1} 3x \quad \therefore \frac{dy}{dx} = \frac{-1}{\sqrt{\{1-(3x)^2\}}} . 3 = \frac{-3}{\sqrt{\{1-9x^2\}}}$

3. $y = \tan^{-1} 2x \quad \therefore \frac{dy}{dx} = \frac{1}{1+(2x)^2} . 2 = \frac{2}{1+4x^2}$

4. $y = \sin^{-1}(x^2) \quad \therefore \frac{dy}{dx} = \frac{1}{\sqrt{\{1-(x^2)^2\}}} . 2x = \frac{2x}{\sqrt{(1-x^4)}}$

5. $y = x^2 . \sin^{-1}\left(\frac{x}{2}\right) \quad \therefore \frac{dy}{dx} = x^2 \frac{1}{\sqrt{\left\{1-\left(\frac{x}{2}\right)^2\right\}}} . \frac{1}{2} + 2x . \sin^{-1}\left(\frac{x}{2}\right)$

$$= \frac{x^2}{2\sqrt{\left\{1-\frac{x^2}{4}\right\}}} + 2x . \sin^{-1}\left(\frac{x}{2}\right)$$

$$= \frac{x^2}{\sqrt{(4-x^2)}} + 2x . \sin^{-1}\left(\frac{x}{2}\right)$$

*Right, now on to the next frame.*

**14**

## Differential coefficients of inverse hyperbolic functions

In just the same way that we have inverse trig. functions, so we have inverse hyperbolic functions and we would not be unduly surprised if their differential coefficients bore some resemblance to those of the inverse trig. functions.

Anyway, let us see what we get. The method is very much as before.

(i) $y = \sinh^{-1} x$ To find $\frac{dy}{dx}$

First express the inverse statement as a direct statement.

$$y = \sinh^{-1} x \quad \therefore x = \sinh y \quad \therefore \frac{dx}{dy} = \cosh y \quad \therefore \frac{dy}{dx} = \frac{1}{\cosh y}$$

We now need to express $\cosh y$ in terms of $x$

We know that $\cosh^2 y - \sinh^2 y = 1 \quad \therefore \cosh^2 y = \sinh^2 y + 1 = x^2 + 1$

$$\cosh y = \sqrt{(x^2+1)}$$

$$\frac{dy}{dx} = \frac{1}{\sqrt{(x^2+1)}} \quad \therefore \frac{d}{dx}\left\{\sinh^{-1} x\right\} = \frac{1}{\sqrt{(x^2+1)}}$$

Let us obtain similar results for $\cosh^{-1} x$ and $\tanh^{-1} x$ and then we will take a look at them.

*So on to the next frame.*

**15**

We have just established $\dfrac{d}{dx}\left\{\sinh^{-1} x\right\} = \dfrac{1}{\sqrt{(x^2+1)}}$

(ii) $y = \cosh^{-1} x \quad \therefore x = \cosh y$

$$\therefore \frac{dx}{dy} = \sinh y \quad \therefore \frac{dy}{dx} = \frac{1}{\sinh y}$$

Now $\cosh^2 y - \sinh^2 y = 1 \quad \therefore \sinh^2 y = \cosh^2 y - 1 = x^2 - 1$

$$\therefore \sinh y = \sqrt{(x^2-1)}$$

$$\therefore \frac{dy}{dx} = \frac{1}{\sqrt{(x^2-1)}} \quad \therefore \frac{d}{dx}\left\{\cosh^{-1} x\right\} = \frac{1}{\sqrt{(x^2-1)}}$$

Now you can deal with the remaining one

$$\text{If} \quad y = \tanh^{-1} x, \frac{dy}{dx} = \ldots\ldots\ldots\ldots$$

Tackle it in much the same way as we did for $\tan^{-1} x$, remembering this time, however, that $\operatorname{sech}^2 x = 1 - \tanh^2 x$. You will find that useful.

*When you have finished, move to frame 16.*

---

**16**

$$y = \tanh^{-1} x \quad \boxed{\frac{dy}{dx} = \frac{1}{1-x^2}}$$

for:

$$y = \tanh^{-1} x \quad \therefore x = \tanh y$$

$$\therefore \frac{dx}{dy} = \operatorname{sech}^2 y = 1 - \tanh^2 y = 1 - x^2 \quad \therefore \frac{dy}{dx} = \frac{1}{1-x^2}$$

$$\frac{d}{dx}\left\{\tanh^{-1} x\right\} = \frac{1}{1-x^2}$$

Now here are the results, all together, so that we can compare them.

$$\frac{d}{dx}\left\{\sinh^{-1} x\right\} = \frac{1}{\sqrt{(x^2+1)}} \quad \ldots\ldots\ldots\ldots\ldots \quad \text{(iv)}$$

$$\frac{d}{dx}\left\{\cosh^{-1} x\right\} = \frac{1}{\sqrt{(x^2-1)}} \quad \ldots\ldots\ldots\ldots\ldots \quad \text{(v)}$$

$$\frac{d}{dx}\left\{\tanh^{-1} x\right\} = \frac{1}{1-x^2} \quad \ldots\ldots\ldots\ldots\ldots \quad \text{(vi)}$$

**Make a note of these in your record book. You will need to remember these results.**

*Now on to frame 17.*

**17** Here are one or two examples, using the last results.

*Example 1.* $y = \cosh^{-1}\left\{3 - 2x\right\}$

$$\therefore \frac{dy}{dx} = \frac{1}{\sqrt{\{(3-2x)^2 - 1\}}}\cdot(-2) = \frac{-2}{\sqrt{(9 - 12x + 4x^2 - 1)}}$$

$$= \frac{-2}{\sqrt{(8 - 12x + 4x^2)}} = \frac{-2}{2\sqrt{(x^2 - 3x + 2)}} = \underline{\frac{-1}{\sqrt{(x^2 - 3x + 2)}}}$$

*Example 2.* $y = \tanh^{-1}\left(\frac{3x}{4}\right)$

$$\therefore \frac{dy}{dx} = \frac{1}{1 - \left(\frac{3x}{4}\right)^2}\cdot\frac{3}{4} = \frac{1}{1 - \frac{9x^2}{16}}\cdot\frac{3}{4}$$

$$= \frac{16}{16 - 9x^2}\cdot\frac{3}{4} = \underline{\frac{12}{16 - 9x^2}}$$

*Example 3.* $y = \sinh^{-1}\{\tan x\}$

$$\therefore \frac{dy}{dx} = \frac{1}{\sqrt{(\tan^2 x + 1)}}\cdot\sec^2 x = \frac{\sec^2 x}{\sqrt{\sec^2 x}}$$

$$= \underline{\sec x}$$

**18** Here are a few for you to do.

**Exercise**

Differentiate:

1. $y = \sinh^{-1} 3x$
2. $y = \cosh^{-1}\left(\frac{5x}{2}\right)$
3. $y = \tanh^{-1}(\tan x)$
4. $y = \sinh^{-1}\sqrt{(x^2 - 1)}$
5. $y = \cosh^{-1}(e^{2x})$

*Finish them all. Then turn on to frame 19 for the results.*

**19**

*Results:*

1. $y = \sinh^{-1} 3x \quad \therefore \frac{dy}{dx} = \frac{1}{\sqrt{\{(3x)^2 + 1\}}} \cdot 3 = \frac{3}{\sqrt{(9x^2 + 1)}}$

2. $y = \cosh^{-1}\left(\frac{5x}{2}\right) \quad \therefore \frac{dy}{dx} = \frac{1}{\sqrt{\left\{\left(\frac{5x}{2}\right)^2 - 1\right\}}} \cdot \frac{5}{2} = \frac{5}{2\sqrt{\left\{\frac{25x^2}{4} - 1\right\}}}$

$$= \frac{5}{2\sqrt{\left(\frac{25x^2 - 4}{4}\right)}} = \frac{5}{\sqrt{(25x^2 - 4)}}$$

3. $y = \tanh^{-1}(\tan x) \quad \therefore \frac{dy}{dx} = \frac{1}{1 - \tan^2 x} \cdot \sec^2 x = \frac{\sec^2 x}{1 - \tan^2 x}$

4. $y = \sinh^{-1}\{\sqrt{(x^2 - 1)}\}$

$$\therefore \frac{dy}{dx} = \frac{1}{\sqrt{(x^2 - 1 + 1)}} \cdot \frac{1}{2}(x^2 - 1)^{-\frac{1}{2}}(2x) = \frac{1}{\sqrt{(x^2 - 1)}}$$

5. $y = \cosh^{-1}(e^{2x}) \quad \therefore \frac{dy}{dx} = \frac{1}{\sqrt{\{(e^{2x})^2 - 1\}}} \cdot 2e^{2x} = \frac{2e^{2x}}{\sqrt{(e^{4x} - 1)}}$

All correct?

*On then to frame 20.*

**20**

Before we leave these inverse trig. and hyperbolic functions, let us look at them all together.

| Inverse Trig. Functions | | Inverse Hyperbolic Functions | |
|---|---|---|---|
| $y$ | $\frac{dy}{dx}$ | $y$ | $\frac{dy}{dx}$ |
| $\sin^{-1} x$ | $\frac{1}{\sqrt{(1 - x^2)}}$ | $\sinh^{-1} x$ | $\frac{1}{\sqrt{(x^2 + 1)}}$ |
| $\cos^{-1} x$ | $\frac{-1}{\sqrt{(1 - x^2)}}$ | $\cosh^{-1} x$ | $\frac{1}{\sqrt{(x^2 - 1)}}$ |
| $\tan^{-1} x$ | $\frac{1}{1 + x^2}$ | $\tanh^{-1} x$ | $\frac{1}{1 - x^2}$ |

It would be a good idea to copy down this combined table, so that you compare and use the results. Do that: it will help you to remember them and to distinguish clearly between them.

**21** Before you do a revision exercise, cover up the table you have just copied and see if you can complete the following correctly.

1. If $y = \sin^{-1} x, \dfrac{dy}{dx} =$ ....................
2. If $y = \cos^{-1} x, \dfrac{dy}{dx} =$ ....................
3. If $y = \tan^{-1} x, \dfrac{dy}{dx} =$ ....................
4. If $y = \sinh^{-1} x, \dfrac{dy}{dx} =$ ....................
5. If $y = \cosh^{-1} x, \dfrac{dy}{dx} =$ ....................
6. If $y = \tanh^{-1} x, \dfrac{dy}{dx} =$ ....................

Now check your results with your table and make a special point of brushing up any of which you are not really sure.

**22** 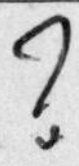

**Revision Exercise**

Differentiate the following with respect to $x$:

1. $\tan^{-1}(\sinh x)$
2. $\sinh^{-1}(\tan x)$
3. $\cosh^{-1}(\sec x)$
4. $\tanh^{-1}(\sin x)$
5. $\sin^{-1}\left(\dfrac{x}{a}\right)$

Take care with these; we have mixed them up to some extent.

*When you have finished them all – and you are sure you have done what was required – check your results with those on frame 23.*

**23**

**Solutions**

1. $y = \tan^{-1}(\sinh x)$ $\qquad \dfrac{d}{dx}\left\{\tan^{-1} x\right\} = \dfrac{1}{1+x^2}$

$$\therefore \frac{dy}{dx} = \frac{1}{1+\sinh^2 x}\,.\,\cosh x = \frac{\cosh x}{\cosh^2 x} = \underline{\operatorname{sech} x}$$

2. $y = \sinh^{-1}(\tan x)$ $\qquad \dfrac{d}{dx}\left\{\sinh^{-1} x\right\} = \dfrac{1}{\sqrt{(x^2+1)}}$

$$\therefore \frac{dy}{dx} = \frac{1}{\sqrt{(\tan^2 x+1)}}\,.\,\sec^2 x = \underline{\frac{\sec^2 x}{\sqrt{\sec^2 x}} = \sec x}$$

3. $y = \cosh^{-1}(\sec x)$ $\qquad \dfrac{d}{dx}\left\{\cosh^{-1} x\right\} = \dfrac{1}{\sqrt{(x^2-1)}}$

$$\therefore \frac{dy}{dx} = \frac{1}{\sqrt{(\sec^2 x-1)}}\,.\sec x\,.\tan x = \frac{\sec x\,.\,\tan x}{\sqrt{\tan^2 x}}$$

$$= \underline{\sec x}$$

4. $y = \tanh^{-1}(\sin x)$ $\qquad \dfrac{d}{dx}\left\{\tanh^{-1} x\right\} = \dfrac{1}{1-x^2}$

$$\therefore \frac{dy}{dx} = \frac{1}{1-\sin^2 x}\,.\,\cos x = \underline{\frac{\cos x}{\cos^2 x} = \sec x}$$

5. $y = \sin^{-1}\left\{\dfrac{x}{a}\right\}$ $\qquad \dfrac{d}{dx}\left\{\sin^{-1} x\right\} = \dfrac{1}{\sqrt{(1-x^2)}}$

$$\therefore \frac{dy}{dx} = \frac{1}{\sqrt{\left\{1-\left(\frac{x}{a}\right)^2\right\}}}\cdot\frac{1}{a} = \frac{1}{a}\cdot\frac{1}{\sqrt{\left(1-\frac{x^2}{a^2}\right)}}$$

$$= \frac{1}{a}\cdot\frac{1}{\sqrt{\left(\frac{a^2-x^2}{a^2}\right)}} = \underline{\frac{1}{\sqrt{(a^2-x^2)}}}$$

If you have got those all correct – or nearly all correct – you now know quite a lot about the differential coefficients of Inverse Trig. and Hyperbolic Functions.

*You are now ready to move on to the next topic of this programme, so off you go to frame 24.*

**24**

## Maximum and minimum values (turning points)

You are already familiar with the basic techniques for finding maximum and minimum values of a function. You have done this kind of operation many times in the past, but just to refresh your memory, let us consider some function, $y = f(x)$ whose graph is shown below.

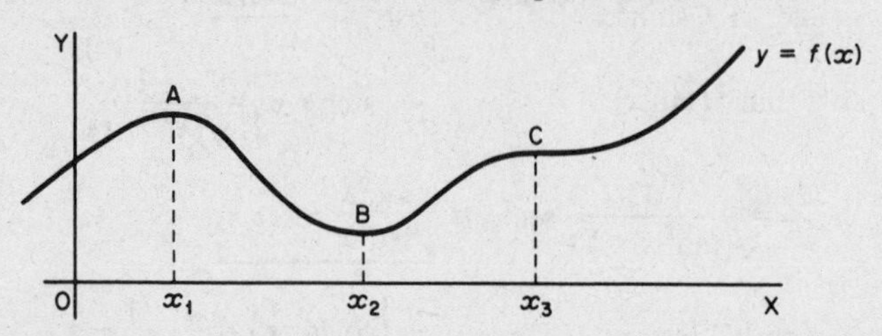


At the point A, i.e. at $x = x_1$, a maximum value of $y$ occurs since at A, the $y$ value is greater than the $y$ values on either side of it and *close to it.*

Similarly, at B, $y$ is a ..................., since the $y$ value at the point B is less than the $y$ values on either side of it and close to it.

**25**

At B, $y$ is a minimum value.

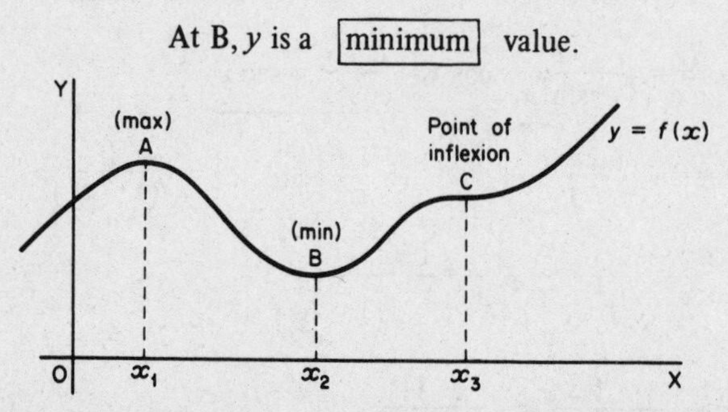


The point C is worth a second consideration. It looks like 'half a max. and half a min.' The curve flattens out at C, but instead of dipping down, it then goes on with an increasingly positive slope. Such a point is an example of a *point of inflexion,* i.e. it is essentially a form of S-bend.

Points A, B and C, are called *turning points* on the graph, or *stationary values of y,* and while you know how to find the positions of A and B, you may know considerably less about points of inflexion. We shall be taking a special look at these.

*On to frame 26.*

**26**

If we consider the slope of the graph as we travel left to right, we can draw a graph to show how this slope varies. We have no actual values for the slope, but we can see whether it is positive or negative, more or less steep. The graph we obtain is the *first derived curve* of the function and we are really plotting the values of $\frac{dy}{dx}$ against values of $x$

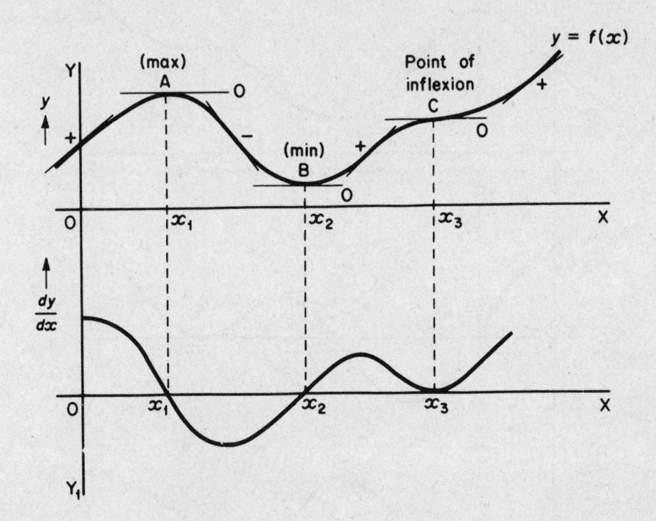


We see that at $x = x_1, x_2, x_3$, (corresponding to our three turning points) the graph of $\frac{dy}{dx}$ is at the $x$-axis – and at no other points.

Therefore, we obtain the first rule, which is that for turning points,

$\frac{dy}{dx} =$ ....................

*Turn on to frame 27.*

**27**

For turning points, A, B, C, $\boxed{\dfrac{dy}{dx} = 0}$

If we now trace the slope of the *first derived curve* and plot this against $x$, we obtain the *second derived curve*, which shows values of $\dfrac{d^2y}{dx^2}$ against $x$.

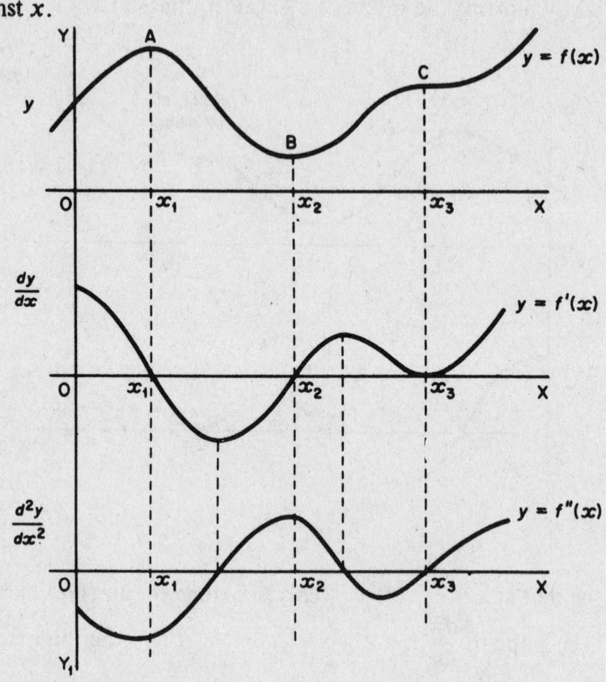


From the first derived curve, we see that for turning points,

$$\frac{dy}{dx} = 0$$

From the second derived curve, we see that

for maximum $y$, $\dfrac{d^2y}{dx^2}$ is negative

for minimum $y$, $\dfrac{d^2y}{dx^2}$ is positive

for P-of-I, $\dfrac{d^2y}{dx^2}$ is zero

*Copy* the diagram into your record book. It summarizes all the facts on max. and min. values so far.

**28**

From the results we have just established, we can now determine

(i) the values of $x$ at which turning points occur, by differentiating the function and then solving the equation $\frac{dy}{dx} = 0$

(ii) the corresponding values of $y$ at these points by merely substituting the $x$ values found, in $y = f(x)$

(iii) the type of each turning point (max., min., or P-of-I) by testing in the expression for $\frac{d^2y}{dx^2}$

With this information, we can go a long way towards drawing a sketch of the curve. So let us apply these results to a straightforward example in the next frame.

---

**29**

*Example.* Find the turning points on the graph of the function $y = \frac{x^3}{3} - \frac{x^2}{2} - 2x + 5$. Distinguish between them and sketch the graph of the function.

There are, of course, two stages:

(i) Turning points are given by $\frac{dy}{dx} = 0$

(ii) The type of each turning point is determined by substituting the roots of the equation $\frac{dy}{dx} = 0$ in the expression for $\frac{d^2y}{dx^2}$

*If* $\frac{d^2y}{dx^2}$ is negative, then $y$ is a maximum,

" " " positive, " " " " minimum,

" " " zero, " " " " point of inflexion.

We shall need both the first and second differential coefficients, so find them ready. If $y = \frac{x^3}{3} - \frac{x^2}{2} - 2x + 5$, then $\frac{dy}{dx} =$ ............. and

$\frac{d^2y}{dx^2} =$ ............. .

**30**

$$\frac{dy}{dx} = x^2 - x - 2; \; \frac{d^2y}{dx^2} = 2x - 1$$

□□□□□□□□□□□□□□□□□□□□□□□□□□□□□□□□□□□□□□□

(i) Turning points occur at $\frac{dy}{dx} = 0$

$\therefore\; x^2 - x - 2 = 0 \;\; \therefore\; (x-2)(x+1) = 0 \;\; \therefore\; x = 2$ and $x = -1$

i.e. turning points occur at $x = 2$ and $x = -1$.

(ii) To determine the type of each turning point, substitute $x = 2$ and then $x = -1$ in the expression for $\frac{d^2y}{dx^2}$

At $x = 2$, $\frac{d^2y}{dx^2} = 4 - 1 = 3$, i.e. positive $\;\therefore\; x = 2$ gives $y_{min}$.

At $x = -1$, $\frac{d^2y}{dx^2} = -2 - 1$, i.e. negative $\;\therefore\; x = -1$ gives $y_{max}$.

Substituting in $y = f(x)$ gives $x = 2, y_{min} = 1\frac{2}{3}$ and $x = -1, y_{max} = 6\frac{1}{6}$

Also, we can see at a glance from the function, that when $x = 0, y = 5$.

*You can now sketch the graph of the function. Do it.*

**31**

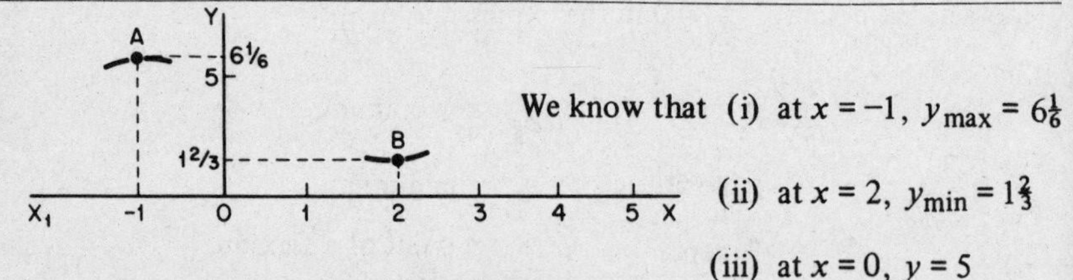


We know that (i) at $x = -1$, $y_{max} = 6\frac{1}{6}$

(ii) at $x = 2$, $y_{min} = 1\frac{2}{3}$

(iii) at $x = 0$, $y = 5$

Joining up with a smooth curve gives:

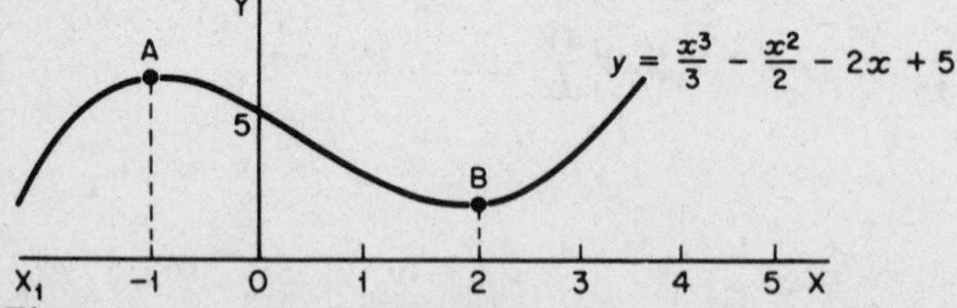


There is no point of inflexion like the point C on this graph. *Move on.*

**32**

All that was just by way of refreshing your memory on work you have done before. Now let us take a wider look at these

**Points of Inflexion**

The point C that we considered on our first diagram was rather a special kind of point of inflexion. In general, it is not necessary for the curve at a P-of-I to have zero slope.

A *point of inflexion* is defined simply as a point on a curve at which the direction of bending changes, i.e. from a right-hand bend to a left-hand bend, or from a left-hand bend to a right-hand bend.

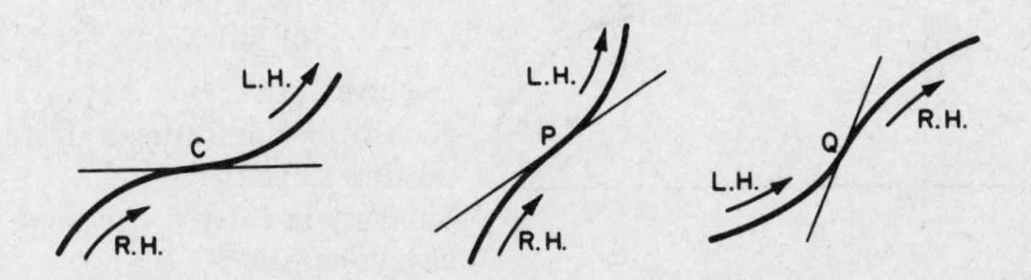


The point C we considered is, of course, a P-of-I, but it is not *essential* at a P-of-I for the slope to be zero. Points P and Q are perfectly good points of inflexion and in fact in these cases the slope is

$$\left\{\begin{array}{l}\text{positive}\\\text{negative}\\\text{zero}\end{array}\right\} \text{Which?}$$

---

**33**

At the points of inflexion, P and Q, the slope is in fact

$$\boxed{\text{positive}}$$

Correct. The slope can of course be positive, negative or zero in any one case, but there is no restriction on its sign.

□□□□□□□□□□□□□□□□□□□□□□□□□□□□□□□□□□□□□□□

A point of inflexion, then, is simply a point on a curve at which there is a change in the *d* .................... of *b* ....................

**34**

Point of inflexion: a point at which there is a change in the

direction of bending

□□□□□□□□□□□□□□□□□□□□□□□□□□□□□□□□□□□□□□□

If the slope at a P-of-I is not zero, it will not appear in our usual max. and min. routine, for $\frac{dy}{dx}$ will not be zero. How, then, are we going to find where such points of inflexion occur? Let us sketch the graphs of the slopes as we did before.

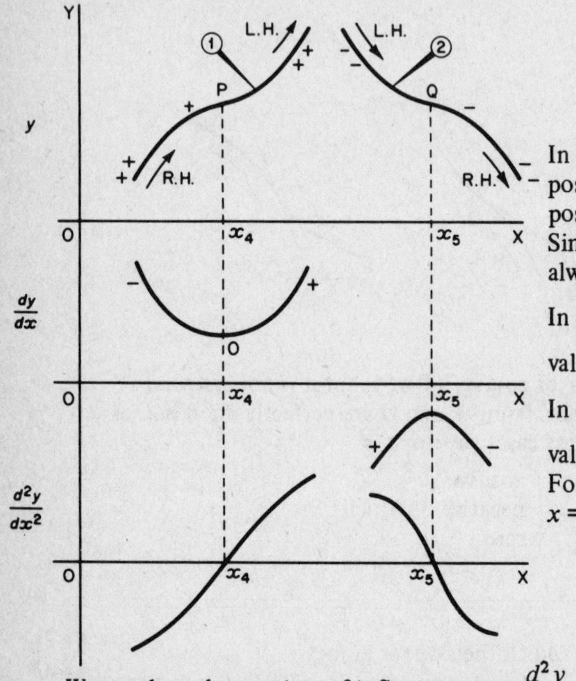


P and Q are points of inflexion.

In curve 1, the slope is always positive, ++ indicating a greater positive slope than +.

Similarly in curve 2, the slope is always negative.

In curve 1, $\frac{dy}{dx}$ reaches a minimum value but not zero.

In curve 2, $\frac{dy}{dx}$ reaches a maximum value but not zero.

For both points of inflexion, i.e. at $x = x_4$ and $x = x_5$

$$\frac{d^2y}{dx^2} = 0$$

We see that where points of inflexion occur, $\frac{d^2y}{dx^2} = 0$

So, is this the clue we have been seeking? If so, it simply means that to find the points of inflexion we differentiate the function of the curve twice and solve the equation $\frac{d^2y}{dx^2} = 0$.

*That sounds easy enough! But turn on to the next frame to see what is involved.*

**35**

We have just found that

$$\boxed{\text{where points of inflexion occur, } \frac{d^2y}{dx^2} = 0}$$

This is perfectly true. Unfortunately, this is not the whole of the story, for it is also possible for $\frac{d^2y}{dx^2}$ to be zero at points other than points of inflexion!

So if we solve $\frac{d^2y}{dx^2} = 0$, we cannot as yet be sure whether the solution $x = a$ gives a point of inflexion or not. How can we decide?

Let us consider just one more set of graphs. This should clear the matter up.

Let S be a true point of inflexion and T a point on $y = f(x)$ as shown. Clearly, T is not a point of inflexion.

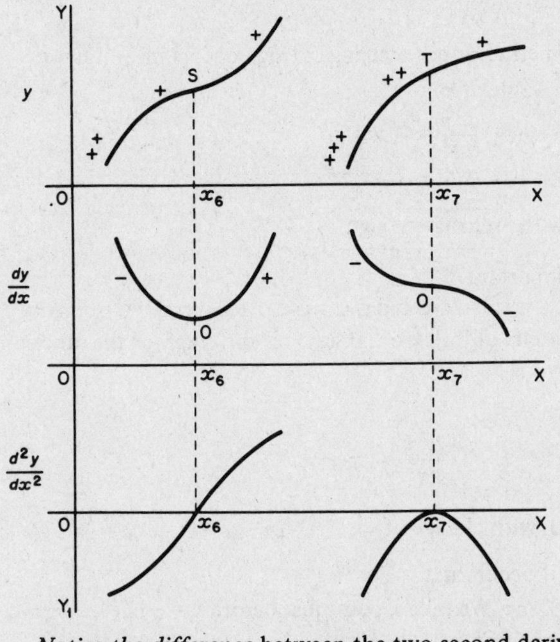


The first derived curves could well look like this.

*Notice the difference* between the two second derived curves.

Although $\frac{d^2y}{dx^2}$ is zero for each (at $x = x_6$ and $x = x_7$), how do they differ?

*When you have discovered the difference, turn on to frame 37.*

**37**

> In the case of the real P-of-I, the graph of $\frac{d^2y}{dx^2}$ crosses the $x$-axis.
>
> In the case of no P-of-I, the graph of $\frac{d^2y}{dx^2}$ only touches the $x$-axis and $\frac{d^2y}{dx^2}$ does not change sign.

□□□□□□□□□□□□□□□□□□□□□□□□□□□□□□□□□□□□□□

This is the clue we have been after, and gives us our final rule.

*For a point of inflexion,* $\frac{d^2y}{dx^2} = 0$ *and there is a change of sign of* $\frac{d^2y}{dx^2}$ *as we go through the point.*

(In the phoney case, there is no change of sign.)

So, to find where points of inflexion occur,

(i) we differentiate $y = f(x)$ twice to get $\frac{d^2y}{dx^2}$

(ii) we solve the equation $\frac{d^2y}{dx^2} = 0$

(iii) we test to see whether or not a change of sign occurs in $\frac{d^2y}{dx^2}$ as we go through this value of $x$.

For points of inflexion, then, $\frac{d^2y}{dx^2} = 0$, with $c$ ............ of $s$............

---

**38**

For a P-of-I, $\frac{d^2y}{dx^2} = 0$ with **change of sign**

This last phrase is all-important.

□□□□□□□□□□□□□□□□□□□□□□□□□□□□□□□□□□□□□□

*Example 1.* Find the points of inflexion, if any, on the graph of the function $y = \frac{x^3}{3} - \frac{x^2}{2} - 2x + 5$.

(i) *Diff. twice.* $\frac{dy}{dx} = x^2 - x - 2, \quad \frac{d^2y}{dx^2} = 2x - 1$

For P-of-I, $\frac{d^2y}{dx^2} = 0$, with change of sign. $\therefore\ 2x - 1 = 0 \quad \therefore\ x = \frac{1}{2}$

If there is a P-of-I, it occurs at $x = \frac{1}{2}$

(ii) *Test for change of sign.* We take a point just before $x = \frac{1}{2}$, i.e. $x = \frac{1}{2} - a$, and a point just after $x = \frac{1}{2}$, i.e. $x = \frac{1}{2} + a$, where $a$ is a small positive quantity, and investigate the sign of $\frac{d^2y}{dx^2}$ at these two values of $x$.

*Turn on.*

**39**

$$\frac{d^2y}{dx^2} = 2x - 1$$

(i) At $x = \frac{1}{2} - a$, $\frac{d^2y}{dx^2} = 2(\frac{1}{2} - a) - 1 = 1 - 2a - 1$

$$= -2a \text{ (negative)}$$

(ii) At $x = \frac{1}{2} + a$, $\frac{d^2y}{dx^2} = 2(\frac{1}{2} + a) - 1 = 1 + 2a - 1$

$$= 2a \text{ (positive)}$$

There *is* a change in sign of $\frac{d^2y}{dx^2}$ as we go through $x = \frac{1}{2}$

$\therefore$ There *is* a point of inflexion at $x = \frac{1}{2}$

If you look at the sketch graph of this function which you have already drawn, you will see the point of inflexion where the right-hand curve changes to the left-hand curve.

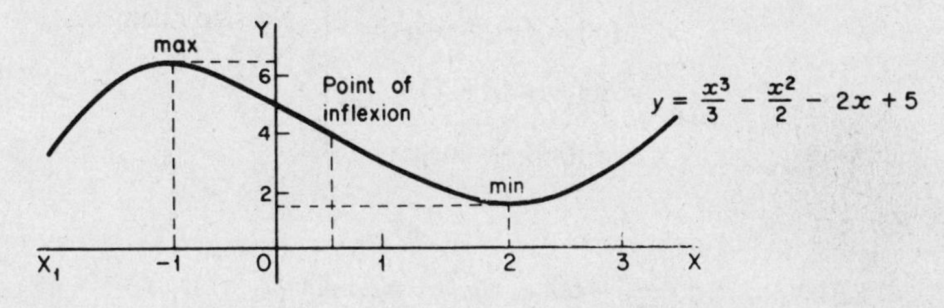


**40**

*Example 2.* Find the points of inflexion on the graph of the function

$$y = 3x^5 - 5x^4 + x + 4$$

First, differentiate twice and solve the equation $\frac{d^2y}{dx^2} = 0$. This will give the values of $x$ at which there are possibly points of inflexion. We cannot be sure until we have then tested for a change of sign in $\frac{d^2y}{dx^2}$. We will do that in due course.

So start off by finding an expression for $\frac{d^2y}{dx^2}$ and solving the equation $\frac{d^2y}{dx^2} = 0$.

*When you have done that, turn on to the next frame.*

**41** *We have:*

$$y = 3x^5 - 5x^4 + x + 4$$

$$\therefore \frac{dy}{dx} = 15x^4 - 20x^3 + 1$$

$$\therefore \frac{d^2y}{dx^2} = 60x^3 - 60x^2 = 60x^2(x-1)$$

For P-of-I, $\frac{d^2y}{dx^2} = 0$, with change of sign.

$$\therefore 60x^2(x-1) = 0 \quad \therefore x = 0 \text{ or } x = 1$$

If there is a point of inflexion, it occurs at $x = 0, x = 1$, or both. Now comes the test for a change of sign. For each of the two values of $x$ we have found, i.e. $x = 0$ and $x = 1$, take points on either side of it, differing from it by a very small amount.

(i) *For* $x = 0$

$$\left.\begin{aligned} \text{At } x = -a, \frac{d^2y}{dx^2} &= 60(-a)^2(-a-1) \\ &= (+)(+)(-) = \text{negative} \\ \text{At } x = +a, \frac{d^2y}{dx^2} &= 60(+a)^2(a-1) \\ &= (+)(+)(-) = \text{negative} \end{aligned}\right\} \begin{array}{l}\text{No sign change.}\\ \text{No P-of-I.}\end{array}$$

(ii) *For* $x = 1$

$$\left.\begin{aligned} \text{At } x = 1-a, \frac{d^2y}{dx^2} &= 60(1-a)^2(1-a-1) \\ &= (+)(+)(-) = \text{negative} \\ \text{At } x = 1+a, \frac{d^2y}{dx^2} &= 60(1+a)^2(1+a-1) \\ &= (+)(+)(+) = \text{positive} \end{aligned}\right\} \begin{array}{l}\text{Change in sign.}\\ \therefore \text{P-of-I.}\end{array}$$

Therefore, the only point of inflexion occurs when $x = 1$, i.e. at the point

$$\underline{x = 1, \ y = 3}$$

That is just about all there is to it. The functions with which we have to deal differ, of course, from problem to problem, but the method remains the same.

*Now turn on to the next frame and complete the Test Exercise awaiting you. The questions are all very straightforward and should not cause you any anxiety.*

**Test Exercise – IX**

42

Answer all the questions.

1. Evaluate (i) $\cos^{-1}(-0{\cdot}6428)$, (ii) $\tan^{-1}(-0{\cdot}7536)$.

2. Differentiate with respect to $x$:

(i) $y = \sin^{-1}(3x + 2)$

(ii) $y = \dfrac{\cos^{-1} x}{x}$

(iii) $y = x^2 \tan^{-1}\left(\dfrac{x}{2}\right)$

(iv) $y = \cosh^{-1}(1 - 3x)$

(v) $y = \sinh^{-1}(\cos x)$

(vi) $y = \tanh^{-1} 5x$

3. Find the stationary values of $y$ and the points of inflexion on the graph of each of the following functions, and in each case, draw a sketch graph of the function.

(i) $y = x^3 - 6x^2 + 9x + 6$

(ii) $y = x + \dfrac{1}{x}$

(iii) $y = x\,\mathrm{e}^{-x}$

*Well done. You are now ready for the next programme.*

**Further Problems – IX**

1. Differentiate (i) $\tan^{-1}\left\{\dfrac{1+\tan x}{1-\tan x}\right\}$

   (ii) $x\sqrt{(1-x^2)}-\sin^{-1}\sqrt{(1-x^2)}$

2. If $y=\dfrac{\sin^{-1}x}{\sqrt{(1-x^2)}}$, prove that

   (i) $(1-x^2)\dfrac{dy}{dx}=xy+1$

   (ii) $(1-x^2)\dfrac{d^2y}{dx^2}-3x\dfrac{dy}{dx}=y$

3. Find $\dfrac{dy}{dx}$ when (i) $y=\tan^{-1}\left\{\dfrac{4\sqrt{x}}{1-4x}\right\}$

   (ii) $y=\tanh^{-1}\left\{\dfrac{2x}{1+x^2}\right\}$

4. Find the co-ordinates of the point of inflexion on the curves

   (i) $y=(x-2)^2(x-7)$

   (ii) $y=4x^3+3x^2-18x-9$

5. Find the values of $x$ for which the function $y=f(x)$, defined by $y(3x-2)=(3x-1)^2$ has maximum and minimum values and distinguish between them. Sketch the graph of the function.

6. Find the values of $x$ at which maximum and minimum values of $y$ and points of inflexion occur on the curve $y=12\ln x+x^2-10x$.

7. If $4x^2+8xy+9y^2-8x-24y+4=0$, show that when $\dfrac{dy}{dx}=0$, $x+y=1$ and $\dfrac{d^2y}{dx^2}=\dfrac{4}{8-5y}$. Hence find the maximum and minimum values of $y$.

8. Determine the smallest positive value of $x$ at which a point of inflexion occurs on the graph of $y=3\,e^{2x}\cos(2x-3)$.

9. If $y^3=6xy-x^3-1$, prove that $\dfrac{dy}{dx}=\dfrac{2y-x^2}{y^2-2x}$ and that the maximum value of $y$ occurs where $x^3=8+2\sqrt{14}$ and the minimum value where $x^3=8-2\sqrt{14}$.

10. For the curve $y = e^{-x} \sin x$, express $\frac{dy}{dx}$ in the form $A e^{-x} \cos(x + a)$ and show that the points of inflexion occur at $x = \frac{\pi}{2} + k\pi$ for any integral value of $k$.

11. Find the turning points and points of inflexion on the following curves, and, in each case, sketch the graph.
    (i) $y = 2x^3 - 5x^2 + 4x - 1$
    (ii) $y = \dfrac{x(x-1)}{x-2}$
    (iii) $y = x + \sin x$ (Take $x$ and $y$ scales as multiples of $\pi$.)

12. Find the values of $x$ at which points of inflexion occur on the following curves.
    (i) $y = e^{-x^2}$ (ii) $y = e^{-2x}(2x^2 + 2x + 1)$
    (iii) $y = x^4 - 10x^2 + 7x + 4$

13. The signalling range ($x$) of a submarine cable is proportional to $r^2 \ln\left(\frac{1}{r}\right)$, where $r$ is the ratio of the radii of the conductor and cable. Find the value of $r$ for maximum range.

14. The power transmitted by a belt drive is proportional to $Tv - \dfrac{wv^3}{g}$, where $v$ = speed of the belt, T = tension on the driving side, and $w$ = weight per unit length of belt. Find the speed at which the transmitted power is a maximum.

15. A right circular cone has a given curved surface A. Show that, when its volume is a maximum, the ratio of the height to the base radius is $\sqrt{2} : 1$.

16. The motion of a particle performing damped vibrations is given by $y = e^{-t} \sin 2t$, $y$ being the displacement from its mean position at time $t$. Show that $y$ is a maximum when $t = \frac{1}{2} \tan^{-1}(2)$ and determine this maximum displacement to three significant figures.

17. The cross-section of an open channel is a trapezium with base 6 cm and sloping sides each 10 cm wide. Calculate the width across the open top so that the cross-sectional area of the channel shall be a maximum.

18. The velocity ($v$) of a piston is related to the angular velocity ($\omega$) of the crank by the relationship $v = \omega r \left\{ \sin\theta + \frac{r}{2l}\sin 2\theta \right\}$ where $r$ = length of crank and $l$ = length of connecting rod. Find the first positive value of $\theta$ for which $v$ is a maximum, for the case when $l = 4r$.

19. A right circular cone of base radius $r$, has a total surface area S and volume V. Prove that $9V^2 = r^2(S^2 - 2\pi r^2 S)$. If S is constant, prove that the vertical angle ($\theta$) of the cone for maximum volume is given by $\theta = 2\sin^{-1}\left(\frac{1}{3}\right)$.

20. Show that the equation $4\frac{d^2x}{dt^2} + 4\mu\frac{dx}{dt} + \mu^2 x = 0$ is satisfied by $x = (At + B)\,e^{-\mu t/2}$, where A and B are arbitrary constants. If $x = 0$ and $\frac{dx}{dt} = C$ when $t = 0$, find A and B and show that the maximum value of $x$ is $\frac{2C}{\mu e}$ and that this occurs when $t = \frac{2}{\mu}$.

# Programme 10

## PARTIAL DIFFERENTIATION

### PART 1

**1** **Partial differentiation**

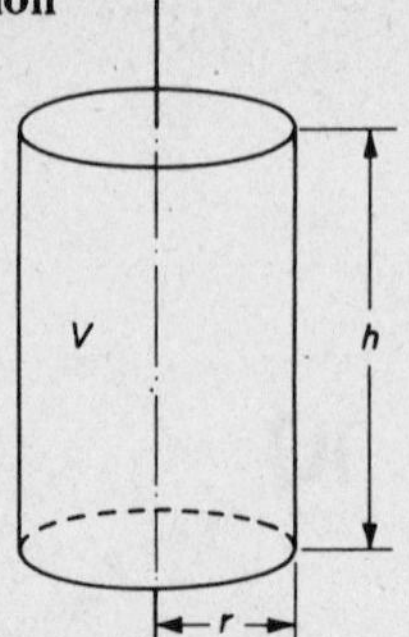


The volume V of a cylinder of radius $r$ and height $h$ is given by

$$V = \pi r^2 h$$

i.e. V depends on two quantities, the values of $r$ and $h$.

If we keep $r$ constant and increase the height $h$, the volume V will increase. In these circumstances, we can consider the differential coefficient of V with respect to $h$ – but only if $r$ is kept constant.

$$\text{i.e.} \left[\frac{dV}{dh}\right]_{r \text{ constant}} \quad \text{is written } \frac{\partial V}{\partial h}$$

Notice the new type of 'delta'. We already know the meaning of $\frac{\delta y}{\delta x}$ and $\frac{dy}{dx}$. Now we have a new one, $\frac{\partial V}{\partial h}$. $\frac{\partial V}{\partial h}$ is called the *partial differential coefficient* of V with respect to $h$ and implies that for our present purpose, the value of $r$ is considered as being kept ....................

**2**

constant

□□□□□□□□□□□□□□□□□□□□□□□□□□□□□□□□□□□□□□

$V = \pi r^2 h$. To find $\frac{\partial V}{\partial h}$, we differentiate the given expression, taking all symbols except V and $h$ as being constant $\therefore \frac{\partial V}{\partial h} = \pi r^2.1 = \pi r^2$

Of course, we could have considered $h$ as being kept constant, in which case, a change in $r$ would also produce a change in V. We can therefore talk about $\frac{\partial V}{\partial r}$ which simply means that we now differentiate $V = \pi r^2 h$ with respect to $r$, taking all symbols except V and $r$ as being constant for the time being.

$$\therefore \frac{\partial V}{\partial r} = \pi\, 2rh = 2\pi rh$$

In the statement, $V = \pi r^2 h$, V is expressed as a function of two variables, $r$ and $h$. It therefore has two partial differential coefficients, one with respect to .................... and one with respect to ....................

One with respect to $r$; one with respect to $h$

**3**

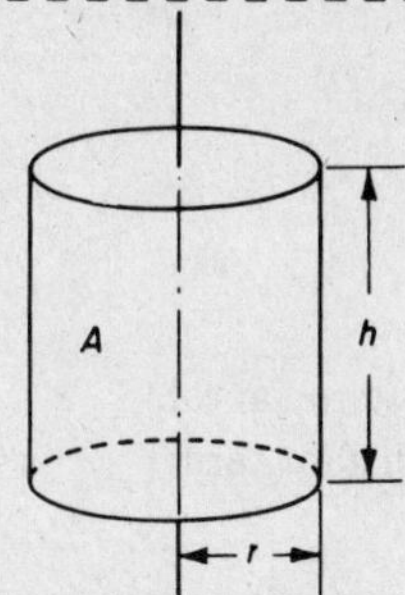


*Another Example*

Let us consider the area of the curved surface of the cylinder.

$$A = 2\pi r h$$

A is a function of $r$ and $h$, so we can find $\frac{\partial A}{\partial r}$ and $\frac{\partial A}{\partial h}$

To find $\frac{\partial A}{\partial r}$ we differentiate the expression for A with respect to $r$, keeping all other symbols constant.

To find $\frac{\partial A}{\partial h}$ we differentiate the expression for A with respect to $h$, keeping all other symbols constant.

So, if $A = 2\pi r h$, then $\frac{\partial A}{\partial r} =$ .................... and $\frac{\partial A}{\partial h} =$ ....................

**4**

$$A = 2\pi r h \qquad \boxed{\frac{\partial A}{\partial r} = 2\pi h} \quad \text{and} \quad \boxed{\frac{\partial A}{\partial h} = 2\pi r}$$

Of course, we are not restricted to the mensuration of the cylinder. The same will happen with any function which is a function of two independent variables. For example, consider $z = x^2 y^3$.

Here $z$ is a function of $x$ and $y$. We can therefore find $\frac{\partial z}{\partial x}$ and $\frac{\partial z}{\partial y}$

(i) To find $\frac{\partial z}{\partial x}$, differentiate w.r.t. $x$, regarding $y$ as a constant.

$$\therefore \frac{\partial z}{\partial x} = 2x \cdot y^3 = \underline{2xy^3}$$

(ii) To find $\frac{\partial z}{\partial y}$, differentiate w.r.t. $y$, regarding $x$ as a constant.

$$\frac{\partial z}{\partial y} = x^2\, 3y^2 = \underline{3x^2y^2}$$

Partial differentiation is easy! For we regard every independent variable, except the one with respect to which we are differentiating, as being for the time being ....................

**5**

constant

Here are one or two examples:

*Example 1.* $u = x^2 + xy + y^2$

(i) To find $\frac{\partial u}{\partial x}$, we regard $y$ as being constant.

Partial diff. w.r.t. $x$ of $x^2 = 2x$

" " " " " $xy = y$ ($y$ is a constant factor)

" " " " " $y^2 = 0$ ($y^2$ is a constant term)

$$\frac{\partial u}{\partial x} = \underline{2x + y}$$

(ii) To find $\frac{\partial u}{\partial y}$, we regard $x$ as being constant.

Partial diff. w.r.t. $y$ of $x^2 = 0$ ($x^2$ is a constant term)

" " " " " $xy = x$ ($x$ is a constant factor)

" " " " " $y^2 = 2y$

$$\frac{\partial u}{\partial y} = \underline{x + 2y}$$

*Another example on frame 6.*

**6**

*Example 2.* $z = x^3 + y^3 - 2x^2y$

$$\frac{\partial z}{\partial x} = 3x^2 + 0 - 4xy = \underline{3x^2 - 4xy}$$

$$\frac{\partial z}{\partial y} = 0 + 3y^2 - 2x^2 = \underline{3y^2 - 2x^2}$$

And it is all just as easy as that.

*Example 3.* $z = (2x - y)(x + 3y)$

This is a product, and the usual product rule applies except that we keep $y$ constant when finding $\frac{\partial z}{\partial x}$, and $x$ constant when finding $\frac{\partial z}{\partial y}$

$$\frac{\partial z}{\partial x} = (2x - y)(1 + 0) + (x + 3y)(2 - 0)$$
$$= 2x - y + 2x + 6y = \underline{4x + 5y}$$

$$\frac{\partial z}{\partial y} = (2x - y)(0 + 3) + (x + 3y)(0 - 1)$$
$$= 6x - 3y - x - 3y = \underline{5x - 6y}$$

Here is one for you to do.

If $z = (4x - 2y)(3x + 5y)$, find $\frac{\partial z}{\partial x}$ and $\frac{\partial z}{\partial y}$

*Find the results and then turn on to frame 7.*

**7**

*Results:* $\dfrac{\partial z}{\partial x} = 24x + 14y$ $\quad$ $\dfrac{\partial z}{\partial y} = 14x - 20y$

For $z = (4x - 2y)(3x + 5y)$, i.e. product

$$\therefore \frac{\partial z}{\partial x} = (4x - 2y)(3 + 0) + (3x + 5y)(4 - 0)$$
$$= 12x - 6y + 12x + 20y = \underline{24x + 14y}$$
$$\frac{\partial z}{\partial y} = (4x - 2y)(0 + 5) + (3x + 5y)(0 - 2)$$
$$= 20x - 10y - 6x - 10y = \underline{14x - 20y}$$

There we are. Now what about this one?

*Example 4.* If $z = \dfrac{2x - y}{x + y}$, find $\dfrac{\partial z}{\partial x}$ and $\dfrac{\partial z}{\partial y}$

Applying the quotient rule, we have

$$\frac{\partial z}{\partial x} = \frac{(x + y)(2 - 0) - (2x - y)(1 + 0)}{(x + y)^2} = \underline{\frac{3y}{(x + y)^2}}$$

and $$\frac{\partial z}{\partial y} = \frac{(x + y)(0 - 1) - (2x - y)(0 + 1)}{(x + y)^2} = \underline{\frac{-3x}{(x + y)^2}}$$

That was not difficult. Now you do this one:

If $z = \dfrac{5x + y}{x - 2y}$, find $\dfrac{\partial z}{\partial x}$ and $\dfrac{\partial z}{\partial y}$

*When you have finished, on to the next frame.*

---

**8**

$\dfrac{\partial z}{\partial x} = \dfrac{-11y}{(x - 2y)^2}$ $\quad$ $\dfrac{\partial z}{\partial y} = \dfrac{11x}{(x - 2y)^2}$

Here is the working:

(i) To find $\dfrac{\partial z}{\partial x}$, we regard $y$ as being constant.

$$\therefore \frac{\partial z}{\partial x} = \frac{(x - 2y)(5 + 0) - (5x + y)(1 - 0)}{(x - 2y)^2}$$
$$= \frac{5x - 10y - 5x - y}{(x - 2y)^2} = \underline{\frac{-11y}{(x - 2y)^2}}$$

(ii) To find $\dfrac{\partial z}{\partial y}$, we regard $x$ as being constant.

$$\therefore \frac{\partial z}{\partial y} = \frac{(x - 2y)(0 + 1) - (5x + y)(0 - 2)}{(x - 2y)^2}$$
$$= \frac{x - 2y + 10x + 2y}{(x - 2y)^2} = \underline{\frac{11x}{(x - 2y)^2}}$$

In practice, we do not write down the zeros that occur in the working, but that is how we think.

*Let us do one more example, so turn on to the next frame.*

**9** *Example 5.* If $z = \sin(3x + 2y)$ find $\dfrac{\partial z}{\partial x}$ and $\dfrac{\partial z}{\partial y}$

Here we have what is clearly a 'function of a function'. So we apply the usual procedure, except to remember that when we are finding

(i) $\dfrac{\partial z}{\partial x}$, we treat $y$ as constant, and

(ii) $\dfrac{\partial z}{\partial y}$, we treat $x$ as constant.

Here goes then.

$$\frac{\partial z}{\partial x} = \cos(3x + 2y) \times \frac{\partial}{\partial x}(3x + 2y)$$
$$= \cos(3x + 2y) \times 3 = \underline{3\cos(3x + 2y)}$$
$$\frac{\partial z}{\partial y} = \cos(3x + 2y) \times \frac{\partial}{\partial y}(3x + 2y)$$
$$= \cos(3x + 2y) \times 2 = \underline{2\cos(3x + 2y)}$$

There it is. So in partial differentiation, we can apply all the ordinary rules of normal differentiation, except that we regard the independent variables other than the one we are using, as being for the time being ....................

**10**

constant

□□□□□□□□□□□□□□□□□□□□□□□□□□□□□□□□□□□□□□□□□□□□

Fine. Now here is a short exercise for you to do by way of revision

***Exercise***

In each of the following cases, find $\dfrac{\partial z}{\partial x}$ and $\dfrac{\partial z}{\partial y}$

1. $z = 4x^2 + 3xy + 5y^2$
2. $z = (3x + 2y)(4x - 5y)$
3. $z = \tan(3x + 4y)$
4. $z = \dfrac{\sin(3x + 2y)}{xy}$

*Finish them all, then turn to frame 11 for the results.*

**11**

Here are the answers:

1. $z = 4x^2 + 3xy + 5y^2$

$$\frac{\partial z}{\partial x} = \underline{8x + 3y} \qquad \frac{\partial z}{\partial y} = \underline{3x + 10y}$$

2. $z = (3x + 2y)(4x - 5y)$

$$\frac{\partial z}{\partial x} = \underline{24x - 7y} \qquad \frac{\partial z}{\partial y} = \underline{-7x - 20y}$$

3. $z = \tan(3x + 4y)$

$$\frac{\partial z}{\partial x} = \underline{3\sec^2(3x + 4y)} \qquad \frac{\partial z}{\partial y} = \underline{4\sec^2(3x + 4y)}$$

4. $z = \dfrac{\sin(3x + 2y)}{xy}$

$$\frac{\partial z}{\partial x} = \underline{\frac{3x\cos(3x + 2y) - \sin(3x + 2y)}{x^2 y}}$$

$$\frac{\partial z}{\partial y} = \underline{\frac{2y\cos(3x + 2y) - \sin(3x + 2y)}{xy^2}}$$

□□□□□□□□□□□□□□□□□□□□□□□□□□□□□□□□□□□□□□

If you have got *all* the answers correct, turn straight on to *frame 15*.
If you have not got all these answers, or are at all uncertain, move to *frame 12*.

**12**

Let us work through these examples in detail.

1. $z = 4x^2 + 3xy + 5y^2$

To find $\dfrac{\partial z}{\partial x}$, regard $y$ as a constant.

$$\therefore \frac{\partial z}{\partial x} = 8x + 3y + 0, \text{ i.e. } 8x + 3y \quad \therefore \frac{\partial z}{\partial x} = \underline{8x + 3y}$$

Similarly, regarding $x$ as constant,

$$\frac{\partial z}{\partial y} = 0 + 3x + 10y, \text{ i.e. } 3x + 10y \quad \therefore \frac{\partial z}{\partial y} = \underline{3x + 10y}$$

2. $z = (3x + 2y)(4x - 5y)$ Product rule.

$$\frac{\partial z}{\partial x} = (3x + 2y)(4) + (4x - 5y)(3)$$
$$= 12x + 8y + 12x - 15y = \underline{24x - 7y}$$

$$\frac{\partial z}{\partial y} = (3x + 2y)(-5) + (4x - 5y)(2)$$
$$= -15x - 10y + 8x - 10y = \underline{-7x - 20y}$$

*Turn on for the solutions to Nos. 3 and 4.*

**13**

3. $z = \tan(3x + 4y)$

$$\frac{\partial z}{\partial x} = \sec^2(3x + 4y)\,(3) = \underline{3\sec^2(3x + 4y)}$$

$$\frac{\partial z}{\partial y} = \sec^2(3x + 4y)\,(4) = \underline{4\sec^2(3x + 4y)}$$

4. $z = \dfrac{\sin(3x + 2y)}{xy}$

$$\frac{\partial z}{\partial x} = \frac{xy\cos(3x + 2y)\,(3) - \sin(3x + 2y)\,(y)}{x^2y^2}$$

$$= \underline{\frac{3x\cos(3x + 2y) - \sin(3x + 2y)}{x^2y}}$$

Now have another go at finding $\dfrac{\partial z}{\partial y}$ in the same way.

*Then check it with frame 14.*

**14**

Here it is:

$$z = \frac{\sin(3x + 2y)}{xy}$$

$$\therefore\ \frac{\partial z}{\partial y} = \frac{xy\cos(3x + 2y).(2) - \sin(3x + 2y).(x)}{x^2y^2}$$

$$= \underline{\frac{2y\cos(3x + 2y) - \sin(3x + 2y)}{xy^2}}$$

That should have cleared up any troubles. This business of partial differentiation is perfectly straightforward. All you have to remember is that for the time being, all the independent variables except the one you are using are kept constant – and behave like constant factors or constant terms according to their positions.

*On you go now to frame 15 and continue the programme.*

**15**

Right. Now let us move on a step.

Consider $z = 3x^2 + 4xy - 5y^2$

$$\text{Then } \frac{\partial z}{\partial x} = 6x + 4y \text{ and } \frac{\partial z}{\partial y} = 4x - 10y$$

The expression $\frac{\partial z}{\partial x} = 6x + 4y$ is itself a function of $x$ and $y$. We could therefore find its partial differential coefficients with respect to $x$ or to $y$.

(i) If we differentiate it partially w.r.t. $x$, we get:

$\frac{\partial}{\partial x}\left\{\frac{\partial z}{\partial x}\right\}$ and this is written $\frac{\partial^2 z}{\partial x^2}$ (much like an ordinary second differential coefficient, but with the partial $\partial$)

$$\therefore \frac{\partial^2 z}{\partial x^2} = \frac{\partial}{\partial x}(6x + 4y) = 6$$

This is called the second partial differential coefficient of $z$ with respect to $x$.

(ii) If we differentiate partially w.r.t. $y$, we get:

$$\frac{\partial}{\partial y}\left\{\frac{\partial z}{\partial x}\right\} \text{ and this is written } \frac{\partial^2 z}{\partial y.\partial x}$$

Note that the operation now being performed is given by the left-hand of the two symbols in the denominator.

$$\frac{\partial^2 z}{\partial y.\partial x} = \frac{\partial}{\partial y}\left\{\frac{\partial z}{\partial x}\right\} = \frac{\partial}{\partial y}\left\{6x + 4y\right\} = 4$$

---

**16**

So we have this:

$$z = 3x^2 + 4xy - 5y^2$$

$$\frac{\partial z}{\partial x} = 6x + 4y \qquad \frac{\partial z}{\partial y} = 4x - 10y$$

$$\frac{\partial^2 z}{\partial x^2} = 6$$

$$\frac{\partial^2 z}{\partial y.\partial x} = 4$$

Of course, we could carry out similar steps with the expression for $\frac{\partial z}{\partial y}$ on the right. This would give us:

$$\frac{\partial^2 z}{\partial y^2} = -10$$

$$\frac{\partial^2 z}{\partial x.\partial y} = 4$$

Note that $\frac{\partial^2 z}{\partial y.\partial x}$ means $\frac{\partial}{\partial y}\left\{\frac{\partial z}{\partial x}\right\}$ so $\frac{\partial^2 z}{\partial x.\partial y}$ means ....................

**17**

$$\frac{\partial^2 z}{\partial x.\partial y} \text{ means } \frac{\partial}{\partial x}\left\{\frac{\partial z}{\partial y}\right\}$$

---

Collecting our previous results together then, we have

$$z = 3x^2 + 4xy - 5y^2$$

$$\frac{\partial z}{\partial x} = 6x + 4y \qquad \frac{\partial z}{\partial y} = 4x - 10y$$

$$\frac{\partial^2 z}{\partial x^2} = 6 \qquad \frac{\partial^2 z}{\partial y^2} = -10$$

$$\frac{\partial^2 z}{\partial y.\partial x} = 4 \qquad \frac{\partial^2 z}{\partial x.\partial y} = 4$$

We see, in this case, that $\dfrac{\partial^2 z}{\partial y.\partial x} = \dfrac{\partial^2 z}{\partial x.\partial y}$

There are then, *two* first differential coefficients, and *four* second differential coefficients, though the last two seem to have the same value.

Here is one for you to do.

If $z = 5x^3 + 3x^2y + 4y^3$, find $\dfrac{\partial z}{\partial x}, \dfrac{\partial z}{\partial y}, \dfrac{\partial^2 z}{\partial x^2}, \dfrac{\partial^2 z}{\partial y^2}, \dfrac{\partial^2 z}{\partial x.\partial y}, \dfrac{\partial^2 z}{\partial y.\partial x}$

*When you have completed all that, turn to frame 18.*

---

**18**

Here are the results:

$$z = 5x^3 + 3x^2y + 4y^3$$

$$\frac{\partial z}{\partial x} = 15x^2 + 6xy \qquad \frac{\partial z}{\partial y} = 3x^2 + 12y^2$$

$$\frac{\partial^2 z}{\partial x^2} = 30x + 6y \qquad \frac{\partial^2 z}{\partial y^2} = 24y$$

$$\frac{\partial^2 z}{\partial y.\partial x} = 6x \qquad \frac{\partial^2 z}{\partial x.\partial y} = 6x$$

Again in this example also, we see that $\dfrac{\partial^2 z}{\partial y.\partial x} = \dfrac{\partial^2 z}{\partial x.\partial y}$. Now do this one.

It looks more complicated, but is done in just the same way. Do not rush at it; take your time and all will be well. Here it is. Find all the first and second partial differential coefficients of $z = x.\cos y - y.\cos x$.

*Then to frame 19.*

**19**

Check your results with these.

$$z = x\cos y - y.\cos x$$

When differentiating w.r.t. $x$, $y$ is constant (and therefore $\cos y$ also)
" " " $y, x$ " " ( " " $\cos x$ " )

So we get:

$$\frac{\partial z}{\partial x} = \cos y + y.\sin x \qquad \frac{\partial z}{\partial y} = -x.\sin y - \cos x$$

$$\frac{\partial^2 z}{\partial x^2} = \text{y}.\cos x \qquad \frac{\partial^2 z}{\partial y^2} = -x.\cos y$$

$$\frac{\partial^2 z}{\partial y.\partial x} = -\sin y + \sin x \qquad \frac{\partial^2 z}{\partial x.\partial y} = -\sin y + \sin x$$

$$\text{And again,} \quad \frac{\partial^2 z}{\partial y.\partial x} = \frac{\partial^2 z}{\partial x.\partial y}$$

In fact this will always be so for the functions you are likely to meet, so that there are really *three* different second partial diff. coeffts. (and not four). In practice, if you have found $\dfrac{\partial^2 z}{\partial y.\partial x}$ it is a useful check to find $\dfrac{\partial^2 z}{\partial x.\partial y}$ separately. They should give the same result, of course.

---

**20**

What about this one?

If $\text{V} = \ln(x^2 + y^2)$, prove that $\dfrac{\partial^2 \text{V}}{\partial x^2} + \dfrac{\partial^2 \text{V}}{\partial y^2} = 0$

This merely entails finding the two second partial diff. coeffts. and substituting them in the left-hand side of the statement. So here goes:

$$\text{V} = \ln(x^2 + y^2)$$

$$\frac{\partial \text{V}}{\partial x} = \frac{1}{(x^2 + y^2)}\,2x = \frac{2x}{x^2 + y^2}$$

$$\frac{\partial^2 \text{V}}{\partial x^2} = \frac{(x^2 + y^2)2 - 2x.2x}{(x^2 + y^2)^2}$$

$$= \frac{2x^2 + 2y^2 - 4x^2}{(x^2 + y^2)^2} = \frac{2y^2 - 2x^2}{(x^2 + y^2)^2} \quad \ldots\ldots\ldots\ldots \text{(i)}$$

Now you find $\dfrac{\partial^2 \text{V}}{\partial y^2}$ in the same way and hence prove the given identity.

*When you are ready, turn on to frame 21.*

**21** We had found that $\dfrac{\partial^2 V}{\partial x^2} = \dfrac{2y^2 - 2x^2}{(x^2 + y^2)^2}$

So making a fresh start from $V = \ln(x^2 + y^2)$, we get

$$\frac{\partial V}{\partial y} = \frac{1}{x^2 + y^2} \cdot 2y = \frac{2y}{x^2 + y^2}$$

$$\frac{\partial^2 V}{\partial y^2} = \frac{(x^2 + y^2)\,2 - 2y.2y}{(x^2 + y^2)^2}$$

$$= \frac{2x^2 + 2y^2 - 4y^2}{(x^2 + y^2)^2} = \frac{2x^2 - 2y^2}{(x^2 + y^2)^2} \quad \text{.............. (ii)}$$

Substituting now the two results in the identity, gives

$$\frac{\partial^2 V}{\partial x^2} + \frac{\partial^2 V}{\partial y^2} = \frac{2y^2 - 2x^2}{(x^2 + y^2)^2} + \frac{2x^2 - 2y^2}{(x^2 + y^2)^2}$$

$$= \frac{2y^2 - 2x^2 + 2x^2 - 2y^2}{(x^2 + y^2)^2} = \underline{0}$$

*Now on to frame 22.*

**22** Here is another kind of example that you should see.

*Example 1.* If $V = f(x^2 + y^2)$, show that $x\dfrac{\partial V}{\partial y} - y\dfrac{\partial V}{\partial x} = 0$

Here we are told that V is a function of $(x^2 + y^2)$ but the precise nature of the function is not given. However, we can treat this as a 'function of a function' and write $f'(x^2 + y^2)$ to represent the diff. coefft. of the function w.r.t. its own combined variable $(x^2 + y^2)$.

$$\therefore \frac{\partial V}{\partial x} = f'(x^2 + y^2) \times \frac{\partial}{\partial x}(x^2 + y^2) = f'(x^2 + y^2).2x$$

$$\frac{\partial V}{\partial y} = f'(x^2 + y^2).\frac{\partial}{\partial y}(x^2 + y^2) = f'(x^2 + y^2).2y$$

$$\therefore x\frac{\partial V}{\partial y} - y\frac{\partial V}{\partial x} = x.f'(x^2 + y^2).2y - y.f'(x^2 + y^2).2x$$

$$= 2xy.f'(x^2 + y^2) - 2xy.f'(x^2 + y^2)$$

$$= \underline{0}$$

*Let us have another one of that kind on the next frame.*

**23**

*Example 2.* If $z = f\left\{\frac{y}{x}\right\}$, show that $x\frac{\partial z}{\partial x} + y\frac{\partial z}{\partial y} = 0$

Much the same as before.

$$\frac{\partial z}{\partial x} = f'\left\{\frac{y}{x}\right\}.\frac{\partial}{\partial x}\left\{\frac{y}{x}\right\} = f'\left\{\frac{y}{x}\right\}\left(-\frac{y}{x^2}\right) = -\frac{y}{x^2}f'\left\{\frac{y}{x}\right\}$$

$$\frac{\partial z}{\partial y} = f'\left\{\frac{y}{x}\right\}.\frac{\partial}{\partial y}\left\{\frac{y}{x}\right\} = f'\left\{\frac{y}{x}\right\}.\frac{1}{x} = \frac{1}{x}f'\left\{\frac{y}{x}\right\}$$

$$\therefore\ x\frac{\partial z}{\partial x} + y\frac{\partial z}{\partial y} = x\left(-\frac{y}{x^2}\right)f'\left\{\frac{y}{x}\right\} + y\frac{1}{x}f'\left\{\frac{y}{x}\right\}$$

$$= -\frac{y}{x}f'\left\{\frac{y}{x}\right\} + \frac{y}{x}f'\left\{\frac{y}{x}\right\}$$

$$\underline{= 0}$$

And one for you, just to get your hand in.

If $V = f(ax + by)$, show that $b\frac{\partial V}{\partial x} - a\frac{\partial V}{\partial y} = 0$

*When you have done it, check your working against that on frame 24.*

**24**

Here is the working; this is how it goes.

$$V = f(ax + by)$$

$$\therefore\ \frac{\partial V}{\partial x} = f'(ax + by).\frac{\partial}{\partial x}(ax + by)$$

$$= f'(ax + by).a = a.f'(ax + by) \quad \text{.............. (i)}$$

$$\frac{\partial V}{\partial y} = f'(ax + by).\frac{\partial}{\partial y}(ax + by)$$

$$= f'(ax + by).b = b.f'(ax + by) \quad \text{..............(ii)}$$

$$\therefore\ b\frac{\partial V}{\partial x} - a\frac{\partial V}{\partial y} = ab.f'(ax + by) - ab.f'(ax + by)$$

$$\underline{= 0}$$

*Turn on to frame 25.*

**25**

So to sum up so far.

Partial differentiation is easy, no matter how complicated the expression to be differentiated may seem.

To differentiate partially w.r.t. $x$, all independent variables other than $x$ are constant for the time being.

To differentiate partially w.r.t. $y$, all independent variables other than $y$ are constant for the time being.

So that, if $z$ is a function of $x$ and $y$, i.e. if $z = f(x, y)$, we can find

$$\frac{\partial z}{\partial x} \qquad \frac{\partial z}{\partial y}$$

$$\frac{\partial^2 z}{\partial x^2} \qquad \frac{\partial^2 z}{\partial y^2}$$

$$\frac{\partial^2 z}{\partial y.\partial x} \qquad \frac{\partial^2 z}{\partial x.\partial y}$$

And also:

$$\frac{\partial^2 z}{\partial y.\partial x} = \frac{\partial^2 z}{\partial x.\partial y}$$

Now for a few revision examples.

**26**

**Revision Exercise**

1. Find all first and second partial differential coefficients for each of the following functions.

   (i) $z = 3x^2 + 2xy + 4y^2$

   (ii) $z = \sin xy$

   (iii) $z = \dfrac{x + y}{x - y}$

2. If $z = \ln(e^x + e^y)$, show that $\dfrac{\partial z}{\partial x} + \dfrac{\partial z}{\partial y} = 1$.

3. If $z = x.f(xy)$, express $x\dfrac{\partial z}{\partial x} - y\dfrac{\partial z}{\partial y}$ in its simplest form.

*When you have finished check with the solutions on frame 27.*

*Results*

**27**

1. (i) $z = 3x^2 + 2xy + 4y^2$

$$\frac{\partial z}{\partial x} = \underline{6x + 2y} \qquad \frac{\partial z}{\partial y} = \underline{2x + 8y}$$

$$\frac{\partial^2 z}{\partial x^2} = \underline{6} \qquad \frac{\partial^2 z}{\partial y^2} = \underline{8}$$

$$\frac{\partial^2 z}{\partial y . \partial x} = \underline{2} \qquad \frac{\partial^2 z}{\partial x . \partial y} = \underline{2}$$

(ii) $z = \sin xy$

$$\frac{\partial z}{\partial x} = \underline{y \cos xy} \qquad \frac{\partial z}{\partial y} = \underline{x \cos xy}$$

$$\frac{\partial^2 z}{\partial x^2} = \underline{-y^2 \sin xy} \qquad \frac{\partial^2 z}{\partial y^2} = \underline{-x^2 \sin xy}$$

$$\frac{\partial^2 z}{\partial y . \partial x} = y(-x \sin xy) + \cos xy = \underline{\cos xy - xy \sin xy}$$

$$\frac{\partial^2 z}{\partial x . \partial y} = x(-y \sin xy) + \cos xy = \underline{\cos xy - xy \sin xy}$$

(iii) $z = \dfrac{x+y}{x-y}$

$$\frac{\partial z}{\partial x} = \frac{(x-y)1 - (x+y)1}{(x-y)^2} = \underline{\frac{-2y}{(x-y)^2}}$$

$$\frac{\partial z}{\partial y} = \frac{(x-y)1 - (x+y)(-1)}{(x-y)^2} = \underline{\frac{2x}{(x-y)^2}}$$

$$\frac{\partial^2 z}{\partial x^2} = (-2y)\frac{(-2)}{(x-y)^3} = \underline{\frac{4y}{(x-y)^3}}.$$

$$\frac{\partial^2 z}{\partial y^2} = 2x \frac{(-2)}{(x-y)^3}(-1) = \underline{\frac{4x}{(x-y)^3}}$$

$$\frac{\partial^2 z}{\partial y . \partial x} = \frac{(x-y)^2(-2) - (-2y)2(x-y)(-1)}{(x-y)^4}$$

$$= \frac{-2(x-y)^2 - 4y(x-y)}{(x-y)^4}$$

$$= \frac{-2}{(x-y)^2} - \frac{4y}{(x-y)^3}$$

$$= \frac{-2x + 2y - 4y}{(x-y)^3} = \underline{\frac{-2x - 2y}{(x-y)^3}}$$

/continued

Continuation of frame 27.

$$\frac{\partial^2 z}{\partial x.\partial y} = \frac{(x-y)^2(2) - 2x.2(x-y)1}{(x-y)^4}$$

$$= \frac{2(x-y)^2 - 4x(x-y)}{(x-y)^4}$$

$$= \frac{2}{(x-y)^2} - \frac{4x}{(x-y)^3}$$

$$= \frac{2x-2y-4x}{(x-y)^3} = \underline{\frac{-2x-2y}{(x-y)^3}}$$

2. $z = \ln(e^x + e^y)$

$$\frac{\partial z}{\partial x} = \frac{1}{e^x + e^y}.e^x \qquad \frac{\partial z}{\partial y} = \frac{1}{e^x + e^y}.e^y$$

$$\frac{\partial z}{\partial x} + \frac{\partial z}{\partial y} = \frac{e^x}{e^x + e^y} + \frac{e^y}{e^x + e^y}$$

$$= \frac{e^x + e^y}{e^x + e^y} = 1$$

$$\underline{\frac{\partial z}{\partial x} + \frac{\partial z}{\partial y} = 1}$$

3. $z = x.f(xy)$

$$\frac{\partial z}{\partial x} = x.f'(xy).y + f(xy)$$

$$\frac{\partial z}{\partial y} = x.f'(xy).x$$

$$x\frac{\partial z}{\partial x} - y\frac{\partial z}{\partial y} = x^2y\,f'(xy) + x\,f(xy) - x^2y\,f'(xy)$$

$$\underline{x\frac{\partial z}{\partial x} - y\frac{\partial z}{\partial y} = x\,f(xy) = z}$$

That was a pretty good revision test. Do not be unduly worried if you made a slip or two in your working. Try to avoid doing so, of course, but you are doing fine. Now on to the next part of the programme.

*Turn on to frame 28.*

**28**

So far we have been concerned with the technique of partial differentiation. Now let us look at one of its applications.

**Small increments**

If we return to the volume of the cylinder with which we started this programme, we have once again that $V = \pi r^2 h$. We have seen that we can

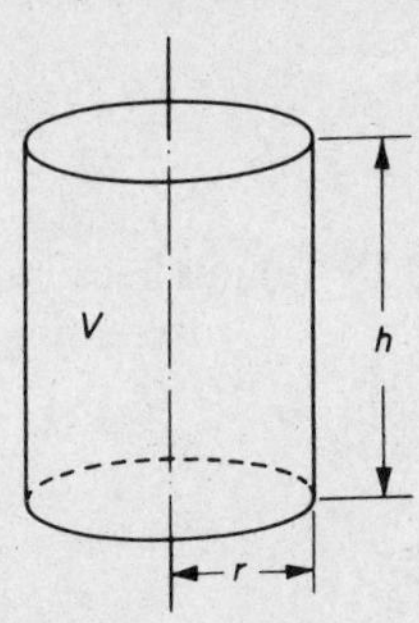


find $\frac{\partial V}{\partial r}$ with $h$ constant, and $\frac{\partial V}{\partial h}$ with $r$ constant.

$$\frac{\partial V}{\partial r} = 2\pi r h; \quad \frac{\partial V}{\partial h} = \pi r^2$$

Now let us see what we get if $r$ and $h$ both change simultaneously.

If $r$ becomes $r + \delta r$, and $h$ becomes $h + \delta h$, let V become $V + \delta V$. Then the new volume is given by

$$\begin{aligned} V + \delta V &= \pi(r + \delta r)^2(h + \delta h) \\ &= \pi(r^2 + 2r.\delta r + \delta r^2)(h + \delta h) \\ &= \pi(r^2 h + 2rh\delta r + h\delta r^2 + r^2\delta h + 2r\delta r\delta h + \delta r^2 .\delta h) \end{aligned}$$

Subtract $V = \pi r^2 h$ from each side, giving

$$\delta V = \pi(2rh.\delta r + h.\delta r^2 + r^2\delta h + 2r\delta r\delta h + \delta r^2.\delta h)$$

$\simeq \pi(2rh\delta r + r^2.\delta h)$ since $\delta r$ and $\delta h$ are small and all the remaining terms are of a higher degree of smallness.

$$\therefore \delta V \simeq 2\pi rh\delta r + \pi r^2\delta h$$

$$\delta V \simeq \frac{\partial V}{\partial r}\delta r + \frac{\partial V}{\partial h}\delta h$$

Let us now do a numerical example to see how it all works out.

*On to frame 29.*

**29**

*Example.*

A cylinder has dimensions $r = 5$ cm, $h = 10$ cm. Find the approximate increase in volume when $r$ increases by 0·2 cm and $h$ decreases by 0·1 cm.

Well now, $$V = \pi r^2 h$$

$$\frac{\partial V}{\partial r} = 2\pi rh \qquad \frac{\partial V}{\partial h} = \pi r^2$$

In this case, when $r = 5$ cm, $h = 10$ cm,

$$\frac{\partial V}{\partial r} = 2\pi 5.10 = 100\pi \qquad \frac{\partial V}{\partial h} = \pi r^2 = \pi 5^2 = 25\pi$$

$\delta r = 0{\cdot}2$ and $\delta h = -0{\cdot}1$ (minus because $h$ is decreasing)

$$\therefore\ \delta V \simeq \frac{\partial V}{\partial r}.\delta r + \frac{\partial V}{\partial h}.\delta h$$

$$\delta V = 100\pi(0{\cdot}2) + 25\pi(-0{\cdot}1)$$

$$= 20\pi - 2{\cdot}5\pi = 17{\cdot}5\pi$$

$$\therefore\ \delta V \simeq 54{\cdot}96 \text{ cm}^3$$

i.e. the volume increases by 54·96 cubic centimetres.

Just like that!

**30**

This kind of result applies not only to the volume of a cylinder, but to any function of two independent variables.

*Example.* If $z$ is a function of $x$ and $y$, i.e. $z = f(x, y)$ and if $x$ and $y$ increase by small amounts $\delta x$ and $\delta y$, the increase $\delta z$ will also be relatively small.

If we expand $\delta z$ in powers of $\delta x$ and $\delta y$, we get

$\delta z = A\delta x + B\,\delta y$ + higher powers of $\delta x$ and $\delta y$, where A and B are functions of $x$ and $y$.

If $y$ remains constant, so that $\delta y = 0$, then

$$\delta z = A\delta x + \text{higher powers of } \delta x$$

$$\therefore\ \frac{\delta z}{\delta x} = A. \text{ So that if } \delta x \to 0, \text{ this becomes } A = \frac{\partial z}{\partial x}$$

Similarly, if $x$ remains constant, making $\delta y \to 0$ gives $B = \dfrac{\partial z}{\partial y}$

$$\therefore\ \delta z = \frac{\partial z}{\partial x}\,\delta x + \frac{\partial z}{\partial y}\,\delta y + \text{higher powers of very small quantities which can be ignored.}$$

$$\underline{\delta z = \frac{\partial z}{\partial x}\,\delta x + \frac{\partial z}{\partial y}\,\delta y}$$

**31**

So, if
$$z = f(x, y)$$
$$\delta z = \frac{\partial z}{\partial x}\delta x + \frac{\partial z}{\partial y}\delta y$$

This is the key to all the forthcoming applications and will be quoted over and over again.

The result is quite general and a similar result applies for a function of three independent variables

e.g. If $z = f(x, y, w)$

$$\text{then } \delta z = \frac{\partial z}{\partial x}\delta x + \frac{\partial z}{\partial y}\delta y + \frac{\partial z}{\partial w}\delta w$$

If we remember the rule for a function of two independent variables, we can easily extend it when necessary.

Here it is once again:

If $z = f(x, y)$ then $\underline{\delta z = \frac{\partial z}{\partial x}\delta x + \frac{\partial z}{\partial y}\delta y}$

Copy this result into your record book in a prominent position, such as it deserves!

**32**

Now for an example or two.

*Example 1.* If $I = \frac{V}{R}$, and V = 250 volts and R = 50 ohms, find the change in I resulting from an increase of 1 volt in V and an increase of 0·5 ohm in R.

$$I = f(V, R) \quad \therefore \delta I = \frac{\partial I}{\partial V}\delta V + \frac{\partial I}{\partial R}\delta R$$
$$\frac{\partial I}{\partial V} = \frac{1}{R} \text{ and } \frac{\partial I}{\partial R} = -\frac{V}{R^2}$$
$$\therefore \delta I = \frac{1}{R}\delta V - \frac{V}{R^2}\delta R$$

So when $R = 50$, $V = 250$, $\delta V = 1$, and $\delta R = 0{\cdot}5$,

$$\delta I = \frac{1}{50}(1) - \frac{250}{2500}(0{\cdot}5)$$
$$= \frac{1}{50} - \frac{1}{20}$$
$$= 0{\cdot}02 - 0{\cdot}05 = -0{\cdot}03$$

i.e. <u>I decreases by 0·03 amperes.</u>

**33** Here is another example.

*Example 2.* If $y = \dfrac{ws^3}{d^4}$, find the percentage increase in $y$, when $w$ increases by 2 per cent, $s$ decreases by 3 per cent and $d$ increases by 1 per cent.

Notice that, in this case, $y$ is a function of three variables, $w$, $s$ and $d$. The formula therefore becomes:

$$\delta y = \frac{\partial y}{\partial w}\delta w + \frac{\partial y}{\partial s}\delta s + \frac{\partial y}{\partial d}\delta d$$

We have $\quad \dfrac{\partial y}{\partial w} = \dfrac{s^3}{d^4}; \quad \dfrac{\partial y}{\partial s} = \dfrac{3ws^2}{d^4}; \quad \dfrac{\partial y}{\partial d} = -\dfrac{4ws^3}{d^5}$

$$\therefore \quad \delta y = \frac{s^3}{d^4}\delta w + \frac{3ws^2}{d^4}\delta s + \frac{-4ws^3}{d^5}\delta d$$

Now then, what are the values of $\delta w$, $\delta s$ and $\delta d$?

Is it true to say that $\quad \delta w = \dfrac{2}{100}; \ \delta s = \dfrac{-3}{100}; \ \delta d = \dfrac{1}{100}$?

If not, why not?

*Next frame.*

**34** No. It is not correct. For $\delta w$ is not $\dfrac{2}{100}$ of a unit, but 2 per cent of $w$,

i.e. $\delta w = \dfrac{2}{100}$ of $w = \dfrac{2w}{100}$

Similarly, $\delta s = \dfrac{-3}{100}$ of $s = \dfrac{-3s}{100}$ and $\delta d = \dfrac{d}{100}$. Now that we have cleared that point up, we can continue with the problem.

$$\begin{aligned}
\delta y &= \frac{s^3}{d^4}\left(\frac{2w}{100}\right) + \frac{3ws^2}{d^4}\left(\frac{-3s}{100}\right) - \frac{4ws^3}{d^5}\left(\frac{d}{100}\right)\\
&= \frac{ws^3}{d^4}\left(\frac{2}{100}\right) - \frac{ws^3}{d^4}\left(\frac{9}{100}\right) - \frac{ws^3}{d^4}\left(\frac{4}{100}\right)\\
&= \frac{ws^3}{d^4}\left\{\frac{2}{100} - \frac{9}{100} - \frac{4}{100}\right\}\\
&= y\left\{-\frac{11}{100}\right\} = -11 \text{ per cent of } y
\end{aligned}$$

i.e. $\underline{y \text{ decreases by } 11 \text{ per cent}}$

Remember that where the increment of $w$ is given as 2 per cent, it is *not* $\dfrac{2}{100}$ of a unit, but $\dfrac{2}{100}$ of $w$, and the symbol $w$ must be included.

*Turn on to frame 35.*

**35**

Now here is one for you to do.

*Exercise*

$P = w^2hd$. If errors of up to 1% (plus or minus) are possible in the measured values of $w$, $h$ and $d$, find the maximum possible percentage error in the calculated value of P.

This is very much like the last example, so you will be able to deal with it without any trouble. Work it right through and then turn on to frame 36 and check your result.

**36**

$$P = w^2hd. \quad \therefore \ \delta P = \frac{\partial P}{\partial w}.\delta w + \frac{\partial P}{\partial h}.\delta h + \frac{\partial P}{\partial d}.\delta d$$

$$\frac{\partial P}{\partial w} = 2whd; \quad \frac{\partial P}{\partial h} = w^2d; \quad \frac{\partial P}{\partial d} = w^2h$$

$$\delta P = 2whd.\delta w + w^2d.\delta h + w^2h.\delta d$$

Now $\quad \delta w = \pm \dfrac{w}{100}, \quad \delta h = \pm \dfrac{h}{100}, \quad \delta d = \pm \dfrac{d}{100}$

$$\delta P = 2whd\left(\pm \frac{w}{100}\right) + w^2d\left(\pm \frac{h}{100}\right) + w^2h\left(\pm \frac{d}{100}\right)$$

$$= \pm \frac{2w^2hd}{100} \pm \frac{w^2dh}{100} \pm \frac{w^2hd}{100}$$

The greatest possible error in P will occur when the signs are chosen so that they are all of the same kind, i.e. all plus or all minus. If they were mixed, they would tend to cancel each other out.

$$\therefore \ \delta P = \pm w^2hd\left\{\frac{2}{100} + \frac{1}{100} + \frac{1}{100}\right\} = \pm P\left(\frac{4}{100}\right)$$

$\therefore$ Maximum possible error in P is 4% of P

Finally, here is one last example for you to do. Work right through it and then check your results with those on frame 37.

*Exercise.* The two sides forming the right-angle of a right-angled triangle are denoted by $a$ and $b$. The hypotenuse is $h$. If there are possible errors of $\pm 0{\cdot}5\%$ in measuring $a$ and $b$, find the maximum possible error in calculating (i) the area of the triangle and (ii) the length of $h$.

**37** *Results:*

(i) $\delta A = 1\%$ of A
(ii) $\delta h = 0{\cdot}5\%$ of $h$

□□□□□□□□□□□□□□□□□□□□□□□□□□□□□□□□□□□□□□□□

Here is the working in detail:

(i) $A = \dfrac{a.b}{2}$ $\qquad \delta A = \dfrac{\partial A}{\partial a}.\delta a + \dfrac{\partial A}{\partial b}.\delta b$

$$\frac{\partial A}{\partial a} = \frac{b}{2}; \quad \frac{\partial A}{\partial b} = \frac{a}{2}; \quad \delta a = \pm\frac{a}{200}; \quad \delta b = \pm\frac{b}{200}$$

h A a b

$$\delta A = \frac{b}{2}\left(\pm\frac{a}{200}\right) + \frac{a}{2}\left(\pm\frac{b}{200}\right)$$

$$= \pm\frac{a.b}{2}\left[\frac{1}{200} + \frac{1}{200}\right] = \pm A.\frac{1}{100}$$

$$\therefore \underline{\delta A = 1\% \text{ of A}}$$

(ii)

$$h = \sqrt{(a^2 + b^2)} = (a^2 + b^2)^{\frac{1}{2}}$$

$$\delta h = \frac{\partial h}{\partial a}\,\delta a + \frac{\partial h}{\partial b}\,\delta b$$

$$\frac{\partial h}{\partial a} = \tfrac{1}{2}(a^2 + b^2)^{-\frac{1}{2}}(2a) = \frac{a}{\sqrt{(a^2 + b^2)}}$$

$$\frac{\partial h}{\partial b} = \tfrac{1}{2}(a^2 + b^2)^{-\frac{1}{2}}(2b) = \frac{b}{\sqrt{(a^2 + b^2)}}$$

Also $\qquad \delta a = \pm\dfrac{a}{200}; \; \delta b = \pm\dfrac{b}{200}$

$$\therefore \delta h = \frac{a}{\sqrt{(a^2 + b^2)}}\left(\pm\frac{a}{200}\right) + \frac{b}{\sqrt{(a^2 + b^2)}}\left(\pm\frac{b}{200}\right)$$

$$= \pm\frac{1}{200}\,\frac{a^2 + b^2}{\sqrt{(a^2 + b^2)}}$$

$$= \pm\frac{1}{200}\sqrt{(a^2 + b^2)} = \pm\frac{1}{200}(h)$$

$$\therefore \underline{\delta h = 0{\cdot}5\% \text{ of } h}$$

That brings us to the end of this particular programme. We shall meet partial differentiation again in a later programme when we shall consider some more of its applications. But for the time being, there remains only the Test Exercise on the next frame. Take your time over the questions; do them carefully.

*So on now to frame 38.*

## Test Exercise – X

**38**

Answer all questions.

1. Find all first and second partial differential coefficients of the following:

$$\text{(i)}\ z = 4x^3 - 5xy^2 + 3y^3$$
$$\text{(ii)}\ z = \cos(2x + 3y)$$
$$\text{(iii)}\ z = e^{(x^2 - y^2)}$$
$$\text{(iv)}\ z = x^2 \sin(2x + 3y)$$

2. (i) If $V = x^2 + y^2 + z^2$, express in its simplest form

$$x\frac{\partial V}{\partial x} + y\frac{\partial V}{\partial y} + z\frac{\partial V}{\partial z}$$

(ii) If $z = f(x + ay) + F(x - ay)$, find $\dfrac{\partial^2 z}{\partial x^2}$ and $\dfrac{\partial^2 z}{\partial y^2}$ and hence prove that $\dfrac{\partial^2 z}{\partial y^2} = a^2 . \dfrac{\partial^2 z}{\partial x^2}$

3. The power P dissipated in a resistor is given by $P = \dfrac{E^2}{R}$. If E = 200 volts and R = 8 ohms, find the change in P resulting from a drop of 5 volts in E and an increase of 0·2 ohm in R.

4. If $\theta = kHLV^{-\frac{1}{2}}$, where $k$ is a constant, and there are possible errors of ± 1% in measuring H, L and V, find the maximum possible error in the calculated value of $\theta$.

That's it.

## Further Problems – X

1. If $z = \dfrac{1}{x^2 + y^2 - 1}$, show that $x\dfrac{\partial z}{\partial x} + y\dfrac{\partial z}{\partial y} = -2z(1 + z)$.

2. Prove that, if $V = \ln(x^2 + y^2)$, then $\dfrac{\partial^2 V}{\partial x^2} + \dfrac{\partial^2 V}{\partial y^2} = 0$.

3. If $z = \sin(3x + 2y)$, verify that $3\dfrac{\partial^2 z}{\partial y^2} - 2\dfrac{\partial^2 z}{\partial x^2} = 6z$.

4. If $u = \dfrac{x + y + z}{(x^2 + y^2 + z^2)^{\frac{1}{2}}}$, show that $x\dfrac{\partial u}{\partial x} + y\dfrac{\partial u}{\partial y} + z\dfrac{\partial u}{\partial z} = 0$.

5. Show that the equation $\dfrac{\partial^2 z}{\partial x^2} + \dfrac{\partial^2 z}{\partial y^2} = 0$, is satisfied by

$$z = \ln \sqrt{(x^2 + y^2)} + \tfrac{1}{2}\tan^{-1}\left(\frac{y}{x}\right)$$

6. If $z = e^x(x\cos y - y\sin y)$, show that $\dfrac{\partial^2 z}{\partial x^2} + \dfrac{\partial^2 z}{\partial y^2} = 0$.

7. If $u = (1 + x)\sinh(5x - 2y)$, verify that

$$4\frac{\partial^2 u}{\partial x^2} + 20\frac{\partial^2 u}{\partial x.\partial y} + 25\frac{\partial^2 u}{\partial y^2} = 0$$

8. If $z = f\left(\dfrac{y}{x}\right)$, show that

$$x^2\frac{\partial^2 z}{\partial x^2} + 2xy\frac{\partial^2 z}{\partial x.\partial y} + y^2\frac{\partial^2 z}{\partial y^2} = 0$$

9. If $z = (x + y).f\left(\dfrac{y}{x}\right)$, where $f$ is an arbitrary function, show that

$$x\frac{\partial z}{\partial x} + y\frac{\partial z}{\partial y} = z$$

10. In the formula $D = \dfrac{Eh^3}{12(1 - \nu^2)}$, $h$ is given as $0{\cdot}1 \pm 0{\cdot}002$ and $\nu$ as $0{\cdot}3 \pm 0{\cdot}02$. Express the approximate maximum error in D in terms of E.

11. The formula $z = \dfrac{a^2}{x^2 + y^2 - a^2}$ is used to calculate $z$ from observed values of $x$ and $y$. If $x$ and $y$ have the same percentage error $p$, show that the percentage error in $z$ is approximately $-2p(1 + z)$.

12. In a balanced bridge circuit, $R_1 = R_2R_3/R_4$. If $R_2$, $R_3$, $R_4$, have known tolerances of $\pm x\%$, $\pm y\%$, $\pm z\%$ respectively, determine the maximum percentage error in $R_1$, expressed in terms of $x$, $y$ and $z$.
13. The deflection $y$ at the centre of a circular plate suspended at the edge and uniformly loaded is given by $y = \dfrac{kwd^4}{t^3}$, where $w$ = total load, $d$ = diameter of plate, $t$ = thickness and $k$ is a constant. Calculate the approximate percentage change in $y$ if $w$ is increased by 3%, $d$ is decreased by 2½% and $t$ is increased by 4%.
14. The coefficient of rigidity ($n$) of a wire of length (L) and uniform diameter ($d$) is given by $n = \dfrac{AL}{d^4}$, where A is a constant. If errors of $\pm 0{\cdot}25\%$ and $\pm 1\%$ are possible in measuring L and $d$ respectively, determine the maximum percentage error in the calculated value of $n$.
15. If $k/k_0 = (T/T_0)^n . p/760$, show that the change in $k$ due to small changes of $a\%$ in T and $b\%$ in $p$ is approximately $(na + b)\%$.
16. The deflection $y$ at the centre of a rod is known to be given by $y = \dfrac{kwl^3}{d^4}$, where $k$ is a constant. If $w$ increases by 2%, $l$ by 3%, and $d$ decreases by 2%, find the percentage increase in $y$.
17. The displacement $y$ of a point on a vibrating stretched string, at a distance $x$ from one end, at time $t$, is given by
$$\frac{\partial^2 y}{\partial t^2} = c^2 . \frac{\partial^2 y}{\partial x^2}$$
Show that one solution of this equation is $y = A \sin \dfrac{px}{c} . \sin(pt + a)$, where A, $p$, $c$ and $a$ are constants.
18. If $y = A \sin(px + a) \cos(qt + b)$, find the error in $y$ due to small errors $\delta x$ and $\delta t$ in $x$ and $t$ respectively.
19. Show that $\phi = Ae^{-kt/2} \sin pt \cos qx$, satisfies the equation $\dfrac{\partial^2 \phi}{\partial x^2} = \dfrac{1}{c^2}\left\{\dfrac{\partial^2 \phi}{\partial t^2} + k\dfrac{\partial \phi}{\partial t}\right\}$, provided that $p^2 = c^2q^2 - \dfrac{k^2}{4}$.
20. Show that (i) the equation $\dfrac{\partial^2 V}{\partial x^2} + \dfrac{\partial^2 V}{\partial y^2} + \dfrac{\partial^2 V}{\partial z^2} = 0$ is satisfied by $V = \dfrac{1}{\sqrt{(x^2 + y^2 + z^2)}}$, and that (ii) the equation $\dfrac{\partial^2 V}{\partial x^2} + \dfrac{\partial^2 V}{\partial y^2} = 0$ is satisfied by $V = \tan^{-1}\left(\dfrac{y}{x}\right)$.

# Programme 11

## PARTIAL DIFFERENTIATION

### PART 2

**1**

**Partial differentiation**

In the first part of the programme on partial differentiation, we established a result which, we said, would be the foundation of most of the applications of partial differentiation to follow.

You surely remember it: it went like this:

If $z$ is a function of two independent variables, $x$ and $y$, i.e. if $z = f(x, y)$, then

$$\delta z = \frac{\partial z}{\partial x}\,\delta x + \frac{\partial z}{\partial y}\,\delta y$$

We were able to use it, just as it stands, to work out certain problems on small increments, errors and tolerances. It is also the key to much of the work of this programme, so copy it down into your record book, thus:

If $z = f(x, y)$, then $\delta z = \dfrac{\partial z}{\partial x}\,\delta x + \dfrac{\partial z}{\partial y}\,\delta y$

---

**2**

If $z = f(x, y)$, then $\delta z = \dfrac{\partial z}{\partial x}\,\delta x + \dfrac{\partial z}{\partial y}\,\delta y$

In this expression, $\dfrac{\partial z}{\partial x}$ and $\dfrac{\partial z}{\partial y}$ are the *partial differential coefficients* of $z$ with respect to $x$ and $y$ respectively, and you will remember that to find

(i) $\dfrac{\partial z}{\partial x}$, we differentiate the function $z$ w.r.t. $x$, keeping all independent variables other than $x$, for the time being, .................. .

(ii) $\dfrac{\partial z}{\partial y}$, we differentiate the function $z$ w.r.t. $y$, keeping all independent variables other than $y$, for the time being, .................. .

**3**

constant constant

An example, just to remind you:

If $z = x^3 + 4x^2y - 3y^3$

then $\dfrac{\partial z}{\partial x} = 3x^2 + 8xy - 0$ ($y$ is constant)

and $\dfrac{\partial z}{\partial y} = 0 + 4x^2 - 9y^2$ ($x$ is constant)

In practice, of course, we do not write down the zero terms.

Before we tackle any further applications, we must be expert at finding partial differential coefficients, so with the reminder above, have a go at this one:

(1) If $z = \tan(x^2 - y^2)$, find $\dfrac{\partial z}{\partial x}$ and $\dfrac{\partial z}{\partial y}$

*When you have finished it, check with the next frame.*

**4**

$$\frac{\partial z}{\partial x} = 2x\sec^2(x^2 - y^2); \quad \frac{\partial z}{\partial y} = -2y\sec^2(x^2 - y^2)$$

for $z = \tan(x^2 - y^2)$

$$\therefore \frac{\partial z}{\partial x} = \sec^2(x^2 - y^2) \times \frac{\partial}{\partial x}(x^2 - y^2)$$

$$= \sec^2(x^2 - y^2)(2x) = \underline{2x\sec^2(x^2 - y^2)}$$

and

$$\frac{\partial z}{\partial y} = \sec^2(x^2 - y^2) \times \frac{\partial}{\partial y}(x^2 - y^2)$$

$$= \sec^2(x^2 - y^2)(-2y) = \underline{-2y\sec^2(x^2 - y^2)}$$

That was easy enough. Now do this one:

(2) If $z = e^{2x - 3y}$, find $\dfrac{\partial^2 z}{\partial x^2}$, $\dfrac{\partial^2 z}{\partial y^2}$, $\dfrac{\partial^2 z}{\partial x . \partial y}$

*Finish them all. Then turn on to frame 5 and check your results.*

**5** Here are the results in detail:

$$z = e^{2x-3y} \quad \therefore \frac{\partial z}{\partial x} = e^{2x-3y}.2 = \underline{2.e^{2x-3y}}$$

$$\frac{\partial z}{\partial y} = e^{2x-3y}(-3) = \underline{-3.e^{2x-3y}}$$

$$\frac{\partial^2 z}{\partial x^2} = 2.e^{2x-3y}.2 = \underline{4.e^{2x-3y}}$$

$$\frac{\partial^2 z}{\partial y^2} = -3.e^{2x-3y}(-3) = \underline{9.e^{2x-3y}}$$

$$\frac{\partial^2 z}{\partial x.\partial y} = -3.e^{2x-3y}.2 = \underline{-6.e^{2x-3y}}$$

All correct?

You remember, too, that in the 'mixed' second partial diff. coefft., the order of differentiating does not matter. So in this case, since

$\dfrac{\partial^2 z}{\partial x.\partial y} = -6.e^{2x-3y}$, then $\dfrac{\partial^2 z}{\partial y.\partial x} =$ ....................

---

**6**

$$\frac{\partial^2 z}{\partial x.\partial y} = \frac{\partial^2 z}{\partial y.\partial x} = \boxed{-6.e^{2x-3y}}$$

---

Well now, before we move on to new work, see what you make of these.

Find all the first and second partial differential coefficients of the following:

(i) $z = x \sin y$

(ii) $z = (x + y) \ln(xy)$

*When you have found all the diff. coefficients, check your work with the solutions in the next frame.*

**7**

Here they are. Check your results carefully.

(i) $z = x \sin y$

$$\therefore \frac{\partial z}{\partial x} = \sin y \qquad \frac{\partial z}{\partial y} = \underline{x \cos y}$$

$$\frac{\partial^2 z}{\partial x^2} = 0 \qquad \frac{\partial^2 z}{\partial y^2} = \underline{-x \sin y}$$

$$\frac{\partial^2 z}{\partial y . \partial x} = \cos y \qquad \frac{\partial^2 z}{\partial x . \partial y} = \underline{\cos y}$$

(ii) $z = (x + y) \ln(xy)$

$$\therefore \frac{\partial z}{\partial x} = (x + y) \frac{1}{xy} . y + \ln(xy) = \underline{\frac{(x + y)}{x} + \ln(xy)}$$

$$\frac{\partial z}{\partial y} = (x + y) \frac{1}{xy} . x + \ln(xy) = \underline{\frac{(x + y)}{y} + \ln(xy)}$$

$$\therefore \frac{\partial^2 z}{\partial x^2} = \frac{x - (x + y)}{x^2} + \frac{1}{xy} . y = \frac{\cancel{x} - \cancel{x} - y}{x^2} + \frac{1}{x}$$

$$= \underline{\frac{x - y}{x^2}}$$

$$\frac{\partial^2 z}{\partial y^2} = \frac{y - (x + y)}{y^2} + \frac{1}{xy} . x = \frac{\cancel{y} - x - \cancel{y}}{y^2} + \frac{1}{y}$$

$$= \underline{\frac{y - x}{y^2}}$$

$$\frac{\partial^2 z}{\partial y . \partial x} = \frac{1}{x} + \frac{1}{xy} . x = \frac{1}{x} + \frac{1}{y}$$

$$= \frac{y + x}{xy}$$

$$\frac{\partial^2 z}{\partial x . \partial y} = \frac{1}{y} + \frac{1}{xy} . y = \frac{1}{y} + \frac{1}{x}$$

$$= \underline{\frac{x + y}{xy}}$$

**8**

Well now, that was just by way of warming up with work you have done before. Let us now move on to the next section of this programme.

**Rates-of-change problems**

Let us consider a cylinder of radius $r$ and height $h$ as before. Then the volume is given by

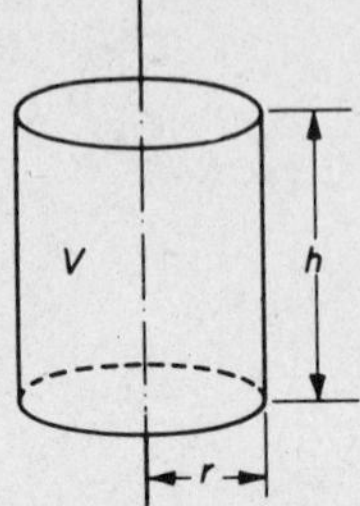


$$V = \pi r^2 h$$

$$\therefore \frac{\partial V}{\partial r} = 2\pi r h \text{ and } \frac{\partial V}{\partial h} = \pi r^2$$

Since V is a function of $r$ and $h$, we also know that

$\delta V = \frac{\partial V}{\partial r}.\delta r + \frac{\partial V}{\partial h}.\delta h$ (Here it is, popping up again!)

Now divide both sides by $\delta t$: $\frac{\delta V}{\delta t} = \frac{\partial V}{\partial r}.\frac{\delta r}{\delta t} + \frac{\partial V}{\partial h}.\frac{\delta h}{\delta t}$

Then if $\delta t \to 0, \frac{\delta V}{\delta t} \to \frac{dV}{dt}, \frac{\delta r}{\delta t} \to \frac{dr}{dt}, \frac{\delta h}{\delta t} \to \frac{dh}{dt}$, but the partial differential coefficients, which do not contain $\delta t$, will remain unchanged.

So our result now becomes $\frac{dV}{dt} =$ ......................

**9**

$$\boxed{\frac{dV}{dt} = \frac{\partial V}{\partial r}.\frac{dr}{dt} + \frac{\partial V}{\partial h}.\frac{dh}{dt}}$$

This result is really the key to problems of the kind we are about to consider. If we know the rate at which $r$ and $h$ are changing, we can now find the corresponding rate of change of V. Like this:

*Example 1.*

The radius of a cylinder increases at the rate of 0·2 cm/sec while the height decreases at the rate of 0·5 cm/sec. Find the rate at which the volume is changing at the instant when $r = 8$ cm and $h = 12$ cm.

*WARNING:* The first inclination is to draw a diagram and to put in the given values for its dimensions, i.e. $r = 8$ cm, $h = 12$ cm. This we *must NOT do*, for the radius and height are changing and the given values are instantaneous values only. Therefore on the diagram we keep the symbols $r$ and $h$ to indicate that they are variables.

**10**

Here it is then:

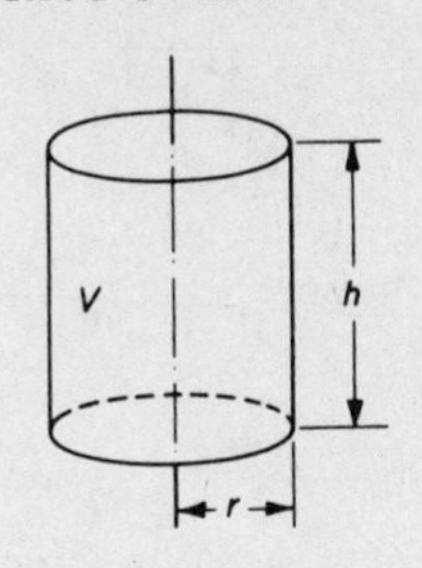


$$V = \pi r^2 h$$

$$\delta V = \frac{\partial V}{\partial r}\delta r + \frac{\partial V}{\partial h}\delta h$$

$$\therefore \frac{dV}{dt} = \frac{\partial V}{\partial r}.\frac{dr}{dt} + \frac{\partial V}{\partial h}.\frac{dh}{dt}$$

$$\frac{\partial V}{\partial r} = 2\pi rh; \quad \frac{\partial V}{\partial h} = \pi r^2$$

$$\therefore \frac{dV}{dt} = 2\pi rh\frac{dr}{dt} + \pi r^2\frac{dh}{dt}$$

Now at the instant we are considering

$r = 8$, $h = 12$, $\frac{dr}{dt} = 0{\cdot}2$, $\frac{dh}{dt} = -0{\cdot}5$ (minus since $h$ is decreasing)

So you can now substitute these values in the last statement and finish off the calculation, giving

$$\frac{dV}{dt} = \ldots\ldots\ldots\ldots$$

---

**11**

$$\boxed{\frac{dV}{dt} = 20{\cdot}1 \text{ cm}^3/\text{sec}}$$

for

$$\frac{dV}{dt} = 2\pi rh.\frac{dr}{dt} + \pi r^2\frac{dh}{dt}$$

$$= 2\pi 8.12.(0{\cdot}2) + \pi 64(-0{\cdot}5)$$

$$= 38{\cdot}4\pi - 32\pi$$

$$= 6{\cdot}4\pi = \underline{20{\cdot}1 \text{ cm}^3/\text{sec.}}$$

Now another one.

*Example 2.*

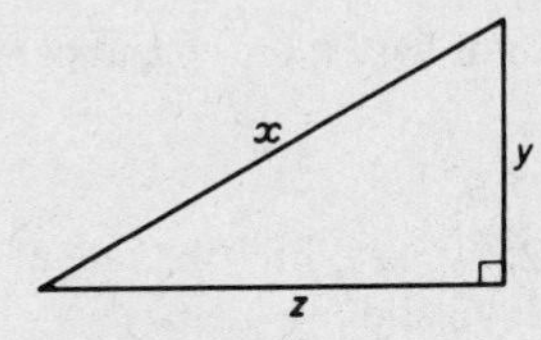


In the right-angled triangle shown, $x$ is increasing at 2 cm/sec while $y$ is decreasing at 3 cm/sec. Calculate the rate at which $z$ is changing when $x = 5$ cm and $y = 3$ cm.

The first thing to do, of course, is to express $z$ in terms of $x$ and $y$. That is not difficult.

$$z = \ldots\ldots\ldots\ldots$$

**12**

$$z = \sqrt{(x^2 - y^2)}$$

□□□□□□□□□□□□□□□□□□□□□□□□□□□□□□□□□□□□□□□

$$z = \sqrt{(x^2 - y^2)} = (x^2 - y^2)^{\frac{1}{2}}$$

$$\therefore \ \delta z = \frac{\partial z}{\partial x}\delta x + \frac{\partial z}{\partial y}\delta y \qquad \text{(The key to the whole business)}$$

$$\therefore \ \frac{dz}{dt} = \frac{\partial z}{\partial x}.\frac{dx}{dt} + \frac{\partial z}{\partial y}.\frac{dy}{dt}$$

In this case

$$\frac{\partial z}{\partial x} = \frac{1}{2}(x^2 - y^2)^{-\frac{1}{2}}(2x) = \frac{x}{\sqrt{(x^2 - y^2)}}$$

$$\frac{\partial z}{\partial y} = \frac{1}{2}(x^2 - y^2)^{-\frac{1}{2}}(-2y) = \frac{-y}{\sqrt{(x^2 - y^2)}}$$

$$\frac{dz}{dt} = \frac{x}{\sqrt{(x^2 - y^2)}}.\frac{dx}{dt} - \frac{y}{\sqrt{(x^2 - y^2)}}.\frac{dy}{dt}$$

So far so good. Now for the numerical values

$$x = 5, \quad y = 3, \quad \frac{dx}{dt} = 2, \quad \frac{dy}{dt} = -3$$

$$\frac{dz}{dt} = \ldots\ldots\ldots\ldots$$

*Finish it off, then move to frame 13.*

**13**

$$\frac{dz}{dt} = 4{\cdot}75 \text{ cm/sec}$$

for we have

$$\frac{dz}{dt} = \frac{5}{\sqrt{(5^2 - 3^2)}}(2) - \frac{3}{\sqrt{(5^2 - 3^2)}}(-3)$$

$$= \frac{5(2)}{4} + \frac{3(3)}{4} \quad = \frac{10}{4} + \frac{9}{4} = \frac{19}{4} = 4{\cdot}75 \text{ cm/sec}$$

$\therefore$ Side $z$ increases at the rate of 4·75 cm/sec

Now here is

*Example 3.* The total surface area S of a cone of base radius $r$ and perpendicular height $h$ is given by

$$S = \pi r^2 + \pi r\sqrt{(r^2 + h^2)}$$

If $r$ and $h$ are each increasing at the rate of 0·25 cm/sec, find the rate at which S is increasing at the instant when $r = 3$ cm and $h = 4$ cm.

Do that one entirely on your own. Take your time: there is no need to hurry. Be quite sure that each step you write down is correct.

*Then turn to frame 14 and check your result.*

**14**

*Solution.* Here it is in detail.

$$S = \pi r^2 + \pi r\sqrt{(r^2 + h^2)} = \pi r^2 + \pi r(r^2 + h^2)^{\frac{1}{2}}$$

$$\delta S = \frac{\partial S}{\partial r}.\delta r + \frac{\partial S}{\partial h}.\delta h \quad \therefore \frac{dS}{dt} = \frac{\partial S}{\partial r}.\frac{dr}{dt} + \frac{\partial S}{\partial h}.\frac{dh}{dt}$$

(i) $$\frac{\partial S}{\partial r} = 2\pi r + \pi r.\frac{1}{2}(r^2 + h^2)^{-\frac{1}{2}}(2r) + \pi(r^2 + h^2)^{\frac{1}{2}}$$

$$= 2\pi r + \frac{\pi r^2}{\sqrt{(r^2 + h^2)}} + \pi\sqrt{(r^2 + h^2)}$$

When $r = 3$ and $h = 4$,

$$\frac{\partial S}{\partial r} = 2\pi 3 + \frac{\pi 9}{5} + \pi 5 = 11\pi + \frac{9\pi}{5} = \frac{64\pi}{5}$$

(ii) $$\frac{\partial S}{\partial h} = \pi r\frac{1}{2}(r^2 + h^2)^{-\frac{1}{2}}(2h) = \frac{\pi rh}{\sqrt{(r^2 + h^2)}}$$

$$= \frac{\pi 3.4}{5} = \frac{12\pi}{5}$$

Also we are given that $\frac{dr}{dt} = 0{\cdot}25$ and $\frac{dh}{dt} = 0{\cdot}25$

$$\therefore \frac{dS}{dt} = \frac{64\pi}{5}.\frac{1}{4} + \frac{12\pi}{5}.\frac{1}{4}$$

$$= \frac{16\pi}{5} + \frac{3\pi}{5} = \frac{19\pi}{5}$$

$$= 3{\cdot}8\pi = \underline{11{\cdot}93 \text{ cm}^2/\text{sec}}$$

**15**

So there we are. Rates-of-change problems are all very much the same. What you must remember is simply this:

(i) The basic statement

If $z = f(x, y)$ then $\delta z = \frac{\partial z}{\partial x}.\delta x + \frac{\partial z}{\partial y}.\delta y$ .................... (i)

(ii) Divide this result by $\delta t$ and make $\delta t \to 0$. This converts the result into the form for rates-of-change problems:

$$\frac{dz}{dt} = \frac{\partial z}{\partial x}.\frac{dx}{dt} + \frac{\partial z}{\partial y}.\frac{dy}{dt} \quad \text{.......................... (ii)}$$

The second result follows directly from the first. Make a note of both of these in your record book for future reference.

*Then for the next part of the work, turn on to frame 16.*

**16**

Partial differentiation can also be used with advantage in finding *differential coefficients of implicit functions.*

For example, suppose we are required to find an expression for $\frac{dy}{dx}$ when we are given that $x^2 + 2xy + y^3 = 0$.

We can set about it in this way:

Let $z$ stand for the function of $x$ and $y$, i.e. $z = x^2 + 2xy + y^3$. Again we use the basic relationship $\delta z = \frac{\partial z}{\partial x}\delta x + \frac{\partial z}{\partial y}\delta y$.

If we divide both sides by $\delta x$, we get

$$\frac{\delta z}{\delta x} = \frac{\partial z}{\partial x} + \frac{\partial z}{\partial y}.\frac{\delta y}{\delta x}$$

Now, if $\delta x \to 0$,

$$\frac{dz}{dx} = \frac{\partial z}{\partial x} + \frac{\partial z}{\partial y}.\frac{dy}{dx}$$

If we now find expressions for $\frac{\partial z}{\partial x}$ and $\frac{\partial z}{\partial y}$, we shall be quite a way towards finding $\frac{dy}{dx}$ (which you see at the end of the expression).

In this particular example, $\frac{\partial z}{\partial x}$ = .................. and $\frac{\partial z}{\partial y}$ = ..................

---

**17**

$$z = x^2 + 2xy + y^3 \qquad \boxed{\frac{\partial z}{\partial x} = 2x + 2y; \quad \frac{\partial z}{\partial y} = 2x + 3y^2}$$

Substituting these in our previous result gives us

$$\frac{dz}{dx} = (2x + 2y) + (2x + 3y^2)\frac{dy}{dx}$$

If only we knew $\frac{dz}{dx}$, we could rearrange this result and obtain an expression for $\frac{dy}{dx}$. So where can we find out something about $\frac{dz}{dx}$?

Refer back to the beginning of the problem. We have used $z$ to stand for $x^2 + 2xy + y^3$ and we were told initially that $x^2 + 2xy + y^3 = 0$.

Therefore $z = 0$, i.e. $z$ is a constant (in this case zero) and hence $\frac{dz}{dx} = 0$.

$$\therefore\ 0 = (2x + 2y) + (2x + 3y^2)\frac{dy}{dx}$$

From this we can find $\frac{dy}{dx}$. So finish it off.

$$\frac{dy}{dx} = \text{......................}$$

*On to frame 18.*

**18**

$$\frac{dy}{dx} = -\frac{2x + 2y}{2x + 3y^2}$$

□□□□□□□□□□□□□□□□□□□□□□□□□□□□□□□□□□□□□□□□

This is almost a routine that always works. In general, we have –

$$\text{If } f(x, y) = 0, \text{ find } \frac{dy}{dx}$$

Let $z = f(x, y)$ then $\delta z = \frac{\partial z}{\partial x}\delta x + \frac{\partial z}{\partial y}\delta y$. Divide by $\delta x$ and make $\delta x \to 0$, in which case

$$\frac{dz}{dx} = \frac{\partial z}{\partial x} + \frac{\partial z}{\partial y}.\frac{dy}{dx}$$

But $z = 0$ (constant) $\therefore \frac{dz}{dx} = 0$ $\qquad \therefore 0 = \frac{\partial z}{\partial x} + \frac{\partial z}{\partial y}.\frac{dy}{dx}$

giving $\qquad \frac{dy}{dx} = -\frac{\partial z}{\partial x}\Big/\frac{\partial z}{\partial y}$

The easiest form to remember is the one that comes direct from the basic result

$$\delta z = \frac{\partial z}{\partial x}\delta x + \frac{\partial z}{\partial y}\delta y$$

Divide by $\delta x$, etc.

$$\frac{dz}{dx} = \frac{\partial z}{\partial x} + \frac{\partial z}{\partial y}.\frac{dy}{dx} \qquad \left\{\frac{dz}{dx} = 0\right\}$$

Make a note of this result.

---

**19**

Now for one or two examples.

*Example 1.* If $e^{xy} + x + y = 1$, evaluate $\frac{dy}{dx}$ at (0,0). The function can be written $e^{xy} + x + y - 1 = 0$.

Let $z = e^{xy} + x + y - 1 \qquad \delta z = \frac{\partial z}{\partial x}.\delta x + \frac{\partial z}{\partial y}.\delta y \quad \therefore \quad \frac{dz}{dx} = \frac{\partial z}{\partial x} + \frac{\partial z}{\partial y}.\frac{dy}{dx}$

$$\frac{\partial z}{\partial x} = e^{xy}.y + 1; \quad \frac{\partial z}{\partial y} = e^{xy}.x + 1 \therefore \quad \frac{dz}{dx} = (y.e^{xy} + 1) + (x.e^{xy} + 1)\frac{dy}{dx}$$

But $z = 0 \quad \therefore \frac{dz}{dx} = 0 \qquad \therefore \quad \frac{dy}{dx} = -\left\{\frac{y.e^{xy} + 1}{x.e^{xy} + 1}\right\}$

At $x = 0, y = 0, \frac{dy}{dx} = -\frac{1}{1} = -1 \quad \therefore \frac{dy}{dx} = -1$

All very easy so long as you can find partial differential coefficients correctly.

*On to frame 20.*

## 20

Now here is:

*Example 2.* If $xy + \sin y = 2$, find $\frac{dy}{dx}$

Let $z = xy + \sin y - 2 = 0$

$$\delta z = \frac{\partial z}{\partial x}\delta x + \frac{\partial z}{\partial y}\delta y$$

$$\frac{dz}{dx} = \frac{\partial z}{\partial x} + \frac{\partial z}{\partial y}.\frac{dy}{dx}$$

$$\frac{\partial z}{\partial x} = y; \quad \frac{\partial z}{\partial y} = x + \cos y$$

$$\therefore \frac{dz}{dx} = y + (x + \cos y)\frac{dy}{dx}$$

But $z = 0 \quad \therefore \frac{dz}{dx} = 0$

$$\therefore \frac{dy}{dx} = \frac{-y}{x + \cos y}$$

Here is one for you to do:

*Example 3.* Find an expression for $\frac{dy}{dx}$ when $x \tan y = y \sin x$. Do it all on your own. *Then check your working with that in frame 21.*

## 21

$$\boxed{\frac{dy}{dx} = -\frac{\tan y - y\cos x}{x\sec^2 y - \sin x}}$$

Did you get that? If so, go straight on to frame 22. If not, here is the working below. Follow it through and see where you have gone astray!

$$x\tan y = y\sin x \quad \therefore x\tan y - y\sin x = 0$$

Let $z = x\tan y - y\sin x = 0$

$$\delta z = \frac{\partial z}{\partial x}\delta x + \frac{\partial z}{\partial y}\delta y$$

$$\frac{dz}{dx} = \frac{\partial z}{\partial x} + \frac{\partial z}{\partial y}.\frac{dy}{dx}$$

$$\frac{\partial z}{\partial x} = \tan y - y\cos x; \quad \frac{\partial z}{\partial y} = x\sec^2 y - \sin x$$

$$\therefore \frac{dz}{dx} = (\tan y - y\cos x) + (x\sec^2 y - \sin x)\frac{dy}{dx}$$

But $z = 0 \quad \therefore \frac{dz}{dx} = 0$

$$\frac{dy}{dx} = -\frac{\tan y - y\cos x}{x\sec^2 y - \sin x}$$

*On now to frame 22.*

**22**

Right. Now here is just one more for you to do. They are really very much the same.

*Example 4.* If $e^{x+y} = x^2y^2$, find an expression for $\dfrac{dy}{dx}$

$$e^{x+y} - x^2y^2 = 0. \quad \text{Let } z = e^{x+y} - x^2y^2 = 0$$

$$\delta z = \frac{\partial z}{\partial x}\,\delta x + \frac{\partial z}{\partial y}\,\delta y$$

$$\frac{dz}{dx} = \frac{\partial z}{\partial x} + \frac{\partial z}{\partial y}\cdot\frac{dy}{dx}$$

So continue with the good work and finish it off, finally getting that

$$\frac{dy}{dx} = \ldots\ldots\ldots\ldots\ldots\ldots$$

*Then move to frame 23.*

**23**

$$\boxed{\frac{dy}{dx} = \frac{2xy^2 - e^{x+y}}{e^{x+y} - 2x^2y}}$$

For $\qquad z = e^{x+y} - x^2y^2 = 0$

$$\frac{\partial z}{\partial x} = e^{x+y} - 2xy^2; \quad \frac{\partial z}{\partial y} = e^{x+y} - 2x^2y$$

$$\therefore \frac{dz}{dx} = (e^{x+y} - 2xy^2) + (e^{x+y} - 2x^2y)\frac{dy}{dx}$$

But $z = 0 \quad \therefore \dfrac{dz}{dx} = 0$

$$\therefore \frac{dy}{dx} = -\frac{(e^{x+y} - 2xy^2)}{(e^{x+y} - 2x^2y)}$$

$$\therefore \underline{\frac{dy}{dx} = \frac{2xy^2 - e^{x+y}}{e^{x+y} - 2x^2y}}$$

That is how they are all done.

*Now on to frame 24.*

**24**

There is one more process that you must know how to tackle.

**Change of variables**

If $z$ is a function of $x$ and $y$, i.e. $z = f(x, y)$, and $x$ and $y$ are themselves functions of two other variables $u$ and $v$, then $z$ is also a function of $u$ and $v$. We may therefore need to find $\frac{\partial z}{\partial u}$ and $\frac{\partial z}{\partial v}$. How do we go about it?

$$z = f(x, y) \quad \therefore \ \delta z = \frac{\partial z}{\partial x}\delta x + \frac{\partial z}{\partial y}\delta y$$

Divide both sides by $\delta u$.

$$\frac{\delta z}{\delta u} = \frac{\partial z}{\partial x}\cdot\frac{\delta x}{\delta u} + \frac{\partial z}{\partial y}\cdot\frac{\delta y}{\delta u}$$

If $v$ is kept constant for the time being, then $\frac{\delta x}{\delta u}$ when $\delta u \to 0$ becomes $\frac{\partial x}{\partial u}$ and $\frac{\delta y}{\delta u}$ becomes $\frac{\partial y}{\partial u}$.

$$\therefore \ \frac{\partial z}{\partial u} = \frac{\partial z}{\partial x}\cdot\frac{\partial x}{\partial u} + \frac{\partial z}{\partial y}\cdot\frac{\partial y}{\partial u}$$

and

$$\frac{\partial z}{\partial v} = \frac{\partial z}{\partial x}\cdot\frac{\partial x}{\partial v} + \frac{\partial z}{\partial y}\cdot\frac{\partial y}{\partial v}$$

Note these

*Next frame.*

**25**

Here is an example on this work.

If $z = x^2 + y^2$, where $x = r\cos\theta$ and $y = r\sin 2\theta$, find $\frac{\partial z}{\partial r}$ and $\frac{\partial z}{\partial \theta}$

$$\frac{\partial z}{\partial r} = \frac{\partial z}{\partial x}\cdot\frac{\partial x}{\partial r} + \frac{\partial z}{\partial y}\cdot\frac{\partial y}{\partial r}$$

and

$$\frac{\partial z}{\partial \theta} = \frac{\partial z}{\partial x}\cdot\frac{\partial x}{\partial \theta} + \frac{\partial z}{\partial y}\cdot\frac{\partial y}{\partial \theta}$$

Now,

$$\frac{\partial z}{\partial x} = 2x \qquad \frac{\partial z}{\partial y} = 2y$$

$$\frac{\partial x}{\partial r} = \cos\theta \qquad \frac{\partial y}{\partial r} = \sin 2\theta$$

$$\therefore \ \frac{\partial z}{\partial r} = \underline{2x\cos\theta + 2y\sin 2\theta}$$

And

$$\frac{\partial x}{\partial \theta} = -r\sin\theta \quad \text{and} \quad \frac{\partial y}{\partial \theta} = 2r\cos 2\theta$$

$$\therefore \ \frac{\partial z}{\partial \theta} = 2x(-r\sin\theta) + 2y(2r\cos 2\theta)$$

$$\frac{\partial z}{\partial \theta} = \underline{4\,yr\cos 2\theta - 2\,xr\sin\theta}$$

And in these two results, the symbols $x$ and $y$ can be replaced by $r\cos\theta$ and $r\sin 2\theta$ respectively.

**26**

One more example.

If $z = e^{xy}$ where $x = \ln(u + v)$ and $y = \sin(u - v)$, find $\frac{\partial z}{\partial u}$ and $\frac{\partial z}{\partial v}$.

We have
$$\frac{\partial z}{\partial u} = \frac{\partial z}{\partial x} \cdot \frac{\partial x}{\partial u} + \frac{\partial z}{\partial y} \cdot \frac{\partial y}{\partial u}$$
$$= y.e^{xy} . \frac{1}{u+v} + x.e^{xy}.\cos(u-v)$$
$$= e^{xy}\left\{\frac{y}{u+v} + x.\cos(u-v)\right\}$$

and
$$\frac{\partial z}{\partial v} = \frac{\partial z}{\partial x} \cdot \frac{\partial x}{\partial v} + \frac{\partial z}{\partial y} \cdot \frac{\partial y}{\partial v}$$
$$= y.e^{xy} . \frac{1}{u+v} + x.e^{xy}\left\{-\cos(u-v)\right\}$$
$$= e^{xy}\left\{\frac{y}{u+v} - x\cos(u-v)\right\}$$

*Now move on to frame 27.*

**27**

Here is one for you to do on your own. All that it entails is to find the various partial differential coefficients and to substitute them in the established results.

$$\frac{\partial z}{\partial u} = \frac{\partial z}{\partial x} \cdot \frac{\partial x}{\partial u} + \frac{\partial z}{\partial y} \cdot \frac{\partial y}{\partial u}$$

and
$$\frac{\partial z}{\partial v} = \frac{\partial z}{\partial x} \cdot \frac{\partial x}{\partial v} + \frac{\partial z}{\partial y} \cdot \frac{\partial y}{\partial v}$$

So you do this one:

If $z = \sin(x + y)$, where $x = u^2 + v^2$ and $y = 2uv$, find $\frac{\partial z}{\partial u}$ and $\frac{\partial z}{\partial v}$

The method is the same as before.

*When you have completed the work, check with the results in frame 28.*

**28**

$$z = \sin(x+y);\quad x = u^2+v^2;\quad y = 2uv$$

$$\frac{\partial z}{\partial x} = \cos(x+y)\ ;\ \frac{\partial z}{\partial y} = \cos(x+y)$$

$$\frac{\partial x}{\partial u} = 2u \qquad \frac{\partial y}{\partial u} = 2v$$

$$\frac{\partial z}{\partial u} = \frac{\partial z}{\partial x}.\frac{\partial x}{\partial u} + \frac{\partial z}{\partial y}.\frac{\partial y}{\partial u}$$

$$= \cos(x+y).2u + \cos(x+y).2v$$

$$= \underline{2(u+v)\cos(x+y)}$$

Also

$$\frac{\partial z}{\partial v} = \frac{\partial z}{\partial x}.\frac{\partial x}{\partial v} + \frac{\partial z}{\partial y}.\frac{\partial y}{\partial v}$$

$$\frac{\partial x}{\partial v} = 2v\ ;\ \frac{\partial y}{\partial v} = 2u$$

$$\frac{\partial z}{\partial v} = \cos(x+y).2v + \cos(x+y).2u$$

$$= \underline{2(u+v)\cos(x+y)}$$

**29** You have now reached the end of this programme and know quite a bit about partial differentiation. We have established some important results during the work, so let us list them once more.

1. *Small increments*

$z = f(x,y)$ $$\delta z = \frac{\partial z}{\partial x}\delta x + \frac{\partial z}{\partial y}\delta y \qquad \text{.............. (i)}$$

2. *Rates of change*

$$\frac{dz}{dt} = \frac{\partial z}{\partial x}.\frac{dx}{dt} + \frac{\partial z}{\partial y}.\frac{dy}{dt} \qquad \text{.............. (ii)}$$

3. *Implicit functions*

$$\frac{dz}{dx} = \frac{\partial z}{\partial x} + \frac{\partial z}{\partial y}.\frac{dy}{dx} \qquad \text{.............. (iii)}$$

4. *Change of variables*

$$\frac{\partial z}{\partial u} = \frac{\partial z}{\partial x}.\frac{\partial x}{\partial u} + \frac{\partial z}{\partial y}.\frac{\partial y}{\partial u}$$

$$\frac{\partial z}{\partial v} = \frac{\partial z}{\partial x}.\frac{\partial x}{\partial v} + \frac{\partial z}{\partial y}.\frac{\partial y}{\partial v} \qquad \text{.............. (iv)}$$

All that now remains is the Test Exercise, so turn on to frame 30 and work through it carefully at your own speed. The questions are just like those you have been doing quite successfully.

**30**

## Test Exercise – XI

Answer *all* the questions. Takc your time over them and work carefully.

1. Use partial differentiation to determine expressions for $\frac{dy}{dx}$ in the following cases:

   (i) $x^3 + y^3 - 2x^2y = 0$

   (ii) $e^x \cos y = e^y \sin x$

   (iii) $\sin^2 x - 5 \sin x \cos y + \tan y = 0$

2. The base radius of a cone, $r$, is decreasing at the rate of 0·1 cm/sec while the perpendicular height, $h$, is increasing at the rate of 0·2 cm/sec. Find the rate at which the volume, V, is changing when $r = 2$ cm and $h = 3$ cm.

3. If $z = 2xy - 3x^2y$ and $x$ is increasing at 2 cm/sec determine at what rate $y$ must be changing in order that $z$ shall be neither increasing nor decreasing at the instant when $x = 3$ cm and $y = 1$ cm.

4. If $z = x^4 + 2x^2y + y^3$ and $x = r\cos\theta$ and $y = r\sin\theta$, find $\frac{\partial z}{\partial r}$ and $\frac{\partial z}{\partial \theta}$ in their simplest forms.

## Further Problems – XI

1. If $F = f(x,y)$ where $x = e^u \cos v$ and $y = e^u \sin v$, show that

$$\frac{\partial F}{\partial u} = x\frac{\partial F}{\partial x} + y\frac{\partial F}{\partial y} \text{ and } \frac{\partial F}{\partial v} = -y\frac{\partial F}{\partial x} + x\frac{\partial F}{\partial y}$$

2. Given that $z = x^3 + y^3$ and $x^2 + y^2 = 1$, determine an expression for $\dfrac{dz}{dx}$ in terms of $x$ and $y$.

3. If $z = f(x, y) = 0$, show that $\dfrac{dy}{dx} = -\dfrac{\partial z}{\partial x}\Big/\dfrac{\partial z}{\partial y}$. The curves $2y^2 + 3x - 8 = 0$ and $x^3 + 2xy^3 + 3y - 1 = 0$ intersect at the point $(2, -1)$. Find the tangent of the angle between the tangents to the curves at this point.

4. If $u = (x^2 - y^2) f(t)$ where $t = xy$ and $f$ denotes an arbitrary function, prove that

$$\frac{\partial^2 u}{\partial x . \partial y} = (x^2 - y^2)\{t.f''(t) + 3f'(t)\}$$

5. If $V = xy/(x^2 + y^2)^2$ and $x = r\cos\theta, y = r\sin\theta$, show that

$$\frac{\partial^2 V}{\partial r^2} + \frac{1}{r}\frac{\partial V}{\partial r} + \frac{1}{r^2}\frac{\partial^2 V}{\partial \theta^2} = 0$$

6. If $u = f(x, y)$ where $x = r^2 - s^2$ and $y = 2rs$, prove that

$$r\frac{\partial u}{\partial r} - s\frac{\partial u}{\partial s} = 2(r^2 + s^2)\frac{\partial u}{\partial x}$$

7. If $f = F(x, y)$ and $x = r e^{\theta}$ and $y = r e^{-\theta}$, prove that

$$2x\frac{\partial f}{\partial x} = r\frac{\partial f}{\partial r} + \frac{\partial f}{\partial \theta} \text{ and } 2y.\frac{\partial f}{\partial y} = r\frac{\partial f}{\partial r} - \frac{\partial f}{\partial \theta}$$

8. If $z = x\ln(x^2 + y^2) - 2y\tan^{-1}\left(\dfrac{y}{x}\right)$ verify that

$$x\frac{\partial z}{\partial x} + y\frac{\partial z}{\partial y} = z + 2x$$

9. By means of partial differentiation, determine $\dfrac{dy}{dx}$ in each of the following cases.

   (i) $xy + 2y - x = 4$

   (ii) $x^3y^2 - 2x^2y + 3xy^2 - 8xy = 5$

   (iii) $\dfrac{4y}{x} + \dfrac{2x}{y} = 3$

10. If $z = 3xy - y^3 + (y^2 - 2x)^{3/2}$, verify that

    (i) $\dfrac{\partial^2 z}{\partial x.\partial y} = \dfrac{\partial^2 z}{\partial y.\partial x}$, and that (ii) $\dfrac{\partial^2 z}{\partial x^2}.\dfrac{\partial^2 z}{\partial y^2} = \left(\dfrac{\partial^2 z}{\partial x.\partial y}\right)^2$

11. If $f = \dfrac{1}{\sqrt{(1 - 2xy + y^2)}}$, show that $y\dfrac{\partial f}{\partial y} = (x - y)\dfrac{\partial f}{\partial x}$

12. If $z = x.f\left(\dfrac{y}{x}\right) + F\left(\dfrac{y}{x}\right)$, prove that

    (i) $x\dfrac{\partial z}{\partial x} + y\dfrac{\partial z}{\partial y} = z - F\left(\dfrac{y}{x}\right)$, (ii) $x^2\dfrac{\partial^2 z}{\partial x^2} + 2xy\dfrac{\partial^2 z}{\partial x.\partial y} + y^2\dfrac{\partial^2 z}{\partial y^2} = 0$

13. If $z = e^{k(r - x)}$, where $k$ is a constant, and $r^2 = x^2 + y^2$, prove

    (i) $\left(\dfrac{\partial z}{\partial x}\right)^2 + \left(\dfrac{\partial z}{\partial y}\right)^2 + 2zk\dfrac{\partial z}{\partial x} = 0$ (ii) $\dfrac{\partial^2 z}{\partial x^2} + \dfrac{\partial^2 z}{\partial y^2} + 2k\dfrac{\partial z}{\partial x} = \dfrac{kz}{r}$

14. If $z = f(x - 2y) + F(3x + y)$, where $f$ and F are arbitrary functions, and if $\dfrac{\partial^2 z}{\partial x^2} + a\dfrac{\partial^2 z}{\partial x.\partial y} + b\dfrac{\partial^2 z}{\partial y^2} = 0$, find the values of $a$ and $b$.

15. If $z = xy/(x^2 + y^2)^2$, verify that $\dfrac{\partial^2 z}{\partial x^2} + \dfrac{\partial^2 z}{\partial y^2} = 0$.

16. If $\sin^2 x - 5\sin x\cos y + \tan y = 0$, find $\dfrac{dy}{dx}$ by using partial differentiation.

17. Find $\dfrac{dy}{dx}$ by partial differentiation, when $x\tan y = y\sin x$.

18. If $V = \tan^{-1}\left\{\dfrac{2xy}{x^2 - y^2}\right\}$, prove that

    (i) $x\dfrac{\partial V}{\partial x} + y\dfrac{\partial V}{\partial y} = 0$, (ii) $\dfrac{\partial^2 V}{\partial x^2} + \dfrac{\partial^2 V}{\partial y^2} = 0$

19. Prove that, if $z = 2xy + x.f\left(\dfrac{y}{x}\right)$ then

$$x\frac{\partial z}{\partial x} + y\frac{\partial z}{\partial y} = z + 2xy$$

20. (i) Find $\dfrac{dy}{dx}$ given that $x^2 y + \sin xy = 0$

    (ii) Find $\dfrac{dy}{dx}$ given that $x\sin xy = 1$

# Programme 12

## CURVES AND CURVE FITTING

**1**

**Introduction**

The purpose of this programme is eventually to devise a reliable method for establishing the relationship between two variables, corresponding values of which have been obtained as a result of tests or experimentation. These results in practice are highly likely to include some errors, however small, due to the imperfect materials used, the limitations of the measuring devices and the shortcomings of the operator conducting the test and recording the results.

There are methods by which we can minimise any further errors or guesswork in processing the results and, indeed, eradicate some of the errors already inherent in the recorded results, but before we consider this important section of the work, some revision of the shape of standard curves and the systematic sketching of curves from their equations would be an advantage.

**2**

**Standard curves**

(a) *Straight line:* The equation is a first-degree relationship and can always be expressed in the form $y = mx + c$, where

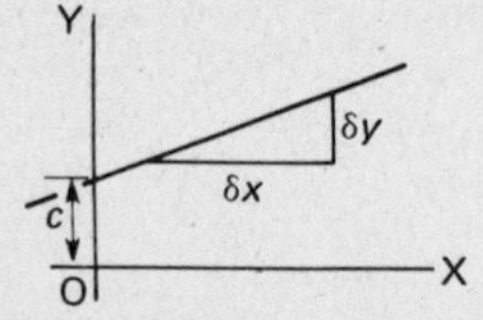


$m$ denotes the slope (gradient), i.e. $\dfrac{\delta y}{\delta x}$

$c$ denotes the intercept on the $y$-axis.

Any first-degree equation gives a straight line graph.

To find where the line crosses the $x$-axis, put $y = 0$.

To find where the line crosses the $y$-axis, put $x = 0$.

Therefore, the line $2y + 3x = 6$ crosses the axes at ..............................

**3**

(2, 0) and (0, 3)

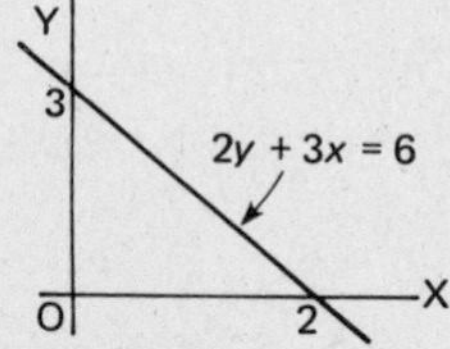


For, when $x = 0$, $2y = 6$ $\therefore\ y = 3$

and when $y = 0$, $3x = 6$ $\therefore\ x = 2$

We can establish the equation of a given straight line by substituting in $y = mx + c$ the $x$- and $y$-coordinates of any three points on the line. Of course, two points are sufficient to determine values of $m$ and $c$, but the third point is taken as a check.

So, if (1, 6), (3, 2) and (5, − 2) lie on a straight line, its equation is ...........

**4**

$$\boxed{y = -2x + 8}$$

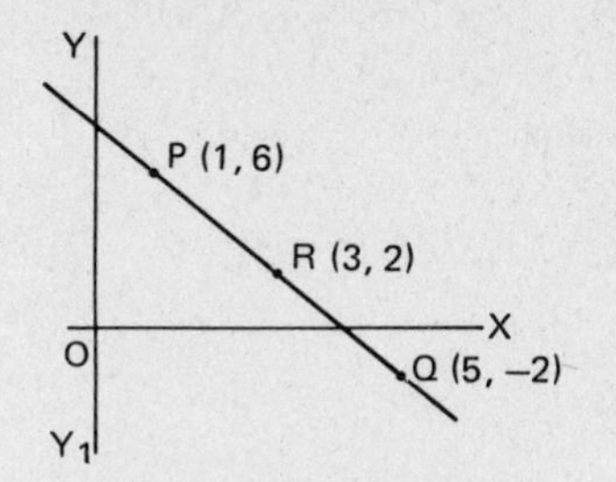


$(1, 6) \qquad 6 = 1m + c$

$(5, -2) \quad -2 = 5m + c$

$\therefore -8 = 4m \quad \therefore m = -2$

$y = -2x + c$

When $\quad x = 1, y = 6 \quad \therefore 6 = -2 + c \quad \therefore c = 8 \qquad \therefore \underline{y = -2x + 8}$

Check: When $x = 3$, $y = -6 + 8 = 2$ which agrees with the third point.

(b) *Second-degree curves:* The basic second-degree curve is $y = x^2$, a parabola symmetrical about the $y$-axis and existing only for $y \geqslant 0$.

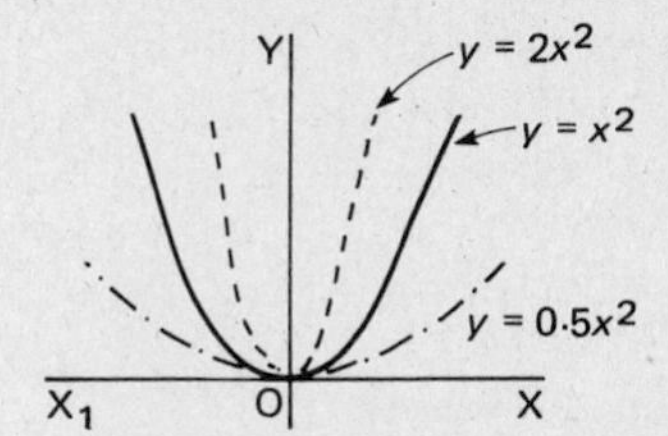


$y = ax^2$ gives a thinner parabola if $a > 1$ and a flatter parabola if $a < 1$.

The general second-degree curve is $y = ax^2 + bx + c$, where the three coefficients, $a$, $b$ and $c$, determine the position of the vertex and the 'width' of the parabola.

*Change of vertex:* If the parabola $y = x^2$ is moved parallel to itself to a vertex position at (2, 3), its equation relative to the new axes is $Y = X^2$.

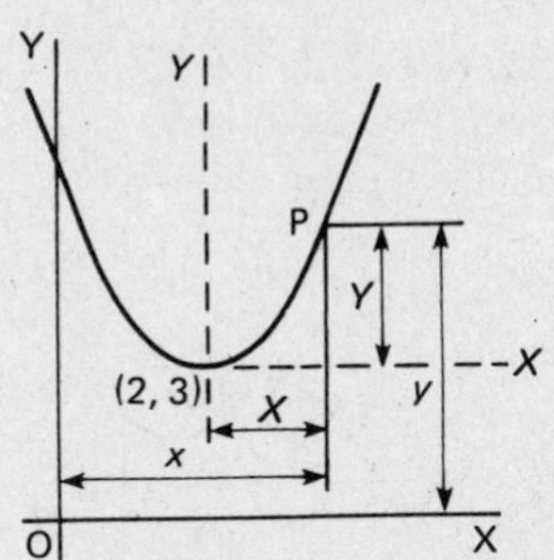


Considering a sample point P, we see that

$$Y = y - 3 \text{ and } X = x - 2$$

So, in terms of the original variables, $x$ and $y$, the equation of the new parabola is ..........................................

**5**

$$\boxed{y = x^2 - 4x + 7}$$

since $Y = X^2$ becomes $y - 3 = (x - 2)^2$ i.e. $y - 3 = x^2 - 4x + 4$ which simplifies to $y = x^2 - 4x + 7$.

*Note:* If the coefficient of $x^2$ is negative, the parabola is inverted.

e.g. $y = -2x^2 + 6x + 5$.

The vertex is at (1·5, 9·5).

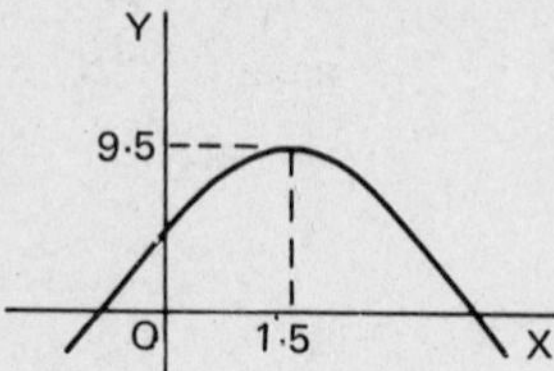


The parabola cuts the $y$-axis at $y$ = ................ and the $x$-axis at $x$ = .............. and $x$ = ....................

**6**

$$\boxed{y = 5;\ x = -0{\cdot}68;\ \text{and } x = 3{\cdot}68}$$

(c) *Third-degree curves:* The basic third-degree curve is $y = x^3$ which passes through the origin. For $x$ positive, $y$ is positive and for $x$ negative, $y$ is negative.

Writing $y = -x^3$, turns the curve upside down.

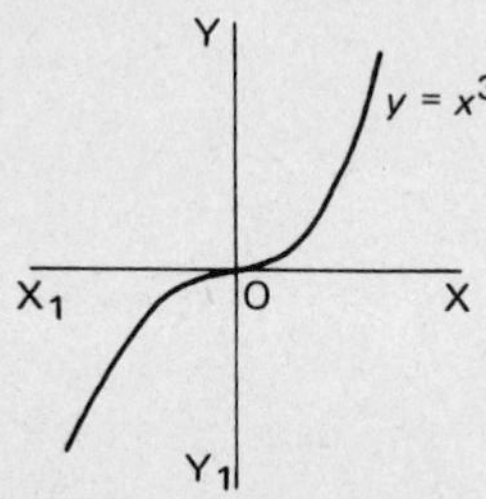


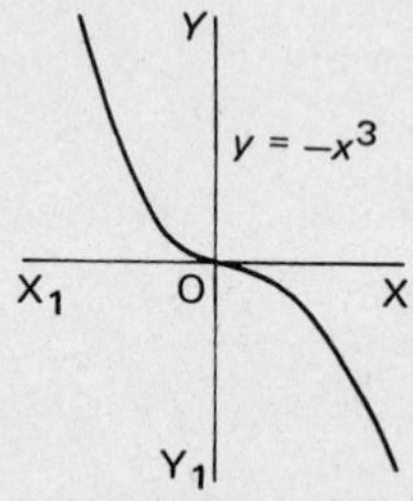


In general, a third-degree curve has a more accentuated double bend and cuts the $x$-axis in three points which may have (i) three real and different values, (ii) two values the same and one different, or (iii) one real value and two complex values.

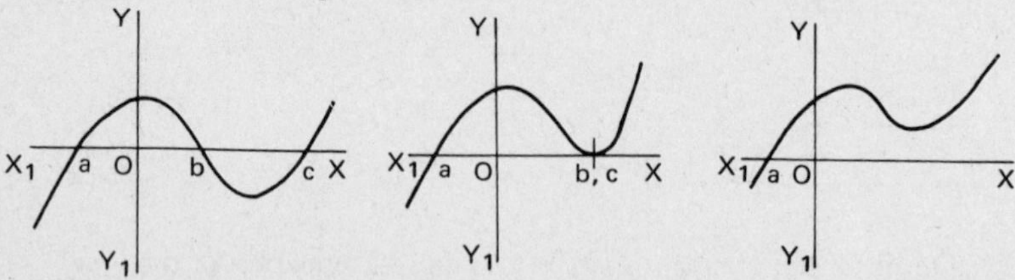


*Now let us collect our ideas so far by working through a short exercise. Move on to the next frame.*

**7**

*Exercise:* Sketch the graphs of the following, indicating relevant information. Do not plot the graphs in detail.

1. $y = 2x - 5$
2. $y = \dfrac{x}{3} + 7$
3. $y = -2x + 4$
4. $2y + 5x - 6 = 0$
5. $y = x^2 + 4$
6. $y = (x - 3)^2$
7. $y = (x + 2)^2 - 4$
8. $y = x - x^2$
9. $y = x^3 - 4$
10. $y = 2 - (x + 3)^3$.

*When you have completed the whole set, check your results with those in the next frame.*

**8** Here are the results

1.

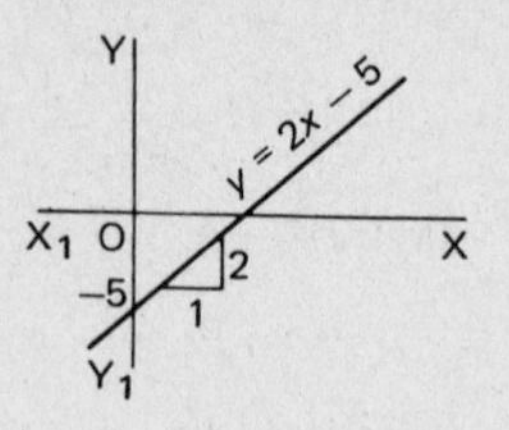


2.

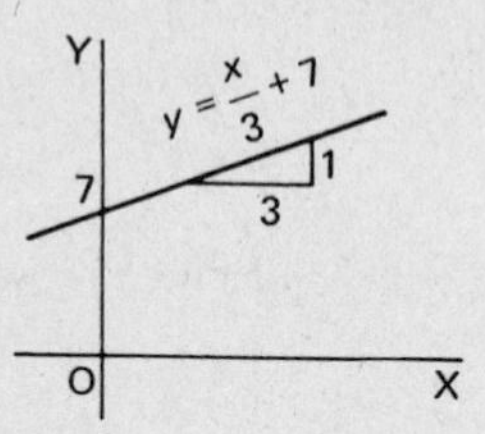


3.

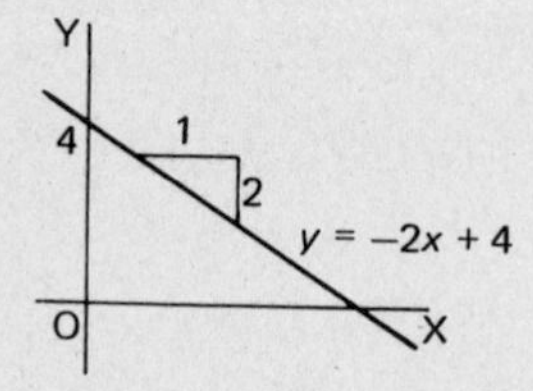


4.

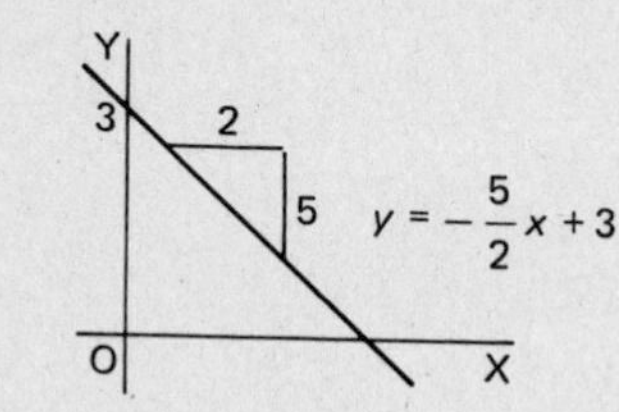


5.

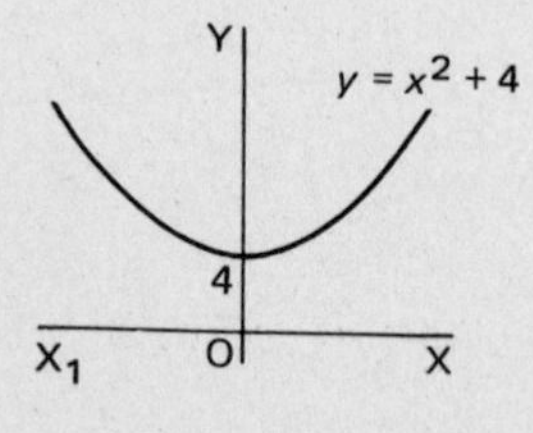


6.

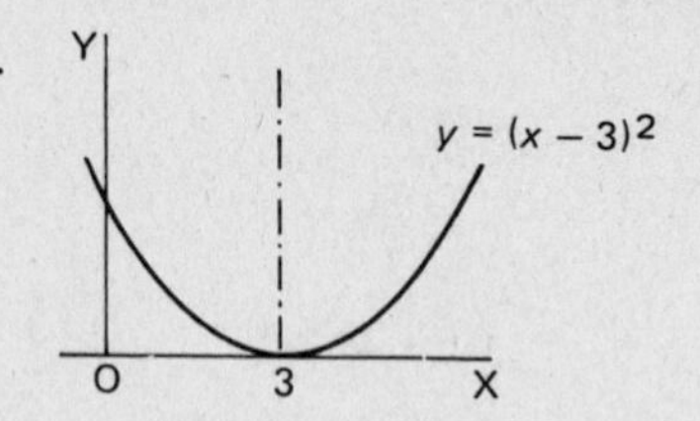


7.

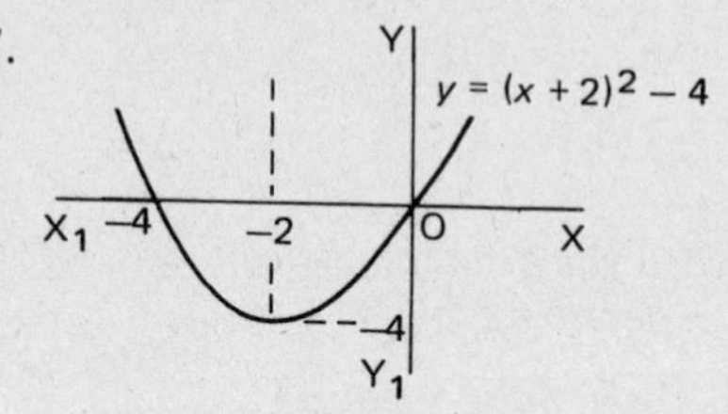


8.

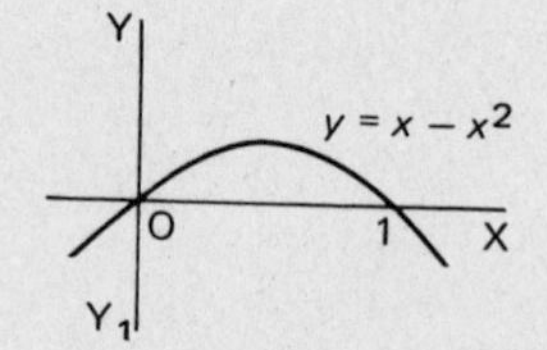


9.

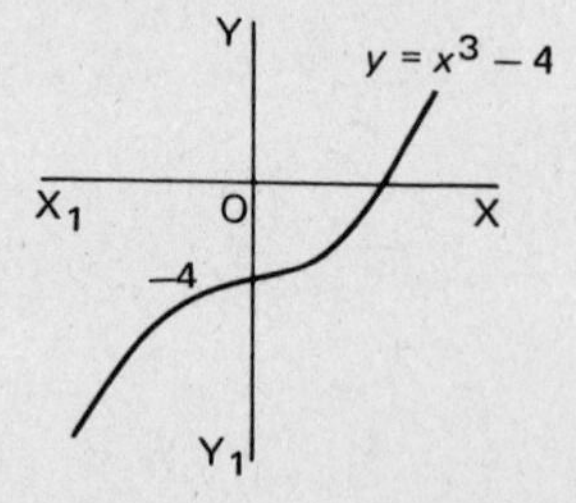


10.

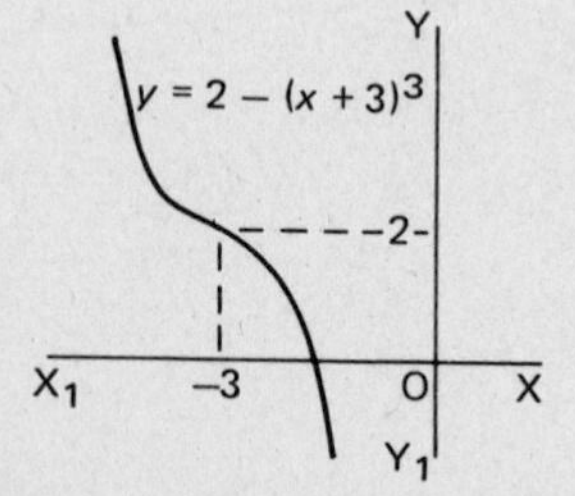


Now we will revise a further set of curves.

(d) *Circle:* The simplest case of the circle is with centre at the origin and radius $r$.

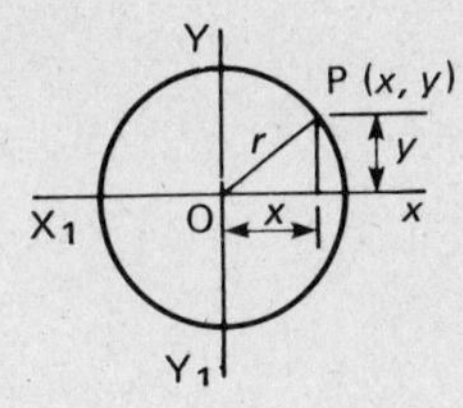


The equation is then

$x^2 + y^2 = r^2$

Moving the centre to a new point (h, k) gives $X^2 + Y^2 = r^2$

where $Y = y - k$

and $X = x - h$

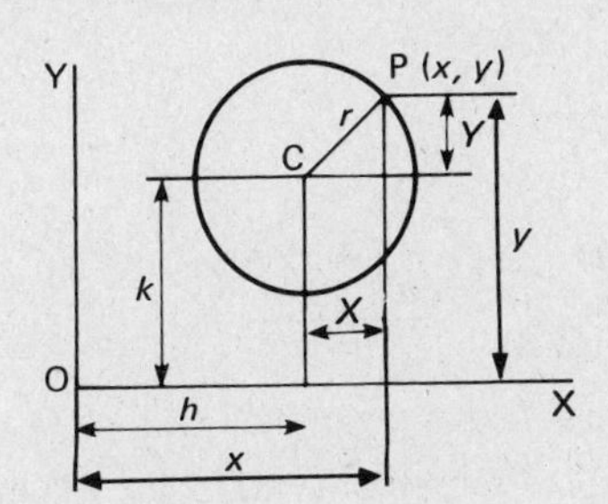


$\therefore \; \underline{(x - h)^2 + (y - k)^2 = r^2}$

The general equation of a circle is

$$x^2 + y^2 + 2gx + 2fy + c = 0$$

where the centre is the point $(-g, -f)$ and radius $= \sqrt{(g^2 + f^2 - c)}$.

Note that, for a second-degree equation to represent a circle,

(i) the coefficients of $x^2$ and $y^2$ are identical,

(ii) there is no product term in $xy$.

So the equation $x^2 + y^2 + 2x - 6y - 15 = 0$ represents a circle with centre ................ and radius ................

**10**

| centre (−1, 3); radius 5 |
|---|

for $2g = 2 \quad \therefore g = 1$
$2f = -6 \quad \therefore f = -3$ } $\therefore$ centre $(-g, -f) = (-1, 3)$

also $c = -15 \quad \therefore$ radius $= \sqrt{(g^2 + f^2 - c)} = \sqrt{(1 + 9 + 15)} = \sqrt{25} = 5.$

(e) *Ellipse:* The equation of an ellipse is

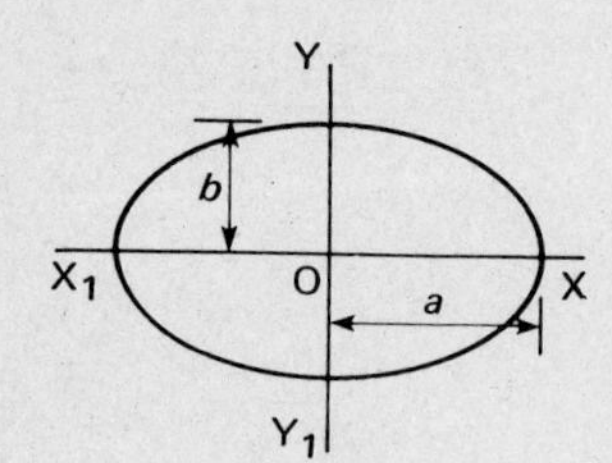


$$\frac{x^2}{a^2} + \frac{y^2}{b^2} = 1$$

where $a$ = semi-major axis $\therefore y = 0, x = \pm a$
and $b$ = semi-minor axis $\therefore x = 0, y = \pm b.$

Of course, when $a^2$ and $b^2$ are equal (say $r^2$) we obtain

........................

**11**

> the equation of a circle, i.e. $x^2 + y^2 = r^2$

(f) *Hyperbola:* The equation of a hyperbola is

$$\frac{x^2}{a^2} - \frac{y^2}{b^2} = 1$$

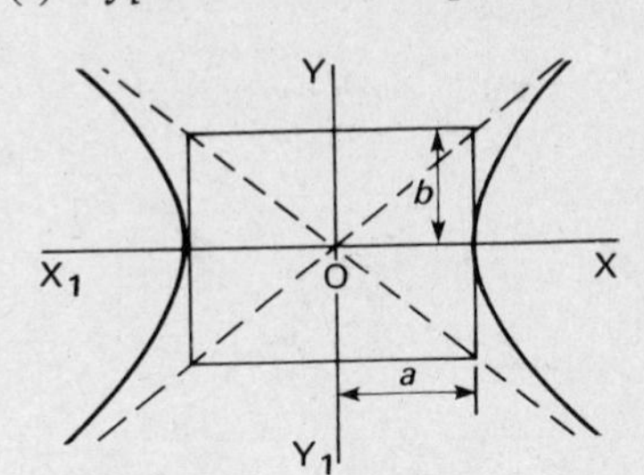


When $y = 0, x = \pm a$

When $x = 0, y^2 = -b^2$ $\therefore$ the curve does not cross the $y$-axis.

Note that the opposite arms of the hyperbola gradually approach two straight lines (asymptotes).

*Rectangular hyperbola:* If the asymptotes are at right-angles to each other, the curve is then a *rectangular hyperbola.* A more usual form of the rectangular hyperbola is to rotate the figure through 45° and to make the asymptotes the axes of $x$ and $y$. The equation of the curve relative to the new axes then becomes

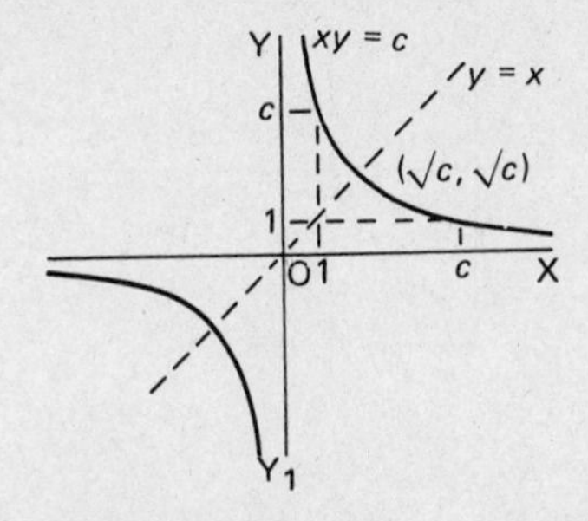


$$xy = c \text{ i.e. } y = \frac{c}{x}$$

Three points are easily located. If $xy = c$

(i) when $x = 1, y = c$
(ii) when $y = 1, x = c$
(iii) the line $y = x$ cuts $xy = c$ at the point $(\pm\sqrt{c}, \pm\sqrt{c})$

These three points are a great help in sketching a rectangular hyperbola. Rectangular hyperbolae frequently occur in practical considerations.

*Now for another short exercise, so move on to the next frame.*

**12**

*Exercise:* Sketch the graphs of the following, showing relevant facts.

1. $x^2 + y^2 = 12{\cdot}25$
2. $x^2 + 4y^2 = 100$
3. $x^2 + y^2 - 4x + 6y - 3 = 0$
4. $2x^2 - 3x + 4y + 2y^2 = 0$
5. $\dfrac{x^2}{36} - \dfrac{y^2}{49} = 1$
6. $xy = 5.$

*When you have sketched all six, compare your results with those in the next frame.*

**13**

The sketch graphs are as follows:

1.

2.

3.

4.

5.

6.

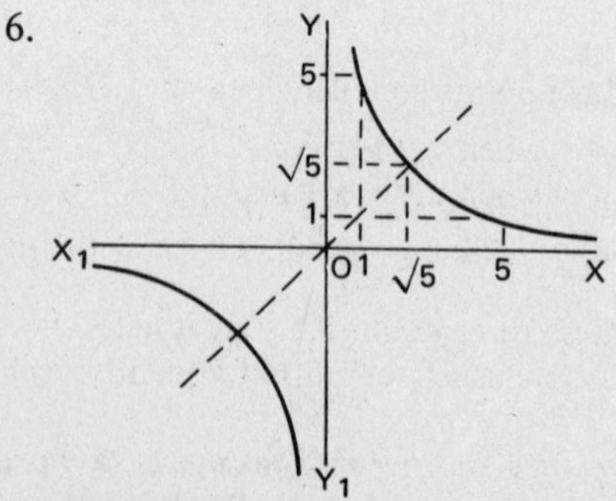


*On to frame 14 for a third set of curves frequently occurring.*

**14**

(g) *Logarithmic curves:* If $y = \log x$, then when $x = 1$, $y = \log 1 = 0$
i.e. the curve crosses the $x$-axis at $x = 1$.

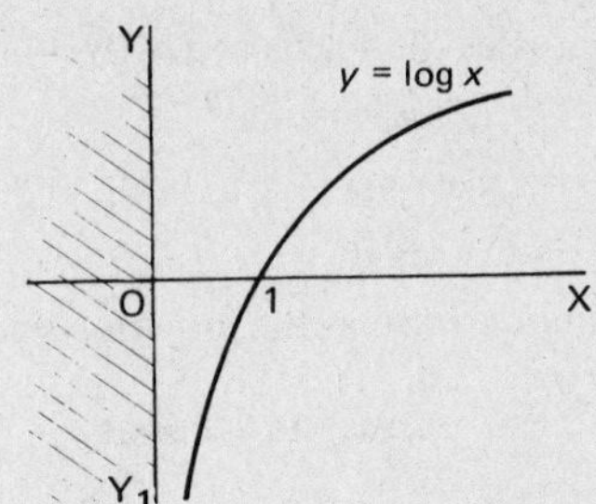


Also, $\log x$ does not exist for $x < 0$.
$y = \log x$ flattens out as $x \to \infty$, but continues to increase at an ever-decreasing rate.

The graph of $y = \ln x$ also has the same shape and crosses the $x$-axis at $x = 1$, but the function values are different.

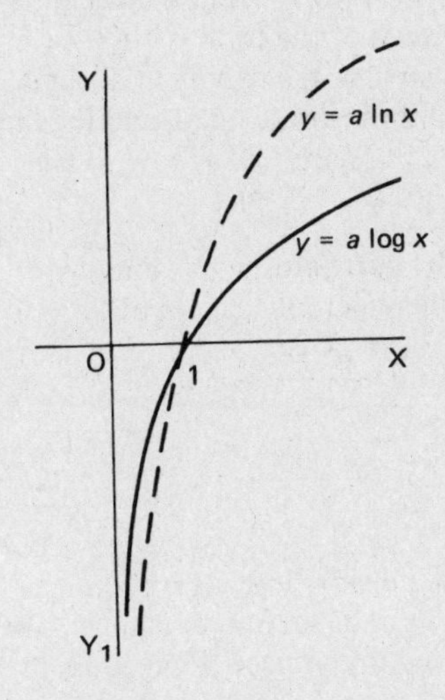


The graphs of $y = a \log x$ and $y = a \ln x$ are similar, but with all ordinates multiplied by the constant factor $a$.

*Continued in the next frame.*

**15** (h) *Exponential curves:*

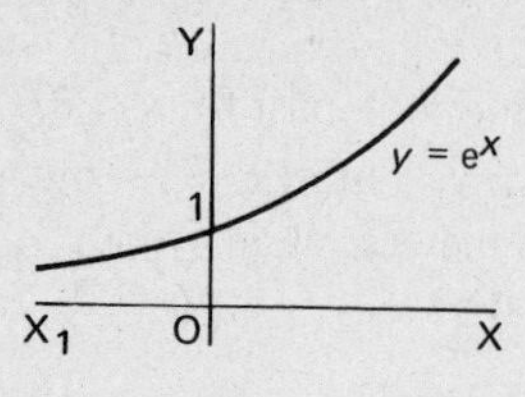


$y = e^x$ crosses the $y$-axis at $x = 0$

i.e. $y = e^0 = 1$.

As $x \rightarrow \infty$, $y \rightarrow \infty$

as $x \rightarrow -\infty$, $y \rightarrow 0$.

Sometimes known as the *growth curve.*

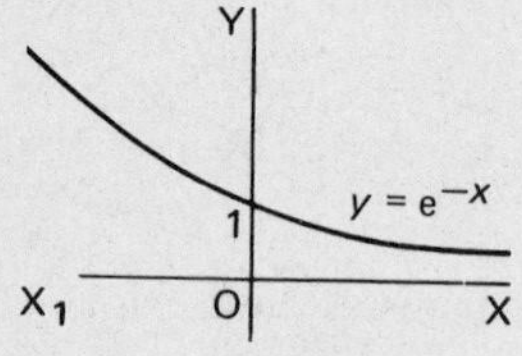


$y = e^{-x}$ also crosses the $y$-axis at $y = 1$.

As $x \rightarrow \infty$, $y \rightarrow 0$

as $x \rightarrow -\infty$, $y \rightarrow \infty$.

Sometimes known as the *decay curve.*

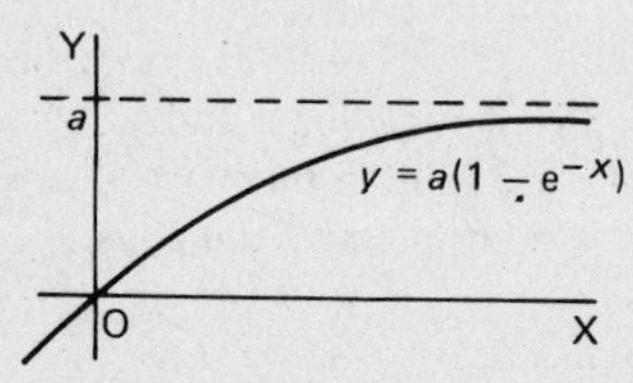


In electrical work, we also frequently have curves of the form $y = a(1 - e^{-x})$. This is an inverted exponential curve, passing through the origin and tending to $y = a$ as asymptote as $x \rightarrow \infty$ (since $e^{-x} \rightarrow 0$ as $x \rightarrow \infty$).

(i) *Hyperbolic curves:* Combination of the curves for $y = e^x$ and $y = e^{-x}$ gives the hyperbolic curves of

$$y = \cosh x = \frac{e^x + e^{-x}}{2} \quad \text{and}$$

$$y = \sinh x = \frac{e^x - e^{-x}}{2}.$$

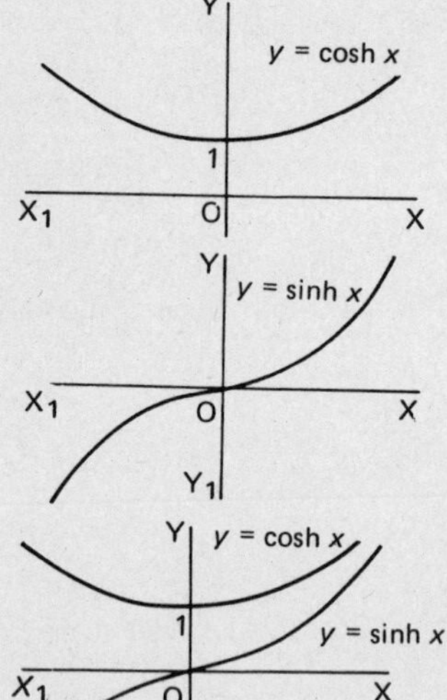


We have already dealt with these functions in detail in programme 3, so refer back if you need to revise them further.

If we draw the two graphs on the same axes, we see that $y = \sinh x$ is always outside $y = \cosh x$, i.e. for any particular value of $x$, $\cosh x > \sinh x$.

*We have one more type of curve to list, so turn on to the next frame.*

(j) *Trigonometrical curves:* The most common occurring in practice is the sine curve.

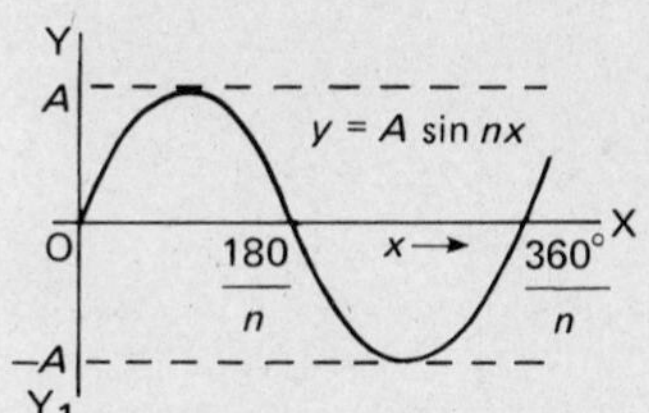


(i) $\underline{y = A \sin nx}$

$$\text{Period} = \frac{360°}{n}$$

$$\text{Amplitude} = A$$

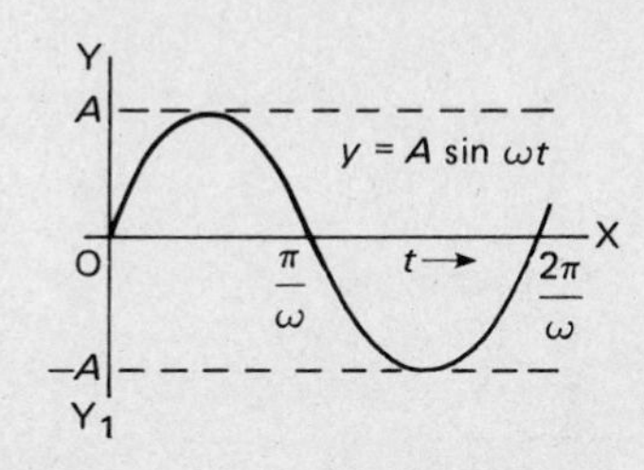


(ii) $\underline{y = A \sin \omega t}$

$$\text{Period} = \frac{2\pi}{\omega}$$

$$\text{Amplitude} = A$$

Now we can sketch a further selection of curves.

*Exercise:* Sketch the following pairs of curves on the same axes. Label each graph clearly and show relevant information.

1. $y = \cosh x$ and $y = 2 \cosh x$
2. $y = \sinh x$ and $y = \sinh 2x$
3. $y = e^x$ and $y = e^{3x}$
4. $y = e^{-x}$ and $y = 2\,e^{-x}$
5. $y = 5 \sin x$ and $y = 3 \sin 2x$
6. $y = 4 \sin \omega t$ and $y = 2 \sin 3\omega t$.

**17** Here they are:

1.

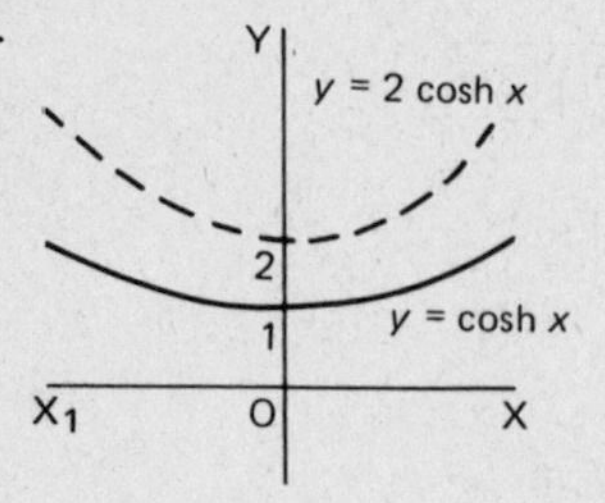


4.

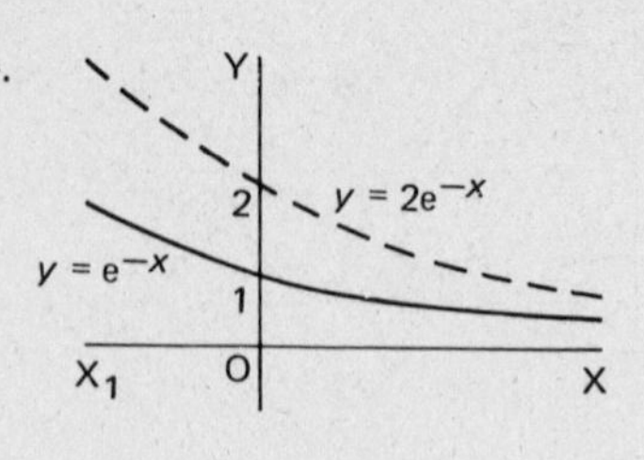


2.

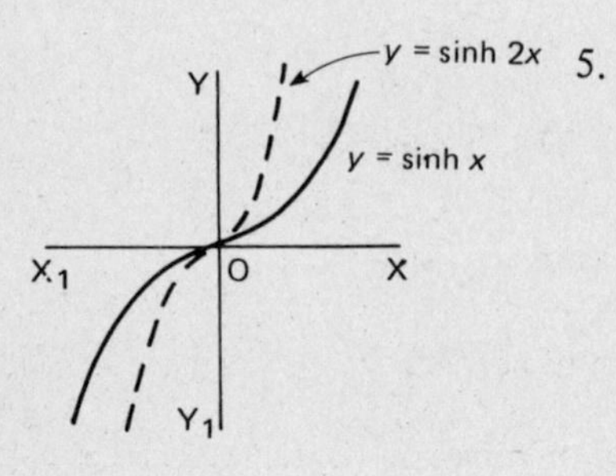


5.

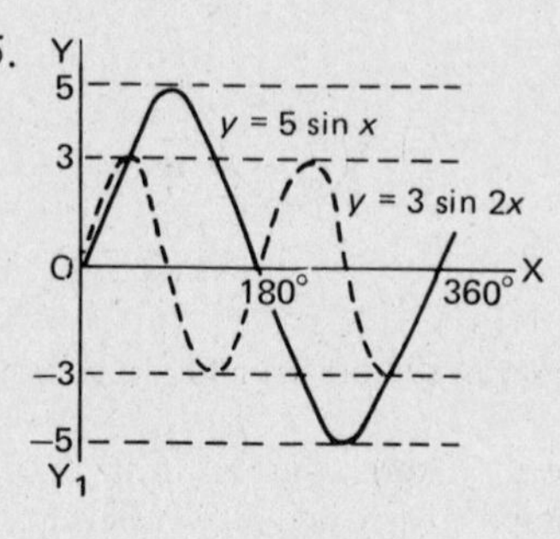


3.

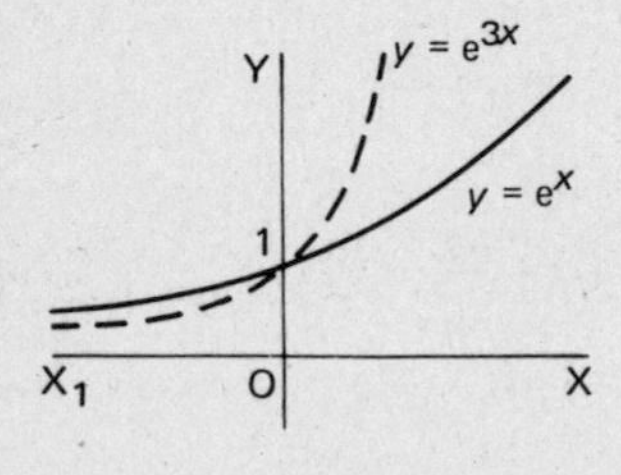


6.

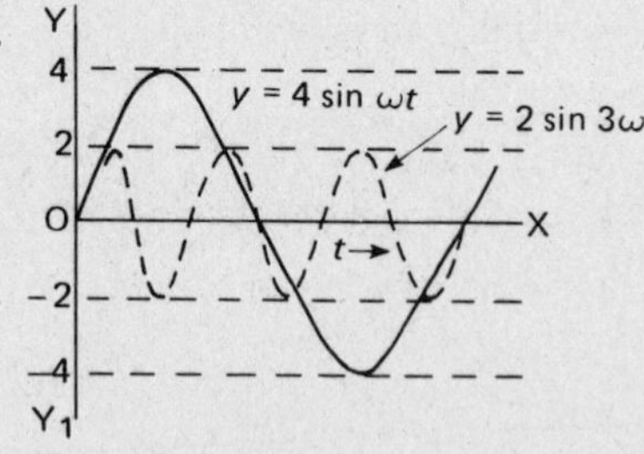

**Asymptotes**

We have already made references to asymptotes in the previous work and it is always helpful to know where asymptotes occur when sketching curves of functions.

*Definition:* An *asymptote* to a curve is a straight line to which the curve approaches as the distance from the origin increases. It can also be thought of as a tangent to the curve at infinity, i.e. the curve touches the asymptote at two coincident points at infinity.

*Condition for infinite roots:* If we consider the equation

$$a_0x^n + a_1x^{n-1} + \ldots + a_{n-2}x^2 + a_{n-1}x + a_n = 0$$

and substitute $x = \dfrac{1}{y}$, the equation becomes

$$a_0\frac{1}{y^n} + a_1\frac{1}{y^{n-1}} + a_2\frac{1}{y^{n-2}} + \ldots + a_{n-1}\frac{1}{y} + a_n = 0$$

If we now multiply through by $y^n$ and reverse the order of the terms, we have

$$a_ny^n + a_{n-1}y^{n-1} + \ldots + a_2y^2 + a_1y + a_0 = 0$$

If $\underline{a_0 = 0}$ and $\underline{a_1 = 0}$, the equation reduces to

$$a_ny^n + a_{n-1}y^{n-1} + \ldots + a_2y^2 = 0$$

$$\therefore\ y^2 = 0 \text{ or } a_ny^{n-2} + a_{n-1}y^{n-3} + \ldots + a_2 = 0.$$

Therefore with the stated condition, $y^2 = 0$ gives two zero roots for $y$,

i.e. two infinite roots for $x$, since $y = \dfrac{1}{x}$ and hence $x = \dfrac{1}{y}$.

*Therefore, the original equation*

$$a_0x^n + a_1x^{n-1} + \ldots + a_{n-1}x + a_n = 0$$

*will have two infinite roots if* $a_0 = 0$ *and* $a_1 = 0$.

**Determination of an asymptote:** From the result we have just established, to find an asymptote to $y = f(x)$,

(a) substitute $y = mx + c$ in the given equation and simplify,
(b) equate to zero ..................................................

**19**

> the coefficients of the two highest powers of $x$ and so determine the values of $m$ and $c$

Let us work through an example to see how it develops.

---

**20**

*Example:* To find the asymptote to the curve $x^2y - 5y - x^3 = 0$.

Substitute $y = mx + c$ in the equation.

$$x^2(mx + c) - 5(mx + c) - x^3 = 0$$

$$mx^3 + cx^2 - 5mx - 5c - x^3 = 0$$

$$(m - 1)x^3 + cx^2 - 5mx - 5c = 0.$$

Equating to zero the coefficients of the two highest powers of $x$,

$$\left.\begin{aligned} m - 1 = 0 \quad \therefore m = 1 \\ c = 0 \quad\quad c = 0 \end{aligned}\right\} \quad \therefore \text{ Asymptote is } \underline{y = x}$$

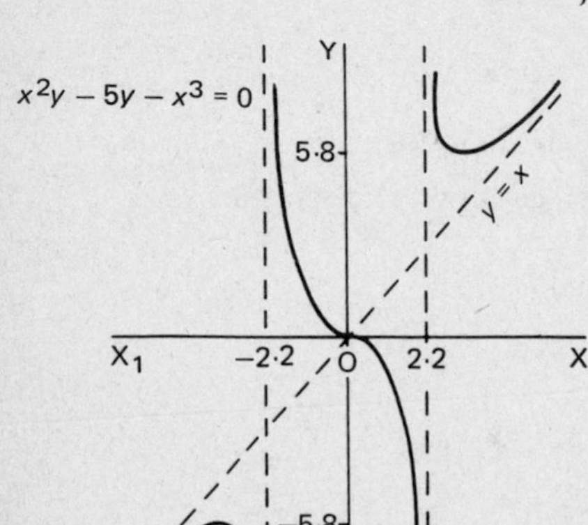


In fact, the graph of $x^2y - 5y - x^3 = 0$ is as shown on the left and we can see that the curve approaches $y = x$ as $x \to \infty$ and as $x \to -\infty$.

From the graph, however, it appears that there are two further asymptotes which are the lines .......................... and ......................

$x = -2{\cdot}2$ and $x = 2{\cdot}2$

These are two lines parallel to the $y$-axis.

**Asymptotes parallel to the x and y axes:** These can be found by a simple rule.

*For the curve $y = f(x)$, the asymptotes parallel to the x-axis can be found by equating the coefficient of the highest power of x to zero. Similarly, the asymptotes parallel to the y-axis can be found by equating the coefficient of the highest power of y to zero.*

*Example 1:* Find the asymptotes, if any, of $y = \dfrac{x-2}{2x+3}$.

First get everything on one line by multiplying by the denominator $(2x + 3)$

$$y(2x+3) = x - 2 \quad \therefore\ 2xy + 3y - x + 2 = 0.$$

(a) Asymptote parallel to $x$-axis: Equate the coefficient of the highest power of $x$ to zero. $(2y - 1)x + 3y + 2 = 0$

$\therefore\ 2y - 1 = 0 \quad \therefore\ 2y = 1 \quad \therefore\ y = 0{\cdot}5 \quad \therefore$ $\underline{y = 0{\cdot}5 \text{ is an asymptote.}}$

(b) Asymptote parallel to y-axis: Equate the coefficient of the highest power of $y$ to zero. $\therefore$ .............................. is also an asymptote.

**22**

$$\boxed{x = -1\cdot5}$$

for, rearranging the equation to obtain the highest power of $y$

$$(2x + 3)y - x + 2 = 0$$

$$\therefore 2x + 3 = 0 \quad \therefore x = -1\cdot5 \quad \therefore \underline{x = -1\cdot5 \text{ is an asymptote.}}$$

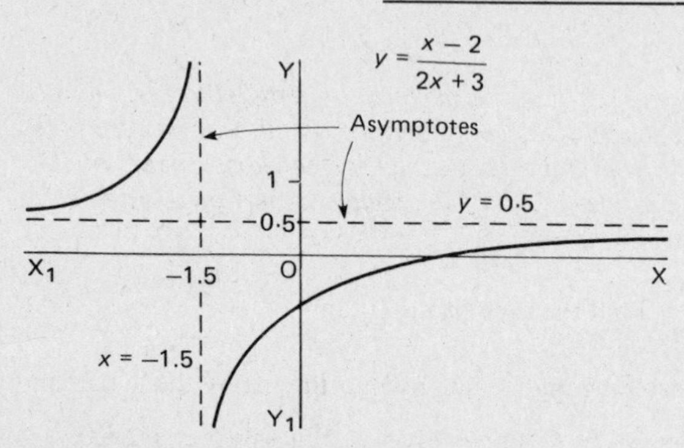


In fact, the graph is positioned as shown above. The only asymptotes are $y = 0\cdot5$ (parallel to the $x$-axis)
and $x = -1\cdot5$ (parallel to the $y$-axis). Let us do another.

*Example 2:* Find the asymptotes of the curve $x^2(x^2 + 2) = y^3(x + 5)$.

(a) Parallel to the $x$-axis: $x^4 + 2x^2 = (x + 5)y^3$.
Equate the coefficient of tne highest power of $x$ to zero.
Highest power of $x$ is $x^4$. Its coefficient is 1, which gives $1 = 0$. This is not the equation of a line. Therefore, there is no asymptote parallel to the $x$-axis.

(b) Parallel to the $y$-axis: This gives ................

**23**

$$\boxed{x = -5 \text{ is an asymptote}}$$

For an asymptote parallel to the $y$-axis, we equate the highest power of $y$ to zero. $\therefore x + 5 = 0 \quad \therefore x = -5$ Therefore, $y = -5$ is an asymptote parallel to the $y$-axis

Now, to find a general asymptote, i.e. not necessarily parallel to either axis, we carry out the method described earlier, which was to ....................
..........................................................................

**24**

> substitute $y = mx + c$ in the equation and equate the coefficients of the two highest powers of $x$ to zero.

If we do that with the equation of this example, we get ............................

*Work it right through to the end.*

---

**25**

> $y = x - \dfrac{5}{3}$ is also an asymptote

For, substituting $y = mx + c$ in $x^2(x^2 + 2) = y^3(x + 5)$ we have

$$x^4 + 2x^2 = (m^3x^3 + 3m^2x^2c + 3mxc^2 + c^3)(x + 5)$$
$$= m^3x^4 + 3m^2x^3c + 3mx^2c^2 + c^3x$$
$$+ 5m^3x^3 + 15m^2x^2c + 15mxc^2 + 5c^3$$
$$\therefore (m^3 - 1)x^4 + (5m^3 + 3m^2c)x^3 + (15m^2c + 3mc^2 - 2)x^2$$
$$+ (15mc^2 + c^3)x + 5c^3 = 0.$$

Equating to zero the coefficients of the two highest powers of $x$,

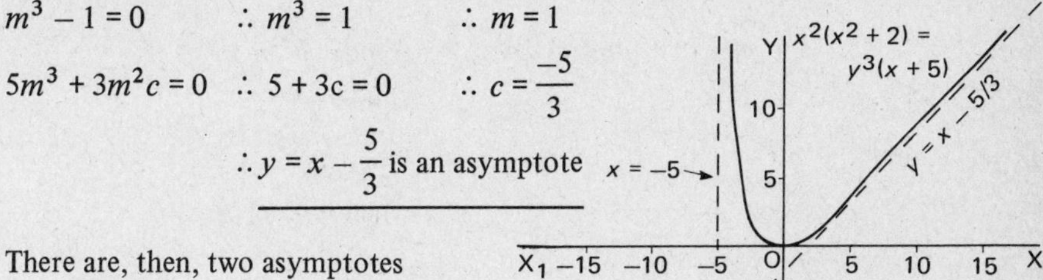

$m^3 - 1 = 0 \qquad \therefore m^3 = 1 \qquad \therefore m = 1$

$5m^3 + 3m^2c = 0 \qquad \therefore 5 + 3c = 0 \qquad \therefore c = \dfrac{-5}{3}$

$\therefore y = x - \dfrac{5}{3}$ is an asymptote

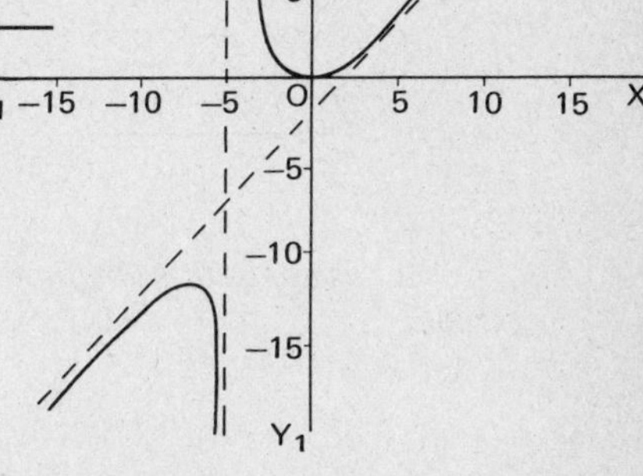


There are, then, two asymptotes

$x = -5$

and $\quad y = x - \dfrac{5}{3}$

In fact, the graph is shown on the right.

Here are two similar problems for your to do on your own. Just apply the rules. There are no tricks.

*Exercise:* Determine the asymptotes, if any, for the following curves

1. $x^4 - 2x^3y + 10x^2 - 7y^2 = 0$
2. $x^3 - xy^2 + 4y^2 - 5 = 0$

**26**

> 1. $y = \dfrac{x}{2}$ is the only asymptote.
>
> 2. $y = x + 2$; $y = -x - 2$; $x = 4$.

*In Question 1*, there are no asymptotes parallel to the axes.
Substituting $y = mx + c$ and collecting up terms gives

$$(1 - 2m)x^4 - 2cx^3 + (10 - 7m^2)x^2 - 14mcx - 7c^2 = 0$$

from which

$$\left.\begin{array}{ll} 1 - 2m = 0 & \therefore\ m = \dfrac{1}{2} \\ c = 0 & \end{array}\right\} \therefore\ \underline{y = \frac{x}{2} \text{ is the only asymptote.}}$$

*In Question 2*, $\underline{x = 4}$ is an asymptote parallel to the $y$-axis.
There is no asymptote parallel to the $x$-axis.
Putting $y = mx + c$ and simplifying produces

$$(1 - m^2)x^3 + (4m^2 - 2mc)x^2 + (8mc - c^2)x + 4c^2 - 5 = 0$$

so that $1 - m^2 = 0 \quad \therefore\ m^2 = 1 \quad \therefore\ m = \pm 1$

and $4m^2 - 2mc = 0 \quad \therefore\ mc = 2.$ When $m = 1, c = 2$

when $m = -1, c = -2$

$\therefore\ \underline{y = x + 2 \text{ and } y = -x - 2 \text{ are asymptotes.}}$

*Now let us apply these methods in a wider context starting in the next frame.*

27

**Systematic curve sketching**, given the equation of the curve.

If, for $y = f(x)$, the function $f(x)$ is known, the graph of the function can be plotted by calculating $x$- and $y$-coordinates of a number of selected points. This, however, can be a tedious occupation and considerable information about the shape and positioning of the curve can be obtained by a systematic analysis of the given equation. There is a list of steps we can take.

1. *Symmetry* Inspect the equation for symmetry.
   (a) If only even powers of $y$ occur, the curve is symmetrical about the $x$-axis.
   (b) If only even powers of $x$ occur, the curve is symmetrical about the $y$-axis.

e.g.

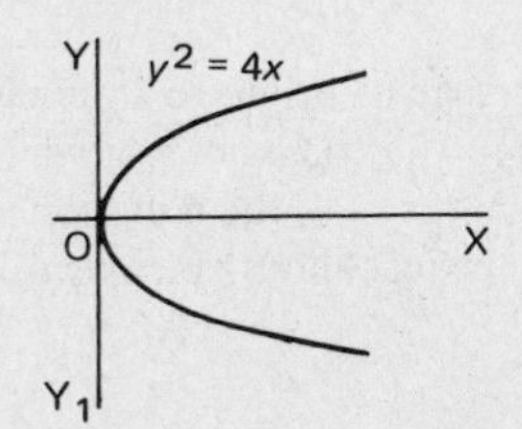


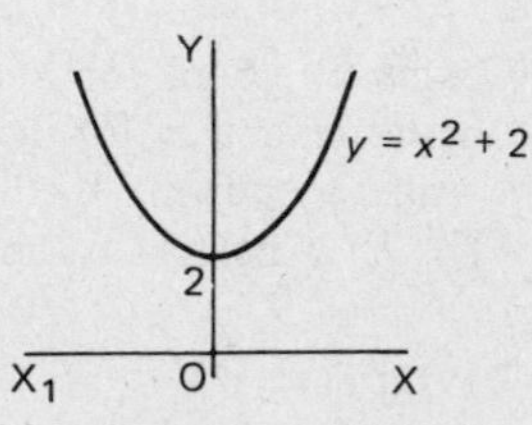


   (c) If only even powers of $y$ and also only even powers of $x$ occur, then ....................................

28

the curve is symmetrical about both axes

For example, $25x^2 + 16y^2 = 400$ is symmetrical about both axes.

$y^2 + 3y - 2 = (x^2 + 7)^2$ is symmetrical about the $y$-axis,

but not about the $x$-axis since ....................

29

both odd and even powers of $y$ occur

2. *Intersection with the axes* Points at which the curve crosses the $x$ and $y$ axes.

   Crosses the $x$-axis: Put $y = 0$ and solve for $x$.
   Crosses the $y$-axis: Put $x = 0$ and solve for $y$.
   So, the curve $y^2 + 3y - 2 = x + 8$ crosses the $x$- and $y$-axes at
   ................................

**30**

| $x$-axis at $x = -10$ |
|---|
| $y$-axis at $y = 2$ and $y = -5$ |

In fact, the curve is

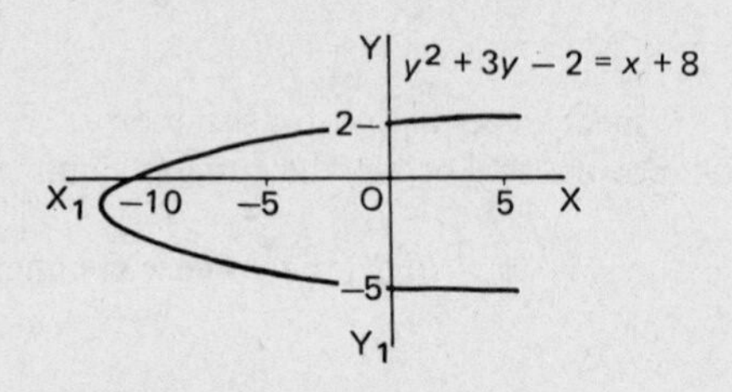


3. *Change of origin* Look for a possible change of origin to simplify the equation. For example, for the curve $4(y + 3) = (x - 4)^2$, if we change the origin by putting $Y = y + 3$ and $X = x - 4$, the equation becomes $4Y = X^2$ which is a parabola symmetrical about the axis of $Y$.

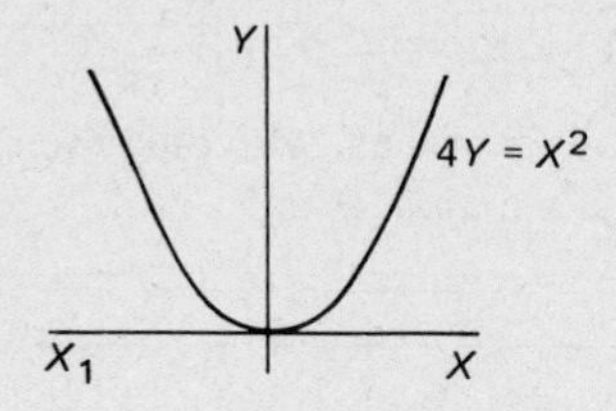


So the curve relative to the original $x$- and $y$-axes is positioned thus:

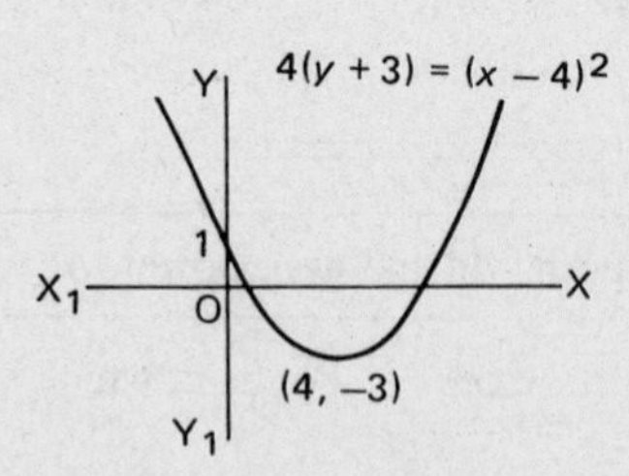

**31**

4. *Asymptotes* We have already dealt with asymptotes in some detail. We investigate (a) asymptotes parallel to the axes, and (b) those of a more general nature.

(a) *Parallel to the axes*

(i) Express the equation 'on one line', i.e. remove algebraic fractions.

(ii) Equate to zero the coefficient of the highest power of $y$ to find the asymptote parallel to the $y$-axis.

(iii) Equate to zero the coefficient of the highest power of $x$ to find the asymptote parallel to the $x$-axis.

As an example, find the asymptotes parallel to the axes for the curve

$$y = \frac{(x-1)(x+6)}{(x+3)(x-4)}.$$

**32**

$$\boxed{x = -3; \quad x = 4; \quad y = 1}$$

since $y(x+3)(x-4) = (x-1)(x+6)$.

Asymptotes parallel to the $y$-axis: $(x+3)(x-4) = 0 \quad \therefore \underline{x = -3}$ and $\underline{x = 4}$.

Re-arranging the equation gives

$$(y-1)x^2 - (y+5)x - 12y + 6 = 0$$

Asymptote parallel to the $x$-axis: $y - 1 = 0 \quad \therefore \underline{y = 1}$

The graph of the function is as shown to the right

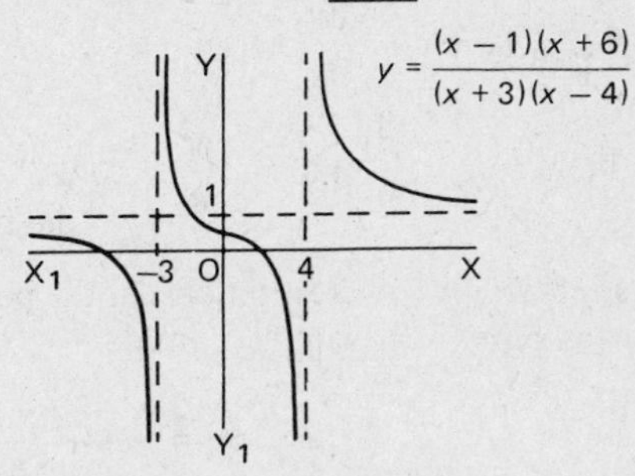


(b) *General asymptotes*

Substitute $y = mx + c$ and equate the coefficients of the two highest powers of $x$ to zero to determine $m$ and $c$.

Thus, for the curve $x^4 - 2x^3y + 5x^2 - 4y^2 = 0$, the asymptote is ...............

**33**

$$\boxed{y = \frac{x}{2}}$$

For, substituting $y = mx + c$ and simplifying the left-hand side, gives

$$(1 - 2m)x^4 - 2cx^3 + (5 - 4m^2)x^2 - 8mcx + 4c^2 = 0.$$

$$\left.\begin{array}{ll} \therefore\ 1 - 2m = 0 & \therefore\ m = \dfrac{1}{2} \\ 2c = 0 & \therefore\ c = 0 \end{array}\right\} \quad \therefore\ y = \frac{x}{2} \text{ is an asymptote.}$$

5. *Large and small values of* x *and* y  If $x$ or $y$ is *small*, higher powers of $x$ or $y$ become negligible and hence only lower powers of $x$ or $y$ appearing in the equation provide an approximate simpler form.

   Similarly, if $x$ or $y$ is *large*, the higher powers have predominance and lower powers can be neglected, i.e. when $x$ is large,
   $y^2 = 2x^2 - 7x + 4$
   approximates to $y^2 = 2x^2$, i.e. $y = \pm x\sqrt{2}$.

6. *Turning points*  Maximum and minimum values; points of inflection. We have dealt at length in a previous programme with this whole topic. We will just summarise the results at this stage:

   For a *maximum*, $\dfrac{dy}{dx} = 0$ and $\dfrac{d^2y}{dx^2}$ is negative: curve concave downwards.

   For a *minimum*, $\dfrac{dy}{dx} = 0$ and $\dfrac{d^2y}{dx^2}$ is positive: curve concave upwards.

   For a *p. of i.*, $\dfrac{dy}{dx} = 0$ and $\dfrac{d^2y}{dx^2} = 0$ with a change of sign through the turning point.

7. *Limitations*  Restrictions on the possible range of values that $x$ or $y$ may have. For example, consider

$$y^2 = \frac{(x+1)(x-3)}{x+4}.$$

For $x < -4$, $y^2$ is negative $\therefore$ no real values of $y$.
For $-4 < x < -1$, $y^2$ is positive $\therefore$ real values of $y$ exist.
For $-1 < x < 3$, $y^2$ is negative $\therefore$ no real values of $y$.
For $3 < x$, $y^2$ is positive $\therefore$ real values of $y$ exist.

The curve finally looks like this:

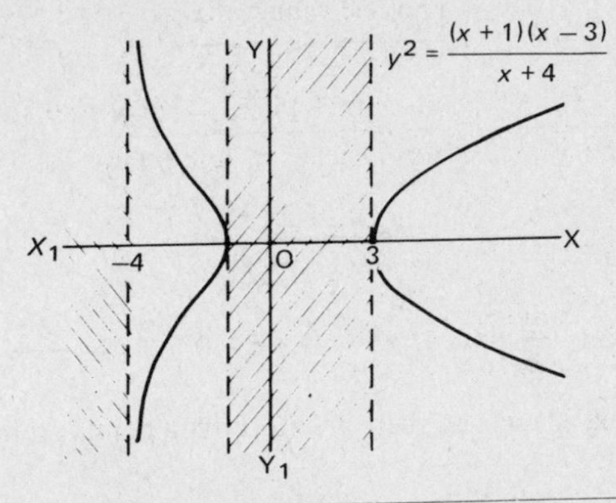


**34**

In practice, not all of these considerations are applicable in any one particular case. Let us work in detail through one such example.

*Example:* Sketch the curve whose equation is $y = \frac{(x+2)(x-3)}{x+1}$.

(a) *Symmetry:* First write the equation 'on the line'.

$$y(x+1) = (x+2)(x-3) = x^2 - x - 6$$

Both odd and even powers of $x$ occur, $\therefore$ no symmetry about $y$-axis.
Only odd powers of $y$ occur, $\therefore$ no symmetry about $x$-axis.

(b) *Crossing the axes:* This is done simply by putting $y = 0$ and $x = 0$, so, in this example, the curve crosses the axes at ................

**35**

| $x$-axis at $x = -2$ and at $x = 3$.<br>$y$-axis at $y = -6$. |
|---|

(c) *Turning points:* The first essential is to find the values of $x$ at which $\frac{dy}{dx} = 0$. Obtain an expression for $\frac{dy}{dx}$ and solve $\frac{dy}{dx} = 0$ for values of $x$.

This gives $\frac{dy}{dx} = 0$ at $x =$ ....................

**36**

no real values of $x$

For, if $y = \dfrac{x^2 - x - 6}{x + 1}$, $\quad \dfrac{dy}{dx} = \dfrac{(x+1)(2x-1) - (x^2 - x - 6)}{(x+1)^2}$

$$= \frac{x^2 + 2x + 5}{(x+1)^2}$$

For turning points, $\dfrac{dy}{dx} = 0 \quad \therefore x^2 + 2x + 5 = 0 \quad \therefore x = \dfrac{-2 \pm \sqrt{-16}}{2}$

i.e. $x$ is complex. Therefore, there are no turning points on the graph.

(d) *When* x *is very small:* $\quad y \simeq -\dfrac{x+6}{x+1} \quad$ i.e. $y \simeq -6$.

*When* x *is very large:* $\quad y \simeq \dfrac{x^2}{x}\quad$ i.e. $x \quad \therefore y \simeq x$.

(e) *Asymptotes:*
(i) First find any asymptotes parallel to the axes.
These are ..................

**37**

parallel to the $y$-axis: $x = -1$

For $y(x + 1) - x^2 + x + 6 = 0 \quad \therefore x + 1 = 0 \quad \therefore x = -1$.

(ii) Now investigate any general asymptote, if any.
This gives ................

**38**

$$\boxed{y = x - 2}$$

This is obtained, as usual, by putting $y = mx + c$ in the equation.

$$(mx + c)(x + 1) = (x + 2)(x - 3)$$

$$\therefore\ mx^2 + mx + cx + c = x^2 - x - 6$$

$$(m - 1)x^2 + (m + c + 1)x + c + 6 = 0.$$

Equating the coefficients of the two highest powers of $x$ to zero,

$$m - 1 = 0 \quad \therefore\ m = 1$$

and $$m + c + 1 = 0 \quad \therefore\ c + 2 = 0 \quad \therefore\ c = -2$$

$$\therefore\ \underline{y = x - 2 \text{ is an asymptote}}.$$

So, collecting our findings together, we have

(a) No symmetry about the $x$- or $y$-axis.
(b) Curve crosses the $x$-axis at $x = -2$ and at $x = 3$.
(c) Curve crosses the $y$-axis at $y = -6$.
(d) There are no turning points on the curve.
(e) Near $x = 0$, the curve approximates to $y = -6$.
(f) For numerically large values of $x$, the curve approximates to $y = x$, i.e. when $x$ is large and positive, $y$ is large and positive and when $x$ is large and negative, $y$ is large and negative.
(g) The only asymptotes are $x = -1$ (parallel to the $y$-axis) and $y = x - 2$.

*With these facts before us, we can now sketch the curve. Do that and then check your result with the result shown in the next frame.*

**39** Here is the graph as it should appear. You can, of course, always plot an odd point or two in critical positions if extra help is needed.

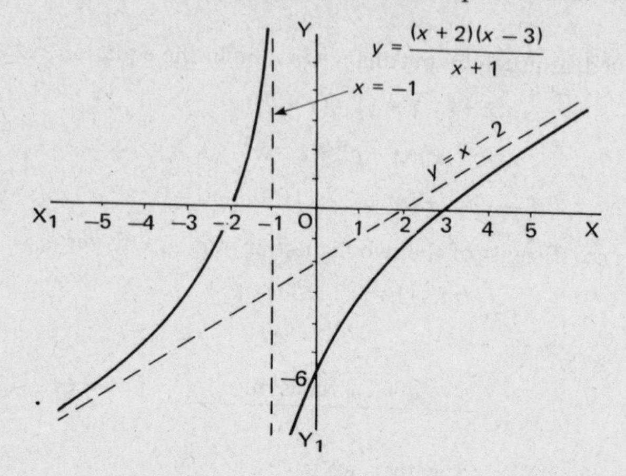


*Now let us move on to something rather different.*

40

## Curve fitting

Readings recorded from a test or experiment normally include errors of various kinds and therefore the points plotted from these data are scattered about the positions they should ideally occupy. Unless very few readings are taken, it can be assumed that the inherent errors will be of a random nature, resulting in some of the values being slightly too high and some slightly too low. Having plotted the points, we then draw as the graph the middle line of this narrow band of points. It may well be that the line drawn does not pass through any of the actual plotted points, but from now on it is the line which is used to determine the relationship between the two variables.

### 1. Straight-line law

*Example:* Values of $V$ and $h$ are recorded in a test.

| $h$ | 6·0 | 10 | 14 | 18 | 21 | 25 |
|---|---|---|---|---|---|---|
| $V$ | 5·5 | 7·0 | 9·5 | 12·5 | 13·5 | 16·5 |

If the law relating $V$ and $h$ is $V = ah + b$, where $a$ and $b$ are constants,

(a) plot the graph of $V$ against $h$,
(b) determine the values of $a$ and $b$.

(a) Plotting the points is quite straightforward. Do it carefully on squared paper.
You get ....................

41

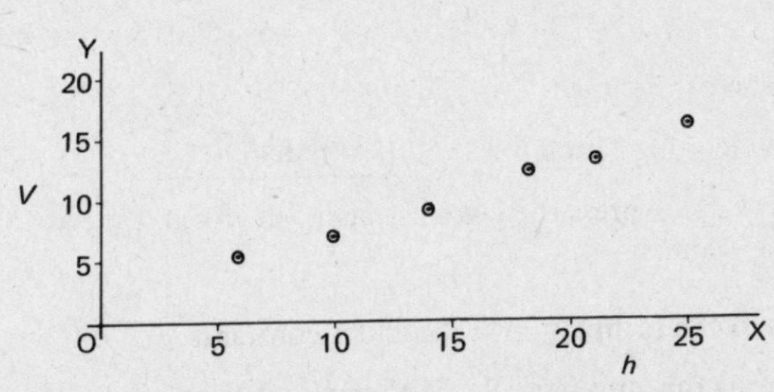


We now estimate by eye the best position for a straight line graph drawn down the middle of this band of points. Draw the line on your graph.

**42**

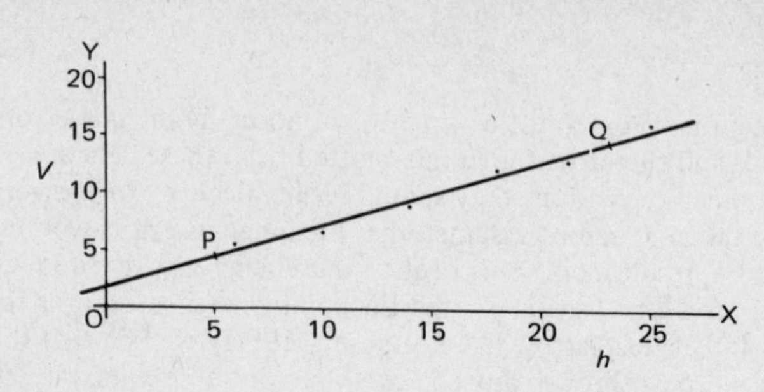


Henceforth, we shall use this line as representing the equation and ignore the actual points we plotted.

(b) The law is first degree, i.e. $V = ah + b$

Compare this with $y = mx + c$

If we therefore find the values of $m$ and $c$ in the normal way, these values will be those of $a$ and $b$.

We now select two convenient points *on the line* and read off their $x$- and $y$-coordinates. For instance, P (5, 4·5) and Q (23, 15). Substituting these values in $y = mx + c$ gives two equations from which we can find the values of $m$ and $c$, which are ................

**43**

$$\boxed{m = 0{\cdot}583; \; c = 1{\cdot}59}$$

For we have $\left.\begin{array}{r} 4{\cdot}5 = 5m + c \\ 15{\cdot}0 = 23m + c \end{array}\right\} \therefore 10{\cdot}5 = 18m \quad \therefore m = \dfrac{10{\cdot}5}{18} = 0{\cdot}583$

and then $4{\cdot}5 = 5(0{\cdot}583) + c \quad \therefore c = 4{\cdot}5 - 2{\cdot}915 \quad \therefore c = 1{\cdot}585$

The equation of the line is $y = 0{\cdot}583x + 1{\cdot}59$

and the law relating $V$ and $h$ is $\underline{V = 0{\cdot}583h + 1{\cdot}59}$

Provided we can express the law in straight line form, they are all tackled in the same manner.

2. **Graphs of the form** $y = ax^n$, $a$ and $n$ constants.

To convert this into 'straight-line form', we take logarithms of both sides. The equation then becomes ................

**44**

$$\boxed{\log y = n \log x + \log a}$$

If we compare this result with $Y = mX + c$ we see we have to plot ................ along the $Y$-axis and ................ along the $X$-axis to obtain a straight-line graph.

**45**

log $y$ along the $Y$-axis
log $x$ along the $X$-axis

$$\left.\begin{array}{l}\log y = n \log x + \log a\\ \quad Y = mX \qquad + c\end{array}\right\}$$ If we then find $m$ and $c$ from the straight line as before, then $m = n$

and $c = \log a \quad \therefore\ a = \text{antilog } c$.

Let us work through an example.

*Example:* Values of $x$ and $y$ are related by the equation $y = ax^n$.

| $x$ | 2 | 5 | 12 | 25 | 32 | 40 |
|---|---|---|---|---|---|---|
| $y$ | 5·62 | 13·8 | 52·5 | 112 | 160 | 200 |

Determine the values of the constants $a$ and $n$.

$$y = ax^n \quad \therefore\ \log y = n \log x + \log a$$
$$Y = mX \qquad + c$$

We must first compile a table showing the corresponding values of log $x$ and log $y$.

*Do that and check with the results shown before moving on.*

**46**

| log $x$ | 0·3010 | 0·6990 | 1·079 | 1·398 | 1·505 | 1·602 |
|---|---|---|---|---|---|---|
| log $y$ | 0·750 | 1·14 | 1·72 | 2·05 | 2·20 | 2·30 |

Now plot these points on graph paper, log $x$ along the $x$-axis
log $y$ along the $y$-axis.
and draw the estimated best straight-line graph.

**47**

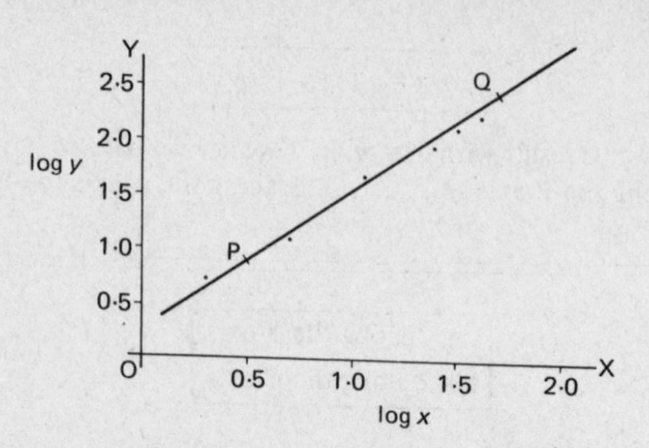


If we select from the graph two points, P (0·50, 0·94) and Q (1·70, 2·45) we can calculate the values of $m$ and $c$ which are $m$ = ................................ and $c$ = ................

**48**

$$m = 1{\cdot}258; \;\; c = 0{\cdot}311$$

Since $\quad Y = mX + c \quad \therefore\ 2{\cdot}45 = 1{\cdot}70m + c$

$$0{\cdot}94 = 0{\cdot}50m + c$$

$$\therefore\ 1{\cdot}51 = 1{\cdot}20m \quad \therefore\ m = 1{\cdot}258; \text{ then } c = 0{\cdot}311$$

Therefore $\quad Y = 1{\cdot}258X + 0{\cdot}311$

$$\log y = n \log x + \log a$$

$$\therefore\ \ n = 1{\cdot}258 \text{ and } \log a = 0{\cdot}311 \qquad \therefore\ a = 2{\cdot}05.$$

Therefore, the equation is $\underline{y = 2{\cdot}05x^{1 \cdot 26}}$

3. **Graphs of the form** $y = ae^{nx}$

Exponential relationships occur frequently in technical situations. As before, our first step is to convert the equation to 'straight-line form' by taking logs of both sides. We could use common logarithms as we did previously, but the work involved is less if we use natural logarithms.

So, taking natural logarithms of both sides, we express the equation in the form ............................

**49**

$$\boxed{\ln y = nx + \ln a}$$

If we compare this with the straight-line equation, we have

$$\ln y = nx + \ln a$$
$$Y = mX + c$$

which shows that we plot values of $\ln y$ along the $Y$-axis,

and values of just $x$ along the $X$-axis

and then the value of $m$ will give the value of $n$

and the value of $c$ will be $\ln a$, hence $a = \text{antiln } c$.

Let us do an example.

*Example:* The following values of $W$ and $T$ are related by the law $W = ae^{nT}$ where $a$ and $n$ are constants.

| $T$ | 3·0 | 10 | 15 | 30 | 50 | 90 |
|---|---|---|---|---|---|---|
| $W$ | 3·857 | 1·974 | 1·733 | 0.4966 | 0.1738 | 0.0091 |

We need values of $\ln W$, so compile a table showing values of $T$ and $\ln W$.

**50**

| $T$ | 3·0 | 10 | 15 | 30 | 50 | 90 |
|---|---|---|---|---|---|---|
| $\ln W$ | 1·35 | 0·68 | 0·55 | –0.70 | –1.75 | –4.70 |

$W = ae^{nT}$ $\quad\therefore\ \ln W = nT + \ln a$

$$Y = mX + c$$

Therefore, we plot $\ln W$ along the $Y$-axis

and $T$ along the $X$-axis

to obtain a straight-line graph, form which $m = n$

and $c = \ln a \ \therefore\ a = \text{antiln } c$.

So, plot the points; draw the best straight-line graph; and from it determine the values of $n$ and $a$. The required law is therefore

$W =$ ................

**51**

$$\boxed{W = 4{\cdot}14\, e^{-0{\cdot}067T}}$$

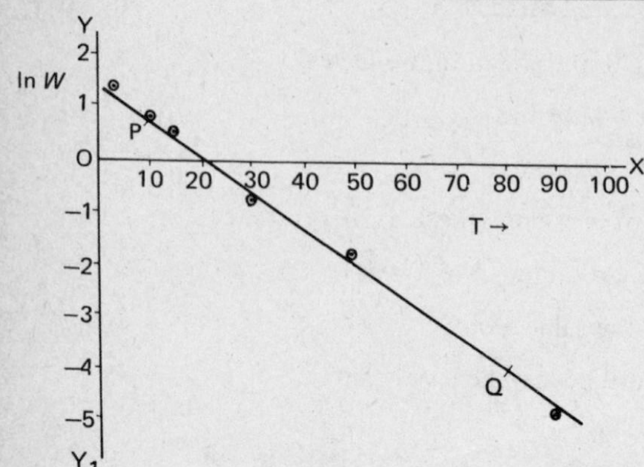


$P = (10, 0{\cdot}75)$; $\quad Q = (80, -3{\cdot}93)$

$$-3{\cdot}93 = 80m + c$$

$$0{\cdot}75 = 10m + c$$

$$\therefore\ -4{\cdot}68 = 70m \quad \therefore\ m = -0{\cdot}067$$

$$\therefore\ c = 0{\cdot}75 + 0{\cdot}67 = 1{\cdot}42$$

$$m = n = -0{\cdot}067$$

$$c = \ln a = 1{\cdot}42 \quad \therefore\ a = 4{\cdot}14$$

$\therefore$ The required law is

$$\underline{W = 4{\cdot}14 e^{-0{\cdot}067T}}$$

If we can express the equation or law in straight line form, the same method can be applied.

How about these? How could you arrange for each of the following to give a straight-line graph?

1. $y = ax^2 + b$
2. $y = ax + b$
3. $y = \dfrac{a}{x} + c$
4. $y = ax^2 + bx$

*Check your suggestions with the next frame.*

**52**

Here they are.

| *Equation* | *Straight-line form* | *Plot* |
|---|---|---|
| $y = ax^2 + b$ | $y = ax^2 + b$ | $y$ against $x^2$ |
| $y = ax + b$ | $y = ax + b$ | $y$ against $x$ |
| $y = \dfrac{a}{x} + b$ | $y = a\dfrac{1}{x} + b$ | $y$ against $\dfrac{1}{x}$ |
| $y = ax^2 + bx$ | $\dfrac{y}{x} = ax + b$ | $\dfrac{y}{x}$ against $x$ |

And now just one more. If we convert $y = \dfrac{a}{x + b}$ to straight-line form, it becomes ..................

53

$$\boxed{y = -\frac{1}{b}xy + \frac{a}{b}}$$

For, if $y = \dfrac{a}{x+b}$, then $xy + by = a \quad \therefore \; by = -xy + a$

$$\therefore \; y = -\frac{1}{b}xy + \frac{a}{b}$$

That is, if we plot values of $y$ against values of the product $xy$, we shall obtain a straight-line graph from which $m = -\dfrac{1}{b}$ and $c = \dfrac{a}{b}$.

From these, $a$ and $b$ can easily be found.

*Finally, to what is perhaps the most important part of this programme. Let us start with a new frame.*

54

**Method of least squares**

All the methods which involve drawing the best straight line by eye are approximate only and depend upon the judgement of the operator. Quite considerable variation can result from various individuals' efforts to draw the 'best straight line'.

*The method of least squares* determines the best straight line entirely by calculation, using the set of recorded results. The form of the equation has to be chosen and this is where the previous revision will be useful.

*Let us start with the case of a linear relationship.*

**55**

*Fitting a straight-line graph* We have to fit a straight line $y = a + bx$ to a set of points $(x_1, y_1), (x_2, y_2) \ldots (x_n, y_n)$ so that the sum of the squares of the distances to this straight line from the given set of points is a minimum. The distance of any point from the line is measured along an ordinate, i.e. in the $y$-direction.

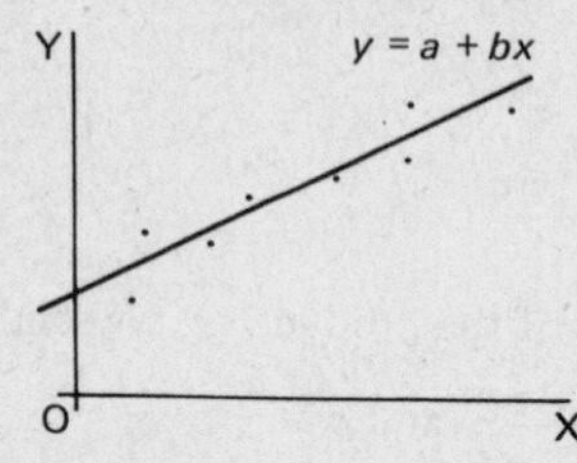


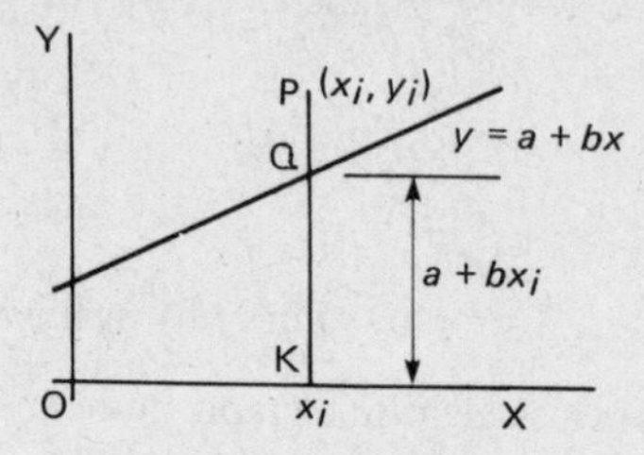


If we take a sample point P $(x_i, y_i)$

QK is the value of $y = a + bx$ at $x = x_i$ i.e. $a + bx_i$.

PQ is the difference between PK and QK, i.e. $y_i - a - bx_i$.

$$\therefore \; \text{PQ}^2 = (y_i - a - bx_i)^2$$

Therefore, the sum $S$ of the squares of these differences for all $n$ such points is given by $S = \sum_{i=1}^{n} (y_i - a - bx_i)^2$.

We have to determine the values $a$ and $b$ so that $S$ shall be a minimum. The right-hand side contains two unknowns $a$ and $b$. Therefore, for the sum of the squares to be a minimum

$$\frac{\partial S}{\partial a} = 0 \quad \text{and} \quad \frac{\partial S}{\partial b} = 0.$$

$$\frac{\partial S}{\partial a} = -2\sum_{i=1}^{n} (y_i - a - bx_i) = 0; \quad \frac{\partial S}{\partial b} = -2\sum_{i=1}^{n} x_i(y_i - abx_i) = 0$$

The first gives $$\sum_1^n y_i - na - b\sum_1^n x_i = 0$$

i.e. $$an + b\sum_1^n x_i = \sum_1^n y_i \quad \ldots\ldots \qquad \text{(A)}$$

The second gives $$\sum_1^n x_i y_i - a\sum_1^n x_i - b\sum_1^n x_i^2 = 0.$$

i.e. $$a\sum_1^n x_i + b\sum_1^n x_i^2 = \sum_1^n x_i y_i \quad \ldots \qquad \text{(B)}$$

Equations (A) and (B) are called the *normal equations* of the problem and we will now write them without the suffixes, remembering that the $x$ and $y$ values are those of the recorded values.

$$\left.\begin{aligned} an + b\Sigma x &= \Sigma y \\ a\Sigma x + b\Sigma x^2 &= \Sigma xy \end{aligned}\right\} \text{ for all the } n \text{ given pairs of values.}$$

From these *normal equations*, the specific values of $a$ and $b$ can be determined.

*We will now work through an example.*

**56**

*Example 1:* Apply the method of least squares to fit a straight-line relationship for the following points

| $x$ | −2·4 | −0·8 | 0·3 | 1·9 | 3·2 |
|---|---|---|---|---|---|
| $y$ | −5·0 | −1·5 | 2·5 | 6·4 | 11·0 |

For this set, $n = 5$ and the normal equations are

$$\left.\begin{aligned} an + b\Sigma x &= \Sigma y \\ a\Sigma x + b\Sigma x^2 &= \Sigma xy \end{aligned}\right\} \text{ where } y = a + bx.$$

Therefore, we need to sum the values of $x$, $y$, $x^2$ and $xy$. This is best done in table form.

| | $x$ | $y$ | $x^2$ | $xy$ |
|---|---|---|---|---|
| | −2.4 | −5·0 | 5·76 | 12·0 |
| | −0·8 | −1·5 | 0·64 | 1·2 |
| | 0·3 | 2·5 | 0·09 | 0·75 |
| | 1·9 | 6·4 | 3·61 | 12·16 |
| | 3·2 | 11·0 | 10·24 | 35·2 |
| $\Sigma$ | 2·2 | 13·4 | 20·34 | 61·31 |

The normal equations now become

................................

and ................................

**57**

$$5a + 2{\cdot}2b = 13{\cdot}4$$
$$2{\cdot}2a + 34{\cdot}20b = 61{\cdot}31$$

Dividing through each equation by the coefficient of $a$ gives

$$a + 0{\cdot}440b = 2{\cdot}68$$
$$a + 9{\cdot}245b = 27{\cdot}87$$

$$\therefore\ 8{\cdot}805b = 25{\cdot}19 \qquad \therefore\ b = 2{\cdot}861$$
$$\therefore\ a = 2{\cdot}68 - 1{\cdot}2588 \qquad \therefore\ a = 1{\cdot}421$$

Therefore, the best straight line for the given values is

$$\underline{y = 1{\cdot}42 + 2{\cdot}86x}$$

To see how well the method works, plot the set of values for $x$ and $y$ and also the straight line whose equation we have just found on the same axes.

**58** The result should look like this:

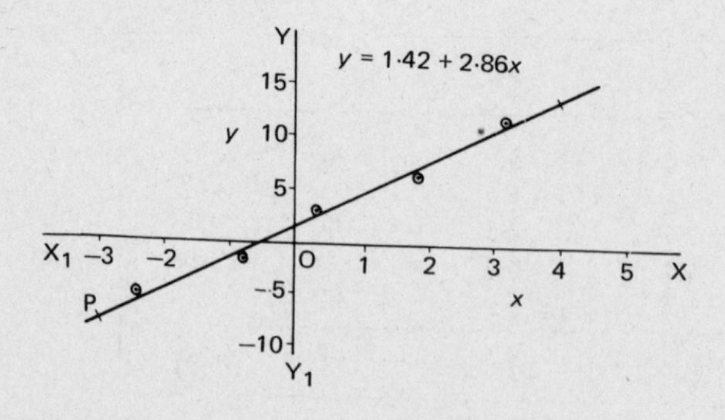

Any relationship that can be expressed in straight-line form can be dealt with in the same way.

*Example 2:* It is required to fit the best rectangular hyperbola $xy = c$ to the set of values given below.

| $x$ | 0·5 | 1·0 | 2·0 | 3·0 | 4·0 | 5·0 |
|---|---|---|---|---|---|---|
| $y$ | 62 | 28 | 17 | 9·0 | 7·0 | 5·0 |

In this case, $n = 6$. Also $y = c \cdot \dfrac{1}{x}$

Ordinate difference between a point and the curve $= y_i - \dfrac{c}{x_i}$

The sum of the squares $S = \sum_{i=1}^{n}\left(y_i - \dfrac{c}{x_i}\right)^2$ and for $S$ to be a minimum,

$\dfrac{\partial S}{\partial c} = 0.$ $\qquad \dfrac{\partial S}{\partial c} = -2\sum_{1}^{n}\dfrac{1}{x_i}\left(y_i - \dfrac{c}{x_i}\right) = 0.$

$$\therefore \sum_{1}^{n}\frac{y_i}{x_i} - c\sum_{1}^{n}\frac{1}{x_i^2} = 0$$

So this time we need values of $x$, $y$, $\dfrac{1}{x}$, $\dfrac{y}{x}$ and $\dfrac{1}{x^2}$.

| $x$ | $y$ | $\dfrac{1}{x}$ | $\dfrac{y}{x}$ | $\dfrac{1}{x^2}$ |
|---|---|---|---|---|
| 0·5 | 62 | 2·0 | 124 | 4·0 |
| 1·0 | 28 | 1·0 | 28 | 1·0 |
| 2·0 | 17 | 0·5 | 8·5 | 0·25 |
| 3·0 | 9 | 0·333 | 3 | 0·111 |
| 4·0 | 7 | 0·25 | 1·75 | 0·0625 |
| 5·0 | 5 | 0·2 | 1 | 0·04 |
| | | $\Sigma$ | 166·25 | 5·4635 |

From this we can find $c$ and the equation of the required hyperbola which is therefore ...................

**59**

$$\boxed{xy = 30{\cdot}4}$$

For $\sum_{1}^{n} \frac{y}{x} - c\sum_{1}^{n} \frac{1}{x^2} = 0.$ $\therefore$ $166{\cdot}25 = 5{\cdot}4635c$ $\therefore$ $c = 30{\cdot}4$

Here is the graph showing the plotted values and also the rectangular hyperbola we have just obtained.

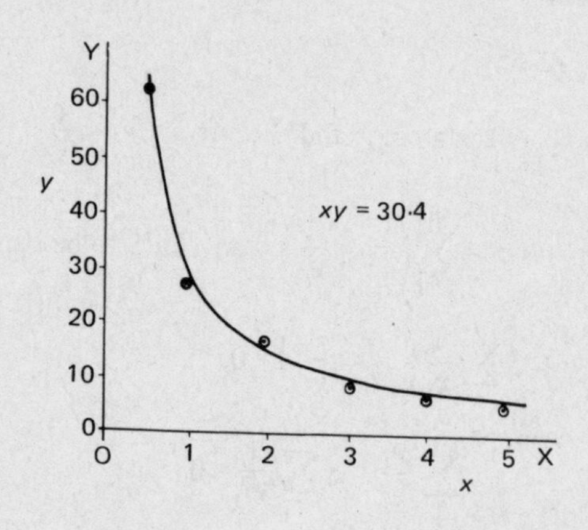

**60**

*Example 3:* Values of $x$ and $y$ are related by the law $y = a \ln x$. Determine the value of $a$ to provide the best fit for the following set of values.

| $x$ | 0·3 | 0·9 | 1·7 | 2·5 | 4·0 | 10·0 |
|---|---|---|---|---|---|---|
| $y$ | –7·54 | –0·672 | 3·63 | 4·41 | 8·22 | 12·2 |

$y = a \ln x$ will give a straight line if we plot $y$ against $\ln x$. Therefore, let $u = \ln x$.

Ordinate difference $= y_i - a \ln x_i = y_i - au_i \quad \therefore S = \sum_{i=1}^{n} (y_i - au_i)^2$

For minimum $S$, $\dfrac{\partial S}{\partial a} = 0. \quad \therefore -2\sum_1^n u_i(y_i - au_i) = 0$

$$\therefore \sum_1^n u_i y_i - a\sum_1^n u_i^2 = 0$$

Therefore we need values of $x$, $\ln x$ (i.e. $u$), $y$, $uy$ and $u^2$.

| $x$ | $u\ (= \ln x)$ | $y$ | $uy$ | $u^2$ |
|---|---|---|---|---|
| 0·3 | –1·204 | –7·54 | 9·078 | 1·450 |
| 0·9 | –0·105 | –0·672 | 0·0708 | 0·0111 |
| 1·7 | 0·531 | 3·63 | 1·9261 | 0·2815 |
| 2·5 | 0·916 | 4·41 | 4·0409 | 0·8396 |
| 4·0 | 1·3863 | 8·22 | 11·3954 | 1·9218 |
| 10·0 | 2·3026 | 12·20 | 28·092 | 5·302 |

Now total up the appropriate columns and finish the problem off.
The equation is finally .....................

**61**

$$\boxed{y = 5{\cdot}57 \ln x}$$

For $\Sigma\, uy = 54{\cdot}603$ and $\Sigma\, u^2 = 9{\cdot}806$

and since $\Sigma\, uy - a\, \Sigma\, u^2 = 0$, then $a = \dfrac{\Sigma\, uy}{\Sigma\, u^2} = 5{\cdot}568$

$$\therefore\ \underline{y = 5{\cdot}57 \ln x}$$

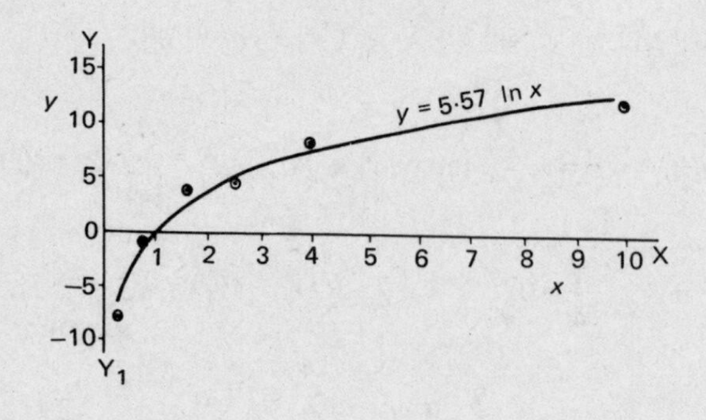

**62**

*Example 4:* A test on the length of tool life at different cutting speeds gave the following results.

| Speed $v$ (m/min) | 120 | 130 | 170 | 200 |
|---|---|---|---|---|
| Life $T$ (min) | 62 | 25 | 7·2 | 2·8 |

If the law relating $v$ and $T$ is $v = aT^k$ determine the constants $a$ and $k$ to give the best fit.

$$v = aT^k \quad \therefore \log v = k \log T + \log a$$

$$Y = mX + c$$

| $X = \log T$ | 1·792 | 1·398 | 0·8573 | 0·4472 |
|---|---|---|---|---|
| $Y = \log v$ | 2·079 | 2·114 | 2·230 | 2·301 |

Arguing as before, $S = \sum_{1}^{n} (Y_i - mX_i - c)^2$

$$\frac{\partial S}{\partial m} = -2\sum_{1}^{n} X(Y - mX - c) = 0 \quad \therefore \sum XY - m\sum X^2 - c\sum X = 0$$

$$\frac{\partial S}{\partial c} = -2\sum_{1}^{n} (Y - mX - c) = 0 \quad \therefore \sum Y - m\sum X - nc = 0$$

So we now need columns for $X$, $Y$, $XY$, and $X^2$.

Compile the appropriate table and finish off the problem, so finding the required law ......................

**63**

$$v = 237T^{-0.173}$$

Here is the working:

| | $X$ | $Y$ | $XY$ | $X^2$ |
|---|---|---|---|---|
| | 1·792 | 2·079 | 3·726 | 3·211 |
| | 1·398 | 2·114 | 2·955 | 1·954 |
| | 0·8573 | 2·230 | 1·912 | 0·735 |
| | 0·4472 | 2·301 | 1·029 | 0·200 |
| $\Sigma$ | 4·4945 | 8·724 | 9·622 | 6·100 |

The normal equations give $9{\cdot}622 - m6{\cdot}100 - c4{\cdot}495 = 0$
and $8{\cdot}724 - m4{\cdot}495 - c4 = 0$.

Dividing through each equation by the coefficient of $m$, we have

$$m + 0{\cdot}7369c = 1{\cdot}5774$$
$$m + 0{\cdot}8899c = 1{\cdot}9408$$

$$\therefore\ 0{\cdot}1530c = 0{\cdot}3634 \quad \therefore\ c = 2{\cdot}375$$

Then $m = 1{\cdot}5774 - 0{\cdot}7369\,(2{\cdot}375) = 1{\cdot}5774 - 1{\cdot}7503 = -0{\cdot}1729$

$$\therefore\ Y = -0{\cdot}173X + 2{\cdot}375$$

$m = n \quad \therefore\ n = -0{\cdot}173$

$c = \log a \quad \therefore\ \log a = 2{\cdot}3752 \quad \therefore\ a = 237{\cdot}25$

$\therefore$ The law is $v = 237T^{-0{\cdot}173}$

The same principles can be applied to fit curves of higher degree to sets of values, but in general fitting a second-degree curve involves three normal equations in three unknowns and curves of higher degree involve an increasing number of simultaneous equations, so that the working tends to become unwieldy without some form of computing facility.

*All that now remains is to check down the Revision Summary that follows and then to work through the Test Exercise as usual.*

**64**

**Revision Summary**

1. *Standard curves* Refer back to frames 2 to 17.
2. *Asymptotes* Rewrite the given equation, if necessary, on one line.
   (a) *Asymptotes parallel to the axes*
   Parallel to the $x$-axis: equate to zero the coefficient of the highest power of $x$.
   Parallel to the $y$-axis: equate to zero the coefficient of the highest power of $y$.
   (b) *General asymptotes*
   Substitute $y = mx + c$ in the equation of the curve and equate to zero the coefficients of the two highest powers of $x$.
3. *Systematic curve sketching*
   (a) *Symmetry*
   Even powers only of $x$: curve symmetrical about $y$-axis.
   Even powers only of $y$: curve symmetrical about $x$-axis.
   Even powers only of both $x$ and $y$: curve symmetrical about both axes.
   (b) *Intersection with the axes* Put $x = 0$ and $y = 0$.
   (c) *Change of origin* to simplify analysis.
   (d) *Asymptotes*
   (e) *Large and small values* of $x$, or $y$
   (f) *Turning points*
   (g) *Limitations* on possible range of values of $x$ or $y$.
4. *Curve fitting*
   (a) Express the law or equation in straight line form.
   (b) Plot values and draw the best straight line as the middle line of the band of points.
   (c) Determine $m$ and $c$ from the $x$ and $y$ coordinates of two points *on the line.* Check with the coordinates of a third point.
   (d) Line of 'best fit' can be calculated by the method of *least squares.* Refer back to frames 54 to 63.

**65**

## Test Exercise – XII

1. Without detailed plotting of points, sketch the graphs of the following, showing relevant information on the graphs.
   (a) $y = (x-3)^2 + 5$
   (b) $y = 4x - x^2$
   (c) $4y^2 + 24y - 14 - 16x + 4x^2 = 0$
   (d) $5xy = 40$
   (e) $y = 6 - e^{-2x}$.

2. Determine the asymptotes of the following curves.
   (a) $x^2y - 9y + x^3 = 0$
   (b) $y^2 = \dfrac{x(x-2)(x+4)}{x-4}$.

3. Analyse and sketch the graph of the function $y = \dfrac{(x-3)(x+5)}{x+2}$

4. Express the following in 'straight line' form and state the variables to be plotted on the $x$ and $y$ axes to give a straight line.
   (a) $y = Ax + Bx^2$
   (b) $y = x + Ae^{kx}$
   (c) $y = \dfrac{A}{B+x}$
   (d) $x^2(y^2 - 1) = k$.

5. The force, $P$ newtons, required to keep an object moving at a speed, $V$ metres per second, was recorded.

| $P$ | 126 | 178 | 263 | 398 | 525 | 724 |
|---|---|---|---|---|---|---|
| $V$ | 1·86 | 2·34 | 2·75 | 3·63 | 4·17 | 4·79 |

If the law connecting $P$ and $V$ is of the form $V = aP^k$, where $a$ and $k$ are constants, apply the method of least squares to obtain the values of $a$ and $k$ that give the best fit to the given set of values.

## Further Problems – XII

1. For each of the following curves, determine the asymptotes parallel to the $x$- and $y$-axes.

(a) $xy^2 + x^2 - 1 = 0$

(b) $x^2y^2 = 4(x^2 + y^2)$

(c) $y = \dfrac{x^2 - 3x + 5}{x - 3}$

(d) $y = \dfrac{x(x + 4)}{(x + 3)(x + 2)}$

(e) $x^2y^2 - x^2 = y^2 + 1$

(f) $y^2 = \dfrac{x}{x - 2}$.

2. Determine all the asymptotes of each of the following curves.

(a) $x^3 - xy^2 + 4x - 16 = 0$

(b) $xy^3 - x^2y + 3x^3 - 4y^3 - 1 = 0$

(c) $y^3 + 2y^2 - x^2y + y - x + 4 = 0$.

3. Analyse and sketch the graphs of the following functions.

(a) $y = x + \dfrac{1}{x}$

(b) $y = \dfrac{1}{x^2 + 1}$

(c) $y^2 = \dfrac{x}{x - 2}$

(d) $y = \dfrac{(x - 1)(x + 4)}{(x - 2)(x - 3)}$

(e) $y(x + 2) = (x + 3)(x - 4)$

(f) $x^2(y^2 - 25) = y$

(g) $xy^2 - x^2y + x + y = 2$.

4. Variables $x$ and $y$ are thought to be related by the law $y = a + bx^2$. Determine the values of $a$ and $b$ that best fit the set of values given.

| $x$ | 5·0 | 7·5 | 12 | 15 | 25 |
|---|---|---|---|---|---|
| $y$ | 13·1 | 28·1 | 70·2 | 109 | 301 |

5. By plotting a suitable graph, show that $P$ and $W$ are related by a law of the form $P = a\sqrt{W} + b$, where $a$ and $b$ are constants, and determine the values of $a$ and $b$.

| $W$ | 7·0 | 10 | 15 | 24 | 40 | 60 |
|---|---|---|---|---|---|---|
| $P$ | 9·76 | 11·0 | 12·8 | 15·4 | 18·9 | 22·4 |

6. If $R = a + \dfrac{b}{d^2}$, find the best values for $a$ and $b$ from the set of corresponding values given below.

| $d$ | 0·1 | 0·2 | 0·3 | 0·5 | 0·8 | 1·5 |
|---|---|---|---|---|---|---|
| $R$ | 5·78 | 2·26 | 1·60 | 1·27 | 1·53 | 1·10 |

7. Two quantities, $x$ and $y$, are related by the law $y = \dfrac{a}{1 - bx^2}$, where $a$ and $b$ are constants. Using the values given below, draw a suitable graph and hence determine the best values of $a$ and $b$.

| $x$ | 4 | 6 | 8 | 10 | 11 | 12 |
|---|---|---|---|---|---|---|
| $y$ | 4·89 | 5·49 | 6·62 | 9·00 | 11·4 | 16·1 |

8. The pressure $p$ and volume $v$ of a mass of gas in a container are related by the law $pv^n = c$, where $n$ and $c$ are constants. From the values given below, plot a suitable graph and hence determine the values of $n$ and $c$.

| $v$ | 4·60 | 7·20 | 10·1 | 15·3 | 20·4 | 30·0 |
|---|---|---|---|---|---|---|
| $p$ | 14·2 | 7·59 | 4·74 | 2·66 | 1·78 | 1·04 |

9. The current, $I$ milliamperes, in a circuit is measured for various values of applied voltage $V$ volts. If the law connecting $I$ and $V$ is $I = aV^n$, where $a$ and $n$ are constants, draw a suitable graph and determine the values of $a$ and $n$ that best fit the set of recorded values.

| $V$ | 8 | 12 | 15 | 20 | 28 | 36 |
|---|---|---|---|---|---|---|
| $I$ | 41·1 | 55·6 | 65·8 | 81·6 | 105 | 127 |

10. Values of $x$ and $y$ are thought to be related by a law of the form $y = ax + b \ln x$, where $a$ and $b$ are constants. By drawing a suitable graph, test whether this is so and determine the values of $a$ and $b$.

| $x$ | 10·4 | 32·0 | 62·8 | 95·7 | 136 | 186 |
|---|---|---|---|---|---|---|
| $y$ | 8·14 | 12·8 | 16·3 | 19·2 | 22·1 | 25·3 |

11. The following pairs of values of $x$ and $y$ are thought to satisfy the law $y = ax^2 + \frac{b}{x}$. Draw a suitable graph to confirm that this is so and determine the values of the constants $a$ and $b$.

| $x$ | 1 | 3 | 4 | 5 | 6 | 7 |
|---|---|---|---|---|---|---|
| $y$ | 5·18 | 15·9 | 27·0 | 41·5 | 59·3 | 80·4 |

12. In a test on breakdown voltages, $V$ kilovolts, for insulation of different thicknesses, $t$ millimetres, the following results were obtained.

| $t$ | 2·0 | 3·0 | 5·0 | 10 | 14 | 18 |
|---|---|---|---|---|---|---|
| $V$ | 153 | 200 | 282 | 449 | 563 | 666 |

If the law connecting $V$ and $t$ is $V = at^n$, draw a suitable graph and determine the values of the constants $a$ and $n$.

13. The torque, $T$ newton metres, required to rotate shafts of different diameters, $D$ millimetres, on a machine is shown below. If the law is $T = aD^n$, where $a$ and $n$ are constants, draw a suitable graph and hence determine the values of $a$ and $n$.

| $D$ | 7·0 | 10 | 18 | 25 | 40 |
|---|---|---|---|---|---|
| $T$ | 0·974 | 1·71 | 4·33 | 7·28 | 15·3 |

# Programme 13

## SERIES

### PART 1

**1**

## Sequences

A *sequence* is a set of quantities, $u_1, u_2, u_3, \ldots,$ stated in a definite order and each term formed according to a fixed pattern, i.e. $u_r = f(r)$.

e.g. $1, 3, 5, 7, \ldots$ is a sequence (the next term would be 9).
$2, 6, 18, 54, \ldots$ is a sequence (the next term would be 162).
$1^2, -2^2, 3^2, -4^2, \ldots$ is a sequence (the next term would be $5^2$).

Also $1, -5, 37, 6, \ldots$ is a sequence, but its pattern is more involved and the next term cannot readily be anticipated.

A *finite* sequence contains only a finite number of terms.
An *infinite* sequence is unending.
So which of the following constitutes a finite sequence:

(i) All the natural numbers, i.e. $1, 2, 3, \ldots$ etc.
(ii) The page numbers of a book.
(iii) The telephone numbers in a telephone directory.

**2**

| The page numbers of a book |
|---|
| The telephone numbers in a directory |

Clearly, the page numbes are in fixed order and terminate at the last page. The telephone numbers form a complicated sequence, ordered by the alphabetical letters of the surnames of the subscribers. The natural numbers form an infinite sequence, since they never come to an end.

## Series

A *series* is formed by the sum of the terms of a sequence.

e.g. $1, 3, 5, 7, \ldots$ is a sequence,
but $1 + 3 + 5 + 7 + \ldots$ is a series.

We shall indicate the terms of a series as follows:
$u_1$ will represent the first term, $u_2$ the second term, $u_3$ the third term, etc., so that $u_r$ will represent the $r^{\text{th}}$ term and $u_{r+1}$ the $(r+1)^{\text{th}}$ term, etc.
Also, the sum of the first 5 terms will be indicated by $S_5$.
So the sum of the first $n$ terms will be stated as .................

**3**

$S_n$

□□□□□□□□□□□□□□□□□□□□□□□□□□□□□□□□□□□□□□□□□□□□□□□□□□

You will already be familiar with two special kinds of series which have many applications. These are (i) *arithmetic series* and (ii) *geometric series.* Just by way of revision, however, we will first review the important results relating to these two series.

1. *Arithmetic series* (or arithmetic progression) denoted by A.P.

An example of an A.P. is the series

$$2 + 5 + 8 + 11 + 14 + \ldots$$

You will note that each term can be written from the previous term by simply adding on a constant value 3. This regular increment is called the *common difference* and is found by selecting any term and subtracting from it the previous term

$$\text{e.g. } 11 - 8 = 3; \quad 5 - 2 = 3; \text{ etc.}$$

*Move on to the next frame.*

---

**4**

The *general arithmetic series* can therefore be written:

$$a + (a + d) + (a + 2d) + (a + 3d) + \ldots \qquad \text{(i)}$$

where $a$ = first term and $d$ = common difference.

You will remember that

(i) the $n^{\text{th}}$ term $= a + (n - 1)d$ .................... (ii)

(ii) the sum of the first $n$ terms is given by

$$S_n = \frac{n}{2}(2a + \overline{n-1}\,d) \qquad \text{(iii)}$$

Make a note of these three items in your record book.

By way of warming up, find the sum of the first 20 terms of the series:

$$10 + 6 + 2 - 2 - 6 \ldots \text{etc.}$$

*Then turn to frame 5.*

**5**

$$\boxed{S_{20} = -560}$$

Since, for the series 10 + 6 + 2 − 2 − 6 . . . etc.

$$a = 10 \text{ and } d = 2 - 6 = -4$$

$$S_n = \frac{n}{2}(2a + \overline{n-1}\,d)$$

$$\therefore S_{20} = \frac{20}{2}(20 + 19[-4])$$

$$= 10(20 - 76) = 10(-56) = \underline{-560}$$

□□□□□□□□□□□□□□□□□□□□□□□□□□□□□□□□□□□□□□

Here is another example:

If the 7th term of an A.P. is 22 and the 12th term is 37, find the series.

We know 7th term = 22 $\therefore a + 6d = 22$
and 12th term = 37 $\therefore a + 11d = 37$ } $5d = 15 \quad \therefore d = 3 \quad \therefore a = 4$

So the series is $\underline{4 + 7 + 10 + 13 + 16 + \ldots \text{etc.}}$

Here is one for you to do:

The 6th term of an A.P. is −5 and the 10th term is −21. Find the sum of the first 30 terms.

**6**

$$\boxed{S_{30} = -1290}$$

since:

6th term = −5 $\quad \therefore a + 5d = -5$
10th term = −21 $\quad \therefore a + 9d = -21$ } $4d = -16 \quad \therefore d = -4 \quad a = 15$

$$\therefore a = 15,\ d = -4,\ n = 30,\ S_n = \frac{n}{2}(2a + \overline{n-1}\,d)$$

$$\therefore S_{30} = \frac{30}{2}(30 + 29[-4])$$

$$= 15(30 - 116) = 15(-86) = \underline{-1290}$$

**Arithmetic mean**

We are sometimes required to find the arith. mean of two numbers, P and Q. This means that we have to insert a number A between P and Q, so that P + A + Q forms an A.P.

$$A - P = d \text{ and } Q - A = d$$

$$\therefore A - P = Q - A \quad 2A = P + Q \quad \therefore \underline{A = \frac{P+Q}{2}}$$

The arithmetic mean of two numbers, then, is simply their average. Therefore, the arithmetic mean of 23 and 58 is ....................

**7**

The arithmetic mean of 23 and 58 is $\boxed{40 \cdot 5}$

If we are required to insert 3 arithmetic means between two given numbers, P and Q, it means that we have to supply three numbers, A, B, C between P and Q, so that P + A + B + C + Q forms an A.P.

*Example.* Insert 3 arithmetic means between 8 and 18.

Let the means be denoted by A, B, C.

Then 8 + A + B + C + 18 forms an A.P.

First term, $a = 8$. fifth term $= a + 4d = 18$

$$\left.\begin{aligned} \therefore \quad a &= 8 \\ a + 4d &= 18 \end{aligned}\right\} 4d = 10 \quad \therefore\ d = 2 \cdot 5$$

$$\left.\begin{aligned} A &= 8 + 2 \cdot 5 = 10 \cdot 5 \\ B &= 8 + 5 \quad\ = 13 \\ C &= 8 + 7 \cdot 5 = 15 \cdot 5 \end{aligned}\right\} \text{Required arith. means are } \underline{10 \cdot 5,\ 13,\ 15 \cdot 5}$$

Now, you find five arithmetic means between 12 and 21·6.

*Then turn to frame 8.*

**8**

Required arith. means: $\boxed{13 \cdot 6,\ 15 \cdot 2,\ 16 \cdot 8,\ 18 \cdot 4,\ 20}$

Here is the working:

Let the 5 arith. means be A, B, C, D, E.

Then 12 + A + B + C + D + E + 21.6 forms an A.P.

$$\therefore\ a = 12; \quad a + 6d = 21 \cdot 6$$

$$\therefore\ 6d = 9 \cdot 6 \quad \therefore\ d = 1 \cdot 6$$

Then

$$\begin{aligned} A &= 12 + 1 \cdot 6 = 13 \cdot 6 & A &= 13 \cdot 6 \\ B &= 12 + 3 \cdot 2 = 15 \cdot 2 & B &= 15 \cdot 2 \\ C &= 12 + 4 \cdot 8 = 16 \cdot 8 & C &= 16 \cdot 8 \\ D &= 12 + 6 \cdot 4 = 18 \cdot 4 & D &= 18 \cdot 4 \\ E &= 12 + 8 \cdot 0 = 20 \cdot 0 & E &= 20. \end{aligned}$$

So that is that! Once you have done one, the others are just like it.

Now we will see how much you remember about Geometric Series.

*So, on to frame 9.*

**9**

2. *Geometric series* (Geometric progression) denoted by G.P.

An example of a G.P. is the series:

$$1 + 3 + 9 + 27 + 81 + \ldots \text{ etc.}$$

Here you see that any term can be written from the previous term by multiplying it by a constant factor 3. This constant factor is called the *common ratio* and is found by selecting any term and dividing it by the previous one.

$$\text{e.g. } 27 \div 9 = 3; \quad 9 \div 3 = 3; \text{ etc.}$$

A G.P. therefore has the form:

$$a + ar + ar^2 + ar^3 + \ldots \text{ etc.}$$

where $a$ = first term, $r$ = common ratio.

So in the geometric series $5 - 10 + 20 - 40 + \ldots$ etc. the common ratio, $r$, is ........................

**10**

$$r = \frac{20}{-10} = -2$$

The general geometric series is therefore:

$$a + ar + ar^2 + ar^3 + \ldots \quad \text{etc.} \quad \text{..........} \quad \text{(iv)}$$

and you will remember that

(i) the $n^{\text{th}}$ term $= ar^{n-1}$ ............................ (v)

(ii) the sum of the first $n$ terms is given by

$$S_n = \frac{a(1 - r^n)}{1 - r} \quad \text{......................} \quad \text{(vi)}$$

Make a note of these items in your record book.

So, now you can do this one:

For the series $8 + 4 + 2 + 1 + \frac{1}{2} + \ldots$ etc., find the sum of the first 8 terms.

*Then on to frame 11.*

**11**

$$\boxed{S_8 = 15\tfrac{15}{16}}$$

Since, for the series 8, 4, 2, 1, . . . etc.

$$a = 8; \quad r = \frac{2}{4} = \frac{1}{2}; \quad S_n = \frac{a(1-r^n)}{1-r}$$

$$\therefore S_8 = \frac{8(1-[\tfrac{1}{2}]^8)}{1-\tfrac{1}{2}}$$

$$= \frac{8(1-\frac{1}{256})}{1-\tfrac{1}{2}} = 16.\frac{255}{256} = \frac{255}{16} = \underline{15\tfrac{15}{16}}$$

Now here is another example.

If the 5th term of a G.P. is 162 and the 8th term is 4374, find the series.

We have

$$5\text{th term} = 162 \quad \therefore a.r^4 = 162$$

$$8\text{th term} = 4374 \quad \therefore a.r^7 = 4374$$

$$\frac{ar^7}{ar^4} = \frac{4374}{162} \quad \therefore r^3 = 27 \quad \therefore r = 3$$

$$\therefore a = \ldots\ldots\ldots\ldots$$

**12**

$$\boxed{a = 2}$$

for $ar^4 = 162$; $ar^7 = 4374$ and $r = 3$

$$\therefore a.3^4 = 162 \quad \therefore a = \frac{162}{81} \quad \therefore a = 2$$

$\therefore$ The series is: $\underline{2 + 6 + 18 + 54 + \ldots \text{ etc.}}$

Of course, now that we know the values of $a$ and $r$, we could calculate the value of any term or the sum of a given number of terms. For this same series, find

(i) the 10th term

(ii) the sum of the first 10 terms.

*When you have finished, turn to frame 13.*

**13**

$$a = 2; \quad r = 3$$

(i) 10th term $= ar^9 = 2.3^9 = 2(19683) = \boxed{39366}$

(ii) $S_{10} = \dfrac{a(1-r^{10})}{1-r} = \dfrac{2(1-3^{10})}{1-3}$

$$= \frac{2(1-59049)}{-2} = \boxed{59048}$$

**Geometric mean**

The geometric mean of two given numbers P and Q is a number A such that P + A + Q forms a G.P.

$$\frac{A}{P} = r \text{ and } \frac{Q}{A} = r$$

$$\therefore \frac{A}{P} = \frac{Q}{A} \quad \therefore A^2 = PQ \quad A = \sqrt{(PQ)}$$

So the geometric mean of 2 numbers is the square root of their product.

Therefore, the geom. mean of 4 and 25 is ....................

**14**

$$A = \sqrt{(4 \times 25)} = \sqrt{100} = \boxed{10}$$

□□□□□□□□□□□□□□□□□□□□□□□□□□□□□□□□□□□□□□

To insert 3 G.M's between two given numbers, P and Q means to insert 3 numbers, A, B, C, such that P + A + B + C + Q forms a G.P.

*Example.* Insert 4 geometric means between 5 and 1215.

Let the means be A, B, C, D. Then 5 + A + B + C + D + 1215 forms a G.P.

$$\text{i.e. } a = 5 \text{ and } ar^5 = 1215$$

$$\therefore r^5 = \frac{1215}{5} = 243 \quad \therefore r = 3$$

$\therefore A = 5.3 = 15$
$B = 5.9 = 45$
$C = 5.27 = 135$
$D = 5.81 = 405$

The required geometric means are:
15, 45, 135, 405

Now here is one for you to do: Insert two geometric means between 5 and 8·64.

*Then on to frame 15.*

**15**

| Required geometric means are 6·0, 7·2 |
|---|

For, let the means be A and B.

Then 5 + A + B + 8·64 forms a G.P.

$$\therefore a = 5; \quad \therefore ar^3 = 8{\cdot}64; \quad \therefore r^3 = 1{\cdot}728; \quad \therefore r = 1{\cdot}2$$

$$\left.\begin{array}{l} A = 5.1{\cdot}2 = 6 \\ B = 5.1{\cdot}44 = 7{\cdot}20 \end{array}\right\} \begin{array}{l}\text{Required means are} \\ \underline{6{\cdot}0 \text{ and } 7{\cdot}2}\end{array}$$

Arithmetic and geometric series are, of course, special kinds of series. There are other special series that are worth knowing. These consist of the series of the powers of the natural numbers. So let us look at these in the next frame.

**16**

**Series of powers of the natural numbers**

1. The series $1 + 2 + 3 + 4 + 5 + \ldots + n$ etc. $= \sum_1^n r$.

   This series, you will see, is an example of an A.P., where $a = 1$ and $d = 1$. The sum of the first $n$ terms is given by:

$$\sum_1^n r = 1 + 2 + 3 + 4 + 5 + \ldots + n$$

$$= \frac{n}{2}(2a + \overline{n-1}\,d) = \frac{n(n+1)}{2}$$

$$\underline{\sum_1^n r = \frac{n(n+1)}{2}}$$

So, the sum of the first 100 natural numbers is ..................

*Then on to frame 17.*

**17**

$$\boxed{\sum_1^{100} r = 5050}$$

for $$r = \frac{100(101)}{2} = 50(101) = 5050$$

□□□□□□□□□□□□□□□□□□□□□□□□□□□□□□□□□□□□□□□□□□□□□□□□□□

2. That was easy enough. Now let us look at this one: To establish the result for the sum of $n$ terms of the series $1^2 + 2^2 + 3^2 + 4^2 + 5^2 + \ldots + n^2$., we make use of the identity

$$(n+1)^3 = n^3 + 3n^2 + 3n + 1$$

We write this as
$$(n+1)^3 - n^3 = 3n^2 + 3n + 1$$

Replacing $n$ by $n - 1$, we get

$$n^3 - (n-1)^3 = 3(n-1)^2 + 3(n-1) + 1$$

and again $$(n-1)^3 - (n-2)^3 = 3(n-2)^2 + 3(n-2) + 1$$

and $$(n-2)^3 - (n-3)^3 = 3(n-3)^2 + 3(n-3) + 1$$

Continuing like this, we should eventually arrive at:

$$3^3 - 2^3 = 3.2^2 + 3.2 + 1$$
$$2^3 - 1^3 = 3.1^2 + 3.1 + 1$$

If we now add all these results together, we find on the left-hand side that all the terms disappear except the first and the last.

$$(n+1)^3 - 1^3 = 3\left\{n^2 + (n-1)^2 + (n-2)^2 + \ldots + 2^2 + 1^2\right\}$$
$$+ 3\left\{n + (n-1) + (n-2) + \ldots + 2 + 1\right\} + n(1)$$
$$= 3.\sum_1^n r^2 + 3\sum_1^n r + n$$

$$\therefore n^3 + 3n^2 + 3n + \not{1} - \not{1} = 3\sum_1^n r^2 + 3\sum_1^n r + n = 3\sum_1^n r^2 + 3\,\frac{n(n+1)}{2} + n$$

$$\therefore n^3 + 3n^2 + 2n = 3\sum_1^n r^2 + \frac{3}{2}(n^2 + n)$$

$$\therefore 2n^3 + 6n^2 + 4n = 6\sum_1^n r^2 + 3n^2 + 3n$$

$$6\sum_1^n r^2 = 2n^3 + 3n^2 + n$$

$$\therefore \sum_1^n r^2 = \frac{n(n+1)(2n+1)}{6}$$

So, the sum of the first 12 terms of the series $1^2 + 2^2 + 3^2 + \ldots$ is ...............

**18**

$$\sum_1^n r^2 = \frac{n(n+1)(2n+1)}{6}$$

$$\therefore \sum_1^{12} r^2 = \frac{12(13)(25)}{6} = 26(25) = \boxed{650}$$

3. The sum of the cubes of the natural numbers is found in much the same way. This time, we use the identity

$$(n+1)^4 = n^4 + 4n^3 + 6n^2 + 4n + 1$$

We rewrite it as before

$$(n+1)^4 - n^4 = 4n^3 + 6n^2 + 4n + 1$$

If we now do the same trick as before and replace $n$ by $(n-1)$ over and over again, and finally total up the results we get the result

$$\sum_1^n r^3 = \left\{\frac{n(n+1)}{2}\right\}^2$$

Note in passing that $\sum_1^n r^3 = \left\{\sum_1^n r\right\}^2$

**19**

Let us collect together these last three results. Here they are:

1. $\sum_1^n r = \dfrac{n(n+1)}{2}$ .............................. (vii)

2. $\sum_1^n r^2 = \dfrac{n(n+1)(2n+1)}{6}$ ................ (viii)

3. $\sum_1^n r^3 = \left\{\dfrac{n(n+1)}{2}\right\}^2$ .......................... (ix)

These are handy results, so copy them into your record book.

*Now turn on to frame 20 and we can see an example of the use of these results.*

**20** *Example:* Find the sum of the series $\sum_{n=1}^{5} n(3 + 2n)$

$$S_5 = \sum_1^5 n(3 + 2n) = \sum_1^5 (3n + 2n^2)$$

$$= \sum_1^5 3n + \sum_1^5 2n^2$$

$$= 3\sum_1^5 n + 2\sum_1^5 n^2$$

$$= \frac{3.5.6}{2} + 2.\frac{5.6.11}{6}$$

$$= 45 + 110$$

$$= \underline{155}$$

It is just a question of using the established results. Here is one for you to do in the same manner.

Find the sum of the series $\sum_{n=1}^{4} (2n + n^3)$

---

**21**

$$S_4 = \sum_1^4 (2n + n^3)$$

$$= 2\sum_1^4 n + \sum_1^4 n^3$$

$$= \frac{2.4.5}{2} + \left\{\frac{4.5}{2}\right\}^2$$

$$= 20 + 100 = \boxed{120}$$

*Remember*

Sum of first $n$ natural numbers $= \dfrac{n(n + 1)}{2}$

Sum of squares of first $n$ natural numbers $= \dfrac{n(n + 1)(2n + 1)}{6}$

Sum of cubes of first $n$ natural numbers $= \left\{\dfrac{n(n + 1)}{2}\right\}^2$

**22**

**Infinite series**

So far, we have been concerned with a finite number of terms of a given series. When we are dealing with the sum of an infinite number of terms of a series, we must be careful about the steps we take.

*Example:* Consider the infinite series $1 + \frac{1}{2} + \frac{1}{4} + \frac{1}{8} + \ldots$

This we recognize as a G.P. in which $a = 1$ and $r = \frac{1}{2}$. The sum of the first $n$ terms is therefore given by

$$S_n = \frac{1(1 - [\frac{1}{2}]^n)}{1 - \frac{1}{2}} = 2(1 - \frac{1}{2^n})$$

Now if $n$ is very large, $2^n$ will be very large and therefore $\frac{1}{2^n}$ will be very small. In fact, as $n \to \infty$, $\frac{1}{2^n} \to 0$. The sum of all the terms in this infinite series is therefore given by $S_\infty$ = the limiting value of $S_n$ as $n \to \infty$.

$$\text{i.e. } S_\infty = \underset{n \to \infty}{\text{Lt}} \{S_n\} = 2(1 - 0) = 2$$

This result means that we can make the sum of the series as near to the value 2 as we please by taking a sufficiently large number of terms.

*Next frame.*

**23**

This is not always possible with an infinite series, for in the case of an A.P. things are very different.

Consider the infinite series $1 + 3 + 5 + 7 + \ldots$

This is an A.P. in which $a = 1$ and $d = 2$.

Then
$$S_n = \frac{n}{2}(2a + \overline{n-1}.d) = \frac{n}{2}(2 + \overline{n-1}.2)$$
$$= \frac{n}{2}(2 + 2n - 2)$$
$$S_n = n^2$$

Of course, in this case, if $n$ is large then the value of $S_n$ is very large. In fact, if $n \to \infty$, then $S_n \to \infty$, which is not a definite numerical value and of little use to us.

This always happens with an A.P.: if we try to find the "sum to infinity", we invariably obtain $+\infty$ or $-\infty$ as the result, depending on the actual series.

*Turn on now to frame 24.*

## 24

In the previous two frames, we made two important points.

(i) We cannot evaluate the sum of an infinite number of terms of an A.P. because the result is always infinite.

(ii) We can sometimes evaluate the sum of an infinite number of terms of a G.P. since, for such a series, $S_n = \dfrac{a(1-r^n)}{1-r}$ and <u>provided $|r| < 1$</u>, then as $n \to \infty$, $r^n \to 0$. In that case $S_\infty = \dfrac{a(1-0)}{1-r} = \dfrac{a}{1-r}$, i.e. $S_\infty = \dfrac{a}{1-r}$

So, find the 'sum to infinity' of the series

$$20 + 4 + 0{\cdot}8 + 0{\cdot}16 + 0{\cdot}032 + \ldots\ldots\ldots\ldots$$

## 25

$$\boxed{S_\infty = 25}$$

For

$$20 + 4 + 0{\cdot}8 + 0{\cdot}16 + 0{\cdot}032 + \ldots$$

$$a = 20; \quad r = \frac{0{\cdot}8}{4} = 0{\cdot}2 = \frac{1}{5}$$

$$\therefore \; S_\infty = \frac{a}{1-r} = \frac{20}{1-\dfrac{1}{5}} = \frac{5}{4}.(20) = \underline{25}$$

□□□□□□□□□□□□□□□□□□□□□□□□□□□□□□□□□□□□□□□

**Limiting values**

In this programme, we have already seen that we have sometimes to determine the limiting value of $S_n$ as $n \to \infty$. Before we leave this topic, let us look a little further into the process of finding limiting values. One or two examples will suffice.

*So turn on to frame 26.*

**26**

*Example 1.* To find the limiting value of $\frac{5n+3}{2n-7}$ as $n \to \infty$

We cannot just substitute $n = \infty$ in the expression and simplify the result, since $\infty$ is not an ordinary number and does not obey the normal rules. So we do it this way:

$$\frac{5n+3}{2n-7} = \frac{5+3/n}{2-7/n} \quad \text{(dividing top and bottom by } n\text{)}$$

$$\underset{n\to\infty}{\text{Limit}} \left\{\frac{5n+3}{2n-7}\right\} = \underset{n\to\infty}{\text{Limit}} \frac{5+3/n}{2-7/n}$$

Now when $n \to \infty$, $3/n \to 0$ and $7/n \to 0$

$$\therefore \underset{n\to\infty}{\text{Lt}} \frac{5n+3}{2n-7} = \underset{n\to\infty}{\text{Lt}} \frac{5+3/n}{2-7/n} = \frac{5+0}{2-0} = \underline{\frac{5}{2}}$$

We can always deal with fractions of the form $\frac{c}{n}, \frac{c}{n^2}, \frac{c}{n^3}$, etc., for when $n \to \infty$, each of these tends to *zero*, which is a precise value.

Let us try another example.

*On to the next frame then.*

**27**

*Example 2.* To find the limiting value of $\frac{2n^2+4n-3}{5n^2-6n+1}$ as $n \to \infty$.

First of all, we divide top and bottom by the highest power of $n$ which is involved, in this case $n^2$.

$$\frac{2n^2+4n-3}{5n^2-6n+1} = \frac{2+4/n-3/n^2}{5-6/n+1/n^2}$$

$$\therefore \underset{n\to\infty}{\text{Lt}} \frac{2n^2+4n-3}{5n^2-6n+1} = \underset{n\to\infty}{\text{Lt}} \frac{2+4/n-3/n^2}{5-6/n+1/n^2}$$

$$= \frac{2+0-0}{5-0+0} = \underline{\frac{2}{5}}$$

*Example 3.* To find $\underset{n\to\infty}{\text{Lt}} \frac{n^3-2}{2n^3+3n-4}$

In this case, the first thing is to ....................

*Turn on to frame 28.*

**28**

Divide top and bottom by $n^3$

Right. So we get

$$\frac{n^3 - 2}{2n^3 + 3n - 4} = \frac{1 - 2/n^3}{2 + 3/n^2 - 4/n^3}$$

$$\therefore \quad \operatorname*{Lt}_{n \to \infty} \frac{n^3 - 2}{2n^3 + 3n - 4} = \dots\dots\dots\dots$$

*Finish it off. Then move on to frame 29.*

**29**

$$\boxed{\frac{1}{2}}$$

□□□□□□□□□□□□□□□□□□□□□□□□□□□□□□□□□□□□□□□□

**Convergent and divergent series**

A series in which the sum ($S_n$) of $n$ terms of the series tends to a definite value, as $n \to \infty$, is called a *convergent* series. If $S_n$ does not tend to a definite value as $n \to \infty$, the series is said to be *divergent.*

*Example:* Consider the G.P. $1 + \frac{1}{3} + \frac{1}{9} + \frac{1}{27} + \frac{1}{81} + \dots$

We know that for a G.P., $S_n = \dfrac{a(1 - r^n)}{1 - r}$, so in this case since $a = 1$ and $r = \frac{1}{3}$, we have:

$$S_n = \frac{1(1 - \frac{1}{3^n})}{1 - \frac{1}{3}} = \frac{1 - \frac{1}{3^n}}{\frac{2}{3}} = \frac{3}{2}(1 - \frac{1}{3^n})$$

$$\therefore \text{ As } n \to \infty, \frac{1}{3^n} \to 0 \quad \therefore \quad \underline{\operatorname*{Lt}_{n \to \infty} S_n = \frac{3}{2}}$$

The sum of $n$ terms of this series tends to the definite value $\frac{3}{2}$ as $n \to \infty$
It is therefore a ................................ series.
(convergent/divergent)

**30**

convergent

If $S_n$ tends to a definite value as $n \to \infty$, the series is *convergent*.
If $S_n$ does not tend to a definite value as $n \to \infty$, the series is *divergent*.

□□□□□□□□□□□□□□□□□□□□□□□□□□□□□□□□□□□□□□

Here is another series. Let us investigate this one.

$$1 + 3 + 9 + 27 + 81 + \ldots$$

This is also a G.P. with $a = 1$ and $r = 3$.

$$\therefore\ S_n = \frac{a(1-r^n)}{1-r} = \frac{1(1-3^n)}{1-3} = \frac{1-3^n}{-2}$$

$$= \frac{3^n - 1}{2}$$

Of course, when $n \to \infty$, $3^n \to \infty$ also.

$$\therefore\ \underset{n\to\infty}{\text{Lt}}\ S_n = \infty \text{ (which is not a definite numerical value)}$$

So in this case, the series is ....................

**31**

divergent

We can make use of infinite series only when they are convergent and it is necessary, therefore, to have some means of testing whether or not a given series is, in fact, convergent.

Of course, we could determine the limiting value of $S_n$ as $n \to \infty$, as we did in the examples a moment ago, and this would tell us directly whether the series in question tended to a definite value (i.e. was convergent) or not.

That is the fundamental test, but unfortunately, it is not always easy to find a formula for $S_n$ and we have therefore to find a test for convergence which uses the terms themselves.

Remember the notation for series in general. We shall denote the terms by $u_1 + u_2 + u_3 + u_4 + \ldots$

*So now turn on to frame 32.*

**32**

**Tests for convergence**

*Test 1. A series cannot be convergent unless its terms ultimately tend to zero, i.e. unless* $\lim_{n\to\infty} u_n = 0$.

If $\lim_{n\to\infty} u_n \neq 0$, the series is divergent.

This is almost just common sense, for if the sum is to approach some definite value as the value of $n$ increases, the numerical value of the individual terms must diminish. For example, we have already seen that

(i) the series $1 + \frac{1}{3} + \frac{1}{9} + \frac{1}{27} + \frac{1}{81} + \ldots$ converges,

while (ii) the series $1 + 3 + 9 + 27 + 81 + \ldots$ diverges.

So what would you say about the series

$$1 + \frac{1}{2} + \frac{1}{3} + \frac{1}{4} + \frac{1}{5} + \frac{1}{6} + \ldots ?$$

Just by looking at it, do you think this series converges or diverges?

**33**

Most likely you said that the series converges since it was clear that the numerical value of the terms decreases as $n$ increases. If so, I am afraid you were wrong, for we shall show later that, in fact, the series

$1 + \frac{1}{2} + \frac{1}{3} + \frac{1}{4} + \frac{1}{5} + \ldots$ diverges.

It was rather a trick question, but be very clear about what the rule states. It says:

*A series cannot be convergent unless its terms ultimately tend to zero, i.e.* $\lim_{n\to\infty} u_n = 0$. It does not say that if the terms tend to zero, then the series is convergent. In fact, it is quite possible for the terms to tend to zero without the series converging – as in the example stated.

In practice, then, we use the rule in the following form:

If $\lim_{n\to\infty} u_n = 0$, the series *may* converge or diverge and we must test further.

If $\lim_{n\to\infty} u_n \neq 0$, we can be sure that the series *diverges.*

*Make a note of these two statements.*

**34**

Before we leave the series

$$1+\frac{1}{2}+\frac{1}{3}+\frac{1}{4}+\frac{1}{5}+\frac{1}{6}+\ldots+\frac{1}{n}+\ldots$$

here is the proof that, although $\underset{n\to\infty}{\text{Lt}}\ u_n = 0$, the series does, in fact, diverge.

We can, of course, if we wish, group the terms as follows:

$$1+\frac{1}{2}+\left\{\frac{1}{3}+\frac{1}{4}\right\}+\left\{\frac{1}{5}+\frac{1}{6}+\frac{1}{7}+\frac{1}{8}\right\}+\ldots$$

Now $$\left\{\frac{1}{3}+\frac{1}{4}\right\}>\left\{\frac{1}{4}+\frac{1}{4}\right\}>\frac{1}{2}$$

and $$\left\{\frac{1}{5}+\frac{1}{6}+\frac{1}{7}+\frac{1}{8}\right\}>\left\{\frac{1}{8}+\frac{1}{8}+\frac{1}{8}+\frac{1}{8}\right\}>\frac{1}{2}\quad\text{etc.}$$

So that $$S_n > 1+\frac{1}{2}+\frac{1}{2}+\frac{1}{2}+\frac{1}{2}+\frac{1}{2}+\ldots$$

$$\therefore\ S_\infty = \infty$$

This is not a definite numerical value, so the series is ....................

**35**

divergent

The best we can get from Test 1, is that a series *may* converge. We must therefore apply a further test.

*Test 2. The comparison test*

*A series of positive terms is convergent if its terms are less than the corresponding terms of a positive series which is known to be convergent. Similarly, the series is divergent if its terms are greater than the corresponding terms of a series which is known to be divergent.*

An example or two will show how we apply this particular test.

*So turn on to the next frame.*

**36** *Example.* To test the series

$$1+\frac{1}{2^2}+\frac{1}{3^3}+\frac{1}{4^4}+\frac{1}{5^5}+\frac{1}{6^6}+\ldots+\frac{1}{n^n}+\ldots$$

we can compare it with the series

$$1+\frac{1}{2^2}+\frac{1}{2^3}+\frac{1}{2^4}+\frac{1}{2^5}+\frac{1}{2^6}+\ \ldots\ \ldots$$

which is known to converge.

If we compare corresponding terms after the first two terms, we see that $\frac{1}{3^3}<\frac{1}{2^3}$; $\frac{1}{4^4}<\frac{1}{2^4}$; and so on for all further terms, so that, after the first two terms, the terms of the first series are each less than the corresponding terms of the series known to converge.

The first series also, therefore, .......................

**37**

converges

The difficulty with the comparison test is knowing which convergent series to use as a standard. A useful series for this purpose is this one:

$$\frac{1}{1^p}+\frac{1}{2^p}+\frac{1}{3^p}+\frac{1}{4^p}+\frac{1}{5^p}+\ldots+\frac{1}{n^p}+\ldots=\sum_{n=1}^{\infty}\frac{1}{n^p}$$

It can be shown that

(i) if $p>1$, the series converges
(ii) if $p\leqslant 1$, the series diverges

So what about the series $\sum_{n=1}^{\infty}\frac{1}{n^2}$?

Does it converge or diverge?

**38**

Converge since the series $\Sigma\frac{1}{n^2}$ is the series $\Sigma\frac{1}{n^p}$ with $p > 1$

□□□□□□□□□□□□□□□□□□□□□□□□□□□□□□□□□□□□□□

Let us look at another example.

To test the series $\frac{1}{1.2}+\frac{1}{2.3}+\frac{1}{3.4}+\frac{1}{4.5}+\ldots$

If we take our standard series

$$\frac{1}{1^p}+\frac{1}{2^p}+\frac{1}{3^p}+\frac{1}{4^p}+\frac{1}{5^p}+\frac{1}{6^p}+\ldots$$

when $p = 2$, we get

$$\frac{1}{1^2}+\frac{1}{2^2}+\frac{1}{3^2}+\frac{1}{4^2}+\frac{1}{5^2}+\frac{1}{6^2}+\ldots$$

which we know to converge.

But $\frac{1}{1.2}<\frac{1}{1^2}$; $\frac{1}{2.3}<\frac{1}{2^2}$; $\frac{1}{3.4}<\frac{1}{3^2}$; etc.

Each term of the given series is less than the corresponding term in the series known to converge.

Therefore ......................

**39**

The given series converges

□□□□□□□□□□□□□□□□□□□□□□□□□□□□□□□□□□□□□□

It is not always easy to devise a suitable comparison series, so we look for yet another test to apply, and here it is:

*Test 3. D'Alembert's ratio test for positive terms*

Let $u_1+u_2+u_3+u_4+\ldots+u_n+\ldots$ be a series of positive terms. Find expressions for $u_n$ and $u_{n+1}$, i.e. the $n^{\text{th}}$ term and the $(n+1)^{\text{th}}$ term, and form the ratio $\frac{u_{n+1}}{u_n}$. Determine the limiting value of this ratio as $n \to \infty$.

If $\underset{n\to\infty}{\text{Lt}}\frac{u_{n+1}}{u_n} < 1$, the series converges

" $> 1$, the series diverges

" $= 1$, the series may converge or diverge and the test gives us no definite information.

*Copy out D'Alembert's ratio test into your record book. Then on to frame 40.*

**40**

Here it is again:

*D'Alembert's ratio test* for positive terms

$$\text{If } \underset{n\to\infty}{\text{Lt}} \frac{u_{n+1}}{u_n} < 1, \text{ the series } converges$$

$$\text{"} \quad > 1, \text{ the series } diverges$$

$$\text{"} \quad = 1, \text{ the result is inconclusive.}$$

□□□□□□□□□□□□□□□□□□□□□□□□□□□□□□□□□□□□□□

*Example:* To test the series $\frac{1}{1} + \frac{3}{2} + \frac{5}{2^2} + \frac{7}{2^3} + \ldots$

We first of all decide on the pattern of the terms and hence write down the $n^{\text{th}}$ term. In this case $u_n = \frac{2n-1}{2^{n-1}}$. The $(n+1)^{\text{th}}$ term will then be the same with $n$ replaced by $(n+1)$

$$\text{i.e. } u_{n+1} = \frac{2n+1}{2^n}$$

$$\therefore \frac{u_{n+1}}{u_n} = \frac{2n+1}{2^n} \cdot \frac{2^{n-1}}{2n-1} = \frac{1}{2} \cdot \frac{2n+1}{2n-1}$$

We now have to find the limiting value of this ratio as $n \to \infty$. From our previous work on limiting values, we know that the next step, then, is to divide top and bottom by ..................

**41**

Divide top and bottom by $n$

$$\text{so } \underset{n\to\infty}{\text{Lt}} \frac{u_{n+1}}{u_n} = \underset{n\to\infty}{\text{Lt}} \frac{1}{2} \cdot \frac{2n+1}{2n-1} = \underset{n\to\infty}{\text{Lt}} \frac{1}{2} \cdot \frac{2+1/n}{2-1/n}$$

$$= \frac{1}{2} \cdot \frac{2+0}{2-0} = \frac{1}{2}$$

Since, in this case, $\underset{n\to\infty}{\text{Lt}} \frac{u_{n+1}}{u_n} < 1$, we know that the given series is *convergent*.

□□□□□□□□□□□□□□□□□□□□□□□□□□□□□□□□□□□□□□

Let us do another one in the same way.

*Example:* Apply D'Alembert's ratio test to the series

$$\frac{1}{2} + \frac{2}{3} + \frac{3}{4} + \frac{4}{5} + \frac{5}{6} + \ldots$$

First of all, we must find an expression for $u_n$.

In this series, $u_n =$ ...................

**42**

$$\frac{1}{2}+\frac{2}{3}+\frac{3}{4}+\frac{4}{5}+\ldots \quad \boxed{u_n = \frac{n}{n+1}}$$

Then $u_{n+1}$ is found by simply replacing $n$ by $(n + 1)$.

$$\therefore u_{n+1} = \frac{n+1}{n+2}$$

So that
$$\frac{u_{n+1}}{u_n} = \frac{n+1}{n+2} \cdot \frac{n+1}{n} = \frac{n^2+2n+1}{n^2+2n}$$

We now have to find $\underset{n\to\infty}{\text{Lt}} \frac{u_{n+1}}{u_n}$ and in order to do that we must divide top and bottom, in this case, by ..................

**43**

$$\boxed{n^2}$$

$$\therefore \underset{n\to\infty}{\text{Lt}} \frac{u_{n+1}}{u_n} = \underset{n\to\infty}{\text{Lt}} \frac{n^2+2n+1}{n^2+2n} = \underset{n\to\infty}{\text{Lt}} \frac{1+2/n+1/n^2}{1+2/n}$$

$$= \frac{1+0+0}{1+0} = 1$$

$\therefore \underset{n\to\infty}{\text{Lt}} \frac{u_{n+1}}{u_n} = 1$, which is inconclusive and which merely tells us that the series may be convergent or divergent. So where do we go from there?

We have, of course, forgotten about Test 1, which states that

(i) if $\underset{n\to\infty}{\text{Lt}} u_n = 0$, the series *may* be convergent

(ii) if $\underset{n\to\infty}{\text{Lt}} u_n \neq 0$, the series is certainly *divergent*

In our present series, $u_n = \dfrac{n}{n+1}$

$$\therefore \underset{n\to\infty}{\text{Lt}} u_n = \underset{n\to\infty}{\text{Lt}} \frac{n}{n+1} = \underset{n\to\infty}{\text{Lt}} \frac{1}{1+1/n} = 1$$

This is *not* zero. Therefore the series is *divergent.*

□□□□□□□□□□□□□□□□□□□□□□□□□□□□□□□□□□□□□□□□□□□□

Now you do this one entirely on your own:

Test the series $\quad \frac{1}{5}+\frac{2}{6}+\frac{2^2}{7}+\frac{2^3}{8}+\frac{2^4}{9}+\ldots$

*When you have finished, check your result with that in frame 44.*

**44** Here is the solution in detail: see if you agree with it.

$$\frac{1}{5}+\frac{2}{6}+\frac{2^2}{7}+\frac{2^3}{8}+\frac{2^4}{9}+\ldots$$

$$u_n=\frac{2^{n-1}}{4+n};\quad u_{n+1}=\frac{2^n}{5+n}$$

$$\therefore \frac{u_{n+1}}{u_n}=\frac{2^n}{5+n}\cdot\frac{4+n}{2^{n-1}}$$

The power $2^{n-1}$ cancels with the power $2^n$ to leave a single factor 2.

$$\therefore \frac{u_{n+1}}{u_n}=\frac{2(4+n)}{5+n}$$

$$\therefore \underset{n\to\infty}{\text{Lt}}\frac{u_{n+1}}{u_n}=\underset{n\to\infty}{\text{Lt}}\frac{2(4+n)}{5+n}=\underset{n\to\infty}{\text{Lt}}\frac{2(4/n+1)}{5/n+1}$$

$$=\frac{2(0+1)}{0+1}=2$$

$$\therefore \underset{n\to\infty}{\text{Lt}}\frac{u_{n+1}}{u_n}=2$$

And since the limiting value is $>1$, we know the series is ......................

**45**

divergent

**Series in general. Absolute convergence**

So far, we have considered series with positive terms only. Some series consist of alternate positive and negative terms.

*Example:* the series $1-\frac{1}{2}+\frac{1}{3}-\frac{1}{4}+\ldots$ is in fact convergent

while the series $1+\frac{1}{2}+\frac{1}{3}+\frac{1}{4}+\ldots$ is divergent.

If $u_n$ denotes the $n^{\text{th}}$ term of a series in general, it may well be positive or negative. But $|u_n|$, or 'mod $u_n$' denotes the numerical value of $u_n$, so that if $u_1+u_2+u_3+u_4+\ldots$ is a series of mixed terms, i.e. some positive, some negative, then the series $|u_1|+|u_2|+|u_3|+|u_4|+\ldots$ will be a series of positive terms.

So if $\Sigma u_n=1-3+5-7+9-\ldots$

Then $\Sigma|u_n|=$ ..............................

**46**

$$\Sigma|u_n| = 1 + 3 + 5 + 7 + 9 + \ldots$$

□□□□□□□□□□□□□□□□□□□□□□□□□□□□□□□□□□□□□□□□

*Note:* If a series $\Sigma u_n$ is convergent, then the series $\Sigma|u_n|$ may very well not be convergent, as in the example stated in the previous frame. But if $\Sigma|u_n|$ is found to be convergent, we can be sure that $\Sigma u_n$ is convergent.

If $\Sigma|u_n|$ converges, the series $\Sigma u_n$ is said to be *absolutely convergent.*

If $\Sigma|u_n|$ is not convergent, but $\Sigma u_n$ does converge, then $\Sigma u_n$ is said to be *conditionally convergent.*

So, if $\Sigma u_n = 1 - \frac{1}{2} + \frac{1}{3} - \frac{1}{4} + \frac{1}{5} - \ldots$ converges

and $\Sigma|u_n| = 1 + \frac{1}{2} + \frac{1}{3} + \frac{1}{4} + \frac{1}{5} + \ldots$ diverges

then $\Sigma u_n$ is ........................................... convergent.
(absolutely or conditionally)

**47**

conditionally

*Example:* Find the range of values of $x$ for which the following series is absolutely convergent.

$$\frac{x}{2.5} - \frac{x^2}{3.5^2} + \frac{x^3}{4.5^3} - \frac{x^4}{5.5^4} + \frac{x^5}{6.5^5} - \ldots$$

$$|u_n| = \frac{x^n}{(n+1)5^n}; \quad |u_{n+1}| = \frac{x^{n+1}}{(n+2)5^{n+1}}$$

$$\therefore \left|\frac{u_{n+1}}{u_n}\right| = \frac{x^{n+1}}{(n+2)5^{n+1}} \cdot \frac{(n+1)5^n}{x^n}$$

$$= \frac{x(n+1)}{5(n+2)} = \frac{x(1+1/n)}{5(1+2/n)}$$

$$\therefore \underset{n\to\infty}{\text{Lt}}\left|\frac{u_{n+1}}{u_n}\right| = \frac{x}{5}$$

For absolute convergence $\underset{n\to\infty}{\text{Lt}}\left|\frac{u_{n+1}}{u_n}\right| < 1$. $\therefore$ Series convergent when $\left|\frac{x}{5}\right| < 1$, i.e. for $|x| < 5$.

*On to frame 48.*

**48**

You have now reached the end of this programme, except for the test exercise which follows in frame 49. Before you work through it, here is a summary of the topics we have covered. Read through it carefully: it will refresh your memory of what we have been doing.

**Revision Sheet**

1. *Arithmetic series:* $a + (a + d) + (a + 2d) + (a + 3d) + \ldots$

$$u_n = a + (n-1)d \qquad S_n = \frac{n}{2}(2a + \overline{n-1}.d)$$

2. *Geometric series:* $a + ar + ar^2 + ar^3 + \ldots$

$$u_n = ar^{n-1} \qquad S_n = \frac{a(1-r^n)}{1-r}$$

$$\text{If } |r| < 1,\ S_\infty = \frac{a}{1-r}$$

3. *Powers of natural numbers:*

$$\sum_1^n r = \frac{n(n+1)}{2} \qquad \sum_1^n r^2 = \frac{n(n+1)(2n+1)}{6}$$

$$\sum_1^n r^3 = \left\{\frac{n(n+1)}{2}\right\}^2$$

4. *Infinite series:* $S_n = u_1 + u_2 + u_3 + u_4 + \ldots + u_n + \ldots$

If $\underset{n \to \infty}{\text{Lt}} S_n$ is a definite value, series is convergent

If " is not a definite value, series is divergent.

5. *Tests for convergence:*

(1) If $\underset{n \to \infty}{\text{Lt}} u_n = 0$, the series *may* be convergent

If " $\neq 0$, the series is certainly divergent.

(2) *Comparison test* – Useful standard series

$$\frac{1}{1^p} + \frac{1}{2^p} + \frac{1}{3^p} + \frac{1}{4^p} + \frac{1}{5^p} + \ldots + \frac{1}{n^p} \ldots$$

For $p > 1$, series converges: for $p < 1$, series diverges.

(3) *D'Alembert's ratio test* for positive terms.

If $\underset{n \to \infty}{\text{Lt}} \dfrac{u_{n+1}}{u_n} < 1$, series converges.

" $> 1$, series diverges.

" $= 1$, inconclusive.

(4) *For general series*

(i) If $\Sigma |u_n|$ converges, $\Sigma u_n$ is absolutely convergent.

(ii) If $\Sigma |u_n|$ diverges, but $\Sigma u_n$ converges, then $\Sigma u_n$ is conditionally convergent.

*Now you are ready for the Test Exercise so turn to frame 49.*

## Test Exercise – XIII

**49**

Answer *all* the questions. Take your time over them and work carefully.

1. The 3rd term of an A.P. is 34 and the 17th term is −8. Find the sum of the first 20 terms.

2. For the series 1 + 1·2 + 1·44 + . . . find the 6th term and the sum of the first 10 terms.

3. Evaluate $\sum_{n=1}^{8} n(3 + 2n + n^2)$.

4. Determine whether each of the following series is convergent.

   (i) $\frac{2}{2.3} + \frac{2}{3.4} + \frac{2}{4.5} + \frac{2}{5.6} + \ldots\ldots$

   (ii) $\frac{2}{1^2} + \frac{2^2}{2^2} + \frac{2^3}{3^2} + \frac{2^4}{4^2} + \ldots + \frac{2^n}{n^2} + \ldots$

   (iii) $u_n = \frac{1 + 2n^2}{1 + n^2}$

   (iv) $u_n = \frac{1}{n!}$

5. Find the range of values of $x$ for which each of the following series is convergent or divergent.

   (i) $1 + x + \frac{x^2}{2!} + \frac{x^3}{3!} + \frac{x^4}{4!} + \ldots\ \ldots$

   (ii) $\frac{x}{1.2} + \frac{x^2}{2.3} + \frac{x^3}{3.4} + \frac{x^4}{4.5} + \ldots\ \ldots$

   (iii) $\sum_{n=1}^{\infty} \frac{(n+1)}{n^3} x^n$

**Further Problems – XIII**

1. Find the sum of $n$ terms of the series

$$S_n = 1^2 + 3^2 + 5^2 + \ldots + (2n-1)^2$$

2. Find the sum to $n$ terms of

$$\frac{1}{1.2.3} + \frac{3}{2.3.4} + \frac{5}{3.4.5} + \frac{7}{4.5.6} + \ldots$$

3. Sum to $n$ terms, the series

$$1.3.5 + 2.4.6 + 3.5.7 + \ldots$$

4. Evaluate the following:

$$\text{(i)}\ \sum_{1}^{n} r(r+3) \qquad \text{(ii)}\ \sum_{1}^{n} (r+1)^3$$

5. Find the sum to infinity of the series

$$1 + \frac{4}{3!} + \frac{6}{4!} + \frac{8}{5!} + \ \ldots$$

6. For the series

$$5 - \frac{5}{2} + \frac{5}{4} - \frac{5}{8} + \ldots + \frac{(-1)^{n-1}5}{2^{n-1}} + \ldots$$

find an expression for $S_n$, the sum of the first $n$ terms. Also, if the series converges, find the sum to infinity.

7. Find the limiting values of

$$\text{(i)}\ \frac{3x^2 + 5x - 4}{5x^2 - x + 7} \quad \text{as} \quad x \to \infty$$

$$\text{(ii)}\ \frac{x^2 + 5x - 4}{2x^2 - 3x + 1} \quad \text{as} \quad x \to \infty$$

8. Determine whether each of the following series converges or diverges.

$$\text{(i)}\ \sum_{1}^{\infty} \frac{n}{n+2} \qquad \text{(ii)}\ \sum_{1}^{\infty} \frac{n}{n^2+1}$$

$$\text{(iii)}\ \sum_{1}^{\infty} \frac{1}{n^2+1} \qquad \text{(iv)}\ \sum_{0}^{\infty} \frac{1}{(2n+1)!}$$

9. Find the range of values of $x$ for which the series

$$\frac{x}{27} + \frac{x^2}{125} + \ldots + \frac{x^n}{(2n+1)^3} + \ldots$$

is absolutely convergent.

10. Show that the series

$$1 + \frac{x}{1.2} + \frac{x^2}{2.3} + \frac{x^3}{3.4} + \ldots$$

is absolutely convergent when $-1 < x < +1$.

11. Determine the range of values of $x$ for which the following series is convergent

$$\frac{x}{1.2.3} + \frac{x^2}{2.3.4} + \frac{x^3}{3.4.5} + \frac{x^4}{4.5.6} + \ldots$$

12. Find the range of values of $x$ for convergence for the series

$$x + \frac{2^4x^2}{2!} + \frac{3^4x^3}{3!} + \frac{4^4x^4}{4!} + \ldots$$

13. Investigate the convergence of the series

$$\frac{1}{1.2} + \frac{x}{2.3} + \frac{x^2}{3.4} + \frac{x^3}{4.5} + \ldots \text{ for } x > 0$$

14. Show that the following series is convergent

$$2 + \frac{3}{2}\cdot\frac{1}{4} + \frac{4}{3}\cdot\frac{1}{4^2} + \frac{5}{4}\cdot\frac{1}{4^3} + \ldots$$

15. Prove that

$$\frac{1}{\sqrt{1}} + \frac{1}{\sqrt{2}} + \frac{1}{\sqrt{3}} + \frac{1}{\sqrt{4}} + \ldots \text{ is divergent}$$

and that

$$\frac{1}{1^2} + \frac{1}{2^2} + \frac{1}{3^2} + \frac{1}{4^2} + \ldots \text{ is convergent.}$$

16. Determine whether each of the following series is convergent or divergent.

(i) $\Sigma \dfrac{1}{2n(2n+1)}$ (ii) $\Sigma \dfrac{1+3n^2}{1+n^2}$

(iii) $\Sigma \dfrac{n}{\sqrt{(4n^2+1)}}$ (iv) $\Sigma \dfrac{3n+1}{3n^2-2}$

17. Show that the series

$$1+\frac{2x}{5}+\frac{3x^2}{25}+\frac{4x^3}{125}+\ldots \text{ is convergent}$$

if $-5<x<5$ and for no other values of $x$.

18. Investigate the convergence of

$$\text{(i)}\quad 1+\frac{3}{2.4}+\frac{7}{4.9}+\frac{15}{8.16}+\frac{31}{16.25}+\ldots$$

$$\text{(ii)}\quad \frac{1}{1.2}+\frac{1}{2.2^2}+\frac{1}{3.2^3}+\frac{1}{4.2^4}+\ldots$$

19. Find the range of values of $x$ for which the following series is convergent.

$$\frac{(x-2)}{1}+\frac{(x-2)^2}{2}+\frac{(x-2)^3}{3}+\ldots+\frac{(x-2)^n}{n}+\ldots$$

20. If $u_r=r(2r+1)+2^{r+1}$, find the value of $\sum_1^n u_r$.

# Programme 14

## SERIES

### PART 2

**1**

## Power series

*Introduction:* In the first programme (No. 11) on series, we saw how important it is to know something of the convergence properties of any infinite series we may wish to use and to appreciate the conditions in which the series is valid.

This is very important, since it is often convenient to represent a function as a series of ascending powers of the variable. This, in fact, is just how a computer finds the value of the sine of a given angle. Instead of storing the whole of the mathematical tables, it sums up the terms of a series representing the sine of an angle.

That is just one example. There are many occasions when we have need to express a function of $x$ as an infinite series of powers of $x$. It is not at all difficult to express a function in this way, as you will soon see in this programme.

*So make a start and turn on to frame 2.*

**2**

Suppose we wish to express sine $x$ as a series of ascending powers of $x$. The series will be of the form

$$\sin x \equiv a + bx + cx^2 + dx^3 + ex^4 + \ldots$$

where $a, b, c$, etc., are constant coefficients, i.e. numerical factors of some kind. Notice that we have used the 'equivalent' sign and not the usual 'equals' sign. The statement is not an equation: it is an identity. The right-hand side does not *equal* the left-hand side: the R.H.S. *is* the L.H.S. expressed in a different form and the expression is therefore true for any value of $x$ that we like to substitute.

Can you pick out an identity from these?

$$(x + 4)^2 = 3x^2 - 2x + 1$$
$$(2x + 1)^2 = 4x^2 + 4x - 3$$
$$(x + 2)^2 = x^2 + 4x + 4$$

*When you have decided, move on to frame 3.*

**3**

$$\boxed{(x + 2)^2 = x^2 + 4x + 4}$$

*Correct.* This is the only identity of the three, since it is the only one in which the R.H.S. is the L.H.S. written in a different form. Right. Now back to our series:

$$\sin x = a + bx + cx^2 + dx^3 + ex^4 + \ldots$$

To establish the series, we have to find the values of the constant coefficients $a, b, c, d$, etc.

Suppose we substitute $x = 0$ on both sides.

Then $$\sin 0 = a + 0 + 0 + 0 + 0 + \ldots$$

and since $\sin 0 = 0$, we immediately get the value of $a$.

$$a = \ldots\ldots\ldots\ldots\ldots\ldots$$

**4**

$$\boxed{a = 0}$$

Now can we substitute some other value for $x$, which will make all the terms disappear except the second? If we could, we should then find the value of $b$. Unfortunately, we cannot find any such substitution, so what is the next step?

Here is the series once again:

$$\sin x = a + bx + cx^2 + dx^3 + ex^4 + \ldots$$

and so far we know that $a = 0$.

The key to the whole business is simply this:

*Differentiate both sides with respect to x.*

On the left, we get $\cos x$.

On the right the terms are simply powers of $x$, so we get

$$\cos x = \ldots\ldots\ldots\ldots\ldots\ldots$$

**5**

$$\boxed{\cos x = b + c.2x + d.3x^2 + e.4x^3 + \ldots}$$

This is still an identity, so we can substitute in it any value for $x$ we like.

Notice that the $a$ has now disappeared from the scene and that the constant term at the beginning of the expression is now $b$.

So what do you suggest that we substitute in the identity as it now stands, in order that all the terms except the first shall vanish?

We substitute $x =$ .................... again.

**6**

$$\boxed{\text{Substitute } x = 0 \text{ again}}$$

*Right:* for then all the terms will disappear except the first and we shall be able to find $b$.

$$\cos x = b + c.2x + d.3x^2 + e.4x^3 + \ldots$$

Put $x = 0$

$$\therefore \cos 0 = 1 = b + 0 + 0 + 0 + 0 + \ldots$$
$$\therefore b = 1$$

So far, so good. We have found the values of $a$ and $b$. To find $c$ and $d$ and all the rest, we merely repeat the process over and over again at each successive stage.

i.e. *Differentiate both sides with respect to x*

*and substitute* ........................

**7**

substitute $x = 0$

So we now get this, from the beginning:

$$\sin x = a + bx + cx^2 + dx^3 + ex^4 + fx^5 + \ldots$$

Put $x = 0$. $\therefore \sin 0 = 0 = a + 0 + 0 + 0 + \ldots \quad \therefore \underline{a = 0}$

Diff. $\cos x = b + c.2x + d.3x^2 + e.4x^3 + f.5x^4 \ldots$

Put $x = 0$. $\therefore \cos 0 = 1 = b + 0 + 0 + 0 + \ldots \quad \therefore \underline{b = 1}$

Diff. $-\sin x = c.2 + d.3.2x + e.4.3x^2 + f.5.4x^3 \ldots$

Put $x = 0$. $\therefore -\sin 0 = 0 = c.2 + 0 + 0 + \ldots \quad \therefore \underline{c = 0}$

Diff. $-\cos x = d.3.2.1 + e.4.3.2x + f.5.4.3x^2 \ldots$

Put $x = 0$. $\therefore -\cos 0 = -1 = d.3! + 0 + 0 + \ldots \quad \therefore \underline{d = -\frac{1}{3!}}$

And again. $\sin x = e.4.3.2.1 + f.5.4.3.2x + \ldots$

Put $x = 0$. $\therefore \sin 0 = 0 = e.4! + 0 + 0 + \quad \therefore \underline{e = 0}$

Once more. $\cos x = f.5.4.3.2.1 + \ldots$

Put $x = 0$. $\therefore \cos 0 = 1 = f.5! + 0 + \quad \therefore \underline{f = \frac{1}{5!}}$

etc. etc.

All that now remains is to put these values for the constant coefficients back into the original series.

$$\sin x = 0 + 1.x + 0.x^2 + -\frac{1}{3!}x^3 + 0.x^4 + \frac{1}{5!}x^5 + \ldots$$

$$\text{i.e.} \quad \underline{\sin x = x - \frac{x^3}{3!} + \frac{x^5}{5!} - \ldots \ldots}$$

Now we have obtained the first few terms of an infinite series representing the function $\sin x$, and you can see how the terms are likely to proceed.

Write down the first six terms of the series for $\sin x$.

*When you have done so, turn on to frame 8.*

**8**

$$\sin x = x - \frac{x^3}{3!} + \frac{x^5}{5!} - \frac{x^7}{7!} + \frac{x^9}{9!} - \frac{x^{11}}{11!} + \ldots$$

Provided we can differentiate a given function over and over again, and find the values of the derivatives when we put $x = 0$, then this method would enable us to express any function as a series of ascending powers of $x$.

However, it entails a considerable amount of writing, so we now establish a general form of such a series, which can be applied to most functions with very much less effort. This general series is known as *Maclaurin's series.*

*So turn on to frame 9 and we will find out all about it.*

**9**

*Maclaurin's series:* To establish the series, we repeat the process of the previous example, but work with a general function, $f(x)$, instead of $\sin x$. The first differential coefficient of $f(x)$ will be denoted by $f'(x)$; the second by $f''(x)$; the third by $f'''(x)$; and so on. Here it is then:

Let $f(x) = a + bx + cx^2 + dx^3 + ex^4 + fx^5 + \ldots$

Put $x = 0$. Then $f(0) = a + 0 + 0 + 0 + \ldots \quad \therefore \underline{a = f(0).}$

i.e. $a$ = the value of the function with $x$ put equal to 0.

Diff. $f'(x) = \quad b + c.2x + d.3x^2 + e.4x^3 + f.5x^4 + \ldots$

Put $x = 0 \quad \therefore f'(0) = \quad b + 0 + 0 + \ldots \quad \therefore \underline{b = f'(0)}$

Diff. $f''(x) = \quad c.2.1 + d.3.2x + e.4.3x^2 + f.5.4x^3 \ldots$

Put $x = 0 \quad \therefore f''(0) = \quad c.2! + 0 + 0 + \ldots \quad \therefore \underline{c = \frac{f''(0)}{2!}}$

Now go on and find $d$ and $e$, remembering that we denote

$$\frac{d}{dx}\left\{f''(x)\right\} \text{ by } f'''(x)$$

and $$\frac{d}{dx}\left\{f'''(x)\right\} \text{ by } f^{\text{iv}}(x), \text{ etc.}$$

So, $d =$ .................... and $e =$ ....................

**10**

$$d = \frac{f'''(0)}{3!}; \quad e = \frac{f^{iv}(0)}{4!}$$

Here it is. We had:

$$f''(x) = c.2.1 + d.3.2x + e.4.3x^2 + f.5.4x^3 + \ldots$$

$$\begin{cases} \text{Diff.} \ \therefore f'''(x) = & d.3.2.1 + e.4.3.2x + f.5.4.3x^2 + \ldots \\ \text{Put } x = 0 \ \therefore f'''(0) = & d.3! + 0 + 0 \ldots \quad \therefore \underline{d = \frac{f'''(0)}{3!}} \end{cases}$$

$$\begin{cases} \text{Diff.} \ \therefore f^{iv}(x) = & e.4.3.2.1 + f.5.4.3.2x + \ldots \\ \text{Put } x = 0 \ \therefore f^{iv}(0) = & e.4! + 0 + 0 + \ldots \quad \therefore \underline{e = \frac{f^{iv}(0)}{4!}} \end{cases}$$

etc. etc.

So $\quad a = f(0); \ b = f'(0); \ c = \dfrac{f''(0)}{2!}; \ d = \dfrac{f'''(0)}{3!}; \ e = \dfrac{f^{iv}(0)}{4!}; \ \ldots$

Now, in just the same way as we did with our series for $\sin x$, we put the expressions for $a, b, c, \ldots$ etc., back into the original series and get:

$$f(x) = \ldots\ldots\ldots\ldots\ldots\ldots$$

**11**

$$f(x) = f(0) + f'(0).x + \frac{f''(0)}{2!}.x^2 + \frac{f'''(0)}{3!}.x^3 + \ldots$$

and this is usually written as

$$\underline{f(x) = f(0) + x.f'(0) + \frac{x^2}{2!}.f''(0) + \frac{x^3}{3!}.f'''(0) \ \ldots} \qquad \text{I}$$

This is *Maclaurin's series* and important!

Notice how tidy each term is.

The term in $x^2$ is divided by 2! and multiplied by $f''(0)$

" " " $x^3$ " " " 3! " " " $f'''(0)$

" " " $x^4$ " " " 4! " " " $f^{iv}(0)$

Copy the series into your record book for future reference.

*Then on to frame 12.*

## 12 Maclaurin's series

$$f(x) = f(0) + x.\, f'(0) + \frac{x^2}{2!}.f''(0) + \frac{x^3}{3!}.f'''(0) + \ldots$$

□□□□□□□□□□□□□□□□□□□□□□□□□□□□□□□□□□□□□□□□□□□□□

Now we will use Maclaurin's series to find a series for $\sinh x$. We have to find the successive differential coefficients of $\sinh x$ and put $x = 0$ in each. Here goes, then:

$$\begin{aligned} f(x) &= \sinh x & f(0) &= \sinh 0 = 0 \\ f'(x) &= \cosh x & f'(0) &= \cosh 0 = 1 \\ f''(x) &= \sinh x & f''(0) &= \sinh 0 = 0 \\ f'''(x) &= \cosh x & f'''(0) &= \cosh 0 = 1 \\ f^{iv}(x) &= \sinh x & f^{iv}(0) &= \sinh 0 = 0 \\ f^{v}(x) &= \cosh x & f^{v}(0) &= \cosh 0 = 1 \quad \text{etc.} \end{aligned}$$

$$\therefore \sinh x = 0 + x.1 + \frac{x^2}{2!}(0) + \frac{x^3}{3!}.(1) + \frac{x^4}{4!}(0) + \frac{x^5}{5!}.(1) + \ldots$$

$$\therefore \sinh x = x + \frac{x^3}{3!} + \frac{x^5}{5!} + \frac{x^7}{7!} + \ldots$$

*Turn on to frame 13.*

## 13

Now let us find a series for $\ln(1 + x)$ in just the same way.

$$\begin{aligned} f(x) &= \ln(1+x) & \therefore f(0) &= \ldots\ldots \\ f'(x) &= \frac{1}{1+x} = (1+x)^{-1} & \therefore f'(0) &= \ldots\ldots \\ f''(x) &= -(1+x)^{-2} = \frac{-1}{(1+x)^2} & \therefore f''(0) &= \ldots\ldots \\ f'''(x) &= 2(1+x)^{-3} = \frac{2}{(1+x)^3} & \therefore f'''(0) &= \ldots\ldots \\ f^{iv}(x) &= -3.2(1+x)^{-4} = -\frac{3.2}{(1+x)^4} & \therefore f^{iv}(0) &= \ldots\ldots \\ f^{v}(x) &= 4.3.2(1+x)^{-5} = \frac{4!}{(1+x)^5} & \therefore f^{v}(0) &= \ldots\ldots \end{aligned}$$

You complete the work. Evaluate the differentials when $x = 0$, remembering that $\ln 1 = 0$, and substitute back into Maclaurin's series to obtain the series for $\ln(1 + x)$.

So, $\ln(1 + x) = \ldots\ldots\ldots$

**14**

$$f(0) = \ln 1 = 0; \quad f'(0) = 1; \quad f''(0) = -1; \quad f'''(0) = 2;$$

$$f^{iv}(0) = -3!; \quad f^{v}(0) = 4!; \ \ldots$$

Also
$$f(x) = f(0) + x.f'(0) + \frac{x^2}{2!}f''(0) + \frac{x^3}{3!}f'''(0) + \ldots$$

$$\ln(1+x) = 0 + x.1 + \frac{x^2}{2!}(-1) + \frac{x^3}{3!}(2) + \frac{x^4}{4!}(-3!) + \ldots$$

$$\underline{\ln(1+x) = x - \frac{x^2}{2} + \frac{x^3}{3} - \frac{x^4}{4} + \frac{x^5}{5} -}$$

Note that in this series, the denominators are the natural numbers, not factorials!

*Another example in frame 15.*

**15**

*Example:* Expand $\sin^2 x$ as a series of ascending powers of $x$.

Maclaurin's series:

$$f(x) = f(0) + x.f'(0) + \frac{x^2}{2!}.f''(0) + \frac{x^3}{3!}.f'''(0) + \ldots$$

| | | |
|---|---|---|
| $\therefore f(x) = \sin^2 x$ | $f(0)$ | = .................. |
| $f'(x) = 2\sin x \cos x = \sin 2x$ | $f'(0)$ | = .................. |
| $f''(x) = 2\cos 2x$ | $f''(0)$ | = .................. |
| $f'''(x) = -4\sin 2x$ | $f'''(0)$ | = .................. |
| $f^{iv}(x) =$ ................ | $f^{iv}(0)$ | = .................. |

There we are! Finish it off: find the first three non-vanishing terms of the series.

*Then move on to frame 16.*

## 16

$$\sin^2 x = x^2 - \frac{x^4}{3} + \frac{2}{45}x^6 \ldots$$

For

$$f(x) = \sin^2 x \qquad \therefore f(0) = 0$$
$$f'(x) = 2\sin x \cos x = \sin 2x \qquad \therefore f'(0) = 0$$
$$f''(x) = 2\cos 2x \qquad \therefore f''(0) = 2$$
$$f'''(x) = -4\sin 2x \qquad \therefore f'''(0) = 0$$
$$f^{iv}(x) = -8\cos 2x \qquad \therefore f^{iv}(0) = -8$$
$$f^{v}(x) = 16\sin 2x \qquad \therefore f^{v}(0) = 0$$
$$f^{vi}(x) = 32\cos 2x \qquad \therefore f^{vi}(0) = 32 \qquad \text{etc.}$$

$$f(x) = f(0) + x.f'(0) + \frac{x^2}{2!}.f''(0) + \frac{x^3}{3!}.f'''(0) + \ldots$$

$$\therefore \sin^2 x = 0 + x(0) + \frac{x^2}{2!}(2) + \frac{x^3}{3!}(0) + \frac{x^4}{4!}(-8) + \frac{x^5}{5!}(0) + \frac{x^6}{6!}(32)$$

$$\therefore \underline{\sin^2 x = x^2 - \frac{x^4}{3} + \frac{2x^6}{45}} \ldots$$

Next we will find the series for $\tan x$. This is a little heavier but the method is always the same.

*Move to frame 17.*

## 17

*Series for tan x*

$$f(x) = \tan x \qquad \therefore f(0) = 0$$
$$\therefore f'(x) = \sec^2 x \qquad \therefore f'(0) = 1$$
$$\therefore f''(x) = 2\sec^2 x \tan x \qquad \therefore f''(0) = 0$$
$$\therefore f'''(x) = 2\sec^4 x + 4\sec^2 x \tan^2 x \qquad \therefore f'''(0) = 2$$
$$= 2\sec^4 x + 4(1 + \tan^2 x)\tan^2 x$$
$$= 2\sec^4 x + 4\tan^2 x + 4\tan^4 x$$
$$\therefore f^{iv}(x) = 8\sec^4 x \tan x + 8\tan x \sec^2 x + 16\tan^3 x \sec^2 x$$
$$= 8(1 + t^2)^2 t + 8t(1 + t^2) + 16t^3(1 + t^2)$$
$$= 8(1 + 2t^2 + t^4)t + 8t + 8t^3 + 16t^3 + 16t^5$$
$$= 16t + 40t^3 + 24t^5 \qquad \therefore f^{iv}(0) = 0$$
$$\therefore f^{v}(x) = 16\sec^2 x + 120t^2.\sec^2 x + 120t^4 \sec^2 x$$
$$\therefore f^{v}(0) = 16$$

$$\therefore \tan x = \ldots\ldots\ldots\ldots\ldots$$

**18**

$$\therefore \tan x = x + \frac{x^3}{3} + \frac{2x^5}{15} + \ldots$$

**Standard series**

By Maclaurin's series, we can build up a list of series representing many of the common functions – we have already found series for $\sin x$, $\sinh x$ and $\ln(1+x)$.

To find a series for $\cos x$, we could apply the same technique all over again. However, let us be crafty about it. Suppose we take the series for $\sin x$ and differentiate both sides with respect to $x$ just once, we get

$$\sin x = x - \frac{x^3}{3!} + \frac{x^5}{5!} - \frac{x^7}{7!} + \ldots$$

$$\text{Diff.} \quad \cos x = 1 - \frac{3x^2}{3!} + \frac{5x^4}{5!} - \frac{7x^6}{7!} \ \ldots \text{ etc.}$$

$$\therefore \cos x = 1 - \frac{x^2}{2!} + \frac{x^4}{4!} - \frac{x^6}{6!} + \ldots$$

In the same way, we can obtain the series for $\cosh x$. We already know that

$$\sinh x = x + \frac{x^3}{3!} + \frac{x^5}{5!} + \frac{x^7}{7!} + \ldots$$

so if we differentiate both sides we shall establish a series for $\cosh x$.

What do we get?

**19**

We get:

$$\sinh x = x + \frac{x^3}{3!} + \frac{x^5}{5!} + \frac{x^7}{7!} + \frac{x^9}{9!} + \ldots$$

Diff.

$$\cosh x = 1 + \frac{3x^2}{3!} + \frac{5x^4}{5!} + \frac{7x^6}{7!} + \frac{9x^8}{9!} \ldots$$

giving:

$$\cosh x = 1 + \frac{x^2}{2!} + \frac{x^4}{4!} + \frac{x^6}{6!} + \frac{x^8}{8!} + \ldots$$

*Let us pause at this point and take stock of the series we have obtained. We will make a list of them, so turn on to frame 20.*

**20** *Summary*

Here are the standard series that we have established so far.

$$\sin x = x - \frac{x^3}{3!} + \frac{x^5}{5!} - \frac{x^7}{7!} + \frac{x^9}{9!} \ldots \qquad \text{II}$$

$$\cos x = 1 - \frac{x^2}{2!} + \frac{x^4}{4!} - \frac{x^6}{6!} + \frac{x^8}{8!} \ldots \qquad \text{III}$$

$$\tan x = x + \frac{x^3}{3} + \frac{2x^5}{15} + \ldots \qquad \text{IV}$$

$$\sinh x = x + \frac{x^3}{3!} + \frac{x^5}{5!} + \frac{x^7}{7!} + \ldots \qquad \text{V}$$

$$\cosh x = 1 + \frac{x^2}{2!} + \frac{x^4}{4!} + \frac{x^6}{6!} + \frac{x^8}{8!} \ldots \qquad \text{VI}$$

$$\ln(1+x) = x - \frac{x^2}{2} + \frac{x^3}{3} - \frac{x^4}{4} + \frac{x^5}{5} \ldots \qquad \text{VII}$$

Make a note of these six series in your record book.

*Then turn on to frame 21.*

**21** **The binomial series**

By the same method, we can apply Maclaurin's series to obtain a power series for $(1+x)^n$. Here it is:

$f(x) = (1+x)^n$ $\qquad f(0) = 1$

$f'(x) = n.(1+x)^{n-1}$ $\qquad f'(0) = n$

$f''(x) = n(n-1).(1+x)^{n-2}$ $\qquad f''(0) = n(n-1)$

$f'''(x) = n(n-1)(n-2).(1+x)^{n-3}$ $\qquad f'''(0) = n(n-1)(n-2)$

$f^{iv}(x) = n(n-1)(n-2)(n-3).(1+x)^{n-4}$ $\qquad f^{iv}(0) = n(n-1)(n-2)(n-3)$

etc. $\qquad$ etc.

General Maclaurin's series:

$$f(x) = f(0) + x.f'(0) + \frac{x^2}{2!}f''(0) + \frac{x^3}{3!}f'''(0) \ldots$$

Therefore, in this case,

$$(1+x)^n = 1 + xn + \frac{x^2}{2!}n(n-1) + \frac{x^3}{3!}n(n-1)(n-2) \ldots$$

$$(1+x)^n = 1 + nx + \frac{n(n-1)}{2!}x^2 + \frac{n(n-1)(n-2)}{3!}x^3 \ldots \qquad \text{VIII}$$

Add this result to your list of series in your record book. Then, by replacing $x$ wherever it occurs by $(-x)$, determine the series for $(1-x)^n$.

*When finished, turn to frame 22.*

**22**

$$(1-x)^n = 1 - nx + \frac{n(n-1)}{2!}x^2 - \frac{n(n-1)(n-2)}{3!}x^3 + \ldots$$

□□□□□□□□□□□□□□□□□□□□□□□□□□□□□□□□□□□□□□

Now we will work through another example. Here it is:

*Example:* To find a series for $\tan^{-1}x$.

As before, we need to know the successive differential coefficients in order to insert them in Maclaurin's series.

$$f(x) = \tan^{-1}x \text{ and } f'(x) = \frac{1}{1+x^2}$$

If we differentiate again, we get $f''(x) = -\frac{2x}{(1+x^2)^2}$, after which the working becomes rather heavy, so let us be crafty and see if we can avoid unnecessary work.

We have $f(x) = \tan^{-1}x$ and $f'(x) = \frac{1}{1+x^2} = (1+x^2)^{-1}$. If we now expand $(1+x^2)^{-1}$ as a binomial series, we shall have a series of powers of $x$ from which we can easily find the higher differential coefficients.

*So see how it works out in the next frame.*

**23**

*To find a series for* $\tan^{-1}x$

$$f(x) = \tan^{-1}x \qquad \therefore f(0) = 0$$

$$\therefore f'(x) = \frac{1}{1+x^2} = (1+x^2)^{-1}$$

$$= 1 - x^2 + \frac{(-1)(-2)}{1.2}x^4 + \frac{(-1)(-2)(-3)}{1.2.3}x^6 + \ldots$$

$$= 1 - x^2 + x^4 - x^6 + x^8 - \ldots \qquad f'(0) = 1$$

$$\therefore f''(x) = -2x + 4x^3 - 6x^5 + 8x^7 - \ldots \qquad f''(0) = 0$$

$$\therefore f'''(x) = -2 + 12x^2 - 30x^4 + 56x^6 - \ldots \qquad f'''(0) = -2$$

$$\therefore f^{iv}(x) = 24x - 120x^3 + 336x^5 - \ldots \qquad f^{iv}(0) = 0$$

$$\therefore f^{v}(x) = 24 - 360x^2 + 1680x^4 - \ldots \qquad f^{v}(0) = 24 \quad \text{etc.}$$

$$\therefore \tan^{-1}x = f(0) + x.f'(0) + \frac{x^2}{2!}f''(0) + \frac{x^3}{3!}f'''(0) + \ldots$$

Substituting the values for the derivatives, gives us that $\tan^{-1}x =$ ...............

*Then on to frame 24.*

**24**

$$\tan^{-1}x = 0 + x(1) + \frac{x^2}{2!}(0) + \frac{x^3}{3!}(-2) + \frac{x^4}{4!}(0) + \frac{x^5}{5!}(24)\ldots$$

$$\boxed{\tan^{-1}x = x - \frac{x^3}{3} + \frac{x^5}{5} - \frac{x^7}{7} + \ldots} \qquad \text{X}$$

This is also a useful series, so make a note of it.

□□□□□□□□□□□□□□□□□□□□□□□□□□□□□□□□□□□□□□□□□

Another series which you already know quite well is the series for $e^x$. Do you remember how it goes? Here it is anyway.

$$e^x = 1 + x + \frac{x^2}{2!} + \frac{x^3}{3!} + \frac{x^4}{4!} + \ldots \qquad \text{XI}$$

and if we simply replace $x$ by $(-x)$, we obtain the series for $e^{-x}$.

$$e^{-x} = 1 - x + \frac{x^2}{2!} - \frac{x^3}{3!} + \frac{x^4}{4!}\ldots \qquad \text{XII}$$

So now we have quite a few. Add the last two to your list.

*And then on to the next frame.*

**25**

*Examples:* Once we have established these standard series, we can of course, combine them as necessary.

*Example 1.* Find the first three terms of the series for $e^x.\ln(1 + x)$.

We know that $$e^x = 1 + x + \frac{x^2}{2!} + \frac{x^3}{3!} + \frac{x^4}{4!} + \ldots$$

and that $$\ln(1 + x) = x - \frac{x^2}{2} + \frac{x^3}{3} - \frac{x^4}{4} + \ldots$$

$$e^x.\ln(1 + x) = \left\{1 + x + \frac{x^2}{2!} + \frac{x^3}{3!} + \frac{x^4}{4!}\ldots\right\}\left\{x - \frac{x^2}{2} + \frac{x^3}{3} - \ldots\right\}$$

Now we have to multiply these series together. There is no constant term in the second series, so the lowest power of $x$ in the product will be $x$ itself. This can only be formed by multiplying the 1 in the first series by the $x$ in the second.

The $x^2$ term is found by multiplying $1 \times \left(-\frac{x^2}{2}\right)$ and $x \times x$ $\Big\}$ $x^2 - \frac{x^2}{2} = \frac{x^2}{2}$

The $x^3$ term is found by multiplying $1 \times \frac{x^3}{3}$ and $x \times \left(-\frac{x^2}{2}\right)$ and $\frac{x^2}{2} \times x$ $\Big\}$ $\frac{x^3}{3} - \frac{x^3}{2} + \frac{x^3}{2} = \frac{x^3}{3}$ and so on.

**26**

$$\therefore\ e^x.\ln(1+x) = x + \frac{x^2}{2} + \frac{x^3}{3} + \ldots$$

It is not at all difficult, provided you are careful to avoid missing any of the products of the terms.

□□□□□□□□□□□□□□□□□□□□□□□□□□□□□□□□□□□□□□

Here is one for you to do in the same way:

*Example 2.* Find the first four terms of the series for $e^x \sinh x$.

*Take your time over it: then check your working with that in frame 27.*

**27**

Here is the solution. Look through it carefully to see if you agree with the result.

$$e^x = 1 + x + \frac{x^2}{2!} + \frac{x^3}{3!} + \frac{x^4}{4!} + \ldots$$

$$\sinh x = x + \frac{x^3}{3!} + \frac{x^5}{5!} + \frac{x^7}{7!} + \ldots$$

$$e^x.\sinh x = \left\{1 + x + \frac{x^2}{2!} + \frac{x^3}{3!}\ldots\right\}\left\{x + \frac{x^3}{3!} + \frac{x^5}{5!} + \ldots\right\}$$

Lowest power is $x$

$$\text{Term in } x = 1.x = x$$

$$\text{" " } x^2 = x.x = x^2$$

$$\text{" " } x^3 = 1.\frac{x^3}{3!} + \frac{x^2}{2!}.x = x^3\left(\frac{1}{6} + \frac{1}{2}\right) = \frac{2x^3}{3}$$

$$\text{" " } x^4 = x.\frac{x^3}{3!} + \frac{x^3}{3!}.x = x^4\left(\frac{1}{6} + \frac{1}{6}\right) = \frac{x^4}{3}$$

$$\therefore\ e^x.\sinh x = x + x^2 + \frac{2x^3}{3} + \frac{x^4}{3} + \ldots$$

*There we are. Now turn on to frame 28.*

## 28

**Approximate values**

This is a very obvious application of series and you will surely have done some examples on this topic some time in the past. Here is just an example or two to refresh your memory.

*Example 1.* Evaluate $\sqrt{1{\cdot}02}$ correct to 5 decimal places.

$$1{\cdot}02 = 1 + 0{\cdot}02$$

$$\sqrt{1{\cdot}02} = (1 + 0{\cdot}02)^{1/2}$$

$$= 1 + \frac{1}{2}(0{\cdot}02) + \frac{\frac{1}{2}(-\frac{1}{2})}{1.2}(0{\cdot}02)^2 + \frac{\frac{1}{2}(-\frac{1}{2})(-\frac{3}{2})}{1.2.3}(0{\cdot}02)^2 \ldots$$

$$= 1 + 0{\cdot}01 - \frac{1}{8}(0{\cdot}0004) + \frac{1}{16}(0{\cdot}000008) - \ldots$$

$$= 1 + 0{\cdot}01 - 0{\cdot}00005 + 0{\cdot}0000005 \ldots$$

$$= 1{\cdot}010001 - 0{\cdot}000050$$

$$= 1{\cdot}009951 \quad \therefore \underline{\sqrt{1{\cdot}02} = 1{\cdot}00995}$$

*Note* that whenever we substitute a value for $x$ in any one of the standard series, we must be satisfied that the substitution value for $x$ is within the range of values of $x$ for which the series is valid.

The present series for $(1 + x)^n$ is valid for $|x| < 1$, so we are safe enough on this occasion.

Here is one for you to do.

*Example 2.* Evaluate $\tan^{-1} 0{\cdot}1$ correct to 4 decimal places.

*Complete the working and then check with the next frame.*

## 29

$$\boxed{\tan^{-1} 0{\cdot}1 = 0{\cdot}0997}$$

$$\tan^{-1} x = x - \frac{x^3}{3} + \frac{x^5}{5} - \frac{x^7}{7} + \ldots$$

$$\therefore \tan^{-1} 0{\cdot}1 = 0{\cdot}1 - \frac{0{\cdot}001}{3} + \frac{0{\cdot}00001}{5} - \frac{0{\cdot}0000001}{7} \ldots$$

$$= 0{\cdot}1 - 0{\cdot}00033 + 0{\cdot}000002 - \ldots$$

$$= \underline{0{\cdot}0997}$$

*We will now consider a further use for series, so* *turn now to frame 30.*

**30**

**Limiting values** – *Indeterminate forms*

In Part I of this programme on series, we had occasion to find the limiting value of $\frac{u_{n+1}}{u_n}$ as $n \to \infty$. Sometimes, we have to find the limiting value of a function of $x$ when $x \to 0$, or perhaps when $x \to a$.

$$\text{e.g.} \quad \lim_{x \to 0}\left\{\frac{x^2 + 5x - 14}{x^2 - 5x + 8}\right\} = \frac{0 + 0 - 14}{0 - 0 + 8} = -\frac{14}{8} = -\frac{7}{4}$$

That is easy enough, but suppose we have to find

$$\lim_{x \to 2}\left\{\frac{x^2 + 5x - 14}{x^2 - 5x + 6}\right\}$$

Putting $x = 2$ in the function, gives $\frac{4 + 10 - 14}{4 - 10 + 6} = \frac{0}{0}$ and what is the value of $\frac{0}{0}$?

Is it zero? Is it 1? Is it indeterminate?

*When you have decided, turn on to frame 31.*

**31**

$\frac{0}{0}$, as it stands, is **indeterminate**

We can sometimes, however, use our knowledge of series to help us out of the difficulty. Let us consider an example or two.

*Example 1.* Find the $\lim_{x \to 0}\left\{\frac{\tan x - x}{x^3}\right\}$

If we just substitute $x = 0$ in the function, we get the result $\frac{0}{0}$ which is indeterminate. So how do we proceed?

Well, we already know that $\tan x = x + \frac{x^3}{3} + \frac{2x^5}{15} + \dots$ So if we replace $\tan x$ by its series in the given function, we get

$$\lim_{x \to 0}\left\{\frac{\tan x - x}{x^3}\right\} = \lim_{x \to 0}\left\{\frac{(\not{x} + \frac{x^3}{3} + \frac{2x^5}{15} + \dots) - \not{x}}{x^3}\right\}$$

$$= \lim_{x \to 0}\left\{\frac{1}{3} + \frac{2x^2}{15} + \dots\right\} = \frac{1}{3}$$

$$\therefore \lim_{x \to 0}\left\{\frac{\tan x - x}{x^3}\right\} = \underline{\frac{1}{3}} \quad \text{– and the job is done!}$$

*Move on to frame 32 for another example.*

**32** *Example 2.* To find $\lim_{x\to 0}\left\{\frac{\sinh x}{x}\right\}$

Direct substitution of $x = 0$ gives $\frac{\sinh 0}{0}$ which is $\frac{0}{0}$ again. So we will express $\sinh x$ by its series, which is

$$\sinh x = \ldots\ldots\ldots\ldots$$

(If you do not remember, you will find it in your list of standard series which you have been compiling. Look it up.)

*Then on to frame 33.*

---

**33**

$$\boxed{\sinh x = x + \frac{x^3}{3!} + \frac{x^5}{5!} + \frac{x^7}{7!} + \ldots}$$

So
$$\lim_{x\to 0}\left\{\frac{\sinh x}{x}\right\} = \lim_{x\to 0}\left\{\frac{x + \frac{x^3}{3!} + \frac{x^5}{5!} + \frac{x^7}{7!}\ldots}{x}\right\}$$

$$= \lim_{x\to 0}\left\{1 + \frac{x^2}{3!} + \frac{x^4}{5!} + \ldots\right\}$$

$$= 1 + 0 + 0 + \ldots = 1$$

$$\therefore \lim_{x\to 0}\left\{\frac{\sinh x}{x}\right\} = 1$$

Now, in very much the same way, you find $\lim_{x\to 0}\left\{\frac{\sin^2 x}{x^2}\right\}$

*Work it through: then check your result with that in the next frame.*

---

**34**

$$\boxed{\lim_{x\to 0}\left\{\frac{\sin^2 x}{x^2}\right\} = 1}$$

Here is the working:

$$\lim_{x\to 0}\left\{\frac{\sin^2 x}{x^2}\right\} = \lim_{x\to 0}\left\{\frac{x^2 - \frac{x^4}{3} + \frac{2x^6}{45} - \ldots}{x^2}\right\}$$

$$= \lim_{x\to 0}\left\{1 - \frac{x^2}{3} + \frac{2x^4}{45}\ \ldots\right\} = 1$$

$$\therefore \lim_{x\to 0}\left\{\frac{\sin^2 x}{x^2}\right\} = 1$$

Here is one more for you to do in like manner.

Find $\lim_{x\to 0}\left\{\frac{\sinh x - x}{x^3}\right\}$

*Then on to frame 35.*

**35**

$$\lim_{x \to 0} \left\{ \frac{\sinh x - x}{x^3} \right\} = \frac{1}{6}$$

Here is the working in detail:

$$\sinh x = x + \frac{x^3}{3!} + \frac{x^5}{5!} + \frac{x^7}{7!} + \ldots$$

$$\therefore \frac{\sinh x - x}{x^3} = \frac{\cancel{x} + \frac{x^3}{3!} + \frac{x^5}{5!} + \frac{x^7}{7!} + \ldots - \cancel{x}}{x^3}$$

$$= \frac{1}{3!} + \frac{x^2}{5!} + \frac{x^4}{7!} + \ldots$$

$$\therefore \lim_{x \to 0} \left\{ \frac{\sinh x - x}{x^3} \right\} = \lim_{x \to 0} \left\{ \frac{1}{3!} + \frac{x^2}{5!} + \frac{x^4}{7!} + \ldots \right\}$$

$$= \frac{1}{3!} = \frac{1}{6}$$

$$\therefore \lim_{x \to 0} \left\{ \frac{\sinh x - x}{x^3} \right\} = \frac{1}{6}$$

So there you are: they are all done the same way.

(i) Express the given function in terms of power series
(ii) Simplify the function as far as possible
(iii) Then determine the limiting value – which should now be possible.

□ □ □ □ □ □ □ □ □ □ □ □ □ □ □ □ □ □ □ □ □ □ □ □ □ □ □ □ □ □ □ □ □ □ □ □ □ □

Of course, there may well be occasions when direct substitution gives the indeterminate form $\frac{0}{0}$ and when we do not know the series expansion of the function concerned. What are we going to do then?

All is not lost! – for we do in fact have another method of finding limiting values which, in many cases, is quicker than the series method. It all depends upon the application of a rule which we must first establish, so turn to the next frame for details thereof.

**36** *L'Hopital's rule* for finding limiting values.

Suppose we have to find the limiting value of a function $F(x) = \dfrac{f(x)}{g(x)}$ at $x = a$, when direct substitution of $x = a$ gives the indeterminate form $\dfrac{0}{0}$, i.e. at $x = a$, $f(x) = 0$ and $g(x) = 0$.

If we represent the circumstances graphically, the diagram would look like this:–

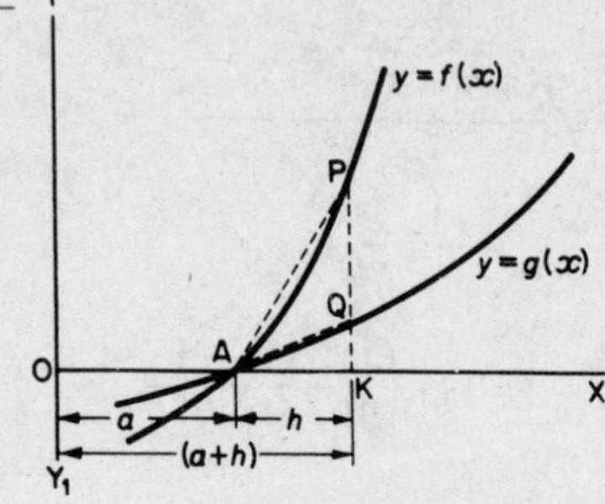


Note that at $x = a$, both of the graphs $y = f(x)$ and $y = g(x)$ cross the $x$-axis, so that at $x = a$, $f(x) =$
and $g(x) =$

At a point K, i.e. $x = (a + h)$, $KP = f(a + h)$ and $KQ = g(a + h)$

$$\frac{f(a+h)}{g(a+h)} = \frac{KP}{KQ}$$

Now divide top and bottom by AK

$$\frac{f(a+h)}{g(a+h)} = \frac{KP/AK}{KQ/AK} = \frac{\tan PAK}{\tan QAK}$$

Now $$\lim_{x \to a} \frac{f(x)}{g(x)} = \lim_{h \to 0} \frac{f(a+h)}{g(a+h)} = \lim_{h \to 0} \frac{\tan PAK}{\tan QAK} = \frac{f'(a)}{g'(a)}$$

i.e. the limiting value of $\dfrac{f(x)}{g(x)}$ as $x \to a$ (at which the function value by direct substitution gives $\dfrac{0}{0}$) is given by the ratio of the differential coefficients of numerator and denominator at $x = a$ (provided, of course, that both $f'(a)$ and $g'(a)$ are not zero themselves)!

$$\therefore \lim_{x \to a} \left\{ \frac{f(x)}{g(x)} \right\} = \frac{f'(a)}{g'(a)} = \lim_{x \to a} \left\{ \frac{f'(x)}{g'(x)} \right\}$$

$$\therefore \lim_{x \to a} \left\{ \frac{f(x)}{g(x)} \right\} = \lim_{x \to a} \left\{ \frac{f'(x)}{g'(x)} \right\}$$

This is known as *l'Hopital's rule* and is extremely useful for finding limiting values when the differential coefficients of the numerator and denominator can easily be found.

*Copy the rule into your record book. Now we will use it.*

37

$$\lim_{x \to a}\left\{\frac{f(x)}{g(x)}\right\} = \lim_{x \to a}\left\{\frac{f'(x)}{g'(x)}\right\}$$

*Example 1.* To find $\lim_{x \to 1}\left\{\frac{x^3 + x^2 - x - 1}{x^2 + 2x - 3}\right\}$

Note first that if we substitute $x = 1$, we get the indeterminate form $\frac{0}{0}$. Therefore we will apply l'Hopital's rule.

We therefore differentiate numerator and denominator separately (*not* as a quotient).

$$\lim_{x \to 1}\left\{\frac{x^3 + x^2 - x - 1}{x^2 + 2x - 3}\right\} = \lim_{x \to 1}\left\{\frac{3x^2 + 2x - 1}{2x + 2}\right\}$$

$$= \frac{3 + 2 - 1}{2 + 2} = \frac{4}{4} = 1$$

$$\therefore \lim_{x \to 1}\left\{\frac{x^3 + x^2 - x - 1}{x^2 - 2x - 3}\right\} = 1$$

and that is all there is to it!

*Let us do another example, so, on to the next frame.*

38

*Example 2.* Determine $\lim_{x \to 0}\left\{\frac{\cosh x - e^x}{x}\right\}$

We first of all try direct substitution, but we find that this leads us to the result $\frac{1 - 1}{0}$, i.e. $\frac{0}{0}$ which is indeterminate. Therefore, apply l'Hopital's rule

$$\lim_{x \to a}\left\{\frac{f(x)}{g(x)}\right\} = \lim_{x \to a}\left\{\frac{f'(x)}{g'(x)}\right\}$$

i.e. differentiate top and bottom *separately* and substitute the given value of $x$ in the differential coefficients.

$$\therefore \lim_{x \to 0}\left\{\frac{\cosh x - e^x}{x}\right\} = \lim_{x \to 0}\left\{\frac{\sinh x - e^x}{1}\right\}$$

$$= \frac{0 - 1}{1} = -1$$

$$\therefore \lim_{x \to 0}\left\{\frac{\cosh x - e^x}{x}\right\} = -1$$

Now you can do this one:

Determine $\lim_{x \to 0}\left\{\frac{x^2 - \sin 3x}{x^2 + 4x}\right\}$

**39**

$$\boxed{\lim_{x \to 0}\left\{\frac{x^2 - \sin 3x}{x^2 + 4x}\right\} = -\frac{3}{4}}$$

The working is simply this:

Direct substitution gives $\frac{0}{0}$, so we apply l'Hopital's rule which gives

$$\lim_{x \to 0}\left\{\frac{x^2 - \sin 3x}{x^2 + 4x}\right\} = \lim_{x \to 0}\left\{\frac{2x - 3\cos 3x}{2x + 4}\right\}$$

$$= \frac{0 - 3}{0 + 4} = -\frac{3}{4}$$

*WARNING:* l'Hopital's rule applies only when the indeterminate form arises. If the limiting value can be found by direct substitution, the rule will not work. An example will soon show this.

Consider $\quad \lim_{x \to 2}\left\{\frac{x^2 + 4x - 3}{5 - 2x}\right\}$

By direct substitution, the limiting value $= \frac{4 + 8 - 3}{5 - 4} = 9$. By l'Hopital's rule $\lim_{x \to 2}\left\{\frac{x^2 + 4x - 3}{5 - 2x}\right\} = \lim_{x \to 2}\left\{\frac{2x + 4}{-2}\right\} = -4$. As you will see, these results do not agree.

*Before using l'Hopital's rule, therefore, you must satisfy yourself that direct substitution gives the indeterminate form* $\frac{0}{0}$. If it does, you may use the rule, but not otherwise.

**40**

Let us look at another example

*Example:* Determine $\lim_{x \to 0}\left\{\frac{x - \sin x}{x^2}\right\}$

By direct substitution, limiting value $= \frac{0 - 0}{0} = \frac{0}{0}$.

Apply l'Hopital's rule:

$$\lim_{x \to 0}\left\{\frac{x - \sin x}{x^2}\right\} = \lim_{x \to 0}\left\{\frac{1 - \cos x}{2x}\right\}$$

We now find, with some horror, that substituting $x = 0$ in the differential coefficients, again produces the indeterminate form $\frac{0}{0}$. So what do you suggest we do now to find $\lim_{x \to 0}\left\{\frac{1 - \cos x}{2x}\right\}$, (without bringing in the use of series)? Any ideas?

We ......................

**41**

We apply the rule a second time.

Correct, for our immediate problem now is to find $\lim_{x \to 0}\left\{\frac{1-\cos x}{2x}\right\}$ If we do that, we get:

$$\underbrace{\lim_{x \to 0}\left\{\frac{x-\sin x}{x^2}\right\} = \lim_{x \to 0}\left\{\frac{1-\cos x}{2x}\right\}}_{\text{First stage}} = \lim_{x \to 0}\left\{\frac{\sin x}{2}\right\} = \frac{0}{2} = 0$$

(Second stage: $\lim_{x \to 0}\left\{\frac{1-\cos x}{2x}\right\} = \lim_{x \to 0}\left\{\frac{\sin x}{2}\right\}$)

$$\therefore \lim_{x \to 0}\left\{\frac{x-\sin x}{x^2}\right\} = 0$$

So now we have the rule complete:

For limiting values when the indeterminate form (i.e. $\frac{0}{0}$) exists, apply l'Hopital's rule

$$\lim_{x \to a}\left\{\frac{f(x)}{g(x)}\right\} = \lim_{x \to a}\left\{\frac{f'(x)}{g'(x)}\right\}$$

and continue to do so until a stage is reached where either the numerator and/or the denominator is not zero.

*Next frame.*

**42**

Just one more example to illustrate the point.

*Example:* Determine $\lim_{x \to 0}\left\{\frac{\sinh x - \sin x}{x^3}\right\}$

Direct substitution gives $\frac{0-0}{0}$, i.e. $\frac{0}{0}$. (indeterminate)

$$\therefore \lim_{x \to 0}\left\{\frac{\sinh x - \sin x}{x^3}\right\} = \lim_{x \to 0}\left\{\frac{\cosh x - \cos x}{3x^2}\right\}, \text{ gives } \frac{1-1}{0} = \frac{0}{0}$$

$$= \lim_{x \to 0}\left\{\frac{\sinh x + \sin x}{6x}\right\}, \text{ gives } \frac{0+0}{0} = \frac{0}{0}$$

$$= \lim_{x \to 0}\left\{\frac{\cosh x + \cos x}{6}\right\} = \frac{1+1}{6} = \frac{1}{3}$$

$$\therefore \lim_{x \to 0}\left\{\frac{\sinh x - \sin x}{x^3}\right\} = \frac{1}{3}$$

*Note* that we apply l'Hopital's rule again and again until we reach the stage where the numerator or the denominator (or both) is *not* zero. We shall then arrive at a definite limiting value of the function.

*Turn on to frame 43.*

**43**

Here are three *Revision Examples* for you to do. Work through all of them and then check your working with the results set out in the next frame. They are all straightforward and easy, so do not peep at the official solutions before you have done them all.

Determine (i) $\lim_{x\to 1}\left\{\frac{x^3-2x^2+4x-3}{4x^2-5x+1}\right\}$

(ii) $\lim_{x\to 0}\left\{\frac{\tan x - x}{\sin x - x}\right\}$ (iii) $\lim_{x\to 0}\left\{\frac{x\cos x - \sin x}{x^3}\right\}$

**44**

*Solutions:*

(i) $\lim_{x\to 1}\left\{\frac{x^3-2x^2+4x-3}{4x^2-5x+1}\right\}$ (Substitution gives $\frac{0}{0}$)

$$= \lim_{x\to 1}\left\{\frac{3x^2-4x+4}{8x-5}\right\} = \frac{3}{3} = 1$$

$$\therefore \lim_{x\to 1}\left\{\frac{x^3-2x^2+4x-3}{4x^2-5x+1}\right\} = 1$$

(ii) $\lim_{x\to 0}\left\{\frac{\tan x - x}{\sin x - x}\right\}$ (Substitution gives $\frac{0}{0}$)

$$= \lim_{x\to 0}\left\{\frac{\sec^2 x - 1}{\cos x - 1}\right\}$$ (still gives $\frac{0}{0}$)

$$= \lim_{x\to 0}\left\{\frac{2\sec^2 x \tan x}{-\sin x}\right\}$$ (and again!)

$$= \lim_{x\to 0}\left\{\frac{2\sec^2 x \sec^2 x + 4\sec^2 x \tan^2 x}{-\cos x}\right\} = \frac{2+0}{-1} = -2$$

$$\therefore \lim_{x\to 0}\left\{\frac{\tan x - x}{\sin x - x}\right\} = -2$$

(iii) $\lim_{x\to 0}\left\{\frac{x\cos x - \sin x}{x^3}\right\}$ (Substitution gives $\frac{0}{0}$)

$$= \lim_{x\to 0}\left\{\frac{-x\sin x + \cos x - \cos x}{3x^2}\right\}$$

$$= \lim_{x\to 0}\left\{\frac{-\sin x}{3x}\right\} = \lim_{x\to 0}\left\{\frac{-\cos x}{3}\right\} = -\frac{1}{3}$$

$$\therefore \lim_{x\to 0}\left\{\frac{x\cos x - \sin x}{x^3}\right\} = -\frac{1}{3}$$

*Next frame.*

**45**

Let us look at another useful series: **Taylor's series.**

Maclaurin's series $f(x) = f(0) + x.f'(0) + \frac{x^2}{2!}f''(0) + \ldots$ expresses a function in terms of its differential coefficients at $x = 0$, i.e. at the point K.

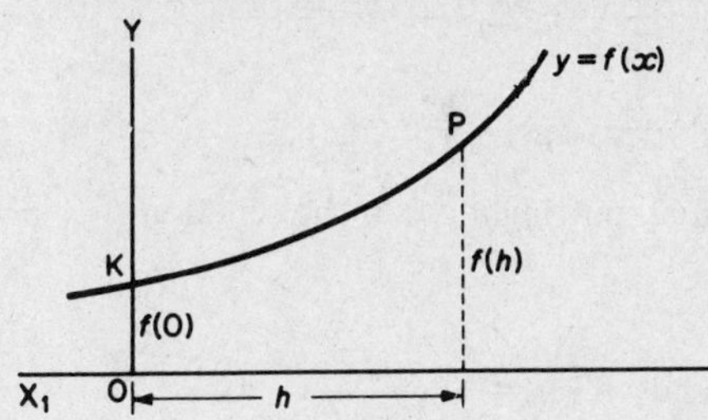


$$\text{At P, } f(h) = f(0) + h.f'(0) + \frac{h^2}{2!}f''(0) + \frac{h^3}{3!}f'''(0) \ldots$$

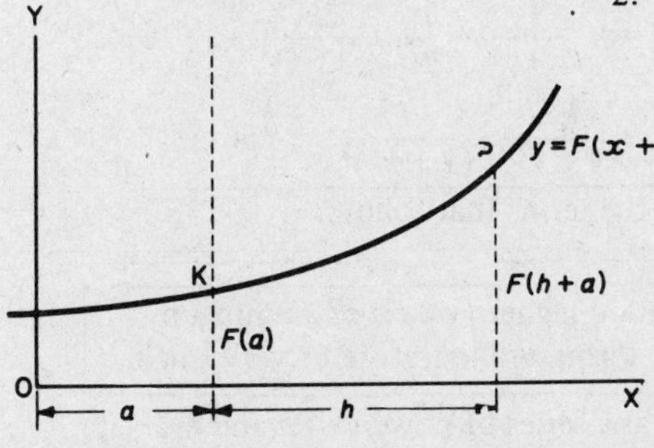


If we now move the $y$-axis $a$ units to the left, the equation of the curve relative to the new axes now becomes $y = F(a + x)$ and the value at K is now $F(a)$

$$\text{At P, } F(a + h) = F(a) + h.F'(a) + \frac{h^2}{2!}F''(a) + \frac{h^3}{3!}F'''(a) + \ldots$$

This is, in fact, a general series and holds good when $a$ and $h$ are both variables. If we write $a = x$ in this result, we obtain

$$f(x + h) = f(x) + h.f'(x) + \frac{h^2}{2!}f''(x) + \frac{h^3}{3!}f'''(x) + \ldots$$

which is the usual form of *Taylor's series.*

**46**

Maclaurin's series and Taylor's series are very much alike in some respects. In fact, Maclaurin's series is really a special case of Taylor's.

*Maclaurin's series:* $f(x) = f(0) + x.f'(0) + \frac{x^2}{2!}f''(0) + \frac{x^3}{3!}f'''(0) + \ldots$

Taylor's *series:* $f(x + h) = f(x) + h.f'(x) + \frac{h^2}{2!}f''(x) + \frac{h^3}{3!}f'''(x) + \ldots$

Copy the two series down together: it will help you learn them.

**47**

*Example 1.* Show that, if $h$ is small, then

$$\tan^{-1}(x+h) = \tan^{-1}x + \frac{h}{1+x^2} - \frac{xh^2}{(1+x^2)^2} \quad \text{approximately.}$$

□□□□□□□□□□□□□□□□□□□□□□□□□□□□□□□□□□□□□□□

Taylor's series states

$$f(x+h) = f(x) + h.f'(x) + \frac{h^2}{2!}f''(x) + \frac{h^3}{3!}f'''(x)\ldots$$

where $f(x)$ is the function obtained by putting $h = 0$ in the function $f(x+h)$.

In this case then, $f(x) = \tan^{-1}x$

$$\therefore f'(x) = \frac{1}{1+x^2} \quad \text{and} \quad f''(x) = -\frac{2x}{(1+x^2)^2}$$

Putting these expressions back into the series, we have

$$\tan^{-1}(x+h) = \tan^{-1}x + h.\frac{1}{1+x^2} - \frac{h^2}{2!}.\frac{2x}{(1+x^2)^2} + \ldots$$

$$= \tan^{-1}x + \frac{h}{1+x^2} - \frac{xh^2}{(1+x^2)^2} \quad \text{approx.}$$

Why are we justified in omitting the terms that follow?

**48**

> The following terms contain higher powers of $h$ which, by definition, is small. These terms will therefore be very small.

*Example 2.* Express $\sin(x+h)$ as a series of powers of $h$ and evaluate $\sin 44°$ correct to 5 decimal places.

$$\sin(x+h) = f(x) + h.f'(x) + \frac{h^2}{2!}f''(x) + \frac{h^3}{3!}f'''(x) + \ldots$$

$$f(x) = \sin x;\quad f'(x) = \cos x;\quad f''(x) = -\sin x;$$

$$f'''(x) = -\cos x;\quad f^{iv}(x) = \sin x;\ \text{etc.}$$

$$\therefore \sin(x+h) = \sin x + h\cos x - \frac{h^2}{2!}\sin x - \frac{h^3}{3!}\cos x + \ldots$$

$$\sin 44° = \sin(45° - 1°) = \sin\left(\frac{\pi}{4} - 0{\cdot}01745\right) \text{ and } \sin\frac{\pi}{4} = \cos\frac{\pi}{4} = \frac{1}{\sqrt{2}}$$

$$\therefore \sin 44° = \frac{1}{\sqrt{2}}\left\{1 + h - \frac{h^2}{2} - \frac{h^3}{6} + \ldots\right\} \quad h = -0{\cdot}01745$$

$$= \frac{1}{\sqrt{2}}\left\{1 - 0{\cdot}01745 - \frac{0{\cdot}0003045}{2} + \frac{0{\cdot}0000053}{6} + \ldots\right\}$$

$$= \frac{1}{\sqrt{2}}\left\{1 - 0{\cdot}01745 - 0{\cdot}0001523 + 0{\cdot}0000009\ldots\right\}$$

$$= 0{\cdot}7071\,(0{\cdot}982399) = 0{\cdot}69466$$

**49**

You have now reached the end of the programme, except for the test exercise which follows. The questions are all straightforward and you will have no trouble with them. Work through all the questions at your own speed. There is no need to hurry.

**Test Exercise – XIV**

1. State Maclaurin's series.
2. Find the first 4 non-zero terms in the expansion of $\cos^2 x$.
3. Find the first 3 non-zero terms in the series for $\sec x$.
4. Show that $\tan^{-1} x = x - \frac{x^3}{3} + \frac{x^5}{5} - \frac{x^7}{7} + \ldots$
5. Assuming the series for $e^x$ and $\tan x$, determine the series for $e^x . \tan x$ up to and including the term in $x^4$.
6. Evaluate $\sqrt{1 \cdot 05}$ correct to 5 significant figures.
7. Find (i) $\lim_{x \to 0} \left\{ \frac{1 - 2\sin^2 x - \cos^3 x}{5x^2} \right\}$

   (ii) $\lim_{x \to 0} \left\{ \frac{\tan x . \tan^{-1} x - x^2}{x^6} \right\}$

   (iii) $\lim_{x \to 0} \left\{ \frac{x - \sin x}{x - \tan x} \right\}$
8. Expand $\cos(x + h)$ as a series of powers of $h$ and hence evaluate $\cos 31°$ correct to 5 decimal places.

*You are now ready to start the next programme.*

## Further Problems – XIV

1. Prove that $\cos x = 1 - \frac{x^2}{2!} + \frac{x^4}{4!} - \frac{x^6}{6!} + \ldots$ and that the series is valid for all values of $x$. Deduce the power series for $\sin^2 x$ and show that, if $x$ is small,

$$\frac{\sin^2 x - x^2 \cos x}{x^4} = \frac{1}{6} + \frac{x^2}{360} \text{ approximately.}$$

2. Apply Maclaurin's series to establish a series for $\ln(1 + x)$. If $1 + x = \frac{b}{a}$, show that

$$(b^2 - a^2)/2ab = x - \frac{x^2}{2} + \frac{x^3}{2} - \ldots$$

Hence show that, if $b$ is nearly equal to $a$, then $(b^2 - a^2)/2ab$ exceeds $\ln\left(\frac{b}{a}\right)$ by approximately $(b - a)^3/6a^3$.

3. Evaluate (i) $\lim_{x\to 0}\left\{\frac{1 - 2\sin^2 x - \cos^3 x}{5x^2}\right\}$

(ii) $\lim_{x\to 0}\left\{\frac{\sin x - x\cos x}{x^3}\right\}$ (iii) $\lim_{x\to 0}\left\{\frac{\tan x - \sin x}{x^3}\right\}$

(iv) $\lim_{x\to 0}\left\{\frac{\sin x - x}{x^3}\right\}$ (v) $\lim_{x\to 0}\left\{\frac{\tan x - x}{x - \sin x}\right\}$

4. Write down the expansions of (i) $\cos x$ and (ii) $\frac{1}{1 + x}$, and hence show that

$$\frac{\cos x}{1 + x} = 1 - x + \frac{x^2}{2} - \frac{x^3}{2} + \frac{13x^4}{24} - \ldots$$

5. State the series for $\ln(1 + x)$ and the range of values of $x$ for which it is valid. Assuming the series for $\sin x$ and for $\cos x$, find the series for $\ln\left(\frac{\sin x}{x}\right)$ and $\ln(\cos x)$ as far as the term in $x^4$. Hence show that, if $x$ is small, $\tan x$ is approximately equal to $x.e^{x^{2}/3}$.

6. Use Maclaurin's series to obtain the expansion of $e^x$ and of $\cos x$ in ascending powers of $x$ and hence determine

$$\lim_{x\to 0}\left\{\frac{e^x + e^{-x} - 2}{2\cos 2x - 2}\right\}$$

7. Find the first four terms in the expansion of $\dfrac{x-3}{(1-x)^2(2+x^2)}$ in ascending powers of $x$.

8. Write down the series for $\ln(1+x)$ in ascending powers of $x$ and state the conditions for convergence.
If $a$ and $b$ are small compared with $x$, show that

$$\ln(x+a) - \ln x = \frac{a}{b}\left(1 + \frac{b-a}{2x}\right)\left\{\ln(x+b) - \ln x\right\}$$

9. Find the value of $k$ for which the expansion of

$$(1+kx)\left(1+\frac{x}{6}\right)^{-1}\ln(1+x)$$

contains no term in $x^2$.

10. Evaluate (i) $\lim\limits_{x\to 0}\left\{\dfrac{\sinh x - \tanh x}{x^3}\right\}$

(ii) $\lim\limits_{x\to 1}\left\{\dfrac{\ln x}{x^2-1}\right\}$ (iii) $\lim\limits_{x\to 0}\left\{\dfrac{x+\sin x}{x^2+x}\right\}$

11. If $u_r$ and $u_{r-1}$ indicate the $r^{th}$ term and the $(r-1)^{th}$ term respectively of the expansion of $(1+x)^n$, determine an expression, in its simplest form, for the ratio $\dfrac{u_r}{u_{r-1}}$. Hence show that in the binomial expansion of $(1+0{\cdot}03)^{12}$, the $r^{th}$ term is less than one-tenth of the $(r-1)^{th}$ term if $r > 4$. Use the expansion to evaluate $(1{\cdot}03)^{12}$ correct to three places of decimals.

12. By the use of Maclaurin's series, show that

$$\sin^{-1}x = x + \frac{x^3}{6} + \frac{3x^5}{40} + \ldots$$

Assuming the series for $e^x$, obtain the expansion of $e^x \sin^{-1}x$, up to and including the term in $x^4$. Hence show that, when $x$ is small, the graph of $y = e^x \sin^{-1}x$ approximates to the parabola $y = x^2 + x$.

13. By application of Maclaurin's series, determine the first two non-vanishing terms of a series for $\ln\cos x$. Express $(1+\cos\theta)$ in terms of $\cos\theta/2$ and show that, if $\theta$ is small,

$$\ln(1+\cos\theta) = \ln 2 - \frac{\theta^2}{4} - \frac{\theta^4}{96} \text{ approximately.}$$

14. If $x$ is small, show that

$$\text{(i)}\quad \sqrt{\left\{\frac{1+x}{1-x}\right\}} \simeq 1 + x + \frac{x^2}{2}$$

$$\text{(ii)}\quad \frac{\sqrt{\{(1+3x^2)e^x\}}}{1-x} \simeq 1 + \frac{3x}{2} + \frac{25x^2}{8}$$

15. Prove that

$$\text{(i)}\quad \frac{x}{e^x - 1} = 1 - \frac{x}{2} + \frac{x^2}{12} - \frac{x^4}{720} + \ldots$$

$$\text{(ii)}\quad \frac{x}{e^x + 1} = \frac{x}{2} - \frac{x^2}{4} + \frac{x^4}{48} - \ldots$$

16. Find (i) $\displaystyle\lim_{x\to 0}\left\{\frac{\sinh^{-1}x - x}{x^3}\right\}$, (ii) $\displaystyle\lim_{x\to 0}\left\{\frac{e^{\sin x} - 1 - x}{x^2}\right\}$.

17. Find the first three terms in the expansion of

$$\frac{\sinh x . \ln(1+x)}{x^2(1+x)^3}$$

18. The field strength of a magnet (H) at a point on the axis, distance $x$ from its centre, is given by

$$H = \frac{M}{2l}\left\{\frac{1}{(x-l)^2} - \frac{1}{(x+l)^2}\right\}$$

where $2l$ = length of magnet and M = moment. Show that, if $l$ is very small compared with $x$, then $H \simeq \dfrac{2M}{x^3}$.

19. Expand $[\ln(1+x)]^2$ in powers of $x$ up to and including the term in $x^4$. Hence determine whether $\cos 2x + [\ln(1+x)]^2$ has a maximum value, minimum value, or point of inflexion at $x = 0$.

20. If $l$ is the length of a circular arc, $a$ is the length of the chord of the whole arc, and $b$ is the length of the chord of half the arc, show that (i) $a = 2r\sin\dfrac{l}{2r}$ and (ii) $b = 2r\sin\dfrac{l}{4r}$, where $r$ is the radius of the circle. By expanding $\sin\dfrac{l}{2r}$ and $\sin\dfrac{l}{4r}$ as series, show that $l = \dfrac{8b - a}{3}$ approximately.

# Programme 15

## INTEGRATION

### PART 1

**1** **Introduction**

You are already familiar with the basic principles of integration and have had plenty of practice at some time in the past. However, that was some time ago, so let us first of all brush up our ideas of the fundamentals.

Integration is the reverse of differentiation. In differentiation, we start with a function and proceed to find its differential coefficient. In integration, we start with the differential coefficient and have to work back to find the function from which it has been derived.

e.g. $\frac{d}{dx}(x^3 + 5) = 3x^2$. Therefore it is true, in this case, to say that the integral of $3x^2$, with respect to $x$, is the function from which it came,

i.e. $\int 3x^2 dx = x^3 + 5$. However, if we had to find $\int 3x^2 dx$, without knowing the past history of the function, we should have no indication of the size of the constant term involved, since all trace of it is lost in the differential coefficient. All we can do is to indicate the constant term by a symbol, e.g. C.

So, in general, $\int 3x^2 dx = x^3 + C$

Although we cannot determine the value of this *constant of integration* without extra information about the function, it is vitally important that we should always include it in our results. There are just one or two occasions when we are permitted to leave it out, not because it is not there, but because in some prescribed situation, it will cancel out in subsequent working. Such occasions, however, are very rare and, in general, the *constant of integration must be included in the result.*

If you omit the constant of integration, your work will be slovenly and, furthermore, it will be completely wrong! So, *do not forget the constant of integration.*

1. **Standard integrals**

Every differential coefficient, when written in reverse, gives us an integral,

$$\text{e.g. } \frac{d}{dx}(\sin x) = \cos x \quad \therefore \int \cos x \, dx = \sin x + C$$

It follows then that our list of standard differential coefficients will form the basis of a list of standard integrals – sometimes slightly modified to give a neater expression.

**2**

Here is a list of basic differential coefficients and the basic integrals that go with them:

1. $\frac{d}{dx}(x^n) = nx^{n-1}$ $\quad \therefore \int x^n\,dx = \frac{x^{n+1}}{n+1} + C \quad \left\{\begin{matrix}\text{provided}\\ n \neq -1\end{matrix}\right\}$
2. $\frac{d}{dx}(\ln x) = \frac{1}{x}$ $\quad \therefore \int \frac{1}{x}\,dx = \ln x + C$
3. $\frac{d}{dx}(e^x) = e^x$ $\quad \therefore \int e^x\,dx = e^x + C$
4. $\frac{d}{dx}(e^{kx}) = ke^{kx}$ $\quad \therefore \int e^{kx}\,dx = \frac{e^{kx}}{k} + C$
5. $\frac{d}{dx}(a^x) = a^x \ln a$ $\quad \therefore \int a^x\,dx = \frac{a^x}{\ln a} + C$
6. $\frac{d}{dx}(\cos x) = -\sin x$ $\quad \therefore \int \sin x\,dx = -\cos x + C$
7. $\frac{d}{dx}(\sin x) = \cos x$ $\quad \therefore \int \cos x\,dx = \sin x + C$
8. $\frac{d}{dx}(\tan x) = \sec^2 x$ $\quad \therefore \int \sec^2 x\,dx = \tan x + C$
9. $\frac{d}{dx}(\cosh x) = \sinh x$ $\quad \therefore \int \sinh x\,dx = \cosh x + C$
10. $\frac{d}{dx}(\sinh x) = \cosh x$ $\quad \therefore \int \cosh x\,dx = \sinh x + C$
11. $\frac{d}{dx}(\sin^{-1} x) = \frac{1}{\sqrt{(1-x^2)}}$ $\quad \therefore \int \frac{1}{\sqrt{(1-x^2)}}\,dx = \sin^{-1} x + C$
12. $\frac{d}{dx}(\cos^{-1} x) = \frac{-1}{\sqrt{(1-x^2)}}$ $\quad \therefore \int \frac{-1}{\sqrt{(1-x^2)}}\,dx = \cos^{-1} x + C$
13. $\frac{d}{dx}(\tan^{-1} x) = \frac{1}{1+x^2}$ $\quad \therefore \int \frac{1}{1+x^2}\,dx = \tan^{-1} x + C$
14. $\frac{d}{dx}(\sinh^{-1} x) = \frac{1}{\sqrt{(x^2+1)}}$ $\quad \therefore \int \frac{1}{\sqrt{(x^2+1)}}\,dx = \sinh^{-1} x + C$
15. $\frac{d}{dx}(\cosh^{-1} x) = \frac{1}{\sqrt{(x^2-1)}}$ $\quad \therefore \int \frac{1}{\sqrt{(x^2-1)}}\,dx = \cosh^{-1} x + C$
16. $\frac{d}{dx}(\tanh^{-1} x) = \frac{1}{1-x^2}$ $\quad \therefore \int \frac{1}{1-x^2}\,dx = \tanh^{-1} x + C$

□ □ □ □ □ □ □ □ □ □ □ □ □ □ □ □ □ □ □ □ □ □ □ □ □ □ □ □ □ □ □ □ □ □ □ □ □ □

Spend a little time copying this list carefully into your record book as a reference list.

**3**

Here is a second look at the last six results, which are less familiar to you than the others.

$$\int \frac{1}{\sqrt{(1-x^2)}} dx = \sin^{-1} x + C \qquad \int \frac{1}{\sqrt{(x^2+1)}} = \sinh^{-1} x + C$$

$$\int \frac{-1}{\sqrt{(1-x^2)}} dx = \cos^{-1} x + C \qquad \int \frac{1}{\sqrt{(x^2-1)}} dx = \cosh^{-1} x + C$$

$$\int \frac{1}{1+x^2} dx = \tan^{-1} x + C \qquad \int \frac{1}{1-x^2} dx = \tanh^{-1} x + C$$

*Notice* (i) How alike the two sets are in shape,
(ii) Where the small, but all important, differences occur.

*On to frame 4.*

**4**

Now cover up the lists you have just copied down and complete the following.

(i) $\int e^{5x}\, dx =$ ................

(ii) $\int x^7\, dx =$ ................

(iii) $\int \sqrt{x}\, dx =$ ................

(iv) $\int \sin x\, dx =$ ................

(v) $\int 2 \sinh x\, dx =$ ................

(vi) $\int \frac{5}{x} dx =$ ................

(vii) $\int \frac{1}{\sqrt{(1-x^2)}} dx =$ ................

(viii) $\int 5^x\, dx =$ ................

(ix) $\int \frac{1}{\sqrt{(x^2-1)}} dx =$ ................

(x) $\int \frac{1}{1+x^2} dx =$ ................

***When you have finished them all, check your results with those given in the next frame.***

**5**

Here they are:

(i) $\int e^{5x}\,dx = \dfrac{e^{5x}}{5} + C$

(ii) $\int x^7\,dx = \dfrac{x^8}{8} + C$

(iii) $\int \sqrt{x}\,dx = \int x^{1/2}\,dx$
$= 2\dfrac{x^{3/2}}{3} + C$

(iv) $\int \sin x\,dx = -\cos x + C$

(v) $\int 2\sinh x\,dx = 2\cosh x + C$

(vi) $\int \dfrac{5}{x}dx = 5\ln x + C$

(vii) $\int \dfrac{1}{\sqrt{(1-x^2)}}dx = \sin^{-1}x + C$

(viii) $\int 5^x\,dx = \dfrac{5^x}{\ln 5} + C$

(ix) $\int \dfrac{1}{\sqrt{(x^2-1)}}dx = \cosh^{-1}x + C$

(x) $\int \dfrac{1}{1+x^2}\,dx = \tan^{-1}x + C$

All correct? – or nearly so? At the moment, these are fresh in your mind, but have a look at your list of standard integrals whenever you have a few minutes to spare. It will help you to remember them.

*Now move on to frame 6.*

**6**

2. **Functions of a linear function of $x$**

We are very often required to integrate functions like those in the standard list, but where $x$ is replaced by a *linear function* of $x$, e.g. $\int (5x-4)^6\,dx$, which is very much like $\int x^6\,dx$ except that $x$ is replaced by $(5x-4)$. If we put $z$ to stand for $(5x-4)$, the integral becomes $\int z^6\,dx$ and before we can complete the operation, we must change the variable, thus

$$\int z^6\,dx = \int z^6\,\frac{dx}{dz}dz$$

Now $\dfrac{dx}{dz}$ can be found from the substitution $z = 5x-4$ for $\dfrac{dz}{dx} = 5$, therefore $\dfrac{dx}{dz} = \dfrac{1}{5}$ and the integral becomes

$$\int z^6 dx = \int z^6\,\frac{dx}{dz}dz = \int z^6\left(\tfrac{1}{5}\right)dz = \frac{1}{5}\int z^6 dz \quad = \frac{1}{5}\cdot\frac{z^7}{7} + C$$

Finally, we must express $z$ in terms of the original variable, $x$, so that

$$\int (5x-4)^6\,dx = \ldots\ldots\ldots\ldots$$

**7**

$$\int (5x-4)^6\,dx = \frac{(5x-4)^7}{5.7} + C$$

$$= \boxed{\frac{(5x-4)^7}{35} + C}$$

The corresponding standard integral is $\int x^6\,dx = \frac{x^7}{7} + C$. We see, therefore, that when $x$ is replaced by $(5x-4)$, the 'power' rule still applies, i.e. $(5x-4)$ replaces the single $x$ in the result, so long as we also *divide by the coefficient of* $x$, in this case 5.

$$\int x^6\,dx = \frac{x^7}{7} + C \quad \therefore \int (5x-4)^6\,dx = \frac{(5x-4)^7}{35} + C$$

This will always happen when we integrate functions of a *linear* function of $x$.

$$\text{e.g.} \int e^x\,dx = \int e^x + C \quad \therefore \int e^{3x+4}\,dx = \frac{e^{3x+4}}{3} + C$$

i.e. $(3x+4)$ replaces $x$ in the integral,
then $(3x+4)$ " " " " result, provided we also divide by the coefficient of $x$.

Similarly, since $\int \cos x\,dx = \sin x + C$,

then $\int \cos(2x+5)\,dx = \ldots\ldots\ldots\ldots$

**8**

Similarly, $\boxed{\int \cos(2x+5)\,dx = \frac{\sin(2x+5)}{2} + C}$

$$\int \sec^2 x\,dx = \tan x + C \quad \therefore \int \sec^2 4x\,dx = \frac{\tan 4x}{4} + C$$

$$\int \frac{1}{x}\,dx = \ln x + C \quad \therefore \int \frac{1}{2x+3}\,dx = \frac{\ln(2x+3)}{2} + C$$

$$\int \sinh x\,dx = \cosh x + C \quad \therefore \int \sinh(3-4x)\,dx = \frac{\cosh(3-4x)}{-4}$$

$$= -\frac{\cosh(3-4x)}{4} + C$$

$$\int \sin x\,dx = -\cos x + C \quad \therefore \int \sin 3x\,dx = -\frac{\cos 3x}{3} + C$$

$$\int e^x\,dx = e^x + C \quad \therefore \int e^{4x}\,dx = \frac{e^{4x}}{4} + C$$

So if a *linear* function of $x$ replaces the single $x$ in the standard integral, the same linear function of $x$ replaces the single $x$ in the result, so long as we also remember to $\ldots\ldots\ldots\ldots$

**9**

... divide by the coefficient of $x$

Now you can do these quite happily – and do not forget the constants of integration!

1. $\int (2x-7)^3\,dx$
2. $\int \cos(7x+2)\,dx$
3. $\int e^{5x+4}\,dx$
4. $\int \sinh 7x\,dx$
5. $\int \frac{1}{4x+3}\,dx$
6. $\int \frac{1}{1+(2x)^2}\,dx$
7. $\int \sec^2(3x+1)\,dx$
8. $\int \sin(2x-5)\,dx$
9. $\int \cosh(1+4x)\,dx$
10. $\int 3^{5x}\,dx$

*Finish them all, then move on to frame 10 and check your results.*

---

**10**

Here are the results:

1. $\int (2x-7)^3\,dx = \frac{(2x-7)^4}{2.4} + C = \frac{(2x-7)^4}{8} + C$
2. $\int \cos(7x+2)\,dx = \frac{\sin(7x+2)}{7} + C$
3. $\int e^{5x+4}\,dx = \frac{e^{5x+4}}{5} + C$
4. $\int \sinh 7x\,dx = \frac{\cosh 7x}{7} + C$
5. $\int \frac{1}{4x+3}\,dx = \frac{\ln(4x+3)}{4} + C$
6. $\int \frac{1}{1+(2x)^2}\,dx = \frac{\tan^{-1}(2x)}{2} + C$
7. $\int \sec^2(3x+1)\,dx = \frac{\tan(3x+1)}{3} + C$
8. $\int \sin(2x-5)\,dx = -\frac{\cos(2x-5)}{2} + C$
9. $\int \cosh(1+4x)\,dx = \frac{\sinh(1+4x)}{4} + C$
10. $\int 3^{5x}\,dx = \frac{3^{5x}}{5\ln 3} + C$

*Now we can start the next section of the programme. So turn on to frame 11.*

**11**

3. **Integrals of the form** $\int \frac{f'(x)}{f(x)}\,dx$ **and** $\int f(x).f'(x)\,dx$.

Consider the integral $\int \frac{2x+3}{x^2+3x-5}\,dx$. This is not one of our standard integrals, so how shall we tackle it? This is an example of a type of integral which is very easy to deal with but which depends largely on how keen your wits are.

You will notice that if we differentiate the denominator, we obtain the expression in the numerator. So, let $z$ stand for the denominator, i.e. $z = x^2 + 3x - 5$

$$\therefore \frac{dz}{dx} = 2x + 3 \quad \therefore dz \equiv (2x+3)\,dx$$

The given integral can then be written in terms of $z$.

$$\int \frac{(2x+3)}{x^2+3x-5}\,dx = \int \frac{dz}{z} \text{ and we know that } \int \frac{1}{z}\,dz = \ln z + C$$

$$= \ln z + C$$

If we now put back what $z$ stands for in terms of $x$, we get

$$\int \frac{(2x+3)}{x^2+3x-5}\,dx = \text{.....................}$$

**12**

$$\boxed{\int \frac{(2x+3)}{x^2+3x-5}\,dx = \ln(x^2+3x-5) + C}$$

Any integral, in which the numerator is the differential coefficient of the denominator, will be of the kind $\int \frac{f'(x)}{f(x)}\,dx = \ln\left\{f(x)\right\} + C$.

e.g. $\int \frac{3x^2}{x^3-4}\,dx$ is of the form $\int \frac{dz}{z}$, since $\frac{d}{dx}(x^3-4) = 3x^2$, i.e. the differential coefficient of the denominator appears as the numerator. Therefore, we can say at once, without any further working

$$\int \frac{3x^2}{x^3-4}\,dx = \ln(x^3-4) + C$$

Similarly, $\int \frac{6x^2}{x^3-4}\,dx = 2\int \frac{3x^2}{x^3-4}\,dx = 2\ln(x^3-4) + C$

and $\int \frac{2x^2}{x^3-4}\,dx = \frac{2}{3}\int \frac{3x^2}{x^3-4}\,dx = \frac{2}{3}\ln(x^3-4) + C$

and $\int \frac{x^2}{x^3-4}\,dx = \text{.................................}$

**13**

$$\int \frac{x^2}{x^3 - 4}\,dx = \frac{1}{3}\int \frac{3x^2}{x^3 - 4}\,dx = \frac{1}{3}\ln(x^3 - 4) + C$$

We shall always get this log form of the result, then, whenever the numerator is the differential coefficient of the denominator, or is a multiple or sub-multiple of it.

*Example:* $\int \cot x\,dx = \int \frac{\cos x}{\sin x}\,dx$ and since we know that $\cos x$ is the differential coefficient of $\sin x$, then

$$\int \cot x\,dx = \int \frac{\cos x}{\sin x}\,dx = \ln \sin x + C$$

In the same way, $\int \tan x\,dx = \int \frac{\sin x}{\cos x}\,dx$

$= \ldots\ldots\ldots\ldots$

**14**

$$\int \tan x\,dx = \int \frac{\sin x}{\cos x}\,dx = -\int \frac{(-\sin x)}{\cos x}\,dx$$

$$= \boxed{-\ln \cos x + C}$$

Whenever we are confronted by an integral in the form of a quotient, our first reaction is to see whether the numerator is the differential coefficient of the denominator. If so, the result is simply the log. of the denominator.

e.g. $\int \frac{4x - 8}{x^2 - 4x + 5}\,dx = \ldots\ldots\ldots\ldots$

## 15

$$\int \frac{4x-8}{x^2-4x+5}\,dx = 2\int \frac{2x-4}{x^2-4x+5}\,ax \boxed{= 2\ln(x^2-4x+5)+C}$$

Here you are: complete the following:

1. $\displaystyle\int \frac{\sec^2 x}{\tan x}\,dx = \ldots\ldots\ldots\ldots\ldots\ldots$

2. $\displaystyle\int \frac{2x+4}{x^2+4x-1}\,dx = \ldots\ldots\ldots\ldots\ldots\ldots$

3. $\displaystyle\int \frac{\sinh x}{\cosh x}\,dx = \ldots\ldots\ldots\ldots\ldots\ldots$

4. $\displaystyle\int \frac{x-3}{x^2-6x+2}\,dx = \ldots\ldots\ldots\ldots\ldots\ldots$

## 16

Here are the results: check yours.

1. $\displaystyle\int \frac{\sec^2 x}{\tan x}\,dx = \ln\tan x + C$

2. $\displaystyle\int \frac{2x+4}{x^2+4x-1}\,dx = \ln(x^2+4x-1)+C$

3. $\displaystyle\int \frac{\sinh x}{\cosh x}\,dx = \ln\cosh x + C$

4. $\displaystyle\int \frac{x-3}{x^2-6x+2}\,dx = \frac{1}{2}\ln(x^2-6x+2)+C$

*Now turn on to frame 17.*

**17**

In very much the same way, we sometimes have integrals such as

$$\int \tan x . \sec^2 x \, dx$$

This, of course, is not a quotient but a product. Nevertheless we notice that one function ($\sec^2 x$) of the product is the differential coefficient of the other function ($\tan x$).

If we put $z = \tan x$, then $dz \equiv \sec^2 x \, dx$ and the integral can then be written $\int z \, dz$ which gives $\frac{z^2}{2} + C$.

$$\therefore \int \tan x . \sec^2 x \, dx = \underline{\frac{\tan^2 x}{2} + C}$$

Here, then, we have a product where one factor is the differential coefficient of the other. We could write it as

$$\int \tan x . d(\tan x)$$

This is just like $\int z \, dz$ which gives $\frac{z^2}{2} + C$

$$\therefore \int \tan x . \sec^2 x \, dx = \int \tan x . d(\tan x) = \underline{\frac{\tan^2 x}{2} + C}$$

*On to the next frame.*

**18**

Here is another example of the same kind:

$$\int \sin x . \cos x \, dx = \int \sin x . d(\sin x) \quad \text{i.e. like} \int z \, dz = \frac{\sin^2 x}{2} + C$$

The only thing you have to spot is that one factor of the product is the differential coefficient of the other, or is some multiple of it.

*Example 1.*

$$\int \frac{\ln x}{x} dx = \int \ln x . \frac{1}{x} dx$$

$$= \int \ln x . d(\ln x) = \frac{(\ln x)^2}{2} + C$$

*Example 2.*

$$\int \frac{\sin^{-1} x}{\sqrt{(1 - x^2)}} dx = \int \sin^{-1} x . \frac{1}{\sqrt{(1 - x^2)}} dx$$

$$= \int \sin^{-1} x . d(\sin^{-1} x)$$

$$= \frac{(\sin^{-1} x)^2}{2} + C$$

*Example 3.*

$$\int \sinh x . \cosh x \, dx = \ldots\ldots\ldots\ldots$$

**19**

$$\int \sinh x \,.\cosh x \, dx = \int \sinh x \,.\, d(\sinh x)$$

$$\boxed{= \frac{\sinh^2 x}{2} + C}$$

Now here is a short revision exercise for you to do. Finish all four and then check your results with those in the next frame.

1. $\displaystyle\int \frac{2x+3}{x^2+3x-7}\,dx$

2. $\displaystyle\int \frac{\cos x}{1+\sin x}\,dx$

2. $\displaystyle\int (x^2+7x-4)(2x+7)\,dx$

4. $\displaystyle\int \frac{4x^2}{x^3-7}\,dx$

**20** *Results:*

1. $\displaystyle\int \frac{2x+3}{x^2+3x-7}\,dx$ Notice that the top is exactly the diff. coefft. of the bottom, i.e. $\displaystyle\int \frac{dz}{z}$

$$\therefore \int \frac{2x+3}{x^2+3x-7}\,dx = \int \frac{d(x^2+3x-7)}{x^2+3x-7}$$

$$= \underline{\ln(x^2+3x-7) + C}$$

2. $$\int \frac{\cos x}{1+\sin x} = \int \frac{d(1+\sin x)}{1+\sin x}$$

$$= \underline{\ln(1+\sin x) + C}$$

3. $$\int (x^2+7x-4)(2x+7)\,dx = \int (x^2+7x-4).d(x^2+7x-4)$$

$$= \underline{\frac{(x^2+7x-4)^2}{2} + C}$$

4. $$\int \frac{4x^2}{x^3-7}\,dx = \frac{4}{3}\int \frac{3x^2}{x^3-7}\,dx$$

$$= \underline{\frac{4}{3}\ln(x^3-7) + C}$$

Always be prepared for these types of integrals. They are often missed, but very easy if you spot them.

*Now on to the next part of the work that starts in frame 21.*

4. **Integration of products – integration by parts**

We often need to integrate a product where either function is *not* the differential coefficient of the other. For example, in the case of

$$\int x^2 . \ln x \, dx,$$

ln $x$ is not the differential coefficient of $x^2$
$x^2$ ” ” ” ” ” ” ln $x$

so in situations like this, we have to find some other method of dealing with the integral. Let us establish the rule for such cases.

If $u$ and $v$ are functions of $x$, then we know that

$$\frac{d}{dx}(uv) = u\frac{dv}{dx} + v\frac{du}{dx}$$

Now integrate both sides with respect to $x$. On the left, we get back to the function from which we started.

$$uv = \int u\frac{dv}{dx}\,dx + \int v\frac{du}{dx}\,dx$$

and rearranging the terms, we have

$$\int u\frac{dv}{dx}\,dx = uv - \int v\frac{du}{dx}\,dx$$

On the left-hand side, we have a product of two factors to integrate. One factor is chosen as the function $u$: the other is thought of as being the differential coefficient of some function $v$. To find $v$, of course, we must integrate this particular factor separately. Then, knowing $u$ and $v$ we can substitute in the right-hand side and so complete the routine.

You will notice that we finish up with another product to integrate on the end of the line, but, unless we are very unfortunate, this product will be easier to tackle than the original one.

This then is the key to the routine:

$$\int u\frac{dv}{dx}\,dx = uv - \int v\frac{du}{dx}\,dx$$

For convenience, this can be memorized as

$$\int u\,dv = uv - \int v\,du$$

In this form it is easier to remember, but the previous line gives its meaning in detail. This method is called *integration by parts.*

**22**

So $\qquad \int u \frac{dv}{dx}\,dx = uv - \int v \frac{du}{dx}\,dx$

i.e. $\qquad \int u\,dv = uv - \int v\,du$

Copy these results into your record book. You will soon learn them. Now for one or two examples involving integration by parts.

*Example 1.* $\int x^2 . \ln x\,dx$

The two factors are $x^2$ and $\ln x$, and we have to decide which to take as $u$ and which as $dv$. If we choose $x^2$ to be $u$ and $\ln x$ to be $dv$, then we shall have to integrate $\ln x$ in order to find $v$. Unfortunately, $\int \ln x\,dx$ is not in our basic list of standard integrals, therefore we must allocate $u$ and $dv$ the other way round, i.e. let $\ln x = u$ and $x^2 = dv$.

$$\therefore \int x^2 . \ln x\,dx = \ln x\left(\frac{x^3}{3}\right) - \frac{1}{3}\int x^3 . \frac{1}{x}\,dx.$$

Notice that we can tidy up the writing of the second integral by writing the constant factors involved, outside the integral.

$$\therefore \int x^2 \ln x\,dx = \ln x\left(\frac{x^3}{3}\right) - \frac{1}{3}\int x^3 . \frac{1}{x}\,dx \quad = \frac{x^3}{3}\ln x - \frac{1}{3}\int x^2\,dx$$

$$= \frac{x^3}{3}.\ln x - \frac{1}{3}.\frac{x^3}{3} + C = \frac{x^3}{3}\left\{\ln x - \frac{1}{3}\right\} + C$$

Note that if one of the factors of the product to be integrated is a log term, this must be chosen as .............. ($u$ or $dv$)

**23**

$u$

*Example 2.* $\int x^2 e^{3x}\,dx \qquad$ Let $u = x^2$ and $dv = e^{3x}$

Then $\int x^2 e^{3x}\,dx = x^2\left(\frac{e^{3x}}{3}\right) - \frac{2}{3}\int e^{3x}\,x\,dx$

$$= \frac{x^2 . e^{3x}}{3} - \frac{2}{3}\left\{x\left(\frac{e^{3x}}{3}\right) - \frac{1}{3}\int e^{3x}\,dx\right\} = \frac{x^2 e^{3x}}{3} - \frac{2x\,e^{3x}}{9} + \frac{2}{9}.\frac{e^{3x}}{3} + C$$

$$= \frac{e^{3x}}{3}\left\{x^2 - \frac{2x}{3} + \frac{2}{9}\right\} + C$$

*On to frame 24.*

**24**

In Example 1 we saw that if one of the factors is a log function, that log function *must* be taken as $u$.

In Example 2 we saw that, provided there is no log term present, the power of $x$ is taken as $u$. (By the way, this method holds good only for positive whole-number powers of $x$. For other powers, a different method must be applied.)

So which of the two factors should we choose to be $u$ in each of the following cases?

(i) $\int x.\ln x\, dx$

(ii) $\int x^3.\sin x\, dx$

**25**

| In $\int x.\ln x\, dx,\quad u = \ln x$ |
|---|
| $\int x^3 \sin x\, dx,\quad u = x^3$ |

Right. Now for a third example.

*Example 3.* $\int e^{3x} \sin x\, dx$. Here we have neither a log factor nor a power of $x$. Let us try putting $u = e^{3x}$ and $dv = \sin x$.

$$\therefore \int e^{3x} \sin x\, dx = e^{3x}(-\cos x) + 3\int \cos x . e^{3x}\, dx$$

$$= -e^{3x} \cos x + 3\int e^{3x} \cos x\, dx$$

$$= -e^{3x} \cos x + 3\left\{e^{3x}(\sin x) - 3\int \sin x . e^{3x}\, dx\right\}$$

and it looks as though we are back where we started. However, let us write I for the integral $\int e^{3x} \sin x\, dx$

$$I = -e^{3x} \cos x + 3e^{3x} \sin x - 9I$$

Then, treating this as a simple equation, we get

$$10I = e^{3x}(3\sin x - \cos x) + C_1$$

$$I = \frac{e^{3x}}{10}(3\sin x - \cos x) + C$$

Whenever we integrate functions of the form $e^{kx} \sin x$ or $e^{kx} \cos x$, we get similar types of results after applying the rule twice.

*Turn on to frame 26.*

**26**

The three examples we have considered enable us to form a priority order for $u$:

(i) $\ln x$
(ii) $x^n$
(iii) $e^{kx}$

i.e. If one factor is a log function, that must be taken as '$u$'.
If there is no log function but a power of $x$, that becomes '$u$'.
If there is neither a log function nor a power of $x$, then the exponential function is taken as '$u$'.

Remembering the priority order will save a lot of false starts.

So which would you choose as '$u$' in the following cases

(i) $\int x^4 \cos 2x\, dx, \qquad u = \ldots\ldots\ldots\ldots$

(ii) $\int x^4 e^{3x}\, dx, \qquad u = \ldots\ldots\ldots\ldots$

(iii) $\int x^3 \ln(x+4)\, dx, \qquad u = \ldots\ldots\ldots\ldots$

(iv) $\int e^{2x} \cos 4x\, dx, \qquad u = \ldots\ldots\ldots\ldots$

**27**

(i) $\int x^4 \cos 2x\, dx, \qquad u = x^4$

(ii) $\int x^4 e^{3x}\, dx, \qquad u = x^4$

(iii) $\int x^3 \ln(x+4)\, dx, \qquad u = \ln(x+4)$

(iv) $\int e^{2x} \cos 4x\, dx, \qquad u = e^{2x}$

Right. Now look at this one.

$$\int e^{5x} \sin 3x\, dx$$

Following our rule for priority for $u$, in this case, we should put

$u = \ldots\ldots\ldots\ldots$

**28**

$$\int e^{5x} \sin 3x \, dx \quad \boxed{\therefore \; u = e^{5x}}$$

Correct. Make a note of that priority list for $u$ in your record book.
Then go ahead and determine the integral given above.

*When you have finished, check your working with that set out in the next frame.*

---

**29**

$$\boxed{\int e^{5x} \sin 3x \, dx = \frac{3\,e^{5x}}{34}\left\{\frac{5}{3}\sin 3x - \cos 3x\right\} + C}$$

Here is the working. Follow it through.

$$\int e^{5x} \sin 3x \, dx = e^{5x}\left(-\frac{\cos 3x}{3}\right) + \frac{5}{3}\int \cos 3x \, . \, e^{5x} \, dx$$

$$= -\frac{e^{5x} \cos 3x}{3} + \frac{5}{3}\left\{e^{5x}\left(\frac{\sin 3x}{3}\right) - \frac{5}{3}\int \sin 3x \, . \, e^{5x} \, dx\right\}$$

$$\therefore \; I = -\frac{e^{5x} \cos 3x}{3} + \frac{5}{9} e^{5x} \sin 3x - \frac{25}{9} I$$

$$\frac{34}{9} I = \frac{e^{5x}}{3}\left\{\frac{5}{3}\sin 3x - \cos 3x\right\} + C_1$$

$$I = \frac{3\,e^{5x}}{34}\left\{\frac{5}{3}\sin 3x - \cos 3x\right\} + C$$

---

There you are. Now do these in much the same way. Finish them both before turning on to the next frame.

(i) $\int x \ln x \, dx$

(ii) $\int x^3 e^{2x} \, dx$

**30**

*Solutions:*

(i) $$\int x \ln x \, dx = \ln x\left(\frac{x^2}{2}\right) - \frac{1}{2}\int x^2 . \frac{1}{x} dx$$

$$= \frac{x^2 \ln x}{2} - \frac{1}{2}\int x \, dx$$

$$= \frac{x^2 \ln x}{2} - \frac{1}{2} . \frac{x^2}{2} + C$$

$$= \frac{x^2}{2}\left\{\ln x - \frac{1}{2}\right\} + C$$

(ii) $$\int x^3 e^{2x} dx = x^3\left(\frac{e^{2x}}{2}\right) - \frac{3}{2}\int e^{2x} x^2 \, dx$$

$$= \frac{x^3 e^{2x}}{2} - \frac{3}{2}\left\{x^2\left(\frac{e^{2x}}{2}\right) - \frac{2}{2}\int e^{2x} x \, dx\right\}$$

$$= \frac{x^3 e^{2x}}{2} - \frac{3 x^2 e^{2x}}{4} + \frac{3}{2}\left\{x\left(\frac{e^{2x}}{2}\right) - \frac{1}{2}\int e^{2x} \, dx\right\}$$

$$= \frac{x^3 e^{2x}}{2} - \frac{3x^2 e^{2x}}{4} + \frac{3x e^{2x}}{4} - \frac{3}{4}\frac{e^{2x}}{2} + C$$

$$= \frac{e^{2x}}{2}\left\{x^3 - \frac{3x^2}{2} + \frac{3x}{2} - \frac{3}{4}\right\} + C.$$

That is all there is to it. You can now deal with the integration of products.
*The next section of the programme begins in frame 31, so turn on now and continue the good work.*

**31**

5. **Integration by partial fractions**

Suppose we have $\int \frac{x+1}{x^2 - 3x + 2} dx$. Clearly this is not one of our standard types, and the numerator is *not* the differential coefficient of the denominator. So how do we go about this one?

In such a case as this, we first of all express the rather cumbersome algebraic fraction in terms of its *partial fractions*, i.e. a number of simpler algebraic fractions which we shall most likely be able to integrate separately without difficulty.

$\frac{x+1}{x^2 - 3x + 2}$ can, in fact, be expressed as $\frac{3}{x-2} - \frac{2}{x-1}$

$$\therefore \int \frac{x+1}{x^2 - 3x + 2} dx = \int \frac{3}{x-2} dx - \int \frac{2}{x-1} dx$$

$$= \text{.............................}$$

**32**

$$\boxed{3 \ln(x-2) - 2 \ln(x-1) + C}$$

The method, of course, hinges on one's being able to express the given function in terms of its partial fractions.

The *rules of partial fractions* are as follows:

(i) The numerator of the given function must be of lower degree than that of the denominator. If it is not, then first of all divide out by long division.

(ii) Factorize the denominator into its prime factors. This is important, since the factors obtained determine the shape of the partial fractions.

(iii) A linear factor $(ax+b)$ gives a partial fraction of the form $\dfrac{A}{ax+b}$

(iv) Factors $(ax+b)^2$ give partial fractions $\dfrac{A}{ax+b} + \dfrac{B}{(ax+b)^2}$

(v) Factors $(ax+b)^3$ give p.f.'s $\dfrac{A}{ax+b} + \dfrac{B}{(ax+b)^2} + \dfrac{C}{(ax+b)^3}$

(vi) A quadratic factor $(ax^2+bx+c)$ gives a p.f. $\dfrac{Ax+B}{ax^2+bx+c}$

Copy down this list of rules into your record book for reference. It will be well worth it.

*Then on to the next frame.*

**33**

Now for some examples.

*Example 1.* $\displaystyle\int \frac{x+1}{x^2-3x+2}\,dx$

$$\frac{x+1}{x^2-3x+2} = \frac{x+1}{(x-1)(x-2)} = \frac{A}{x-1} + \frac{B}{x-2}$$

Multiply both sides by the denominator $(x-1)(x-2)$.

$$x+1 = A(x-2) + B(x-1)$$

This is an identity and true for any value of $x$ we like to substitute. Where possible, choose a value of $x$ which will make one of the brackets zero.

Let $(x-1)=0$, i.e. substitute $x=1$

$$\therefore\ 2 = A(-1) + B(0) \quad \therefore\ A = -2$$

Let $(x-2)=0$, i.e. substitute $x=2$

$$\therefore\ 3 = A(0) + B(1) \quad \therefore\ B = 3$$

So the integral can now be written ..........................

**34**

$$\int \frac{x+1}{x^2-3x+2}\,dx = \int \frac{3}{x-2}\,dx - \int \frac{2}{x-1}\,dx$$

Now the rest is easy.

$$\int \frac{x+1}{x^2-3x+2}\,dx = 3\int \frac{1}{x-2}\,dx - 2\int \frac{1}{x-1}\,dx$$

$= 3\ln(x-2) - 2\ln(x-1) + C$ (Do not forget the constant of integration!)

And now another one.

*Example 2.* To determine $\displaystyle\int \frac{x^2}{(x+1)(x-1)^2}\,dx$

Numerator = 2nd degree; denominator = 3rd degree. Rule 1 is satisfied. Denominator already factorized into its prime factors. Rule 2 is satisfied.

$$\frac{x^2}{(x+1)(x-1)^2} = \frac{A}{x+1} + \frac{B}{x-1} + \frac{C}{(x-1)^2}$$

Clear the denominators $\quad x^2 = A(x-1)^2 + B(x+1)(x-1) + C(x+1)$

Put $(x-1) = 0$, i.e. $x = 1 \quad \therefore\ 1 = A(0) + B(0) + C(2) \quad \therefore\ C = \frac{1}{2}$

Put $(x+1) = 0$, i.e. $x = -1 \quad \therefore\ 1 = A(4) + B(0) + C(0) \quad \therefore\ A = \frac{1}{4}$

When the crafty substitution has come to an end, we can find the remaining constants (in this case, just B) by equating coefficients. Choose the highest power involved, i.e. $x^2$ in this example.

$$[x^2] \quad \therefore\ 1 = A + B \quad \therefore\ B = 1 - A = 1 - \tfrac{1}{4} \ \therefore\ B = \tfrac{3}{4}$$

$$\therefore\ \frac{x^2}{(x+1)(x-1)^2} = \frac{1}{4}\cdot\frac{1}{x+1} + \frac{3}{4}\cdot\frac{1}{x-1} + \frac{1}{2}\cdot\frac{1}{(x-1)^2}$$

$$\therefore \int \frac{x^2}{(x+1)(x-1)^2}\,dx = \frac{1}{4}\int\frac{1}{x+1}\,dx + \frac{3}{4}\int\frac{1}{x-1}\,dx + \frac{1}{2}\int(x-1)^{-2}\,dx$$

$$= \ldots\ldots\ldots\ldots\ldots\ldots$$

**35**

$$\int \frac{x^2}{(x+1)(x-1)^2}\,dx = \frac{1}{4}\ln(x+1) + \frac{3}{4}\ln(x-1) - \frac{1}{2(x-1)} + C$$

*Example 3.* To determine $\displaystyle\int \frac{x^2+1}{(x+2)^3}\,dx$

Rules 1 and 2 of partial fractions are satisfied. The next stage is to write down the form of the partial fractions.

$$\frac{x^2+1}{(x+2)^3} = \ldots\ldots\ldots\ldots\ldots$$

**36**

$$\boxed{\frac{x^2+1}{(x+2)^3} = \frac{A}{x+2} + \frac{B}{(x+2)^2} + \frac{C}{(x+2)^3}}$$

Now clear the denominators by multiplying both sides by $(x+2)^3$. So we get

$$x^2 + 1 = \ldots\ldots\ldots\ldots\ldots\ldots$$

---

**37**

$$\boxed{x^2 + 1 = A(x+2)^2 + B(x+2) + C}$$

We now put $(x + 2) = 0$, i.e. $x = -2$

$$\therefore\ 4 + 1 = A(0) + B(0) + C \quad \therefore\ C = 5$$

There are no other brackets in this identity so we now equate coefficients, starting with the highest power involved, i.e. $x^2$. What does that give us?

---

**38**

$$x^2 + 1 = A(x+2)^2 + B(x+2) + C. \quad C = 5$$

$$[x^2]\ \therefore\ 1 = A \qquad \boxed{\therefore\ A = 1}$$

We now go to the other extreme and equate the lowest power involved, i.e. the constant terms (or absolute terms) on each side.

$$[\text{C.T.}]\quad \therefore\ 1 = 4A + 2B + C$$

$$\therefore\ 1 = 4 \quad + 2B + 5 \quad \therefore\ 2B = -8 \quad \therefore\ B = -4$$

$$\therefore\ \frac{x^2+1}{(x+2)^3} = \frac{1}{x+2} - \frac{4}{(x+2)^2} + \frac{5}{(x+2)^3}$$

$$\therefore \int \frac{x^2+1}{(x+2)^3}\,dx = \ldots\ldots\ldots\ldots\ldots\ldots$$

---

**39**

$$\int \frac{x^2+1}{(x+2)^3}\,dx = \ln(x+2) - 4\,\frac{(x+2)^{-1}}{-1} + 5\,\frac{(x+2)^{-2}}{-2} + C$$

$$= \underline{\ln(x+2) + \frac{4}{x+2} - \frac{5}{2(x+2)^2} + C}$$

*Now for another example, turn on to frame 40.*

**40**

*Example 4.* To find $\int \frac{x^2}{(x-2)(x^2+1)}\,dx$

In this example, we have a quadratic factor which will not factorize any further.

$$\therefore \quad \frac{x^2}{(x-2)(x^2+1)} = \frac{A}{x-2} + \frac{Bx+C}{x^2+1}$$

$$\therefore \quad x^2 = A(x^2+1) + (x-2)(Bx+C)$$

Put $(x-2) = 0$, i.e. $x = 2$

$$\therefore \quad 4 = A(5) + 0 \qquad\qquad \therefore \ A = \frac{4}{5}$$

Equate coefficients

$$[x^2] \quad 1 = A + B \quad \therefore \ B = 1 - A = 1 - \frac{4}{5} \quad \therefore \ B = \frac{1}{5}$$

$$[\text{C.T.}] \quad 0 = A - 2C \quad \therefore \ C = A/2 \qquad \therefore \ C = \frac{2}{5}$$

$$\therefore \quad \frac{x^2}{(x-2)(x^2+1)} = \frac{4}{5}\cdot\frac{1}{x-2} + \frac{\frac{1}{5}x + \frac{2}{5}}{x^2+1}$$

$$= \frac{4}{5}\cdot\frac{1}{x-2} + \frac{1}{5}\cdot\frac{x}{x^2+1} + \frac{2}{5}\cdot\frac{1}{x^2+1}$$

$$\therefore \int \frac{x^2}{(x-2)(x^2+1)}\,dx = \ldots\ldots\ldots\ldots$$

**41**

$$\boxed{\int \frac{x^2}{(x-2)(x^2+1)}\,dx = \frac{4}{5}\ln(x-2) + \frac{1}{10}\ln(x^2+1) + \frac{2}{5}\tan^{-1}x + C}$$

Here is one for you to do on your own.

*Example 5.* Determine $\int \frac{4x^2+1}{x(2x-1)^2}\,dx$

Rules 1 and 2 are satisfied, and the form of the partial fractions will be

$$\frac{4x^2+1}{x(2x-1)^2} = \frac{A}{x} + \frac{B}{2x-1} + \frac{C}{(2x-1)^2}$$

*Off you go then. When you have finished it completely, turn on to frame 42.*

**42**

$$\int \frac{4x^2+1}{x(2x-1)^2}\,dx = \ln x - \frac{2}{2x-1} + C$$

Check through your working in detail.

$$\frac{4x^2+1}{x(2x-1)^2} = \frac{A}{x} + \frac{B}{2x-1} + \frac{C}{(2x-1)^2}$$

$$\therefore\ 4x^2+1 = A(2x-1)^2 + Bx(2x-1) + Cx$$

Put $(2x-1) = 0$, i.e. $x = 1/2$

$$\therefore\ 2 = A(0) + B(0) + \frac{C}{2} \qquad \therefore\ C = 4$$

$$\left.\begin{aligned} &[x^2] \quad 4 = 4A + 2B \quad \therefore\ 2A + B = 2 \\ &[\text{C.T.}] \quad 1 = A \end{aligned}\right\} \quad \begin{aligned} A &= 1 \\ B &= 0 \end{aligned}$$

$$\therefore\ \frac{4x^2+1}{x(2x-1)^2} = \frac{1}{x} + \frac{4}{(2x-1)^2}$$

$$\therefore \int \frac{4x^2+1}{x(2x-1)^2}\,dx = \int \frac{1}{x}\,dx + 4\int (2x-1)^{-2}\,dx$$

$$= \ln x + \frac{4.(2x-1)^{-1}}{-1.\,2} + C$$

$$= \ln x - \frac{2}{2x-1} + C$$

*Move on to frame 43.*

**43**

We have done quite a number of integrals of one type or another in our work so far. We have covered:

1. the basic standard integrals,
2. functions of a linear function of $x$,
3. integrals in which one part is the differential coefficient of the other part,
4. integration by parts, i.e. integration of products,
5. integration by partial fractions.

Before we finish this part of the programme on integration, let us look particularly at some types of integrals involving trig. functions.

*So, on we go to frame 44.*

**44**

**6. Integration of trigonometrical functions**

(a) *Powers of sin x and of cos x*

(i) We already know that

$$\int \sin x \, dx = -\cos x + C$$
$$\int \cos x \, dx = \sin x + C$$

(ii) To integrate $\sin^2 x$ and $\cos^2 x$, we express the function in terms of the cosine of the double angle.

$$\cos 2x = 1 - 2\sin^2 x \quad \text{and} \quad \cos 2x = 2\cos^2 x - 1$$

$$\therefore \sin^2 x = \frac{1}{2}(1 - \cos 2x) \quad \text{and} \quad \cos^2 x = \frac{1}{2}(1 + \cos 2x)$$

$$\therefore \int \sin^2 x \, dx = \frac{1}{2}\int (1 - \cos 2x) \, dx = \underline{\frac{x}{2} - \frac{\sin 2x}{4} + C}$$

$$\therefore \int \cos^2 x \, dx = \frac{1}{2}\int (1 + \cos 2x) \, dx = \underline{\frac{x}{2} + \frac{\sin 2x}{4} + C}$$

Notice how nearly alike these two results are. One must be careful to distinguish between them, so make a note of them in your record book for future reference.

*Then move on to frame 45.*

**45**

(iii) To integrate $\sin^3 x$ and $\cos^3 x$.

To integrate $\sin^3 x$, we release one of the factors $\sin x$ from the power and convert the remaining $\sin^2 x$ into $(1 - \cos^2 x)$, thus:

$$\int \sin^3 x \, dx = \int \sin^2 x . \sin x \, dx = \int (1 - \cos^2 x) \sin x \, dx$$
$$= \int \sin x \, dx - \int \cos^2 x . \sin x \, dx$$
$$= \underline{-\cos x + \frac{\cos^3 x}{3} + C}$$

We do not normally remember this as a standard result, but we certainly do remember the method by which we can find $\int \sin^3 x \, dx$ when necessary.

So, in a similar way, you can now find $\int \cos^3 x \, dx$.

*When you have done it, turn on to frame 46.*

**46**

$$\int \cos^3 x\, dx = \sin x - \frac{\sin^3 x}{3} + C$$

For:

$$\int \cos^3 x\, dx = \int \cos^2 x \,.\cos x\, dx = \int (1 - \sin^2 x) \cos x\, dx$$

$$= \int \cos x\, dx - \int \sin^2 x \,.\cos x\, dx \;=\; \underline{\sin x - \frac{\sin^3 x}{3} + C}$$

Now what about this one?

(iv) To integrate $\sin^4 x$ and $\cos^4 x$.

$$\int \sin^4 x\, dx = \int (\sin^2 x)^2\, dx = \int \frac{(1 - \cos 2x)^2}{4}\, dx$$

$$= \int \frac{1 - 2\cos 2x + \cos^2 2x}{4}\, dx \quad \text{N.B.} \begin{cases} \cos^2 x = \frac{1}{2}(1 + \cos 2x) \\ \cos^2 2x = \frac{1}{2}(1 + \cos 4x) \end{cases}$$

$$= \frac{1}{4}\int (1 - 2\cos 2x + \frac{1}{2} + \frac{1}{2}.\cos 4x)\, dx$$

$$= \frac{1}{4}\int (\frac{3}{2} - 2\cos 2x + \frac{1}{2}\cos 4x)\, dx$$

$$= \frac{1}{4}\left\{\frac{3x}{2} - \frac{2\sin 2x}{2} + \frac{1}{2}.\frac{\sin 4x}{4}\right\} + C \;=\; \frac{3x}{8} - \frac{\sin 2x}{4} + \frac{\sin 4x}{32} + C$$

Remember not this result, but the *method.*

Now you find $\int \cos^4 x\, dx$ in much the same way.

**47**

$$\int \cos^4 x\, dx = \frac{3x}{8} + \frac{\sin 2x}{4} + \frac{\sin 4x}{32} + C$$

The working is very much like that of the last example.

$$\int \cos^4 x\, dx = \int (\cos^2 x)^2\, dx = \int \frac{(1 + \cos 2x)^2}{4}\, dx$$

$$= \int \frac{(1 + 2\cos 2x + \cos^2 2x)}{4}\, dx \;=\; \frac{1}{4}\int (1 + 2\cos 2x + \frac{1}{2} + \frac{1}{2}.\cos 4x)\, dx$$

$$= \frac{1}{4}\int (\frac{3}{2} + 2\cos 2x + \frac{1}{2}.\cos 4x)\, dx \;=\; \frac{1}{4}\left\{\frac{3x}{2} + \sin 2x + \frac{\sin 4x}{8}\right\} + C$$

$$= \frac{3x}{8} + \frac{\sin 2x}{4} + \frac{\sin 4x}{32} + C$$

*On to the next frame.*

**48** (v) To integrate $\sin^5 x$ and $\cos^5 x$

We can integrate $\sin^5 x$ in very much the same way as we found the integral of $\sin^3 x$.

$$\int \sin^5 x\,dx = \int \sin^4 x.\sin x\,dx = \int (1-\cos^2 x)^2 \sin x\,dx$$
$$= \int (1 - 2\cos^2 x + \cos^4 x)\sin x\,dx$$
$$= \int \sin x\,dx - 2\int \cos^2 x.\sin x\,dx + \int \cos^4 x.\sin x\,dx$$
$$= -\cos x + \frac{2\cos^3 x}{3} - \frac{\cos^5 x}{5} + C$$

Similarly,

$$\int \cos^5 x\,dx = \int \cos^4 x.\cos x\,dx = \int (1-\sin^2 x)^2 \cos x\,dx$$
$$= \int (1 - 2\sin^2 x + \sin^4 x)\cos x\,dx$$
$$= \int \cos x\,dx - 2\int \sin^2 x.\cos x\,dx + \int \sin^4 x.\cos x\,dx$$
$$= \sin x - \frac{2\sin^3 x}{3} + \frac{\sin^5 x}{5} + C$$

Note the method, but do not try to memorize these results. Sometimes we need to integrate higher powers of $\sin x$ and $\cos x$ than those we have considered. In those cases, we make use of a different approach which we shall deal with in due course.

**49** (b) *Products of sines and cosines*

Finally, while we are dealing with the integrals of trig. functions, let us consider one further type. Here is an example:

$$\int \sin 4x.\cos 2x\,dx$$

To determine this, we make use of the identity

$$2\sin A\cos B = \sin(A+B) + \sin(A-B)$$

$$\therefore\ \sin 4x.\cos 2x = \frac{1}{2}(2\sin 4x\cos 2x)$$
$$= \frac{1}{2}\Big\{\sin(4x+2x) + \sin(4x-2x)\Big\}$$
$$= \frac{1}{2}\Big\{\sin 6x + \sin 2x\Big\}$$

$$\therefore\ \int \sin 4x\cos 2x\,dx = \frac{1}{2}\int(\sin 6x + \sin 2x)\,dx = -\frac{\cos 6x}{12} - \frac{\cos 2x}{4} + C$$

**50**

This type of integral means, of course, that you must know your trig. identities. Do they need polishing up? Now is the chance to revise some of them, anyway.

There are four identities very like the one we have just used.

$$2\sin A\cos B = \sin(A+B) + \sin(A-B)$$
$$2\cos A\sin B = \sin(A+B) - \sin(A-B)$$
$$2\cos A\cos B = \cos(A+B) + \cos(A-B)$$
$$2\sin A\sin B = \cos(A-B) - \cos(A+B)$$

Remember that the compound angles are interchanged in the last line. These are important and very useful, so copy them down into your record book and learn them.

*Now move to frame 51.*

**51**

Now another example of the same kind.

*Example:* $\int \cos 5x \sin 3x\, dx$

$$= \frac{1}{2}\int (2\cos 5x \sin 3x)\, dx$$
$$= \frac{1}{2}\int \left\{\sin(5x+3x) - \sin(5x-3x)\right\} dx$$
$$= \frac{1}{2}\int \left\{\sin 8x - \sin 2x\right\} dx$$
$$= \frac{1}{2}\left\{-\frac{\cos 8x}{8} + \frac{\cos 2x}{2}\right\} + C$$
$$= \frac{\cos 2x}{4} - \frac{\cos 8x}{16} + C$$

And now here is one for you to do:

$$\int \cos 6x \cos 4x\, dx = \ldots\ldots\ldots\ldots$$

*Off you go. Finish it, then turn on to frame 52.*

**52**

$$\boxed{\int \cos 6x \cos 4x\, dx = \frac{\sin 10x}{20} + \frac{\sin 2x}{4} + C}$$

For
$$\int \cos 6x \cos 4x\, dx = \frac{1}{2}\int 2\cos 6x \cos 4x\, dx$$
$$= \frac{1}{2}\int \left\{\cos 10x + \cos 2x\right\} dx$$
$$= \frac{1}{2}\left\{\frac{\sin 10x}{10} + \frac{\sin 2x}{2}\right\} + C$$
$$= \frac{\sin 10x}{20} + \frac{\sin 2x}{4} + C$$

Well, there you are. They are all done in the same basic way. Here is one last one for you to do. Take care!

$$\int \sin 5x \sin x\, dx = \ldots\ldots\ldots\ldots\ldots$$

This will use the last of our four trig. identities, the one in which the compound angles are interchanged, so do not get caught.

*When you have finished, move on to frame 53.*

**53** Well, here it is, worked out in detail. Check your result.

$$\int \sin 5x \sin x\, dx = \frac{1}{2}\int 2\sin 5x \sin x\, dx$$
$$= \frac{1}{2}\int \left\{\cos(5x - x) - \cos(5x + x)\right\} dx$$
$$= \frac{1}{2}\int \left\{\cos 4x - \cos 6x\right\} dx$$
$$= \frac{1}{2}\left\{\frac{\sin 4x}{4} - \frac{\sin 6x}{6}\right\} + C$$
$$= \frac{\sin 4x}{8} - \frac{\sin 6x}{12} + C$$

□□□□□□□□□□□□□□□□□□□□□□□□□□□□□□□□□□□□□

This brings us to the end of Part 1 of the programme on integration, except for the Test Exercise which follows in the next frame. Before you work the exercise, look back through the notes you have made in your record book, and brush up any points on which you are not perfectly clear.

*When you are ready, turn on to the next frame.*

**54**

Here is the Test Exercise on the work you have been doing in this programme. The integrals are all quite straightforward so you will have no trouble with them. Take your time: there is no need to hurry – and no extra marks for speed!

**Test Exercise – XV**

Answer *all* the questions.

Determine the following integrals:

1. $\int e^{\cos x} \sin x \, dx$
2. $\int \frac{\ln x}{\sqrt{x}} dx$
3. $\int \tan^2 x \, dx$
4. $\int x^2 \sin 2x \, dx$
5. $\int e^{-3x} \cos 2x \, dx$
6. $\int \sin^5 x \, dx$
7. $\int \cos^4 x \, dx$
8. $\int \frac{4x + 2}{x^2 + x + 5} dx$
9. $\int x\sqrt{(1 + x^2)} \, dx$
10. $\int \frac{2x - 1}{x^2 - 8x + 15} dx$
11. $\int \frac{2x^2 + x + 1}{(x - 1)(x^2 + 1)} dx$
12. $\int \sin 5x \cos 3x \, dx$

*You are now ready to start Part 2 of the programme on integration.*

## Further Problems – XV

Determine the following:

1. $\displaystyle\int \frac{3x^2}{(x-1)(x^2+x+1)}\,dx$

2. $\displaystyle\int_0^{\pi/2} \sin 7x \cos 5x \, dx$

3. $\displaystyle\int \frac{\sin 2x}{1+\cos^2 x}\,dx$

4. $\displaystyle\int_0^{a/2} x^2(a^2-x^2)^{-3/2}\,dx$

5. $\displaystyle\int_0^{\pi} x \sin^2 x \, dx$

6. $\displaystyle\int \frac{2x+1}{(x^2+x+1)^{3/2}}\,dx$

7. $\displaystyle\int \frac{x+1}{(x-1)(x^2+x+1)}\,dx$

8. $\displaystyle\int \frac{x^2}{x+1}\,dx$

9. $\displaystyle\int \frac{2x^2+x+1}{(x-1)(x^2+1)}\,dx$

10. $\displaystyle\int_0^{\pi} (\pi - x)\cos x \, dx$

11. $\displaystyle\int_0^{n} x^2 (n-x)^p \, dx$, for $p > 0$

12. $\displaystyle\int \frac{4x^2-7x+13}{(x-2)(x^2+1)}\,dx$

13. $\displaystyle\int_0^{\pi/2} \sin 5x \cos 3x \, dx$

14. $\displaystyle\int \frac{\sin^{-1} x}{\sqrt{(1-x^2)}}\,dx$

15. $\displaystyle\int_0^{1} \frac{x^2-2x}{(2x+1)(x^2+1)}\,dx$

16. $\displaystyle\int_0^{\pi} x^2 \sin x \, dx$

17. $\displaystyle\int_0^{\pi} x^2 \sin^2 x \, dx$

18. $\displaystyle\int_0^{1} x \tan^{-1} x \, dx$

19. $\displaystyle\int \frac{dx}{x^2(1+x^2)}$

20. $\displaystyle\int x\sqrt{(1+x^2)}\,dx$

21. $\displaystyle\int \frac{8-x}{(x-2)^2(x+1)}\,dx$

22. $\displaystyle\int_0^{\pi} e^{2x} \cos 4x \, dx$

23. $\displaystyle\int_0^{\pi/2} \sin^5 x \cos^3 x \, dx$

24. $\displaystyle\int_0^{\pi/6} e^{2\theta} \cos 3\theta \, d\theta$

25. $\displaystyle\int_0^{\pi/\omega} \sin \omega t \cos 2\omega t \, dt$

26. $\displaystyle\int \tan^2 x \sec^2 x \, dx$

27. $\displaystyle\int \frac{2x+3}{(x-4)(5x+2)}\,dx$

28. $\displaystyle\int \frac{dx}{\sqrt{x^2+4x+4}}$

29. $\displaystyle\int \frac{5x^2+11x-2}{(x+5)(x^2+9)}\,dx$

30. $\displaystyle\int \frac{x-1}{9x^2-18x+17}\,dx$

31. $\int \frac{4x^5}{x^4 - 1}\, dx$

32. $\int x^2 \ln(1 + x^2)\, dx$

33. $\int \frac{\cos\theta - \sin\theta}{\cos\theta + \sin\theta}\, d\theta$

34. $\int \frac{1 - \sin\theta}{\cos^2\theta}\, d\theta$

35. $\int \frac{2x - 3}{(x - 1)(x - 2)(x + 3)}\, dx$

36. $\int_0^{\pi/3} \frac{\sin x}{(1 + \cos x)^2}\, dx$

37. $\int_1^2 (x - 1)^2 \ln x\, dx$

38. $\int \frac{4x^2 - x + 12}{x(x^2 + 4)}\, dx$

39. $\int \frac{x^3 + x + 1}{x^4 + x^2}\, dx$

40. If $L\frac{di}{dt} + Ri = E$, where L, R and E are constants, and it is known that $i = 0$ at $t = 0$, show that

$$\int_0^t (Ei - Ri^2)\, dt = \frac{Li^2}{2}$$

*Note.* Some of the integrals above are *definite integrals,* so here is a reminder.

In $\int_a^b f(x)\, dx$, the values of $a$ and $b$ are called the *limits* of the integral.

If $\int f(x)\, dx = F(x) + C$

then $\int_a^b f(x)\, dx = [F(x)]_{x=b} - [F(x)]_{x=a}$

# Programme 16

## INTEGRATION

### PART 2

**1** I. Consider the integral $\int \frac{dZ}{Z^2 - A^2}$

From our work in Part 1 of this programme on integration, you will recognize that the denominator can be factorized and that the function can therefore be expressed in its *partial fractions*.

$$\frac{1}{Z^2 - A^2} \equiv \frac{1}{(Z-A)(Z+A)} \equiv \frac{P}{Z-A} + \frac{Q}{Z+A}$$

where P and Q are constants.

$$\therefore\ 1 \equiv P(Z+A) + Q(Z-A)$$

Put $Z = A$ $\quad \therefore\ 1 = P(2A) + Q(0) \quad \therefore\ P = \frac{1}{2A}$

Put $Z = -A$ $\quad \therefore\ 1 = P(0) + Q(-2A) \quad \therefore\ Q = -\frac{1}{2A}$

$$\therefore\ \frac{1}{Z^2 - A^2} = \frac{1}{2A} \cdot \frac{1}{Z-A} - \frac{1}{2A} \cdot \frac{1}{Z+A}$$

$$\therefore \int \frac{1}{Z^2 - A^2}\, dZ = \frac{1}{2A} \int \frac{1}{Z-A}\, dZ - \frac{1}{2A} \int \frac{1}{Z+A}\, dZ$$

$$= \ldots\ldots\ldots\ldots\ldots\ldots$$

**2**

$$\int \frac{1}{Z^2 - A^2}\, dZ = \frac{1}{2A} . \ln(Z-A) - \frac{1}{2A} . \ln(Z+A) + C$$

$$= \boxed{\frac{1}{2A} . \ln\left\{\frac{Z-A}{Z+A}\right\} + C}$$

This is the first of nine standard results which we are going to establish in this programme. They are useful to remember since the standard results will remove the need to work each example in detail, as you will see.

We have $\int \frac{1}{Z^2 - A^2}\, dZ = \frac{1}{2A} \ln\left\{\frac{Z-A}{Z+A}\right\} + C$

$$\therefore \int \frac{1}{Z^2 - 16}\, dZ = \int \frac{1}{Z^2 - 4^2}\, dZ = \frac{1}{8} \ln\left\{\frac{Z-4}{Z+4}\right\} + C$$

and $\int \frac{1}{x^2 - 5}\, dx = \int \frac{1}{x^2 - (\sqrt{5})^2}\, dx = \frac{1}{2\sqrt{5}} \ln\left\{\frac{x - \sqrt{5}}{x + \sqrt{5}}\right\} + C$

(Note that 5 can be written as the square of its own square root.)

So $\quad \int \frac{1}{Z^2 - A^2}\, dZ = \frac{1}{2A} \ln\left\{\frac{Z-A}{Z+A}\right\} + C \quad \ldots\ldots\ldots\ldots \text{(i)}$

*Copy this result into your record book and move on to frame 3.*

**3**

We had $$\boxed{\int \frac{dZ}{Z^2 - A^2} = \frac{1}{2A} \ln\left\{\frac{Z-A}{Z+A}\right\} + C}$$

So therefore:

$$\int \frac{dZ}{Z^2 - 25} = \ldots\ldots\ldots\ldots$$

$$\int \frac{dZ}{Z^2 - 7} = \ldots\ldots\ldots\ldots$$

**4**

$$\int \frac{dZ}{Z^2 - 25} = \int \frac{dZ}{Z^2 - 5^2} = \boxed{\frac{1}{10} . \ln\left\{\frac{Z-5}{Z+5}\right\} + C}$$

$$\int \frac{dZ}{Z^2 - 7} = \int \frac{dZ}{Z^2 - (\sqrt{7})^2} = \boxed{\frac{1}{2\sqrt{7}} . \ln\left\{\frac{Z-\sqrt{7}}{Z+\sqrt{7}}\right\} + C}$$

□□□□□□□□□□□□□□□□□□□□□□□□□□□□□□□□□□□□□□□□

Now what about this one?

$$\int \frac{1}{x^2 + 4x + 2} dx$$

At first sight, this seems to have little to do with the standard result, or to the examples we have done so far. However, let us re-write the denominator, thus:

$$x^2 + 4x + 2 = x^2 + 4x \qquad\qquad + 2. \text{ (Nobody will argue with that!)}$$

Now we complete the square with the first two terms, by adding on the square of half the coefficient of $x$.

$$x^2 + 4x + 2 = x^2 + 4x + 2^2 \quad + 2$$

and of course we must subtract an equal amount, i.e. 4, to keep the identity true.

$$\therefore\ x^2 + 4x + 2 = \underbrace{x^2 + 4x + 2^2}_{} + 2 - 4$$

$$= \quad (x+2)^2 \quad - 2$$

So $\displaystyle\int \frac{1}{x^2 + 4x + 2} dx$ can be written $\displaystyle\int \frac{1}{\ldots\ldots\ldots\ldots} dx$

*Turn on to frame 5.*

**5**

$$\int \frac{1}{x^2 + 4x + 2}\, dx = \boxed{\int \frac{1}{(x+2)^2 - 2}\, dx}$$

Then we can express the constant 2 as the square of its own square root.

$$\therefore \int \frac{1}{x^2 + 4x + 2}\, dx = \int \frac{1}{(x+2)^2 - (\sqrt{2})^2}\, dx$$

You will see that we have re-written the given integral in the form $\int \frac{1}{Z^2 - A^2}\, dZ$ where, in this case, $Z = (x + 2)$ and $A = \sqrt{2}$. Now the standard result was

$$\int \frac{1}{Z^2 - A^2}\, dZ = \frac{1}{2A} \ln\left\{\frac{Z - A}{Z + A}\right\} + C$$

Substituting our expressions for Z and A in this result, gives

$$\int \frac{1}{x^2 + 4x + 2}\, dx = \int \frac{1}{(x+2)^2 - (\sqrt{2})^2}\, dx$$

$$= \underline{\frac{1}{2\sqrt{2}} \ln\left\{\frac{x + 2 - \sqrt{2}}{x + 2 + \sqrt{2}}\right\} + C}$$

Once we have found our particular expressions for Z and A, all that remains is to substitute these expressions in the standard result.

*On now to frame 6.*

---

**6**

Here is another example.

$$\int \frac{1}{x^2 + 6x + 4}\, dx$$

First complete the square with the first two terms of the given denominator and subtract an equal amount.

$$\begin{aligned} x^2 + 6x + 4 &= x^2 + 6x \qquad + 4 \\ &= \underbrace{x^2 + 6x + 3^2} + 4 - 9 \\ &= \quad (x+3)^2 \quad - 5 \\ &= \quad (x+3)^2 \quad - (\sqrt{5})^2 \end{aligned}$$

So $$\int \frac{1}{x^2 + 6x + 4}\, dx = \int \frac{1}{(x+3)^2 - (\sqrt{5})^2}\, dx$$

$$= \dots\dots\dots\dots\dots\dots$$

**7**

$$\int \frac{1}{x^2+6x+4}\,dx = \frac{1}{2\sqrt{5}} \ln \frac{x+3-\sqrt{5}}{x+3+\sqrt{5}} + C$$

□□□□□□□□□□□□□□□□□□□□□□□□□□□□□□□□□□□□□□

And another on your own:

Find $\displaystyle\int \frac{1}{x^2-10x+18}\,dx$

*When you have finished, move on to frame 8.*

**8**

$$\int \frac{1}{x^2-10x+18}\,dx = \frac{1}{2\sqrt{7}} \ln\left\{\frac{x-5-\sqrt{7}}{x-5+\sqrt{7}}\right\} + C$$

For:

$$\begin{aligned} x^2-10x+18 &= x^2-10x \qquad +18 \\ &= x^2-10x+5^2+18-25 \\ &= (x-5)^2-7 \\ &= (x-5)^2-(\sqrt{7})^2 \end{aligned}$$

$$\therefore \int \frac{1}{x^2-10x+18}\,dx = \frac{1}{2\sqrt{7}} \ln\left\{\frac{x-5-\sqrt{7}}{x-5+\sqrt{7}}\right\} + C$$

*Now on to frame 9.*

**9**

Now what about this one? $\displaystyle\int \frac{1}{5x^2-2x-4}\,dx$

In order to complete the square, as we have done before, the coefficient of $x$ must be 1. Therefore, in the denominator, we must first of all take out a factor 5 to reduce the second degree term to a single $x^2$.

$$\therefore \int \frac{1}{5x^2-2x-4}\,dx = \frac{1}{5}\int \frac{1}{x^2-\frac{2}{5}x-\frac{4}{5}}\,dx$$

Now we can proceed as in the previous examples.

$$\begin{aligned} x^2-\tfrac{2}{5}x-\tfrac{4}{5} &= x^2-\tfrac{2}{5}x-\tfrac{4}{5} \\ &= x^2-\tfrac{2}{5}x+\left(\tfrac{1}{5}\right)^2-\tfrac{4}{5}-\tfrac{1}{25} \\ &= \left(x-\tfrac{1}{5}\right)^2-\tfrac{21}{25} \\ &= \left(x-\tfrac{1}{5}\right)^2-\left(\tfrac{\sqrt{21}}{5}\right)^2 \end{aligned}$$

$$\therefore \int \frac{1}{5x^2-2x-4}\,dx = \ldots\ldots\ldots\ldots\ldots$$

(Remember the factor 1/5 in the front!)

**10**

$$\int \frac{1}{5x^2 - 2x - 4}\,dx = \frac{1}{2\sqrt{21}} \ln\left\{\frac{5x - 1 - \sqrt{21}}{5x - 1 + \sqrt{21}}\right\} + C$$

Here is the working: follow it through.

$$\int \frac{1}{5x^2 - 2x - 4}\,dx = \frac{1}{5}\int \frac{1}{\left(x - \frac{1}{5}\right)^2 - \left(\frac{\sqrt{21}}{5}\right)^2}\,dx$$

$$= \frac{1}{5} \cdot \frac{5}{2\sqrt{21}} \ln\left\{\frac{x - 1/5 - \sqrt{21}/5}{x - 1/5 + \sqrt{21}/5}\right\} + C$$

$$= \frac{1}{2\sqrt{21}} \ln\left\{\frac{5x - 1 - \sqrt{21}}{5x - 1 + \sqrt{21}}\right\} + C$$

□□□□□□□□□□□□□□□□□□□□□□□□□□□□□□□□□□□□□□□

II. Now, in very much the same way, let us establish the second standard result by considering $\int \frac{dZ}{A^2 - Z^2}$

This looks rather like the last one and can be determined again by partial fractions.

Work through it on your own and determine the general result.

*Then turn on to frame 11 and check your working.*

**11**

$$\int \frac{dZ}{A^2 - Z^2} = \frac{1}{2A} \ln\left\{\frac{A + Z}{A - Z}\right\} + C$$

For: $\frac{1}{A^2 - Z^2} = \frac{1}{(A - Z)(A + Z)} = \frac{P}{A - Z} + \frac{Q}{A + Z}$

$$\therefore\ 1 = P(A + Z) + Q(A - Z)$$

Put $Z = A$ $\quad \therefore\ 1 = P(2A) + Q(0)$ $\quad \therefore\ P = \frac{1}{2A}$

Put $Z = -A$ $\quad \therefore\ 1 = P(0) + Q(2A)$ $\quad \therefore\ Q = \frac{1}{2A}$

$$\therefore \int \frac{1}{A^2 - Z^2}\,dZ = \frac{1}{2A} \cdot \int \frac{1}{A + Z}\,dZ + \frac{1}{2A}\int \frac{1}{A - Z}\,dZ$$

$$= \frac{1}{2A} . \ln(A + Z) - \frac{1}{2A} . \ln(A - Z) + C$$

$$\therefore \int \frac{1}{A^2 - Z^2}\,dZ = \frac{1}{2A} \ln\left\{\frac{A + Z}{A - Z}\right\} + C \quad \text{................ (ii)}$$

Copy this second standard form into your record book and compare it with the first result. They are very much alike. *Turn to frame 12.*

**12**

So we have:

$$\int \frac{dZ}{Z^2 - A^2} = \frac{1}{2A} \ln\left\{\frac{Z - A}{Z + A}\right\} + C$$

$$\int \frac{dZ}{A^2 - Z^2} = \frac{1}{2A} \ln\left\{\frac{A + Z}{A - Z}\right\} + C$$

Note how nearly alike these two results are.

Now for some examples on the second standard form.

*Example 1.* $\int \frac{1}{9 - x^2} dx = \int \frac{1}{3^2 - x^2} dx = \frac{1}{6} \ln\left\{\frac{3 + x}{3 - x}\right\} + C$

*Example 2.* $\int \frac{1}{5 - x^2} dx = \int \frac{1}{(\sqrt{5})^2 - x^2} dx = \frac{1}{2\sqrt{5}} \ln\left\{\frac{\sqrt{5} + x}{\sqrt{5} - x}\right\} + C$

*Example 3.* $\int \frac{1}{3 - x^2} dx = \ldots\ldots\ldots\ldots$

**13**

$$\frac{1}{2\sqrt{3}} \ln\left\{\frac{\sqrt{3} + x}{\sqrt{3} - x}\right\} + C$$

*Example 4.* $\int \frac{1}{3 + 6x - x^2} dx$

We complete the square in the denominator as before, but we must be careful of the signs – and, do not forget, the coefficient of $x^2$ must be 1. So we do it like this:

$$3 + 6x - x^2 = 3 - (x^2 - 6x \qquad )$$

Note that we put the $x^2$ term and the $x$ term inside brackets with a minus sign outside. Naturally, the $6x$ becomes $-6x$ inside the brackets. Now we can complete the square inside the brackets and *add* on a similar amount outside the brackets (since everything inside the brackets is negative).

So
$$\begin{aligned} 3 + 6x - x^2 &= 3 - (x^2 - 6x + 3^2) + 9 \\ &= 12 - (x - 3)^2 \\ &= (2\sqrt{3})^2 - (x - 3)^2 \end{aligned}$$

In this case, then, $A = 2\sqrt{3}$ and $Z = (x - 3)$

$$\therefore \int \frac{1}{3 + 6x - x^2} dx = \int \frac{1}{(2\sqrt{3})^2 - (x - 3)^2} dx$$

$$= \ldots\ldots\ldots\ldots\ldots\ldots$$

*Finish it off.*

**14**

$$\boxed{\frac{1}{4\sqrt{3}}\ln\left\{\frac{2\sqrt{3}+x-3}{2\sqrt{3}-x+3}\right\}+C}$$

□□□□□□□□□□□□□□□□□□□□□□□□□□□□□□□□□□□□□□□□□□□

Here is another example of the same type.

*Example 5.* $\displaystyle\int\frac{1}{9-4x-x^2}\,dx$

First of all we carry out the 'completing the square' routine.

$$\begin{aligned}9-4x-x^2 &= 9-(x^2+4x\qquad)\\ &= 9-(x^2+4x+2^2)+4\\ &= 13-(x+2)^2\\ &= (\sqrt{13})^2-(x+2)^2\end{aligned}$$

In this case, $A=\sqrt{13}$ and $Z=(x+2)$

Now we know that $\displaystyle\int\frac{dZ}{A^2-Z^2}=\frac{1}{2A}\ln\left\{\frac{A+Z}{A-Z}\right\}+C$

So that, in this example $\displaystyle\int\frac{1}{9-4x-x^2}\,dx=$ ....................

**15**

$$\boxed{\frac{1}{2\sqrt{13}}\ln\left\{\frac{\sqrt{13}+x+2}{\sqrt{13}-x-2}\right\}+C}$$

□□□□□□□□□□□□□□□□□□□□□□□□□□□□□□□□□□□□□□□□□□□

*Example 6.* $\displaystyle\int\frac{1}{5+4x-2x^2}\,dx$

Remember that we must first remove the factor 2 from the denominator to reduce the coefficient of $x^2$ to 1.

$$\therefore\int\frac{1}{5+4x-2x^2}\,dx=\frac{1}{2}\int\frac{1}{\frac{5}{2}+2x-x^2}\,dx$$

Now we proceed as before.

$$\begin{aligned}\frac{5}{2}+2x-x^2 &= \frac{5}{2}-(x^2-2x\qquad)\\ &= \frac{5}{2}-(x^2-2x+1^2)+1\\ &= \frac{7}{2}-(x-1)^2\\ &= (\sqrt{3{\cdot}5})^2-(x-1)^2\end{aligned}$$

$$\therefore\int\frac{1}{5+4x-2x^2}\,dx= \text{....................}$$

(Do not forget the factor 2 we took out of the denominator.)

**16**

$$\frac{1}{4\sqrt{3.5}}\ln\left\{\frac{\sqrt{3.5}+x-1}{\sqrt{3.5}-x+1}\right\}+C$$

□□□□□□□□□□□□□□□□□□□□□□□□□□□□□□□□□□□□□□□□□□□□□

Right. Now just one more.

*Example 7.* Determine $\int\frac{1}{6-6x-5x^2}\,dx$.

What is the first thing to do?

**17**

Reduce the coefficient of $x^2$ to 1,
i.e. take out a factor 5 from the denominator.

Correct. Let us do it then.

$$\int\frac{1}{6-6x-5x^2}\,dx=\frac{1}{5}\int\frac{1}{\frac{6}{5}-\frac{6}{5}x-x^2}\,dx$$

Now you can complete the square as usual and finish it off.

*Then move to frame 18.*

**18**

$$\int\frac{1}{6-6x-5x^2}\,dx=\frac{1}{2\sqrt{39}}\ln\left\{\frac{\sqrt{39}+5x+3}{\sqrt{39}-5x-3}\right\}+C$$

For:

$$\int\frac{1}{6-6x-5x^2}\,dx=\frac{1}{5}\int\frac{1}{\frac{6}{5}-\frac{6}{5}x-x^2}\,dx$$

$$\begin{aligned}\frac{6}{5}-\frac{6}{5}x-x^2&=\frac{6}{5}-\left(x^2+\frac{6}{5}x\qquad\right)\\&=\frac{6}{5}-\left\{x^2+\frac{6}{5}x+\left(\frac{3}{5}\right)^2\right\}+\frac{9}{25}\\&=\frac{39}{25}-\left(x+\frac{3}{5}\right)^2\\&=\left(\frac{\sqrt{39}}{5}\right)^2-\left(x+\frac{3}{5}\right)^2\end{aligned}$$

So that $A=\frac{\sqrt{39}}{5}$ and $Z=(x+\frac{3}{5})$

Now $\qquad\int\frac{1}{A^2-Z^2}\,dZ=\frac{1}{2A}\ln\left\{\frac{A+Z}{A-Z}\right\}+C$

$$\therefore\int\frac{1}{6-6x-x^2}\,dx=\frac{1}{5}\cdot\frac{1}{\frac{2\sqrt{39}}{5}}\ln\left\{\frac{\sqrt{39/5}+x+3/5}{\sqrt{39/5}-x-3/5}\right\}+C$$

$$=\frac{1}{2\sqrt{39}}\ln\left\{\frac{\sqrt{39}+5x+3}{\sqrt{39}-5x-3}\right\}+C$$

*Now turn to frame 19.*

**19**

By way of revision, cover up your notes and complete the following. Do not work out the integrals in detail; just quote the results.

$$\text{(i)} \int \frac{dZ}{Z^2 - A^2} = \dots\dots\dots\dots\dots$$

$$\text{(ii)} \int \frac{dZ}{A^2 - Z^2} = \dots\dots\dots\dots\dots$$

*Check your results with frame 20.*

**20**

$$\int \frac{dZ}{Z^2 - A^2} = \frac{1}{2A} \ln\left\{\frac{Z-A}{Z+A}\right\} + C$$

$$\int \frac{dZ}{A^2 - Z^2} = \frac{1}{2A} \ln\left\{\frac{A+Z}{A-Z}\right\} + C$$

□□□□□□□□□□□□□□□□□□□□□□□□□□□□□□□□□□□□□□

III. Now for the third standard form.

Consider $\displaystyle\int \frac{dZ}{Z^2 + A^2}$

Here, the denominator will not factorize, so we cannot apply the rules of partial fractions. We now turn to substitution, i.e. we try to find a substitution for Z which will enable us to write the integral in a form which we already know how to tackle.

Suppose we put $Z = A \tan\theta$.

Then $\quad Z^2 + A^2 = A^2 \tan^2\theta + A^2 = A^2(1 + \tan^2\theta) = A^2 \sec^2\theta$

Also $\quad \dfrac{dZ}{d\theta} = A \sec^2\theta \quad$ i.e. $dZ \equiv A \sec^2\theta \, d\theta$

The integral now becomes

$$\int \frac{1}{Z^2 + A^2}\, dZ = \int \frac{1}{A^2 \sec^2\theta} . A \sec^2\theta \, d\theta = \int \frac{1}{A}\, d\theta$$

$$= \frac{1}{A}.\theta + C$$

This is a nice simple result, but we cannot leave it like that, for $\theta$ is a variable we introduced in the working. We must express $\theta$ in terms of the original variable Z.

$$Z = A \tan\theta, \quad \therefore \frac{Z}{A} = \tan\theta \quad \therefore \theta = \tan^{-1}\frac{Z}{A}$$

$$\therefore \int \frac{1}{Z^2 + A^2}\, dZ = \frac{1}{A} \tan^{-1}\left\{\frac{Z}{A}\right\} + C \quad \dots\dots\dots\dots \text{(iii)}$$

*Add this one to your growing list of standard forms.*

**21**

$$\int \frac{1}{Z^2 + A^2}\, dZ = \frac{1}{A}\tan^{-1}\left\{\frac{Z}{A}\right\} + C$$

*Example 1.* $\displaystyle\int \frac{1}{x^2 + 16}\, dx = \int \frac{1}{x^2 + 4^2}\, dx = \frac{1}{4}\tan^{-1}\left\{\frac{x}{4}\right\} + C$

*Example 2.* $\displaystyle\int \frac{1}{x^2 + 10x + 30}\, dx$

As usual, we complete the square in the denominator

$$\begin{aligned} x^2 + 10x + 30 &= x^2 + 10x \qquad + 30 \\ &= x^2 + 10x + 5^2 + 30 - 25 \\ &= (x+5)^2 + 5 \\ &= (x+5)^2 + (\sqrt{5})^2 \end{aligned}$$

$$\therefore \int \frac{1}{x^2 + 10x + 30}\, dx = \int \frac{1}{(x+5)^2 + (\sqrt{5})^2}\, dx$$

$$= \text{.............................}$$

**22**

$$\frac{1}{\sqrt{5}}.\tan^{-1}\left\{\frac{x+5}{\sqrt{5}}\right\} + C$$

□□□□□□□□□□□□□□□□□□□□□□□□□□□□□□□□□□□□□□

Once you know the standard form, you can find the expressions for Z and A in any example and then substitute these in the result. Here you are; do this one on your own:

*Example 3.* Determine $\displaystyle\int \frac{1}{2x^2 + 12x + 32}\, dx$

Take your time over it. Remember all the rules we have used and then you cannot go wrong.

*When you have completed it, turn to frame 23 and check your working.*

**23**

$$\boxed{\int \frac{1}{2x^2 + 12x + 32}\,dx = \frac{1}{2\sqrt{7}}\tan^{-1}\left\{\frac{x+3}{\sqrt{7}}\right\} + C}$$

Check your working.

$$\int \frac{1}{2x^2 + 12x + 32}\,dx = \frac{1}{2}\int \frac{1}{x^2 + 6x + 16}\,dx$$

$$\begin{aligned} x^2 + 6x + 16 &= x^2 + 6x \qquad + 16 \\ &= \underbrace{x^2 + 6x + 3^2}\, + 16 - 9 \\ &= \quad (x+3)^2 + 7 \\ &= \quad (x+3)^2 + (\sqrt{7})^2 \end{aligned}$$

So $Z = (x + 3)$ and $A = \sqrt{7}$

$$\int \frac{1}{Z^2 + A^2}\,dZ = \frac{1}{A}\tan^{-1}\left\{\frac{Z}{A}\right\} + C$$

$$\therefore \int \frac{1}{2x^2 + 12x + 32}\,dx = \frac{1}{2}\cdot\frac{1}{\sqrt{7}}\tan^{-1}\left\{\frac{x+3}{\sqrt{7}}\right\} + C$$

*Now move to frame 24.*

**24**

IV. Let us now consider a different integral.

$$\int \frac{1}{\sqrt{(A^2 - Z^2)}}\,dZ$$

We clearly cannot employ partial fractions, because of the root sign. So we must find a suitable substitution.

Put $\qquad Z = A \sin\theta$

Then $\quad A^2 - Z^2 = A^2 - A^2\sin^2\theta = A^2(1 - \sin^2\theta) = A^2\cos^2\theta$

$$\sqrt{(A^2 - Z^2)} = A\cos\theta$$

Also $\qquad \dfrac{dZ}{d\theta} = A\cos\theta \quad \therefore\ dZ \equiv A\cos\theta\,.\,d\theta$

So the integral becomes

$$\int \frac{1}{\sqrt{(A^2 - Z^2)}}\,dZ = \int \frac{1}{A\cos\theta}\,.\,A\cos\theta\,.\,d\theta = \int d\theta = \theta + C$$

Expressing $\theta$ in terms of the original variable:

$$Z = A\sin\theta \quad \therefore\ \sin\theta = \frac{Z}{A} \quad \therefore\ \theta = \sin^{-1}\frac{Z}{A}$$

$$\therefore \int \frac{1}{\sqrt{(A^2 - Z^2)}}\,dZ = \sin^{-1}\left\{\frac{Z}{A}\right\} + C \quad \text{............ (iv)}$$

***This is our next standard form, so add it to the list in your record book. Then move on to frame 25.***

**25**

$$\int \frac{1}{\sqrt{(A^2 - Z^2)}} dZ = \sin^{-1}\left\{\frac{Z}{A}\right\} + C$$

*Example 1.* $\int \frac{1}{\sqrt{(25 - x^2)}} dx = \int \frac{1}{\sqrt{(5^2 - x^2)}} dx = \sin^{-1}\left\{\frac{x}{5}\right\} + C$

*Example 2.* $\int \frac{1}{\sqrt{(3 - 2x - x^2)}} dx$

As usual

$$\begin{aligned} 3 - 2x - x^2 &= 3 - (x^2 + 2x \qquad) \\ &= 3 - (x^2 + 2x + 1^2) + 1 \\ &= 4 - (x + 1)^2 \\ &= 2^2 - (x + 1)^2 \end{aligned}$$

So, in this case, A = 2 and Z = $(x + 1)$

$$\int \frac{1}{\sqrt{(3 - 2x - x^2)}} dx = \int \frac{1}{\sqrt{\{2^2 - (x + 1)^2\}}} dx$$

$$= \sin^{-1}\left\{\frac{x + 1}{2}\right\} + C$$

Similarly,

*Example 3.* $\int \frac{1}{\sqrt{(5 - 4x - x^2)}} dx = \ldots\ldots\ldots\ldots\ldots\ldots\ldots\ldots$

**26**

$$\int \frac{1}{\sqrt{(5 - 4x - x^2)}} dx = \sin^{-1}\left\{\frac{x + 2}{3}\right\} + C$$

For:

$$\begin{aligned} 5 - 4x - x^2 &= 5 - (x^2 + 4x \qquad) \\ &= 5 - (x^2 + 4x + 2^2) + 4 \\ &= 9 - (x + 2)^2 \\ &= 3^2 - (x + 2)^2 \end{aligned}$$

$$\therefore \int \frac{1}{\sqrt{(5 - 4x - x^2)}} dx = \sin^{-1}\left\{\frac{x + 2}{3}\right\} + C$$

Now this one:

*Example 4.* Determine $\int \frac{1}{\sqrt{(14 - 12x - 2x^2)}} dx.$

Before we can complete the square, we must reduce the coefficient of $x^2$ to 1, i.e. we must divide the expression $14 - 12x - 2x^2$ by 2, but note that this becomes $\sqrt{2}$ when brought outside the root sign.

$$\int \frac{1}{\sqrt{(14 - 12x - 2x^2)}} dx = \frac{1}{\sqrt{2}} \int \frac{1}{\sqrt{(7 - 6x - x^2)}} dx$$

*Now finish that as in the last example.*

**27**

$$\boxed{\int \frac{1}{\sqrt{(14-12x-2x^2)}}dx = \frac{1}{\sqrt{2}}\sin^{-1}\left\{\frac{x+3}{4}\right\} + C}$$

For:

$$\int \frac{1}{\sqrt{(14-12x-2x^2)}}\,dx = \frac{1}{\sqrt{2}}\int \frac{1}{\sqrt{(7-6x-x^2)}}\,dx$$

$$\begin{aligned} 7-6x-x^2 &= 7-(x^2+6x\qquad) \\ &= 7-(x^2+6x+3^2)+9 \\ &= 16-(x+3)^2 \\ &= 4^2-(x+3)^2 \end{aligned}$$

So A = 4 and Z = $(x + 3)$

$$\int \frac{1}{\sqrt{(A^2-Z^2)}}dZ = \sin^{-1}\left\{\frac{Z}{A}\right\} + C$$

$$\therefore \int \frac{1}{\sqrt{(14-12x-2x^2)}}\,dx = \frac{1}{\sqrt{2}}\sin^{-1}\left\{\frac{x+3}{4}\right\} + C$$

---

**28**

V. Let us now look at the next standard integral in the same way.

To determine $\int \frac{dZ}{\sqrt{(Z^2+A^2)}}$. Again we try to find a convenient substitution for Z, but no trig. substitution converts the function into a form that we can manage.

We therefore have to turn to the *hyperbolic identities* and put $Z = A\sinh\theta$.

Then $\qquad Z^2 + A^2 = A^2\sinh^2\theta + A^2 = A^2(\sinh^2\theta + 1)$

Remember $\qquad \cosh^2\theta - \sinh^2\theta = 1 \quad \therefore \cosh^2\theta = \sinh^2\theta + 1$

$$\therefore Z^2 + A^2 = A^2\cosh^2\theta \quad \therefore \sqrt{(Z^2+A^2)} = A\cosh\theta$$

Also $\qquad \frac{dZ}{d\theta} = A\cosh\theta \therefore dZ \equiv A\cosh\theta\,.\,d\theta$

So $\qquad \int \frac{dZ}{\sqrt{(Z^2+A^2)}} = \int \frac{1}{A\cosh\theta}\,.A\cosh\theta\,d\theta = \int d\theta = \theta + C$

But $\qquad Z = A\sinh\theta \quad \therefore \sinh\theta = \frac{Z}{A} \quad \therefore \theta = \sinh^{-1}\left\{\frac{Z}{A}\right\}$

$$\therefore \int \frac{dZ}{\sqrt{(Z^2+A^2)}} = \sinh^{-1}\left\{\frac{Z}{A}\right\} + C \quad \text{.............. (v)}$$

Copy this result into your record book for future reference.

Then $\qquad \int \frac{1}{\sqrt{(x^2+4)}}dx = $ ......................

**29**

$$\int \frac{1}{\sqrt{(x^2+4)}}\,dx = \sinh^{-1}\left\{\frac{x}{2}\right\} + C$$

□□□□□□□□□□□□□□□□□□□□□□□□□□□□□□□□□□□□□

Once again, all we have to do is to find the expressions for Z and A in any particular example and substitute in the standard form.

Now you can do this one all on your own.

Determine $\int \frac{1}{\sqrt{(x^2+5x+12)}}\,dx$

*Complete the working: then check with frame 30.*

**30**

$$\int \frac{1}{\sqrt{(x^2+5x+12)}}\,dx = \sinh^{-1}\left\{\frac{2x+5}{\sqrt{23}}\right\} + C$$

Here is the working set out in detail:

$$\begin{aligned} x^2+5x+12 &= x^2+5x \qquad\qquad +12 \\ &= x^2+5x+\left(\frac{5}{2}\right)^2 +12-\frac{25}{4} \\ &= \left(x+\frac{5}{2}\right)^2+\frac{23}{4} \\ &= \left(x+\frac{5}{2}\right)^2+\left(\frac{\sqrt{23}}{2}\right)^2 \end{aligned}$$

So that $Z = x+\frac{5}{2}$ and $A = \frac{\sqrt{23}}{2}$

$$\begin{aligned} \therefore \int \frac{1}{\sqrt{(x^2+5x+12)}}\,dx &= \sinh^{-1}\left\{\frac{x+\frac{5}{2}}{\sqrt{23}/2}\right\} + C \\ &= \sinh^{-1}\left\{\frac{2x+5}{\sqrt{23}}\right\} + C \end{aligned}$$

Now do one more.

$$\int \frac{1}{\sqrt{(2x^2+8x+15)}}\,dx = \ldots\ldots\ldots\ldots\ldots$$

**31**

$$\boxed{\frac{1}{\sqrt{2}}\sinh^{-1}\left\{\frac{(x+2)\sqrt{2}}{\sqrt{7}}\right\}+C}$$

Here is the working:

$$\int\frac{1}{\sqrt{(2x^2+8x+15)}}\,dx=\frac{1}{\sqrt{2}}\int\frac{1}{\sqrt{\left(x^2+4x+\frac{15}{2}\right)}}\,dx$$

$$\begin{aligned}x^2+4x+\frac{15}{2}&=x^2+4x\qquad+\frac{15}{2}\\&=x^2+4x+2^2+\frac{15}{2}-4\\&=(x+2)^2+\frac{7}{2}\\&=(x+2)^2+\left(\sqrt{\frac{7}{2}}\right)^2\end{aligned}$$

So that $Z=(x+2)$ and $A=\sqrt{\frac{7}{2}}$

$$\therefore\int\frac{1}{\sqrt{(2x^2+8x+15)}}\,dx=\frac{1}{\sqrt{2}}\sinh^{-1}\left\{\frac{x+2}{\sqrt{\frac{7}{2}}}\right\}+C$$

$$=\frac{1}{\sqrt{2}}\sinh^{-1}\frac{(x+2)\sqrt{2}}{\sqrt{7}}+C$$

*Fine. Now on to frame 32.*

**32**

Now we will establish another standard result.

VI. Consider $\int\frac{dZ}{\sqrt{(Z^2-A^2)}}$

The substitution here is to put $Z=A\cosh\theta$.

$$Z^2-A^2=A^2\cosh^2\theta-A^2=A^2(\cosh^2\theta-1)=A^2\sinh^2\theta$$

$$\therefore\sqrt{(Z^2-A^2)}=A\sinh\theta$$

Also $\qquad Z=A\cosh\theta\quad\therefore dZ=A\sinh\theta\,d\theta$

$$\therefore\int\frac{dZ}{\sqrt{(Z^2-A^2)}}=\int\frac{1}{A\sinh\theta}.A\sinh\theta\,d\theta=\int d\theta=\theta+C$$

$$Z=A\cosh\theta\quad\therefore\cosh\theta=\frac{Z}{A}\quad\therefore\theta=\cosh^{-1}\left\{\frac{Z}{A}\right\}+C$$

$$\therefore\int\frac{dZ}{\sqrt{(Z^2-A^2)}}=\cosh^{-1}\left\{\frac{Z}{A}\right\}+C\quad\text{............ (vi)}$$

This makes the sixth standard result we have established. Add it to your list.

*Then move on to frame 33.*

**33**

$$\int \frac{dZ}{\sqrt{(Z^2 - A^2)}} = \cosh^{-1}\left\{\frac{Z}{A}\right\} + C$$

*Example 1.* $\displaystyle\int \frac{1}{\sqrt{(x^2 - 9)}} dx = \cosh^{-1}\left\{\frac{x}{3}\right\} + C$

*Example 2.* $\displaystyle\int \frac{1}{\sqrt{(x^2 + 6x + 1)}} dx = \ldots\ldots\ldots\ldots\ldots$

You can do that one on your own. The method is the same as before: just complete the square and find out what Z and A are in this case and then substitute in the standard result.

**34**

$$\int \frac{1}{\sqrt{(x^2 + 6x + 1)}} dx = \cosh^{-1}\left\{\frac{x + 3}{2\sqrt{2}}\right\} + C$$

Here it is:

$$\begin{aligned} x^2 + 6x + 1 &= x^2 + 6x \qquad + 1 \\ &= x^2 + 6x + 3^2 + 1 - 9 \\ &= (x + 3)^2 - 8 \\ &= (x + 3)^2 - (2\sqrt{2})^2 \end{aligned}$$

So that $Z = (x + 3)$ and $A = 2\sqrt{2}$

$$\begin{aligned} \therefore \int \frac{1}{\sqrt{(x^2 + 6x + 1)}} dx &= \int \frac{1}{\sqrt{\{(x + 3)^2 - (2\sqrt{2})^2\}}} dx \\ &= \cosh^{-1}\left\{\frac{x + 3}{2\sqrt{2}}\right\} + C \end{aligned}$$

Let us now collect together the results we have established so far so that we can compare them.

*So turn on to frame 35.*

**35** Here are our standard forms so far, with the method indicated in each case.

| | | |
|---|---|---|
| 1. | $\int \frac{dZ}{Z^2 - A^2} = \frac{1}{2A} \ln\left\{\frac{Z-A}{Z+A}\right\} + C$ | Partial fractions |
| 2. | $\int \frac{dZ}{A^2 - Z^2} = \frac{1}{2A} \ln\left\{\frac{A+Z}{A-Z}\right\} + C$ | " " |
| 3. | $\int \frac{dZ}{Z^2 + A^2} = \frac{1}{A} \tan^{-1}\left\{\frac{Z}{A}\right\} + C$ | Put $Z = A \tan\theta$ |
| 4. | $\int \frac{dZ}{\sqrt{(A^2 - Z^2)}} = \sin^{-1}\left\{\frac{Z}{A}\right\} + C$ | Put $Z = A \sin\theta$ |
| 5. | $\int \frac{dZ}{\sqrt{(Z^2 + A^2)}} = \sinh^{-1}\left\{\frac{Z}{A}\right\} + C$ | Put $Z = A \sinh\theta$ |
| 6. | $\int \frac{dZ}{\sqrt{(Z^2 - A^2)}} = \cosh^{-1}\left\{\frac{Z}{A}\right\} + C$ | Put $Z = A \cosh\theta$ |

Note that the first three make one group (without square roots).
Note that the second three make a group with the square roots in the denominators.

You should make an effort to memorize these six results for you will be expected to know them and to be able to quote them and use them in various examples.

---

**36** You will remember that in the programme on hyperbolic functions, we obtained the result $\sinh^{-1}x = \ln\left\{x + \sqrt{(x^2 + 1)}\right\}$

$$\therefore \sinh^{-1}\left\{\frac{Z}{A}\right\} = \ln\left\{\frac{Z}{A} + \sqrt{\left(\frac{Z^2}{A^2} + 1\right)}\right\}$$

$$= \ln\left\{\frac{Z}{A} + \sqrt{\frac{(Z^2 + A^2)}{A^2}}\right\}$$

$$= \ln\left\{\frac{Z}{A} + \frac{\sqrt{(Z^2 + A^2)}}{A}\right\}$$

$$\sinh^{-1}\left\{\frac{Z}{A}\right\} = \ln\left\{\frac{Z + \sqrt{(Z^2 + A^2)}}{A}\right\}$$

Similarly

$$\cosh^{-1}\left\{\frac{Z}{A}\right\} = \ln\left\{\frac{Z + \sqrt{(Z^2 - A^2)}}{A}\right\}$$

This means that the results of standard integrals 5 and 6 can be expressed either as inverse hyperbolic functions or in log form according to the needs of the exercise.

*Turn on now to frame 37.*

**37**

The remaining three standard integrals in our list are:

$$7.\int\sqrt{(A^2-Z^2)}.dZ \quad 8.\int\sqrt{(Z^2+A^2)}.dZ \quad 9.\int\sqrt{(Z^2-A^2)}.dZ$$

In each case, the appropriate substitution is the same as with the corresponding integral in which the same expression occurred in the denominator.

i.e. for $\int\sqrt{(A^2-Z^2)}.dZ$ put $Z = A\sin\theta$

$\int\sqrt{(Z^2+A^2)}.dZ$ " $Z = A\sinh\theta$

$\int\sqrt{(Z^2-A^2)}.\,dZ$ " $Z = A\cosh\theta$

Making these substitutions, gives the following results.

$$\int\sqrt{(A^2-Z^2)}.\,dZ = \frac{A^2}{2}\left\{\sin^{-1}\left(\frac{Z}{A}\right)+\frac{Z\sqrt{(A^2-Z^2)}}{A^2}\right\}+C \quad\text{...... (vii)}$$

$$\int\sqrt{(Z^2+A^2)}dZ = \frac{A^2}{2}\left\{\sinh^{-1}\left(\frac{Z}{A}\right)+\frac{Z\sqrt{(Z^2+A^2)}}{A^2}\right\}+C \quad\text{...... (viii)}$$

$$\int\sqrt{(Z^2-A^2)}.\,dZ = \frac{A^2}{2}\left\{\frac{Z\sqrt{(Z^2-A^2)}}{A^2}-\cosh^{-1}\left(\frac{Z}{A}\right)\right\}+C \quad\text{...... (ix)}$$

These results are more complicated and difficult to remember but the method of using them is much the same as before. Copy them down.

---

**38**

Let us see how the first of these results is obtained.

$$\int\sqrt{(A^2-Z^2)}.dZ \qquad \text{Put } Z = A\sin\theta$$

$$\therefore\ A^2-Z^2 = A^2-A^2\sin^2\theta = A^2(1-\sin^2\theta) = A^2\cos^2\theta$$

$$\therefore\ \sqrt{(A^2-Z^2)} = A\cos\theta \qquad \text{Also } dZ \equiv A\cos\theta\, d\theta$$

$$\int\sqrt{(A^2-Z^2)}.\,dZ = \int A\cos\theta\,.\,A\cos\theta\,d\theta = A^2\int\cos^2\theta\,d\theta$$

$$= A^2\left[\frac{\theta}{2}+\frac{\sin 2\theta}{4}\right]+C = \frac{A^2}{2}\left\{\theta+\frac{2\sin\theta\cos\theta}{2}\right\}+C$$

Now $\sin\theta = \dfrac{Z}{A}$ and $\cos^2\theta = 1-\dfrac{Z^2}{A^2} = \dfrac{A^2-Z^2}{A^2}$ $\therefore\ \cos\theta = \dfrac{\sqrt{(A^2-Z^2)}}{A}$

$$\therefore\ \int\sqrt{(A^2-Z^2)}.\,dZ = \frac{A^2}{2}\left\{\sin^{-1}\left(\frac{Z}{A}\right)+\frac{Z}{A}.\frac{\sqrt{(A^2-Z^2)}}{A}\right\}$$

$$= \frac{A^2}{2}\left\{\sin^{-1}\left(\frac{Z}{A}\right)+\frac{Z\sqrt{(A^2-Z^2)}}{A^2}\right\}+C$$

The other two are proved in a similar manner. *Now on to frame 39.*

**39** Here is an example

$$\int \sqrt{(x^2 + 4x + 13)}.dx$$

First of all complete the square and find Z and A as before. Right. Do that.

**40**

$$x^2 + 4x + 13 = (x + 2)^2 + 3^2$$

So that, in this case $\boxed{Z = x + 2}$ and $\boxed{A = 3}$

$$\therefore \int \sqrt{(x^2 + 4x + 13)}.dx = \int \sqrt{\{(x + 2)^2 + 3^2\}}.dx$$

This is of the form

$$\int \sqrt{(Z^2 + A^2)}.dZ = \frac{A^2}{2}\left\{\sinh^{-1}\left(\frac{Z}{A}\right) + \frac{Z\sqrt{(Z^2 + A^2)}}{A^2}\right\} + C$$

So, substituting our expressions for Z and A, we get

$$\int \sqrt{(x^2 + 4x + 13)}.dx = \dots\dots\dots\dots$$

**41**

$$\boxed{\int \sqrt{(x^2 + 4x + 13)}.\,dx = \frac{9}{2}\left\{\sinh^{-1}\left(\frac{x + 2}{3}\right) + \frac{(x + 2)\sqrt{(x^2 + 4x + 13)}}{9}\right\} + C}$$

We see then that the use of any of these standard forms merely involves completing the square as we have done on many occasions, finding the expressions for Z and A, and substituting these in the appropriate result. This means that you can now tackle a wide range of integrals which were beyond your ability before you worked through this programme.

□□□□□□□□□□□□□□□□□□□□□□□□□□□□□□□□□□□□□□□

Now, by way of revision, *without looking at your notes*, complete the following:

(i) $\int \frac{dZ}{Z^2 - A^2} = \dots\dots\dots\dots$

(ii) $\int \frac{dZ}{A^2 - Z^2} = \dots\dots\dots\dots$

(iii) $\int \frac{dZ}{Z^2 + A^2} = \dots\dots\dots\dots$

**42**

$$\int \frac{dZ}{Z^2 - A^2} = \frac{1}{2A} \cdot \ln\left\{\frac{Z - A}{Z + A}\right\} + C$$
$$\int \frac{dZ}{A^2 - Z^2} = \frac{1}{2A} \cdot \ln\left\{\frac{A + Z}{A - Z}\right\} + C$$
$$\int \frac{dZ}{Z^2 + A^2} = \frac{1}{A} \cdot \tan^{-1}\left\{\frac{Z}{A}\right\} + C$$

And now the second group:

$$\int \frac{dZ}{\sqrt{(A^2 - Z^2)}} = \ldots\ldots\ldots\ldots$$
$$\int \frac{dZ}{\sqrt{(Z^2 + A^2)}} = \ldots\ldots\ldots\ldots$$
$$\int \frac{dZ}{\sqrt{(Z^2 - A^2)}} = \ldots\ldots\ldots\ldots$$

**43**

$$\int \frac{dZ}{\sqrt{(A^2 - Z^2)}} = \sin^{-1}\left\{\frac{Z}{A}\right\} + C$$
$$\int \frac{dZ}{\sqrt{(Z^2 + A^2)}} = \sinh^{-1}\left\{\frac{Z}{A}\right\} + C$$
$$\int \frac{dZ}{\sqrt{(Z^2 - A^2)}} = \cosh^{-1}\left\{\frac{Z}{A}\right\} + C$$

You will not have remembered the third group, but here they are again. Take another look at them.

$$\int \sqrt{(A^2 - Z^2)}.\,dZ = \frac{A^2}{2}\left\{\sin^{-1}\left(\frac{Z}{A}\right) + \frac{Z\sqrt{(A^2 - Z^2)}}{A^2}\right\} + C$$
$$\int \sqrt{(Z^2 + A^2)}.\,dZ = \frac{A^2}{2}\left\{\sinh^{-1}\left(\frac{Z}{A}\right) + \frac{Z\sqrt{(Z^2 + A^2)}}{A^2}\right\} + C$$
$$\int \sqrt{(Z^2 - A^2)}.\,dZ = \frac{A^2}{2}\left\{\frac{Z\sqrt{(Z^2 - A^2)}}{A^2} - \cosh^{-1}\left(\frac{Z}{A}\right)\right\} + C$$

Notice that the square root in the result is the same root as that in the integral in each case.

□□□□□□□□□□□□□□□□□□□□□□□□□□□□□□□□□□□□□□

That ends that particular section of the programme, but there are other integrals that require substitution of some kind, so we will now deal with one or two of these.

*Turn on to frame 44.*

**44**

Integrals of the form $\displaystyle\int \frac{1}{a + b\sin^2 x + c\cos^2 x}\,dx$

*Example 1.* Consider $\displaystyle\int \frac{1}{3 + \cos^2 x}\,dx$, which is different from any we have had before. It is certainly not one of the standard forms.

The key to the method is to substitute $t = \tan x$ in the integral. Of course, $\tan x$ is not mentioned in the integral, but if $\tan x = t$, we can soon find corresponding expressions for $\sin x$ and $\cos x$. Draw a sketch diagram, thus:

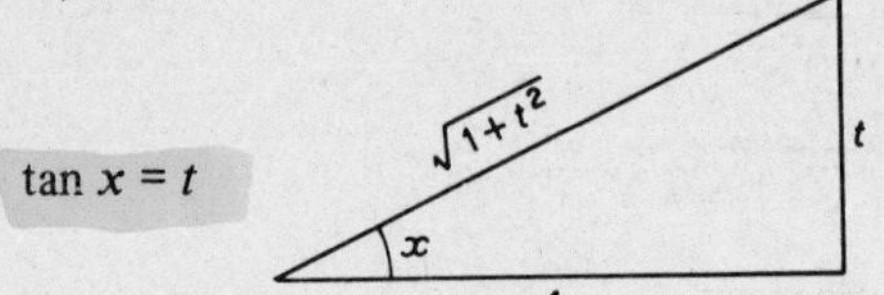


$$\therefore\ \sin x = \frac{t}{\sqrt{(1 + t^2)}}$$

$$\therefore\ \cos x = \frac{1}{\sqrt{(1 + t^2)}}$$

Also, since $t = \tan x$, $\dfrac{dt}{dx} = \sec^2 x = 1 + \tan^2 x = 1 + t^2$

$$\therefore\ \frac{dx}{dt} = \frac{1}{1 + t^2} \qquad \therefore dx \equiv \frac{dt}{1 + t^2}$$

Then $3 + \cos^2 x = 3 + \dfrac{1}{1 + t^2} = \dfrac{3 + 3t^2 + 1}{1 + t^2} = \dfrac{4 + 3t^2}{1 + t^2}$

So the integral now becomes:

$$\int \frac{1}{3 + \cos^2 x}\,.\,dx = \int \frac{1 + t^2}{4 + 3t^2}\,.\,\frac{dt}{1 + t^2}$$

$$= \int \frac{1}{4 + 3t^2}\,dt = \frac{1}{3}\int \frac{1}{\frac{4}{3} + t^2}\,dt$$

and from what we have done in the earlier part of this programme, this is ...................

**45**

$$\frac{1}{3}\int \frac{1}{\frac{4}{3} + t^2}\,dt = \frac{1}{3}.\frac{\sqrt{3}}{2}\tan^{-1}\left\{\frac{t}{2/\sqrt{3}}\right\} + C$$

$$= \frac{1}{3}\,\frac{\sqrt{3}}{2}\tan^{-1}\left\{\frac{t\sqrt{3}}{2}\right\} + C$$

Finally, since $t = \tan x$, we can return to the original variable and obtain

$$\int \frac{1}{3 + \cos^2 x}\,dx = \frac{1}{2\sqrt{3}}\tan^{-1}\left\{\frac{\sqrt{3}.\tan x}{2}\right\} + C$$

*Turn to frame 46.*

**46**

The method is the same for all integrals of the type

$$\int \frac{1}{a + b\sin^2 x + c\cos^2 x}\,dx$$

In practice, some of the coefficients may be zero and those terms missing from the function. But the routine remains the same.

Use the substitution $t = \tan x$. That is all there is to it.

From the diagram

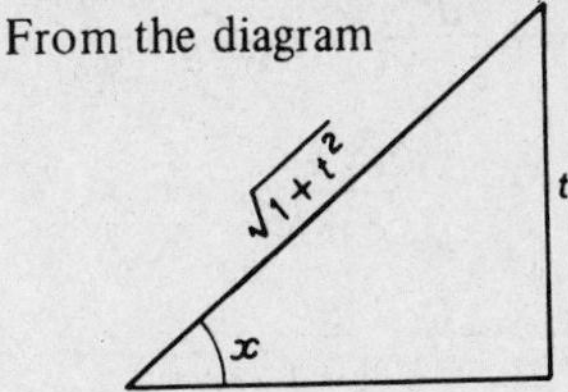


we get

$\sin x = \dots\dots\dots\dots$

$\cos x = \dots\dots\dots\dots$

**47**

$$\sin x = \frac{t}{\sqrt{(1+t^2)}} \qquad \cos x = \frac{1}{\sqrt{(1+t^2)}}$$

We also have to change the variable.

$$t = \tan x \ \therefore\ \frac{dt}{dx} = \sec^2 x = 1 + \tan^2 x = 1 + t^2$$

$$\therefore\ \frac{dx}{dt} = \frac{1}{1+t^2}\,;\ dx \equiv \dots\dots\dots\dots$$

**48**

$$dx \equiv \frac{dt}{1+t^2}$$

Armed with these substitutions we can deal with any integral of the present type. This does not give us a standard result, but provides us with a standard method.

We will work through another example in the next frame, but first of all, what were those substitutions?

$\sin x = \dots\dots\dots\dots$

$\cos x = \dots\dots\dots\dots$

**49**

$$\sin x = \frac{t}{\sqrt{(1+t^2)}} \qquad \cos x = \frac{1}{\sqrt{(1+t^2)}}$$

Right. Now for an example.

*Example 2.* Determine $\int \frac{1}{2\sin^2 x + 4\cos^2 x}\,dx$

Using the substitution above, and that $dx \equiv \frac{dt}{1+t^2}$, we have

$$2\sin^2 x + 4\cos^2 x = \frac{2t^2}{1+t^2} + \frac{4}{1+t^2} = \frac{2t^2+4}{1+t^2}$$

$$\therefore \int \frac{1}{2\sin^2 x + 4\cos^2 x}\,dx = \int \frac{1+t^2}{2t^2+4} \cdot \frac{dt}{1+t^2}$$

$$= \frac{1}{2}\int \frac{1}{t^2+2}\,dt$$

$$= \ldots\ldots\ldots\ldots\ldots\ldots$$

**50**

$$\frac{1}{2\sqrt{2}}\tan^{-1}\left\{\frac{t}{\sqrt{2}}\right\} + C$$

and since $t = \tan x$, we can return to the original variable, so that

$$\int \frac{1}{2\sin^2 x + 4\cos^2 x}.dx = \frac{1}{2\sqrt{2}}\tan^{-1}\left\{\frac{\tan x}{\sqrt{2}}\right\} + C$$

Now here is one for you to do on your own.
Remember the substitutions:

$$t = \tan x \qquad \sin x = \frac{t}{\sqrt{(1+t^2)}}$$

$$\cos x = \frac{1}{\sqrt{(1+t^2)}}$$

$$dx \equiv \frac{dt}{(1+t^2)}$$

Right, then here it is:

*Example 3.* $\int \frac{1}{2\cos^2 x + 1}\,dx = \ldots\ldots\ldots\ldots$

*Work it right through to the end and then check your result and your working with that in the next frame.*

**51**

$$\int \frac{1}{2\cos^2 x + 1}\,dx = \frac{1}{\sqrt{3}}\tan^{-1}\left\{\frac{\tan x}{\sqrt{3}}\right\} + C$$

Here is the working:

$$2\cos^2 x + 1 = \frac{2}{1+t^2} + 1 = \frac{2+1+t^2}{1+t^2}$$

$$= \frac{3+t^2}{1+t^2}$$

$$\therefore \int \frac{1}{2\cos^2 x + 1}\,dx = \int \frac{1+t^2}{3+t^2}\cdot\frac{dt}{1+t^2}$$

$$= \int \frac{1}{3+t^2}\,dt = \frac{1}{\sqrt{3}}\tan^{-1}\left(\frac{t}{\sqrt{3}}\right) + C$$

$$= \frac{1}{\sqrt{3}}\tan^{-1}\left\{\frac{\tan x}{\sqrt{3}}\right\} + C$$

So whenever we have an integral of this type, with $\sin^2 x$ and/or $\cos^2 x$ in the denominator, the key to the whole business is to make the substitution $t =$ ..................

**52**

$$t = \tan x$$

Let us now consider the integral $\int \frac{1}{5 + 4\cos x}\,dx$

This is clearly not one of the last type, for the trig. function in the denominator is $\cos x$ and not $\cos^2 x$.

In fact, this is an example of a further group of integrals that we are going to cover in this programme. In general they are of the form $\int \frac{1}{a + b\sin x + c\cos x}\,dx$, i.e. sines and cosines in the denominator but not squared.

*So turn on to frame 53 and we will start to find out something about these integrals.*

**53**

Integrals of the type $\int \frac{1}{a + b \sin x + c \cos x}\, dx$

The key this time is to substitute $t = \tan \frac{x}{2}$

From this, we can find corresponding expressions for $\sin \frac{x}{2}$ and $\cos \frac{x}{2}$ from a simple diagram as before, but it also means that we must express $\sin x$ and $\cos x$ in terms of the trig. ratios of the half-angle – so it will entail a little more work, but only a little, so do not give up. It is a lot easier than it sounds.

First of all let us establish the substitutions in detail.

$t = \tan \frac{x}{2}$

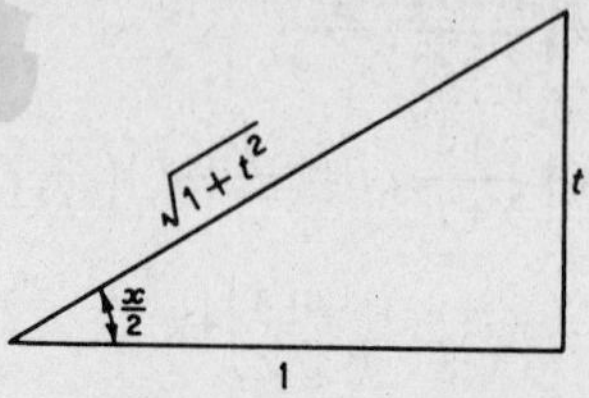

$\therefore\ \sin \frac{x}{2} = \frac{t}{\sqrt{(1 + t^2)}}$

$\therefore\ \cos \frac{x}{2} = \frac{1}{\sqrt{(1 + t^2)}}$

$$\sin x = 2 \sin \frac{x}{2} \cos \frac{x}{2} = 2 . \frac{t}{\sqrt{(1 + t^2)}} . \frac{1}{\sqrt{(1 + t^2)}} = \frac{2t}{1 + t^2}$$

$$\cos x = \cos^2 \frac{x}{2} - \sin^2 \frac{x}{2} = \frac{1}{1 + t^2} - \frac{t^2}{1 + t^2} = \frac{1 - t^2}{1 + t^2}$$

Also, since $t = \tan \frac{x}{2}$, $\frac{dt}{dx} = \frac{1}{2} \sec^2 \frac{x}{2} = \frac{1}{2}(1 + \tan^2 \frac{x}{2})$

$$= \frac{1 + t^2}{2}$$

$$\frac{dx}{dt} = \frac{2}{1 + t^2} \qquad \therefore\ dx \equiv \frac{2\, dt}{1 + t^2}$$

So we have:

If $t = \tan \frac{x}{2}$

$$\sin x = \frac{2t}{1 + t^2}$$

$$\cos x = \frac{1 - t^2}{1 + t^2}$$

$$dx \equiv \frac{2\, dt}{1 + t^2}$$

It is worth remembering these substitutions for use in examples. So copy them down into your record book for future reference. Then we shall be ready to use them.

*On to frame 54.*

**54**

*Example 1.* $\displaystyle\int \frac{dx}{5 + 4\cos x}$

Using the substitution $t = \tan\dfrac{x}{2}$, we have

$$5 + 4\cos x = 5 + 4\,\frac{(1 - t^2)}{1 + t^2}$$

$$= \frac{5 + 5t^2 + 4 - 4t^2}{1 + t^2} = \frac{9 + t^2}{1 + t^2}$$

$$\therefore \int \frac{dx}{5 + 4\cos x} = \int \frac{1 + t^2}{9 + t^2} \cdot \frac{2\,dt}{1 + t^2} = 2\int \frac{dt}{9 + t^2}$$

$$= \ldots\ldots\ldots\ldots\ldots\ldots$$

**55**

$$\frac{2}{3}\tan^{-1}\left\{\frac{t}{3}\right\} + C$$

$$= \frac{2}{3}\tan^{-1}\left\{\frac{\tan x/2}{3}\right\} + C$$

Here is another.

*Example 2.* $\displaystyle\int \frac{dx}{3\sin x + 4\cos x}$

Using the substitution $t = \tan\dfrac{x}{2}$

$$3\sin x + 4\cos x = \frac{6t}{1 + t^2} + \frac{4(1 - t^2)}{1 + t^2}$$

$$= \frac{4 + 6t - 4t^2}{1 + t^2}$$

$$\therefore \int \frac{dx}{3\sin x + 4\cos x} = \int \frac{1 + t^2}{4 + 6t - 4t^2} \cdot \frac{2\,dt}{1 + t^2}$$

$$= \int \frac{1}{2 + 3t - 2t^2}\,dt$$

$$= \frac{1}{2}\int \frac{1}{1 + \frac{3}{2}t - t^2}\,dt$$

Now complete the square in the denominator as we were doing earlier in the programme and finish it off.

*Then on to frame 56.*

**56**

$$\boxed{\frac{1}{5}.\ln\left\{\frac{1+2\tan x/2}{4-2\tan x/2}\right\}+C}$$

For
$$1+\frac{3}{2}t-t^2=1-(t^2-\frac{3}{2}t \qquad )$$
$$=1-\left(t^2-\frac{3}{2}t+\left[\frac{3}{4}\right]^2\right)+\frac{9}{16}$$
$$=\frac{25}{16}-\left(t-\frac{3}{4}\right)^2$$
$$=\left(\frac{5}{4}\right)^2-\left(t-\frac{3}{4}\right)^2$$

Integral, $$I=\frac{1}{2}\int\frac{1}{\left(\frac{5}{4}\right)^2-\left(t-\frac{3}{4}\right)^2}\,dt$$

$$=\frac{1}{2.\frac{5}{2}}\ln\left\{\frac{5/4+t-3/4}{5/4-t+3/4}\right\}+C=\frac{1}{5}\ln\left\{\frac{1/2+t}{2-t}\right\}+C$$

$$=\frac{1}{5}\ln\left\{\frac{1+2t}{4-2t}\right\}+C=\frac{1}{5}\ln\left\{\frac{1+2\tan x/2}{4+2\tan x/2}\right\}+C$$

And here is one more for you, all on your own. Finish it: then check your working with that in the next frame. Here it is.

*Example 3.* $\int\frac{1}{1+\sin x-\cos x}\,dx=$ ....................

**57**

$$\boxed{\ln\left\{\frac{\tan x/2}{1+\tan x/2}\right\}+C}$$

Here is the working.

$$1+\sin x-\cos x=1+\frac{2t}{1+t^2}-\frac{1-t^2}{1+t^2}$$
$$=\frac{1+t^2+2t-1+t^2}{1+t^2}=\frac{2(t^2+t)}{1+t^2}$$
$$I=\int\frac{1+t^2}{2(t^2+t)}.\frac{2\,dt}{1+t^2}=\int\frac{1}{t^2+t}\,dt$$
$$=\int\left(\frac{1}{t}-\frac{1}{1+t}\right)dt$$
$$=\ln\left\{\frac{t}{1+t}\right\}+C=\ln\left\{\frac{\tan x/2}{1+\tan x/2}\right\}+C$$

**58**

You have now reached the end of this programme except for the Test Exercise which follows. Before you work through the questions, brush up any parts of the programme about which you are not perfectly clear. Look back through the programme if you want to do so. There is no hurry. Your success is all that matters.

When you are ready, work all the questions in the Test Exercise. The integrals in the Test are just like those we have been doing in the programme, so you will find them quite straightforward.

## Test Exercise – XVI

Determine the following:

1. $\int \frac{1}{\sqrt{(49-x^2)}}dx$
2. $\int \frac{dx}{x^2+3x-5}$
3. $\int \frac{dx}{2x^2+8x+9}$
4. $\int \frac{1}{\sqrt{(3x^2+16)}}dx$
5. $\int \frac{dx}{9-8x-x^2}$
6. $\int \sqrt{(1-x-x^2)}.dx$
7. $\int \frac{dx}{\sqrt{(5x^2+10x-16)}}$
8. $\int \frac{dx}{1+2\sin^2 x}$
9. $\int \frac{dx}{2\cos x+3\sin x}$
10. $\int \sec x\, dx$

*You are now ready for your next programme. Well done!*

## Further Problems – XVI

Determine the following:

1. $\displaystyle\int \frac{dx}{x^2 + 12x + 15}$

2. $\displaystyle\int \frac{dx}{8 - 12x - x^2}$

3. $\displaystyle\int \frac{dx}{x^2 + 14x + 60}$

4. $\displaystyle\int \frac{x - 8}{x^2 + 4x + 16}\, dx$

5. $\displaystyle\int \frac{dx}{\sqrt{(x^2 + 12x + 48)}}$

6. $\displaystyle\int \frac{dx}{\sqrt{(17 - 14x - x^2)}}$

7. $\displaystyle\int \frac{dx}{\sqrt{(x^2 + 16x + 36)}}$

8. $\displaystyle\int \frac{6x - 5}{\sqrt{(x^2 - 12x + 52)}}\, dx$

9. $\displaystyle\int \frac{dx}{2 + \cos x}$

10. $\displaystyle\int_0^{\pi/2} \frac{dx}{4 \sin^2 x + 5 \cos^2 x}$

11. $\displaystyle\int \frac{dx}{x^2 + 5x + 5}$

12. $\displaystyle\int \frac{3x^3 - 4x^2 + 3x}{x^2 + 1}\, dx$

13. $\displaystyle\int \sqrt{(3 - 2x - x^2)}\, dx$

14. $\displaystyle\int_2^4 \frac{dx}{\sqrt{(6x - 8 - x^2)}}$

15. $\displaystyle\int \frac{dx}{\sqrt{(x^2 - 4x - 21)}}$

16. $\displaystyle\int \frac{dx}{4 \sin^2 x + 9 \cos^2 x}$

17. $\displaystyle\int \frac{dx}{3 \sin x - 4 \cos x}$

18. $\displaystyle\int_0^1 \sqrt{\frac{x}{2 - x}}\, dx$
    (Put $x = 2 \sin^2\theta$)

19. $\displaystyle\int \frac{x + 3}{\sqrt{(1 - x^2)}}\, dx$

20. $\displaystyle\int \frac{\cos x}{2 - \cos x}\, dx$

21. $\displaystyle\int \frac{x^2 - x + 14}{(x + 2)(x^2 + 4)}\, dx$

22. $\displaystyle\int \frac{dx}{5 + 4 \cos^2 x}$

23. $\displaystyle\int \frac{x + 2}{\sqrt{(x^2 + 9)}}\, dx$

24. $\displaystyle\int \frac{dx}{\sqrt{(2x^2 - 7x + 5)}}$

25. $\displaystyle\int_1^4 \frac{dx}{\sqrt{\{(x + 2)(4 - x)\}}}$

26. $\displaystyle\int \frac{d\theta}{2 \sin^2\theta - \cos^2\theta}$

27. $\displaystyle\int \frac{x + 3}{\sqrt{(x^2 + 2x + 10)}}\, dx$

28. $\displaystyle\int \sqrt{(15 - 2x - x^2 dx)}$

29. $\displaystyle\int_0^a \frac{dx}{(a^2 + x^2)^2}$
    (Put $x = a \tan\theta$)

30. $\displaystyle\int \frac{a^2 dx}{(x + a)(x^2 + 2a^2)}$

# Programme 17

## REDUCTION FORMULAE

**1** In an earlier programme on integration, we dealt with the method of *integration by parts*, and you have had plenty of practice in that since that time. You remember that it can be stated thus:

$$\int u\,dv = u\,v - \int v\,du$$

So just to refresh your memory, do this one to start with.

$$\int x^2 e^x\,dx = \text{..................}$$

*When you have finished, move on to frame 2.*

**2**

$$\boxed{\int x^2 e^x\,dx = e^x[x^2 - 2x + 2] + C}$$

Here is the working, so that you can check your solution.

$$\int x^2 e^x\,dx = x^2(e^x) - 2\int e^x\,x\,dx$$
$$= x^2 e^x - 2[x(e^x) - \int e^x\,dx]$$
$$= x^2 e^x - 2x e^x + 2e^x + C$$
$$= \underline{e^x\,[x^2 - 2x + 2] + C}$$

*On to frame 3.*

**3** Now let us try the same thing with this one –

$$\int x^n e^x\,dx = x^n(e^x) - n\int e^x x^{n-1}\,dx$$
$$= x^n e^x - n\int e^x x^{n-1}\,dx$$

Now you will see that the integral on the right, i.e. $\int e^x x^{n-1}\,dx$, is of exactly the same form as the one we started with, i.e. $\int e^x x^n\,dx$, except for the fact that $n$ has now been replaced by $(n-1)$

Then, if we denote $\int x^n e^x\,dx$ by $I_n$

we can denote $\int x^{n-1} e^x\,dx$ by $I_{n-1}$

So our result

$$\int x^n e^x\,dx = x^n e^x - n\int e^x x^{n-1}\,dx$$

can be written

$$I_n = x^n e^x - \text{..................}$$

*Then on to frame 4.*

**4**

$$I_n = x^n e^x - n\, I_{n-1}$$

This relationship is called a *reduction formula* since it expresses an integral in $n$ in terms of the same integral in $(n-1)$. Here it is again.

$$\text{If } I_n = \int x^n e^x \, dx$$

$$\text{then } I_n = x^n e^x - n.I_{n-1}$$

Make a note of this result in your record book, since we shall be using it in the examples that follow.

*Then to frame 5.*

**5**

*Example* Consider $\int x^2 e^x \, dx$

This is, of course, the case of $I_n = \int x^n e^x \, dx$ in which $n = 2$.

We know that $I_n = x^n e^x - n\, I_{n-1}$ applies to this integral, so, putting $n = 2$, we get

$$I_2 = x^2 e^x - 2.I_1$$

and then

$$I_1 = x^1 e^x - 1.I_0$$

Now we can easily evaluate $I_0$ in the normal manner –

$$I_0 = \int x^0 e^x \, dx = \int 1\, e^x \, dx = \int e^x \, dx = e^x + C$$

$$\text{So} \quad I_2 = x^2 e^x - 2.I_1$$

$$\text{and} \quad I_1 = x e^x - e^x + C_1$$

$$\therefore\ I_2 = x^2 e^x - 2x e^x + 2e^x + C$$

$$= e^x \left[x^2 - 2x + 2\right] + C$$

And that is it. Once you have established the reduction formula for a particular type of integral, its use is very simple.

In just the same way, using the same reduction formula, determine the integral $\int x^3 e^x \, dx$.

*Then check with the next frame.*

**6**

$$\int x^3 e^x \, dx = e^x [x^3 - 3x^2 + 6x - 6] + C$$

Here is the working. Check yours. $I_n = x^n e^x - n I_{n-1}$

$n = 3$ $\quad I_3 = x^3 e^x - 3.I_2$

$n = 2$ $\quad I_2 = x^2 e^x - 2.I_1$

$n = 1$ $\quad I_1 = x e^x - 1.I_0$

$$\text{and } I_0 = \int x^0 e^x \, dx = \int e^x \, dx = e^x + C$$

$$\begin{aligned} \therefore \; I_3 &= x^3 e^x - 3.I_2 \\ &= x^3 e^x - 3x^2 e^x + 6.I_1 \\ &= x^3 e^x - 3x^2 e^x + 6x e^x - 6e^x + C \\ &= \underline{e^x [x^3 - 3x^2 + 6x - 6] + C} \end{aligned}$$

*Now move on to frame 7.*

---

**7**

Let us now find a reduction formula for the integral $\int x^n \cos x \, dx$.

$$\begin{aligned} I_n &= \int x^n \cos x \, dx \\ &= x^n(\sin x) - n \int \sin x \, x^{n-1} \, dx \\ &= x^n \sin x - n \int x^{n-1} \sin x \, dx. \end{aligned}$$

Note that this is *not* a reduction formula yet, since the integral on the right is *not* of the same form as that of the original integral. So let us apply the integration-by-parts routine again.

$$\begin{aligned} I_n &= x^n \sin x - n \int x^{n-1} \sin x \, dx \\ &= x^n \sin x - n \ldots\ldots\ldots\ldots\ldots\ldots \end{aligned}$$

---

**8**

$$I_n = x^n \sin x + n \, x^{n-1} \cos x - n(n-1) \int x^{n-2} \cos x \, dx$$

Now you will see that the integral $\int x^{n-2} \cos x \, dx$ is the same as the integral $\int x^n \cos x \, dx$, with $n$ replaced by ...............

**9**

$$\boxed{n-2}$$

i.e. $\underline{I_n = x^n \sin x + n\,x^{n-1} \cos x - n(n-1)\,I_{n-2}}$

So this is the reduction formula for $I_n = \int x^n \cos x\, dx$

Copy the result down in your record book and then use it to find $\int x^2 \cos x\, dx$. First of all, put $n = 2$ in the result, which then gives ..................................................

**10**

$$\boxed{I_2 = x^2 \sin x + 2x \cos x - 2.1.\ I_0}$$

Now $\quad I_0 = \int x^0 \cos x\, dx = \int \cos x\, dx = \sin x + C_1$

And so $\quad \underline{I_2 = x^2 \sin x + 2x \cos x - 2 \sin x + C}$

Now you know what it is all about, how about this one?

Find a reduction formula for $\int x^n \sin x\, dx$.

Apply the integration-by-parts routine: it is very much like the last one.

*When you have finished, move on to frame 11.*

**11**

$$\boxed{I_n = -x^n \cos x + n\,x^{n-1} \sin x - n(n-1)\,I_{n-2}}$$

For: $\quad I_n = \int x^n \sin x\, dx$

$$= x^n(-\cos x) + n \int \cos x\, x^{n-1}\, dx$$

$$= -x^n \cos x + n\left\{x^{n-1}(\sin x) - (n-1)\int \sin x\, x^{n-2}\, dx\right\}$$

$$\therefore\ \underline{I_n = -x^n \cos x + n\,x^{n-1} \sin x - n(n-1)\,I_{n-2}}$$

Make a note of the result, and then let us find $\int x^3 \sin x\, dx$.

Putting $n = 3$, $I_3 = -x^3 \cos x + 3\,x^2 \sin x - 3.2.\ I_1$

and then $\quad I_1 = \int x \sin x\, dx$

$\quad = \ldots\ldots\ldots\ldots\ldots\ldots\ldots\ldots$

*Find this and then finish the result – then on to frame 12.*

**12**

$$I_1 = -x\cos x + \sin x + C$$

So that $I_3 = -x^3 \cos x + 3x^2 \sin x - 6 I_1$

$\therefore\ I_3 = -x^3 \cos x + 3x^2 \sin x + 6x \cos x - 6\sin x + C$

Note that a reduction formula can be repeated until the value of $n$ decreases to $n = 1$ or $n = 0$, when the final integral is determined by normal methods.

*Now move on to frame 13 for the next section of the work.*

**13** Let us now see what complications there are when the integral has limits.

*Example.* To determine $\int_0^{\pi} x^n \cos x\, dx$.

Now we have already established that, if $I_n = \int x^n \cos x\, dx$, then

$$I_n = x^n \sin x + n\,x^{n-1} \cos x - n(n-1)\, I_{n-2}$$

If we now define $I_n = \int_0^{\pi} x^n \cos x\, dx$, all we have to do is to apply the limits to the calculated terms on the right-hand side of our result

$$I_n = \Big[x^n \sin x + n\,x^{n-1} \cos x\Big]_0^{\pi} - n(n-1)\, I_{n-2}$$

$$= \big[0 + n\,\pi^{n-1}(-1)\big] - \big[0 + 0\big] - n(n-1)\, I_{n-2}$$

$$\therefore\ I_n = -n\,\pi^{n-1} - n(n-1)\, I_{n-2}$$

This, of course, often simplifies the reduction formula and is much quicker than obtaining the complete general result and then having to substitute the limits.

Use the result above to evaluate $\int_0^{\pi} x^4 \cos x\, dx$.

First put $n = 4$, giving $I_4 =$ ................................

**14**

$$I_4 = -4\pi^3 - 4.3.\, I_2$$

Now put $n = 2$ to find $I_2$, which is $I_2 =$ ...................

**15**

$$\boxed{I_2 = -2.\pi - 2.1.I_0}$$

and $\quad I_0 = \int_0^\pi x^0 \cos x \, dx = \int_0^\pi \cos x \, dx = \Big[\sin x\Big]_0^\pi = 0$

So we have

$$I_4 = -4\pi^3 - 12\,I_2$$
$$I_2 = -2\pi$$

and $\quad \therefore \; I_4 = \ldots\ldots\ldots\ldots\ldots\ldots$

---

**16**

$$\boxed{\int_0^\pi x^4 \cos x \, dx = I_4 = -4\pi^3 + 24\pi}$$

Now here is one for you to do in very much the same way.

Evaluate $\quad \int_0^\pi x^5 \cos x \, dx.$

*Work it right through and then check your working with frame 17.*

---

**17**

$$\boxed{I_5 = -5\pi^4 + 60\pi^2 - 240}$$

Working:

$$I_n = -n\pi^{n-1} - n(n-1)\,I_{n-2}$$
$$\therefore\; I_5 = -5\pi^4 - 5.4.\,I_3$$
$$I_3 = -3\pi^2 - 3.2.\,I_1$$

and

$$I_1 = \int_0^\pi x \cos x \, dx = \Big[x(\sin x)\Big]_0^\pi - \int_0^\pi \sin x \, dx$$
$$= [0-0] - \Big[-\cos x\Big]_0^\pi$$
$$= \Big[\cos x\Big]_0^\pi = (-1) - (1) = -2$$

$$\therefore\; I_5 = -5\pi^4 - 20\,I_3$$
$$I_3 = -3\pi^2 - 6(-2)$$
$$\therefore\; \underline{I_5 = -5\pi^4 + 60\pi^2 - 240}$$

*Turn on to frame 18.*

**18**

*Reduction formulae for* (i) $\int \sin^n x\, dx$ *and* (ii) $\int \cos^n x\, dx$.

(i) $\int \sin^n x\, dx$.

Let $\quad I_n = \int \sin^n x\, dx = \int \sin^{n-1} x . \sin x\, dx = \int \sin^{n-1} x . d(-\cos x)$

Then, integration by parts, gives

$$I_n = \sin^{n-1} x.(-\cos x) + (n-1)\int \cos x . \sin^{n-2} x . \cos x\, dx$$

$$= -\sin^{n-1} x . \cos x + (n-1)\int \cos^2 x . \sin^{n-2} x\, dx$$

$$= -\sin^{n-1} x . \cos x + (n-1)\int (1-\sin^2 x)\sin^{n-2} x\, dx$$

$$= -\sin^{n-1} x . \cos x + (n-1)\left\{\int \sin^{n-2} x\, dx - \int \sin^n x\, dx\right\}$$

$$\therefore\ I_n = -\sin^{n-1} x . \cos x + (n-1)\, I_{n-2} - (n-1)\, I_n$$

Now bring the last term over to the left-hand side, and we have

$$n\, I_n = -\sin^{n-1} x . \cos x + (n-1)\, I_{n-2}$$

So finally, if $I_n = \int \sin^n x\, dx$, $\quad \underline{I_n = -\frac{1}{n}\sin^{n-1} x . \cos x + \frac{n-1}{n}\, I_{n-2}}$

Make a note of this result, and then use it to find $\int \sin^6 x\, dx$

**19**

$$\boxed{I_6 = -\frac{1}{6}\sin^5 x . \cos x - \frac{5}{24}\sin^3 x.. \cos x - \frac{5}{16}\sin x \cos x + \frac{5x}{16} + C}$$

For $\quad I_6 = -\frac{1}{6}\sin^5 x \cos x + \frac{5}{6} . I_4 .$

$$I_4 = -\frac{1}{4}\sin^3 x \cos x + \frac{3}{4} . I_2 .$$

$$I_2 = -\frac{1}{2}\sin x \cos x + \frac{1}{2} . I_0 . \qquad I_0 = \int dx = x + C$$

$$\therefore\ I_6 = -\frac{1}{6}\sin^5 x \cos x + \frac{5}{6}\left[-\frac{1}{4}\sin^3 x \cos x + \frac{3}{4} . I_2\right]$$

$$= -\frac{1}{6}\sin^5 x \cos x - \frac{5}{24}\sin^3 x \cos x + \frac{5}{8}\left[-\frac{1}{2}\sin x \cos x + \frac{x}{2}\right] + C$$

$$= -\frac{1}{6}\sin^5 x \cos x - \frac{5}{24}\sin^3 x \cos x - \frac{5}{16}\sin x \cos x + \frac{5x}{16} + C$$

**20**

(ii) $\int \cos^n x\, dx$.

Let $\quad I_n = \int \cos^n x\, dx = \int \cos^{n-1} x \cos x\, dx = \int \cos^{n-1} x\, d(\sin x)$

$$= \cos^{n-1} x . \sin x - (n-1) \int \sin x . \cos^{n-2} x\, (-\sin x)\, dx$$

$$= \cos^{n-1} x . \sin x + (n-1) \int \sin^2 x \cos^{n-2} x\, dx$$

$$= \cos^{n-1} x . \sin x + (n-1) \int (1 - \cos^2 x) \cos^{n-2} x\, dx$$

$$= \cos^{n-1} x . \sin x + (n-1) \left\{ \int \cos^{n-2} x\, dx - \int \cos^n x\, dx \right\}$$

Now finish it off, so that $I_n =$ ....................................

**21**

$$\boxed{I_n = \frac{1}{n} \cos^{n-1} x . \sin x + \frac{n-1}{n} . I_{n-2}}$$

For $\quad I_n = \cos^{n-1} x . \sin x + (n-1) I_{n-2} - (n-1) I_n$

$$n\, I_n = \cos^{n-1} x . \sin x + (n-1) I_{n-2}.$$

$$\therefore\ I_n = \frac{1}{n} \cos^{n-1} x . \sin x + \frac{n-1}{n} I_{n-2}$$

Add this result to your list and then apply it to find $\int \cos^5 x\, dx$

*When you have finished it, move to frame 22.*

**22**

$$\boxed{\int \cos^5 x\, dx = \frac{1}{5} \cos^4 x \sin x + \frac{4}{15} \cos^2 x \sin x + \frac{8}{15} \sin x + C}$$

Here it is:

$$I_5 = \frac{1}{5} \cos^4 x \sin x + \frac{4}{5} I_3$$

$$I_3 = \frac{1}{3} \cos^2 x \sin x + \frac{2}{3} I_1$$

$$\text{And } I_1 = \int \cos x\, dx = \sin x + C_1$$

$$\therefore\ I_5 = \frac{1}{5} \cos^4 x \sin x + \frac{4}{5} \left[ \frac{1}{3} \cos^2 x \sin x + \frac{2}{3} \sin x \right] + C$$

$$= \frac{1}{5} \cos^4 x \sin x + \frac{4}{15} \cos^2 x \sin x + \frac{8}{15} \sin x + C$$

*On to frame 23.*

**23** The integrals $\int \sin^n x\, dx$ and $\int \cos^n x\, dx$ with limits $x = 0$ and $x = \pi/2$, give some interesting and useful results.
We already know the reduction formula

$$\int \sin^n x\, dx = I_n = -\frac{1}{n}\sin^{n-1}x\,.\cos x + \frac{n-1}{n}I_{n-2}$$

Inserting the limits

$$I_n = \left[-\frac{1}{n}\sin^{n-1}x\cos x\right]_0^{\pi/2} + \frac{n-1}{n}I_{n-2}$$

$$= \left[0 - 0\right] + \frac{n-1}{n}I_{n-2}$$

$$\therefore\ I_n = \frac{n-1}{n}I_{n-2}$$

And if you do the same with the reduction formula for $\int \cos^n x\, dx$, you get exactly the same result.

So for $\int_0^{\pi/2} \sin^n x\, dx$ and $\int_0^{\pi/2} \cos^n x\, dx$, we have

$$\underline{I_n = \frac{n-1}{n}I_{n-2}}$$

Also

(i) If $n$ is even, the formula eventually reduces to $I_0$

i.e. $\int_0^{\pi/2} 1\, dx = \left[x\right]_0^{\pi/2} = \pi/2 \quad \therefore\ \underline{I_0 = \pi/2}$

(ii) If $n$ is odd, the formula eventually reduces to $I_1$

i.e. $\int_0^{\pi/2} \sin x\, dx = \left[-\cos x\right]_0^{\pi/2} = -(-1) \quad \therefore\ \underline{I_1 = 1}$

So now, all on your own, evaluate $\int_0^{\pi/2} \sin^5 x\, dx$. What do you get?

---

**24**

$$\boxed{I_5 = \frac{4}{5}\cdot\frac{2}{3}\cdot 1 = \frac{8}{15}}$$

For $\quad I_5 = \frac{4}{5}\,.\,I_3$

$I_3 = \frac{2}{3}\,.\,I_1 \qquad$ and we know that $I_1 = 1$

$$\therefore\ I_5 = \frac{4}{5}\cdot\frac{2}{3}\cdot 1 = \underline{\frac{8}{15}}$$

In the same way, find $\int_0^{\pi/2} \cos^6 x\, dx$.

*Then to frame 25.*

**25**

$$\boxed{I_6 = \frac{5\pi}{32}}$$

For
$$I_6 = \frac{5}{6} . I_4$$
$$I_4 = \frac{3}{4} . I_2$$
$$I_2 = \frac{1}{2} . I_0 \quad \text{and } I_0 = \frac{\pi}{2}$$
$$\therefore \; I_6 = \frac{5}{6} . \frac{3}{4} . \frac{1}{2} . \frac{\pi}{2} = \underline{\frac{5\pi}{32}}$$

*Note* that all the natural numbers from $n$ down to 1 appear alternately on the bottom or top of the expression. In fact, if we start writing the numbers with the value of $n$ on the bottom, we can obtain the result with very little working.

$$\frac{(n-1) \qquad (n-3) \qquad (n-5) \; ....}{n \qquad (n-2) \qquad (n-4) \qquad ....} \text{ etc.}$$

If $n$ is odd, the factors end with 1 on the bottom

e.g. $\dfrac{6 . 4 . 2}{7 . 5 . 3 . 1}$ and that is all there is to it.

If $n$ is even, the factor 1 comes on top and then we add the factor $\pi/2$

e.g. $\dfrac{7 . 5 . 3 . 1}{8 . 6 . 4 . 2} . \dfrac{\pi}{2}$

So (i) $\int \sin^4 x \, dx =$ ....................

and (ii) $\int \cos^5 x \, dx =$ ....................

**26**

$$\boxed{\int \sin^4 x \, dx = \frac{3\pi}{16}, \int \cos^5 x \, dx = \frac{8}{15}}$$

This result for evaluating $\int \sin^n x \, dx$ or $\int \cos^n x \, dx$ between the limits $x = 0$ and $x = \pi/2$, is known as *Wallis's formula.* It is well worth remembering, so make a few notes on it.

*Then on to frame 27 for a further example.*

**27** Here is another example on the same theme.

*Example.* Evaluate $\int_0^{\pi/2} \sin^5 x \cos^2 x \, dx$.

We can write

$$\int_0^{\pi/2} \sin^5 x \cos^2 x \, dx = \int_0^{\pi/2} \sin^5 x(1 - \sin^2 x) \, dx$$

$$= \int_0^{\pi/2} (\sin^5 x - \sin^7 x) \, dx$$

$$= I_5 - I_7$$

$$= \ldots\ldots\ldots\ldots\ldots\ldots$$

*Finish it off.*

**28**

$$\boxed{\frac{8}{105}}$$

$$I_5 = \frac{4 \,.\, 2}{5 \,.\, 3 \,.\, 1} = \frac{8}{15}; \; I_7 = \frac{6 \,.\, 4 \,.\, 2}{7 \,.\, 5 \,.\, 3 \,.\, 1} = \frac{16}{35}$$

$$\therefore I_5 - I_7 = \frac{8}{15} - \frac{16}{35} = \underline{\frac{8}{105}}$$

**29** All that now remains is the Test Exercise. The examples are all very straightforward and should cause no difficulty.

Before you work the exercise, look back through your notes and revise any points on which you are not absolutely certain: there should not be many.

*On then to frame 30.*

30

**Test Exercise – XVII**

Work through all the questions. Take your time over the exercise: there are no prizes for speed!

Here they are then.

1. If $I_n = \int x^n e^{2x}\,dx$, show that

$$I_n = \frac{x^n e^{2x}}{2} - \frac{n}{2}.\,I_{n-1}$$

and hence evaluate $\int x^3 e^{2x}\,dx$.

2. Evaluate (i) $\displaystyle\int_0^{\pi/2} \sin^2 x \cos^6 x\,dx$

(ii) $\displaystyle\int_0^{\pi/2} \sin^4 x \cos^5 x\,dx$

3. By the substitution $x = a \sin\theta$, determine

$$\int_0^a x^3 (a^2 - x^2)^{3/2}\,dx$$

4. By writing $\tan^n x$ as $\tan^{n-2} x\,.(\sec^2 x - 1)$, obtain a reduction formula for $\int \tan^n x\,dx$.

Hence show that $I_n = \displaystyle\int_0^{\pi/4} \tan^n x\,dx = \frac{1}{n-1} - I_{n-2}$

5. By the substitution $x = \sin^2\theta$, determine a reduction formula for the integral

$$\int x^{5/2}(1-x)^{3/2}\,dx$$

Hence evaluate

$$\int_0^1 x^{5/2}(1-x)^{3/2}\,dx$$

**Further Problems – XVII**

1. If $I_n = \int_0^{\pi/2} x \cos^n x \, dx$, when $n > 1$, show that

$$I_n = \frac{n(n-1)}{n^2} I_{n-2} - 1$$

2. Establish a reduction formula for $\int \sin^n x \, dx$ in the form

$$I_n = -\frac{1}{n} \sin^{n-1} x \cos x + \frac{n-1}{n} I_{n-2}$$

and hence determine $\int \sin^7 x \, dx$.

3. If $I_n = \int_0^{\infty} x^n e^{-ax} \, dx$, show that $I_n = \frac{n}{a} . I_{n-1}$. Hence evaluate $\int_0^{\infty} x^9 e^{-2x} dx$

4. If $I_n = \int_0^{\pi} e^{-x} \sin^n x \, dx$, show that $I_n = \frac{n(n-1)}{n^2+1} I_{n-2}$.

5. If $I_n = \int_0^{\pi/2} x^n \sin x \, dx$, prove that, for $n \geqslant 2$,

$$I_n = n\left(\frac{\pi}{2}\right)^{n-1} - n(n-1) I_{n-2}$$

Hence evaluate $I_3$ and $I_4$.

6. If $I_n = \int x^n e^x \, dx$, obtain a reduction formula for $I_n$ in terms of $I_{n-1}$ and hence determine $\int x^4 e^x \, dx$.

7. If $I_n = \int \sec^n x \, dx$, prove that

$$I_n = \frac{1}{n-1} \tan x \sec^{n-2} x + \frac{n-2}{n-1} I_{n-2} \quad (n \geqslant 2)$$

Hence evaluate $\int_0^{\pi/6} \sec^8 x \, dx$.

8. If $I_n = \int_0^{\pi/2} e^{-x} \cos^n x \, dx$, where $n \geqslant 2$, prove that

$$\text{(i)} \quad I_n = 1 - n \int_0^{\pi/2} e^{-x} \sin x \cos^{n-1} x \, dx$$

$$\text{(ii)} \quad (n^2 + 1) I_n = 1 + n(n-1) I_{n-2}$$

Show that $I_6 = \dfrac{263 - 144\, e^{-\pi/2}}{629}$

9. If $I_n = \int (x^2 + a^2)^n \, dx$, show that

$$I_n = \frac{1}{2n+1} \left[ x(x^2 + a^2)^n + 2na^2 \, I_{n-1} \right]$$

10. If $I_n = \int \cot^n x \, dx$, $(n > 1)$, show that

$$I_n = -\frac{\cot^{n-1} x}{(n-1)} - I_{n-2}$$

Hence determine $I_6$.

11. If $I_n = \int (\ln x)^n \, dx$, show that

$$I_n = x(\ln x)^n - n . I_{n-1}$$

Hence find $\int (\ln x)^3 \, dx$.

12. If $I_n = \int \cosh^n x \, dx$, prove that

$$I_n = \frac{1}{n} \cosh^{n-1} x \sinh x + \frac{n-1}{n} I_{n-2}$$

Hence evaluate $\int_0^a \cosh^3 x \, dx$, where $a = \cosh^{-1} (\sqrt{2})$.

# Programme 18

## INTEGRATION APPLICATIONS

### PART 1

**1**

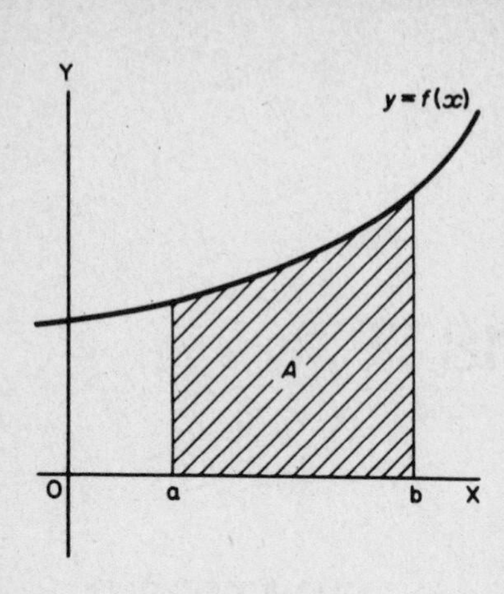


We now look at some of the applications to which integration can be put. Some you already know from earlier work: others will be new to you. So let us start with one you first met long ago.

*Areas under curves*

*To find the area bounded by the curve $y = f(x)$, the x-axis and the ordinates at $x = a$ and $x = b$*

There is, of course, no mensuration formula for this, since its shape depends on the function $f(x)$. Do you remember how you established the method for finding this area?

*Move on to frame 2.*

**2**

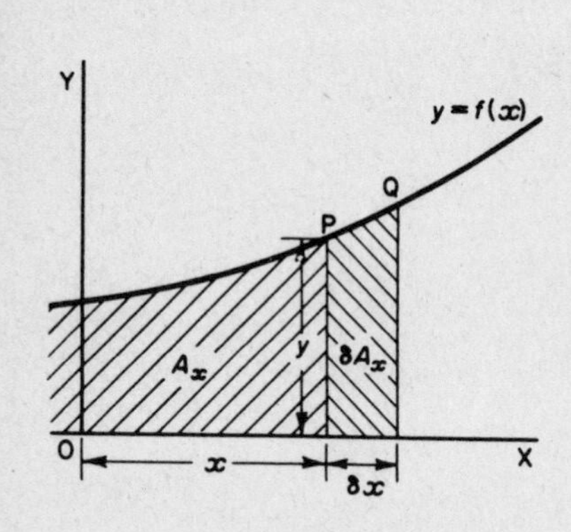


Let us revise this, for the same principles are applied in many other cases.

Let $P(x, y)$ be a point on the curve $y = f(x)$ and let $A_x$ denote the area under the curve measured from some point away to the left of the diagram.

The point Q, near to P, will have co-ordinates $(x + \delta x, y + \delta y)$ and the area is increased by the extent of the shaded strip. Denote this by $\delta A_x$.

If we 'square off' the strip at the level of P, then we can say that the area of the strip is approximately equal to that of the rectangle (omitting PQR).

i.e. area of strip $= \delta A_x \triangleq$ ..........................

*Turn to frame 3.*

**3**

$$\delta A_x \simeq y\delta x$$

Therefore, $\dfrac{\delta A_x}{\delta x} \simeq y$

i.e. the total area of the strip divided by the width, $\delta x$, of the strip gives approximately the value $y$.

The area above the rectangle represents the error in our stated approximation, but if we reduce the width of the strips, the total error is very much reduced.

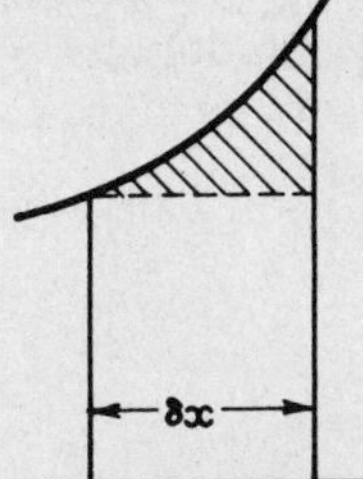


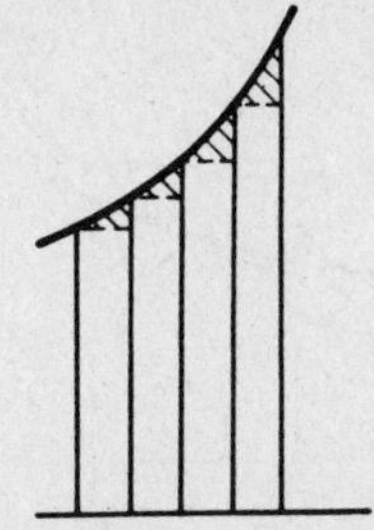

If we continue this process and make $\delta x \to 0$, then in the end the error will vanish, and, at the same time, $\dfrac{\delta A_x}{\delta x} \to$ ..........................

**4**

$$\frac{\delta A_x}{\delta x} \to \frac{dA_x}{dx}$$

Correct. So we have $\dfrac{dA_x}{dx} = y$ (no longer an approximation)

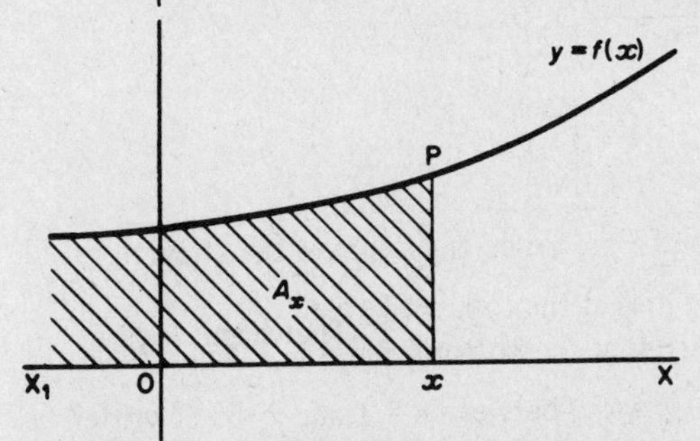


$$\therefore\ A_x = \int y\,dx$$

$$= \int f(x)\,dx$$

$$A_x = F(x) + C$$

and this represents the area under the curve up to the point P.

*Note* that, as it stands, this result would not give us a numerical value for the area, because we do not know from what point the measurement of the area began (somewhere off to the left of the figure). Nevertheless, we can make good use of the result, so turn on now to frame 5.

**5** $A_x = \int y\,dx$ gives the area up to the point $P(x, y)$.

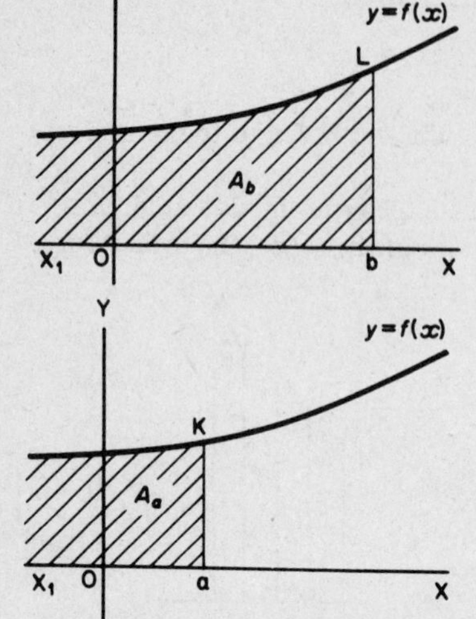


So:

(i) If we substitute $x = b$, we have the area up to the point L

i.e. $A_b = \int y\,dx$ with $x = b$.

(ii) If we substitute $x = a$, we have the area up to the point K

i.e. $A_a = \int y\,dx$ with $x = a$.

If we now subtract the second result from the first, we have the area under the curve between the ordinates at $x = a$ and $x = b$.

i.e. $A = \int y\,dx_{(x=b)} - \int y\,dx_{(x=a)}$

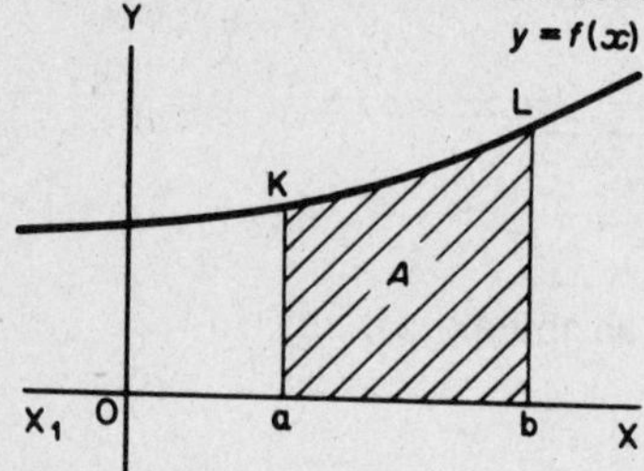


This is written

$$A = \int_a^b y\,dx$$

and the boundary values $a$ and $b$ are called the *limits* of the integral.

Remember: the higher limit goes at the top.
the lower limit goes at the bottom. } That seems logical.

So, the area under the curve $y = f(x)$ between $x = 1$ and $x = 5$ is written

$$A = \int_1^5 y\,dx.$$

Similarly, the area under the curve $y = f(x)$ between $x = -5$ and $x = -1$ is written A = ..................

*On to frame 6.*

**6**

$$A = \int_{-5}^{-1} y\,dx$$

Let us do a simple example.
Find the area under the curve $y = x^2 + 2x + 1$ between $x = 1$ and $x = 2$.

$$A = \int_1^2 y\,dx = \int_1^2 (x^2 + 2x + 1)\,dx$$

$$= \left[\frac{x^3}{3} + x^2 + x + C\right]_1^2$$

$$= \left[\frac{8}{3} + 4 + 2 + C\right] - \left[\frac{1}{3} + 1 + 1 + C\right]$$

(putting $x = 2$) (putting $x = 1$)

$$= \left[8\tfrac{2}{3} + C\right] - \left[2\tfrac{1}{3} + C\right]$$

$$= \underline{6\tfrac{1}{3}\text{ units}^2}$$

*Note:* When we have limits to substitute, the constant of integration appears in each bracket and will therefore always disappear. In practice therefore, we may leave out the constant of integration when we have limits, since we know it will always vanish in the next line of working.
Now you do this one:

Find the area under the curve $y = 3x^2 + 4x - 5$ between $x = 1$ and $x = 3$.

*Then move on to frame 7.*

**7**

$$A = 32\text{ units}^2$$

For
$$A = \int_1^3 (3x^2 + 4x - 5)\,dx = \left[x^3 + 2x^2 - 5x\right]_1^3$$

$$= \left[27 + 18 - 15\right] - \left[1 + 2 - 5\right]$$

$$= \left[30\right] - \left[-2\right] = \underline{32\text{ units}^2}$$

*Definite integrals*

An integral with limits is called a *definite integral.*

With a definite integral, the constant of integration may be omitted, not because it is not there, but because ...............................................

...............................................

*On to frame 8.*

**8**

... it occurs in both brackets and disappears in subsequent working.

So, to evaluate a definite integral

(i) Integrate the function (omitting the constant of integration) and enclose within square brackets with the limits at the right-hand end.
(ii) Substitute the upper limit.
(iii) Substitute the lower limit.
(iv) Subtract the second result from the first result.

$$\int_a^b y\,dx = \Big[F(x)\Big]_a^b = F(b) - F(a)$$

Now, you evaluate this one.

$$\int_{-1}^{\frac{1}{2}} 4\,e^{2x}\,dx = \ldots\ldots\ldots\ldots\ldots\ldots$$

**9**

5·166

Here it is:

$$\int_{-1}^{\frac{1}{2}} 4e^{2x}\,dx = 4\left[\frac{e^{2x}}{2}\right]_{-1}^{\frac{1}{2}} = 2\Big[e^{2x}\Big]_{-1}^{\frac{1}{2}}$$

$$= 2\Big[e - e^{-2}\Big]$$

$$= 2\left[e - \frac{1}{e^2}\right]$$

$$= 5{\cdot}166$$

Now, what about this one: $\displaystyle\int_0^{\pi/2} x\cos x\,dx.$

First of all, forget about the limits.

$$\int x\cos x\,dx = \ldots\ldots\ldots\ldots\ldots\ldots$$

*When you have done that part, turn to frame 10.*

**10**

$$\int x \cos x \, dx = x(\sin x) - \int \sin x \, dx$$
$$= x \sin x + \cos x + C$$

$$\therefore \int_0^{\pi/2} x \cos x \, dx = \left[ x \sin x + \cos x \right]_0^{\pi/2}$$

= .................... You finish it off.

**11**

$$\frac{\pi}{2} - 1$$

for
$$\int_0^{\pi/2} x \cos x \, dx = \left[ x \sin x + \cos x \right]_0^{\pi/2}$$
$$= \left[ \frac{\pi}{2} + 0 \right] - \left[ 0 + 1 \right]$$
$$= \frac{\pi}{2} - 1$$

If you can integrate the given function, the rest is easy.

*So move to the next frame and work one or two on your own.*

**12**

**Exercise**

Evaluate:

(1) $\displaystyle\int_1^2 (2x - 3)^4 \, dx$

(2) $\displaystyle\int_0^5 \frac{1}{x + 5} \, dx$

(3) $\displaystyle\int_{-3}^3 \frac{dx}{x^2 + 9}$

(4) $\displaystyle\int_1^e x^2 \ln x \, dx$

*When you have finished them all, check your results with the solutions given in the next frame.*

**13** *Solutions*

$$(1)\ \int_1^2 (2x-3)^4\,dx = \left[\frac{(2x-3)^5}{10}\right]_1^2 = \frac{1}{10}\left\{(1)^5 - (-1)^5\right\}$$

$$= \frac{1}{10}\left\{(1) - (-1)\right\} = \frac{2}{10} = \underline{\frac{1}{5}}$$

$$(2)\ \int_0^5 \frac{1}{x+5}\,dx = \Big[\ln(x+5)\Big]_0^5$$

$$= \ln 10 - \ln 5 = \ln\frac{10}{5} = \underline{\ln 2}$$

$$(3)\ \int_{-3}^{3} \frac{dx}{x^2+9} = \left[\frac{1}{3}\tan^{-1}\frac{x}{3}\right]_{-3}^{3}$$

$$= \frac{1}{3}\left[(\tan^{-1} 1) - (\tan^{-1}[-1])\right]$$

$$= \frac{1}{3}\left[\frac{\pi}{4} - \left(-\frac{\pi}{4}\right)\right] = \underline{\frac{\pi}{6}}$$

$$(4)\ \int x^2 \ln x\,dx = \ln x\left(\frac{x^3}{3}\right) - \frac{1}{3}\int x^3 \cdot \frac{1}{x}\,dx$$

$$= \frac{x^3 \ln x}{3} - \frac{x^3}{9} + C$$

$$\therefore \int_1^e x^2 \ln x\,dx = \left[\frac{x^3 \ln x}{3} - \frac{x^3}{9}\right]_1^e$$

$$= \left(\frac{e^3}{3} - \frac{e^3}{9}\right) - \left(0 - \frac{1}{9}\right)$$

$$= \frac{2e^3}{9} + \frac{1}{9} = \underline{\frac{1}{9}(2e^3+1)}$$

*On to frame 14.*

**14** In very many practical applications we shall be using definite integrals, so let us practise a few more.

Do these:

$$(5)\ \int_0^{\pi/2} \frac{\sin 2x}{1+\cos^2 x}\,dx$$

$$(6)\ \int_1^2 x\,e^x\,dx$$

$$(7)\ \int_0^{\pi} x^2 \sin x\,dx$$

*Finish them off and then check with the next frame.*

**15**

*Solutions*

(5) $$\int_0^{\pi/2} \frac{\sin 2x}{1+\cos^2 x}\,dx = \left[-\ln(1+\cos^2 x)\right]_0^{\pi/2}$$
$$= \left[-\ln(1+0)\right] - \left[-\ln(1+1)\right]$$
$$= \left[-\ln 1 + \ln 2\right] = \underline{\ln 2}$$

(6) $$\int x\,e^x\,dx = x(e^x) - \int e^x\,dx$$
$$= x\,e^x - e^x + C$$
$$\therefore \int_1^2 x\,e^x\,dx = \left[e^x(x-1)\right]_1^2$$
$$= e^2 - 0 = \underline{e^2}$$

(7) $$\int x^2 \sin x\,dx = x^2(-\cos x) + 2\int x\cos x\,dx$$
$$= -x^2\cos x + 2\left\{x(\sin x) - \int \sin x\,dx\right\}$$
$$= -x^2\cos x + 2x\sin x + 2\cos x + C$$
$$\therefore \int_0^{\pi} x^2 \sin x\,dx = \left[(2-x^2)\cos x + 2x\sin x\right]_0^{\pi}$$
$$= \left[(2-\pi^2)(-1)+0\right] - \left[2+0\right]$$
$$= \pi^2 - 2 - 2 = \underline{\pi^2 - 4}$$

*Now move on to frame 16.*

**16**

Before we move on to the next piece of work, here is just one more example for you to do on areas.

*Example.* Find the area bounded by the curve $y = x^2 - 6x + 5$, the $x$-axis, and the ordinates at $x = 1$ and $x = 3$.

*Work it through and then turn on to frame 17.*

**17**

$$A = -5\frac{1}{3} \text{ units}^2$$

Here is the working:

$$A = \int_1^3 y\,dx = \int_1^3 (x^2 - 6x + 5)\,dx = \left[\frac{x^3}{3} - 3x^2 + 5x\right]_1^3$$

$$= (9 - 27 + 15) - (\frac{1}{3} - 3 + 5)$$

$$= (-3) - (2\frac{1}{3}) = \underline{-5\frac{1}{3} \text{ units}^2}$$

If you are concerned about the negative sign of the result, let us sketch the graph of the function. Here it is:

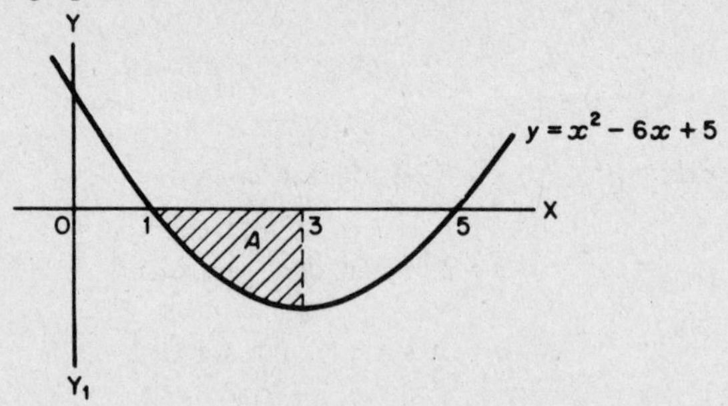


We find that between the limits we are given, the area lies below the $x$-axis.

For such an area, $y$ is negative

$\therefore$ $y\delta x$ is negative

$\therefore$ $\delta A$ is negative $\therefore$ A is negative.

So remember,

Areas below the $x$-axis are *negative.*

*Next frame.*

**18**

The danger comes when we are integrating between limits and part of the area is above the $x$-axis and part below it. In that case, the integral will give the algebraic sum of the area, i.e. the negative area will partly or wholly cancel out the positive area. If this is likely to happen, sketch the curve and perform the integration in two parts.

*Now turn to frame 19.*

**19**

**Parametric equations**

*Example.* A curve has parametric equations $x = at^2, y = 2at$. Find the area bounded by the curve, the $x$-axis, and the ordinates at $t = 1$ and $t = 2$.

We know that $A = \int_a^b y\,dx$ where $a$ and $b$ are the limits or boundary values of the variable.

Replacing $y$ by $2at$, gives

$$A = \int_a^b 2at\,dx$$

but we cannot integrate a function of $t$ with respect to $x$ directly. We therefore have to change the variable of the integral and we do it thus –

We are given $\quad x = at^2 \quad \therefore \dfrac{dx}{dt} = 2at \quad \therefore dx \equiv 2at\,dt$

We now have $\quad A = \int_1^2 2at.2at\,dt = \int_1^2 4a^2t^2\,dt$

$= \ldots\ldots\ldots\ldots\ldots\ldots$ Finish it off.

**20**

$$A = \int_1^2 4a^2t^2\,dt = 4a^2\left[\frac{t^3}{3}\right]_1^2$$

$$= 4a^2\left\{\frac{8}{3} - \frac{1}{3}\right\} = \underline{\frac{28a^2}{3}}$$

*The method* is always the same –

(i) Express $x$ and $y$ in terms of the parameter,
(ii) Change the variable,
(iii) Insert limits of the parameter.

*Example.* If $x = a\sin\theta, y = b\cos\theta$, find the area under the curve between $\theta = 0$ and $\theta = \pi$.

$$A = \int_a^b y\,dx = \int_0^\pi b\cos\theta . a\cos\theta . d\theta \qquad x = a\sin\theta, \quad dx \equiv a\cos\theta\,d\theta$$

$$= ab\int_0^\pi \cos^2\theta\,d\theta$$

$= \ldots\ldots\ldots\ldots\ldots\ldots\ldots\ldots\ldots\ldots$

**21**

$$\boxed{A = \frac{\pi ab}{2}}$$

For
$$A = ab\int_0^{\pi} \cos^2\theta\, d\theta = ab\left[\frac{\theta}{2} + \frac{\sin 2\theta}{4}\right]_0^{\pi}$$
$$= ab\left[\frac{\pi}{2}\right] = \underline{\frac{\pi ab}{2}}$$

Now do this one on your own:

*Example.* If $x = \theta - \sin\theta$, $y = 1 - \cos\theta$, find the area under the curve between $\theta = 0$ and $\theta = \pi$.

*When you have finished it, move on to frame 22.*

**22**

$$\boxed{A = \frac{3\pi}{2}\text{ units}^2}$$

Working:

$$A = \int_a^b y\, dx \qquad\qquad y = (1 - \cos\theta)$$
$$x = (\theta - \sin\theta)$$
$$dx \equiv (1 - \cos\theta)\, d\theta$$
$$= \int_0^{\pi} (1 - \cos\theta)(1 - \cos\theta)\, d\theta$$
$$= \int_0^{\pi} (1 - 2\cos\theta + \cos^2\theta)\, d\theta$$
$$= \left[\theta - 2\sin\theta + \frac{\theta}{2} + \frac{\sin 2\theta}{4}\right]_0^{\pi}$$
$$= \left[\frac{3\pi}{2}\right] - \left[0\right] = \underline{\frac{3\pi}{2}\text{ units}^2}$$

**23**

**Mean values**

To find the mean height of the students in a class, we could measure their individual heights, total the results and divide by the number of subjects. That is, in such cases, the *mean value* is simply the *average* of the separate values we were considering.

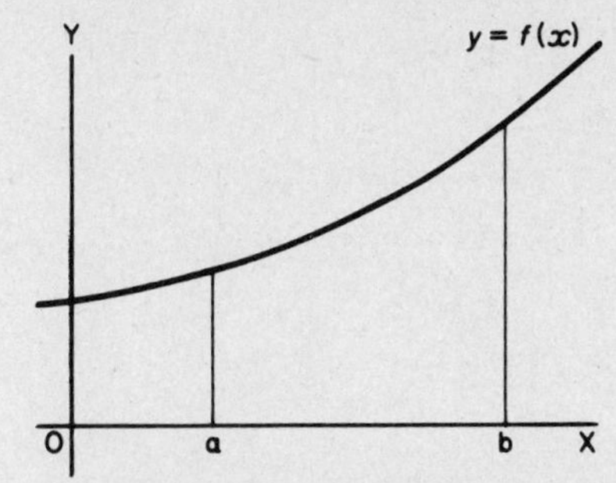


To find the mean value of a continuous function, however, requires further consideration.

When we set out to find the mean value of the function $y = f(x)$ between $x = a$ and $x = b$, we are no longer talking about separate items but a quantity which is continuously changing from $x = a$ to $x = b$. If we estimate the mean height of the figure in the diagram, over the given range, we are selecting a value M such that the part of the figure cut off would fill in the space below.

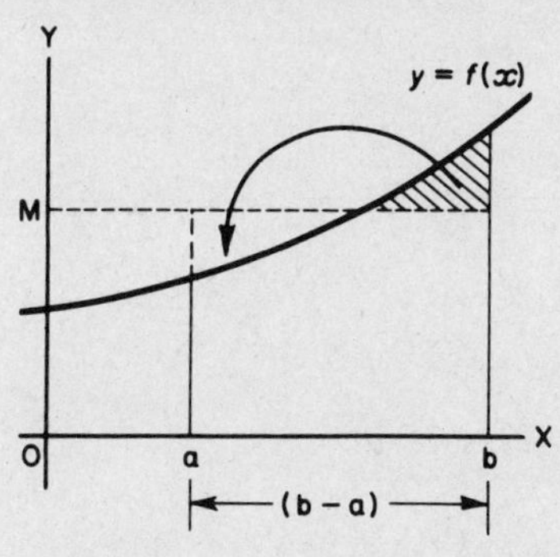


In other words, the area of the figure between $x = a$ and $x = b$ is shared out equally along the base line of the figure to produce the rectangle.

$$\therefore \ \text{M} = \frac{\text{Area}}{\text{Base line}} = \frac{\text{A}}{b-a}$$

$$\therefore \ \text{M} = \frac{1}{b-a}\int_a^b y\,dx$$

So, to find the mean value of a function between two limits, find the area under the curve between those limits and divide by ...........................

*On to frame 24.*

**24**

| length of the base line |
|---|

So it is really an application of areas.

*Example.* To find the mean value of $y = 3x^2 + 4x + 1$ between $x = -1$ and $x = 2$.

$$M = \frac{1}{b-a}\int_a^b y\,dx$$

$$= \frac{1}{2-(-1)}\int_{-1}^{2}(3x^2 + 4x + 1)\,dx$$

$$= \frac{1}{3}\left[x^3 + 2x^2 + x\right]_{-1}^{2}$$

$$= \frac{1}{3}\left[(8 + 8 + 2) - (-1 + 2 - 1)\right]$$

$$= \frac{1}{3}\left[18\right] = 6 \quad \therefore \underline{M = 6}$$

Here is one for you:

*Example.* Find the mean value of $y = 3 \sin 5t + 2 \cos 3t$ between $t = 0$ and $t = \pi$.

*Check your result with frame 25.*

**25**

Here is the working in full:

$$M = \frac{1}{\pi - 0}\int_0^{\pi}(3 \sin 5t + 2 \cos 3t)\,dt$$

$$= \frac{1}{\pi}\left[\frac{-3 \cos 5t}{5} + \frac{2 \sin 3t}{3}\right]_0^{\pi}$$

$$= \frac{1}{\pi}\left\{\left[\frac{-3 \cos 5\pi}{5} + \frac{2 \sin 3\pi}{3}\right] - \left[-\frac{3}{5} + 0\right]\right\}$$

$$= \frac{1}{\pi}\left\{\frac{3}{5} + \frac{3}{5}\right\} \qquad \underline{M = \frac{6}{5\pi}}$$

**26**

**R.M.S. values**

The phrase 'r.m.s. value of $y$' stands for 'the square *r*oot of the *m*ean value of the *s*quares of $y$' between some stated limits.

*Example.* If we are asked to find the r.m.s. value of $y = x^2 + 3$ between $x = 1$ and $x = 3$, we have –

$$\text{r.m.s.} = \sqrt{(\text{Mean value of } y^2 \text{ between } x = 1 \text{ and } x = 3)}$$

$$\therefore (\text{r.m.s.})^2 = \text{Mean value of } y^2 \text{ between } x = 1 \text{ and } x = 3$$

$$= \frac{1}{3-1}\int_1^3 y^2\,dx$$

$$= \ldots\ldots\ldots\ldots\ldots\ldots$$

**27**

$$(\text{r.m.s.})^2 = \frac{1}{2}\int_1^3 (x^4 + 6x^2 + 9)\,dx$$

$$= \frac{1}{2}\left[\frac{x^5}{5} + 2x^3 + 9x\right]_1^3$$

$$= \frac{1}{2}\left\{\left[\frac{243}{5} + 54 + 27\right] - \left[\frac{1}{5} + 2 + 9\right]\right\}$$

$$= \frac{1}{2}\left\{48{\cdot}6 + 81 - 11{\cdot}2\right\}$$

$$= \frac{1}{2}\left\{129{\cdot}6 - 11{\cdot}2\right\}$$

$$= \frac{1}{2}\left\{118{\cdot}4\right\} = 59{\cdot}2$$

$$\text{r.m.s.} = \sqrt{59{\cdot}2} = 7{\cdot}694 \quad \therefore \underline{\text{r.m.s.} = 7{\cdot}69}$$

So, *in words*, the r.m.s. value of $y$ between $x = a$ and $x = b$ means

..............................................................................

(Write it out)

*Then to the next frame.*

**28**

> '.. the square root of the mean value of the squares of $y$ between $x = a$ and $x = b$ '

There are three distinct steps:

(1) Square the given function.
(2) Find the mean value of the result over the interval given.
(3) Take the square root of the mean value.

So here is one for you to do:

*Example.* Find the r.m.s. value of $y = 400 \sin 200\pi t$ between $t = 0$ and $t = \dfrac{1}{100}$

*When you have the result, move on to frame 29.*

**29**

See if you agree with this –

$$y^2 = 160000 \sin^2 200\pi t$$

$$= 160000.\frac{1}{2}(1 - \cos 400\pi t)$$

$$= 80000\,(1 - \cos 400\pi t)$$

$$\therefore\ (\text{r.m.s.})^2 = \frac{1}{\frac{1}{100} - 0}\int_0^{1/100} 80000\,(1 - \cos 400\pi t)\,dt$$

$$= 100.\,80000\left[t - \frac{\sin 400\pi t}{400\pi}\right]_0^{1/100}$$

$$= 8.10^6\left[\frac{1}{100} - 0\right]$$

$$= 8.10^4$$

$$\therefore\ \text{r.m.s.} = \sqrt{(8.10^4)} = 200\sqrt{2} = \underline{282.8}$$

*Now on to frame 30.*

**30**

Before we come to the end of this particular programme, let us think back once again to the beginning of the work. We were, of course, considering the area bounded by the curve $y = f(x)$, the $x$-axis, and the ordinates at $x = a$ and $x = b$.

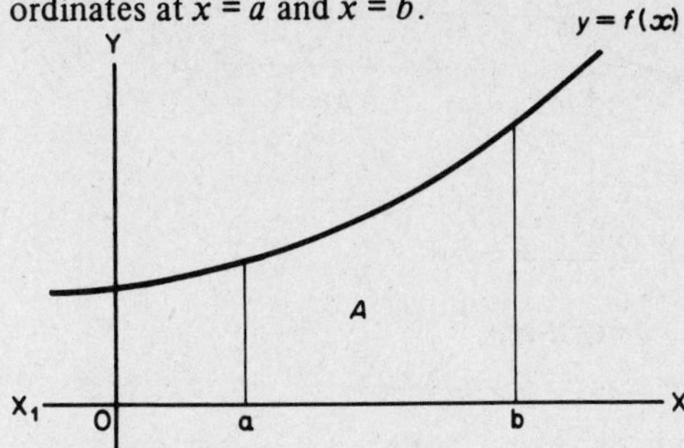


We found that

$$A = \int_a^b y\,dx$$

Let us look at the figure again.

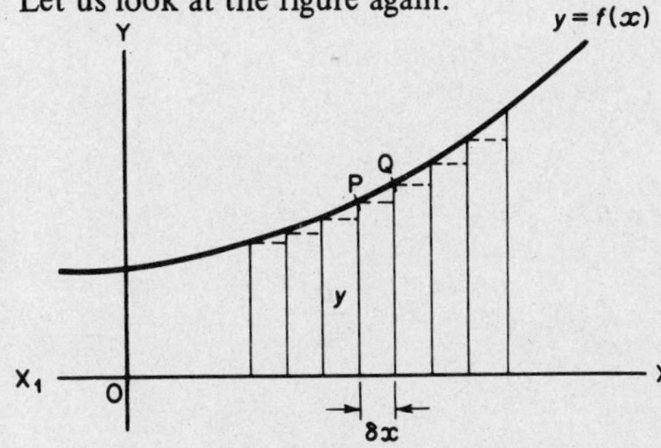


If P is the point $(x, y)$ then the area of the strip $\delta A$ is given by

$$\delta A \simeq y.\delta x$$

If we divide the complete figure up into a series of such strips, then the total area is given approximately by the sum of the areas of these strips.

i.e. A = sum of the strips between $x = a$ and $x = b$

i.e. $A \simeq \sum_{x=a}^{x=b} y.\delta x$ $\qquad \Sigma \equiv$ 'the sum of all terms like..'

The error in our approximation is caused by ignoring the area over each rectangle. But if the strips are made narrower, this error progressively decreases and, at the same time, the number of strips required to cover the figure increases. Finally, when $\delta x \to 0$,

A = sum of an infinite number of minutely thin rectangles

$$\therefore \quad A = \int_a^b y\,dx = \sum_{x=a}^{x=b} y.\delta x \text{ when } x \to 0$$

It is sometimes convenient, therefore, to regard integration as a summing up process of an infinite number of minutely small quantities each of which is too small to exist alone.

We shall make use of this idea at a later date.

*Next frame.*

**31**

## Summary Sheet

1. ***Areas under curves***

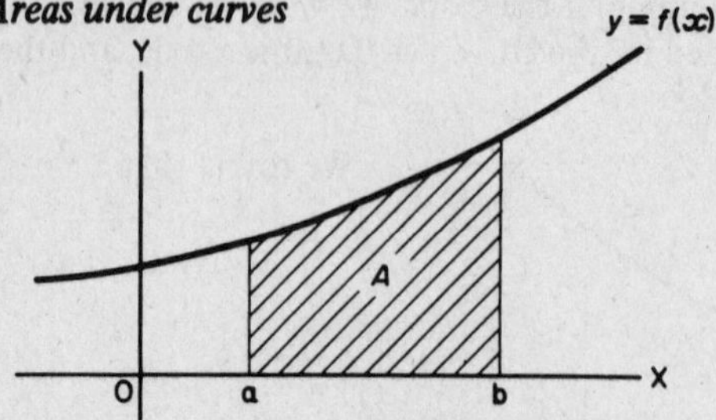


$$A = \int_a^b y\,dx$$

Areas below the $x$-axis are *negative.*

2. ***Definite integrals***

A definite integral is an integral with *limits.*

$$\int_a^b y\,dx = \Big[F(x)\Big]_a^b = F(b) - F(a)$$

3. ***Parametric equations***

$$x = f(t),\ y = F(t)$$

$$\int_{x_1}^{x_2} y\,dx = \int_{x_1}^{x_2} F(t)\,dx = \int_{t_1}^{t_2} F(t).\frac{dx}{dt}\,dt$$

4. ***Mean values***

$$M = \frac{1}{b-a}\int_a^b y\,dx$$

5. ***R.M.S. values***

$$(\text{r.m.s.})^2 = \frac{1}{b-a}\int_a^b y^2\,dx$$

6. ***Integration as a summing process***

$$\text{When } \delta x \to 0,\ \sum_{x=a}^{x=b} y.\delta x = \int_a^b y\,dx$$

All that now remains is the Test Exercise set out in the next frame. Before you work through it, be sure there is nothing that you wish to brush up. It is all very straightforward, so take your time.

*On then to frame 32.*

**Test Exercise – XVIII**

Work all the questions.

1. Find the area bounded by the curves $y = 3\,e^{2x}$ and $y = 3\,e^{-x}$ and the ordinates at $x = 1$ and $x = 2$.

2. The parametric equations of a curve are

$$y = 2\sin\frac{\pi}{10}t,\ x = 2 + 2t - 2\cos\frac{\pi}{10}t$$

Find the area under the curve between $t = 0$ and $t = 10$.

3. Find the mean value of $y = \dfrac{5}{2 - x - 3x^2}$ between $x = -\dfrac{1}{3}$ and $x = +\dfrac{1}{3}$.

4. Calculate the r.m.s. value of $i = 20 + 100\sin 100\pi t$ between $t = 0$ and $t = 1/50$.

5. If $i = \mathrm{I}\sin\omega t$ and $v = \mathrm{L}\dfrac{di}{dt} + \mathrm{R}i$, find the mean value of the product $vi$ between $t = 0$ and $t = \dfrac{2\pi}{\omega}$.

6. If $i = 300\sin 100\pi t + \mathrm{I}$, and the r.m.s. value of $i$ between $t = 0$ and $t = 0{\cdot}02$ is 250, determine the value of I.

**Further Problems–XVIII**

1. Find the mean height of the curve $y = 3x^2 + 5x - 7$ above the $x$-axis between $x = -2$ and $x = 3$.

2. Find the r.m.s. value of $i = \cos x + \sin x$ over the range $x = 0$ to $x = \dfrac{3\pi}{4}$.

3. Determine the area of one arch of the cycloid $x = \theta - \sin\theta$, $y = 1 - \cos\theta$, i.e. find the area of the plane figure bounded by the curve and the $x$-axis between $\theta = 0$ and $\theta = 2\pi$.

4. Find the area enclosed by the curves $y = \sin x$ and $y = \sin 2x$, between $x = 0$ and $x = \pi/3$.

5. If $i = 0{\cdot}2 \sin 10\pi t + 0{\cdot}01 \sin 30\pi t$, find the mean value of $i$ between $t = 0$ and $t = 0{\cdot}2$.

6. If $i = i_1 \sin pt + i_2 \sin 2pt$, show that the mean value of $i^2$ over a period is $\frac{1}{2}(i_1^2 + i_2^2)$.

7. Sketch the curves $y = 4e^x$ and $y = 9\sinh x$, and show that they intersect when $x = \ln 3$. Find the area bounded by the two curves and the $y$-axis.

8. If $v = v_0 \sin\omega t$ and $i = i_0 \sin(\omega t - a)$, find the mean value of $vi$ between $t = 0$ and $t = \dfrac{2\pi}{\omega}$.

9. If $i = \dfrac{E}{R} + I\sin\omega t$, where E, R, I, $\omega$ are constants, find the r.m.s. value of $i$ over the range $t = 0$ to $t = \dfrac{2\pi}{\omega}$.

10. The parametric equations of a curve are

$$x = a\cos^2 t \sin t, y = a\cos t \sin^2 t$$

Show that the area enclosed by the curve between $t = 0$ and $t = \dfrac{\pi}{2}$ is $\dfrac{\pi a^2}{32}$ units$^2$.

11. Find the area bounded by the curve $(1 - x^2) y = (x - 2)(x - 3)$, the $x$-axis and the ordinates at $x = 2$ and $x = 3$.

12. Find the area enclosed by the curve $a(a - x) y = x^3$, the $x$-axis and the line $2x = a$.

13. Prove that the area bounded by the curve $y = \tanh x$ and the straight line $y = 1$ between $x = 0$ and $x = \infty$, is ln 2.

14. Prove that the curve defined by $x = \cos^3 t$, $y = 2 \sin^3 t$, encloses an area $\frac{3\pi}{4}$ units$^2$.

15. Find the mean value of $y = x e^{-x/a}$ between $x = 0$ and $x = a$.

16. A plane figure is bounded by the curves $2y = x^2$ and $x^3 y = 16$, the $x$-axis and the ordinate at $x = 4$. Calculate the area enclosed.

17. Find the area of the loop of the curve $y^2 = x^4 (4 + x)$.

18. If $i = I_1 \sin(\omega t + a) + I_2 \sin(2\omega t + \beta)$, where $I_1$, $I_2$, $\omega$, $a$, and $\beta$ are constants, find the r.m.s. value of $i$ over a period, i.e. from $t = 0$ to $t = \frac{2\pi}{\omega}$.

19. Show that the area enclosed by the curve $x = a(2t - \sin 2t)$, $y = 2a \sin^2 t$, and the $x$-axis between $t = 0$ and $t = \pi$ is $3\pi a^2$ units$^2$.

20. A plane figure is bounded by the curves $y = 1/x^2$, $y = e^{x/2} - 3$ and the lines $x = 1$ and $x = 2$. Determine the extent of the area of the figure.

# Programme 19

## INTEGRATION APPLICATIONS

### PART 2

**1** **Introduction**

In the previous programme, we saw how integration could be used

(a) to calculate areas under plane curves,
(b) to find mean values of functions,
(c) to find r.m.s. values of functions.

We are now going to deal with a few more applications of integration: with some of these you will already be familiar and the work will serve as revision; others may be new to you. Anyway, let us make a start, so move on to frame 2.

**2** **Volumes of solids of revolution**

If the plane figure bounded by the curve $y = f(x)$, the $x$-axis, and the ordinates at $x = a$ and $x = b$, rotates through a complete revolution about the $x$-axis, it will generate a solid symmetrical about OX.

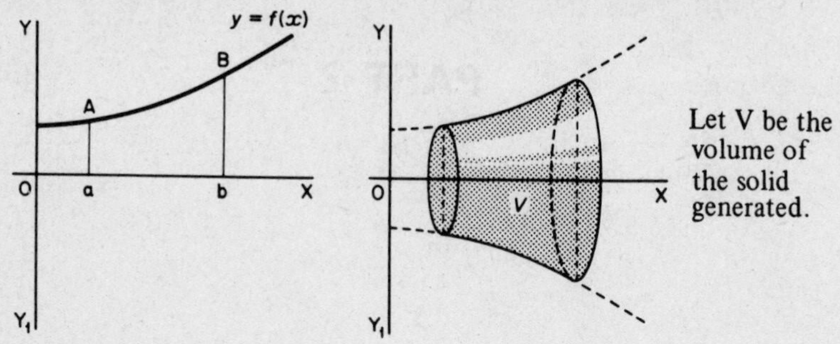


Let V be the volume of the solid generated.

To find V, let us first consider a thin strip of the original plane figure.

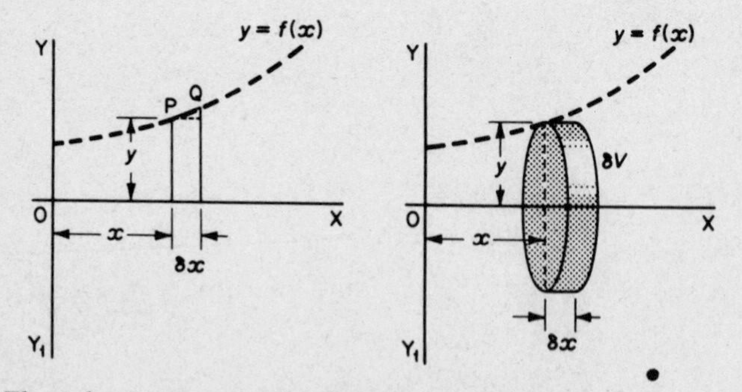


The volume generated by the strip ≏ the volume generated by the rectangle.

i.e. $\delta V \simeq$ ............

**3**

$$\delta V \simeq \pi y^2 . \delta x$$

Correct, since the solid generated is a flat cylinder.

If we divide the whole plane figure up into a number of such strips, each will contribute its own flat disc with volume $\pi y^2 . \delta x$.

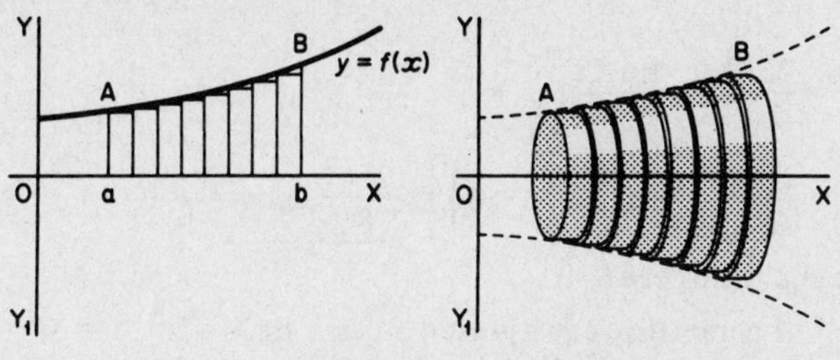


$$\therefore \text{ Total volume, } V \simeq \sum_{x=a}^{x=b} \pi y^2 . \delta x$$

The error in the approximation is due to the areas above the rectangles, which cause the step formation in the solid. However, if $\delta x \to 0$, the error disappears, so that finally V = ..........................

**4**

$$V = \int_a^b \pi y^2 . dx$$

This is a standard result, which you have doubtless seen many times before, so make a note of it in your record book and move on to frame 5.

**5**

Here is an example:

*Example.* Find the volume generated when the plane figure bounded by $y = 5 \cos 2x$, the $x$-axis, and ordinates at $x = 0$ and $x = \frac{\pi}{4}$, rotates about the $x$-axis through a complete revolution.

We have: $$V = \int_0^{\pi/4} \pi y^2 dx = 25\pi \int_0^{\pi/4} \cos^2 2x \, dx$$

Express this in terms of the double angle (i.e. $4x$) and finish it off.

*Then turn on to frame 6.*

**6**

$$V = \frac{25\pi^2}{8} \text{ units}^3$$

For: $V = \pi\int_0^{\pi/4} y^2\,dx = 25\pi\int_0^{\pi/4} \cos^2 2x\,dx$

$$= \frac{25\pi}{2}\int_0^{\pi/4} (1 + \cos 4x)\,dx$$

$\cos 2\theta = 2\cos^2\theta - 1$

$\cos^2\theta = \frac{1}{2}(1 + \cos 2\theta)$

$$= \frac{25\pi}{2}\left[x + \frac{\sin 4x}{4}\right]_0^{\pi/4}$$

$$= \frac{25\pi}{2}\left[\left\{\frac{\pi}{4} + 0\right\} - \left\{0 + 0\right\}\right] = \frac{25\pi^2}{8} \text{ units}^3$$

Now what about this one?

*Example.* The parametric equations of a curve are $x = 3t^2$, $y = 3t - t^2$. Find the volume generated when the plane figure bounded by the curve, the $x$-axis and the ordinates corresponding to $t = 0$ and $t = 2$, rotates about the $x$-axis. [Remember to change the variable of the integral!]

*Work it right through and then check with the next frame.*

**7**

$$V = 49{\cdot}62\pi = 156 \text{ units}^3$$

Here is the solution. Follow it through.

$$V = \int_a^b \pi y^2\,dx$$

$x = 3t^2$, $y = 3t - t^2$

$$V = \int_{t=0}^{t=2} \pi(3t - t^2)^2\,dx$$

$x = 3t^2$

$dx = 6t\,dt$

$$= \pi\int_0^2 (9t^2 - 6t^3 + t^4)\,6t\,dt$$

$$= 6\pi\int_0^2 (9t^3 - 6t^4 + t^5)\,dt$$

$$= 6\pi\left[\frac{9t^4}{4} - \frac{6t^5}{5} + \frac{t^6}{6}\right]_0^2$$

$$= 6\pi\left[36 - 38{\cdot}4 + 10{\cdot}67\right] = 6\pi\left[46{\cdot}67 - 38{\cdot}4\right]$$

$$= 6\pi(8{\cdot}27) = 49{\cdot}62\pi = 156 \text{ units}^3$$

So they are all done in very much the same way.

*Turn on now to frame 8.*

**8**

Here is a slightly different example.

*Example.* Find the volume generated when the plane figure bounded by the curve $y = x^2 + 5$, the $x$-axis, and the ordinates $x = 1$ and $x = 3$, rotates about the *y-axis* through a complete revolution.

Note that this time the figure rotates about the axis of $y$.

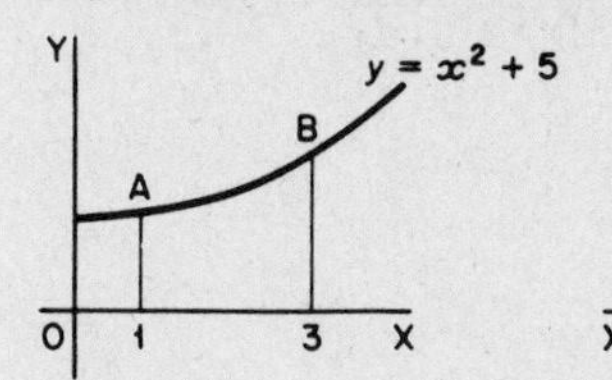


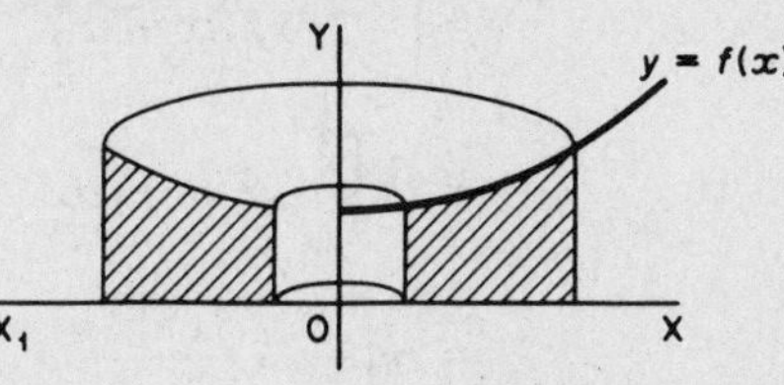


Half of the solid formed, is shown in the right-hand diagram. We have no standard formula for this case. $\left[ V = \int_a^b \pi y^2 \, dx \right.$ refers to rotation about the $x$-axis.$\left.\right]$ In all such cases, we build up the integral from first principles.

*To see how we go about this, move on to frame 9.*

**9**

Here it is: note the general method.

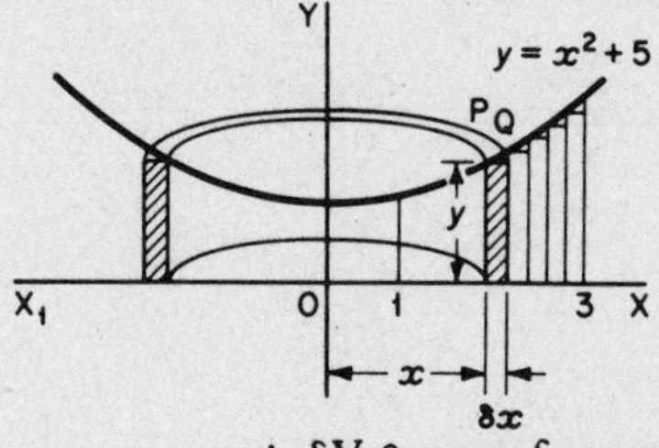


If we rotate an elementary strip PQ, we can say –

Vol. generated by the strip $\triangleq$ vol. generated by rectangle (i.e. hollow thin cylinder)

$$\therefore \; \delta V \triangleq \text{area of cross section} \times \text{circumference}$$

$$\delta V \triangleq y\delta x \,.\, 2\pi x \triangleq 2\pi xy \, \delta x$$

For all such strips between $x = 1$ and $x = 3$

$$V \triangleq \Sigma \delta V \triangleq \sum_{x=1}^{x=3} 2\pi xy . \delta x$$

As usual, if $\delta x \to 0$, the error disappears and we finally obtain

$$V = 2\int_1^3 \pi xy \, dx$$

Since $y = x^2 + 5$, we can now substitute for $y$ and finish the calculation.

*Do that, and then on to the next frame.*

**10**

$$V = 80\pi \text{ units}^3$$

Here is the working: check yours.

$$V = \int_1^3 2\pi xy\, dx = 2\pi \int_1^3 x(x^2 + 5)\, dx$$

$$= 2\pi \int_1^3 (x^3 + 5x)\, dx$$

$$= 2\pi \left[\frac{x^4}{4} + \frac{5x^2}{2}\right]_1^3$$

$$= 2\pi \left[\left\{\frac{81}{4} + \frac{45}{2}\right\} - \left\{\frac{1}{4} + \frac{5}{2}\right\}\right]$$

$$= 2\pi \left[\frac{80}{4} + \frac{40}{2}\right]$$

$$= 2\pi \left[20 + 20\right] = \underline{80\pi \text{ units}^3}$$

Whenever we have a problem not covered by our standard results, we build up the integral from first principles.

**11**

This last result is often required, so let us write it out again.

The volume generated when the plane figure bounded by the curve $y = f(x)$, the $x$-axis and the ordinates $x = a$ and $x = b$ rotates completely about the *y-axis* is given by:

$$V = 2\pi \int_a^b xy\, dx$$

Copy this into your record book for future reference.

*Then on to frame 12, where we will deal with another application of integration.*

## Centroid of a plane figure



The position of the centroid of a plane figure depends not only on the extent of the area but also on how the area is distributed. It is very much like the idea of the centre of gravity of a thin plate, but we cannot call it a centre of gravity, since a plane figure has no mass.

We can find its position, however, by taking an elementary strip and then taking moments (i) about OY to find $\bar{x}$, and (ii) about OX to find $\bar{y}$. No doubt, you remember the results. Here they are:

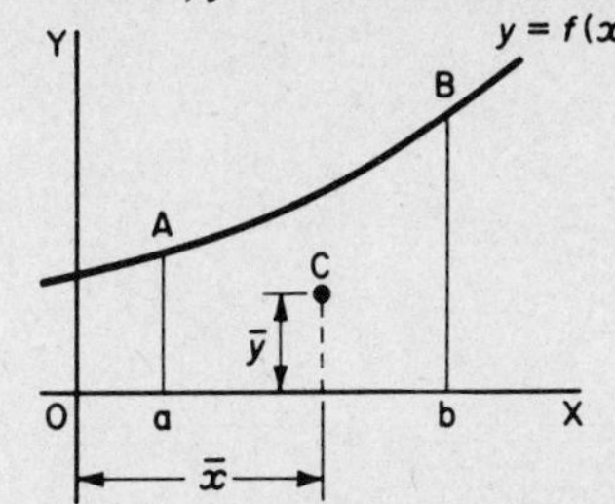


$$A\bar{x} \simeq \sum_{x=a}^{x=b} x.y\delta x$$

$$A\bar{y} \simeq \sum_{x=a}^{x=b} \frac{y}{2}.y\delta x.$$

Which give $\bar{x} = \dfrac{\int_a^b xy\,dx}{\int_a^b y\,dx}$, $\bar{y} = \dfrac{\frac{1}{2}\int_a^b y^2\,dx}{\int_a^b y\,dx}$

*Add these to your list of results.*

---



Now let us do one example. Here goes.

Find the position of the centroid of the figure bounded by $y = e^{2x}$, the $x$-axis, the $y$-axis, and the ordinate at $x = 2$.

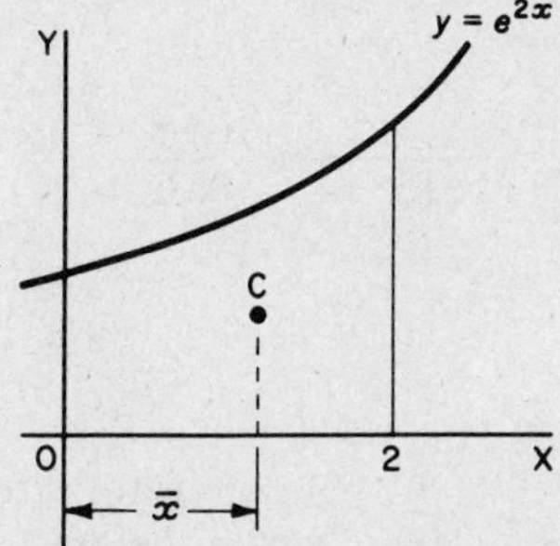


First, to find $\bar{x}$

$$\bar{x} = \frac{\int_0^2 xy\,dx}{\int_0^2 y\,dx}$$

We evaluate the two integrals quite separately, so let $\bar{x} = \dfrac{I_1}{I_2}$

Then $\qquad I_1 = \int_0^2 x\,e^{2x}\,dx = \ldots\ldots\ldots\ldots$

**14**

$$I_1 = \frac{3e^4 + 1}{4}$$

For: $I_1 = \int_0^2 x\, e^{2x}\, dx = \left[ x\left(\frac{e^{2x}}{2}\right) - \frac{1}{2}\int e^{2x}\, dx \right]_0^2$

$$= \left[ \frac{x\, e^{2x}}{2} - \frac{e^{2x}}{4} \right]_0^2$$

$$= \left(e^4 - \frac{e^4}{4}\right) - \left(-\frac{1}{4}\right)$$

$$= \frac{3\, e^4}{4} + \frac{1}{4} = \frac{3\, e^4 + 1}{4}$$

Similarly, $I_2 = \int_0^2 e^{2x}\, dx$ which gives $I_2 =$ ....................

**15**

$$I_2 = \frac{e^4 - 1}{2}$$

For: $I_2 = \int_0^2 e^{2x}\, dx = \left[\frac{e^{2x}}{2}\right]_0^2 = \frac{e^4}{2} - \frac{1}{2} = \frac{e^4 - 1}{2}$

So, therefore, $\bar{x} = \frac{I_1}{I_2} = \frac{3\, e^4 + 1}{4} \times \frac{2}{e^4 - 1}$

$=$ ................................

**16**

$$\bar{x} = 1{\cdot}523$$

$$\bar{x} = \frac{3e^4 + 1}{2(e^4 - 1)} = \frac{3(54{\cdot}60) + 1}{2(54{\cdot}60 - 1)} = \frac{163{\cdot}8 + 1}{109{\cdot}2 - 1} = \frac{164{\cdot}8}{108{\cdot}2}$$

$$\therefore \bar{x} = 1{\cdot}523$$

Now we have to find $\bar{y}$

$$\bar{y} = \frac{\int_0^2 \frac{1}{2} y^2\, dx}{\int_0^2 y\, dx} = \frac{I_3}{I_2}$$

Note that the denominator is the same as before.

$I_3 = \frac{1}{2}\int_0^2 y^2\, dx =$ ....................

**17**

$$\boxed{I_3 = \frac{1}{8}\left[e^8 - 1\right] \quad \therefore \bar{y} = \frac{1}{4}\left[e^4 + 1\right]}$$

$$I_3 = \frac{1}{2}\int_0^2 y^2\,dx = \frac{1}{2}\int_0^2 e^{4x}\,dx = \frac{1}{2}\left[\frac{e^{4x}}{4}\right]_0^2$$

$$= \frac{1}{8}\left[e^8 - 1\right]$$

$$\therefore \bar{y} = \frac{I_3}{I_2} = \frac{\frac{1}{8}(e^8 - 1)}{\frac{1}{2}(e^4 - 1)} = \frac{1}{4}(e^4 + 1) = \frac{1}{4}(54{\cdot}60 + 1)$$

$$= \frac{55{\cdot}60}{4} = 13{\cdot}9$$

So the results are:

$$\bar{x} = 1{\cdot}523; \ \bar{y} = 13{\cdot}9$$

Now do this one on your own in just the same way.

*Example.* Find the position of the centroid of the figure bounded by the curve $y = 5 \sin 2x$, the $x$-axis, and the ordinates at $x = 0$ and $x = \frac{\pi}{6}$

(First of all find $\bar{x}$ and check your result before going on to find $\bar{y}$)

**18**

$$\boxed{\bar{x} = 0{\cdot}3424}$$

$$I_1 = \int_0^{\pi/6} xy\,dx = 5\int_0^{\pi/6} x \sin 2x\,dx$$

$$= 5\left[x\frac{(-\cos 2x)}{2} + \frac{1}{2}\int_0^{\pi/6} \cos 2x\,dx\right]$$

$$= 5\left[-\frac{x \cos 2x}{2} + \frac{\sin 2x}{4}\right]_0^{\pi/6}$$

$$= 5\left[-\frac{\pi}{6}.\frac{1}{2}.\frac{1}{2} + \frac{\sqrt{3}}{8}\right]$$

$$= 5\left[\frac{\sqrt{3}}{8} - \frac{\pi}{24}\right] = \frac{5}{4}\left[\frac{\sqrt{3}}{2} - \frac{\pi}{6}\right]$$

Also' $I_2 = \int_0^{\pi/6} 5 \sin 2x\,dx = 5\left[-\frac{\cos 2x}{2}\right]_0^{\pi/6} = -\frac{5}{2}\left[\frac{1}{2} - 1\right] = \frac{5}{4}$

$$\therefore \bar{x} = \frac{5}{4}\left[\frac{\sqrt{3}}{2} - \frac{\pi}{6}\right].\frac{4}{5} = \left[\frac{\sqrt{3}}{2} - \frac{\pi}{6}\right]$$

$$= 0{\cdot}8660 - 0{\cdot}5236 \quad \therefore \bar{x} = 0{\cdot}3424$$

Do you agree with that? If so, push on and find $\bar{y}$.

*When you have finished, move on to frame 19.*

**19**

$$\bar{y} = 1{\cdot}542$$

Here is the working in detail.

$$I_3 = \frac{1}{2}\int_0^{\pi/6} 25 \sin^2 2x \, dx$$

$$= \frac{25}{2}\int_0^{\pi/6} \frac{1}{2}(1 - \cos 4x)\, dx$$

$$= \frac{25}{4}\left[x - \frac{\sin 4x}{4}\right]_0^{\pi/6}$$

$$= \frac{25}{4}\left[\frac{\pi}{6} - \frac{\sin(2\pi/3)}{4}\right] \qquad \sin\frac{2\pi}{3} = \sin\frac{\pi}{3} = \frac{\sqrt{3}}{2}$$

$$= \frac{25}{4}\left[\frac{\pi}{6} - \frac{\sqrt{3}}{8}\right]$$

$$= \frac{25}{4}\left[0{\cdot}5236 - 0{\cdot}2153\right]$$

$$= \frac{25}{4}\left[0{\cdot}3083\right] = 25(0{\cdot}07708) = \underline{1{\cdot}927}$$

Therefore

$$\bar{y} = \frac{I_3}{I_2} = \frac{1{\cdot}927}{5/4} = \frac{(1{\cdot}927)4}{5} = \underline{1{\cdot}542}$$

So the final results are

$$\underline{\bar{x} = 0{\cdot}342, \ \bar{y} = 1{\cdot}542}$$

*Now to frame 20.*

**20**

Here is another application of integration not very different from the last.

**Centre of gravity of a solid of revolution**

To find the position of the centre of gravity of the solid formed when the plane figure bounded by the curve $y = f(x)$, the $x$-axis, and the ordinates at $x = a$ and $x = b$ rotates about the $x$-axis.

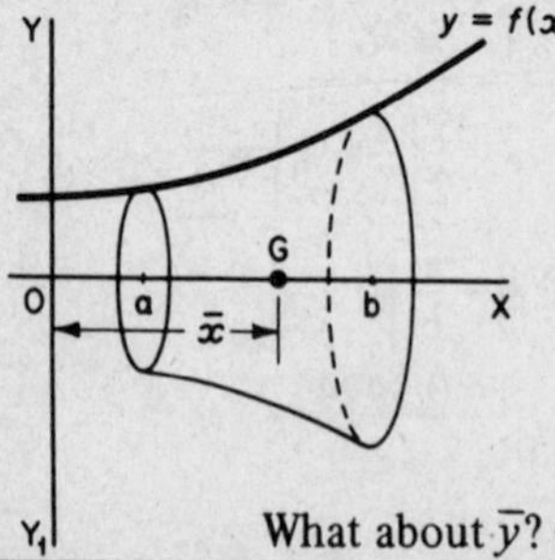


If we take elementary discs and sum the moments of volume (or mass) about OY, we can calculate $\bar{x}$.

This gives $\bar{x} = \dfrac{\displaystyle\int_a^b xy^2\, dx}{\displaystyle\int_a^b y^2\, dx}$

What about $\bar{y}$? Clearly, $\bar{y}$ = ..............................

**21**

$\bar{y} = 0$

Correct, since the solid generated is symmetrical about OX and therefore the centre of gravity lies on this axis, i.e. $\bar{y} = 0$.

So we have to find only $\bar{x}$, using

$$\bar{x} = \frac{\int_a^b xy^2\,dx}{\int_a^b y^2\,dx} = \frac{I_1}{I_2}$$

and we proceed in much the same way as we did for centroids.

Do this example, all on your own:

*Example.* Find the position of the centre of gravity of the solid formed when the plane figure bounded by the curve $x^2 + y^2 = 16$, the $x$-axis, and the ordinates $x = 1$ and $x = 3$ rotates about the $x$-axis.

*When you have finished, move to frame 22.*

**22**

$\bar{x} = 1{\cdot}89,\ \bar{y} = 0$

Check your working.

$$I_1 = \int_1^3 x(16 - x^2)\,dx = \int_1^3 (16x - x^3)\,dx = \left[8x^2 - \frac{x^4}{4}\right]_1^3$$

$$= \left(72 - \frac{81}{4}\right) - \left(8 - \frac{1}{4}\right)$$

$$= 64 - 20 = 44 \quad \therefore\ \underline{I_1 = 44}$$

$$I_2 = \int_1^3 (16 - x^2)\,dx = \left[16x - \frac{x^3}{3}\right]_1^3$$

$$= (48 - 9) - \left(16 - \frac{1}{3}\right)$$

$$= 23\tfrac{1}{3} \quad \therefore\ \underline{I_2 = 23\tfrac{1}{3}}$$

$$\therefore\ \bar{x} = \frac{I_1}{I_2} = \frac{44}{1} \cdot \frac{3}{70} = \frac{132}{70} = \underline{1{\cdot}89}$$

So $\underline{\bar{x} = 1{\cdot}89,\ \bar{y} = 0}$

They are all done in the same manner.

Now for something that may be new to you.

*Turn on to frame 23.*

## 23

**Lengths of curves**

To find the length of the arc of the curve $y = f(x)$ between $x = a$ and $x = b$.

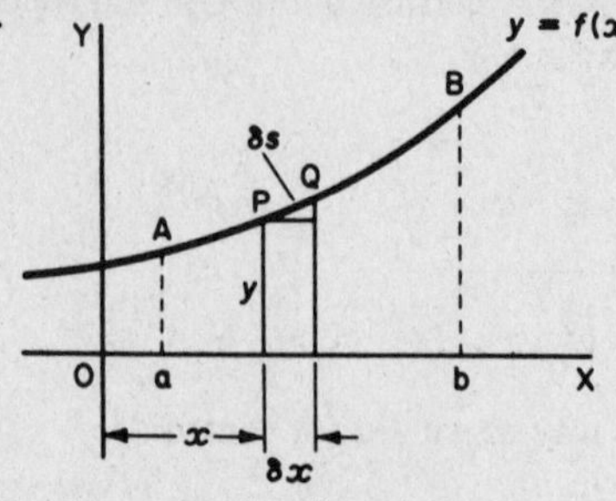


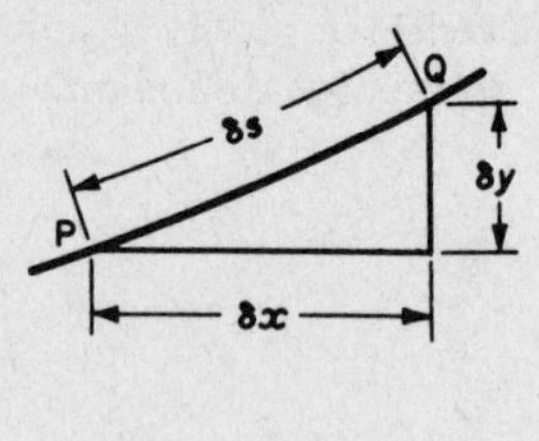


Let P be the point $(x, y)$ and Q a point on the curve near to P.
Let $\delta s$ = length of the small arc PQ.

Then $(\delta s)^2 \simeq (\delta x)^2 + (\delta y)^2 \qquad \therefore \dfrac{(\delta s)^2}{(\delta x)^2} \simeq 1 + \dfrac{(\delta y)^2}{(\delta x)^2}$

$$\left(\frac{\delta s}{\delta x}\right)^2 \simeq 1 + \left(\frac{\delta y}{\delta x}\right)^2 \qquad \therefore \frac{\delta s}{\delta x} \simeq \sqrt{\left\{1 + \left(\frac{\delta y}{\delta x}\right)^2\right\}}$$

If $\delta x \to 0 \quad \dfrac{ds}{dx} = \sqrt{\left\{1 + \left(\dfrac{dy}{dx}\right)^2\right\}} \qquad \therefore s = \displaystyle\int_a^b \sqrt{\left\{1 + \left(\frac{dy}{dx}\right)^2\right\}}\,dx$

Make a note of this result.
*Then on to the next frame.*

## 24

*Example.* Find the length of the curve $y^2 = x^3$ between $x = 0$ and $x = 4$.

$$y^2 = x^3 \quad \therefore y = x^{3/2} \quad \therefore \frac{dy}{dx} = \frac{3}{2}x^{\frac{1}{2}} \quad \therefore 1 + \left(\frac{dy}{dx}\right)^2 = 1 + \frac{9x}{4}$$

$$\therefore s = \int_0^4 \left(1 + \frac{9x}{4}\right)^{\frac{1}{2}} dx = \left[\frac{2}{3} \cdot \frac{4}{9}\left(1 + \frac{9x}{4}\right)^{3/2}\right]_0^4$$

$$= \frac{8}{27}\left[10\sqrt{10} - 1\right] = \frac{8}{27}\left[31{\cdot}62 - 1\right]$$

$$= \frac{8}{27}(30{\cdot}62) = \underline{9{\cdot}07 \text{ units}}$$

That is all there is to it. Now here is one for you:

*Example.* Find the length of the curve $y = 10\cosh\dfrac{x}{10}$ between $x = -1$ and $x = 2$.

*Finish it, then turn to frame 25.*

**25**

$$\boxed{s = 3{\cdot}015 \text{ units}}$$

Here is the working set out.

$$y = 10\cosh\frac{x}{10} \qquad s = \int_{-1}^{2}\sqrt{\left\{1+\left(\frac{dy}{dx}\right)^2\right\}}dx$$

$$\frac{dy}{dx} = \sinh\frac{x}{10} \quad \therefore\ 1+\left(\frac{dy}{dx}\right)^2 = 1+\sinh^2\frac{x}{10} = \cosh^2\frac{x}{10}$$

$$\therefore\ s = \int_{-1}^{2}\sqrt{\left\{\cosh^2\frac{x}{10}\right\}}.\,dx = \int_{-1}^{2}\cosh\frac{x}{10}\,dx = \left[10\sinh\frac{x}{10}\right]_{-1}^{2}$$

$$= 10\left[\sinh 0{\cdot}2 - \sinh(-0{\cdot}1)\right] \qquad \sinh(-x) = -\sinh x$$

$$= 10\left[\sinh 0{\cdot}2 + \sinh 0{\cdot}1\right]$$

$$= 10\left[0{\cdot}2013 + 0{\cdot}1002\right]$$

$$= 10\left[0{\cdot}3015\right] = \underline{3{\cdot}015 \text{ units}}$$

*Now to frame 26.*

**26**

**Lengths of curves – parametric equations**

Instead of changing the variable of the integral as we have done before when the curve is defined in terms of parametric equations, we establish a special form of the result which saves a deal of working when we use it. Here it is.

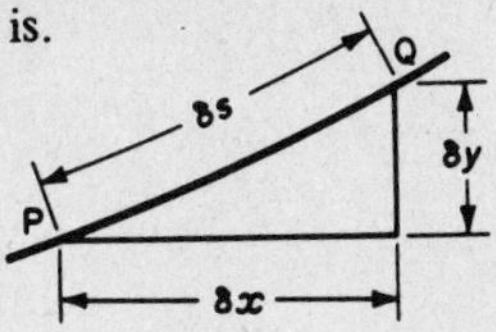

Let $y = f(t),\ x = F(t)$

As before

$$(\delta s)^2 \triangleq (\delta x)^2 + (\delta y)^2$$

Divide by $(\delta t)^2$

$$\therefore\ \left(\frac{\delta s}{\delta t}\right)^2 \triangleq \left(\frac{\delta x}{\delta t}\right)^2 + \left(\frac{\delta y}{\delta t}\right)^2$$

If $\delta t \to 0$, this becomes

$$\left(\frac{ds}{dt}\right)^2 = \left(\frac{dx}{dt}\right)^2 + \left(\frac{dy}{dt}\right)^2$$

$$\therefore\ \frac{ds}{dt} = \sqrt{\left\{\left(\frac{dx}{dt}\right)^2 + \left(\frac{dy}{dt}\right)^2\right\}}$$

$$\therefore\ s = \int_{t=t_1}^{t=t_2}\sqrt{\left\{\left(\frac{dx}{dt}\right)^2 + \left(\frac{dy}{dt}\right)^2\right\}}dt$$

*This is a very useful result. Make a note of it in your book and then turn on to the next frame.*

**27**

*Example.* Find the length of the curve $x = 2\cos^3\theta, y = 2\sin^3\theta$ between the points corresponding to $\theta = 0$ and $\theta = \pi/2$.

Remember $$s = \int_0^{\pi/2} \sqrt{\left\{\left(\frac{dx}{d\theta}\right)^2 + \left(\frac{dy}{d\theta}\right)^2\right\}}.\,d\theta$$

We have $$\frac{dx}{d\theta} = 6\cos^2\theta\,(-\sin\theta) = -6\cos^2\theta\sin\theta$$

$$\frac{dy}{d\theta} = 6\sin^2\theta\cos\theta$$

$$\therefore \left(\frac{dx}{d\theta}\right)^2 + \left(\frac{dy}{d\theta}\right)^2 = 36\cos^4\theta\sin^2\theta + 36\sin^4\theta\cos^2\theta$$

$$= 36\sin^2\theta\cos^2\theta\,(\cos^2\theta + \sin^2\theta)$$

$$= 36\sin^2\theta\cos^2\theta$$

$$\therefore \sqrt{\left\{\left(\frac{dx}{d\theta}\right)^2 + \left(\frac{dy}{d\theta}\right)^2\right\}} = 6\sin\theta\cos\theta = 3\sin 2\theta$$

$$\therefore s = \int_0^{\pi/2} 3\sin 2\theta\,d\theta$$

$= \ldots\ldots\ldots\ldots\ldots\ldots$ Finish it off.

**28**

$$\boxed{s = 3 \text{ units}}$$

For we had $$s = \int_0^{\pi/2} 3\sin 2\theta\,d\theta$$

$$= 3\left[\frac{-\cos 2\theta}{2}\right]_0^{\pi/2}$$

$$= 3\left[\left(\frac{1}{2}\right) - \left(-\frac{1}{2}\right)\right] = \underline{3 \text{ units}}$$

It is all very straightforward and not at all difficult. Just take care not to make any silly slips that would wreck the results.

Here is one for you to do in much the same way.

*Example.* Find the length of the curve $x = 5(2t - \sin 2t), y = 10\sin^2 t$ between $t = 0$ and $t = \pi$.

*When you have completed it, turn on to frame 29.*

**29**

$$\boxed{s = 40 \text{ units}}$$

For: $x = 5(2t - \sin 2t), \ y = 10 \sin^2 t$

$$\therefore \frac{dx}{dt} = 5(2 - 2\cos 2t) = 10(1 - \cos 2t)$$

$$\frac{dy}{dt} = 20 \sin t \cos t = 10 \sin 2t.$$

$$\left(\frac{dx}{dt}\right)^2 + \left(\frac{dy}{dt}\right)^2 = 100(1 - 2\cos 2t + \cos^2 2t) + 100 \sin^2 2t$$

$$= 100(1 - 2\cos 2t + \cos^2 2t + \sin^2 2t)$$

$$= 200(1 - \cos 2t) \qquad \text{But } \cos 2t = 1 - 2\sin^2 t$$

$$= 400 \sin^2 t$$

$$\therefore \sqrt{\left\{\left(\frac{dx}{dt}\right)^2 + \left(\frac{dy}{dt}\right)^2\right\}} = 20 \sin t$$

$$\therefore s = \int_0^\pi 20 \sin t \, dt = 20\left[-\cos t\right]_0^\pi$$

$$= 20\left[(1) - (-1)\right] = \underline{40 \text{ units}}$$

*Next frame.*

**30**

So, for the lengths of curves, there are two forms:

(i) $s = \int_{x_1}^{x_2} \sqrt{\left\{1 + \left(\frac{dy}{dx}\right)^2\right\}} dx$ when $y = F(x)$

(ii) $s = \int_{\theta_1}^{\theta_2} \sqrt{\left\{\left(\frac{dx}{d\theta}\right)^2 + \left(\frac{dy}{d\theta}\right)^2\right\}} d\theta$ for parametric equations.

Just check that you have made a note of these in your record book.

*Now turn on to frame 31 and we will consider a further application of integration. This will be the last for this programme.*

# 31 Surfaces of revolution

If an arc of a curve rotates about an axis, it will generate a surface. Let us take the general case.

Find the area of the surface generated when the arc of the curve $y = f(x)$ between $x = x_1$ and $x = x_2$ rotates about the $x$-axis through a complete revolution.

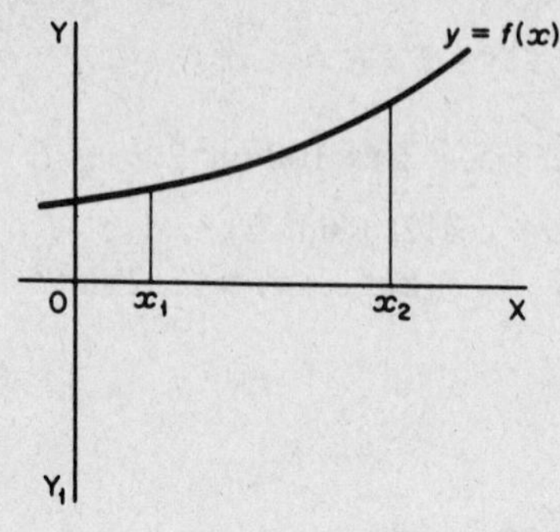


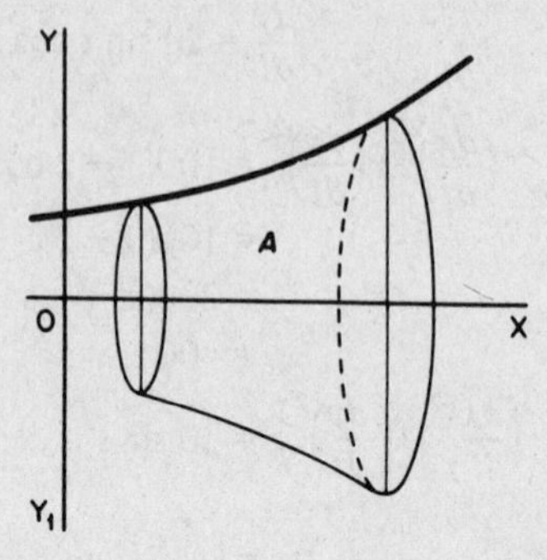


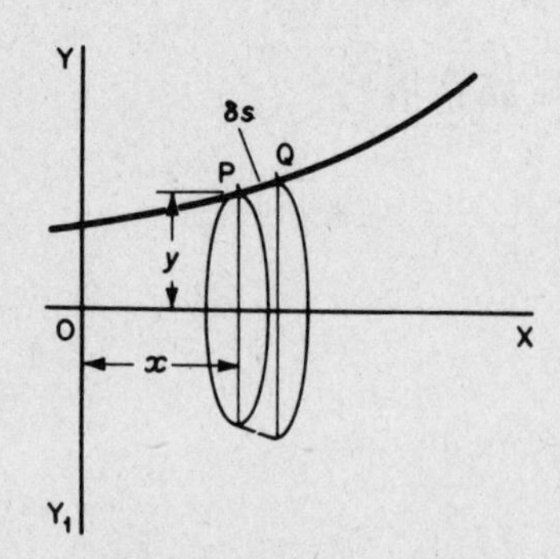


If we rotate a small element of arc $\delta s$ units long, it will generate a thin band of area $\delta A$.

Then $\delta A \simeq 2\pi y . \delta s$

Dividing by $\delta x$, gives
$$\frac{\delta A}{\delta x} \simeq 2\pi y \frac{\delta s}{\delta x}$$

and if $\delta x \to 0$,
$$\frac{dA}{dx} = 2\pi y \frac{ds}{dx}$$

Now we have previously seen that $\dfrac{ds}{dx} = \sqrt{\left\{1 + \left(\dfrac{dy}{dx}\right)^2\right\}}$

$$\therefore \frac{dA}{dx} = 2\pi y \sqrt{\left\{1 + \left(\frac{dy}{dx}\right)^2\right\}}$$

So that A = ......................

**32**

$$A = \int_{x_1}^{x_2} 2\pi y \sqrt{\left\{1 + \left(\frac{dy}{dx}\right)^2\right\}} dx$$

This is another standard result, so copy it down into your record book.

*Then on to the next frame.*

**33**

Here is an example requiring the last result.

*Example.* Find the area generated when the arc of the parabola $y^2 = 8x$ between $x = 0$ and $x = 2$ rotates about the $x$-axis.

We have $$A = \int_0^2 2\pi y \sqrt{\left\{1 + \left(\frac{dy}{dx}\right)^2\right\}} dx$$

$$y^2 = 8x \quad \therefore\ y = 2\sqrt{2}\,x^{\frac{1}{2}} \quad \therefore\ \frac{dy}{dx} = \sqrt{2}\,x^{-\frac{1}{2}} \quad \therefore\ \left(\frac{dy}{dx}\right)^2 = \frac{2}{x}$$

$$\therefore\ 1 + \left(\frac{dy}{dx}\right)^2 = 1 + \frac{2}{x} = \frac{x+2}{x}$$

$$\therefore\ A = \int_0^2 2\pi\, 2\sqrt{2}\, x^{\frac{1}{2}} \sqrt{\left\{\frac{x+2}{x}\right\}} dx$$

$$= \int_0^2 4\sqrt{2}.\pi.x^{\frac{1}{2}} \frac{(x+2)^{\frac{1}{2}}}{x^{\frac{1}{2}}} dx$$

$$= 4\sqrt{2}.\pi \int_0^2 (x+2)^{\frac{1}{2}} dx$$

= .............................. Finish it off: then move on.

**34**

$$A = 19{\cdot}5\pi = 61{\cdot}3 \text{ units}^2$$

For we had $$A = 4\sqrt{2}.\pi \int_0^2 (x+2)^{\frac{1}{2}} dx$$

$$= 4\sqrt{2}.\pi \left[\frac{(x+2)^{3/2}}{3/2}\right]_0^2$$

$$= \frac{8\sqrt{2}\,\pi}{3}\left[(8) - (2\sqrt{2})\right]$$

$$= \frac{8\pi}{3}\left[8\sqrt{2} - 4\right] = \frac{8\pi}{3}\left[7{\cdot}312\right]$$

$$= 19{\cdot}5\pi = \underline{61{\cdot}3 \text{ units}^2}$$

*Now continue the good work by moving on to frame 35.*

## 35 Surfaces of revolution – parametric equations

We have already seen that if we rotate a small arc $\delta s$, the area $\delta A$ of the thin band generated is given by

$$\delta A \simeq 2\pi y\,.\,\delta s$$

If we divide by $\delta\theta$, we get

$$\frac{\delta A}{\delta\theta} \simeq 2\pi y.\frac{\delta s}{\delta\theta}$$

and if $\delta\theta \to 0$, this becomes

$$\frac{dA}{d\theta} = 2\pi y.\frac{ds}{d\theta}$$

We already have established in our work on lengths of curves that

$$\frac{ds}{d\theta} = \sqrt{\left\{\left(\frac{dx}{d\theta}\right)^2 + \left(\frac{dy}{d\theta}\right)^2\right\}}$$

$$\therefore \frac{dA}{d\theta} = 2\pi y\sqrt{\left\{\left(\frac{dx}{d\theta}\right)^2 + \left(\frac{dy}{d\theta}\right)^2\right\}}$$

$$\therefore A = \int_{\theta_1}^{\theta_2} 2\pi y\sqrt{\left\{\left(\frac{dx}{d\theta}\right)^2 + \left(\frac{dy}{d\theta}\right)^2\right\}}\,d\theta$$

This is a special form of the result for use when the curve is defined as a pair of parametric equations.

*On to frame 36.*

## 36

*Example.* Find the area generated when the curve $x = a(\theta - \sin\theta)$, $y = a(1 - \cos\theta)$ between $\theta = 0$ and $\theta = \pi$, rotates about the $x$-axis through a complete revolution.

Here $\quad \dfrac{dx}{d\theta} = a(1 - \cos\theta) \quad \therefore \left(\dfrac{dx}{d\theta}\right)^2 = a^2(1 - 2\cos\theta + \cos^2\theta)$

$\quad\quad\quad \dfrac{dy}{d\theta} = a\sin\theta \quad\quad \therefore \left(\dfrac{dy}{d\theta}\right)^2 = a^2\sin^2\theta$

$$\therefore \left(\frac{dx}{d\theta}\right)^2 + \left(\frac{dy}{d\theta}\right)^2 = a^2(1 - 2\cos\theta + \cos^2\theta + \sin^2\theta)$$

$$= 2a^2(1 - \cos\theta) \quad \text{But } \cos\theta = 1 - 2\sin^2\frac{\theta}{2}$$

$$= 4a^2\sin^2\frac{\theta}{2}$$

$$\therefore \sqrt{\left\{\left(\frac{dx}{d\theta}\right)^2 + \left(\frac{dy}{d\theta}\right)^2\right\}} = \ldots\ldots\ldots\ldots$$

Finish the integral and so find the area of the surface generated.

**37**

$$\sqrt{\left\{\left(\frac{dx}{d\theta}\right)^2 + \left(\frac{dy}{d\theta}\right)^2\right\}} = 2a\sin\frac{\theta}{2}$$

$$A = \int_0^\pi 2\pi y \sqrt{\left\{\left(\frac{dx}{d\theta}\right)^2 + \left(\frac{dy}{d\theta}\right)^2\right\}}.\,d\theta$$

$$= 2\pi \int_0^\pi a(1 - \cos\theta).\,2a\sin\frac{\theta}{2}.\,d\theta$$

$$= 2\pi \int_0^\pi a\left(2\sin^2\frac{\theta}{2}\right).\,2a\sin\frac{\theta}{2}\,d\theta$$

$$= 8\pi a^2 \int_0^\pi \left(1 - \cos^2\frac{\theta}{2}\right)\sin\frac{\theta}{2}.\,d\theta$$

$$= 8\pi a^2 \int_0^\pi \left(\sin\frac{\theta}{2} - \cos^2\frac{\theta}{2}\sin\frac{\theta}{2}\right)d\theta$$

$$= 8\pi a^2 \left[-2\cos\frac{\theta}{2} + \frac{2\cos^3\theta/2}{3}\right]_0^\pi$$

$$= 8\pi a^2 \left[(0) - (-2 + 2/3)\right]$$

$$= 8\pi a^2 \left[4/3\right] = \underline{\frac{32\pi a^2}{3}\ \text{units}^2}$$

Here is one final one for you to do.

*Example.* Find the surface area generated when the arc of the curve $y = 3t^2$, $x = 3t - t^3$ between $t = 0$ and $t = 1$ rotates about OX through $2\pi$ radians.

*When you have finished – next frame.*

---

**38**

Here it is in full. $y = 3t^2 \quad \therefore \frac{dy}{dt} = 6t \quad \therefore \left(\frac{dy}{dt}\right)^2 = 36t^2$

$$x = 3t - t^3 \quad \therefore \frac{dx}{dt} = 3 - 3t^2 = 3(1 - t^2) \quad \therefore \left(\frac{dx}{dt}\right)^2 = 9(1 - 2t^2 + t^4)$$

$$\left(\frac{dx}{dt}\right)^2 + \left(\frac{dy}{dt}\right)^2 = 9 - 18t^2 + 9t^4 + 36t^2$$

$$= 9 + 18t^2 + 9t^4 = 9(1 + t^2)^2$$

$$\therefore A = \int_0^1 2\pi\, 3t^2 \sqrt{9(1 + t^2)^2}.\,dt$$

$$= 18\pi \int_0^1 t^2(1 + t^2)\,dt = 18\pi \int_0^1 (t^2 + t^4)\,dt$$

$$= 18\pi \left[\frac{t^3}{3} + \frac{t^5}{5}\right]_0^1 = 18\pi\left[\frac{1}{3} + \frac{1}{5}\right] = 18\pi\,\frac{8}{15} = \underline{\frac{48\pi}{5}\ \text{units}^2}$$

---

**39** **Rules of Pappus**

There are two useful rules worth knowing which can well be included with this stage of the work. In fact we have used them already in our work just by common sense. Here they are:

1. If an arc of a plane curve rotates about an axis in its plane, the area of the surface generated is equal to the length of the line multiplied by the distance travelled by its centroid.
2. If a plane figure rotates about an axis in its plane, the volume generated is equal to the area of the figure multiplied by the distance travelled by its centroid.

You can see how much alike they are.

By the way, there is just one proviso in using the rules of Pappus: the axis of rotation must not cut the rotating arc or plane figure.

So copy the rules down into your record book. You may need to refer to them at some future time.

*Now on to frame 40.*

**40** **Revision Summary**

1. *Volumes of solids of revolution*

(a) *about x-axis*

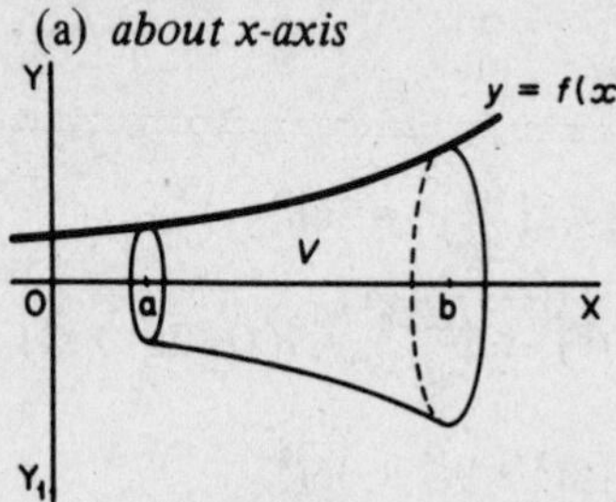


$$V = \int_a^b \pi y^2 \, dx \qquad \text{............ (i)}$$

Parametric equations $$V = \int_{\theta_1}^{\theta_2} \pi y^2 . \frac{dx}{d\theta} . \, d\theta \qquad \text{............ (ii)}$$

(b) *about y-axis*

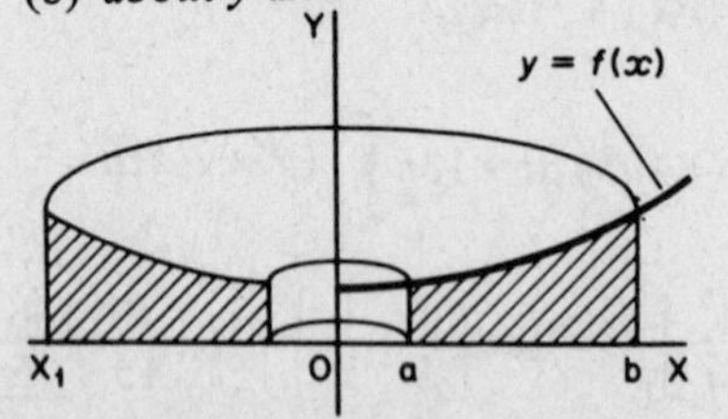


$$V = \int_a^b 2\pi xy \, dx \qquad \text{............ (iii)}$$

2. *Centroids of plane figures*

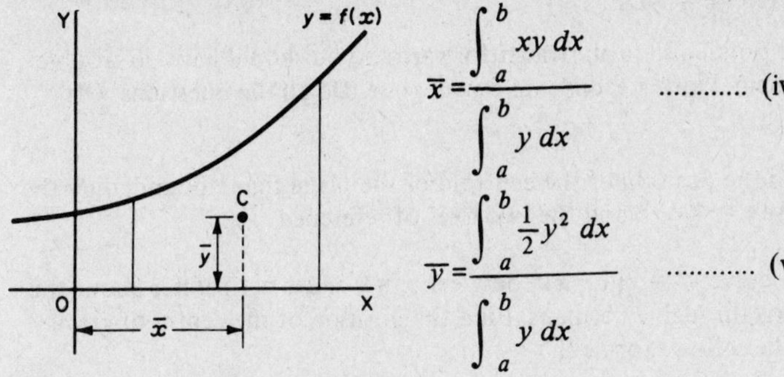


$$\bar{x} = \frac{\int_a^b xy\,dx}{\int_a^b y\,dx} \quad \text{............ (iv)}$$

$$\bar{y} = \frac{\int_a^b \frac{1}{2}y^2\,dx}{\int_a^b y\,dx} \quad \text{............ (v)}$$

3. *Centres of gravity of solids of revolution*

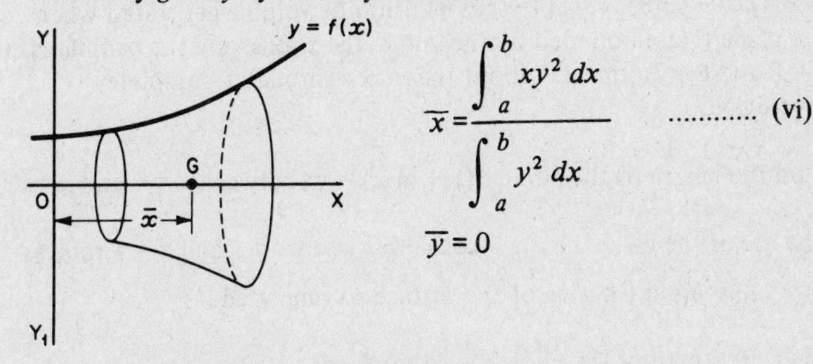


$$\bar{x} = \frac{\int_a^b xy^2\,dx}{\int_a^b y^2\,dx} \quad \text{............ (vi)}$$

$$\bar{y} = 0$$

4. *Lengths of curves*

$y = f(x)$
$$s = \int_{x_1}^{x_2} \sqrt{\left\{1 + \left(\frac{dy}{dx}\right)^2\right\}}.\,dx \quad \text{............ (vii)}$$

Parametric equations
$$s = \int_{\theta_1}^{\theta_2} \sqrt{\left\{\left(\frac{dx}{d\theta}\right)^2 + \left(\frac{dy}{d\theta}\right)^2\right\}}.\,d\theta \quad \text{............ (viii)}$$

5. *Surfaces of revolution*

$y = f(x)$
$$A = \int_{x_1}^{x_2} 2\pi y \sqrt{\left\{1 + \left(\frac{dy}{dx}\right)^2\right\}}.\,dx \quad \text{............ (ix)}$$

Parametric equations
$$A = \int_{\theta_1}^{\theta_2} 2\pi y \sqrt{\left\{\left(\frac{dx}{d\theta}\right)^2 + \left(\frac{dy}{d\theta}\right)^2\right\}}.\,d\theta \quad \text{............ (x)}$$

10.

*All that now remains is the Test Exercise in frame 41, so when you are ready, turn on and work through it.*

**41** **Test Exercise – XIX**

The problems are all straightforward so you should have no trouble with them. Work steadily: take your time. Do all the questions. Off you go.

1. Find the position of the centroid of the plane figure bounded by the curve $y = 4 - x^2$ and the two axes of reference.

2. The curve $y^2 = x(1 - x)^2$ between $x = 0$ and $x = 1$ rotates about the $x$-axis through $2\pi$ radians. Find the position of the centre of gravity of the solid so formed.

3. If $x = a(\theta - \sin\theta), y = a(1 - \cos\theta)$, find the volume generated when the plane figure bounded by the curve, the $x$-axis, and the ordinates at $\theta = 0$ and $\theta = 2\pi$, rotates about the $x$-axis through a complete revolution.

4. Find the length of the curve $8(y + \ln x) = x^2$ between $x = 1$ and $x = e$.

5. The arc of the catenary $y = 5\cosh\dfrac{x}{5}$ between $x = 0$ and $x = 5$ rotates about OX. Find the area of the surface so generated.

6. Find the length of the curve $x = 5(\cos\theta + \theta\sin\theta)$, $y = 5(\sin\theta - \theta\cos\theta)$ between $\theta = 0$ and $\theta = \pi/2$.

7. The parametric equations of a curve are $x = e^t \sin t, y = e^t \cos t$. If the arc of this curve between $t = 0$ and $t = \pi/2$ rotates through a complete revolution about the $x$-axis, calculate the area of the surface generated.

Now you are all ready for the next programme. Well done, keep it up!

**Further Problems – XIX**

1. Find the length of the curve $y = \frac{x}{2} - \frac{x^2}{4} + \frac{1}{2}\ln(1 - x)$ between $x = 0$ and $x = \frac{1}{2}$.

2 For the catenary $y = 5\cosh\frac{x}{5}$, calculate
   (i) the length of arc of the curve between $x = 0$ and $x = 2$.
   (ii) the surface area generated when this arc rotates about the $x$-axis through a complete revolution.

3. The plane figure bounded by the parabola $y^2 = 4ax$, the $x$-axis and the ordinate at $x = a$, is rotated through a complete revolution about the line $x = -a$. Find the volume of the solid generated.

4. A plane figure is enclosed by the parabola $y^2 = 4x$ and the line $y = 2x$. Determine (i) the position of the centroid of the figure, and (ii) the centre of gravity of the solid formed when the plane figure rotates completely about the $x$-axis.

5. The area bounded by $y^2x = 4a^2(2a - x)$, the $x$-axis and the ordinates $x = a$, $x = 2a$, is rotated through a complete revolution about the $x$-axis. Show that the volume generated is $4\pi a^3(2\ln 2 - 1)$.

6. Find the length of the curve $x^{2/3} + y^{2/3} = 4$ between $x = 0$ and $x = 8$.

7. Find the length of the arc of the curve $6xy = x^4 + 3$, between $x = 1$ and $x = 2$.

8. A solid is formed by the rotation about the $y$-axis of the area bounded by the $y$-axis, the lines $y = -5$ and $y = 4$, and an arc of the curve $2x^2 - y^2 = 8$. Given that the volume of the solid is $\frac{135\pi}{2}$, find the distance of the centre of gravity from the $x$-axis.

9. The line $y = x - 1$ is a tangent to the curve $y = x^3 - 5x^2 + 8x - 4$ at $x = 1$ and cuts the curve again at $x = 3$. Find the $x$ coordinate of the centroid of the plane figure so formed.

10. Find by integration, the area of the minor segment of the circle $x^2 + y^2 = 4$ cut off by the line $y = 1$. If this plane figure rotates about the $x$-axis through $2\pi$ radians, calculate the volume of the solid generated and hence obtain the distance of the centroid of the minor segment from the $x$-axis.

11. If the parametric equations of a curve are $x = 3a \cos\theta - a\cos 3\theta$, $y = 3a \sin\theta - a \sin 3\theta$, show that the length of arc between points corresponding to $\theta = 0$ and $\theta = \phi$ is $6a(1 - \cos\phi)$.

12. A curve is defined by the parametric equations

$$x = \theta - \sin\theta,\ \ y = 1 - \cos\theta$$

(i) Determine the length of the curve between $\theta = 0$ and $\theta = 2\pi$.
(ii) If the arc in (i) rotates through a complete revolution about the $x$-axis, determine the area of the surface generated.
(iii) Deduce the distance of the centroid of the arc from the $x$-axis.

13. Find the length of the curve $y = \cosh x$ between $x = 0$ and $x = 1$. Show that the area of the surface of revolution obtained by rotating the arc through four right-angles about the $y$-axis is $\dfrac{2\pi(e-1)}{e}$ units.

14. A parabolic reflector is formed by revolving the arc of the parabola $y^2 = 4ax$ from $x = 0$ to $x = h$ about the $x$-axis. If the diameter of the reflector is $2l$, show that the area of the reflecting surface is

$$\frac{\pi l}{6h^2}\left\{(l^2 + 4h^2)^{3/2} - l^3\right\}$$

15. A segment of a sphere has a base radius $r$ and maximum height $h$. Prove that its volume is $\dfrac{\pi h}{6}\left\{h^2 + 3r^2\right\}$

16. A groove, semi-circular in section and 1 cm deep, is turned in a solid cylindrical shaft of diameter 6 cm. Find the volume of material removed and the surface area of the groove.

17. Prove that the length of arc of the parabola $y^2 = 4ax$, between the points where $y = 0$ and $y = 2a$, is $a\left\{\sqrt{2} + \ln(1 + \sqrt{2})\right\}$ This arc is rotated about the $x$-axis through $2\pi$ radians. Find the area of the surface generated. Hence find the distance of the centroid of the arc from the line $y = 0$.

18. A cylindrical hole of length $2a$ is bored centrally through a sphere. Prove that the volume of material remaining is $\frac{4\pi a^3}{3}$.

19. Prove that the centre of gravity of the zone of a thin uniform spherical shell, cut off by two parallel planes is halfway between the centres of the two circular end sections.

20. Sketch the curve $3ay^2 = x(x-a)^2$, when $a > 0$. Show that $\frac{dy}{dx} = \pm\frac{3x-a}{2\sqrt{(3ax)}}$ and hence prove that the perimeter of the loop is $4a/\sqrt{3}$ units.

# Programme 20

## INTEGRATION APPLICATIONS

### PART 3

**1**

1. **Moments of inertia**

The amount of work that an object of mass $m$, moving with velocity $v$, will do against a resistance before coming to rest, depends on the values of these two quantities: its mass and its velocity.

The store of energy possessed by the object, due to its movement, is called its *kinetic energy*, and it can be shown experimentally that the kinetic energy of a moving object is proportional

(i) to its mass,

and (ii) to the square of its ................................

**2**

velocity

That is,

$$\text{K.E.} \propto m v^2 \quad \therefore \text{K.E.} = k m v^2$$

and if standard units of mass and velocity are used, the value of the constant $k$ is $\frac{1}{2}$.

$$\therefore \text{K.E.} = \tfrac{1}{2} m v^2$$

No doubt, you have met and used that result elsewhere.

It is important, so make a note of it.

**3**

$$\text{K.E.} = \tfrac{1}{2} m v^2$$

In many applications in engineering, we are concerned with objects that are rotating – wheels, cams, shafts, armatures, etc. – and we often refer to their movement in terms of 'revolutions per second'. Each particle of the rotating object, however, has a linear velocity, and so has its own store of K.E. – and it is the K.E. of rotating objects that we are concerned with in this part of the programme.

*So turn on to frame 4.*

**4**

Let us first consider a single particle P of mass $m$ rotating about an axis X with constant angular velocity $\omega$ radians per second.

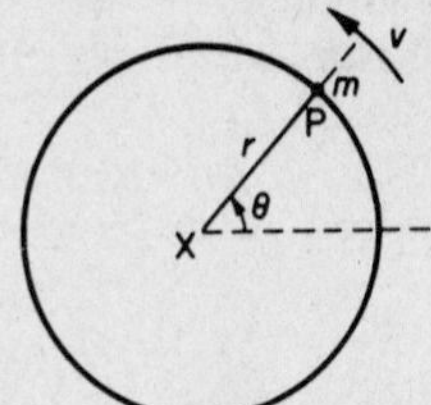


This means that the angle $\theta$ at the centre is increasing at the rate of $\omega$ radians/ per second.

Of course, the linear velocity of P, $v$ cm/s, depends upon two quantities

(i) the angular velocity ($\omega$ rad/s)

and also (ii) ..........................................

**5**

how far P is from the centre

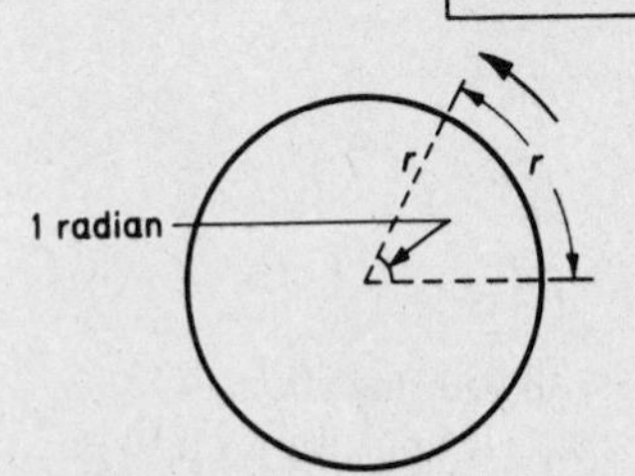


To generate an angle of 1 radian in a second, P must move round the circle a distance equal to 1 radius length, i.e. $r$ (cm).

If $\theta$ is increasing at 1 rad/s, P is moving at $r$ cm/s,
" " " " " 2 " P is moving at $2r$ cm/s,
" " " " " 3 " P is moving at $3r$ cm/s, etc.

So, in general,

if $\theta$ is increasing at $\omega$ rad/s, P is moving at $\omega r$ cm/s.

Therefore, if the angular velocity of P is $\omega$ rad/s, the linear velocity,

$v$, of P is ..................

**6**

$v = \omega r$

We have already established that the kinetic energy of an object of mass $m$ moving with velocity $v$ is given by

K.E. = ..................

**7**

$$\boxed{\text{K.E.} = \tfrac{1}{2}mv^2}$$

So, for our rotating particle, we have

$$\begin{aligned}\text{K.E.} &= \tfrac{1}{2}mv^2\\ &= \tfrac{1}{2}m(\omega r)^2\\ &= \tfrac{1}{2}m\omega^2 r^2\end{aligned}$$

and changing the order of the factors we can write

$$\underline{\text{K.E.} = \tfrac{1}{2}\omega^2 . mr^2}$$

where $\omega$ = the angular velocity of the particle P about the axis (rad/s)
$m$ = mass of P
$r$ = distance of P from the axis of rotation

*Make a note of that result: we shall certainly need that again.*

**8**

$$\boxed{\text{K.E.} = \tfrac{1}{2}\,\omega^2 . mr^2}$$

If we now have a whole system of particles, all rotating about XX with the same angular velocity $\omega$ rad/s, each particle contributes its own store of energy.

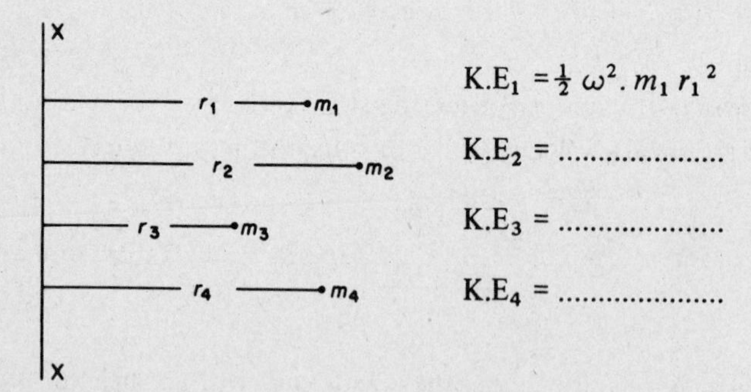

**9**

$$\boxed{\begin{aligned} K.E_1 &= \tfrac{1}{2}\,\omega^2 . m_1 r_1^{\,2} \\ K.E_2 &= \tfrac{1}{2}\,\omega^2 . m_2 r_2^{\,2} \\ K.E_3 &= \tfrac{1}{2}\,\omega^2 . m_3 r_3^{\,2} \\ K.E_4 &= \tfrac{1}{2}\,\omega^2 . m_4 r_4^{\,2} \end{aligned}}$$

So that, the total energy of the system (or solid object) is given by

$$\begin{aligned} K.E. &= K.E_1 + K.E_2 + K.E_3 + K.E_4 + \ldots \\ &= \tfrac{1}{2}\,\omega^2 . m_1 r_1^{\,2} + \tfrac{1}{2}\,\omega^2 . m_2 r_2^{\,2} + \tfrac{1}{2}\,\omega^2 . m_3 r_3^{\,2} + \ldots \\ K.E. &= \Sigma \tfrac{1}{2}\,\omega^2 . m r^2 \\ \underline{K.E.} &\underline{= \tfrac{1}{2}\,\omega^2 . \Sigma m r^2} \qquad \text{(since } \omega \text{ is a constant)} \end{aligned}$$

This is another result to note.

---

**10**

$$\boxed{K.E. = \tfrac{1}{2}\,\omega^2 . \Sigma m r^2}$$

This result is the product of two distinct factors:

(i) $\frac{1}{2}\omega^2$ can be varied by speeding up or slowing down the rate of rotation,

but (ii) $\Sigma m r^2$ is a property of the rotating object. It depends on the total mass but also on where that mass is distributed in relation to the axis XX. It is a physical property of the object and is called its *second moment of mass*, or its *moment of inertia* (denoted by the symbol I).

$$\therefore \underline{\text{I} = \Sigma m r^2} \qquad \text{(for all the particles)}$$

*Example:* For the system of particles shown, find its moment of inertia about the axis XX.

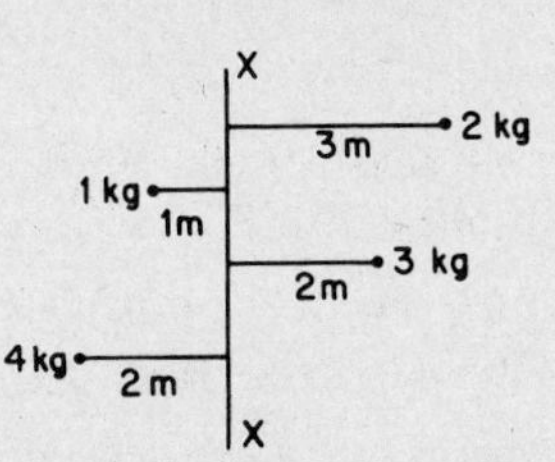


I = ....................

**11**

$$\boxed{I = 21 \text{ kg m}^2}$$

Since $\quad I = \Sigma m r^2$

$= 2.3^2 + 1.1^2 + 3.2^2 + 4.2^2$

$= 18 + 1 + 12 + 16 = \underline{47 \text{ kg m}^2}$

*Move on to frame 12.*

**12**

2. **Radius of gyration**

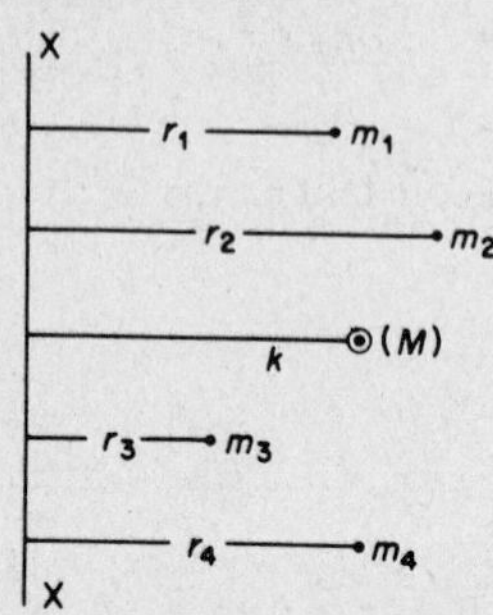


If we imagine the total mass M of the system arranged at a distance $k$ from the axis, so that the K.E. of M would be the same as the total K.E. of the distributed particles,

then $\quad \frac{1}{2}\omega^2 . Mk^2 = \frac{1}{2}\omega^2 . \Sigma m r^2$

$\therefore \quad Mk^2 = \Sigma m r^2$

and $k$ is called the *radius of gyration* of the object about the particular axis of rotation.

So, we have $\quad \underline{I = \Sigma m r^2}; \quad \underline{Mk^2 = I}$

I = moment of inertia (or second moment of mass)

$k$ = radius of gyration about the given axis.

*Now let us apply some of these results, so on you go to frame 13.*

**13**

*Example 1.* To find the moment of inertia (I) and the radius of gyration ($k$) of a uniform thin rod about an axis through one end perpendicular to the length of the rod.

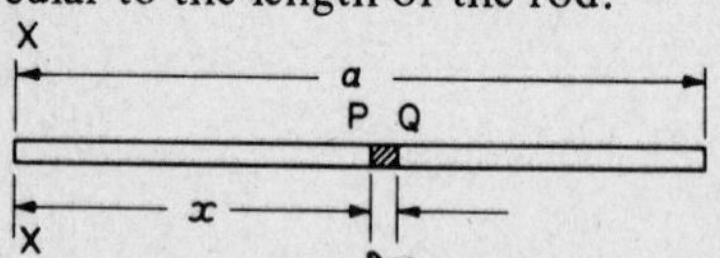


Let $\rho$ = mass per unit length of rod

Mass of element PQ = $\rho.\delta x$.

$\therefore$ Second moment of mass of PQ about XX = mass × (distance)$^2$

$= \rho.\delta x.\, x^2 = \rho x^2.\delta x$.

$\therefore$ Total second moment for all such elements can be written

$I \simeq$ ..................

**14**

$$I \simeq \sum_{0}^{a} \rho x^2.\delta x$$

The approximation sign is included since $x$ is the distance up to the left-hand side of the element PQ. But, if $\delta x \to 0$, this becomes

$$I = \int_0^a \rho x^2.dx = \rho\left[\frac{x^3}{3}\right]_0^a = \frac{\rho a^3}{3} \quad \therefore \underline{I = \frac{\rho a^3}{3}}$$

Now, to find $k$, we shall use $Mk^2 = I$, so we must first determine the total mass M.

Since $\rho$ = mass per unit length of rod, and the rod is $a$ units long, the total mass, M = ....................

**15**

$$M = a\rho$$

$$Mk^2 = I \quad \therefore a\rho.k^2 = \frac{\rho a^3}{3}$$

$$\therefore k^2 = \frac{a^2}{3} \quad \therefore k = \frac{a}{\sqrt{3}}$$

$$\therefore \underline{I = \frac{\rho a^3}{3}} \quad \text{and} \quad \underline{k = \frac{a}{\sqrt{3}}}$$

Now for another:

*Example 2.* Find I for a rectangular plate about an axis through its c.g. parallel to one side, as shown.

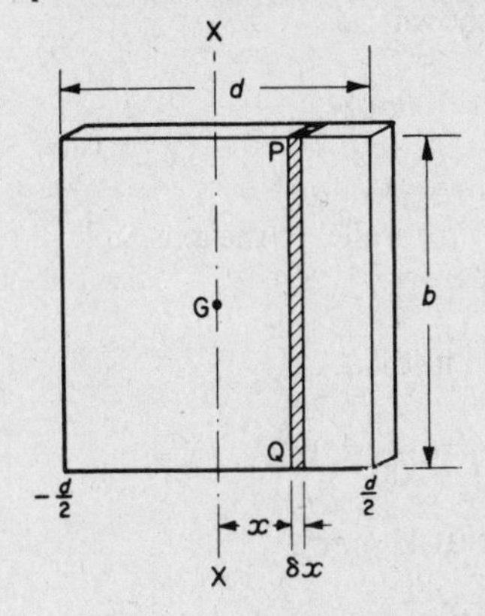


Let $\rho$ = mass per unit area of plate.

Mass of strip PQ = $b.\delta x.\rho$

Second moment of mass of strip about XX

$$\simeq b\,\delta x\,\rho.x^2$$

(i.e. mass × distance²)

$\therefore$ Total second moment for all strips covering the figure

$$I \simeq \sum_{x=}^{x=} \text{....................}$$

**16**

$$I \simeq \sum_{x=-d/2}^{x=d/2} b\rho x^2.\delta x$$

Did you remember the limits?

So now, if $\delta x \to 0$,

$$I = \int_{-d/2}^{d/2} b\rho x^2.dx = b\rho\left[\frac{x^3}{3}\right]_{-d/2}^{d/2}$$

$$= b\rho\left\{\left(\frac{d^3}{24}\right) - \left(-\frac{d^3}{24}\right)\right\} = \frac{b\rho d^3}{12}$$

$$\therefore I = \frac{bd^3\rho}{12}$$

and since the total mass $M = bd\rho$, $I = \dfrac{Md^2}{12}$

$$\therefore I = \frac{bd^3\rho}{12} = \frac{Md^2}{12}$$

*This is a useful standard result for a rectangular plate, so make a note of it for future use.*

**17**

Here is an example, very much like the last, for you to do.

*Example 3.* Find I for a rectangular plate, 20 cm × 10 cm, of mass 2 kg, about an axis 5 cm from one 20-cm side as shown.

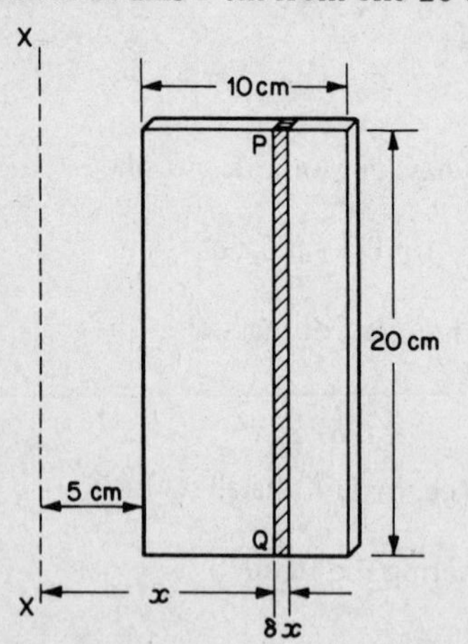


Take a strip parallel to the axis and argue as before.

*Note* that, in this case,

$$\rho = \frac{2}{10.20} = \frac{2}{200} = 0{\cdot}01$$

i.e. $\rho = 0{\cdot}01$ kg/cm$^2$

*Finish it off and then turn on to the next frame.*

**18**

$$\boxed{I = 217 \text{ kg cm}^2}$$

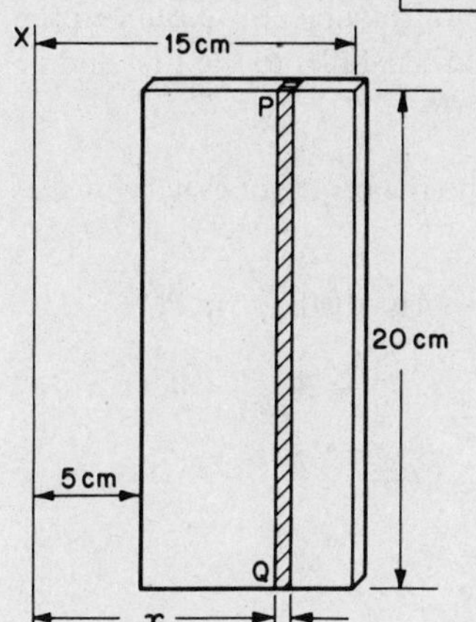


Here is the working in full:

$$\rho = 0{\cdot}01 \text{ kg cm}^2$$

Area of strip = $20.\delta x$
$\therefore$ Mass of strip = $20.\delta x.\rho$
$\therefore$ 2nd moment of mass of strip about XX $\triangleq 20.\delta x.\rho.x^2$

$$\therefore \text{ Total 2nd moment of mass} = I \triangleq \sum_{x=5}^{x=15} 20\,\rho\, x^2.\delta x.$$

If $\delta x \to 0$, $\quad I = \int_5^{15} 20\,\rho\, x^2.dx = 20\,\rho\left[\frac{x^3}{3}\right]_5^{15} = \frac{20\rho}{3}\left\{3375 - 125\right\}$

$$= \frac{20}{3}\left\{3250\right\}\frac{1}{100} = \frac{650}{3} = \underline{217 \text{ kg cm}^2}$$

Now, for the same problem, find the value of $k$.

---

**19**

$$\boxed{k = 10{\cdot}4 \text{ cm}}$$

for $Mk^2 = I$ and $M = 2$ kg

$$\therefore\ 2k^2 = 217 \qquad \therefore\ k^2 = 108{\cdot}5$$

$$\therefore\ k = \sqrt{108{\cdot}5} = \underline{10{\cdot}4 \text{ cm}}$$

Normally, then, we find I this way:

(i) Take an elementary strip parallel to the axis of rotation at a distance $x$ from it.
(ii) Form an expression for its second moment of mass about the axis.
(iii) Sum for all such strips.
(iv) Convert to integral form and evaluate.

It is just as easy as that!

**20**

3. **Parallel axes theorem**

If I is known about an axis through the c.g. of the object, we can easily write down the value of I about any other axis parallel to the first and a known distance from it.

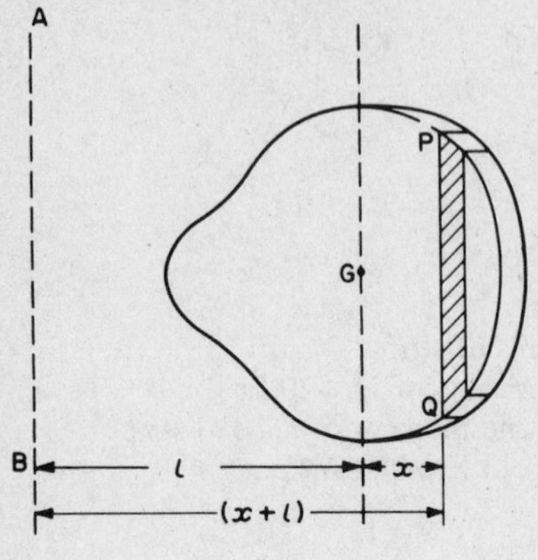


Let G be the centre of gravity of the object

Let $m$ = mass of the strip PQ

Then $\quad I_G = \Sigma m x^2$

and $\quad I_{AB} = \Sigma m(x + l)^2$

$$\therefore \; I_{AB} = \Sigma m(x^2 + 2lx + l^2)$$
$$= \Sigma mx^2 + \Sigma 2mxl + \Sigma ml^2$$
$$= \Sigma mx^2 + 2l\Sigma mx + l^2\Sigma m \quad \text{(since } l \text{ is a constant)}$$

Now, $\quad \Sigma mx^2 = \ldots\ldots\ldots\ldots\ldots\ldots$

and $\quad \Sigma m = \ldots\ldots\ldots\ldots\ldots\ldots$

---

**21**

$$\boxed{\Sigma mx^2 = I_G \;;\; \Sigma m = M}$$

Right. In the middle term we have $\Sigma mx$. This equals 0, since the axis XX by definition passes through the c.g. of the solid.

In our previous result, then,

$$\Sigma mx^2 = I_G; \;\; \Sigma mx = 0; \;\; \Sigma m = M$$

and substituting these in, we get

$$\underline{I_{AB} = I_G + Ml^2}$$

Thus, if we know $I_G$, we can obtain $I_{AB}$ by simply adding on the product of the total mass × square of the distance of transfer.

*This result is important: make a note of it in your book.*

**22**

*Example 1.* To find I about the axis AB for the rectangular plate shown below.

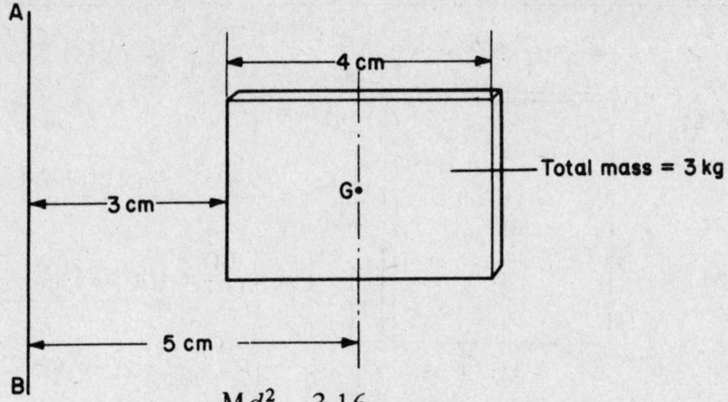


We have: $I_G = \frac{Md^2}{12} = \frac{3.16}{12} = 4 \text{ kg cm}^2$

$$\therefore I_{AB} = I_G + Ml^2$$

$$= 4 + 3.25 = 4 + 75 = 79 \text{ kg cm}^2$$

$$\therefore \underline{I_{AB} = 79 \text{ kg cm}^2}$$

As easy as that!

*Next frame.*

**23**

You do this one:

*Example 2.* A metal door, 40 cm × 60 cm, has a mass of 8 kg and is hinged along one 60-cm side.

Here is the figure:

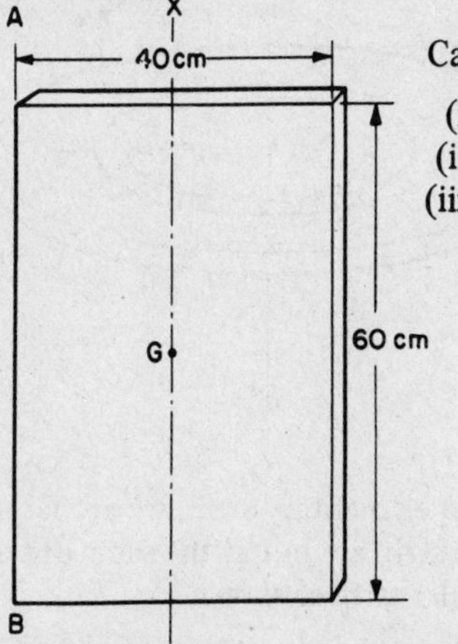


Calculate

(i) I about XX, the axis through the c.g.
(ii) I about the line of hinge, AB.
(iii) $k$ about AB.

*Find all three results: then turn on to frame 24 and check your working.*

**24**

$I_{XX} = 1067 \text{ kg cm}^2$; $I_{AB} = 4267 \text{ kg cm}^2$; $k_{AB} = 23{\cdot}1 \text{ cm}$

*Solutions:*

(i) $I_G = \dfrac{Md^2}{12}$

$= \dfrac{8.\,40^2}{12} = \dfrac{8.\,1600}{12}$

$= \dfrac{3200}{3} = \underline{1067 \text{ kg cm}^2}$

(ii) $I_{AB} = I_G + Ml^2 \quad = 1067 + 8.\,20^2 = 1067 + 3200$

$= \underline{4267 \text{ kg cm}^2}$

(iii) $Mk^2 = I_{AB} \quad \therefore 8k^2 = 4267 \quad \therefore k^2 = 533{\cdot}4 \quad \therefore \underline{k = 23{\cdot}1 \text{ cm}}$

If you made any slips, be sure to clear up any difficulties.

*Then move on to the next example.*

**25**

Let us now consider wheels, cams, etc. – basically rotating discs.

To find the moment of inertia of a circular plate about an axis through its centre, perpendicular to the plane of the plate.

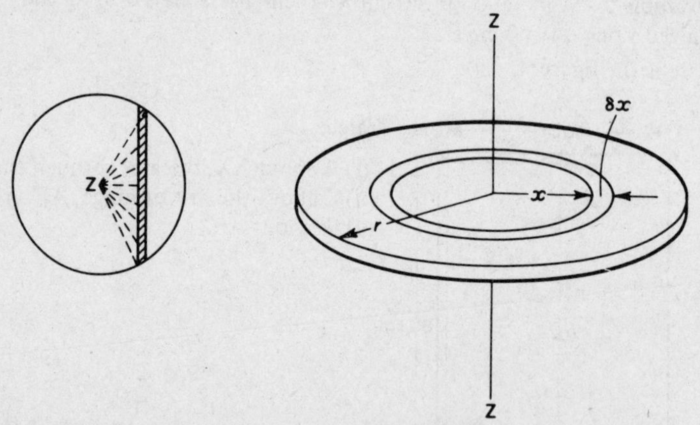


If we take a slice across the disc as an elementary strip, we are faced with the difficulty that all points in the strip are not at the same distance from the axis. We therefore take a circular strip as shown.

Mass of strip $\simeq 2\pi x.\delta x.\rho$ ($\rho$ = mass per unit area of plate)

$\therefore$ 2nd moment of strip about ZZ $\simeq$ ...............................

**26**

$$2\pi x.\delta x.\rho.x^2$$

$\therefore$ 2nd moment of strip about ZZ $= 2\pi\rho x^3.\delta x$

$\therefore$ Total 2nd moment for all such circular strips about ZZ, is given by

$$I_Z \triangleq \sum_{x=0}^{x=r} 2\pi\rho x^3.\delta x$$

If $\delta x \to 0$,

$$I_Z = \int_0^r 2\pi\rho x^3.dx = 2\pi\rho\left[\frac{x^4}{4}\right]_0^r$$

$$= \frac{2\pi\rho r^4}{4} = \frac{\pi r^4\rho}{2}$$

Total mass, $M = \pi r^2\rho$

$$\therefore\ I_Z = \frac{\pi r^4\rho}{2} = \frac{M.r^2}{2}$$

This is another standard result, so note it down.

*Next frame.*

**27**

$$I_Z = \frac{\pi r^4\rho}{2} = \frac{M.r^2}{2}$$

*Example 1.* Find the radius of gyration of a metal disc of radius 6 cm and total mass 0·5 kg.

We know that, for a circular disc,

$$I_Z = \frac{M.r^2}{2} \quad \text{and, of course,} \quad Mk^2 = I$$

so off you go and find the value of $k$.

**28**

$$k = 4{\cdot}24 \text{ cm}$$

$$I_Z = \frac{M.r^2}{2} = \frac{0{\cdot}5.\,36}{2} = 9 \text{ kg cm}^2$$

$$Mk^2 = I \quad \therefore\ \tfrac{1}{2}k^2 = 9 \quad \therefore\ k^2 = 18$$

$$\therefore\ k = 4{\cdot}24 \text{ cm}$$

They are all done in very much the same way.

*Turn to frame 29.*

**29** 4 **Perpendicular axes theorem** (for thin plates)

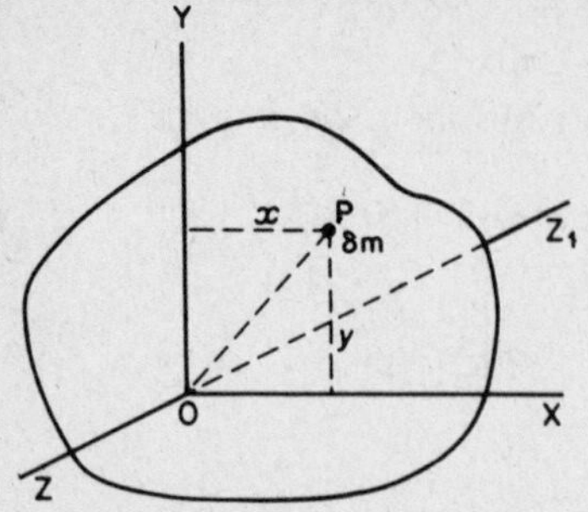


Let $\delta m$ be a small mass at P.

Then $I_X \simeq \Sigma \delta m . y^2$

and $I_Y \simeq \Sigma \delta m . x^2$

Let ZZ be the axis perpendicular to both XX and YY.

Then $$I_Z = \Sigma \delta m.(OP)^2 = \Sigma \delta m.(x^2 + y^2)$$
$$= \Sigma \delta m.y^2 + \Sigma \delta m.x^2$$
$$\therefore \; I_Z = I_X + I_Y$$

$\therefore$ If we know the second moment about two perpendicular axes in the plane of the plate, the second moment about a third axis, perpendicular to both (through the point of intersection) is given by

$$\underline{I_Z = I_X + I_Y}$$

And that is another result to note.

**30** To find I for a circular disc about a diameter as axis.

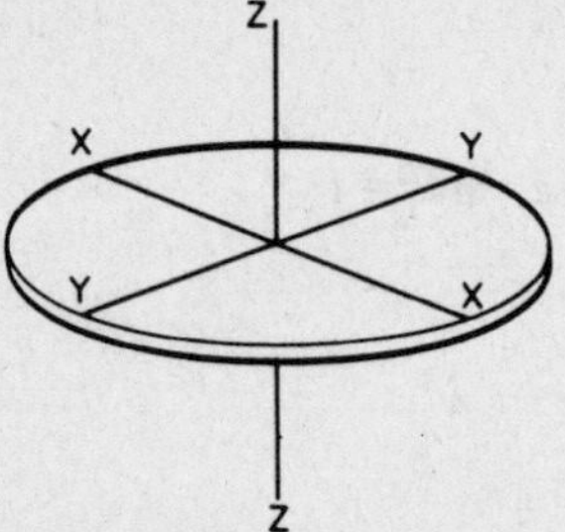


We have already established that

$$I_Z = \frac{\pi r^4 \rho}{2} = \frac{M.r^2}{2}$$

Let XX and YY be two diameters perpendicular to each other.

Then we know $$I_X + I_Y = I_Z = \frac{M.r^2}{2}$$

But all diameters are identical

$$\therefore \; I_X = I_Y \quad \therefore \; 2\,I_X = \frac{M.r^2}{2} \quad \therefore \; I_X = \frac{M.r^2}{4}$$

$\therefore$ For a circular disc:

$$\underline{I_Z = \frac{\pi r^4 \rho}{2} = \frac{M.r^2}{2}} \quad \text{and} \quad \underline{I_X = \frac{\pi r^4 \rho}{4} = \frac{M.r^2}{4}}$$

Make a note of these too.

31

*Example.* Find I for a circular disc, 40 cm diameter, and of mass 12 kg,

(i) about the normal axis (Z axis),
(ii) about a diameter as axis,
(iii) about a tangent as axis.

Work it through on your own. When you have obtained (ii) you can find (iii) by applying the parallel axes theorem.

*Then check with the next frame.*

32

$$I_Z = 2400 \text{ kg cm}^2; \quad I_X = 1200 \text{ kg cm}^2; \quad I_T = 6000 \text{ kg cm}^2$$

For:

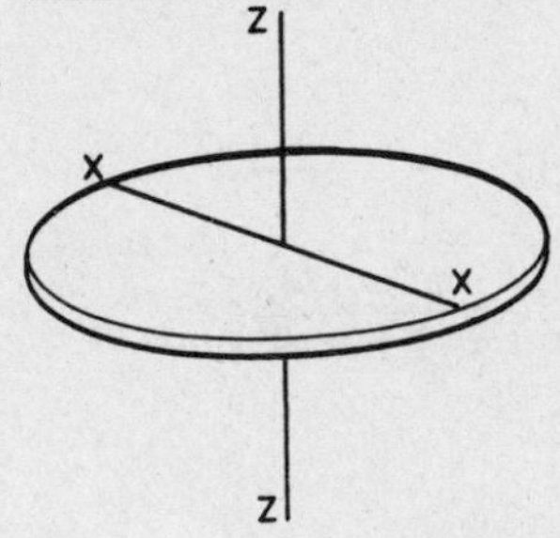


(i) $I_Z = \dfrac{M.r^2}{2} = \dfrac{12.20^2}{2}$

$= \underline{2400 \text{ kg cm}^2}$

(ii) $I_X = \dfrac{M.r^2}{4} = \dfrac{1}{2}I_Z = \underline{1200 \text{ kg cm}^2}$

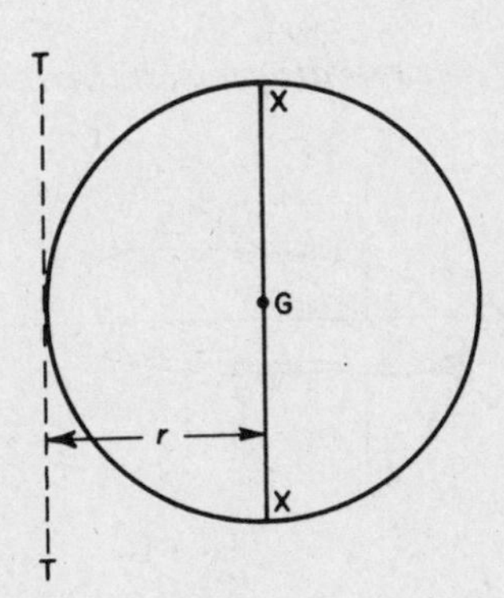


(iii) $I_X = 1200 \text{ kg cm}^2$

By the parallel axes theorem

$$I_T = I_X + Ml^2$$
$$= 1200 + 12.20^2$$
$$= 1200 + 4800$$
$$= \underline{6000 \text{ kg cm}^2}$$

In the course of our work, we have established a number of important results, so, at this point, let us collect them together, so that we can see them as a whole.

*On then to the next frame.*

**33** **Useful standard results**, so far.

1. $I = \Sigma mr^2$; $M.k^2 = I$
2. *Rectangular plate* ($\rho$ = mass/unit area)

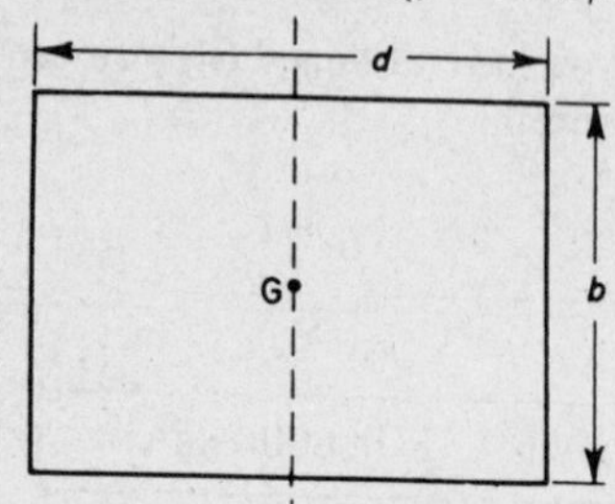


$$I_G = \frac{bd^3\rho}{12} = \frac{M.d^2}{12}$$

3. *Circular disc*

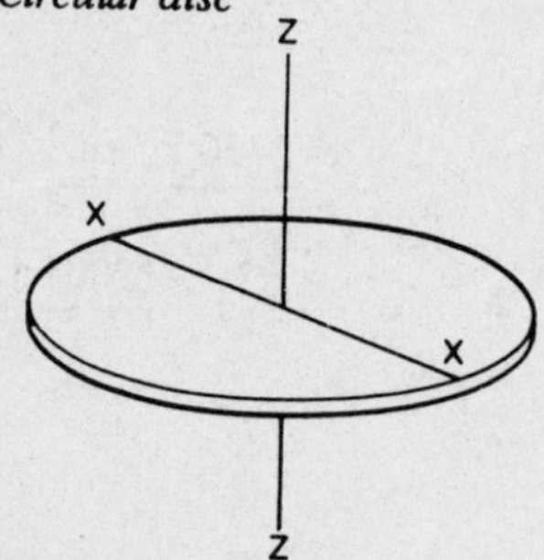


$$I_Z = \frac{\pi r^4 \rho}{2} = \frac{M.r^2}{2}$$

$$I_X = \frac{\pi r^4 \rho}{4} = \frac{M.r^2}{4}$$

4. *Parallel axes theorem*

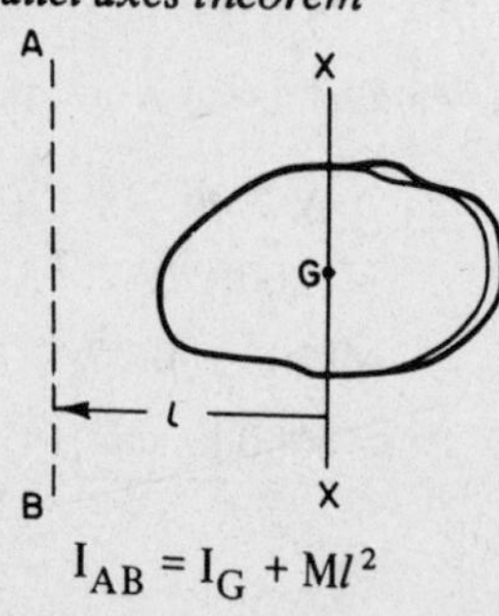


$$I_{AB} = I_G + Ml^2$$

5. *Perpendicular axes theorem*

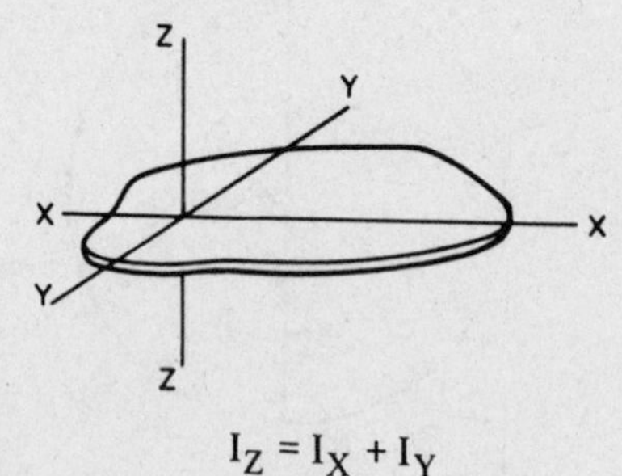


$$I_Z = I_X + I_Y$$

These standard results cover a large number of problems, but sometimes it is better to build up expressions in particular cases from first principles. *Let us see an example using that method.*

**34**

*Example 1.* Find I for the hollow shaft shown, about its natural axis. Density of material = 0·008 kg/cm³.

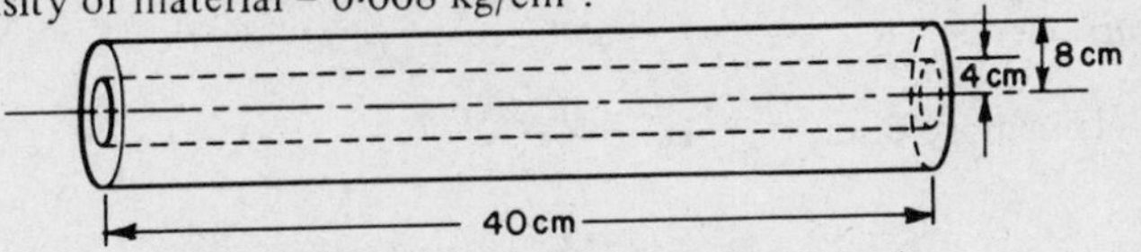


First consider a thin shell, distance $x$ from the axis

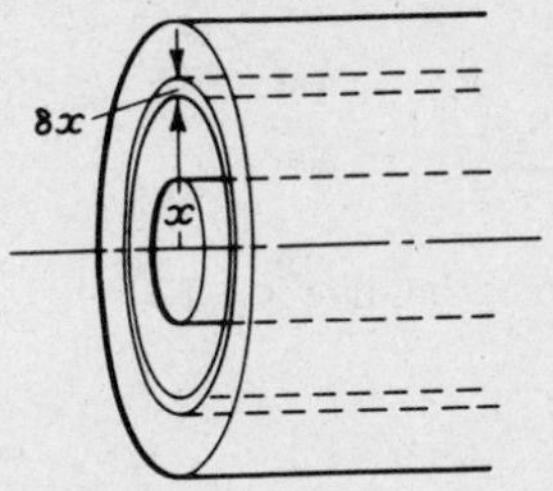


Mass of shell $\simeq 2\pi x . \delta x . 40\rho$. kg

$\therefore$ 2nd mt. about XX $\simeq 2\pi x . \delta x . 40\rho . x^2$

$\simeq 80\pi\rho x^3 . \delta x$

$\therefore$ Total 2nd mt. $\sum_{x=4}^{x=8} 80\pi\rho x^3 . \delta x$

Now, if $\delta x \to 0$, I = ........................

*and finish it off, then check with the next frame.*

**35**

$$I = 1931 \text{ kg cm}^2$$

For

$$I = 80\pi\rho \int_4^8 x^3\, dx = 80\pi\rho \left[\frac{x^4}{4}\right]_4^8$$

$$= \frac{80\pi\rho}{4} [64^2 - 16^2]$$

$$= 20\pi\rho \,.\, 48.\, 80 = 20\pi \,.48.80.0{\cdot}008$$

$$= 614{\cdot}4\pi = \underline{1931 \text{ kg cm}^2}$$

Here is another:

*Example 2.* Find I and $k$ for the solid cone shown, about its natural axis of symmetry.

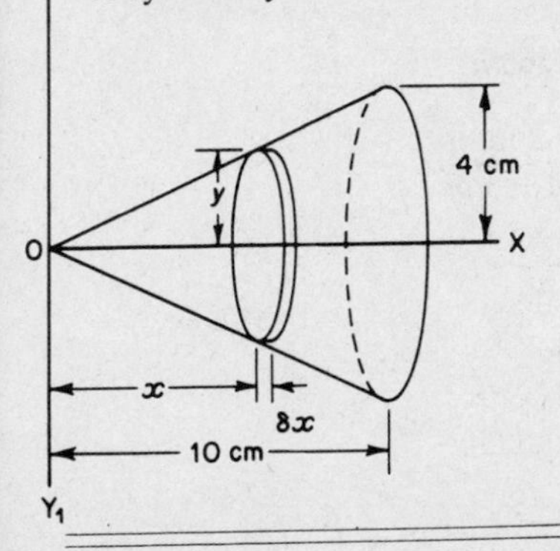


First take an elementary disc at distance $x$ from the origin. For this disc, OX is the normal axis, so

$I_X$ = ...........................

Then sum for all the discs, etc.

*Finish it off.*

# 36

$$I_X = 256\,\pi\rho \; ; \; k = 2{\cdot}19 \text{ cm}$$

*Solution:*

For elementary disc: $\quad I_X = \dfrac{\pi y^4.\delta x \rho}{2}$

$$\therefore \text{ Total } I_X \triangleq \sum_{x=0}^{x=10} \frac{\pi y^4 \delta x \rho}{2}$$

If $\delta x \to 0$, $\quad I_X = \displaystyle\int_0^{10} \frac{\pi \rho y^4}{2}\, dx = \frac{\pi\rho}{2}\int_0^{10} y^4\, dx$

Now, from the figure, the slope of the generating line is 4/10.

$$\therefore y = \frac{4x}{10}$$

$$\therefore I_X = \frac{\pi\rho}{2}\int_0^{10}\left(\frac{4x}{10}\right)^4 dx$$

$$= \frac{\pi\rho}{2}\, 0{\cdot}16^2 \left[\frac{x^5}{5}\right]_0^{10}$$

$$= \frac{\pi\rho\, 0{\cdot}0256}{2}\left[\frac{10^5}{5}\right]$$

$$= \pi\rho\, 0{\cdot}0256.\, 10^4 = \underline{256\,\pi\rho}$$

Now we proceed to find $k$.

$$\text{Total mass} = M = \frac{1}{3}\pi 4^2\, 10\rho = \frac{160\,\pi\rho}{3}$$

$$Mk^2 = I$$

$$\therefore \frac{160\,\pi\rho}{3} k^2 = 256\,\pi\rho$$

$$\therefore k^2 = \frac{3.256.\pi\rho}{160.\pi\rho}$$

$$= \frac{3.64}{40}$$

$$= 4{\cdot}8$$

$$\therefore k = \sqrt{4{\cdot}8} = \underline{2{\cdot}19 \text{ cm}}$$

*Turn now to frame 37.*

37

## 5. Second moments of area

In the theory of bending of beams, the expression $\Sigma ar^2$, relating to the cross-section of the beam, has to be evaluated. This expression is called the *second moment of area* of the section and although it has nothing to do with kinetic energy of rotation, the mathematics involved is clearly very much akin to that for moments of inertia, i.e. $\Sigma mr^2$.

Indeed, all the results we have obtained for thin plates, could apply to plane figures, provided always that 'mass' is replaced by 'area'. In fact, the mathematical processes are so nearly alike that the same symbol (I) is used in practice both for *moment of inertia* and for *second moment of area.*

38

| **Moments of inertia** | **Second moments of area** |
|---|---|
| $I = \Sigma mr^2$ | $I = \Sigma ar^2$ |
| $Mk^2 = I$ | $Ak^2 = I$ |
| *Rectangular plate* | *Rectangle* |
| $I_G = \dfrac{bd^3\rho}{12} = \dfrac{M.d^2}{12}$ | $I_C = \dfrac{bd^3}{12} = \dfrac{A.d^2}{12}$ |
| *Circular plate* | *Circle* |
| $I_Z = \dfrac{\pi r^4 \rho}{2} = \dfrac{M.r^2}{2}$ | $I_Z = \dfrac{\pi r^4}{2} = \dfrac{A.r^2}{2}$ |
| $I_X = \dfrac{\pi r^4 \rho}{4} = \dfrac{M.r^2}{4}$ | $I_X = \dfrac{\pi r^4}{4} = \dfrac{A.r^2}{4}$ |

*Parallel axes theorem* – applies to both.

$I_{AB} = I_G + Ml^2$ $\qquad$ $I_{AB} = I_C + Al^2$

*Perpendicular axes theorem* – applies to thin plates and plane figures only.

$$I_Z = I_X + I_Y$$

*Turn on.*

**39**

There is really nothing new about this: all we do is replace 'mass' by 'area'.

*Example 1.* Find the second moment of area of a rectangle about an axis through one corner perpendicular to the plane of the figure.

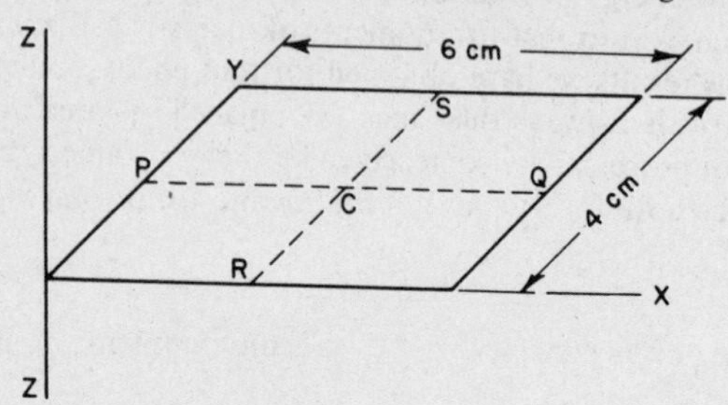


$$I_{PQ} = \frac{bd^3}{12} = \frac{6.\,4^3}{12} = 32\text{ cm}^4$$

By the parallel axes theorem, $I_X$ = ..................

**40**

$$\boxed{I_X = 128\text{ cm}^4}$$

for

$$I_X = 32 + 24.2^2 = 32 + 24.4$$
$$= 32 + 96 = \underline{128\text{ cm}^4}$$

Also $I_{RS} = \dfrac{bd^3}{12}$ = ....................

**41**

$$\boxed{I_{RS} = 72\text{ cm}^4}$$

for

$$I_{RS} = \frac{4.\,6^3}{12} = 72\text{ cm}^4$$

$\therefore I_Y$ = ..................

**42**

$$\boxed{I_Y = 288\text{ cm}^4}$$

For, again by the parallel axes theorem,

$$I_Y = 72 + 24.3^2 = 72 + 216 = \underline{288\text{ cm}^4}$$

So we have therefore: $I_X = 128\text{ cm}^4$

and $I_Y = 288\text{ cm}^4$

$\therefore I_Z$ (which is perpendicular to both $I_X$ and $I_Y$) = ..................

$$\boxed{I_Z = 416 \text{ cm}^4}$$

**43**

□□□□□□□□□□□□□□□□□□□□□□□□□□□□□□□□□□□□□□□□□□□□□□□□□□□□

When the plane figure is bounded by an analytical curve, we proceed in much the same way.

*Example 2.* Find the second moment of area of the plane figure bounded by the curve $y = x^2 + 3$, the $x$-axis, and the ordinates at $x = 1$ and $x = 3$, about the $y$-axis.

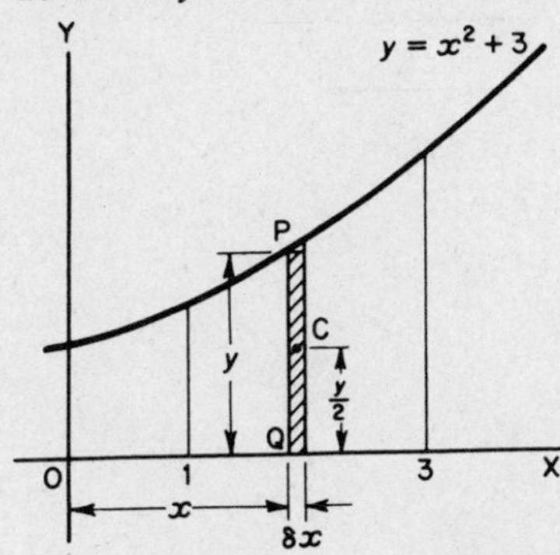


Area of strip PQ $= y.\delta x$

$\therefore$ 2nd mt. of strip about OY $= y.\delta x.x^2$

$= x^2.y.\delta x$

$$\therefore\ I_Y \simeq \sum_{x=1}^{x=3} x^2\, y\, \delta x$$

If $\delta x \to 0$, $\qquad I_Y = \int_1^3 x^2 y\, dx = \ldots\ldots\ldots\ldots\ldots\ldots$

*Finish it off.*

$$\boxed{I_Y = 74{\cdot}4 \text{ units}^4}$$

**44**

for
$$I_Y = \int_1^3 x^2(x^2+3)\, dx = \int_1^3 (x^4 + 3x^2)\, dx$$
$$= \left[\frac{x^5}{5} + x^3\right]_1^3$$
$$= \left(\frac{243}{5} + 27\right) - \left(\frac{1}{5} + 1\right)$$
$$= \frac{242}{5} + 26 = 48{\cdot}4 + 26 = \underline{74{\cdot}4 \text{ units}^4}$$

*Note:* Had we been asked to find $I_X$, we should take second moment of the strip about OX, i.e. $\dfrac{y^3}{3}\,\delta x$ ; sum for all strips $\displaystyle\sum_{x=1}^{x=3} \frac{y^3}{3}\,\delta x$; and then evaluate the integral.

*Now, one further example, so turn on to the next frame.*

**45**

*Example 3.* For the triangle PQR shown, find the second moment of area and $k$ about an axis AB through the vertex and parallel to the base.

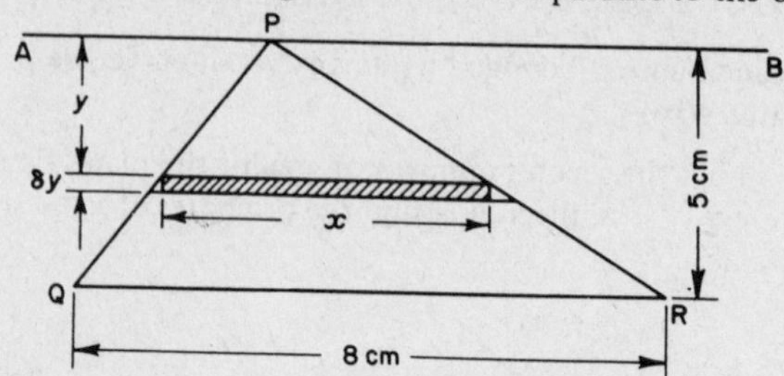


First consider an elementary strip. Area of strip = $x.\delta y$

$\therefore$ 2nd mt. of strip about AB = $x.\delta y.y^2 = xy^2.\delta y$

$\therefore$ Total 2nd mt. about AB for all such strips

$$\simeq \sum_{y=0}^{y=5} xy^2.\delta y$$

If $\delta y \to 0$, $$I_{AB} = \int_0^5 xy^2\,dy$$

We must now write $x$ in terms of $y$ – and we can obtain this from the figure by similar triangles.

Finish the work off, so that $I_{AB}$ = ........................

**46**

$$\boxed{I = 250\text{ cm}^4\ ;\ k = 3{\cdot}536\text{ cm}}$$

For we have $$\frac{y}{5} = \frac{x}{8} \quad \therefore\ x = \frac{8y}{5}$$

$$\therefore\ I_{AB} = \int_0^5 xy^2\,dy = \frac{8}{5}\int_0^5 y^3\,dy = \frac{8}{5}\left[\frac{y^4}{4}\right]_0^5$$

$$= \frac{8}{20}\,[5^4 - 0] = \frac{8}{20}(625) = \underline{250\text{ cm}^4}$$

Also, total area, $A = \frac{5.8}{2} = 20\text{ cm}^2$

$$\therefore\ Ak^2 = I \quad \therefore\ 20k^2 = 250$$

$$\therefore\ k^2 = 12{\cdot}5$$

$$\therefore\ \underline{k = 3{\cdot}536\text{ cm}}$$

*Next frame.*

47

**Composite figures**

If a figure is made up of a number of standard figures whose individual second moments about a given axis are $I_1$, $I_2$, $I_3$, etc., then the second moment of the composite figure about the same axis is simply the sum of $I_1$, $I_2$, $I_3$, etc.

Similarly, if a figure whose second moment about a given axis is $I_2$ is removed from a larger figure with second moment $I_1$ about the same axis, the second moment of the remaining figure is $I = I_1 - I_2$.

Now for something new.

48

6. **Centres of pressure**

*Pressure at a point P depth z below the surface of a liquid.*

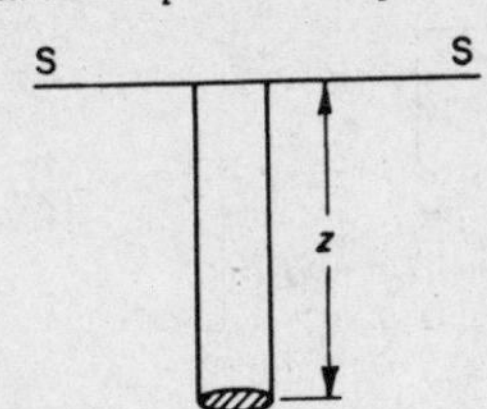


If we have a perfect liquid, the pressure at P, i.e. the thrust on unit area at P, is due to the weight of the column of liquid of height $z$ above it.

Pressure at P = $p = wz$, where $w$ = weight of unit volume of the liquid.

Also, the pressure at P operates equally in all directions.

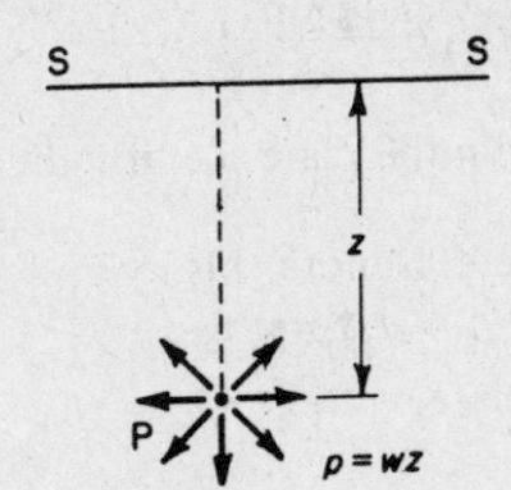


Note that, in our considerations, we shall ignore the atmospheric pressure which is also acting on the surface of the liquid.

The pressure, then, at any point in a liquid is proportional to the .................... of the point below the surface.

**49**

depth

*Total thrust on a vertical plate* immersed in liquid.

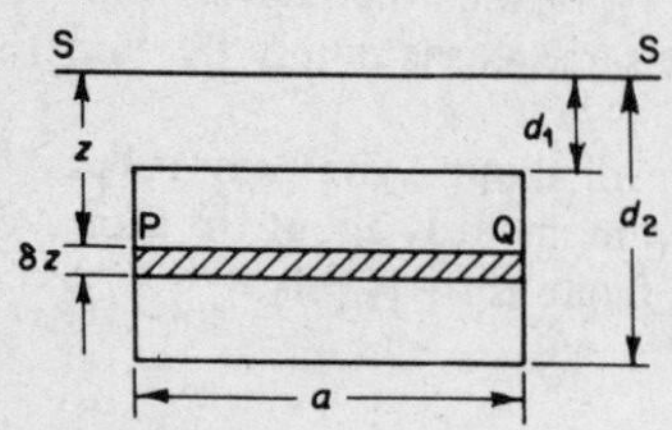


Consider a thin strip at a depth $z$ below the surface of the liquid.

Pressure at P $= wz$.

$\therefore$ Thrust on strip PQ $\simeq wz$ (area of strip)

$\simeq w.z.a.\delta z$

Then the total thrust on the whole plate

$$\simeq \sum_{z=d_1}^{z=d_2} a\,w\,z\,\delta z$$

If $\delta z \to 0$, total thrust $= \int_{d_1}^{d_2} a\,w\,z\,dz =$ ....................

**50**

$$\frac{a\,w}{2}\left[d_2^{\,2} - d_1^{\,2}\right]$$

for: total thrust $= a\,w\left[\frac{z^2}{2}\right]_{d_1}^{d_2} = \frac{a\,w}{2}\left[d_2^{\,2} - d_1^{\,2}\right]$

This can be written

$$\text{Total thrust} = \frac{a\,w}{2}(d_2 - d_1)(d_2 + d_1)$$

$$= w\,a\,(d_2 - d_1)\left(\frac{d_2 + d_1}{2}\right)$$

Now, $\left(\frac{d_2 + d_1}{2}\right)$ is the depth half way down the plate, i.e. it indicates the depth of the centre of gravity of the plate. Denote this by $\bar{z}$.

Then, total thrust $= w\,a(d_2 - d_1)\bar{z} = a(d_2 - d_1)\,w\,\bar{z}$.

Also $a(d_2 - d_1)$ is the total area of the plate.

So we finally obtain the fact that

total thrust = area of plate $\times$ pressure at the c.g. of the plate.

In fact, this result applies whatever the shape of the plate, so copy the result down for future use.

*On to the next frame.*

**51**

Total thrust = area of plate × pressure at the c.g. of plate

So, if $w$ is the weight per unit volume of liquid, determine the total thrust on the following plates, immersed as shown.

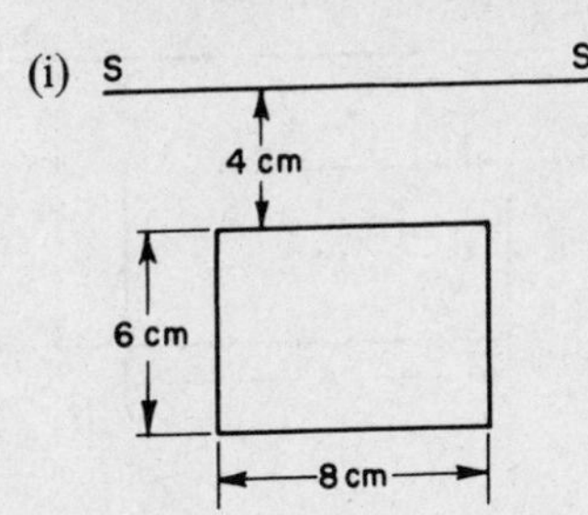


(ii) S S 4 cm 10 cm 10 cm

So, thrust (i) = .................... and thrust (ii) = ....................

**52**

thrust (i) = 336 $w$ : thrust (ii) = 180 $w$

For, in each case,

total thrust = area of surface × pressure at the c.g.

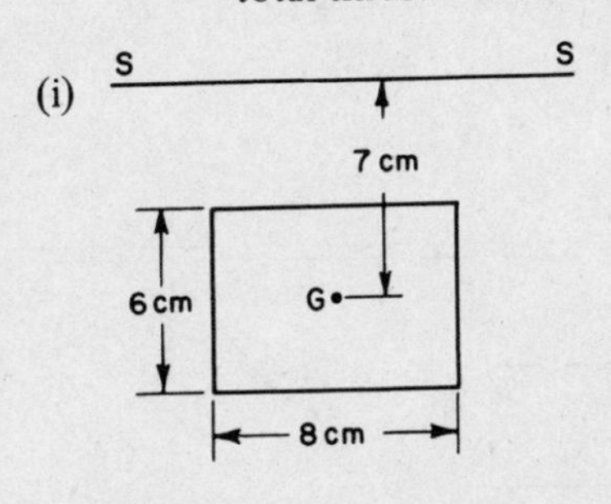


Area = $6 \times 8 = 48 \text{ cm}^2$

Pressure at G = $7w$

$\therefore$ Total thrust = $48.7w$

$= \underline{336w}$

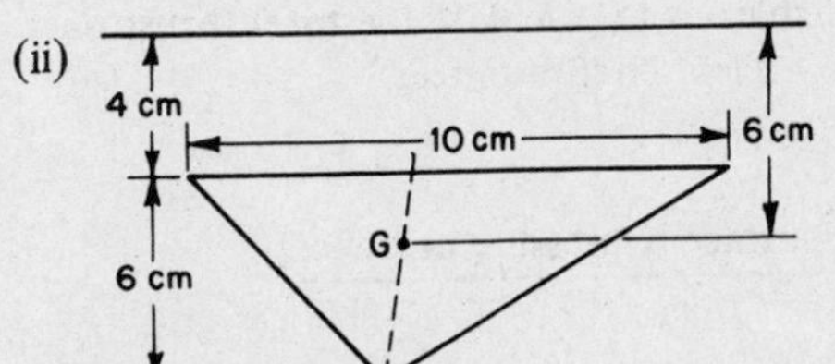


Area = $\dfrac{10 \times 6}{2} = 30 \text{ cm}^2$

Pressure at G = $6w$

$\therefore$ Total thrust = $30.6w$

$= \underline{180w}$

*On to the next frame.*

**53**

If the plate is not vertical, but inclined at an angle $\theta$ to the horizontal, the rule still holds good.

*Example:*

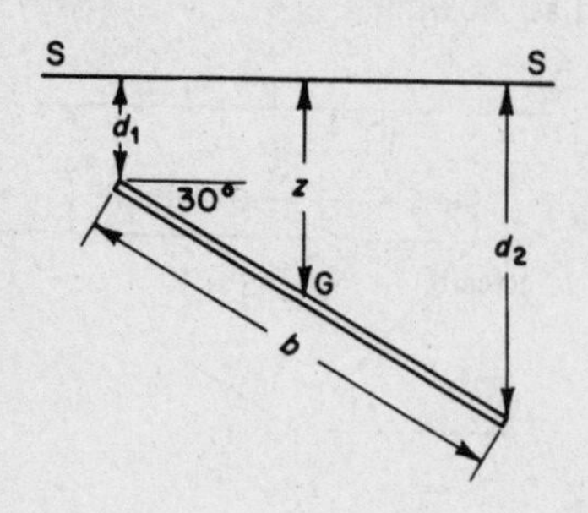


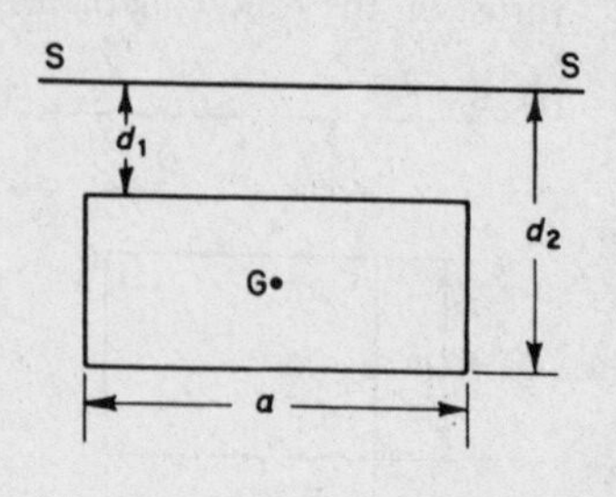


$$\text{Depth of G} = d_1 + \frac{b}{2}\sin 30° = d_1 + \frac{b}{4}$$

$$\text{Pressure at G} = \left(d_1 + \frac{b}{4}\right)w$$

$$\text{Total area} = a\,b$$

$$\therefore \quad \text{Total thrust} = \text{.........................}$$

**54**

$$\boxed{a\,b\left(d_1 + \frac{b}{4}\right)w}$$

Remember this general rule enables us to calculate the total thrust on an immersed surface in almost any set of circumstances.

So make a note of it:

<u>total thrust = area of surface × pressure at the c.g.</u>

*Then on to frame 55.*

**55**

**Depth of centre of pressure**

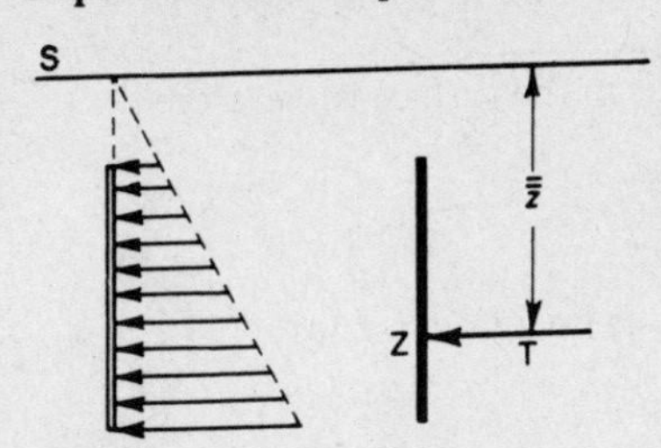


The pressure on an immersed plate increases with depth and we have seen how to find the total thrust T on the plate.

The resultant of these forces is a single force equal to the total thrust, T, in magnitude and acting at a point Z called the *centre of pressure* of the plate. Let $\bar{\bar{z}}$ denote the depth of the centre of pressure.

To find $\bar{\bar{z}}$ we take moments of forces about the axis where the plane of the plate cuts the surface of the liquid. Let us consider our same rectangular plate again.

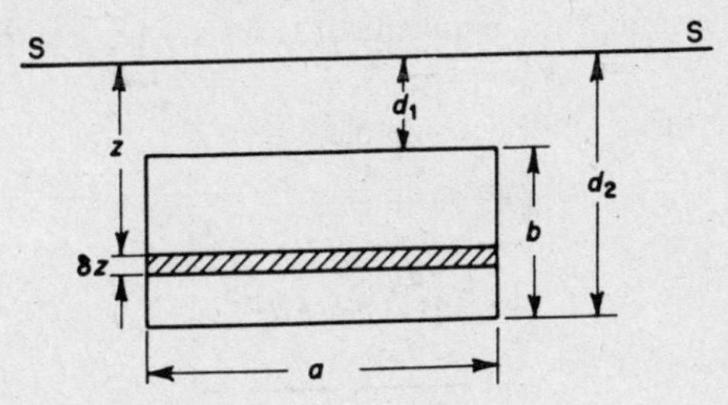


The area of the strip PQ = ........................

**56**

$$\boxed{a.\delta z}$$

The pressure at the level of PQ = ........................

**57**

$$\boxed{z\,w}$$

So the thrust on the strip PQ = ........................

**58**

$$\boxed{a.\delta z.w.z \quad \text{i.e.} \quad a\,w\,z\,\delta z}$$

The moment of this thrust about the axis in the surface is therefore

$$= a\,w\,z\,\delta z.\,z$$
$$= a\,w\,z^2.\,\delta z$$

So that the sum of the moments of thrusts on all such strips

= ................

**59**

$$\boxed{\sum_{d_1}^{d_2} a\,w\,z^2\,\delta z}$$

Now, if $\delta z \to 0$,

$$\text{the sum of the moments of thrusts} = \int_{d_1}^{d_2} a\,w\,z^2.\,dz$$

Also, the total thrust on the whole plate = ........................

**60**

$$\boxed{\int_{d_1}^{d_2} a\,w\,z\,dz}$$

Right. Now the total thrust $\times\ \bar{\bar{z}}$ = sum of moments of all individual thrusts

$$\therefore \int_{d_1}^{d_2} a\,w\,z\,dz \times \bar{\bar{z}} = \int_{d_1}^{d_2} a\,w\,z^2\,dz$$

$$\therefore \text{Total thrust} \times \bar{\bar{z}} = w\int_{d_1}^{d_2} a\,z^2\,dz$$

$$= w\mathrm{I}$$

Therefore, we have

$$\bar{\bar{z}} = \frac{w\mathrm{I}}{\text{total thrust}} = \frac{w\,\mathrm{A}\,k^2}{\mathrm{A}\,w\,\bar{z}}$$

$$\therefore\ \bar{\bar{z}} = \frac{k^2}{\bar{z}}$$

*Make a note of that and then turn on.*

61

So we have these two important results:

(i) The total thrust on a submerged surface
= total area of face × pressure at its centroid (depth $\bar{z}$)

(ii) The resultant thrust acts at the centre of pressure, the depth of which, $\bar{\bar{z}}$, is given by $\bar{\bar{z}} = \dfrac{k^2}{\bar{z}}$.

Now for an example on this.

62

*Example 1.* For a vertical rectangular dam, 40 m × 20 m, the top edge of the dam coincides with the surface level. Find the depth of the centre of pressure.

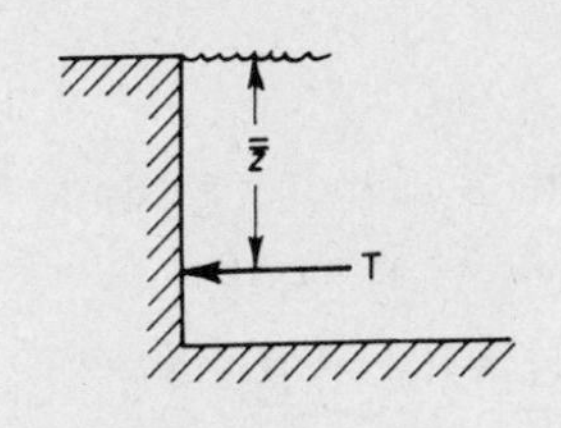


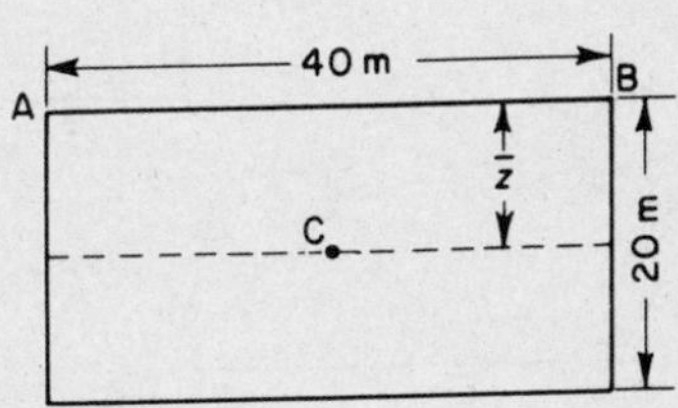


In this case, $\bar{z} = 10$ m.
To find $k^2$ about AB

$$I_c = \frac{A d^2}{12} = \frac{40 \,.\, 20 \,.\, 400}{12} = \frac{80\,000}{3}\ \text{m}^4$$

$$I_{AB} = I_C + A l^2 = \frac{80\,000}{3} + 800 \,.\, 100$$

$$= \frac{4}{3}.(80\,000)$$

$$Ak^2 = I \quad \therefore\ k^2 = \frac{4}{3}.\frac{80\,000}{800} = \frac{400}{3}$$

$$\therefore \quad \bar{\bar{z}} = \frac{k^2}{\bar{z}} = \frac{400}{3.10} = \frac{40}{3} = \underline{13{\cdot}33\ \text{m}}$$

*Note* that, in this case,

(i) the centroid is half-way down the rectangle,

but (ii) the centre of pressure is two-thirds of the way down the rectangle.

**63**

Here is one for you.

*Example 2.* An outlet from a storage tank is closed by a circular cover hung vertically. The diameter of the cover = 1 m and the top of the cover is 2·5 m below the surface of the liquid. Determine the depth of the centre of pressure of the cover.

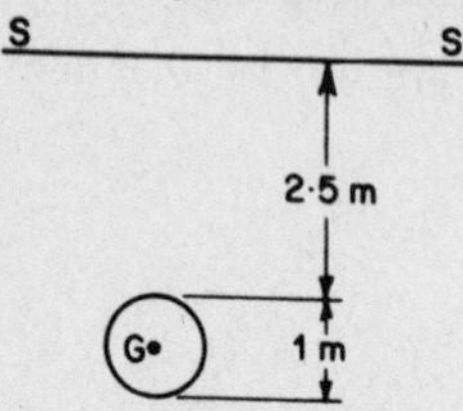


*Work completely through it: then check your working with the next frame.*

**64**

$$\bar{\bar{z}} = 3{\cdot}02 \text{ m}$$

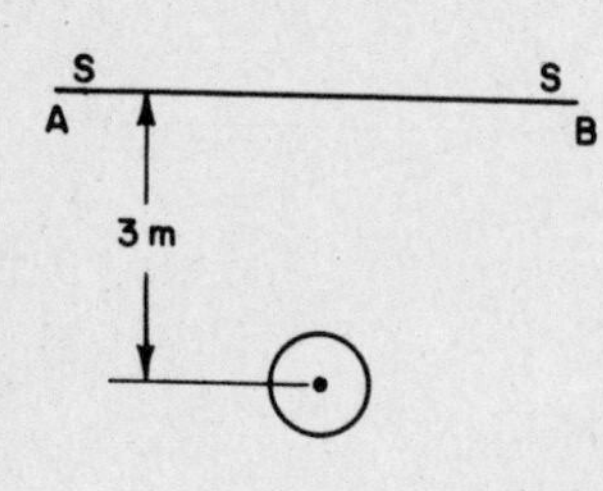


We have:

(i) Depth of centroid = $\bar{z}$ = 3 m

(ii) To find $k^2$ about AB

$$I_C = \frac{Ar^2}{4} = \frac{\pi(\frac{1}{2})^2 . (\frac{1}{2})^2}{4} = \frac{\pi}{64}$$

$$I_{AB} = \frac{\pi}{64} + A.3^2$$

$$= \frac{\pi}{64} + \pi(\tfrac{1}{2})^2.9$$

$$= \frac{\pi}{64} + \frac{9\pi}{4} = \frac{145\pi}{64}$$

For AB $\quad k^2 = \frac{I_{AB}}{A} = \frac{145\pi}{64}.\frac{4}{\pi} = \frac{145}{16}$

$$\therefore \bar{\bar{z}} = \frac{k^2}{\bar{z}} = \frac{145}{16}.\frac{1}{3} = \frac{145}{48} = \underline{3{\cdot}02 \text{ m}}$$

□□□□□□□□□□□□□□□□□□□□□□□□□□□□□□□□□□□□□□□□□

And that brings us to the end of this piece of work. Before you work through the Test Exercise, check down the revision sheet that follows in frame 65 and brush up any part of the programme about which you may not be absolutely clear.

*Then, when you are ready, turn on to the Test Exercise.*

**65**

**Revision Sheet**

1. SECOND MOMENTS

| | *Mts. of Inertia* | | *2nd Mts. of Area* |
|---|---|---|---|
| (i) | $I = \Sigma m r^2$ <br> $M k^2 = I$ | (i) | $I = \Sigma a r^2$ <br> $A k^2 = I$ |
| (ii) | Rectangular plate: <br> $I_G = \frac{b d^3 \rho}{12} = \frac{M.d^2}{12}$ | (ii) | Rectangle: <br> $I_C = \frac{b.d^3}{12} = \frac{A.d^2}{12}$ |
| (iii) | Circular disc: <br> $I_Z = \frac{\pi r^4 \rho}{2} = \frac{M.r^2}{2}$ <br> $I_X = \frac{\pi r^4 \rho}{4} = \frac{M.r^2}{4}$ | (iii) | Circle: <br> $I_Z = \frac{\pi r^4}{2} = \frac{A.r^2}{2}$ <br> $I_X = \frac{\pi r^4}{4} = \frac{A.r^2}{4}$ |
| (iv) | Parallel axes theorem: <br> $I_{AB} = I_G + M l^2$ | | $I_{AB} = I_C + A l^2$ |

(v) Perpendicular axes theorem (thin plates and plane figures only):

$$I_Z = I_X + I_Y$$

2. CENTRES OF PRESSURE

(i) Pressure at depth $z = w z$ ($w$ = weight of unit volume of liquid)

(ii) Total thrust on plane surface
= area of surface × pressure at the centroid.

(iii) Depth of centre of pressure ($\bar{\bar{z}}$):

Total thrust × $\bar{\bar{z}}$ = sum of moments of distributed thrust

$$\bar{\bar{z}} = \frac{k^2}{\bar{z}}$$

where $k$ = radius of gyration of figure about axis in surface of liquid,
$\bar{z}$ = depth of centroid.

Note: The magnitude of the total thrust
= (area × pressure at the centroid)
but it acts through the centre of pressure.

□□□□□□□□□□□□□□□□□□□□□□□□□□□□□□□□□□□□□□□□□□□□□□

*Now for the Test Exercise, on to frame 66.*

**66**

Work through all the questions in the Test Exercise. They are very much like those we have been doing, so will cause you no difficulty: there are no tricks. Take your time and work carefully.

**Test Exercise – XX**

1. (i) Find the moment of inertia of a rectangular plate, of sides $a$ and $b$, about an axis through the mid-point of the plate and perpendicular to the plane of the plate. (ii) Hence find also the moment of inertia about an axis parallel to the first axis and passing through one corner of the plate. (iii) Find the radius of gyration about the second axis.

2. Show that the radius of gyration of a thin rod of length $l$ about an axis through its centre and perpendicular to the rod is $\dfrac{l}{2\sqrt{3}}$.

   An equilateral triangle ABC is made of three identical thin rods each of length $l$. Find the radius of gyration of the triangle about an axis through A, perpendicular to the plane of ABC.

3. A plane figure is bounded by the curve $xy = 4$, the $x$-axis, and the ordinates at $x = 2$ and $x = 4$. Calculate the square of the radius of gyration of the figure (i) about OX, and (ii) about OY.

4. Prove that the radius of gyration of a uniform solid cone with base radius $r$ about its natural axis is $\sqrt{\dfrac{3r^2}{10}}$.

5. An equilateral triangular plate is immersed in water vertically with one edge in the surface. If the length of each side is $a$, find the total thrust on the plate and the depth of the centre of pressure.

**Further Problems – XX**

1. A plane figure is enclosed by the curve $y = a \sin x$ and the $x$-axis between $x = 0$ and $x = \pi$. Show that the radius of gyration of the figure about the $x$-axis is $\dfrac{a\sqrt{2}}{3}$.

2. A length of thin uniform wire of mass M is made into a circle of radius $a$. Find the moment of inertia of the wire about a diameter as axis.

3. A solid cylinder of mass M has a length $l$ and radius $r$. Show that its moment of inertia about a diameter of the base is $M\left[\dfrac{r^2}{4} + \dfrac{l^2}{3}\right]$.

4. Show that the moment of inertia of a solid sphere of radius $r$ and mass M, about a diameter as axis, is $\dfrac{2}{5}Mr^2$.

5. Prove that, if $k$ is the radius of gyration of an object about an axis through its centre of gravity, and $k_1$ is the radius of gyration about another axis parallel to the first and at a distance $l$ from it, then $k_1 = \sqrt{(k^2 + l^2)}$.

6. A plane figure is bounded by the parabola $y^2 = 4ax$, the $x$-axis and the ordinate $x = c$. Find the radius of gyration of the figure (i) about the $x$-axis, and (ii) about the $y$-axis.

7. Prove that the moment of inertia of a hollow cylinder of length $l$, with inner and outer radii $r$ and R respectively, and total mass M, about its natural axis, is given by $I = \frac{1}{2}M(R^2 + r^2)$.

8. Show that the depth of the centre of pressure of a vertical triangle with one side in the surface is $\frac{1}{2}h$, if $h$ is the perpendicular height of the triangle.

9. Calculate the second moment of area of a square of side $a$ about a diagonal as axis.

10. Find the moment of inertia of a solid cone of mass M and base radius $r$ and height $h$, about a diameter of the base as axis. Find also the radius of gyration.

11. A thin plate in the form of a trapezium with parallel sides of length $a$ and $b$, distance $d$ apart, is immersed vertically in water with the side of length $a$ in the surface. Prove that the depth of the centre of pressure ($\bar{\bar{z}}$) is given by

$$\bar{\bar{z}} = \frac{d(a + 3b)}{2(a + 2b)}$$

12. Find the second moment of area of an ellipse about its major axis.

13. A square plate of side $a$ is immersed vertically in water with its upper side horizontal and at a depth $d$ below the surface. Prove that the centre of pressure is at a distance $\dfrac{a^2}{6(a + 2d)}$ below the centre of the square.

14. Find the total thrust and the depth of the centre of pressure when a semicircle of radius $a$ is immersed vertically in liquid with its diameter in the surface.

15. A plane figure is bounded by the curve $y = e^x$, the $x$-axis, the $y$-axis and the ordinate at $x = 1$. Calculate the radius of gyration of the figure (i) about OX as axis, and (ii) about OY as axis.

16. A vertical dam is a parabolic segment of width 12 m and maximum depth 4 m at the centre. If the water reaches the top of the dam, find the total thrust on the face.

17. A circle of diameter 6 cm is removed from the centre of a rectangle measuring 10 cm by 16 cm. For the figure that remains, calculate the radius of gyration about one 10-cm side as axis.

18. Prove that the moment of inertia of a thin hollow spherical shell of mass M and radius $r$, about a diameter as axis is $\frac{2}{3}\text{M}r^2$.

19. A semicircular plate of radius $a$ is immersed vertically in water, with its diameter horizontal and the centre of the arc just touching the surface. Find the depth of the centre of pressure.

20. A thin plate of uniform thickness and total mass M, is bounded by the curve $y = c\cosh\dfrac{x}{c}$, the $x$-axis, the $y$-axis, and the ordinate $x = a$. Show that the moment of inertia of the plate about the $y$-axis is $\text{M}\left\{a^2 - 2ca\coth^{-1}\left(\dfrac{a}{c}\right) + 2c^2\right\}$.

# Programme 21

## APPROXIMATE INTEGRATION

**1**

**Introduction**

In previous programmes, we have seen how to deal with various types of integral, but there are still some integrals that look simple enough, but which cannot be determined by any of the standard methods we have studied.

For instance, $\int_0^{\frac{1}{2}} x\,e^x\,dx$ can be evaluated by the method of integration by parts.

$$\int_0^{\frac{1}{2}} x\,e^x\,dx = \ldots\ldots\ldots\ldots$$

What do you get?

**2**

$$1 - \tfrac{1}{2}\sqrt{e}$$

for:

$$\int_0^{\frac{1}{2}} x\,e^x\,dx = \Big[x(e^x)\Big]_0^{\frac{1}{2}} - \int_0^{\frac{1}{2}} e^x\,dx$$

$$= \Big[x e^x - e^x\Big]_0^{\frac{1}{2}} = \Big[e^x(x-1)\Big]_0^{\frac{1}{2}}$$

$$= e^{\frac{1}{2}}(-\tfrac{1}{2}) - e^0(-1) = 1 - \tfrac{1}{2}\sqrt{e}$$

That was easy enough, and this method depends, of course, on the fact that on each application of the routine, the power of $x$ decreases by 1, until it disappears, leaving $\int e^x\,dx$ to be completed without difficulty.

But suppose we try to evaluate $\int_0^{\frac{1}{2}} x^{\frac{1}{2}} e^x\,dx$ by the same method. The process now breaks down. Work through it and see if you can decide why

*When you have come to a conclusion, move on to the next frame.*

**3**

Reducing the power of $x$ by 1 at each application of the method, will never give $x^0$, i.e. the power of $x$ will never disappear and so the resulting integral will always be a product.

For we get:

$$\int_0^{\frac{1}{2}} x^{\frac{1}{2}} e^x\,dx = \Big[x^{\frac{1}{2}}(e^x)\Big]_0^{\frac{1}{2}} - \tfrac{1}{2}\int_0^{\frac{1}{2}} e^x\,x^{-\frac{1}{2}}\,dx$$

and in the process, we have hopped over $x^0$.

So here is a complication. The present programme will show you how to deal with this and similar integrals that do not fit in to our normal patterns.

*So on, then, to frame 4.*

4

**Approximate integration**

First of all, the results we shall get will be approximate in value, but like many other 'approximate' methods in mathematics, this does not imply that they are 'rough and ready' and of little significance.

The word 'approximate' in this context simply means that the numerical value cannot be completely defined, but that we can state the value to as many decimal places as we wish.

e.g. To say $x = \sqrt{3}$ is exact, but to say $x = 1{\cdot}732$ is an approximate result since, in fact, $\sqrt{3}$ has a value $1{\cdot}7321\ldots$ with an infinite number of decimal places.

Let us not be worried, then, by approximate values: we use them whenever we quote a result correct to a stated number of decimal places, or significant figures.

$$\pi = 3\tfrac{1}{7}; \quad \pi = 3{\cdot}142 \ : \ \pi = 3{\cdot}14159$$

are all ........................ values

5

approximate

We note, of course, that an approximate value can be made nearer and nearer to the real value by taking a larger number of decimal places – and that usually means more work!

Evaluation of definite integrals is often required in science and engineering problems: a numerical approximation of the result is then quite satisfactory.

Let us see two methods that we can apply when the standard routines fail.

*On to frame 6.*

**6** **Method 1.** *By series*

Consider the integral $\int_0^{\frac{1}{2}} x^{\frac{1}{2}} e^x \, dx$, which we have already seen cannot be evaluated by the normal means. We have to convert this into some other form that we can deal with.

Now we know that

$$e^x = 1 + x + \frac{x^2}{2!} + \frac{x^3}{3!} + \frac{x^4}{4!} + \ldots$$

$$\therefore x^{\frac{1}{2}} e^x = x^{\frac{1}{2}}\left\{1 + x + \frac{x^2}{2!} + \frac{x^3}{3!} + \frac{x^4}{4!} + \ldots\right\}$$

$$\therefore \int_0^{\frac{1}{2}} x^{\frac{1}{2}} e^x \, dx = \int_0^{\frac{1}{2}} x^{\frac{1}{2}}\left\{1 + x + \frac{x^2}{2!} + \frac{x^3}{3!} + \ldots\right\} dx$$

$$= \int_0^{\frac{1}{2}} \left\{x^{1/2} + x^{3/2} + \frac{x^{5/2}}{2!} + \frac{x^{7/2}}{3!} + \ldots\right\} dx$$

Now these are simply powers of $x$, so, on the next line, we have

= ...................

**7**

$$I = \left[\frac{2\,x^{3/2}}{3} + \frac{2\,x^{5/2}}{5} + \frac{2\,x^{7/2}}{7.2} + \frac{2\,x^{9/2}}{9.6} + \ldots\right]_0^{\frac{1}{2}}$$

$$= \left[\frac{2\,x^{3/2}}{3} + \frac{2\,x^{5/2}}{5} + \frac{x^{7/2}}{7} + \frac{x^{9/2}}{27} + \ldots\right]_0^{\frac{1}{2}}$$

To ease the calculation, take out the factor $x^{\frac{1}{2}}$

$$I = \left[x^{\frac{1}{2}}\left\{\frac{2\,x}{3} + \frac{2\,x^2}{5} + \frac{x^3}{7} + \frac{x^4}{27} + \frac{x^5}{132} + \ldots\right\}\right]_0^{\frac{1}{2}}$$

$$= \frac{1}{\sqrt{2}}\left\{\frac{1}{3} + \frac{2}{4\,.\,5} + \frac{1}{8\,.\,7} + \frac{1}{16\,.\,27} + \ldots\right\}$$

$$= \frac{\sqrt{2}}{2}\left\{0{\cdot}3333 + 0{\cdot}1000 + 0{\cdot}0179 + 0{\cdot}0023 + 0{\cdot}0002 \ldots\right\}$$

$$= \frac{\sqrt{2}}{2}\left\{0{\cdot}4537\right\}$$

$$= 1{\cdot}414\,(0{\cdot}2269)$$

$$= \underline{0{\cdot}3207}$$

All we do is to express the function as a series and integrate the powers of $x$ one at a time.

*Let us see another example, so turn on to frame 8.*

**8**

Here is another.

To evaluate $\int_0^{\frac{1}{2}} \frac{\ln(1+x)}{\sqrt{x}}\,dx$

First we expand $\ln(1+x)$ as a power series. Do you remember what it is?

$$\ln(1+x) = \text{....................}$$

**9**

$$\boxed{\ln(1+x) \equiv x - \frac{x^2}{2} + \frac{x^3}{3} - \frac{x^4}{4} + \frac{x^5}{5} + \dots}$$

$$\therefore \frac{\ln(1+x)}{\sqrt{x}} = x^{-\frac{1}{2}}\left\{x - \frac{x^2}{2} + \frac{x^3}{3} - \frac{x^4}{4} + \frac{x^5}{5} - \dots\right\}$$

$$= x^{1/2} - \frac{x^{3/2}}{2} + \frac{x^{5/2}}{3} - \frac{x^{7/2}}{4} + \frac{x^{9/2}}{5} - \dots$$

$$\therefore \int \frac{\ln(1+x)}{\sqrt{x}}\,dx = \text{........................}$$

**10**

$$\boxed{\int \frac{\ln(1+x)}{\sqrt{x}}\,dx = \frac{2}{3}x^{3/2} - \frac{x^{5/2}}{5} + \frac{2x^{7/2}}{21} - \frac{x^{9/2}}{18} \dots}$$

So that, applying the limits, we get

$$\int_0^{\frac{1}{2}} \frac{\ln(1+x)}{\sqrt{x}}\,dx = \left[x^{1/2}\left\{\frac{2x}{3} - \frac{x^2}{5} + \frac{2x^3}{21} - \frac{x^4}{18} \dots\right\}\right]_0^{\frac{1}{2}}$$

$$= \frac{1}{\sqrt{2}}\left\{\frac{1}{3} - \frac{1}{20} + \frac{1}{84} - \frac{1}{288} + \frac{1}{880} - \frac{1}{2496} \dots\right\}$$

$$= 0{\cdot}7071\left\{0{\cdot}3333 - 0{\cdot}0500 + 0{\cdot}0119 - 0{\cdot}0035 + 0{\cdot}0011 - 0{\cdot}0004 \dots\right\}$$

$$= 0{\cdot}7071\ (0{\cdot}2924)$$

$$= \underline{0{\cdot}2067}$$

Here is one for you to do in very much the same way.

$$\text{Evaluate} \quad \int_0^1 \sqrt{x}.\cos x\,dx$$

*Complete the working and then check your result with that given in the next frame.*

**11**

0·531 to 3 decimal places

*Solution:*

$$\cos x = 1 - \frac{x^2}{2!} + \frac{x^4}{4!} - \frac{x^6}{6!} + \frac{x^8}{8!} - \dots$$

$$\therefore \sqrt{x}\cos x = x^{1/2} - \frac{x^{5/2}}{2} + \frac{x^{9/2}}{24} - \frac{x^{13/2}}{720} + \dots$$

$$\therefore \int_0^1 \sqrt{x}\cos x\,dx = \left[\frac{2x^{3/2}}{3} - \frac{x^{7/2}}{7} + \frac{x^{11/2}}{132} - \frac{x^{15/2}}{5400} + \dots\right]_0^1$$

$$= \left\{\frac{2}{3} - \frac{1}{7} + \frac{1}{132} - \frac{1}{5400} + \dots\right\}$$

$$= 0{\cdot}6667 - 0{\cdot}1429 + 0{\cdot}007576 - 0{\cdot}000185 + \dots$$

$$= \underline{0{\cdot}531 \text{ to 3 dec. pl.}}$$

*Check carefully if you made a slip. Then on to frame 12.*

**12**

The method, then, is really very simple, providing the function can readily be expressed in the form of a series.

But we must use this method with caution. Remember that we are dealing with infinite series which are valid only for values of $x$ for which the series converges. In many cases, if the limits are less than 1 we are safe, but with limits greater than 1 we must be extra careful. For instance, the integral $\int_2^4 \frac{1}{1+x^3}\,dx$ would give a divergent series when the limits were substituted. So what tricks can we employ in a case such as this?

*On to the next frame, and we will find out.*

**13**

To evaluate $\int_2^4 \frac{1}{1+x^3}\,dx$

We first of all take out the factor $x^3$ from the denominator

$$\frac{1}{1+x^3} \equiv \frac{1}{x^3}\left\{\frac{1}{\frac{1}{x^3}+1}\right\} = \frac{1}{x^3}\left\{1+\frac{1}{x^3}\right\}^{-1}$$

This is better, for if $x^3$ is going to be greater than 1 when we substitute the limits, $\frac{1}{x^3}$ will be ..............................

---

**14**

less than 1

Right. So in this form we can expand without further trouble.

$$\text{I} = \int_2^4 x^{-3}\left\{1 - \frac{1}{x^3} + \frac{1}{x^6} - \frac{1}{x^9} + \ldots\right\} dx$$

$$= \int_2^4 x^{-3}\left\{1 - x^{-3} + x^{-6} - x^{-9} + \ldots\right\} dx$$

$$= \int_2^4 \left\{x^{-3} - x^{-6} + x^{-9} - x^{-12} + \ldots\right\} dx$$

$=$ .......................... Now finish it off.

---

**15**

0·088 to 3 decimal places

for $\text{I} = \int_2^4 \left\{x^{-3} - x^{-6} + x^{-9} - x^{-12} + \ldots\right\} dx$

$$= \left[-\frac{x^{-2}}{2} + \frac{x^{-5}}{5} - \frac{x^{-8}}{8} + \frac{x^{-11}}{11} \ldots\right]_2^4$$

$$= \left[-\frac{1}{2x^2} + \frac{1}{5x^5} - \frac{1}{8x^8} + \frac{1}{11x^{11}} \ldots\right]_2^4$$

$$= \left\{-\frac{1}{32} + \frac{1}{5120} - \frac{1}{524288} + \ldots\right\}$$

$$- \left\{-\frac{1}{8} + \frac{1}{160} - \frac{1}{2048} + \ldots\right\}$$

$= -0{\cdot}03125 + 0{\cdot}00020 - 0{\cdot}00000 + 0{\cdot}12500 - 0{\cdot}00625 + 0{\cdot}00049$

$= 0{\cdot}12569 - 0{\cdot}03750$

$= 0{\cdot}08819$

$= \underline{0{\cdot}088 \text{ to 3 dec. pl.}}$

**16**

**Method 2.** *By Simpson's rule*

Integration by series is rather tedious and cannot always be applied, so let us start afresh and try to discover some other method of obtaining the approximate value of a definite integral.

We know, of course, that integration can be used to calculate the area under a curve $y = f(x)$ between two given points $x = a$ and $x = b$.

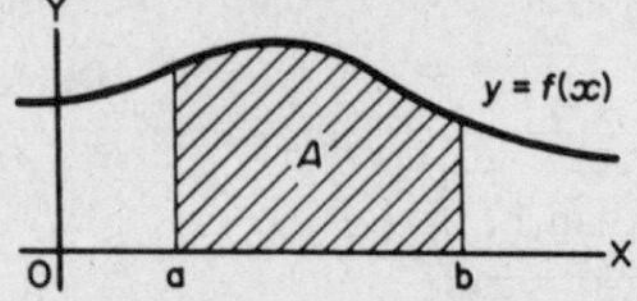


$$A = \int_a^b y\,dx = \int_a^b f(x)\,dx$$

So, if only we could find the area A by some other means, this would give us the numerical value of the integral we have to evaluate. There are various practical ways of doing this and the one we shall choose is to apply Simpson's rule.

*So on to frame 17.*

**17**

*Simpson's rule*

To find the area under the curve $y = f(x)$ between $x = a$ and $x = b$.

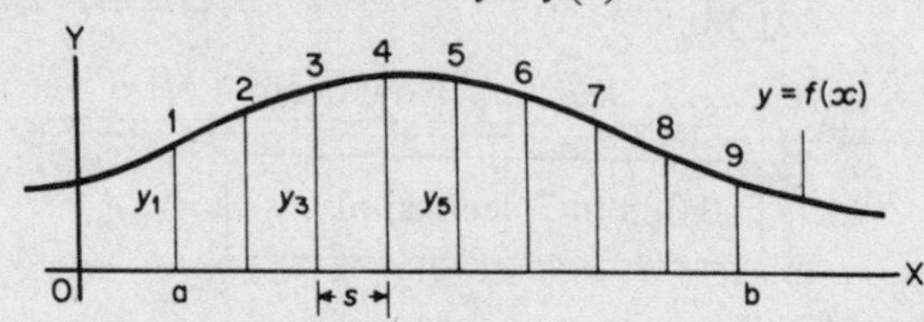


(a) Divide the figure into any even number ($n$) of equal-width strips (width $s$)

(b) Number and measure each ordinate: $y_1, y_2, y_3, \ldots, y_{n+1}$. The number of ordinates will be one more than the number of strips.

(c) The area A of the figure is then given by:

$$A \simeq \frac{s}{3}\left[(F + L) + 4E + 2R\right]$$

Where $s$ = width of each strip,
F + L = sum of the first and last ordinates,
4E = 4 × the sum of the even-numbered ordinates,
2R = 2 × the sum of the remaining odd-numbered ordinates.

*Note* that each ordinate is used once – and only once.

Make a note of this result in your record book for future reference.

$$A \simeq \frac{s}{3}\left[(F+L)+4E+2R\right]$$

The symbols themselves remind you of what they represent.

*Example:* To evaluate $\int_2^6 y\,dx$ for the function $y = f(x)$, the graph of which is shown.

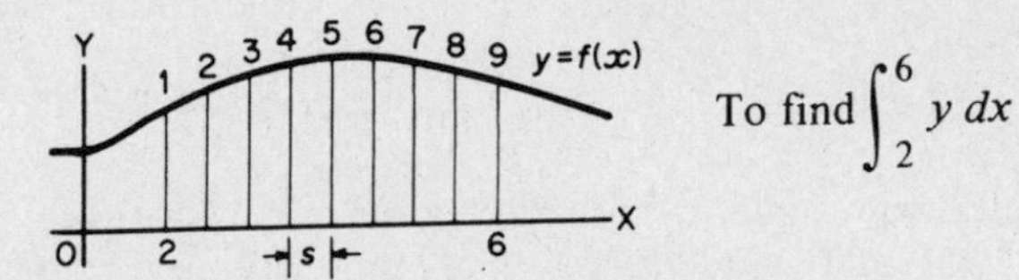


To find $\int_2^6 y\,dx$

If we take 8 strips, then $s = \frac{6-2}{8} = \frac{4}{8} = \frac{1}{2}$. $\quad s = \frac{1}{2}$

Suppose we find the lengths of the ordinates to be as follows:

| Ord. No. | 1 | 2 | 3 | 4 | 5 | 6 | 7 | 8 | 9 |
|---|---|---|---|---|---|---|---|---|---|
| Length | 7·5 | 8·2 | 10·3 | 11·5 | 12·4 | 12·8 | 12·3 | 11·7 | 11·5 |

Then we have

$$F + L = 7{\cdot}5 + 11{\cdot}5 = 19$$
$$4E = 4(8{\cdot}2 + 11{\cdot}5 + 12{\cdot}8 + 11{\cdot}7) = 4(44{\cdot}2) = 176{\cdot}8$$
$$2R = 2(10{\cdot}3 + 12{\cdot}4 + 12{\cdot}3) = 2(35) = 70$$

So that

$$A \simeq \frac{1/2}{3}\,[19 + 176{\cdot}8 + 70]$$
$$= \frac{1}{6}[265{\cdot}8] = 44{\cdot}3 \quad \therefore\ A = 44{\cdot}3 \text{ units}^2$$
$$\therefore \int_2^6 f(x)\,dx \simeq 44{\cdot}3$$

The accuracy of the result depends on the number of strips into which we divide the figure. A larger number of thinner strips gives a more accurate result.

Simpson's rule is important: it is well worth remembering.

Here it is again: write it out, but replace the query marks with the appropriate coefficients.

$$A \simeq \frac{s}{?}\left[(F+L) + ?E + ?R\right]$$

**19**

$$A \simeq \frac{s}{3}\left[(F + L) + 4E + 2R\right]$$

In practice, we do not have to plot the curve in order to measure the ordinates. We calculate them at regular intervals. Here is an example.

*Example:* To evaluate $\int_0^{\pi/3} \sqrt{\sin x}\, dx$, using six intervals.

(a) Find the value of $s$:

$$s = \frac{\pi/3 - 0}{6} = \frac{\pi}{18} \quad (= 10^\circ \text{ intervals})$$

(b) Calculate the values of $y$(i.e. $\sqrt{\sin x}$) at intervals of $\pi/18$ between $x = 0$ (lower limit) and $x = \pi/3$ (upper limit), and set your work out in the form of the table below.

| $x$ | $\sin x$ | $\sqrt{\sin x}$ |
|---|---|---|
| 0 $(0^\circ)$ | 0·0000 | 0·0000 |
| $\pi/18$ $(10^\circ)$ | 0·1736 | 0·4166 |
| $\pi/9$ $(20^\circ)$ | 0·3420 | ....... |
| $\pi/6$ $(30^\circ)$ | 0·5000 | ....... |
| $2\pi/9$ $(40^\circ)$ | ........ | ....... |
| $5\pi/18$ $(50^\circ)$ | ........ | ....... |
| $\pi/3$ $(60^\circ)$ | ........ | ....... |

Leave the right-hand side of your page blank for the moment.

Copy and complete the table as shown on the left-hand side above.

**20**

Here it is: check your results so far.

| $x$ | $\sin x$ | $\sqrt{\sin x}$ | (i) F + L | (ii) E | (iii) R |
|---|---|---|---|---|---|
| 0 $(0^\circ)$ | 0·0000 | 0·0000 | → | | |
| $\pi/18$ $(10^\circ)$ | 0·1736 | 0·4166 | | → | |
| $\pi/9$ $(20^\circ)$ | 0·3420 | 0·5848 | | | → |
| $\pi/6$ $(30^\circ)$ | 0·5000 | 0·7071 | | → | |
| $2\pi/9$ $(40^\circ)$ | 0·6428 | 0·8016 | | | → |
| $5\pi/18$ $(50^\circ)$ | 0·7660 | 0·8752 | | → | |
| $\pi/3$ $(60^\circ)$ | 0·8660 | 0·9306 | → | | |

Now form three more columns on the right-hand side, headed as shown, and transfer the final results across as indicated. This will automatically sort out the ordinates into their correct groups.

*Then on to frame 21.*

**21**

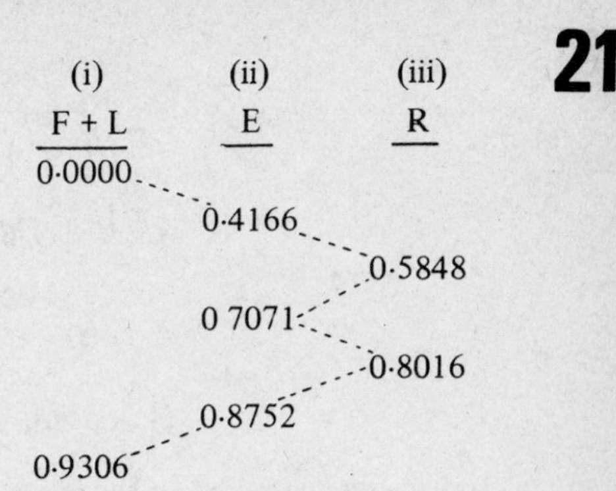

| (i) F + L | (ii) E | (iii) R |
|---|---|---|
| 0·0000 | | |
| | 0·4166 | |
| | | 0·5848 |
| | 0 7071 | |
| | | 0·8016 |
| | 0·8752 | |
| 0·9306 | | |

*Note that*

(a) You start in column 1

(b) You then zig-zag down the two right-hand columns

(c) You finish back in column 1.

Now total up each of the three columns.

**22**

Your results should be:

| (i) F + L | (ii) E | (iii) R |
|---|---|---|
| 0·9306 | 1·9989 | 1·3864 |

Now (a) Multiply column (ii) by 4 so as to give 4E,

(b) Multiply column (iii) by 2 so as to give 2R,

(c) Transfer the result in columns (ii) and (iii) to column (i) and total column (i) to obtain (F + L) + 4E + 2R.

Now do that.

**23**

This gives:

| | F + L | E | R |
|---|---|---|---|
| F + L ⟶ | 0·9306 | 1·9989 | 1·3864 |
| 4E ⟶ | 7·9956 | 4 | 2 |
| 2R ⟶ | 2·7728 | 7·9956 | 2·7728 |
| (F + L) + 4E + 2R ⟶ | 11·6990 | | |

The formula is $A \simeq \frac{s}{3}[(F+L)+4E+2R]$ so to find A we simply need to multiply our last result by $\frac{s}{3}$. Remember $s = \pi/18$.

So now you can finish it off.

$$\int_0^{\pi/3} \sqrt{\sin x}\, dx = \dots\dots\dots\dots$$

**24**

$$\boxed{0.681}$$

For:
$$A \simeq \frac{s}{3}[(F+L)+4E+2R]$$
$$\simeq \frac{\pi/18}{3}[11.6990]$$
$$\simeq \pi/54\,[11.6990]$$
$$\simeq 0.6806$$
$$\therefore \int_0^{\pi/3} \sqrt{\sin x}\,dx \simeq 0.681$$

Before we do another example, let us see the last solution complete.

To evaluate $\int_0^{\pi/3} \sqrt{\sin x}\,dx$ by Simpson's rule, using 6 intervals.

$$s = \frac{\pi/3 - 0}{6} = \pi/18 \quad (= 10^\circ \text{ intervals})$$

| $x$ | $\sin x$ | $\sqrt{\sin x}$ | F + L | E | R |
|---|---|---|---|---|---|
| 0 (0°) | 0·0000 | 0·0000 | 0·0000 | | |
| $\pi/18$ (10°) | 0·1736 | 0·4166 | | 0·4166 | |
| $\pi/9$ (20°) | 0·3420 | 0·5848 | | | 0·5848 |
| $\pi/6$ (30°) | 0·5000 | 0·7071 | | 0·7071 | |
| $2\pi/9$ (40°) | 0·6428 | 0·8016 | | | 0·8016 |
| $5\pi/18$ (50°) | 0·7660 | 0·8752 | | 0·8752 | |
| $\pi/3$ (60°) | 0·8660 | 0·9306 | 0·9306 | | |
| | F + L ⟶ | | 0·9306 | 1·9989 | 1·3864 |
| | 4E ⟶ | | 7·9956 | 4 | 2 |
| | 2R ⟶ | | 2·7728 | 7·9956 | 2·7728 |
| | (F + L) + 4E + 2R ⟶ | | 11·6990 | | |

$$I \simeq \frac{s}{3}[(F+L)+4E+2R]$$
$$\simeq \frac{\pi}{54}[11.6990]$$
$$\simeq 0.6806$$
$$\therefore \int_0^{\pi/3} \sqrt{\sin x}\,dx \simeq 0.681$$

Now we will tackle example 2 and set it out in much the same way.
*Turn to frame 25.*

**25**

*Example 2.* To evaluate $\int_{0\cdot2}^{1\cdot0} \sqrt{(1+x^3)}dx$, using 8 intervals.

First of all, find the value of $s$ in this case.

$s =$ ....................

**26**

0·1

For $s = \dfrac{1\cdot0 - 0\cdot2}{8} = \dfrac{0\cdot8}{8} = 0\cdot1$ $\qquad s = 0\cdot1$

Now write the column headings required to build up the function values. What will they be on this occasion?

**27**

| $x$ | $x^3$ | $1+x^3$ | $\sqrt{(1+x^3)}$ | F + L | E | R |
|---|---|---|---|---|---|---|

Right. So your table will look like this, with $x$ ranging from 0·2 to 1·0.

| $x$ | $x^3$ | $1+x^3$ | $\sqrt{(1+x^3)}$ | F + L | E | R |
|---|---|---|---|---|---|---|
| 0·2 | 0·008 | 1·008 | 1·0039 | | | |
| 0·3 | 0·027 | 1·027 | 1·0134 | | | |
| 0·4 | 0·064 | | | | | |
| 0·5 | 0·125 | | | | | |
| 0·6 | 0·216 | | | | | |
| 0·7 | 0·343 | | | | | |
| 0·8 | | | | | | |
| 0·9 | | | | | | |
| 1·0 | | | | | | |
| | | | F + L → | | | |
| | | | 4E → | | 4 | 2 |
| | | | 2R → | | | |
| | | | (F + L) + 4E + 2R → | | | |

Copy down and complete the table above and finish off the working to

evaluate $\int_{0\cdot2}^{1\cdot0} \sqrt{(1+x^3)}dx$.

*Check with the next frame.*

**28**

$$\int_{0\cdot2}^{1\cdot0} \sqrt{(1+x^3)}\,dx = 0\cdot911$$

| $x$ | $x^3$ | $1+x^3$ | $\sqrt{(1+x^3)}$ | F + L | E | R |
|---|---|---|---|---|---|---|
| 0·2 | 0·008 | 1·008 | 1·0039 | 1·0039 | | |
| 0·3 | 0·027 | 1·027 | 1·0134 | | 1·0134 | |
| 0·4 | 0·064 | 1·064 | 1·0316 | | | 1·0316 |
| 0·5 | 0·125 | 1·125 | 1·0607 | | 1·0607 | |
| 0·6 | 0·216 | 1·216 | 1·1027 | | | 1·1027 |
| 0·7 | 0·343 | 1·343 | 1·1589 | | 1·1589 | |
| 0·8 | 0·512 | 1·512 | 1·2296 | | | 1·2296 |
| 0·9 | 0·729 | 1·729 | 1·3149 | | 1·3149 | |
| 1·0 | 1·000 | 2·000 | 1·4142 | 1·4142 | | |
| | | | F + L ⟶ | 2·4181 | 4·5479 | 3·3639 |
| | | | 4E ⟶ | 18·1916 | 4 | 2 |
| | | | 2R ⟶ | 6·7278 | 18·1916 | 6·7278 |
| | | | (F + L) + 4E + 2R ⟶ | 27·3375 | | |

$$I = \frac{s}{3}\,[(F+L) + 4E + 2R]$$

$$= \frac{0\cdot1}{3}\,[27\cdot3375] = \frac{1}{3}\,[2\cdot73375] = 0\cdot9113$$

$$\therefore \int_{0\cdot2}^{1\cdot0} \sqrt{(1+x^3)}\,dx \simeq 0\cdot911$$

There it is. *Next frame.*

**29**

Here is another one: let us work through it together.

*Example 3.* Using Simpson's rule with 8 intervals, evaluate $\int_1^3 y\,dx$, where the values of $y$ at regular intervals of $x$ are given.

| $x$ | 1·0 | 1·25 | 1·50 | 1·75 | 2·00 | 2·25 | 2·50 | 2·75 | 3·00 |
|---|---|---|---|---|---|---|---|---|---|
| $y$ | 2·45 | 2·80 | 3·44 | 4·20 | 4·33 | 3·97 | 3·12 | 2·38 | 1·80 |

If these function values are to be used as they stand, they must satisfy the requirements for Simpson's rule, which are:

(i) the function values must be spaced at .................. intervals of $x$, and

(ii) there must be an .................. number of strips and therefore an .................. number of ordinates.

**30**

regular; even; odd

These conditions are satisfied in this case, so we can go ahead and evaluate the integral. In fact, the working will be a good deal easier for we are told the function values and there is no need to build them up as we had to do before.

In this example, $s =$ ....................

**31**

$s = 0{\cdot}25$

For $s = \frac{3-1}{8} = \frac{2}{8} = 0{\cdot}25$

Off you go, then. Set out your table and evaluate the integral defined by the values given in frame 29. When you have finished, move on to frame 32 to check your working.

**32**

6·62

| $x$ | $y$ | F + L | E | R |
|---|---|---|---|---|
| 1·0 | 2·45 | 2·45 | | |
| 1·25 | 2·80 | | 2·80 | |
| 1·50 | 3·44 | | | 3·44 |
| 1·75 | 4·20 | | 4·20 | |
| 2·00 | 4·33 | | | 4·33 |
| 2·25 | 3·97 | | 3·97 | |
| 2·50 | 3·12 | | | 3·12 |
| 2·75 | 2·38 | | 2·38 | |
| 3·00 | 1·80 | 1·80 | | |
| | F + L → | 4·25 | 13·35 | 10·89 |
| | 4E → | 53·40 | 4 | 2 |
| | 2R → | 21·78 | 53·40 | 21·78 |
| | (F + L) + 4E + 2R → | 79·43 | | |

$$I = \frac{s}{3}\left[(F + L) + 4E + 2R\right] = \frac{0{\cdot}25}{3}\left[79{\cdot}43\right]$$

$$= \frac{1}{12}\left[79{\cdot}43\right] = 6{\cdot}62$$

$$\therefore \int_1^3 y\,dx \simeq 6{\cdot}62$$

**33**

Here is one further example.

*Example 4.* A pin moves along a straight guide so that its velocity $v$ (cm/s) when it is a distance $x$ (cm) from the beginning of the guide at time $t$ (s), is as given in the table below.

| $t$ (s) | 0 | 0·5 | 1·0 | 1·5 | 2·0 | 2·5 | 3·0 | 3·5 | 4·0 |
|---|---|---|---|---|---|---|---|---|---|
| $v$ (cm/s) | 0 | 4·00 | 7·94 | 11·68 | 14·97 | 17·39 | 18·25 | 16·08 | 0 |

Apply Simpson's rule, using 8 intervals, to find the approximate total distance travelled by the pin between $t = 0$ and $t = 4$.

We must first interpret the problem, thus:

$$v = \frac{dx}{dt} \quad \therefore \; x = \int_0^4 v\,dt$$

and since we are given values of the function $v$ at regular intervals of $t$, and there is an even number of intervals, then we are all set to apply Simpson's rule.

Complete the problem, then, entirely on your own.

*When you have finished it, check with frame 34.*

**34**

46·5 cm

| $t$ | $v$ | F + L | E | R |
|---|---|---|---|---|
| 0 | 0·00 | 0·00 | | |
| 0·5 | 4·00 | | 4·00 | |
| 1·0 | 7·94 | | | 7·94 |
| 1·5 | 11·68 | | 11·68 | |
| 2·0 | 14·97 | | | 14·97 |
| 2·5 | 17·39 | | 17·39 | |
| 3·0 | 18·25 | | | 18·25 |
| 3·5 | 16·08 | | 16·08 | |
| 4·0 | 0·00 | 0·00 | | |
| | F + L → | 0·00 | 49·15 | 41·16 |
| | 4E → | 196·60 | 4 | 2 |
| | 2R → | 82·32 | 196·60 | 82·32 |
| | (F + L) + 4E + 2R → | 278·92 | | |

$$x = \frac{s}{3}\,[(F + L) + 4E + 2R] \quad \text{and} \quad s = 0{\cdot}5$$

$$\therefore \; x = \frac{1}{6}\,[278{\cdot}92] = 46{\cdot}49 \quad \therefore \; \underline{\text{Total distance} \simeq 46{\cdot}5 \text{ cm}}$$

**35**

**Proof of Simpson's rule**

So far, we have been using Simpson's rule, but we have not seen how it is established. You are not likely to be asked to prove it, but in case you are interested here is one proof.

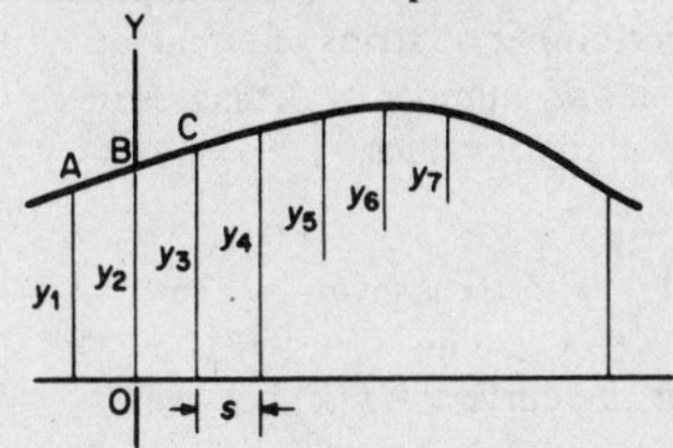


Divide into an even number of strips ($2n$) of equal width ($s$). Let the ordinates be $y_1, y_2, y_3, \ldots y_{2n+1}$. Take OX and OY as axes in the position shown.

Then $A = (-s, y_1)$; $B = (0, y_2)$; $C = (s, y_3)$

Let the curve through A, B, C be represented by $y = a + bx + cx^2$

$$y_1 = a + b(-s) + cs^2 \quad \text{(i)}$$
$$y_2 = a \quad \text{(ii)}$$
$$y_3 = a + bs + cs^2 \quad \text{(iii)}$$

(iii) – (i) $y_3 - y_1 = 2\,bs \quad \therefore\ b = \frac{1}{2s}(y_3 - y_1)$

(i) + (iii) – 2(ii) $y_1 + y_3 - 2y_2 = 2\,cs^2 \quad \therefore c = \frac{1}{2s^2}(y_1 - 2y_2 + y_3)$

Let $A_1$ = area of the first pair of strips.

$$A_1 = \int_{-s}^{s} y\,dx \doteq \int_{-s}^{s} (a + bx + cx^2)\,dx \doteq \left[ax + \frac{bx^2}{2} + \frac{cx^3}{3}\right]_{-s}^{s}$$

$$\doteq 2as + \frac{2cs^3}{3} \doteq 2sy_2 + \frac{2s^3}{3} \cdot \frac{1}{2s^2}(y_1 - 2y_2 + y_3)$$

$$\doteq \frac{s}{3}(6y_2 + y_1 - 2y_2 + y_3) \doteq \frac{s}{3}(y_1 + 4y_2 + y_3)$$

So $\quad A_1 \doteq \frac{s}{3}(y_1 + 4y_2 + y_3)$

Similarly $\quad A_2 \doteq \frac{s}{3}(y_3 + 4y_4 + y_5)$

$$A_3 \doteq \frac{s}{3}(y_5 + 4y_6 + y_7)$$

..................................

$$A_n \doteq \frac{s}{3}(y_{2n-1} + 4y_{2n} + y_{2n+1})$$

Total area $A = A_1 + A_2 + A_3 + \ldots\ldots\ldots + A_n$.

$$\therefore A \doteq \frac{s}{3}\Big[(y_1 + y_{2n+1}) + 4(y_2 + y_4 + \ldots + y_{2n}) + 2(y_3 + y_5 + \ldots + y_{2n-1})\Big]$$

$$\underline{A = \frac{s}{3}[(F + L) + 4E + 2R]}$$

*On to frame 36.*

**36** We have almost reached the end of the programme, except for the usual Test Exercise that awaits you. Before we turn to that, let us revise once again the requirements for applying Simpson's rule.

(a) The figure is divided into an *even* number of strips of equal width $s$. There will therefore be an *odd* number of ordinates or function values, including both boundary values.

(b) The value of the definite integral $\int_a^b f(x)\,dx$ is given by the numerical value of the area under the curve $y = f(x)$ between $x = a$ and $x = b$

$$I = A \simeq \frac{s}{3}\left[(F + L) + 4E + 2R\right]$$

where $s$ = width of strip (or interval),
F + L = sum of the first and last ordinates,
4E = 4 × sum of the even-numbered ordinates,
2R = 2 × sum of remaining odd-numbered ordinates.

(c) A practical hint to finish with:

*Always* set your work out in the form of a table, as we have done in the examples. It prevents your making slips in method and calculation, and enables you to check without difficulty.

Now for the Test Exercise. The problems are similar to those we have been considering in the programme, so you will find them quite straightforward.

*On then to frame 37.*

**37**

**Test Exercise – XXI**

Work through all the questions in the exercise. Set the solutions out neatly. Take your time: it is very easy to make numerical slips with work of this kind.

1. Express $\sin x$ as a power series and hence evaluate

$$\int_0^1 \frac{\sin x}{x}\,dx \text{ to 3 places of decimals.}$$

2. Evaluate $\int_{0\cdot 1}^{0\cdot 2} x^{-1}\, e^{2x}\, dx$ correct to 3 decimal places.

3. The values of a function $y = f(x)$ at stated values of $x$ are given below.

| $x$ | 2·0 | 2·5 | 3·0 | 3·5 | 4·0 | 4·5 | 5·0 | 5·5 | 6·0 |
|---|---|---|---|---|---|---|---|---|---|
| $y$ | 3·50 | 6·20 | 7·22 | 6·80 | 5·74 | 5·03 | 6·21 | 8·72 | 11·10 |

Using Simpson's rule, with 8 intervals, find an approximate value of $\int_2^6 y\,dx$.

4. Evaluate $\int_0^{\pi/2} \sqrt{\cos\theta}\, d\theta$, using 6 intervals.

5. Find an approximate value of $\int_0^{\pi/2} \sqrt{(1 - 0\cdot 5\sin^2\theta)}\,d\theta$ using Simpson's rule with 6 intervals.

*Now you are ready for the next programme.*

**Further Problems – XXI**

1. Evaluate $\int_0^{\frac{1}{2}} \sqrt{(1-x^2)}dx$ (i) by direct integration,
   (ii) by expanding as a power series,
   (iii) by Simpson's rule (8 intervals).

2. State the series for $\ln(1+x)$ and for $\ln(1-x)$ and hence obtain a series for $\ln\left\{\frac{1+x}{1-x}\right\}$.

   Evaluate $\int_0^{0{\cdot}3} \ln\left\{\frac{1+x}{1-x}\right\}dx$, correct to 3 decimal places.

3. In each of the following cases, apply Simpson's rule (6 intervals) to obtain an approximate value of the integral.

   (a) $\int_0^{\pi/2} \frac{dx}{1+3\cos x}$ (b) $\int_0^{\pi} (5-4\cos\theta)^{\frac{1}{2}}\,d\theta$

   (c) $\int_0^{\pi/2} \frac{d\theta}{\sqrt{(1-\frac{1}{2}\sin^2\theta)}}$

4. The coordinates of a point on a curve are given below.

| $x$ | 0 | 1 | 2 | 3 | 4 | 5 | 6 | 7 | 8 |
|---|---|---|---|---|---|---|---|---|---|
| $y$ | 4 | 5·9 | 7·0 | 6·4 | 4·8 | 3·4 | 2·5 | 1·7 | 1 |

   The plane figure bounded by the curve, the $x$-axis and the ordinates at $x = 0$ and $x = 8$, rotates through a complete revolution about the $x$-axis. Use Simpson's rule (8 intervals) to obtain an approximate value of the volume generated.

5. The perimeter of an ellipse with parametric equations $x = 3\cos\theta$, $y = 2\sin\theta$, is $2\sqrt{2}\int_0^{\pi/2} (13-5\cos 2\theta)^{\frac{1}{2}}\,d\theta$. Evaluate this integral using Simpson's rule with 6 intervals.

6. Calculate the area bounded by the curve $y = e^{-x^2}$, the $x$-axis, and the ordinates at $x = 0$ and $x = 1$. Use Simpson's rule with 6 intervals.

7. The voltage of a supply at regular intervals of 0·01 s, over a half-cycle, is found to be: 0, 19·5, 35, 45, 40·5, 25, 20·5, 29, 27, 12·5, 0. By Simpson's rule (10 intervals) find the r.m.s. value of the voltage over the half-cycle.

8. Show that the length of arc of the curve $x = 3\theta - 4 \sin \theta$, $y = 3 - 4 \cos \theta$, between $\theta = 0$ and $\theta = 2\pi$, is given by the integral $\int_0^{2\pi} \sqrt{(25 - 24 \cos \theta)}\,d\theta$. Evaluate the integral, using Simpson's rule with 8 intervals.

9. Obtain the first four terms of the expansion of $(1 + x^3)^{\frac{1}{2}}$ and use them to determine the approximate value of $\int_0^{\frac{1}{2}} \sqrt{(1 + x^3)}\,dx$, correct to three decimal places.

10. Establish the integral in its simplest form representing the length of the curve $y = \frac{1}{2} \sin \theta$ between $\theta = 0$ and $\theta = \pi/2$. Apply Simpson's rule, using 6 intervals, to find an approximate value of this integral.

11. Determine the first four non-zero terms of the series for $\tan^{-1}x$ and hence evaluate $\int_0^{\frac{1}{2}} \sqrt{x}.\tan^{-1}x\,dx$ correct to 3 decimal places.

12. Evaluate, correct to three decimal places,

    (i) $\int_0^1 \sqrt{x}.\cos x\,dx$, (ii) $\int_0^1 \sqrt{x}.\sin x\,dx$.

13. Evaluate $\int_0^{\pi/2} \sqrt{(2{\cdot}5 - 1{\cdot}5 \cos 2\theta)}\,d\theta$ by Simpson's rule, using 6 intervals.

14. Determine the approximate value of $\int_0^1 (4 + x^4)^{\frac{1}{2}}\,dx$

    (i) by first expanding the expression in powers of $x$,
    (ii) by applying Simpson's rule, using 4 intervals.

    In each case, give the result to 2 places of decimals.

# Programme 22

## POLAR CO-ORDINATES SYSTEM

**1**

## Introduction to polar co-ordinates

We already know that there are two main ways in which the position of a point in a plane can be represented.

(i) by Cartesian co-ordinates, i.e. $(x, y)$
(ii) by polar co-ordinates, i.e. $(r, \theta)$.

The relationship between the two systems can be seen from a diagram.

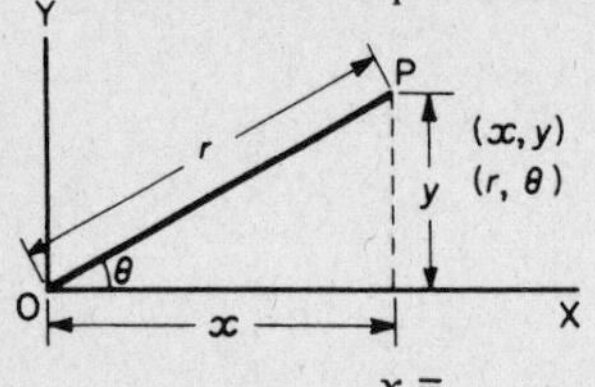


For instance, $x$ and $y$ can be expressed in terms of $r$ and $\theta$.

$x =$ .................... ; $y =$ ....................

**2**

$$x = r\cos\theta;\; y = r\sin\theta$$

Or, working in the reverse direction, the co-ordinates $r$ and $\theta$ can be found if we know the values of $x$ and $y$.

$r =$ .................... ; $\theta =$ ....................

**3**

$$r = \sqrt{(x^2 + y^2)};\; \theta = \tan^{-1}\left(\frac{y}{x}\right)$$

This is just by way of revision. We first met polar co-ordinates in an earlier programme on complex numbers. In this programme, we are going to direct a little more attention to the *polar co-ordinates system* and its applications.

First of all, an easy example or two to warm up.

*Example 1.* Express in polar co-ordinates the position $(-5, 2)$.
Important hint: *always* draw a diagram; it will enable you to see which quadrant you are dealing with and prevent your making an initial slip.

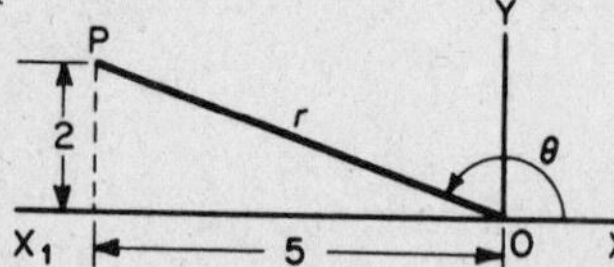


Remember that $\theta$ is measured from the positive OX direction.

In this case, the polar co-ordinates of P are ....................

## 4

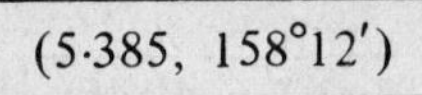


For:

(i) $r^2 = 2^2 + 5^2 = 4 + 25 = 29$

$\therefore r = \sqrt{29} = 5{\cdot}385$ .

(ii) $\tan E = \frac{2}{5} = 0{\cdot}4 \quad \therefore E = 21°48'$

$\therefore \theta = 158°12'$

Position of P is $(5{\cdot}385, 158°12')$

A sketch diagram will help you to check that $\theta$ is in the correct quadrant.

*Example 2.* Express $(4, -3)$ in polar co-ordinates. Draw a sketch and you cannot go wrong!

*When you are ready, move to frame 5.*

## 5

(5, 323°8′)

Here it is.

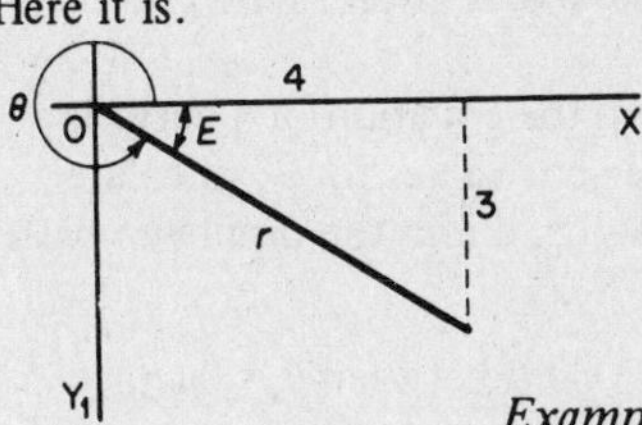


(i) $r^2 = 3^2 + 4^2 = 25 \quad \therefore r = 5$

(ii) $\tan E = \frac{3}{4} = 0{\cdot}75 \quad \therefore E = 36°52'$

$\therefore \theta = 323°8'$

$(4, -3) = (5, 323°8')$

*Example 3.* Express in polar co-ordinates $(-2, -3)$.

*Finish it off and then move to frame 6.*

## 6

3·606, 236°19′

Check your result.

(i) $r^2 = 2^2 + 3^2 = 4 + 9 = 13$

$r = \sqrt{13} = 3{\cdot}606$

(ii) $\tan E = \frac{3}{2} = 1{\cdot}5 \quad \therefore E = 56°19'$

$\therefore \theta = 236°19'$

$(-2, -3) = (3{\cdot}606, 236°19')$

Of course, conversion in the opposite direction is just a matter of evaluating $x = r\cos\theta$ and $y = r\sin\theta$. Here is an example.

*Example 4.* Express $(5, 124°)$ in Cartesian co-ordinates.

*Do that, and then move on to frame 7.*

**7**

$(-2{\cdot}796, 4{\cdot}145)$

Working:

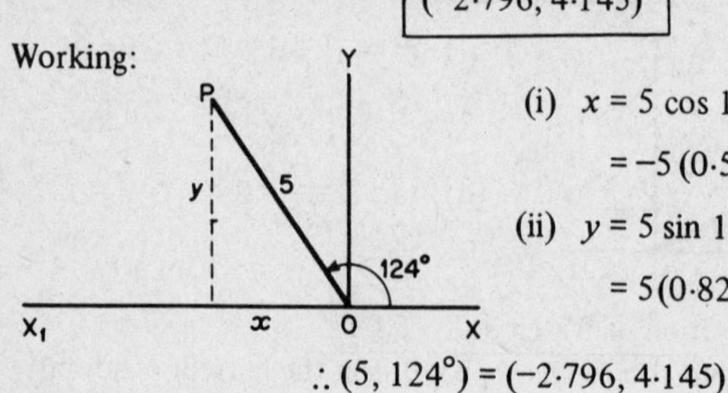


(i) $x = 5\cos 124° = -5\cos 56°$

$= -5\,(0{\cdot}5592) = -2{\cdot}7960$

(ii) $y = 5\sin 124° = 5\sin 56°$

$= 5(0{\cdot}8290) = 4{\cdot}1450$

$\therefore (5, 124°) = \underline{(-2{\cdot}796, 4{\cdot}145)}$

That was all very easy.

*Now, on to the next frame.*

**8**

## Polar curves

In Cartesian co-ordinates, the equation of a curve is given as the general relationship between $x$ and $y$, i.e. $y = f(x)$.

Similarly, in the polar co-ordinate system, the equation of a curve is given in the form $r = f(\theta)$. We can then take spot values for $\theta$, calculate the corresponding values of $r$, plot $r$ against $\theta$, and join the points up with a smooth curve to obtain the graph of $r = f(\theta)$.

*Example 1.* To plot the polar graph of $r = 2\sin\theta$ between $\theta = 0$ and $\theta = 2\pi$.

We take values of $\theta$ at convenient intervals and build up a table of values giving the corresponding values of $r$.

| $\theta°$ | 0 | 30 | 60 | 90 | 120 | 150 | 180 |
|---|---|---|---|---|---|---|---|
| $\sin\theta$ | 0 | 0·5 | 0·866 | 1 | 0·866 | 0·5 | 0 |
| $r = 2\sin\theta$ | 0 | 1·0 | 1·732 | 2 | 1·732 | 1·0 | 0 |

| $\theta°$ | 210 | 240 | 270 | 300 | 330 | 360 |
|---|---|---|---|---|---|---|
| $\sin\theta$ | | | | | | |
| $r = 2\sin\theta$ | | | | | | |

Complete the table, being careful of signs.

*When you have finished, turn on to frame 9.*

**9**

Here is the complete table.

| $\theta°$ | 0 | 30 | 60 | 90 | 120 | 150 | 180 |
|---|---|---|---|---|---|---|---|
| $\sin\theta$ | 0 | 0·5 | 0·866 | 1 | 0·866 | 0·5 | 0 |
| $r = 2\sin\theta$ | 0 | 1·0 | 1·732 | 2 | 1·732 | 1·0 | 0 |

| $\theta°$ | 210 | 240 | 270 | 300 | 330 | 360 |
|---|---|---|---|---|---|---|
| $\sin\theta$ | −0·5 | −0·866 | −1 | −0·866 | −0·5 | 0 |
| $r = 2\sin\theta$ | −1·0 | −1·732 | −2 | −1·732 | −1·0 | 0 |

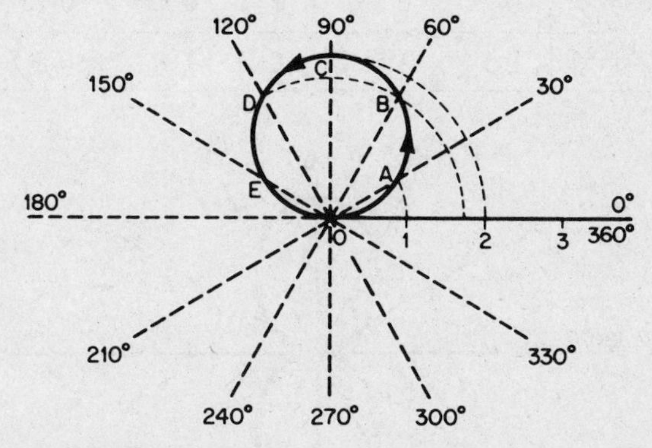


(i) We choose a linear scale for $r$ and indicate it along the initial line.

(ii) The value of $r$ is then laid off along each direction in turn, points plotted, and finally joined up with a smooth curve. The resulting graph is as shown above.

*Note* that when we are dealing with the 210° direction, the value of $r$ is negative (−1) and this distance is therefore laid off in the reverse direction which once again brings us to the point A. So for values of $\theta$ between $\theta = 180°$ and $\theta = 360°$, $r$ is negative and the first circle is retraced exactly. The graph, therefore, looks like one circle, but consists, in fact, of two circles, one on top of the other.

Now, in the same way, you can plot the graph of $r = 2\sin^2\theta$.

Compile a table of values at 30° intervals between $\theta = 0°$ and $\theta = 360°$ and proceed as we did above.

Take a little time over it.

*When you have finished, move on to frame 10.*

**10**

Here is the result in detail.

| $\theta$ | 0 | 30 | 60 | 90 | 120 | 150 | 180 |
|---|---|---|---|---|---|---|---|
| $\sin\theta$ | 0 | 0·5 | 0·866 | 1 | 0·866 | 0·5 | 0 |
| $\sin^2\theta$ | 0 | 0·25 | 0·75 | 1 | 0·75 | 0·25 | 0 |
| $r = 2\sin^2\theta$ | 0 | 0·5 | 1·5 | 2 | 1·5 | 0·5 | 0 |

| $\theta$ | 210 | 240 | 270 | 300 | 330 | 360 |
|---|---|---|---|---|---|---|
| $\sin\theta$ | −0·5 | −0·866 | −1 | −0·866 | −0·5 | 0 |
| $\sin^2\theta$ | 0·25 | 0·75 | 1 | 0·75 | 0·25 | 0 |
| $r = 2\sin^2\theta$ | 0·5 | 1·5 | 2 | 1·5 | 0·5 | 0 |

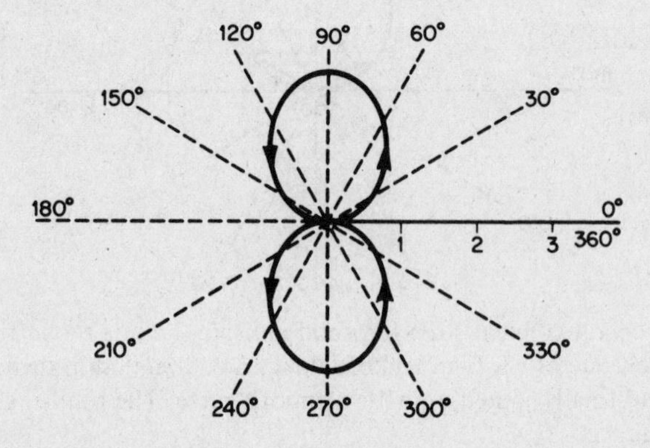


This time, $r$ is always positive and so there are, in fact, two distinct loops.

*Now on to the next frame.*

**11**

**Standard polar curves**

Polar curves can always be plotted from sample points as we have done above. However, it is often useful to know something of the shape of the curve without the rather tedious task of plotting points in detail.

In the next few frames, we will look at some of the more common polar curves.

*So on to frame 12.*

*Typical polar curves* **12**

1. $r = a \sin \theta$

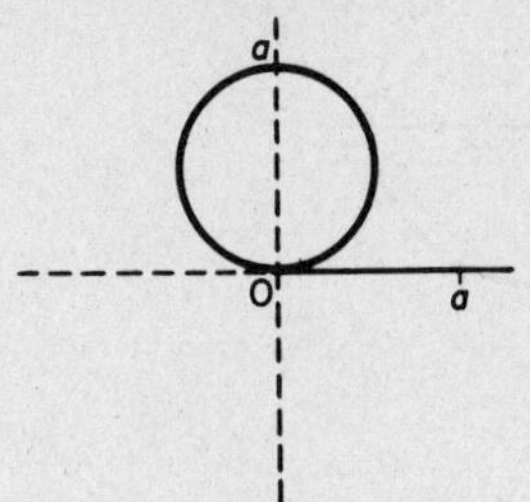


2. $r = a \sin^2\theta$

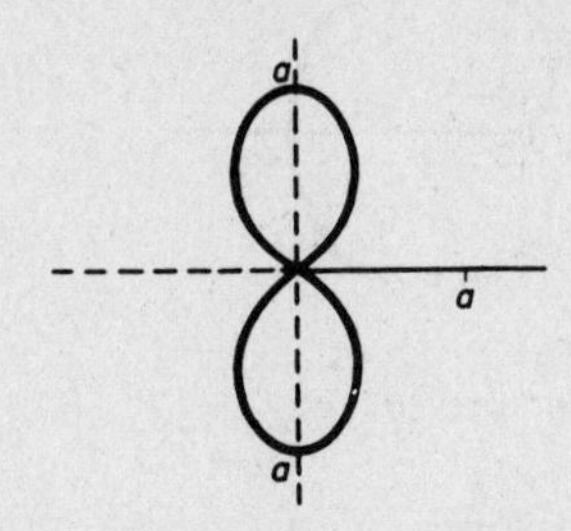


3. $r = a \cos \theta$

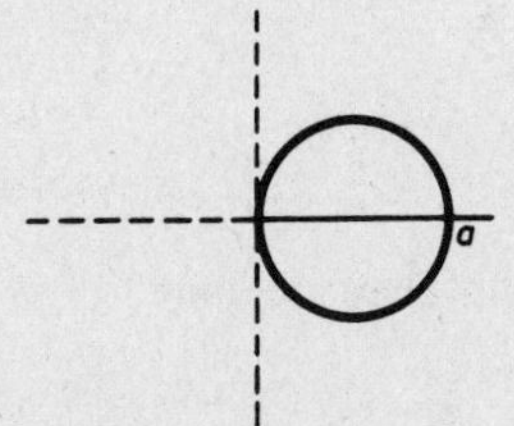


4. $r = a \cos^2\theta$

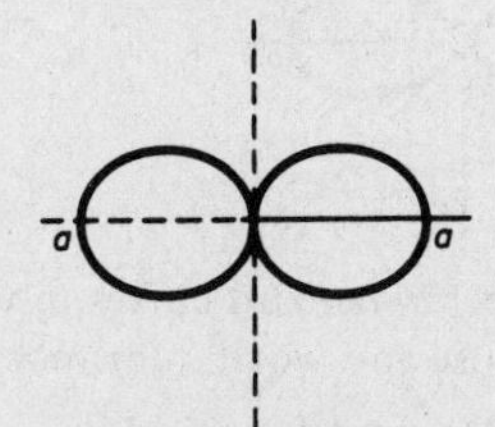


5. $r = a \sin 2\theta$

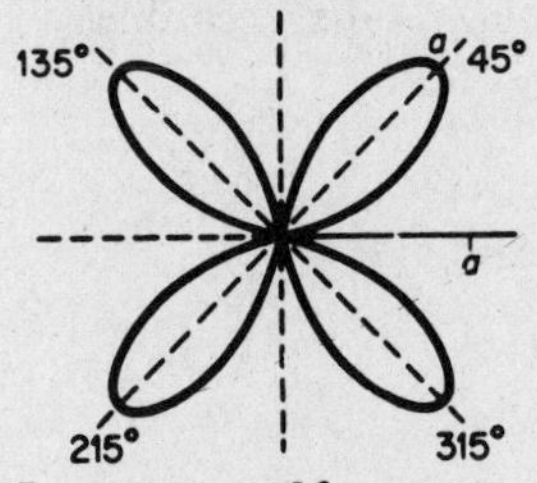


6. $r = a \sin 3\theta$

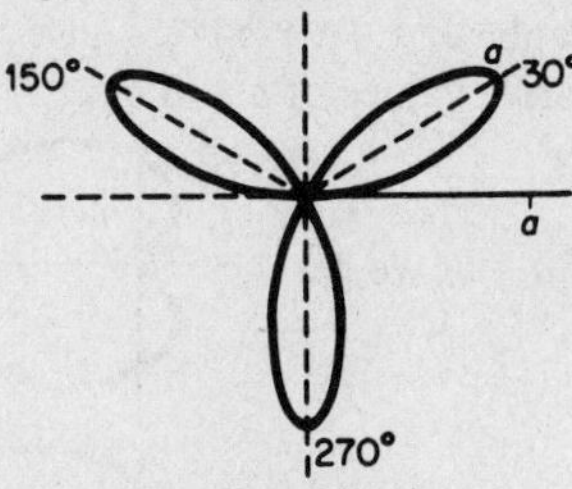


7. $r = a \cos 2\theta$

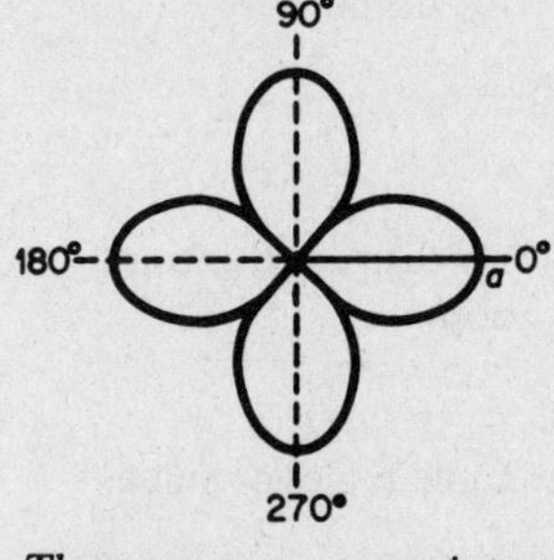


8. $r = a \cos 3\theta$

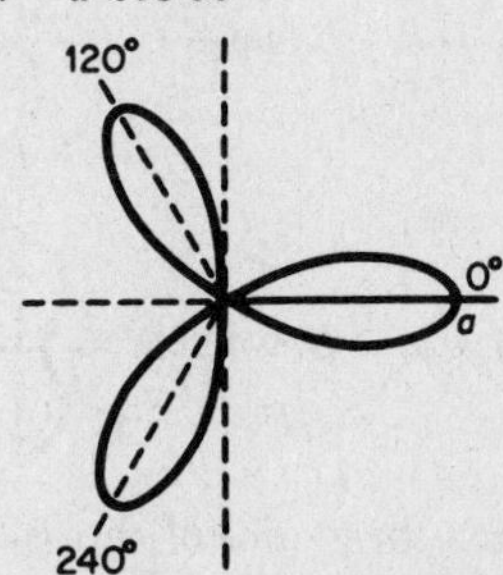


*There are some more interesting polar curves worth seeing, so turn on to frame 13.*

**13**

9. $r = a(1 + \cos\theta)$

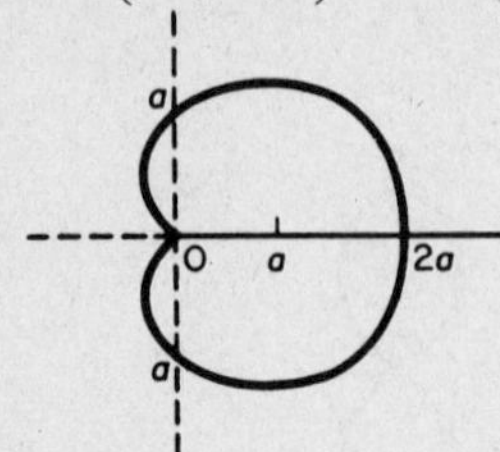


10. $r = a(1 + 2\cos\theta)$

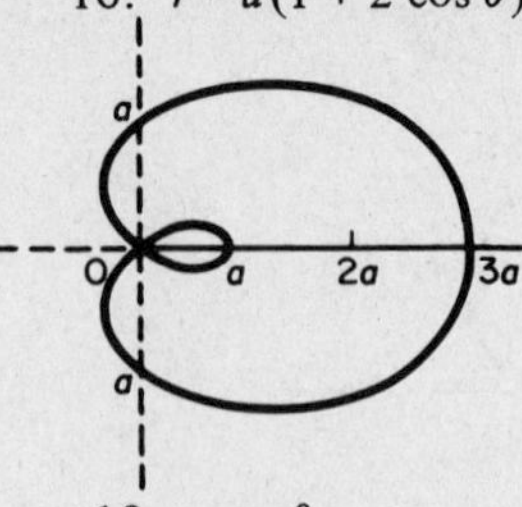


11. $r^2 = a^2 \cos 2\theta$

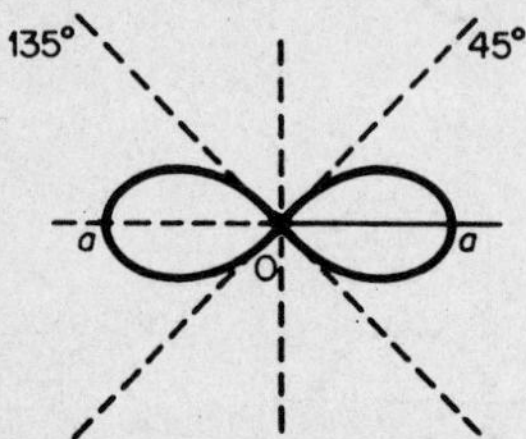


12. $r = a\theta$

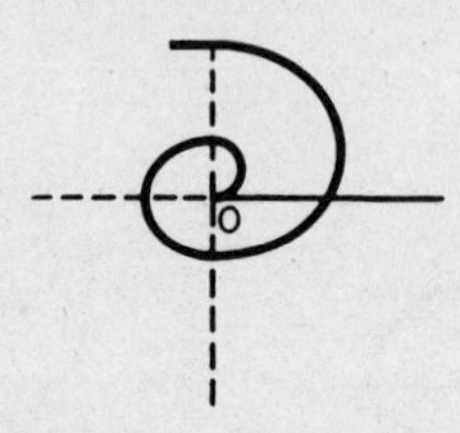


Sketch these 12 standard curves in your record book. They are quite common in use and worth remembering.

*Then on to the next frame.*

**14**

The graphs of $r = a + b\cos\theta$ give three interesting results, according to the relative values of $a$ and $b$.

(i) If $a = b$, we get

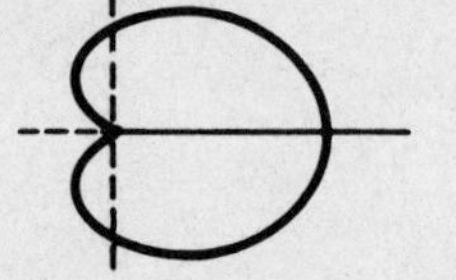

(cardioid)

(ii) If $a < b$, we get

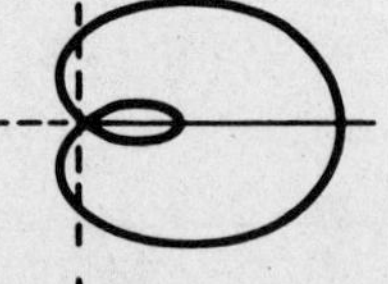

(re-entrant loop)

(iii) If $a > b$, we get

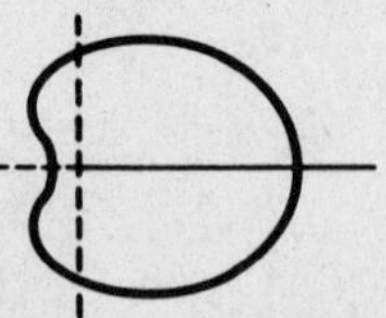

(no cusp or re-entrant loop)

So sketch the graphs of the following. Do *not* compile tables of values.

(i) $r = 2 + 2\cos\theta$ (iii) $r = 1 + 2\cos\theta$

(ii) $r = 5 + 3\cos\theta$ (iv) $r = 2 + \cos\theta$

**15**

Here they are. See how closely you agree.

(i) $r = 2 + 2\cos\theta$ $(a = b)$

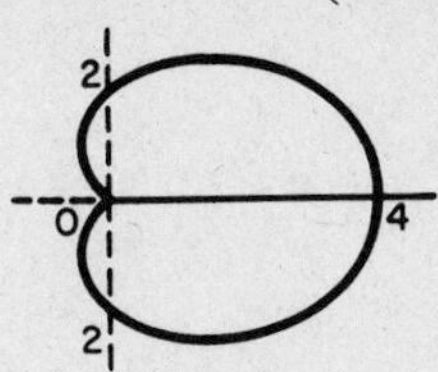


(ii) $r = 5 + 3\cos\theta$ $(a > b)$

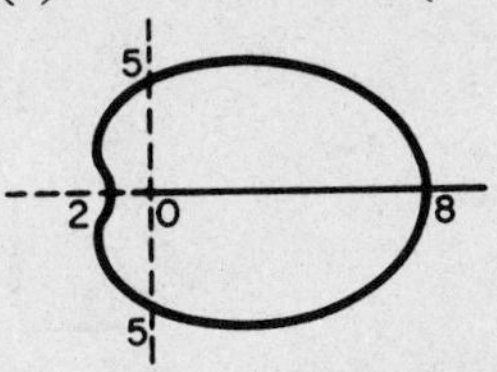


(iii) $r = 1 + 2\cos\theta$ $(a < b)$

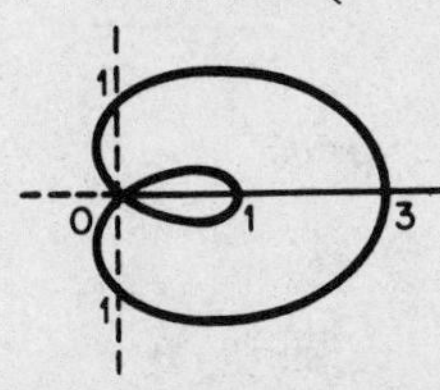


(iv) $r = 2 + \cos\theta$ $(a > b)$

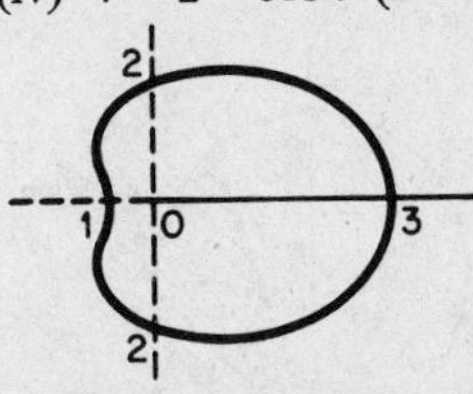


If you have slipped up with any of them, it would be worth while to plot a few points to confirm how the curve goes.

*On to frame 16.*

**16**

*To find the area of the plane figure bounded by the polar curve* $r = f(\theta)$ *and the radius vectors at* $\theta = \theta_1$ *and* $\theta = \theta_2$.

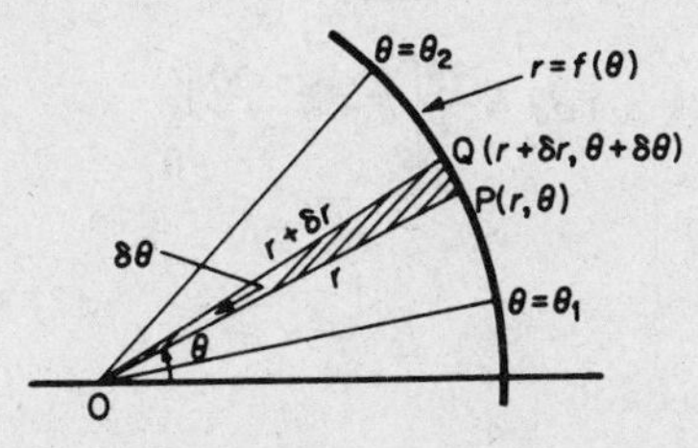


$$\text{Area of sector OPQ} = \delta A \simeq \tfrac{1}{2}r(r + \delta r)\sin\delta\theta$$

$$\therefore \frac{\delta A}{\delta\theta} \simeq \tfrac{1}{2}r(r + \delta r)\frac{\sin\delta\theta}{\delta\theta}$$

If $\delta\theta \to 0$, $\dfrac{\delta A}{\delta\theta} \to \dfrac{dA}{d\theta}$, $\delta r \to 0$, $\dfrac{\sin\delta\theta}{\delta\theta} \to$ ............................

*Next frame.*

**17**

$$\boxed{\frac{\sin \delta\theta}{\delta\theta} \to 1}$$

$$\therefore \frac{dA}{d\theta} = \tfrac{1}{2}r(r + 0)\,1 = \tfrac{1}{2}r^2$$

$$\therefore A = \int_{\theta_1}^{\theta_2} \tfrac{1}{2}r^2\,d\theta$$

*Example 1.* To find the area enclosed by the curve $r = 5 \sin \theta$ and the radius vectors at $\theta = 0$ and $\theta = \pi/3$.

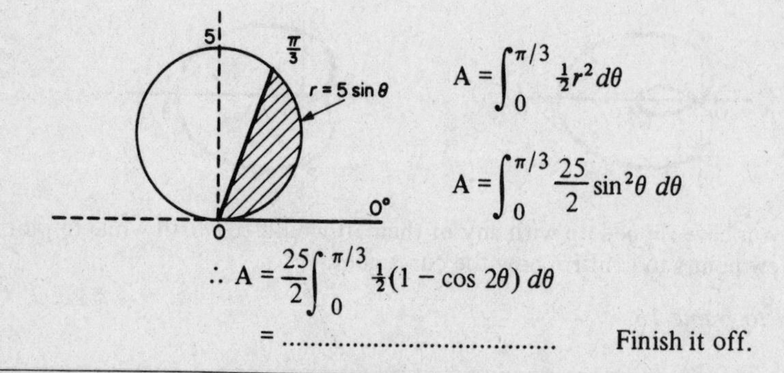


$$A = \int_0^{\pi/3} \tfrac{1}{2}r^2\,d\theta$$

$$A = \int_0^{\pi/3} \frac{25}{2}\sin^2\theta\,d\theta$$

$$\therefore A = \frac{25}{2}\int_0^{\pi/3} \tfrac{1}{2}(1 - \cos 2\theta)\,d\theta$$

$= \ldots\ldots\ldots\ldots\ldots\ldots\ldots\ldots$ Finish it off.

**18**

$$\boxed{A = \frac{25}{4}\left[\frac{\pi}{3} - \frac{\sqrt{3}}{4}\right] = 3{\cdot}84}$$

For:
$$A = \frac{25}{4}\int_0^{\pi/3} (1 - \cos 2\theta)\,d\theta = \frac{25}{4}\left[\theta - \frac{\sin 2\theta}{2}\right]_0^{\pi/3}$$

$$= \frac{25}{4}\left[\frac{\pi}{3} - \frac{\sin 2\pi/3}{2}\right]$$

$$= \frac{25}{4}\left[\frac{\pi}{3} - \frac{\sqrt{3}}{4}\right] = 3{\cdot}8388$$

$\underline{A = 3{\cdot}84}$ to 2 decimal places

Now this one:

*Example 2.* Find the area enclosed by the curve $r = 1 + \cos \theta$ and the radius vectors at $\theta = 0$ and $\theta = \pi/2$.

First of all, what does the curve look like?

19

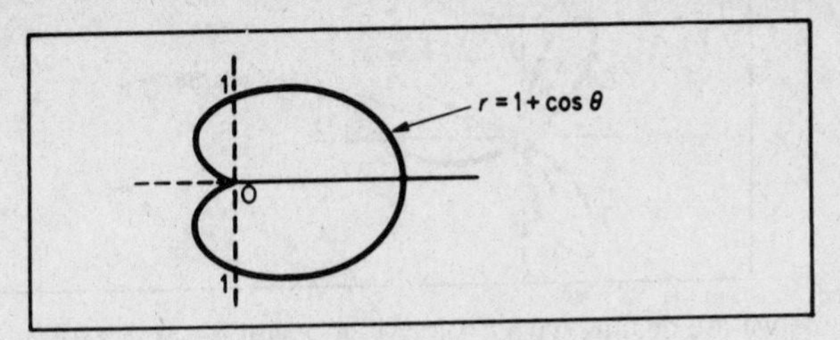


Right. So now calculate the value of A between $\theta = 0$ and $\theta = \pi/2$.

*When you have finished, move on to frame 20.*

20

$$A = \frac{3\pi}{8} + 1 = 2{\cdot}178$$

For:
$$A = \tfrac{1}{2}\int_0^{\pi/2} r^2 d\theta = \tfrac{1}{2}\int_0^{\pi/2} (1 + 2\cos\theta + \cos^2\theta)\, d\theta$$

$$= \tfrac{1}{2}\left[\theta + 2\sin\theta + \frac{\theta}{2} + \frac{\sin 2\theta}{4}\right]_0^{\pi/2}$$

$$= \tfrac{1}{2}\left\{\left(\frac{3\pi}{4} + 2 + 0\right) - \left(0\right)\right\}$$

$$\therefore A = \frac{3\pi}{8} + 1 = 2{\cdot}178$$

So the area of a polar sector is easy enough to obtain. It is simply

$$A = \int_{\theta_1}^{\theta_2} \tfrac{1}{2} r^2\, d\theta$$

Make a note of this general result in your record book, if you have not already done so.

*Next frame.*

21

*Example 3.* Find the total area enclosed by the curve $r = 2\cos 3\theta$.

Notice that no limits are given, so we had better sketch the curve to see what is implied.

This was in fact one of the standard polar curves that we listed earlier in this programme. Do you remember how it goes? If not, refer to your notes: it should be there.

*Then on to frame 22.*

**22**

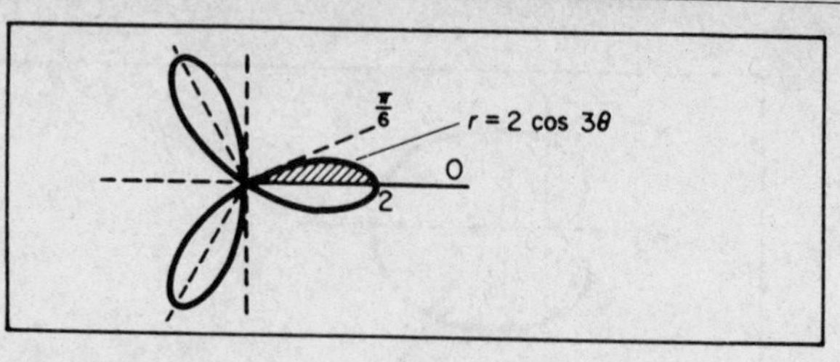


Since we are dealing with $r = 2\cos 3\theta$, $r$ will become zero when $\cos 3\theta = 0$, i.e. when $3\theta = \pi/2$, i.e. when $\theta = \pi/6$.

We see that the figure consists of 3 equal loops, so that the total area, A, is given by

$$A = 3\text{ (area of one loop)}$$
$$= 6\text{ (area between } \theta = 0 \text{ and } \theta = \pi/6.)$$
$$A = 6\int_0^{\pi/6} \tfrac{1}{2}r^2\,d\theta = 3\int_0^{\pi/6} 4\cos^2 3\theta\,d\theta$$
$$= \ldots\ldots\ldots\ldots$$

**23**

$\pi$ units$^2$

since
$$A = 12\int_0^{\pi/6} \tfrac{1}{2}(1 + \cos 6\theta)\,d\theta$$
$$= 6\left[\theta + \frac{\sin 6\theta}{6}\right]_0^{\pi/6} = \underline{\pi \text{ units}^2}$$

Now here is one for you to do on your own.

*Example 4.* Find the area enclosed by one loop of the curve $r = a\sin 2\theta$.

First sketch the graph.

**24**

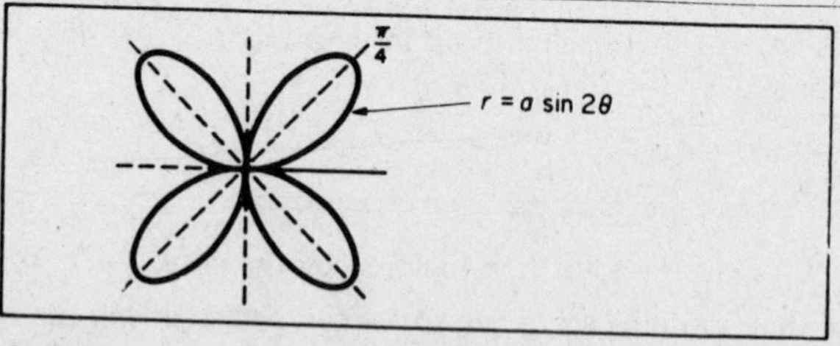


Arguing as before, $r = 0$ when $a\sin 2\theta = 0$, i.e. $\sin 2\theta = 0$, i.e. $2\theta = 0$, so that $2\theta = 0, \pi, 2\pi$, etc.

$$\therefore\ \theta = 0, \pi/2, \pi, \text{ etc.}$$

So the integral denoting the area of the loop in the first quadrant will be

$$A = \ldots\ldots\ldots\ldots$$

**25**

$$A = \frac{1}{2}\int_0^{\pi/2} r^2\, d\theta$$

Correct. Now go ahead and calculate the area.

---

**26**

$$A = \pi a^2/8 \text{ units}^2$$

Here is the working: check yours.

$$A = \frac{1}{2}\int_0^{\pi/2} r^2\, d\theta = \frac{a^2}{2}\int_0^{\pi/2} \sin^2 2\theta\, d\theta$$

$$= \frac{a^2}{4}\int_0^{\pi/2} (1 - \cos 4\theta)\, d\theta$$

$$= \frac{a^2}{4}\left[\theta - \frac{\sin 4\theta}{4}\right]_0^{\pi/2} = \underline{\frac{\pi a^2}{8} \text{ units}^2}.$$

*Now on to frame 27.*

---

**27**

*To find the volume generated when the plane figure bounded by $r = f(\theta)$ and the radius vectors at $\theta = \theta_1$ and $\theta = \theta_2$ rotates about the initial line.*

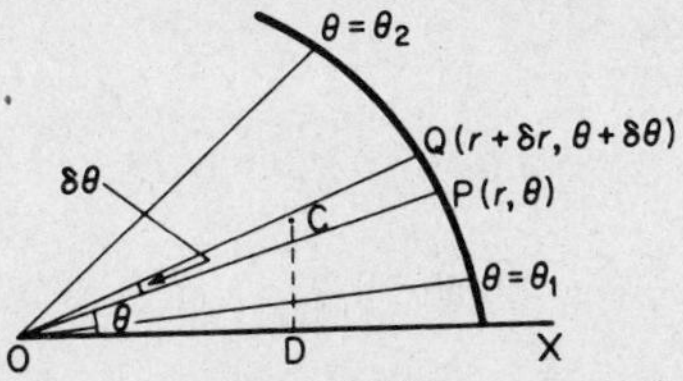

If we regard the elementary sector OPQ as approximately equal to the $\Delta$ OPQ, then the centroid C is distance $\frac{2r}{3}$ from 0.

We have: Area OPQ $\simeq \frac{1}{2}r(r + \delta r) \sin \delta\theta$

Volume generated when OPQ rotates about OX = $\delta V$

$\therefore$ $\delta V$ = area OPQ $\times$ distance travelled by its centroid (Pappus)

$$= \frac{1}{2}r(r + \delta r) \sin \delta\theta \,.\, 2\pi\, \text{CD}$$

$$= \frac{1}{2}r(r + \delta r) \sin \delta\theta \,.\, 2\pi . \frac{2}{3}r \sin \theta$$

$$= \frac{2}{3}\pi r^2 (r + \delta r) \sin \delta\theta \,.\, \sin \theta$$

$$\therefore \frac{\delta V}{\delta\theta} = \frac{2}{3}\pi r^2 (r + \delta r)\frac{\sin \delta\theta}{\delta\theta} . \sin \theta$$

Then when $\delta\theta \to 0$, $\frac{dV}{d\theta}$ = .........................

**28**

$$\frac{dV}{dx} = \frac{2}{3}\pi r^3 \sin\theta$$

and $\quad \therefore \; V = \ldots\ldots\ldots\ldots$

**29**

$$V = \int_{\theta_1}^{\theta_2} \frac{2}{3}\pi r^3 \sin\theta \, d\theta$$

Correct. This is another standard result, so add it to your notes.

*Then move to the next frame for an example.*

**30** *Example 1.* Find the volume of the solid formed when the plane figure bounded by $r = 2\sin\theta$ and the radius vectors at $\theta = 0$ and $\theta = \pi/2$, rotates about the initial line.

Well now, $\quad V = \int_0^{\pi/2} \frac{2}{3}\pi r^3 \sin\theta \, d\theta$

$$= \int_0^{\pi/2} \frac{2}{3}.\pi.(2\sin\theta)^3.\sin\theta \, d\theta = \int_0^{\pi/2} \frac{16}{3}\pi \sin^4\theta \, d\theta$$

Since the limits are between 0 and $\pi/2$, we can use Wallis's formula for this. (Remember?)

So $\quad V = \ldots\ldots\ldots\ldots$

**31**

$$V = \pi^2 \text{ units}^3$$

For $\quad V = \frac{16\pi}{3}\int_0^{\pi/2} \sin^4\theta \, d\theta$

$$= \frac{16\pi}{3}.\frac{3.1}{4.2}.\frac{\pi}{2} = \underline{\pi^2 \text{ units}^3}$$

*Example 2.* Find the volume of the solid formed when the plane figure bounded by $r = 2a\cos\theta$ and the radius vectors at $\theta = 0$ and $\theta = \pi/2$, rotates about the initial line.

Do that one entirely on your own.

*When you have finished it, move on to the next frame.*

**32**

$$\boxed{V = \frac{4\pi a^3}{3} \text{ units}^3}$$

For
$$V = \int_0^{\pi/2} \frac{2}{3}.\pi.r^3 \sin\theta \, d\theta \quad \text{and} \quad r = 2a\cos\theta$$

$$= \int_0^{\pi/2} \frac{2}{3}.\pi.8a^3\cos^3\theta.\sin\theta \, d\theta$$

$$= -\frac{16\pi a^3}{3}\int_0^{\pi/2} \cos^3\theta\,(-\sin\theta)\,d\theta$$

$$= -\frac{16\pi a^3}{3}\left[\frac{\cos^4\theta}{4}\right]_0^{\pi/2} = -\frac{16\pi a^3}{3}\left[-\frac{1}{4}\right]$$

$$V = \frac{4\pi a^3}{3} \text{ units}^3$$

So far, then, we have had

(i) $A = \int_{\theta_1}^{\theta_2} \frac{1}{2}r^2\,d\theta$

(ii) $V = \int_{\theta_1}^{\theta_2} \frac{2}{3}\pi r^3 \sin\theta\,d\theta$

Check that you have noted these results in your book.

**33**

*To find the length of arc of the polar curve $r = f(\theta)$, between $\theta = \theta_1$ and $\theta = \theta_2$.*

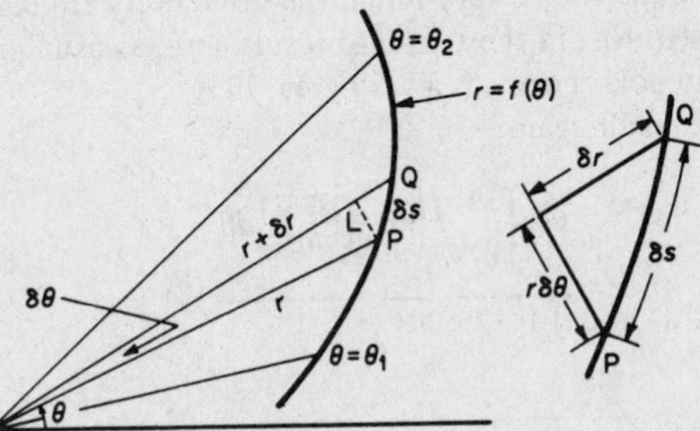


With the usual figure $\quad \delta s^2 \simeq r^2.\delta\theta^2 + \delta r^2 \quad \therefore \frac{\delta s^2}{\delta\theta^2} \simeq r^2 + \frac{\delta r^2}{\delta\theta^2}$

If $\delta\theta \to 0$, $\left(\frac{ds}{d\theta}\right)^2 = r^2 + \left(\frac{dr}{d\theta}\right)^2 \quad \therefore \frac{ds}{d\theta} = \sqrt{\left\{r^2 + \left(\frac{dr}{d\theta}\right)^2\right\}} \therefore s = \text{....................}$

**34**

$$s = \int_{\theta_1}^{\theta_2} \sqrt{\left\{r^2 + \left(\frac{dr}{d\theta}\right)^2\right\}} d\theta$$

*Example 1.* Find the length of arc of the spiral $r = a\,e^{3\theta}$ from $\theta = 0$ to $\theta = 2\pi$.

Now, $$r = a\,e^{3\theta} \quad \therefore \frac{dr}{d\theta} = 3a\,e^{3\theta}$$

$$\therefore r^2 + \left(\frac{dr}{d\theta}\right)^2 = a^2 e^{6\theta} + 9a^2 e^{6\theta} = 10a^2 e^{6\theta}$$

$$\therefore s = \int_0^{2\pi} \sqrt{\left\{r^2 + \left(\frac{dr}{d\theta}\right)^2\right\}} d\theta = \int_0^{2\pi} \sqrt{10}.a\,e^{3\theta}\, d\theta$$

$$= \ldots\ldots\ldots\ldots\ldots\ldots\ldots$$

**35**

$$s = \frac{a\sqrt{10}}{3}\left\{e^{6\pi} - 1\right\}$$

Since $$\int_0^{2\pi} \sqrt{10}.a.e^{3\theta}\, d\theta = \frac{\sqrt{10}\,a}{3}\Big[e^{3\theta}\Big]_0^{2\pi} = \frac{a\sqrt{10}}{3}\left\{e^{6\pi} - 1\right\}$$

As you can see, the method is very much the same every time. It is merely a question of substituting in the standard result, and, as usual, a knowledge of the shape of the polar curves is a very great help.

Here is our last result again.

$$s = \int_{\theta_1}^{\theta_2} \sqrt{\left\{r^2 + \left(\frac{dr}{d\theta}\right)^2\right\}} d\theta$$

Make a note of it: add it to the list.

**36**

Now here is an example for you to do.

*Example 2.* Find the length of the cardioid $r = a(1 + \cos\theta)$ between $\theta = 0$ and $\theta = \pi$.

*Finish it completely, and then check with the next frame.*

**37**

$$\boxed{s = 4a \text{ units}}$$

Here is the working:

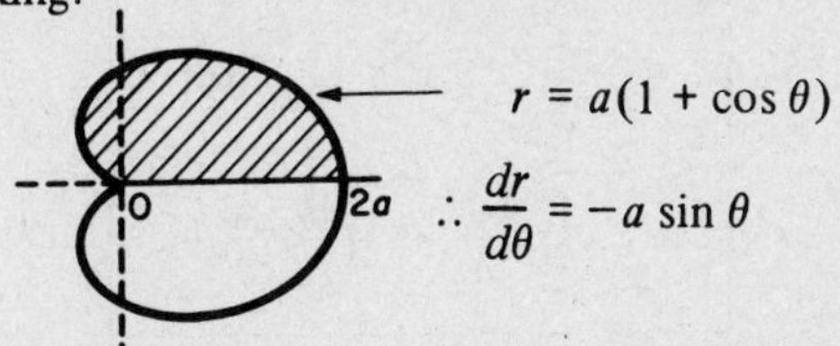


$$r = a(1 + \cos\theta)$$

$$\therefore \frac{dr}{d\theta} = -a\sin\theta$$

$$\therefore r^2 + \left(\frac{dr}{d\theta}\right)^2 = a^2\left\{1 + 2\cos\theta + \cos^2\theta + \sin^2\theta\right\}$$
$$= a^2\left\{2 + 2\cos\theta\right\} = 2a^2(1 + \cos\theta)$$

Now $\cos\theta$ can be re-written as $\left(2\cos^2\frac{\theta}{2} - 1\right)$

$$\therefore r^2 + \left(\frac{dr}{d\theta}\right)^2 = 2a^2.2\cos^2\frac{\theta}{2}$$

$$\therefore \sqrt{\left\{r^2 + \left(\frac{dr}{d\theta}\right)^2\right\}} = 2a\cos\frac{\theta}{2}$$

$$\therefore s = \int_0^\pi 2a\cos\frac{\theta}{2}\,d\theta = 2a\left[2\sin\frac{\theta}{2}\right]_0^\pi$$
$$= 4a\left[1 - 0\right] = \underline{4a \text{ units}}$$

*Next frame.*

**38**

Let us pause a moment and think back. So far we have established three useful results relating to polar curves. Without looking back in this programme, or at your notes, complete the following.

If $r = f(\theta)$, (i) A = ....................

(ii) V = ....................

(iii) $s$ = ....................

*To see how well you have got on, turn on to frame 39.*

**39**

$$A = \int_{\theta_1}^{\theta_2} \frac{1}{2} r^2 \, d\theta$$

$$V = \int_{\theta_1}^{\theta_2} \frac{2}{3} . \pi . r^3 \sin\theta \, d\theta$$

$$s = \int_{\theta_1}^{\theta_2} \sqrt{\left\{ r^2 + \left( \frac{dr}{d\theta} \right)^2 \right\}} d\theta$$

If you were uncertain of any of them, be sure to revise that particular result now. When you are ready, move on to the next section of the programme.

**40**

Finally, we come to this topic.

*To find the area of the surface generated when the arc of the curve $r = f(\theta)$ between $\theta = \theta_1$ and $\theta = \theta_2$ rotates about the initial line.*

Once again, we refer to our usual figure.

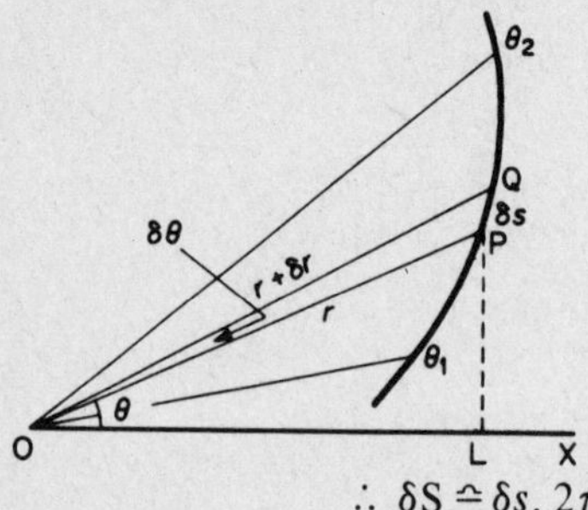


If the elementary arc PQ rotates about OX, then, by the theorem of Pappus, the surface generated, $\delta S$, is given by (length of arc) × (distance travelled by its centroid).

$$\therefore \delta S \simeq \delta s . 2\pi PL \simeq \delta s . 2\pi r \sin\theta$$

$$\therefore \frac{\delta S}{\delta\theta} \simeq 2\pi r \sin\theta \frac{\delta s}{\delta\theta}$$

From our previous work, we know that $\frac{\delta s}{\delta\theta} \simeq \sqrt{\left\{ r^2 + \left( \frac{\delta r}{\delta\theta} \right)^2 \right\}}$

so that $$\frac{\delta S}{\delta\theta} \simeq 2\pi r \sin\theta \sqrt{\left\{ r^2 + \left( \frac{\delta r}{\delta\theta} \right)^2 \right\}}$$

And now, if $\delta\theta \to 0$, $$\frac{dS}{d\theta} = 2\pi r \sin\theta \sqrt{\left\{ r^2 + \left( \frac{dr}{d\theta} \right)^2 \right\}}$$

$$\therefore S = \int_{\theta_1}^{\theta_2} 2\pi r \sin\theta \sqrt{\left\{ r^2 + \left( \frac{dr}{d\theta} \right)^2 \right\}} d\theta$$

This is also an important result, so add it to your list.

**41**

$$\boxed{S = \int_{\theta_1}^{\theta_2} 2\pi r \sin\theta \sqrt{\left\{r^2 + \left(\frac{dr}{d\theta}\right)^2\right\}}\, d\theta}$$

This looks a little more involved, but the method of attack is much the same. An example will show.

*Example 1.* Find the surface area generated when the arc of the curve $r = 5(1 + \cos\theta)$ between $\theta = 0$ and $\theta = \pi$, rotates completely about the initial line.

Now, $\quad r = 5(1 + \cos\theta) \quad \therefore \frac{dr}{d\theta} = -5\sin\theta$

$$\therefore r^2 + \left(\frac{dr}{d\theta}\right)^2 = \ldots\ldots\ldots\ldots\ldots\ldots$$

**42**

$$\boxed{50(1 + \cos\theta)}$$

for $\quad r^2 + \left(\frac{dr}{d\theta}\right)^2 = 25(1 + 2\cos\theta + \cos^2\theta + \sin^2\theta)$

$$= 25(2 + 2\cos\theta)$$

$$= 50(1 + \cos\theta)$$

We would like to express this as a square, since we have to take its root, so we now write $\cos\theta$ in terms of its half angle.

$$\therefore r^2 + \left(\frac{dr}{d\theta}\right)^2 = 50\left(1 + 2\cos^2\frac{\theta}{2} - 1\right)$$

$$= 100\cos^2\frac{\theta}{2}$$

$$\therefore \sqrt{\left\{r^2 + \left(\frac{dr}{d\theta}\right)^2\right\}} = 10\cos\frac{\theta}{2}$$

So the formula in this case now becomes

$$S = \ldots\ldots\ldots\ldots\ldots\ldots$$

**43**

$$S = \int_0^{\pi} 2\pi.5(1 + \cos\theta)\sin\theta.10\cos\frac{\theta}{2}.d\theta$$

$$\therefore S = 100\pi\int_0^{\pi}(1 + \cos\theta)\sin\theta\cos\frac{\theta}{2}\,d\theta$$

We can make this more convenient if we express $(1 + \cos\theta)$ and $\sin\theta$ also in terms of $\frac{\theta}{2}$.

What do we get?

**44**

$$S = 400\pi\int_0^{\pi}\cos^4\frac{\theta}{2}\sin\frac{\theta}{2}\,d\theta$$

For:

$$S = 100\pi\int_0^{\pi}(1 + \cos\theta)\sin\theta\cos\frac{\theta}{2}\,d\theta$$

$$= 100\pi\int_0^{\pi} 2\cos^2\frac{\theta}{2}.\,2\sin\frac{\theta}{2}\cos\frac{\theta}{2}.\cos\frac{\theta}{2}\,d\theta$$

$$= 400\pi\int_0^{\pi}\cos^4\frac{\theta}{2}\sin\frac{\theta}{2}\,d\theta.$$

Now the differential coefficient of $\cos\frac{\theta}{2}$ is $\left\{-\frac{\sin\frac{\theta}{2}}{2}\right\}$

$$\therefore S = -800\pi\int_0^{\pi}\cos^4\frac{\theta}{2}\left\{-\frac{\sin\frac{\theta}{2}}{2}\right\}d\theta$$

$= \ldots\ldots\ldots\ldots\ldots\ldots\ldots\ldots\ldots\ldots$ Finish it off.

**45**

$$S = 160\pi \text{ units}^2$$

Since

$$S = -800\pi\int_0^{\pi}\cos^4\frac{\theta}{2}\left\{-\frac{\sin\frac{\theta}{2}}{2}\right\}d\theta$$

$$= -800\pi\left[\frac{\cos^5\frac{\theta}{2}}{5}\right]_0^{\pi} = \frac{-800\pi}{5}[0 - 1]$$

$$\underline{S = 160\pi \text{ units}^2}$$

And finally, here is one for you to do.

*Example 2.* Find the area of the surface generated when the arc of the curve $r = a\,e^{\theta}$ between $\theta = 0$ and $\theta = \pi/2$ rotates about the initial line.

*Finish it completely and then check with the next frame.*

**46**

$$\boxed{S = \frac{2\sqrt{2}}{5}.\pi a^2(2e^{\pi} + 1)}$$

For, we have:

$$S = \int_0^{\pi/2} 2\pi r \sin\theta \sqrt{\left\{r^2 + \left(\frac{dr}{d\theta}\right)^2\right\}} d\theta$$

And, in this case,

$$r = a e^{\theta} \quad \therefore \frac{dr}{d\theta} = a e^{\theta}$$

$$\therefore r^2 + \left(\frac{dr}{d\theta}\right)^2 = a^2 e^{2\theta} + a^2 e^{2\theta} = 2a^2 e^{2\theta}$$

$$\therefore \sqrt{\left\{r^2 + \left(\frac{dr}{d\theta}\right)^2\right\}} = \sqrt{2}.a.e^{\theta}$$

$$\therefore S = \int_0^{\pi/2} 2\pi a e^{\theta} \sin\theta . \sqrt{2} a e^{\theta} d\theta$$

$$= 2\sqrt{2}\pi a^2 \int_0^{\pi/2} e^{2\theta} \sin\theta \, d\theta$$

Let $I = \int e^{2\theta} \sin\theta \, d\theta = e^{2\theta}(-\cos\theta) + 2\int \cos\theta \, e^{2\theta} d\theta$

$$= -e^{2\theta}\cos\theta + 2\left\{e^{2\theta}\sin\theta - 2\int \sin\theta \, e^{2\theta} d\theta\right\}$$

$$I = -e^{2\theta}\cos\theta + 2e^{2\theta}\sin\theta - 4I$$

$$\therefore 5I = e^{2\theta}\left\{2\sin\theta - \cos\theta\right\}$$

$$I = \frac{e^{2\theta}}{5}\left\{2\sin\theta - \cos\theta\right\}$$

$$\therefore S = 2\sqrt{2}.\pi.a^2\left[\frac{e^{2\theta}}{5}\left\{2\sin\theta - \cos\theta\right\}\right]_0^{\pi/2}$$

$$= \frac{2\sqrt{2}.\pi.a^2}{5}\left\{e^{\pi}(2-0) - 1(0-1)\right\}$$

$$\underline{S = \frac{2\sqrt{2}.\pi.a^2}{5}(2e^{\pi} + 1) \text{ units}^2}$$

We are almost at the end, but before we finish the programme, let us collect our results together.

*So turn on to frame 47.*

## 47 Revision Sheet

*Polar curves* – applications.

1. *Area* $$A = \int_{\theta_1}^{\theta_2} \frac{1}{2} r^2 \, d\theta$$

2. *Volume* $$V = \int_{\theta_1}^{\theta_2} \frac{2}{3} \pi r^3 \sin\theta \, d\theta$$

3. *Length of arc* $$s = \int_{\theta_1}^{\theta_2} \sqrt{\left\{ r^2 + \left( \frac{dr}{d\theta} \right)^2 \right\}} \, d\theta$$

4. *Surface of revolution* $$S = \int_{\theta_1}^{\theta_2} 2\pi r \sin\theta \sqrt{\left\{ r^2 + \left( \frac{dr}{d\theta} \right)^2 \right\}} \, d\theta$$

It is important to know these. The detailed working will depend on the particular form of the function $r = f(\theta)$, but, as you have seen, the method of approach is mainly consistent.

The Test Exercise now remains to be worked. First brush up any points on which you are not perfectly clear; then, when you are ready, turn on to the next frame.

## Test Exercise – XXII

Answer all the questions. They are quite straightforward: there are no tricks. But take your time and work carefully.

1. Calculate the area enclosed by the curve $r\theta^2 = 4$ and the radius vectors at $\theta = \pi/2$ and $\theta = \pi$.

2. Sketch the polar curves:

   (i) $r = 2 \sin \theta$ (ii) $r = 5 \cos^2\theta$ (iii) $r = \sin 2\theta$

   (iv) $r = 1 + \cos \theta$ (v) $r = 1 + 3 \cos \theta$ (vi) $r = 3 + \cos \theta$

3. The plane figure bounded by the curve $r = 2 + \cos \theta$ and the radius vectors at $\theta = 0$ and $\theta = \pi$, rotates about the initial line through a complete revolution. Determine the volume of the solid generated.

4. Find the length of the polar curve $r = 4 \sin^2 \dfrac{\theta}{2}$ between $\theta = 0$ and $\theta = \pi$.

5. Find the area of the surface generated when the arc of the curve $r = a(1 - \cos \theta)$ between $\theta = 0$ and $\theta = \pi$, rotates about the initial line.

*That completes the work on polar curves. You are now ready for the next programme.*

## Further Problems – XXII

1. Sketch the curve $r = \cos^2\theta$. Find (i) the area of one loop and (ii) the volume of the solid formed by rotating it about the initial line.

2. Show that $\sin^4\theta = \frac{3}{8} - \frac{1}{2}\cos 2\theta + \frac{1}{8}\cos 4\theta$. Hence find the area bounded by the curve $r = 4\sin^2\theta$ and the radius vectors at $\theta = 0$ and $\theta = \pi$.

3. Find the area of the plane figure enclosed by the curve $r = a\sec^2\left(\frac{\theta}{2}\right)$ and the radius vectors at $\theta = 0$ and $\theta = \pi/2$.

4. Determine the area bounded by the curve $r = 2\sin\theta + 3\cos\theta$ and the radius vectors at $\theta = 0$ and $\theta = \pi/2$.

5. Find the area enclosed by the curve $r = \dfrac{2}{1 + \cos 2\theta}$ and the radius vectors at $\theta = 0$ and $\theta = \pi/4$.

6. Plot the graph of $r = 1 + 2\cos\theta$ at intervals of 30° and show that it consists of a small loop within a larger loop. The area between the two loops is rotated about the initial line through two right-angles. Find the volume generated.

7. Find the volume generated when the plane figure enclosed by the curve $r = 2a\sin^2\left(\frac{\theta}{2}\right)$ between $\theta = 0$ and $\theta = \pi$, rotates around the initial line.

8. The plane figure bounded by the cardioid $r = 2a(1 + \cos\theta)$ and the parabola $r(1 + \cos\theta) = 2a$ rotates around the initial line. Show that the volume generated is $18\pi a^3$.

9. Find the length of the arc of the curve $r = a\cos^3\left(\frac{\theta}{3}\right)$ between $\theta = 0$ and $\theta = 3\pi$.

10. Find the length of the arc of the curve $r = 3\sin\theta + 4\cos\theta$ between $\theta = 0$ and $\theta = \pi/2$.

11. Find the length of the spiral $r = a\theta$ between $\theta = 0$ and $\theta = 2\pi$.
12. Sketch the curve $r = a \sin^3\left(\frac{\theta}{3}\right)$ and calculate its total length.
13. Show that the length of arc of the curve $r = a \cos^2\theta$ between $\theta = 0$ and $\theta = \pi/2$ is $a[2\sqrt{3} + \ln(2 + \sqrt{3})]/(2\sqrt{3})$.
14. Find the length of the spiral $r = a\,e^{b\theta}$ between $\theta = 0$ and $\theta = \theta_1$, and the area swept out by the radius vector between these two limits.
15. Find the area of the surface generated when the arc of the curve $r^2 = a^2 \cos 2\theta$ between $\theta = 0$ and $\theta = \pi/4$, rotates about the initial line.

# Programme 23

## MULTIPLE INTEGRALS

**1**

**Summation in two directions**

Let us consider the rectangle bounded by the straight lines, $x = r$, $x = s, y = k, y = m$, as shown.

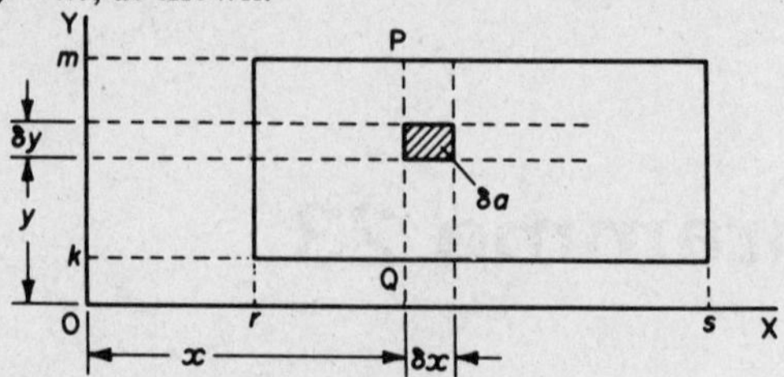


Then the area of the shaded element, $\delta a$ = ...................

**2**

$$\delta a = \delta y . \delta x$$

If we add together all the elements of area, like $\delta a$, to form the vertical strip PQ, then $\delta A$, the area of the strip, can be expressed as

$\delta A$ = ...........

**3**

$$\delta A = \sum_{y=k}^{y=m} \delta y . \delta x$$

Did you remember to include the limits?

Note that during this summation in the $y$-direction, $\delta x$ is constant.

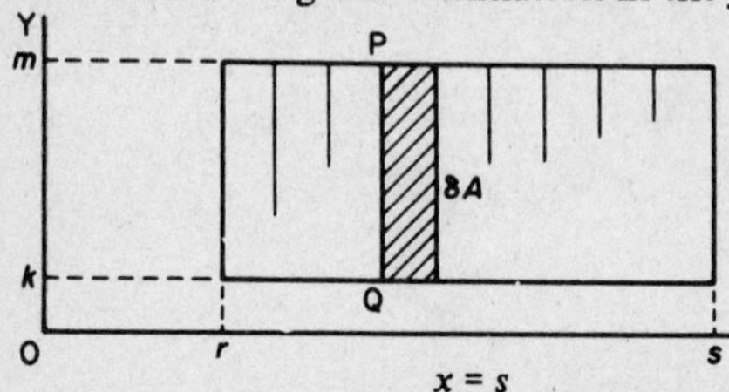


If we now sum all the strips across the figure from $x = r$ to $x = s$, we shall obtain the total area of the rectangle, A.

$$\therefore A = \sum_{x=r}^{x=s} \text{(all vertical strips like PQ)}$$

$$= \sum_{x=r}^{x=s} \left\{ \sum_{y=k}^{y=m} \delta y . \delta x \right\}$$

Removing the brackets, this becomes

$$A = \sum_{x=r}^{x=s} \sum_{y=k}^{y=m} \delta y . \delta x.$$

If now $\delta y \to 0$ and $\delta x \to 0$, the finite summations become integrals, so the expression becomes A = ...................

**4**

$$A = \int_{x=r}^{x=s} \int_{y=k}^{y=m} dy \cdot dx$$

To evaluate this expression, we start from the inside and work outwards.

$$A = \int_{x=r}^{x=s} \left\{ \int_{y=k}^{y=m} dy \right\} dx = \int_{x=r}^{x=s} \Big[ y \Big]_{y=k}^{y=m} dx$$

$$= \int_{x=r}^{x=s} (m-k)\, dx$$

and since $m$ and $k$ are constants, this gives A = ...................

**5**

$$A = (m-k) \,.\, (s-r)$$

for $$A = \Big[ (m-k)\, x \Big]_{x=r}^{x=s} = (m-k) \Big[ x \Big]_{x=r}^{x=s}$$

$$\underline{A = (m-k) \,.\, (s-r)}$$

which we know is correct, for it is merely A = length × breadth.

That may seem a tedious way to find the area of a rectangle, but we have done it to introduce the method we are going to use.

First we define an element of area $\delta y \,.\, \delta x$.

Then we sum in the $y$-direction to obtain the area of a ...................

Finally, we sum the result in the $x$-direction to obtain the area of the ...................

**6**

vertical strip; whole figure

We could have worked slightly differently:

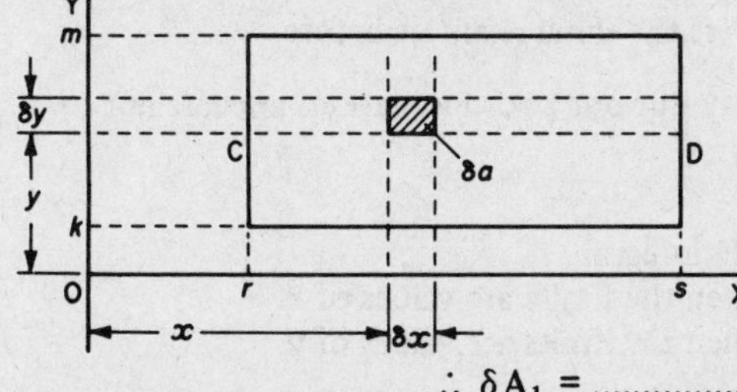


As before $\delta a = \delta x \,.\, \delta y$.
If we sum the elements in the *x-direction* this time, we get the area $\delta A_1$ of the horizontal strip CD

$\therefore\ \delta A_1 =$ ...................

## 7

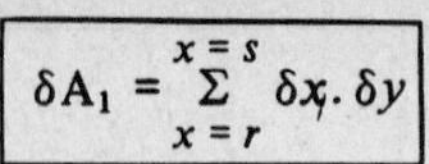

$$\delta A_1 = \sum_{x=r}^{x=s} \delta x . \delta y$$

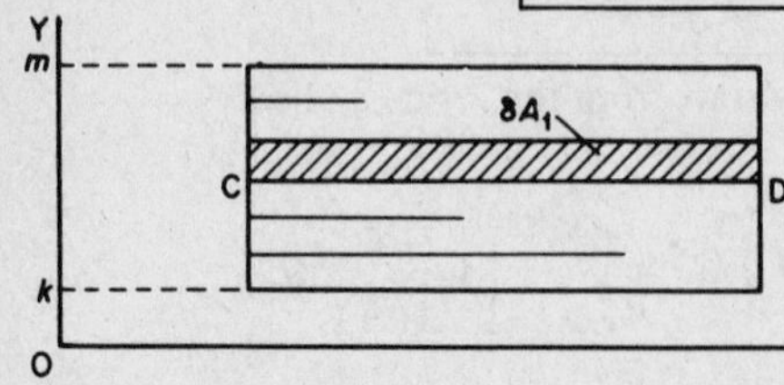


Now sum the strips vertically and we obtain once again the area of the whole rectangle.

$$A_1 = \sum_{y=k}^{y=m} \text{(all horizontal strips like CD)} = \sum_{y=k}^{y=m} \left\{ \sum_{x=r}^{x=s} \delta x . \delta y \right\}$$

As before, if we now remove the brackets and consider what this becomes when $\delta x \rightarrow 0$ and $\delta y \rightarrow 0$, we get

$$A_1 = \ldots\ldots\ldots\ldots\ldots$$

## 8

$$A_1 = \int_{y=k}^{y=m} \int_{x=r}^{x=s} dx . dy$$

To evaluate this we start from the centre

$$A_1 = \int_{y=k}^{y=m} \boxed{\int_{x=r}^{x=s} dx} \; dy$$

$$= \ldots\ldots\ldots\ldots\ldots\ldots$$

*Complete the working to find* $A_1$ *and then move on to frame 9.*

## 9

$$A_1 = (s-r) . (m-k)$$

For $\quad A_1 = \int_{y=k}^{y=m} \Big[ x \Big]_r^s dy = \int_k^m (s-r)\, dy = (s-r) \Big[ y \Big]_k^m$

$\therefore$ $\underline{A_1 = (s-r) . (m-k)}$ which is the same result as before.

So the order in which we carry out our two summations appears not to matter.

***Remember***

(i) We work from the inside integral.
(ii) We integrate w.r.t. $x$ when the limits are values of $x$.
(iii) We integrate w.r.t. $y$ when the limits are values of $y$.

*Turn to the next frame.*

**Double integrals** **10**

The expression $\int_{y_1}^{y_2} \int_{x_1}^{x_2} f(x, y)\, dx\, dy$ is called a *double integral* (for obvious reasons!) and indicates that

(i) $f(x, y)$ is first integrated with respect to $x$ (regarding $y$ as being constant) between the limits $x = x_1$ and $x = x_2$,

(ii) the result is then integrated with respect to $y$ between the limits $y = y_1$ and $y = y_2$.

*Example 1*

$$\text{Evaluate } I = \int_1^2 \int_2^4 (x + 2y)\, dx\, dy$$

So $(x + 2y)$ is first integrated w.r.t. $x$ between $x = 2$ and $x = 4$, with $y$ regarded as constant for the time being.

$$I = \int_1^2 \boxed{\int_2^4 (x + 2y)\, dx}\; dy.$$

$$= \int_1^2 \left[\frac{x^2}{2} + 2xy\right]_2^4 . dy$$

$$= \int_1^2 \left\{(8 + 8y) - (2 + 4y)\right\} dy$$

$$= \int_1^2 (6 + 4y)\, dy = \ldots\ldots\ldots\ldots$$

*Finish it off*

**11**

$$\boxed{I = 12}$$

For
$$I = \int_1^2 (6 + 4y)\, dy = \left[6y + 2y^2\right]_1^2$$
$$= (12 + 8) - (6 + 2) = 20 - 8 = \underline{12}$$

Here is another.

*Example 2*

$$\text{Evaluate } I = \int_1^2 \int_0^3 x^2 y\, dx\, dy$$

Do this one on your own. Remember to start with $\int_0^3 x^2 y\, dx$ with $y$ constant.

*Finish the double integral completely and then turn on to frame 12.*

**12**

$$\boxed{I = 13{\cdot}5}$$

Check your working:

$$I = \int_1^2 \int_0^3 x^2 y \, dx \, dy = \int_1^2 \left[ \int_0^3 x^2 y \, dx \right] dy$$

$$= \int_1^2 \left[ \frac{x^3}{3} \cdot y \right]_{x=0}^{x=3} dy$$

$$= \int_1^2 (9y) \, dy = \left[ \frac{9y^2}{2} \right]_1^2$$

$$= 18 - 4{\cdot}5 = \underline{13{\cdot}5}$$

Now do this one in just the same way.

*Example 3*

$$\text{Evaluate } I = \int_1^2 \int_0^\pi (3 + \sin\theta) \, d\theta \, dr$$

*When you have finished, check with the next frame.*

**13**

$$\boxed{I = 3\pi + 2}$$

Here it is:

$$I = \int_1^2 \int_0^\pi (3 + \sin\theta) \, d\theta \, dr.$$

$$= \int_1^2 \Big[ 3\theta - \cos\theta \Big]_0^\pi dr$$

$$= \int_1^2 \Big\{ (3\pi + 1) - (-1) \Big\} dr$$

$$= \int_1^2 (3\pi + 2) \, dr$$

$$= \Big[ (3\pi + 2) \, r \Big]_1^2$$

$$= (3\pi + 2)(2 - 1) = \underline{3\pi + 2}$$

*On to the next frame.*

**14**

**Triple integrals.** Sometimes we have to deal with expressions such as

$$I = \int_a^b \int_c^d \int_e^f f(x, y, z)\, dx \,.\, dy \,.\, dz$$

but the rules are as before. Start with the innermost integral and work outwards.

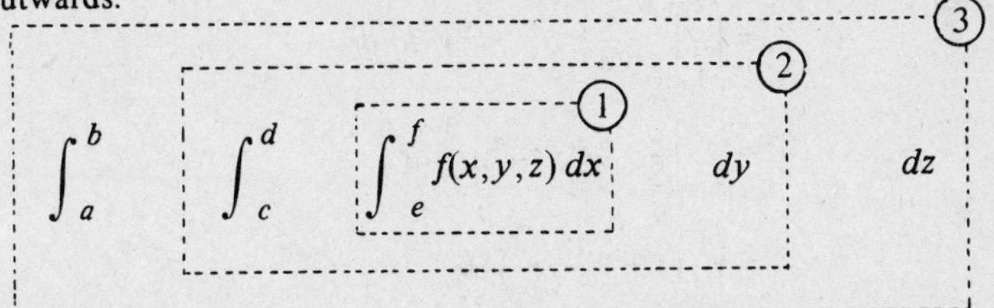

All symbols are regarded as constant for the time being, except the one variable with respect to which the stage of integration is taking place. So try this one on your own straight away.

*Example 1.* Evaluate $I = \int_1^3 \int_{-1}^1 \int_0^2 (x + 2y - z)\, dx.\, dy.\, dz$

**15**

$$\boxed{I = -8}$$

Did you manage it first time? Here is the working in detail.

$$I = \int_1^3 \int_{-1}^1 \int_0^2 (x + 2y - z)\, dx \,.\, dy \,.\, dz$$

$$= \int_1^3 \int_{-1}^1 \left[\frac{x^2}{2} + 2xy - xz\right]_0^2 dy.dz$$

$$= \int_1^3 \int_{-1}^1 (2 + 4y - 2z)\, dy.\, dz = \int_1^3 \left[2y + 2y^2 - 2yz\right]_{-1}^1 dz$$

$$= \int_1^3 \left\{(2 + 2 - 2z) - (-2 + 2 + 2z)\right\} dz = \int_1^3 (4 - 4z)\, dz$$

$$= \left[4z - 2z^2\right]_1^3 = (12 - 18) - (4 - 2) \underline{= -8}$$

*Example 2.* Evaluate $\int_1^2 \int_0^3 \int_0^1 (p^2 + q^2 - r^2)\, dp \,.\, dq \,.\, dr$

*When you have finished it, turn on to frame 16.*

**16**

$$\boxed{I = 3}$$

For
$$I = \int_1^2 \int_0^3 \int_0^1 (p^2 + q^2 - r^2)\, dp\, dq\, dr$$
$$= \int_1^2 \int_0^3 \left[\frac{p^3}{3} + pq^2 - pr^2\right]_0^1 dq\, dr$$
$$= \int_1^2 \int_0^3 \left\{\frac{1}{3} + q^2 - r^2\right\} dq\, dr$$
$$= \int_1^2 \left[\frac{q}{3} + \frac{q^3}{3} - qr^2\right]_0^3 dr$$
$$= \int_1^2 (1 + 9 - 3r^2)\, dr$$
$$= \left[10r - r^3\right]_1^2 = (20 - 8) - (10 - 1)$$
$$= 12 - \underline{9 = 3}$$

It is all very easy if you take it steadily, step by step.

Now two quickies for revision:

Evaluate (i) $\int_1^2 \int_3^5 dy\, dx$, (ii) $\int_0^4 \int_1^{3x} 2y\, dy\, dx$.

*Finish them both and then move on to the next frame.*

**17**

$$\boxed{\text{(i) } I = 2; \quad \text{(ii) } I = 188}$$

Here they are.

(i) $I = \int_1^2 \int_3^5 dy\, dx = \int_1^2 \left[y\right]_3^5 . \, dx = \int_1^2 (5 - 3)\, dx = \int_1^2 2\, dx = \left[2x\right]_1^2$

$= 4 - 2 = 2$

(ii) $I = \int_0^4 \int_1^{3x} 2y\, dy\, dx = \int_0^4 \left[y^2\right]_1^{3x} dx = \int_0^4 (9x^2 - 1)\, dx$

$= \left[3x^3 - x\right]_0^4 = 192 - 4 = \underline{188}$

And finally, do this one.

$$I = \int_0^5 \int_1^2 (3x^2 - 4)\, dx\, dy = \text{....................}$$

**18**

$$\boxed{I = 15}$$

Check the working.

$$I = \int_0^5 \int_1^2 (3x^2 - 4)\, dx\, dy$$

$$= \int_0^5 \left[ x^3 - 4x \right]_1^2 dy$$

$$= \int_0^5 \left\{ (8 - 8) - (1 - 4) \right\} dy$$

$$= \int_0^5 3\, dy = \left[ 3y \right]_0^5 = \underline{15}$$

Now let us see a few applications of multiple integrals.

*Move on then to the next frame.*

**19**

**Applications**

*Example 1.* Find the area bounded by $y = \frac{4x}{5}$, the $x$-axis and the ordinate at $x = 5$.

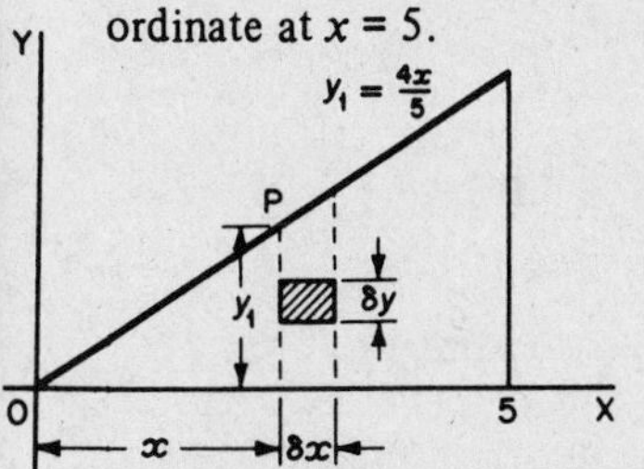


Area of element = $\delta y . \delta x$

$\therefore$ Area of strip $\sum_{y=0}^{y=y_1} \delta y . \delta x$

The sum of all such strips across the figure gives us

$$A \simeq \sum_{x=0}^{x=5} \left\{ \sum_{y=0}^{y=y_1} \delta y . \delta x \right\}$$

$$\simeq \sum_{x=0}^{x=5} \sum_{y=0}^{y=y_1} \delta y . \delta x.$$

Now, if $\delta y \to 0$ and $\delta x \to 0$, then

$$A = \int_0^5 \int_0^{y_1} dy\, dx$$

$$= \int_0^5 \left[ y \right]_0^{y_1} dx = \int_0^5 y_1\, dx$$

But $y_1 = \frac{4x}{5}$

So A = ............

*Finish it off.*

**20**

$$\boxed{A = 10 \text{ units}^2}$$

For
$$A = \int_0^5 \frac{4x}{5}\,dx = \left[\frac{2x^2}{5}\right]_0^5 = \underline{10}$$

Right. Now what about this one?

*Example 2.* Find the area under the curve $y = 4\sin\dfrac{x}{2}$ between $x = \dfrac{\pi}{3}$ and $x = \pi$, by double integral method.

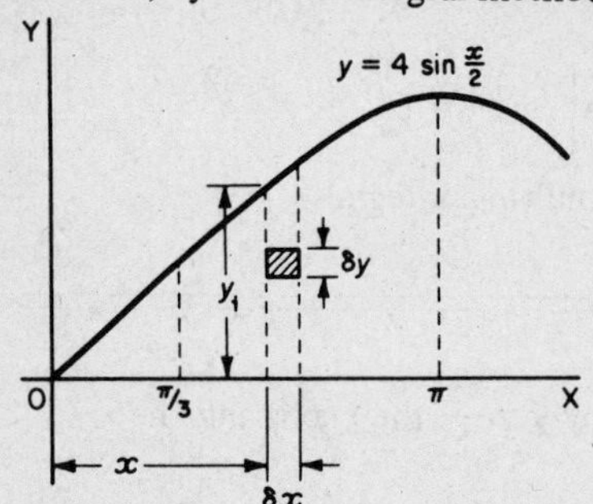


Steps as before.
Area of element $= \delta y . \delta x$
Area of vertical strip
$$\sum_{y=0}^{y=y_1} \delta y . \delta x$$

Total area of figure:
$$A \triangleq \sum_{x=\pi/3}^{x=\pi} \left\{ \sum_{y=0}^{y=y_1} \delta y . \delta x \right\}$$

If $\delta y \to 0$ and $\delta x \to 0$, then
$$A = \int_{\pi/3}^{\pi} \int_0^{y_1} dy\,dx = \ldots\ldots\ldots\ldots\ldots$$

Complete it, remembering that $y_1 = 4\sin\dfrac{x}{2}$

**21**

$$\boxed{A = 4\sqrt{3} \text{ units}^2}$$

For you get
$$A = \int_{\pi/3}^{\pi}\int_0^{y_1} dy\,dx = \int_{\pi/3}^{\pi}\Big[y\Big]_0^{y_1} dx = \int_{\pi/3}^{\pi} y_1\,dx$$
$$= \int_{\pi/3}^{\pi} 4\sin\frac{x}{2}\,dx = \left[-8\cos\frac{x}{2}\right]_{\pi/3}^{\pi}$$
$$= (-8\cos\pi/2) - (-8\cos\pi/6)$$
$$= 0 - 8.\frac{\sqrt{3}}{2} = \underline{4\sqrt{3} \text{ units}^2}$$

*Now for a rather more worthwhile example – on to frame 22.*

**22**

*Example 3.* Find the area enclosed by the curves

$$y_1{}^2 = 9x \text{ and } y_2 = \frac{x^2}{9}$$

First we must find the points of intersection. For that, $y_1 = y_2$.

$$\therefore\ 9x = \frac{x^4}{81} \quad \therefore\ x = 0 \text{ or } x^3 = 729, \text{ i.e. } x = 9$$

So we have a diagram like this:

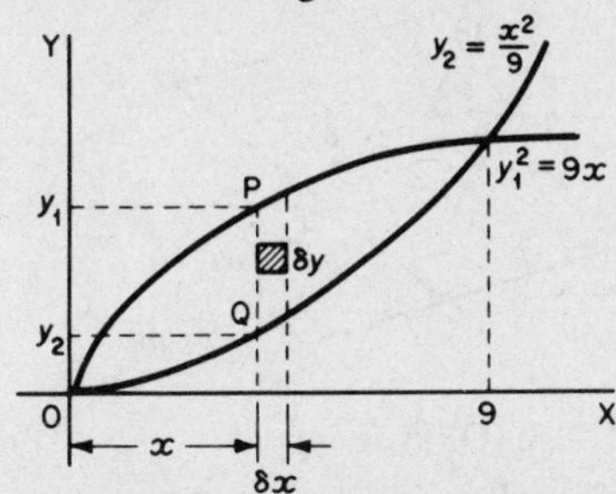


As usual,
Area of element = $\delta y . \delta x$
$\therefore$ Area of strip PQ

$$\sum_{y=y_2}^{y=y_1} \delta y . \delta x$$

Summing all strips between $x = 0$ and $x = 9$,

$$A \simeq \sum_{x=0}^{x=9} \left\{ \sum_{y=y_2}^{y=y_1} \delta y . \delta x \right\} = \sum_{x=0}^{x=9} \sum_{y=y_2}^{y=y_1} \delta y . \delta x$$

If $\delta y \to 0$ and $\delta x \to 0$, $\quad A = \int_0^9 \int_{y_2}^{y_1} dy\, dx$

Now finish it off, remembering that $y_1{}^2 = 9x$ and $y_2 = \dfrac{x^2}{9}$

**23**

$$\boxed{A = 27 \text{ units}^2}$$

Here it is.

$$A = \int_0^9 \int_{y_2}^{y_1} dy\, dx = \int_0^9 \Big[ y \Big]_{y_2}^{y_1} dx$$

$$= \int_0^9 (y_1 - y_2)\, dx$$

$$= \int_0^9 \left\{ 3x^{\frac{1}{2}} - \frac{x^2}{9} \right\} dx$$

$$= \left[ 2x^{3/2} - \frac{x^3}{27} \right]_0^9$$

$$= 54 - 27 = \underline{27 \text{ units}^2}$$

*Now for a different one. So turn on to the next frame.*

**24** Double integrals can conveniently be used for finding other values besides areas.

*Example 4.* Find the second moment of area of a rectangle 6 cm X 4 cm about an axis through one corner perpendicular to the plane of the figure.

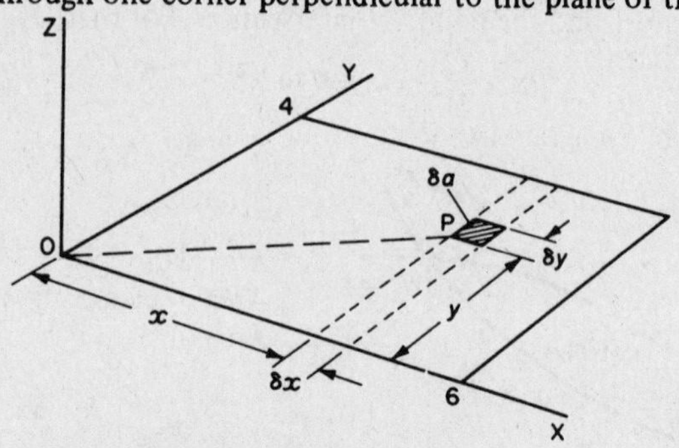


Second moment of element P about OZ $\simeq \delta a\,(OP)^2$

$$\simeq \delta y\,.\,\delta x\,.\,(x^2 + y^2)$$

Total second moment about OZ

$$I \simeq \sum_{x=0}^{x=6}\ \sum_{y=0}^{y=4} (x^2 + y^2)\,dy\,dx$$

If $\delta x \to 0$ and $\delta y \to 0$, this becomes

$$I = \int_0^6\int_0^4 (x^2 + y^2)\,dy\,dx$$

Now complete the working. I = ..................

**25**

$$\boxed{I = 416\ \text{cm}^4}$$

For:

$$I = \int_0^6\int_0^4 (x^2 + y^2)\,dy\,dx = \int_0^6 \left[x^2y + \frac{y^3}{3}\right]_0^4 dx$$

$$= \int_0^6 \left\{4x^2 + \frac{64}{3}\right\} dx$$

$$= \left[\frac{4x^3}{3} + \frac{64x}{3}\right]_0^6 = \underline{288 + 128 = 416\ \text{cm}^4}$$

Now here is one for you to do on your own.

*Example 5.* Find the second moment of area of a rectangle 5 cm X 3 cm about one 5 cm side as axis.

*Complete it and then turn on to frame 26.*

**26**

$$\boxed{I = 45 \text{ cm}^4}$$

Here it is: check through the working.

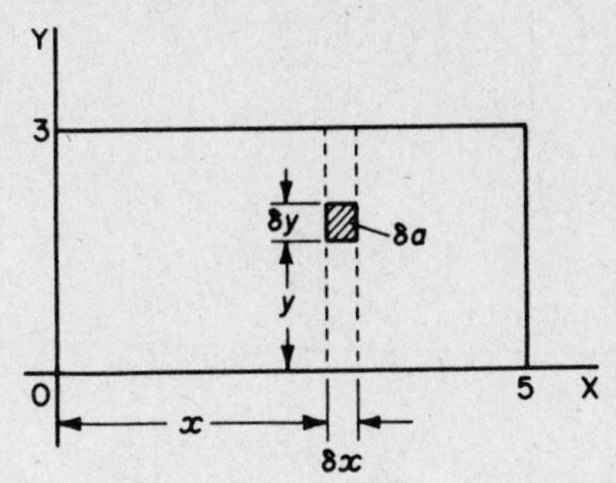


Area of element = $\delta a = \delta y.\, \delta x$
Second moment of area of $\delta a$ about OX = $\delta a\, y^2$
$= y^2\, \delta y\, \delta x$

Second moment of strip $\triangleq \sum_{y=0}^{y=3} y^2 . \delta y . \delta x$

Second moment of whole figure $\triangleq \sum_{x=0}^{x=5} \sum_{y=0}^{y=3} y^2 . \delta y . \delta x$

If $\delta y \to 0$ and $\delta x \to 0$

$$I = \int_0^5 \int_0^3 y^2 \, dy \, dx$$

$$\therefore\ I = \int_0^5 \left[\frac{y^3}{3}\right]_0^3 dx = \int_0^5 9\, dx = \left[9x\right]_0^5$$

$$\underline{I = 45 \text{ cm}^4}$$

*On to frame 27.*

**27**

Now a short revision exercise. Finish both integrals, before turning on to the next frame. Here they are.

**Revision**

Evaluate the following:

$$\text{(i)} \quad \int_0^2 \int_1^3 (y^2 - xy)\, dy\, dx$$

$$\text{(ii)} \quad \int_0^3 \int_1^2 (x^2 + y^2)\, dy\, dx.$$

*When you have finished both, turn on.*

**28**

$$\text{(i) } I = 9\tfrac{1}{3}; \quad \text{(ii) } I = 16$$

Here they are in detail.

(i)
$$I = \int_0^2 \int_1^3 (y^2 - xy)\,dy\,dx = \int_0^2 \left[\frac{y^3}{3} - \frac{xy^2}{2}\right]_1^3 dx$$
$$= \int_0^2 \left\{\left(9 - \frac{9x}{2}\right) - \left(\frac{1}{3} - \frac{x}{2}\right)\right\} dx$$
$$= \int_0^2 \left(\frac{26}{3} - 4x\right) dx = \left[\frac{26x}{3} - 2x^2\right]_0^2$$
$$= 17\tfrac{1}{3} - 8 = \underline{9\tfrac{1}{3}}$$

(ii)
$$I = \int_0^3 \int_1^2 (x^2 + y^2)\,dy\,dx = \int_0^3 \left[x^2 y + \frac{y^3}{3}\right]_1^2 dx$$
$$= \int_0^3 \left\{\left(2x^2 + \frac{8}{3}\right) - \left(x^2 + \frac{1}{3}\right)\right\} dx$$
$$= \int_0^3 \left(x^2 + \frac{7}{3}\right) dx = \left[\frac{x^3}{3} + \frac{7x}{3}\right]_0^3$$
$$= 9 + 7 = \underline{16}$$

*Now on to frame 29.*

**29** **Alternative notation**

Sometimes, double integrals are written in a slightly different way. For example, the last double integral $I = \int_0^3 \int_1^2 (x^2 + y^2)\,dy\,dx$ could have been written

$$\int_0^3 dx \int_1^2 (x^2 + y^2)\,dy$$

The key now is that we start working from the *right-hand* side integral and gradually work back towards the front. Of course, we get the same result and the working is identical.

Let us have an example or two, to get used to this notation.

*Move on then to frame 30.*

**30**

*Example 1.*

$$I = \int_0^2 dx \int_0^{\pi/2} 5 \cos\theta \, d\theta$$
$$= \int_0^2 dx \left[ 5 \sin\theta \right]_0^{\pi/2}$$
$$= \int_0^2 dx \left[ 5 \right] = \int_0^2 5 \, dx = \left[ 5x \right]_0^2$$
$$= \underline{10}$$

It is all very easy, once you have seen the method. You try this one.

*Example 2.* Evaluate $I = \int_3^6 dy \int_0^{\pi/2} 4 \sin 3x \, dx$

**31**

$$\boxed{I = 4}$$

Here it is.

$$I = \int_3^6 dy \int_0^{\pi/2} 4 \sin 3x \, dx$$
$$= \int_3^6 dy \left[ \frac{-4 \cos 3x}{3} \right]_0^{\pi/2}$$
$$= \int_3^6 dy \left\{ (0) - \left( -\frac{4}{3} \right) \right\} = \int_3^6 dy \, \frac{4}{3}$$
$$= \left[ \frac{4y}{3} \right]_3^6 = (8) - (4) = \underline{4}$$

Now do these two.

*Example 3.*
$$\int_2^3 dx \int_0^1 (x - x^2) \, dy$$

*Example 4.*
$$\int_1^2 dy \int_y^{2y} (x - y) \, dx$$

(Take care with the second one)

***When you have finished them both, turn on to the next frame.***

## 32

$$Ex.\ 3.\ \mathrm{I} = -4{\cdot}5, \quad Ex.\ 4.\ \mathrm{I} = \frac{7}{6}$$

***Results:***

*Example 3.*

$$\mathrm{I} = \int_0^3 dx \int_0^1 (x - x^2)\,dy$$

$$= \int_0^3 dx \Big[ xy - x^2 y \Big]_0^1$$

$$= \int_0^3 dx\,(x - x^2) = \int_0^3 (x - x^2)\,dx$$

$$= \left[\frac{x^2}{2} - \frac{x^3}{3}\right]_0^3 = \frac{9}{2} - 9 = \underline{-4{\cdot}5}$$

*Example 4*

$$\mathrm{I} = \int_1^2 dy \int_y^{2y} (x - y)\,dx$$

$$= \int_1^2 dy \left[\frac{x^2}{2} - xy\right]_{x=y}^{x=2y}$$

$$= \int_1^2 dy \left\{(2y^2 - 2y^2) - \left(\frac{y^2}{2} - y^2\right)\right\}$$

$$= \int_1^2 dy\,\frac{y^2}{2} = \int_1^2 \frac{y^2}{2}\,dy$$

$$= \left[\frac{y^3}{6}\right]_1^2 = \frac{8}{6} - \frac{1}{6} = \underline{\frac{7}{6}}$$

*Next frame.*

## 33

Now, by way of revision, evaluate these.

$$\text{(i)} \quad \int_0^4 \int_y^{2y} (2x + 3y)\,dx\,dy$$

$$\text{(ii)} \quad \int_1^4 dx \int_0^{\sqrt{x}} (2y - 5x)\,dy.$$

***When you have completed both of them, turn on to frame 34.***

**34**

(i) 128, (ii) −54·5

*Working*

$$\text{(i)}\quad I = \int_0^4 \int_y^{2y} (2x + 3y)\, dx\, dy$$

$$= \int_0^4 \left[ x^2 + 3xy \right]_{x=y}^{x=2y} .\, dy$$

$$= \int_0^4 \left\{ (4y^2 + 6y^2) - (y^2 + 3y^2) \right\} dy$$

$$= \int_0^4 \left\{ 10y^2 - 4y^2 \right\} dy = \int_0^4 6y^2\, dy$$

$$= \left[ \frac{6y^3}{3} \right]_0^4 = \left[ 2y^3 \right]_0^4 = \underline{128}$$

$$\text{(ii)}\quad I = \int_1^4 dx \int_0^{\sqrt{x}} (2y - 5x)\, dy$$

$$= \int_1^4 dx \left[ y^2 - 5xy \right]_{y=0}^{y=\sqrt{x}}$$

$$= \int_1^4 dx \left\{ x - 5x^{3/2} \right\}$$

$$= \int_1^4 (x - 5x^{3/2})\, dx = \left[ \frac{x^2}{2} - 2x^{5/2} \right]_1^4$$

$$= (8 - 64) - (\tfrac{1}{2} - 2)$$

$$= -56 + 1{\cdot}5 = \underline{-54{\cdot}5}$$

So it is just a question of being able to recognize and to interpret the two notations.

Now let us look at one or two further examples of the use of multiple integrals.

*Turn on then to frame 35.*

**35** **Example**

To find the area of the plane figure bounded by the polar curve $r = f(\theta)$, and the radius vectors at $\theta = \theta_1$ and $\theta = \theta_2$.

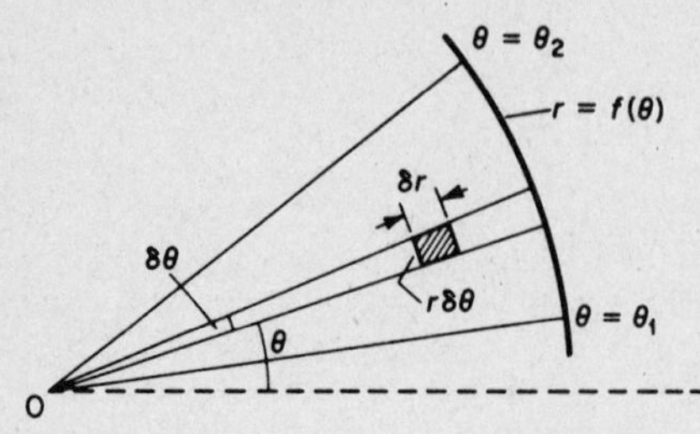


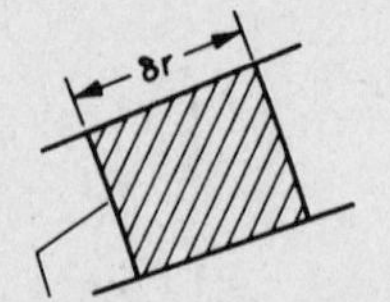


Small arc of a circle of radius $r$, subtending an angle $\delta\theta$ at the centre.

$\therefore$ arc $= r.\delta\theta$

We proceed very much as before.

$$\text{Area of element} \simeq \delta r.\, r\delta\theta$$

$$\text{Area of thin sector} \simeq \sum_{r=0}^{r=r_1} \delta r.\, r\delta\theta$$

$$\text{Total area} \simeq \sum_{\theta=\theta_1}^{\theta=\theta_2} \text{(all such thin sectors)}$$

$$\simeq \sum_{\theta=\theta_1}^{\theta=\theta_2} \left\{ \sum_{r=0}^{r=r_1} r.\delta r.\delta\theta \right\}$$

$$\simeq \sum_{\theta=\theta_1}^{\theta=\theta_2} \sum_{r=0}^{r=r_1} r.\delta r.\delta\theta$$

Then if $\delta\theta \to 0$ and $\delta r \to 0$,

$$A = \int_{\theta_1}^{\theta_2} \int_0^{r_1} r.dr.d\theta$$

$$= \ldots\ldots\ldots\ldots\ldots\ldots \quad \text{Finish it off.}$$

---

**36** The working continues:

$$A = \int_{\theta_1}^{\theta_2} \left[\frac{r^2}{2}\right]_0^{r_1} d\theta$$

$$= \int_{\theta_1}^{\theta_2} \left(\frac{r_1^{\,2}}{2}\right) d\theta$$

i.e. in general,

$$A = \int_{\theta_1}^{\theta_2} \frac{1}{2} r^2 \, d\theta$$

Which is the result we have met before.

*Let us work an actual example of this, so turn on to frame 37.*

**37**

*Example.* By the use of double integrals, find the area enclosed by the polar curve $r = 4(1 + \cos\theta)$ and the radius vectors at $\theta = 0$ and $\theta = \pi$.

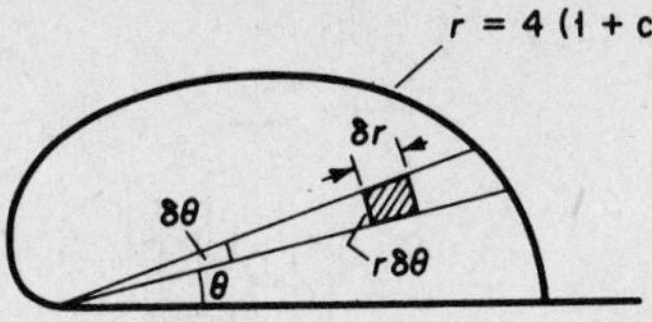

r = 4 (1 + cos θ)
δr
δθ
rδθ
θ

$$A \simeq \sum_{\theta=0}^{\theta=\pi} \sum_{r=0}^{r=r_1} r\,\delta r.\delta\theta$$

$$A = \int_0^{\pi}\int_0^{r_1} r\,dr\,d\theta$$

$$= \int_0^{\pi}\left[\frac{r^2}{2}\right]_0^{r_1}.\,d\theta$$

$$= \int_0^{\pi}\left[\frac{{r_1}^2}{2}\right]d\theta \qquad \text{But } r_1 = f(\theta) = 4(1 + \cos\theta)$$

$$\therefore\ A = \int_0^{\pi} 8(1 + \cos\theta)^2\,d\theta$$

$$= \int_0^{\pi} 8(1 + 2\cos\theta + \cos^2\theta)\,d\theta$$

$= \ldots\ldots\ldots\ldots\ldots\ldots\ldots\ldots$

---

**38**

$$\boxed{A = 12\pi \text{ units}^2}$$

For

$$A = 8\int_0^{\pi}(1 + 2\cos\theta + \cos^2\theta)\,d\theta$$

$$= 8\left[\theta + 2\sin\theta + \frac{\theta}{2} + \frac{\sin 2\theta}{4}\right]_0^{\pi}$$

$$= 8\left(\pi + \frac{\pi}{2}\right) - (0)$$

$$= 8\pi + 4\pi = \underline{12\pi \text{ units}^2}$$

*Now let us deal with volumes by the same method, so move on to the next frame.*

## 39 Determination of volumes by multiple integrals

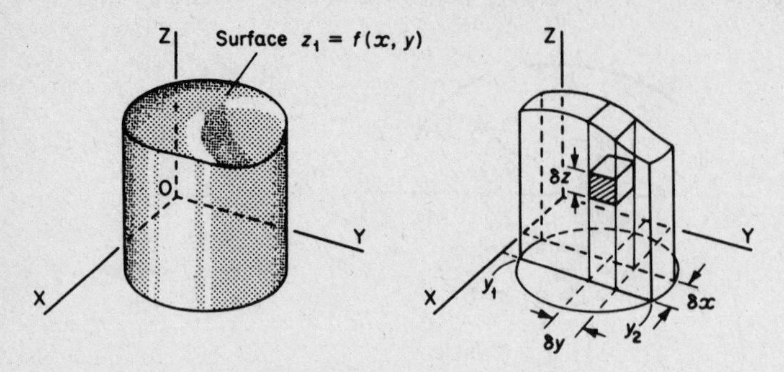


Element of volume $\delta v = \delta x . \delta y . \delta z$.
Summing the elements up the column, we have

$$\delta V_c = \sum_{z=0}^{z=z_1} \delta x . \delta y . \delta z$$

If we now sum the columns between $y = y_1$ and $y = y_2$, we obtain the volume of the slice.

$$\delta V_s = \sum_{y=y_1}^{y=y_2} \sum_{z=0}^{z=z_1} \delta x . \delta y . \delta z$$

Then, summing all slices between $x = x_1$ and $x = x_2$, we have the total volume.

$$V = \sum_{x=x_1}^{x=x_2} \sum_{x=x_1}^{y=y_2} \sum_{z=0}^{z=z_1} \delta x . \delta y . \delta z$$

Then, as usual, if $\delta x \to 0$, $\delta y \to 0$ and $\delta z \to 0$

$$V = \int_{x_1}^{x_2} \int_{y_1}^{y_2} \int_{0}^{z_1} dx . dy . dz$$

The result this time is a triple integral, but the development is very much the same as in our previous examples.

Let us see this in operation in the following examples.

*Next frame.*

**40**

*Example 1.* A solid is enclosed by the plane $z = 0$, the planes $x = 1, x = 4$, $y = 2, y = 5$ and the surface $z = x + y$. Find the volume of the solid.

First of all, what does the figure look like?

The plane $z = 0$ is the $x$-$y$ plane and the plane $x = 1$ is positioned thus:

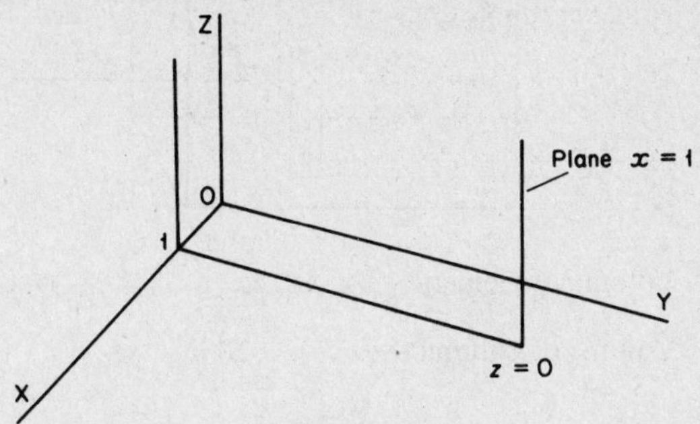


Working on the same lines, draw a sketch of the vertical sides.

The figure so far now looks like this:

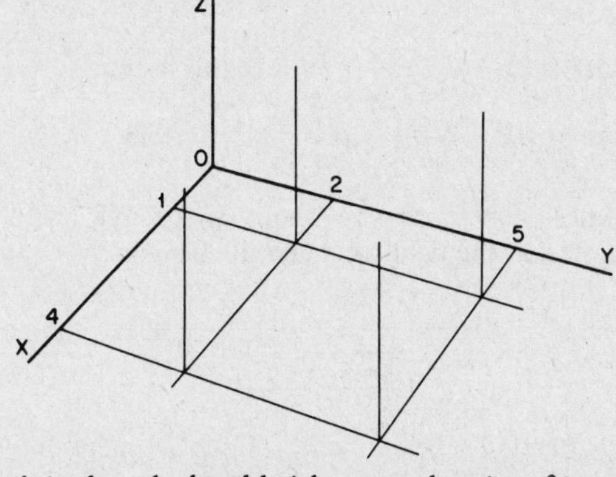


If we now mark in the calculated heights at each point of intersection ($z = x + y$), we get

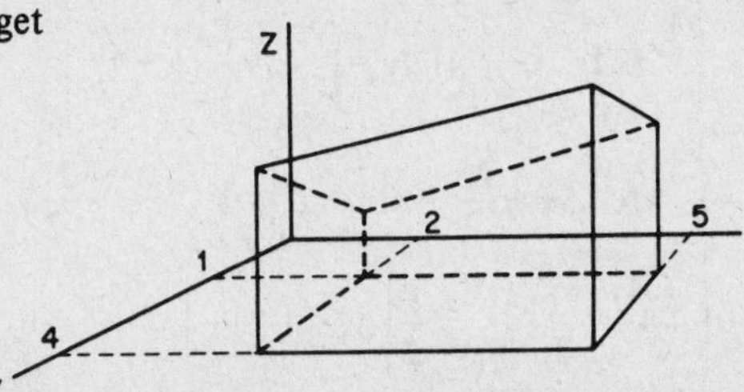


This is just preparing the problem, so that we can see how to develop the integral. *For the calculation stage, turn on to the next frame.*

**42**

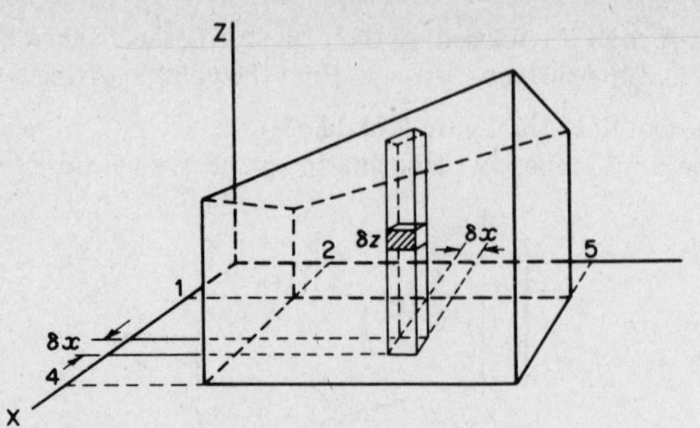


Volume of element $\simeq \delta x \,.\, \delta y \,.\, \delta z$

Volume of column $\simeq \delta x \,.\, \delta y \sum_{z=0}^{z=(x+y)} \delta z$

Volume of slice $\simeq \delta x \sum_{y=2}^{y=5} \delta y \sum_{z=0}^{z=x+y} \delta z$

Volume of total solid $\simeq \sum_{x=1}^{x=4} \delta x \sum_{y=2}^{y=5} \delta y \sum_{z=0}^{z=x+y} \delta z$

Then, as usual, if $\delta x \to 0$, $\delta y \to 0$, $\delta z \to 0$, this becomes

$$V = \int_1^4 dx \int_2^5 dy \int_0^{x+y} dz$$

And this you can now finish off without any trouble. (With this form of notation, start at the right-hand end. Remember?)

So V = ....................

**43**

$$\boxed{V = 54 \text{ units}^3}$$

$$V = \int_1^4 dx \int_2^5 dy \int_0^{x+y} dz = \int_1^4 dx \int_2^5 dy\,(x+y)$$

$$= \int_1^4 dx \int_2^5 (x+y)\,dy = \int_1^4 dx \left[xy + \frac{y^2}{2}\right]_2^5$$

$$= \int_1^4 dx \left[5x + \frac{25}{2} - 2x - 2\right] = \int_1^4 \left(3x + \frac{21}{2}\right) dx$$

$$= \left[\frac{3x^2}{2} + \frac{21x}{2}\right]_1^4 = \frac{1}{2}\left[3x^2 + 21x\right]_1^4$$

$$= \frac{1}{2}\left\{(48 + 84) - (3 + 21)\right\} = \frac{1}{2}\left\{132 - 24\right\} = \underline{54 \text{ units}^3}$$

**44**

*Example 2.* Find the volume of the solid bounded by the planes, $z = 0, x = 1, x = 2, y = -1, y = 1$ and the surface $z = x^2 + y^2$.

In the light of the last example, can you conjure up a mental picture of what this solid looks like? As before it will give rise to a triple integral.

$$V = \int_1^2 dx \int_{-1}^1 dy \int_0^{x^2+y^2} dz$$

Evaluate this and so find V. V = ....................

**45**

$$\boxed{V = \frac{16}{3} \text{ units}^3}$$

For we have:

$$\begin{aligned} V &= \int_1^2 dx \int_{-1}^1 dy \int_0^{x^2+y^2} dz \\ &= \int_1^2 dx \int_{-1}^1 dy \,(x^2 + y^2) \\ &= \int_1^2 dx \left[ x^2 y + \frac{y^3}{3} \right]_{-1}^1 \\ &= \int_1^2 \left\{ \left(x^2 + \frac{1}{3}\right) - \left(-x^2 - \frac{1}{3}\right) \right\} dx \\ &= \int_1^2 \left\{ 2x^2 + \frac{2}{3} \right\} dx \\ &= \frac{2}{3} \left[ x^3 + x \right]_1^2 \\ &= \frac{2}{3} \left\{ (8 + 2) - (1 + 1) \right\} \\ &= \underline{\frac{16}{3} \text{ units}^3} \end{aligned}$$

*Next frame.*

**46**

That brings us almost to the end of this programme.

In our work on multiple integrals, we have been developing a form of approach rather than compiling a catalogue of formulae. There is little therefore that we can list by way of revision on this occasion, except perhaps to remind you, once again, of the two forms of notation.

Remember:

(i) For integrals written $\int_c^d \int_a^b f(x, y)\, dx \,.\, dy$, work from the centre outwards.

(ii) For integrals written $\int_c^d dy \int_a^b f(x, y)\, dx$ work from the right-hand side.

Now there is the Test Exercise to follow. Before working through it, turn back into the programme and revise any points on which you are not perfectly clear. If you have followed all the directions you will have no trouble with the test.

*So when you are ready, move on to the Test Exercise.*

**47**

## Test Exercise – XXIII

Answer all questions. They are all quite straightforward and should cause you no trouble.

1. Evaluate (i) $\displaystyle\int_1^3\int_0^2 (y^3 - xy)\,dy\,dx$

   (ii) $\displaystyle\int_0^a dx\int_0^{y_1}(x - y)\,dy$, where $y_1 = \sqrt{(a^2 - x^2)}$

2. Determine

   (i) $\displaystyle\int_0^{\sqrt{3}+2}\int_0^{\pi/3}(2\cos\theta - 3\sin 3\theta)\,d\theta\,.\,dr$

   (ii) $\displaystyle\int_2^4\int_1^2\int_0^4 xy(z+2)\,dx\,dy\,dz$

   (iii) $\displaystyle\int_0^1 dz\int_1^2 dx\int_0^x (x + y + z)\,dy$

3. The line $y = 2x$ and the parabola $y^2 = 16x$ intersect at $x = 4$. Find by a double integral, the area enclosed by $y = 2x$, $y^2 = 16x$ and the ordinate at $x = 1$.

4. A triangle is bounded by the $x$-axis, the line $y = 2x$ and the ordinate at $x = 4$. Build up a double integral representing the second moment of area of this triangle about the $x$-axis and evaluate the integral.

5. Form a double integral to represent the area of the plane figure bounded by the polar curve $r = 3 + 2\cos\theta$ and the radius vectors at $\theta = 0$ and $\theta = \pi/2$, and evaluate it.

6. A solid is enclosed by the planes $z = 0, y = 1, y = 3, x = 0, x = 3$, and the surface $z = x^2 + xy$. Calculate the volume of the solid.

*That's it!*

## Further Problems – XXIII

1. Evaluate $\displaystyle\int_0^{\pi}\int_0^{\cos\theta} r\sin\theta\, dr\, d\theta$

2. " $\displaystyle\int_0^{2\pi}\int_0^{3} r^3(9-r^2)\, dr\, d\theta$

3. " $\displaystyle\int_{-2}^{1}\int_{x^2+4x}^{3x+2} dy\, dx$

4. " $\displaystyle\int_0^{a}\int_0^{b}\int_0^{c} (x^2+y^2)\, dx\, dy\, dz$

5. " $\displaystyle\int_0^{\pi}\int_0^{\pi/2}\int_0^{r} x^2\sin\theta\, dx\, d\theta\, d\phi$

6. Find the area bounded by the curve $y = x^2$ and the line $y = x + 2$.

7. Find the area of the polar figure enclosed by the circle $r = 2$ and the cardioid $r = 2(1 + \cos\theta)$.

8. Evaluate $\displaystyle\int_0^{2} dx\int_1^{3} dy\int_1^{2} xy^2z\, dz$

9. " $\displaystyle\int_0^{2} dx\int_1^{2} (x^2+y^2)\, dy$

10. " $\displaystyle\int_0^{1} dr\int_0^{\pi/4} r\cos^2\theta\, d\theta$

11. Determine the area bounded by the curves $x = y^2$ and $x = 2y - y^2$.

12. Express as a double integral, the area contained by one loop of the curve $r = 2\cos 3\theta$ and evaluate the integral.

13. Evaluate $\displaystyle\int_0^{\pi/2}\int_{\pi/4}^{\tan^{-1}(2)}\int_0^{4} x\sin y\, dx\, dy\, dz$

14. Evaluate $\displaystyle\int_0^{\pi}\int_0^{4\cos z}\int_0^{\sqrt{(16-y^2)}} y\, dx\, dy\, dz$

15. A plane figure is bounded by the polar curve $r = a(1 + \cos\theta)$ between $\theta = 0$ and $\theta = \pi$, and the initial line OA. Express as a double integral the first moment of area of the figure about OA and evaluate the integral. If the area of the figure is known to be $\frac{3\pi a^2}{4}$ units$^2$, find the distance (h) of the centroid of the figure from OA.

16. Using double integrals, find (i) the area and (ii) the second moment about OX of the plane figure bounded by the $x$-axis and that part of the ellipse $\frac{x^2}{a^2} + \frac{y^2}{b^2} = 1$ which lies above OX. Find also the position of the centroid.

17. The base of a solid is the plane figure in the $xy$-plane bounded by $x = 0, x = 2, y = x$, and $y = x^2 + 1$. The sides are vertical and the top is the surface $z = x^2 + y^2$. Calculate the volume of the solid so formed.

18. A solid consists of vertical sides standing on the plane figure enclosed by $x = 0, x = b, y = a$ and $y = c$. The top is the surface $z = xy$. Find the volume of the solid so defined.

19. Show that the area outside the circle $r = a$ and inside the circle $r = 2a\cos\theta$ is given by

$$A = 2\int_0^{\pi/3}\int_a^{2a\cos\theta} r\,dr\,d\theta$$

Evaluate the integral.

20. A rectangular block is bounded by the co-ordinate planes of reference and by the planes $x = 3, y = 4, z = 2$. Its density at any point is numerically equal to the square of its distance from the origin. Find the total mass of the solid.

# Programme 24

## FIRST ORDER DIFFERENTIAL EQUATIONS

**1**

## Introduction

A *differential equation* is a relationship between an independent variable, $x$, a dependent variable, $y$, and one or more differential coefficients of $y$ with respect to $x$.

$$\text{e.g.}\quad x^2\frac{dy}{dx} + y\sin x = 0$$

$$xy\frac{d^2y}{dx^2} + y\frac{dy}{dx} + e^{3x} = 0$$

Differential equations represent dynamic relationships, i.e. quantities that change, and are thus frequently occurring in scientific and engineering problems.

The *order* of a differential equation is given by the highest derivative involved in the equation.

$x\frac{dy}{dx} - y^2 = 0$ is an equation of the 1st order

$xy\frac{d^2y}{dx^2} - y^2\sin x = 0$ " " " " 2nd "

$\frac{d^3y}{dx^3} - y\frac{dy}{dx} + e^{4x} = 0$ " " " " 3rd "

So that $\frac{d^2y}{dx^2} + 2\frac{dy}{dx} + 10y = \sin 2x$ is an equation of the .......... order.

**2**

second

Since in the equation $\frac{d^2y}{dx^2} + 2\frac{dy}{dx} + 10y = \sin 2x$, the highest derivative involved is $\frac{d^2y}{dx^2}$.

Similarly,

(i) $x\frac{dy}{dx} = y^2 + 1$ is a ...... order equation

(ii) $\cos^2 x\frac{dy}{dx} + y = 1$ is a ...... order equation

(iii) $\frac{d^2y}{dx^2} - 3\frac{dy}{dx} + 2y = x^2$ is a ...... order equation

(iv) $(y^3 + 1)\frac{dy}{dx} - xy^2 = x$ is a ...... order equation

*On to frame 3.*

3

(i) first, (ii) first, (iii) second, (iv) first.

**Formation of differential equations**

Differential equations may be formed in practice from a consideration of the physical problems to which they refer. Mathematically, they can occur when arbitrary constants are eliminated from a given function. Here are a few examples:

*Example 1.* Consider $y = A \sin x + B \cos x$, where A and B are two arbitrary constants.

If we differentiate, we get

$$\frac{dy}{dx} = A \cos x - B \sin x$$

and $$\frac{d^2y}{dx^2} = -A \sin x - B \cos x$$

which is identical to the original equation, but with the sign changed.

i.e. $$\frac{d^2y}{dx^2} = -y \quad \therefore \frac{d^2y}{dx^2} + y = 0$$

This is a differential equation of the ...... order.

4

second

*Example 2.* Form a differential equation from the function $y = x + \frac{A}{x}$

We have $$y = x + \frac{A}{x} = x + A\,x^{-1}$$

$$\therefore \frac{dy}{dx} = 1 - A\,x^{-2} = 1 - \frac{A}{x^2}$$

From the given equation, $\frac{A}{x} = y - x \quad \therefore A = x(y - x)$

$$\therefore \frac{dy}{dx} = 1 - \frac{x(y-x)}{x^2}$$

$$= 1 - \frac{y-x}{x} = \frac{x-y+x}{x} = \frac{2x-y}{x}$$

$$\therefore x\frac{dy}{dx} = 2x - y$$

This is an equation of the ...... order.

**5**

first

Now one more.

*Example 3.* Form the diff. equation for $y = Ax^2 + Bx$.

We have $y = Ax^2 + Bx$ (i)

$\therefore \frac{dy}{dx} = 2Ax + B$ (ii)

$\therefore \frac{d^2y}{dx^2} = 2A$ (iii) $A = \frac{1}{2}\frac{d^2y}{dx^2}$

Substitute for 2A in (ii) $\frac{dy}{dx} = x\frac{d^2y}{dx^2} + B$

$$\therefore B = \frac{dy}{dx} - x\frac{d^2y}{dx^2}$$

Substituting for A and B in (i), we have

$$y = x^2 . \frac{1}{2}\frac{d^2y}{dx^2} + x\left(\frac{dy}{dx} - x\frac{d^2y}{dx^2}\right)$$

$$= \frac{x^2}{2} . \frac{d^2y}{dx^2} + x . \frac{dy}{dx} - x^2 . \frac{d^2y}{dx^2}$$

$$\therefore y = x\frac{dy}{dx} - \frac{x^2}{2} . \frac{d^2y}{dx^2}$$

and this is an equation of the .......... order.

**6**

second

If we collect our last few results together, we have:

$y = A \sin x + B \cos x$ gives the equation $\frac{d^2y}{dx^2} + y = 0$ (2nd order)

$y = Ax^2 + Bx$ " " " $y = x\frac{dy}{dx} - \frac{x^2}{2} . \frac{d^2y}{dx^2}$ (2nd order)

$y = x + \frac{A}{x}$ " " " $x\frac{dy}{dx} = 2x - y$ (1st order)

If we were to investigate the following, we should also find that

$y = Axe^x$ gives the diff. equation $x\frac{dy}{dx} - y(1 + x) = 0$ (1st order)

$y = Ae^{-4x} + Be^{-6x}$ " " " " $\frac{d^2y}{dx^2} + 10\frac{dy}{dx} + 24y = 0$ (2nd order)

Some of the functions give 1st order equations: some give 2nd order equations. Now look at these five results and see if you can find any distinguishing features in the functions which decide whether we obtain a 1st order equation or a 2nd order equation in any particular case.

*When you have come to a conclusion, turn on to frame 7.*

**7**

A function with 1 arbitrary constant gives a 1st order equation.
" " " 2 arbitrary constants " " 2nd order "

Correct, and in the same way,

A function with 3 arbitrary constants would give a 3rd order equation.

So, without working each out in detail, we can say that

(i) $y = e^{-2x}(A + Bx)$ would give a diff. equation of .......... order.

(ii) $y = A\dfrac{x-1}{x+1}$ " " " " " " .......... "

(iii) $y = e^{3x}(A\cos 3x + B\sin 3x)$ " " " " .......... "

**8**

(i) 2nd, (ii) 1st, (iii) 2nd

since (i) and (iii) each have 2 arbitrary constants, while (ii) has only 1 arbitrary constant.

Similarly,

(i) $x^2\dfrac{dy}{dx} + y = 1$ is derived from a function having ........ arbitrary constants.

(ii) $\cos^2 x\dfrac{dy}{dx} = 1 - y$ " " a function having ........ arbitrary constants.

(iii) $\dfrac{d^2y}{dx^2} + 4\dfrac{dy}{dx} + y = e^{2x}$ " a function having ........ arbitrary constants.

**9**

| (i) 1, (ii) 1, (iii) 2 |
|---|

So from all this, the following rule emerges:

A 1st order diff. equation is derived from a function having 1 arbitrary constant.

A 2nd " " " " " " " " " 2 arbitrary constants.

An $n$th order differential equation is derived from a function having $n$ arbitrary constants.

Copy this last statement into your record book. It is important to remember this rule and we shall make use of it at various times in the future.

*Then on to frame 10.*

**10**

**Solution of differential equations**

To solve a differential equation, we have to find the function for which the equation is true. This means that we have to manipulate the equation so as to eliminate all the differential coefficients and leave a relationship between $y$ and $x$.

The rest of this particular programme is devoted to the various methods of solving *first order differential equations*. Second order equations will be dealt with in a subsequent programme.

*So, for the first method, turn on to frame 11.*

**11**

**Method 1** *By direct integration*

If the equation can be arranged in the form $\frac{dy}{dx} = f(x)$, then the equation can be solved by simple integration.

*Example 1.* $\frac{dy}{dx} = 3x^2 - 6x + 5$

Then $y = \int (3x^2 - 6x + 5)\,dx = x^3 - 3x^2 + 5x + C$

i.e. $\underline{y = x^3 - 3x^2 + 5x + C}$

As always, of course, the constant of integration must be included. Here it provides the one arbitrary constant which we always get when solving a first order differential equation.

*Example 2.* Solve $x\frac{dy}{dx} = 5x^3 + 4$

In this case, $\frac{dy}{dx} = 5x^2 + \frac{4}{x}$

So, $y = \ldots\ldots\ldots\ldots\ldots$

**12**

$$\boxed{y = \frac{5x^3}{3} + 4\ln x + C}$$

As you already know from your work on integration, the value of C cannot be determined unless further information about the function is given. In its present form, the function is called the *general solution* (or *primitive*) of the given equation.

If we are told the value of $y$ for a given value of $x$, C can be evaluated and the result is then a *particular solution* of the equation.

*Example 3.* Find the particular solution of the equation $e^x \frac{dy}{dx} = 4$, given that $y = 3$ when $x = 0$.

First re-write the equation in the form $\frac{dy}{dx} = \frac{4}{e^x} = 4e^{-x}$

Then $y = \int 4e^{-x}\,dx = -4e^{-x} + C$

Knowing that when $x = 0, y = 3$, we can evaluate C in this case, so that the required particular solution is

$y = \ldots\ldots\ldots\ldots\ldots$

$$y = -4e^{-x} + 7$$

**13**

**Method 2** *By separating the variables*

If the given equation is of the form $\frac{dy}{dx} = f(x, y)$, the variable $y$ on the right-hand side, prevents solving by direct integration. We therefore have to devise some other method of solution.

Let us consider equations of the form $\frac{dy}{dx} = f(x).F(y)$ and of the form $\frac{dy}{dx} = \frac{f(x)}{F(y)}$, i.e. equations in which the right-hand side can be expressed as products or quotients of functions of $x$ or of $y$.

A few examples will show how we proceed.

*Example 1.* Solve $\frac{dy}{dx} = \frac{2x}{y+1}$

We can re-write this as $(y+1)\frac{dy}{dx} = 2x$

Now integrate both sides with respect to $x$

$$\int (y+1)\frac{dy}{dx}\,dx = \int 2x\,dx \quad \text{i.e.} \quad \int (y+1)\,dy = \int 2x\,dx$$

and this gives $\frac{y^2}{2} + y = x^2 + C$

**14**

*Example 2.* Solve $\frac{dy}{dx} = (1+x)(1+y)$

$$\frac{1}{1+y}\frac{dy}{dx} = 1 + x$$

Integrate both sides with respect to $x$

$$\int \frac{1}{1+y}\frac{dy}{dx}\,dx = \int (1+x)\,dx \quad \therefore \quad \int \frac{1}{1+y}\,dy = \int (1+x)\,dx$$

$$\ln(1+y) = x + \frac{x^2}{2} + C$$

The method depends on our being able to express the given equation in the form $F(y).\frac{dy}{dx} = f(x)$. If this can be done, the rest is then easy, for we have $\int F(y)\frac{dy}{dx}\,dx = \int f(x)\,dx \quad \therefore \quad \int F(y)\,dy = \int f(x)\,dx$

and we then continue as in the examples.

*Let us see another example, so turn on to frame 15.*

*Example 3.* Solve $\dfrac{dy}{dx} = \dfrac{1+y}{2+x}$ (i)

This can be written as $\dfrac{1}{1+y}\dfrac{dy}{dx} = \dfrac{1}{2+x}$

Integrate both sides with respect to $x$

$$\int \frac{1}{1+y}\frac{dy}{dx}\,dx = \int \frac{1}{2+x}\,dx$$

$$\therefore \int \frac{1}{1+y}\,dy = \int \frac{1}{2+x}\,dx \qquad \text{(ii)}$$

$$\therefore \ln(1+y) = \ln(2+x) + C$$

It is convenient to write the constant C as the logarithm of some other constant A

$$\ln(1+y) = \ln(2+x) + \ln A$$

$$\therefore 1 + y = A(2+x)$$

*Note:* We can, in practice, get from the given equation (i) to the form of the equation in (ii) by a simple routine, thus:

$$\frac{dy}{dx} = \frac{1+y}{2+x}$$

First, multiply across by the $dx$

$$dy = \frac{1+y}{2+x}\,dx$$

Now collect the '$y$-factor' with the $dy$ on the left, i.e. divide by $(1+y)$

$$\frac{1}{1+y}\,dy = \frac{1}{2+x}\,dx$$

Finally, add the integral signs

$$\int \frac{1}{1+y}\,dy = \int \frac{1}{2+x}\,dx$$

and then continue as before.

This is purely a routine which enables us to sort out the equation algebraically, the whole of the work being done in one line. Notice, however, that the R.H.S. of the given equation must first be expressed as '$x$-factors' and '$y$-factors'.

Now for another example, using this routine.

*Example 4.* Solve $\dfrac{dy}{dx} = \dfrac{y^2 + xy^2}{x^2y - x^2}$

First express the R.H.S. in '$x$-factors' and '$y$-factors'

$$\frac{dy}{dx} = \frac{y^2(1+x)}{x^2(y-1)}$$

Now re-arrange the equation so that we have the '$y$-factors' and $dy$ on the L.H.S. and the '$x$-factors' and $dx$ on the R.H.S.

So we get ...........................

## 16

$$\boxed{\frac{y-1}{y^2}\,dy = \frac{1+x}{x^2}\,dx}$$

We now add the integral signs

$$\int \frac{y-1}{y^2}\,dy = \int \frac{1+x}{x^2}\,dx$$

and complete the solution

$$\int \left\{\frac{1}{y} - y^{-2}\right\} dy = \int \left\{x^{-2} + \frac{1}{x}\right\} dx$$

$$\therefore\ \ln y + y^{-1} = \ln x - x^{-1} + C$$

$$\therefore\ \ln y + \frac{1}{y} = \ln x - \frac{1}{x} + C$$

Here is another.

*Example 5.* Solve $\dfrac{dy}{dx} = \dfrac{y^2 - 1}{x}$

Re-arranging, we have $dy = \dfrac{y^2 - 1}{x}\,dx$

$$\frac{1}{y^2 - 1}\,dy = \frac{1}{x}\,dx$$

$$\therefore\ \int \frac{1}{y^2 - 1}\,dy = \int \frac{1}{x}\,dx$$

Which gives ..............................................

## 17

$$\boxed{\frac{1}{2}\ln\frac{y-1}{y+1} = \ln x + C}$$

$$\therefore\ \ln\frac{y-1}{y+1} = 2\ln x + \ln A$$

$$\therefore\ \frac{y-1}{y+1} = A\,x^2$$

$$y - 1 = A\,x^2\,(y+1)$$

You see they are all done in the same way. Now here is one for you to do:

*Example 6.* Solve $xy\dfrac{dy}{dx} = \dfrac{x^2 + 1}{y + 1}$

First of all, re-arrange the equation into the form

$$F(y)\,dy = f(x)\,dx$$

i.e. arrange the '$y$-factors' and $dy$ on the L.H.S. and the '$x$-factors' and $dx$ on the R.H.S.

*What do you get?*

**18**

$$y(y+1)\,dy = \frac{x^2+1}{x}\,dx$$

for

$$xy\frac{dy}{dx} = \frac{x^2+1}{y+1}$$

$$\therefore\ xy\,dy = \frac{x^2+1}{y+1}\,dx$$

$$\therefore\ y(y+1)\,dy = \frac{x^2+1}{x}\,dx$$

So we now have

$$\int (y^2+y)\,dy = \int \left(x+\frac{1}{x}\right)dx$$

*Now finish it off, then move on to the next frame.*

---

**19**

$$\frac{y^3}{3}+\frac{y^2}{2} = \frac{x^2}{2} + \ln x + C$$

□□□□□□□□□□□□□□□□□□□□□□□□□□□□□□□□□□□□□□

Provided that the R.H.S. of the equation $\frac{dy}{dx} = f(x, y)$ can be separated into '$x$-factors' and '$y$-factors', the equation can be solved by the method of *separating the variables.*

Now do this one entirely on your own.

*Example 6.* Solve $x\frac{dy}{dx} = y + xy$

*When you have finished it completely, turn to frame 20 and check your solution.*

**20**

Here is the result. Follow it through carefully, even if your own answer is correct.

$$x\frac{dy}{dx} = y + xy \quad \therefore\ x\frac{dy}{dx} = y(1 + x)$$

$$x\,dy = y(1 + x)\,dx$$

$$\therefore\ \frac{dy}{y} = \frac{1 + x}{x}\,dx$$

$$\therefore\ \int\frac{1}{y}dy = \int\left(\frac{1}{x} + 1\right)dx$$

$$\therefore\ \ln y = \ln x + x + C$$

At this stage, we have eliminated the differential coefficients and so we have solved the equation. However, we can express the result in a neater form, thus:

$$\ln y - \ln x = x + C$$

$$\therefore\ \ln\left\{\frac{y}{x}\right\} = x + C$$

$$\therefore\ \frac{y}{x} = e^{x + c} = e^x.e^c$$

Now $e^c$ is a constant; call it A.

$$\therefore\ \frac{y}{x} = \mathrm{A}e^x \quad \therefore\ y = \mathrm{A}xe^x$$

*Next frame.*

**21**

This final example looks more complicated, but it is solved in just the same way. We go through the same steps as before. Here it is.

*Example 7.* Solve $y\tan x\dfrac{dy}{dx} = (4 + y^2)\sec^2 x$

First separate the variables, i.e. arrange the '$y$-factors' and $dy$ on one side and the '$x$-factors' and $dx$ on the other.

So we get ..............................

**22**

$$\boxed{\frac{y}{4 + y^2}\,dy = \frac{\sec^2 x}{\tan x}\,dx}$$

Adding the integral signs, we get

$$\int\frac{y}{4 + y^2}\,dy = \int\frac{\sec^2 x}{\tan x}\,dx$$

Now determine the integrals, so that we have ..............................

**23**

$$\boxed{\frac{1}{2}\ln(4+y^2) = \ln\tan x + C}$$

This result can now be simplified into:

$$\ln(4+y^2) = 2\ln\tan x + \ln A \quad \text{(expressing the constant 2C as ln A)}$$

$$\therefore\ 4+y^2 = A\tan^2 x$$

$$\therefore\ \underline{y^2 = A\tan^2 x - 4}$$

So there we are. Provided we can factorize the equation in the way we have indicated, solution by separating the variables is not at all difficult. So now for a short revision exercise to wind up this part of the programme.

*Move on to frame 24.*

**24**

**Revision Exercise**

Work all the exercise before checking your results.

Find the general solutions of the following equations:

1. $\dfrac{dy}{dx} = \dfrac{y}{x}$
2. $\dfrac{dy}{dx} = (y+2)(x+1)$
3. $\cos^2 x \dfrac{dy}{dx} = y + 3$
4. $\dfrac{dy}{dx} = xy - y$
5. $\dfrac{\sin x}{1+y} \cdot \dfrac{dy}{dx} = \cos x$

***When you have finished them all, turn to frame 25 and check your solutions.***

**25**

**Solutions**

1. $$\frac{dy}{dx} = \frac{y}{x} \quad \therefore \int \frac{1}{y}\,dy = \int \frac{1}{x}\,dx$$
$$\therefore \ln y = \ln x + C$$
$$= \ln x + \ln A$$
$$\therefore \underline{y = Ax}$$

2. $$\frac{dy}{dx} = (y+2)(x+1)$$
$$\therefore \int \frac{1}{y+2}\,dy = \int (x+1)\,dx$$
$$\therefore \underline{\ln(y+2) = \frac{x^2}{2} + x + C}$$

3. $$\cos^2 x\,\frac{dy}{dx} = y + 3$$
$$\therefore \int \frac{1}{y+3}\,dy = \int \frac{1}{\cos^2 x}\,dx$$
$$= \int \sec^2 x\,dx$$
$$\underline{\ln(y+3) = \tan x + C}$$

4. $$\frac{dy}{dx} = xy - y \quad \therefore \frac{dy}{dx} = y(x-1)$$
$$\therefore \int \frac{1}{y}\,dy = \int (x-1)\,dx$$
$$\therefore \underline{\ln y = \frac{x^2}{2} - x + C}$$

5. $$\frac{\sin x}{1+y} \cdot \frac{dy}{dx} = \cos x$$
$$\int \frac{1}{1+y}\,dy = \int \frac{\cos x}{\sin x}\,dx$$
$$\therefore \ln(1+y) = \ln \sin x + C$$
$$= \ln \sin x + \ln A$$
$$1 + y = A \sin x$$
$$\therefore \underline{y = A \sin x - 1}$$

□□□□□□□□□□□□□□□□□□□□□□□□□□□□□□□□□□□□□□□□□□□□□□□

**If you are quite happy about those, we can start the next part of the programme, so turn on now to frame 26.**

**26**

**Method 3** *Homogeneous equations – by substituting y = vx*

Here is an equation: $\dfrac{dy}{dx} = \dfrac{x + 3y}{2x}$

This looks simple enough, but we find that we cannot express the R.H.S. in the form of '$x$-factors' and '$y$-factors', so we cannot solve by the method of separating the variables.

In this case we make the substitution $y = vx$, where $v$ is a function of $x$.

$$\text{So} \quad y = vx$$

Differentiate with respect to $x$ (using the product rule).

$$\therefore \frac{dy}{dx} = v.1 + x\frac{dv}{dx} = v + x\frac{dv}{dx}$$

Also $$\frac{x + 3y}{2x} = \frac{x + 3vx}{2x} = \frac{1 + 3v}{2}$$

The equation now becomes $v + x\dfrac{dv}{dx} = \dfrac{1 + 3v}{2}$

$$\therefore x\frac{dv}{dx} = \frac{1 + 3v}{2} - v$$

$$= \frac{1 + 3v - 2v}{2} = \frac{1 + v}{2}$$

$$\therefore x\frac{dv}{dx} = \frac{1 + v}{2}$$

The given equation is now expressed in terms of $v$ and $x$, and in this form we find that we can solve by separating the variables. Here goes:

$$\int \frac{2}{1 + v}\,dv = \int \frac{1}{x}\,dx$$

$$\therefore 2 \ln (1 + v) = \ln x + C = \ln x + \ln A$$

$$(1 + v)^2 = Ax$$

But $\quad y = vx \therefore v = \left\{\dfrac{y}{x}\right\} \therefore \left(1 + \dfrac{y}{x}\right)^2 = Ax$

which gives $\quad \underline{(x + y)^2 = Ax^3}$

*Note.* $\dfrac{dy}{dx} = \dfrac{x + 3y}{2x}$ is an example of a *homogeneous diff. equation.*

This is determined by the fact that the total degree in $x$ and $y$ for each of the terms involved is the same (in this case, of degree 1). The key to solving every homogeneous equation is to substitute $y = vx$ where $v$ is a function of $x$. This converts the equation into a form in which we can solve by separating the variables.

*Let us work another example, so turn on to frame 27.*

**27** *Example 2.* Solve $\dfrac{dy}{dx} = \dfrac{x^2 + y^2}{xy}$

Here, all terms on the R.H.S. are of degree 2, i.e. the equation is homogeneous. ∴ We substitute $y = vx$ (where $v$ is a function of $x$)

$$\therefore \frac{dy}{dx} = v + x\frac{dv}{dx}$$

and $$\frac{x^2 + y^2}{xy} = \frac{x^2 + v^2x^2}{vx^2} = \frac{1 + v^2}{v}$$

The equation now becomes

$$v + x\frac{dv}{dx} = \frac{1 + v^2}{v}$$

$$\therefore x\frac{dv}{dx} = \frac{1 + v^2}{v} - v$$

$$= \frac{1 + v^2 - v^2}{v} = \frac{1}{v}$$

$$\therefore x\frac{dv}{dx} = \frac{1}{v}$$

Now you can separate the variables and get the result in terms of $v$ and $x$.

*Off you go: when you have finished, move to frame 28.*

---

**28**

$$\boxed{\frac{v^2}{2} = \ln x + C}$$

for $$\int v\,dv = \int \frac{1}{x}\,dx$$

$$\therefore \frac{v^2}{2} = \ln x + C$$

All that now remains is to express $v$ back in terms of $x$ and $y$. The substitution we used was $y = vx \quad \therefore v = \dfrac{y}{x}$

$$\therefore \frac{1}{2}\left(\frac{y}{x}\right)^2 = \ln x + C$$

$$\underline{y^2 = 2x^2(\ln x + C)}$$

Now, what about this one?

*Example 3.* Solve $\dfrac{dy}{dx} = \dfrac{2xy + 3y^2}{x^2 + 2xy}$

Is this a homogeneous equation? If you think so, what are your reasons?

*When you have decided, turn on to frame 29.*

**29**

Yes, because the degree of each term is the same

Correct. They are all, of course, of degree 2.
So we now make the substitution, $y =$ ......................

**30**

$y = vx$, where $v$ is a function of $x$

Right. That is the key to the whole process.

$$\frac{dy}{dx} = \frac{2xy + 3y^2}{x^2 + 2xy}$$

So express each side of the equation in terms of $v$ and $x$.

$$\frac{dy}{dx} = \ldots\ldots\ldots\ldots$$

and
$$\frac{2xy + 3y^2}{x^2 + 2xy} = \ldots\ldots\ldots\ldots$$

*When you have finished, move on to the next frame.*

**31**

$$\frac{dy}{dx} = v + x\frac{dv}{dx}$$

$$\frac{2xy + 3y^2}{x^2 + 2xy} = \frac{2vx^2 + 3v^2x^2}{x^2 + 2vx^2} = \frac{2v + 3v^2}{1 + 2v}$$

So that
$$v + x\frac{dv}{dx} = \frac{2v + 3v^2}{1 + 2v}$$

Now take the single $v$ over to the R.H.S. and simplify, giving

$$x\frac{dv}{dx} = \ldots\ldots\ldots\ldots$$

**32**

$$x\frac{dv}{dx}=\frac{2v+3v^2}{1+2v}-v$$
$$=\frac{2v+3v^2-v-2v^2}{1+2v}$$
$$x\frac{dv}{dx}=\frac{v+v^2}{1+2v}$$

Now you can separate the variables, giving ..................

**33**

$$\int\frac{1+2v}{v+v^2}\,dv=\int\frac{1}{x}\,dx$$

Integrating both sides, we can now obtain the solution in terms of $v$ and $x$. What do you get?

**34**

$$\ln(v+v^2)=\ln x+C$$
$$=\ln x+\ln A$$
$$\therefore v+v^2=Ax$$

We have almost finished the solution. All that remains is to express $v$ back in terms of $x$ and $y$.

Remember the substitution was $y = vx$, so that $v=\frac{y}{x}$

So finish it off.

*Then move on.*

**35**

$$xy+y^2=Ax^3$$

for $$v+v^2=Ax \text{ and } v=\frac{y}{x}$$

$$\therefore \frac{y}{x}+\frac{y^2}{x^2}=Ax$$

$$\underline{xy+y^2=Ax^3}$$

And that is all there is to it.

*Turn to frame 36.*

**36**

Here is the solution of the last equation, all in one piece. Follow it through again.

To solve $$\frac{dy}{dx} = \frac{2xy + 3y^2}{x^2 + 2xy}$$

This is homogeneous, all terms of degree 2. Put $y = vx$

$$\therefore \frac{dy}{dx} = v + x\frac{dv}{dx}$$

$$\frac{2xy + 3y^2}{x^2 + 2xy} = \frac{2vx^2 + 3v^2x^2}{x^2 + 2vx^2} = \frac{2v + 3v^2}{1 + 2v}$$

$$\therefore v + x\frac{dv}{dx} = \frac{2v + 3v^2}{1 + 2v}$$

$$x\frac{dv}{dx} = \frac{2v + 3v^2}{1 + 2v} - v$$

$$= \frac{2v + 3v^2 - v - 2v^2}{1 + 2v}$$

$$\therefore x\frac{dv}{dx} = \frac{v + v^2}{1 + 2v}$$

$$\therefore \int \frac{1 + 2v}{v + v^2}\,dv = \int \frac{1}{x}\,dx$$

$$\therefore \ln(v + v^2) = \ln x + C = \ln x + \ln A$$

$$v + v^2 = Ax$$

But $$y = vx \quad \therefore v = \frac{y}{x}$$

$$\therefore \frac{y}{x} + \frac{y^2}{x^2} = Ax$$

$$\therefore xy + y^2 = Ax^3$$

Now, in the same way, you do this one. Take your time and be sure that you understand each step.

*Example 4.* Solve $$(x^2 + y^2)\frac{dy}{dx} = xy$$

*When you have completely finished it, turn to frame 37 and check your solution.*

**37** Here is the solution in full.

$$(x^2 + y^2)\frac{dy}{dx} = xy \quad \therefore \frac{dy}{dx} = \frac{xy}{x^2 + y^2}$$

Put $$y = vx \quad \therefore \frac{dy}{dx} = v + x\frac{dv}{dx}$$

and $$\frac{xy}{x^2 + y^2} = \frac{vx^2}{x^2 + v^2x^2} = \frac{v}{1 + v^2}$$

$$\therefore v + x\frac{dv}{dx} = \frac{v}{1 + v^2}$$

$$x\frac{dv}{dx} = \frac{v}{1 + v^2} - v$$

$$x\frac{dv}{dx} = \frac{v - v - v^3}{1 + v^2} = \frac{-v^3}{1 + v^2}$$

$$\therefore \int \frac{1 + v^2}{v^3}\,dv = -\int \frac{1}{x}\,dx$$

$$\therefore \int \left(v^{-3} + \frac{1}{v}\right) dv = -\ln x + C$$

$$\therefore \frac{-v^{-2}}{2} + \ln v = -\ln x + \ln A$$

$$\ln v + \ln x + \ln K = \frac{1}{2v^2}$$

$$\ln Kvx = \frac{1}{2v^2}$$

But $$v = \frac{y}{x} \quad \therefore \ln Ky = \frac{x^2}{2y^2}$$

$$\underline{2y^2 \ln Ky = x^2}$$

This is one form of the solution: there are of course other ways of expressing it.

*Now for a short revision exercise on this part of the work, move on to frame 38.*

---

**38** **Revision Exercise**

Solve the following:

1. $(x - y)\dfrac{dy}{dx} = x + y$

2. $2x^2 \dfrac{dy}{dx} = x^2 + y^2$

3. $(x^2 + xy)\dfrac{dy}{dx} = xy - y^2$

*When you have finished all three, turn on and check your results.*

**39**

The solution of equation 1 can be written as

$$\tan^{-1}\left\{\frac{y}{x}\right\} = \ln A + \ln x + \frac{1}{2}\ln\left\{1 + \frac{y^2}{x^2}\right\}$$

Did you get that? If so, move straight on to frame 40. If not, check your working with the following.

1. $$(x - y)\frac{dy}{dx} = x + y \quad \therefore \frac{dy}{dx} = \frac{x + y}{x - y}$$

Put $$y = vx \quad \therefore \frac{dy}{dx} = v + x\frac{dv}{dx} \qquad \frac{x + y}{x - y} = \frac{1 + v}{1 - v}$$

$$\therefore v + x\frac{dv}{dx} = \frac{1 + v}{1 - v} \qquad \therefore x\frac{dv}{dx} = \frac{1 + v}{1 - v} - v = \frac{1 + v - v + v^2}{1 - v} = \frac{1 + v^2}{1 - v}$$

$$\therefore \int \frac{1 - v}{1 + v^2}\,dv = \int \frac{1}{x}\,dx \qquad \therefore \int\left\{\frac{1}{1 + v^2} - \frac{v}{1 + v^2}\right\}dv = \ln x + C$$

$$\therefore \tan^{-1} v - \frac{1}{2}\ln(1 + v^2) = \ln x + \ln A$$

But $$v = \frac{y}{x} \quad \therefore \tan^{-1}\left\{\frac{y}{x}\right\} = \ln A + \ln x + \frac{1}{2}\ln\left(1 + \frac{y^2}{x^2}\right)$$

This result can, in fact, be simplified further.

*Now on to frame 40.*

**40**

Equation 2 gives the solution

$$\frac{2x}{x - y} = \ln x + C$$

If you agree, move straight on to frame 41. Otherwise, follow through the working. Here it is.

2. $$2x^2\frac{dy}{dx} = x^2 + y^2 \quad \therefore \frac{dy}{dx} = \frac{x^2 + y^2}{2x^2}$$

Put $$y = vx \quad \therefore \frac{dy}{dx} = v + x\frac{dv}{dx}; \qquad \frac{x^2 + y^2}{2x^2} = \frac{x^2 + v^2x^2}{2x^2} = \frac{1 + v^2}{2}$$

$$\therefore v + x\frac{dv}{dx} = \frac{1 + v^2}{2} \qquad \therefore x\frac{dv}{dx} = \frac{1 + v^2}{2} - v = \frac{1 - 2v + v^2}{2} = \frac{(v - 1)^2}{2}$$

$$\therefore \int \frac{2}{(v - 1)^2}\,dv = \int \frac{1}{x}\,dx \quad \therefore -2\,\frac{1}{v - 1} = \ln x + C$$

But $$v = \frac{y}{x} \text{ and } \frac{2}{1 - v} = \ln x + C \quad \therefore \frac{2x}{x - y} = \ln x + C$$

*On to frame 41.*

**41** One form of the result for equation 3 is $\boxed{xy = A\,e^{x/y}}$ Follow through the working and check yours.

3. $$(x^2 + xy)\frac{dy}{dx} = xy - y^2 \quad \therefore \frac{dy}{dx} = \frac{xy - y^2}{x^2 + xy}$$

Put $y = vx \quad \therefore \dfrac{dy}{dx} = v + x\dfrac{dv}{dx}; \quad \dfrac{xy - y^2}{x^2 + xy} = \dfrac{vx^2 - v^2x^2}{x^2 + vx^2} = \dfrac{v - v^2}{1 + v}$

$$\therefore v + x\frac{dv}{dx} = \frac{v - v^2}{1 + v}$$

$$x\frac{dv}{dx} = \frac{v - v^2}{1 + v} - v = \frac{v - v^2 - v - v^2}{1 + v} = \frac{-2v^2}{1 + v}$$

$$\therefore \int \frac{1 + v}{v^2}\,dv = \int \frac{-2}{x}\,dx$$

$$\int \left(v^{-2} + \frac{1}{v}\right)dv = -\int \frac{2}{x}\,dx$$

$$\therefore \ln v - \frac{1}{v} = -2\ln x + C. \quad \text{Let } C = \ln A$$

$$\ln v + 2\ln x = \ln A + \frac{1}{v}$$

$$\ln\left\{\frac{y}{x}.\,x^2\right\} = \ln A + \frac{x}{y} \qquad \therefore \underline{xy = A\,e^{x/y}}$$

*Now move to the next frame.*

**42** **Method 4** *Linear equations – use of integrating factor*

Consider the equation $\dfrac{dy}{dx} + 5y = e^{2x}$

This is clearly an equation of the first order, but different from those we have dealt with so far. In fact, none of our previous methods could be used to solve this one, so we have to find a further method of attack.

In this case, we begin by multiplying both sides by $e^{5x}$. This gives

$$e^{5x}\frac{dy}{dx} + y\,5\,e^{5x} = e^{2x}.e^{5x} \quad = e^{7x}$$

We now find that the L.H.S. is, in fact, the differential coefficient of $y.e^{5x}$. $\quad \therefore \dfrac{d}{dx}\left\{y.e^{5x}\right\} = e^{7x}$

Now, of course, the rest is easy. Integrate both sides w.r.t. $x$.

$$\therefore y.e^{5x} = \int e^{7x}\,dx \quad = \frac{e^{7x}}{7} + C \qquad \therefore \; y = \ldots\ldots\ldots\ldots\ldots\ldots$$

**43**

$$y = \frac{e^{2x}}{7} + Ce^{-5x}$$

Did you forget to divide the C by the $e^{5x}$? It is a common error so watch out for it.

□□□□□□□□□□□□□□□□□□□□□□□□□□□□□□□□□□□□□□□

The equation we have just solved is an example of a set of equations of the form $\frac{dy}{dx} + Py = Q$, where P and Q are functions of $x$ (or constants). This equation is called a *linear equation of the first order* and to solve any such equation, we multiply both sides by an *integrating factor* which is always $e^{\int P\,dx}$. This converts the L.H.S. into a complete differential coefficient.

In our last example, $\frac{dy}{dx} + 5y = e^{2x}$, P = 5. $\therefore \int P\,dx = 5x$ and the integrating factor was therefore $e^{5x}$. *Note* that in determining $\int P\,dx$, we do not include a constant of integration. This omission is purely for convenience, for a constant of integration here would in fact give a constant factor on both sides of the equation, which would subsequently cancel. This is one of the rare occasions when we do not write down the constant of integration.

So: *To solve a differential equation of the form*

$$\frac{dy}{dx} + Py = Q$$

*where* P *and* Q *are constants or functions of x, multiply both sides by the integrating factor* $e^{\int P\,dx}$.

This is important, so copy this rule down into your record book.

*Then move on to frame 44.*

**44**

*Example 1.* To solve $\frac{dy}{dx} - y = x$.

If we compare this with $\frac{dy}{dx} + Py = Q$, we see that in this case

$$P = -1 \text{ and } Q = x$$

The integrating factor is always $e^{\int P\,dx}$ and here P = −1.

$\therefore \int P\,dx = -x$ and the integrating factor is therefore ....................

**45**

$$\boxed{e^{-x}}$$

We therefore multiply both sides by $e^{-x}$.

$$\therefore\ e^{-x}\frac{dy}{dx} - y\,e^{-x} = x\,e^{-x}$$

$$\frac{d}{dx}\left\{e^{-x}\,y\right\} = x\,e^{-x} \qquad \therefore\ y\,e^{-x} = \int x\,e^{-x}\,dx$$

The R.H.S. integral can now be determined by integrating by parts.

$$y\,e^{-x} = x(-e^{-x}) + \int e^{-x}\,dx \qquad = -x\,e^{-x} - e^{-x} + C$$

$$\therefore\ y = -x - 1 + C\,e^{x} \qquad \therefore\ \underline{y = C\,e^{x} - x - 1}$$

The whole method really depends on

(i) being able to find the integrating factor,
(ii) being able to deal with the integral that emerges on the R.H.S.

Let us consider the general case.

**46**

Consider $\dfrac{dy}{dx} + Py = Q$ where P and Q are functions of $x$. Integrating factor, IF $= e^{\int P\,dx}$ $\quad \therefore\ \dfrac{dy}{dx}.\,e^{\int P\,dx} + Py\,e^{\int P\,dx} = Q\,e^{\int P\,dx}$

You will now see that the L.H.S. is the differential coefficient of $y\,e^{\int P\,dx}$

$$\therefore\ \frac{d}{dx}\left\{y\,e^{\int P\,dx}\right\} = Q\,e^{\int P\,dx}$$

Integrate both sides with respect to $x$

$$y\,e^{\int P\,dx} = \int Q\,e^{\int P\,dx}.\,dx$$

This result looks far more complicated than it really is. If we indicate the integrating factor by IF, this result becomes

$$y.\text{IF} = \int Q.\text{IF}\,dx$$

and, in fact, we remember it in that way.

*So, the solution of an equation of the form*

$$\frac{dy}{dx} + Py = Q \text{ (where P and Q are functions of } x)$$

is given by $\quad \underline{y.\text{IF} = \int Q.\text{IF}\,dx, \text{ where IF} = e^{\int P\,dx}}$

Copy this into your record book. *Then turn to frame 47.*

**47**

So if we have the equation

$$\frac{dy}{dx} + 3y = \sin x$$

$$\left[\frac{dy}{dx} + \mathrm{P}y = \mathrm{Q}\right]$$

then in this case

(i) $\mathrm{P} =$ ........ (ii) $\int \mathrm{P}\,dx =$ ........ (iii) IF = ........

---

**48**

(i) $\mathrm{P} = 3$; (ii) $\int \mathrm{P}\,dx = 3x$; (iii) $\mathrm{IF} = e^{3x}$

□□□□□□□□□□□□□□□□□□□□□□□□□□□□□□□□□□□□□□

Before we work through any further examples, let us establish a very useful piece of simplification, which we can make good use of when we are finding integrating factors. We want to simplify $e^{\ln \mathrm{F}}$, where F is a function of $x$.

$$\text{Let} \quad y = e^{\ln \mathrm{F}}$$

Then, by the very definition of a logarithm, $\ln y = \ln \mathrm{F}$

$$\therefore\ y = \mathrm{F} \quad \therefore\ \mathrm{F} = e^{\ln \mathrm{F}} \quad \text{i.e.}\ \underline{e^{\ln \mathrm{F}} = \mathrm{F}}$$

This means that $e^{\ln (\text{function})} = \text{function}$. Always!

$$e^{\ln x} = x$$

$$e^{\ln \sin x} = \sin x$$

$$e^{\ln \tanh x} = \tanh x$$

$$e^{\ln (x^2)} = \ldots\ldots\ldots\ldots$$

---

**49**

$x^2$

Similarly, what about $e^{k \ln \mathrm{F}}$? If the log in the index is multiplied by any external coefficient, this coefficient must be taken inside the log as a power.

$$\text{e.g.} \quad e^{2 \ln x} = e^{\ln (x^2)} = x^2$$

$$e^{3 \ln \sin x} = e^{\ln (\sin^3 x)} = \sin^3 x$$

$$e^{-\ln x} = e^{\ln (x^{-1})} = x^{-1} = \frac{1}{x}$$

$$\text{and} \quad e^{-2 \ln x} = \ldots\ldots\ldots\ldots\ldots\ldots$$

**50**

$\boxed{\dfrac{1}{x^2}}$ for $e^{-2\ln x} = e^{\ln(x^{-2})} = x^{-2} = \dfrac{1}{x^2}$

So here is the rule once again: $e^{\ln F} = F$

Make a note of this rule in your record book.

*Then on to frame 51.*

**51**

Now let us see how we can apply this result to our working.

*Example 2.* Solve $x\dfrac{dy}{dx} + y = x^3$

First we divide through by $x$ to reduce the first term to a single $\dfrac{dy}{dx}$

$$\text{i.e.}\quad \frac{dy}{dx} + \frac{1}{x}.y = x^2$$

$$\text{Compare with}\quad \left[\frac{dy}{dx} + Py = Q\right] \quad \therefore \; P = \frac{1}{x} \text{ and } Q = x^2$$

$$IF = e^{\int P\,dx} \qquad \int P\,dx = \int \frac{1}{x}dx = \ln x$$

$$\therefore IF = e^{\ln x} = x \quad \therefore IF = x$$

$$\text{The solution is}\quad y.IF = \int Q.IF\,dx$$

so $yx = \int x^2.x\,dx = \int x^3\,dx = \dfrac{x^4}{4} + C \quad \therefore \; \underline{xy = \dfrac{x^4}{4} + C}$

*Move to frame 52.*

**52**

*Example 3.* Solve $\dfrac{dy}{dx} + y\cot x = \cos x$

$$\text{Compare with}\quad \left[\frac{dy}{dx} + Py = Q\right] \quad \therefore \; \begin{cases} P = \cot x \\ Q = \cos x \end{cases}$$

$$IF = e^{\int P\,dx} \quad \int P\,dx = \int \cot x\,dx = \int \frac{\cos x}{\sin x}dx = \ln \sin x$$

$$\therefore IF = e^{\ln \sin x} = \sin x$$

$$y.IF = \int Q.IF\,dx \quad \therefore y\sin x = \int \sin x\cos x\,dx = \frac{\sin^2 x}{2} + C$$

$$\therefore \; \underline{y = \frac{\sin x}{2} + C\,\mathrm{cosec}\,x}$$

Now here is another.

*Example 4.* Solve $(x+1)\dfrac{dy}{dx} + y = (x+1)^2$

The first thing is to ..........................................................

**53**

Divide through by $(x + 1)$

Correct, since we must reduce the coefficient of $\dfrac{dy}{dx}$ to 1.

$$\therefore \frac{dy}{dx} + \frac{1}{x+1}.y = x + 1$$

Compare with $\qquad \dfrac{dy}{dx} + \text{P}y = \text{Q}$

In this case $\qquad \text{P} = \dfrac{1}{x+1}$ and $\text{Q} = x + 1$

Now determine the integrating factor, which simplifies to

IF = ......................

**54**

$\text{IF} = x + 1$

for $\qquad \displaystyle\int \text{P}\,dx = \int \frac{1}{x+1}\,dx = \ln(x+1)$

$$\therefore \text{IF} = e^{\ln(x+1)} = (x+1)$$

The solution is always $\qquad y.\text{IF} = \displaystyle\int \text{Q}.\text{IF}\,dx$

and we know that, in this case, $\text{IF} = x + 1$ and $\text{Q} = x + 1$.

*So finish off the solution and then move on to frame 55.*

**55**

$$y = \frac{(x+1)^2}{3} + \frac{\text{C}}{x+1}$$

Here is the solution in detail:

$$y.(x+1) = \int (x+1)(x+1)\,dx$$
$$= \int (x+1)^2\,dx$$
$$= \frac{(x+1)^3}{3} + \text{C}$$
$$\therefore y = \frac{(x+1)^2}{3} + \frac{\text{C}}{x+1}$$

Now let us do another one.

*Example 5.* Solve $\qquad x\dfrac{dy}{dx} - 5y = x^7$

In this case, P = ................ Q = ................

**56**

$$P = -\frac{5}{x};\ Q = x^6$$

for if

$$x\frac{dy}{dx} - 5y = x^7$$

$$\therefore \frac{dy}{dx} - \frac{5}{x}.y = x^6$$

Compare with $\left[\frac{dy}{dx} + Py = Q\right]$ $\therefore P = -\frac{5}{x};\ Q = x^6$

So the integrating factor, IF = ........................

**57**

$$IF = x^{-5} = \frac{1}{x^5}$$

for $\quad IF = e^{\int P\,dx} \quad \int P\,dx = -\int \frac{5}{x}dx = -5\ln x$

$$\therefore IF = e^{-5\ln x} = e^{\ln(x^{-5})} = x^{-5} = \frac{1}{x^5}$$

So the solution is

$$y.\frac{1}{x^5} = \int x^6.\frac{1}{x^5}\,dx$$

$$\frac{y}{x^5} = \int x\,dx = \frac{x^2}{2} + C$$

$$y = \ldots\ldots\ldots\ldots\ldots\ldots$$

**58**

$$y = \frac{x^7}{2} + Cx^5$$

Did you remember to multiply the C by $x^5$?

□□□□□□□□□□□□□□□□□□□□□□□□□□□□□□□□□□□□□□

Fine. Now you do this one entirely on your own.

*Example 6.* Solve $\quad (1 - x^2)\frac{dy}{dx} - xy = 1.$

*When you have finished it, turn to frame 59.*

**59**

$$y\sqrt{(1-x^2)} = \sin^{-1}x + C$$

Here is the working in detail. Follow it through.

$$(1-x^2)\frac{dy}{dx} - xy = 1$$

$$\therefore \frac{dy}{dx} - \frac{x}{1-x^2}.y = \frac{1}{1-x^2}$$

$$\text{IF} = e^{\int P\,dx} \qquad \int P\,dx = \int \frac{-x}{1-x^2}\,dx = \frac{1}{2}\ln(1-x^2)$$

$$\therefore \text{IF} = e^{\frac{1}{2}\ln(1-x^2)} = e^{\ln\{(1-x^2)^{\frac{1}{2}}\}} = (1-x^2)^{\frac{1}{2}}$$

$$\text{Now} \qquad y.\text{IF} = \int Q.\text{IF}\,dx$$

$$\therefore y\sqrt{(1-x^2)} = \int \frac{1}{1-x^2}\sqrt{(1-x^2)}.dx$$

$$= \int \frac{1}{\sqrt{(1-x^2)}}\,dx = \sin^{-1}x + C$$

$$y\sqrt{(1-x^2)} = \underline{\sin^{-1}x + C}$$

*Now on to frame 60.*

**60**

In practically all the examples so far, we have been concerned with finding the general solutions. If further information is available, of course, particular solutions can be obtained. Here is one final example for you to do.

*Example 7.* Solve the equation

$$(x-2)\frac{dy}{dx} - y = (x-2)^3$$

given that $y = 10$ when $x = 4$.

Off you go then. It is quite straightforward

*When you have finished it, turn on to frame 61 and check your solution.*

**61**

$$2y = (x-2)^3 + 6(x-2)$$

Here it is:

$$(x-2)\frac{dy}{dx} - y = (x-2)^3$$

$$\frac{dy}{dx} - \frac{1}{x-2}\,.\,y = (x-2)^2$$

$$\int \mathrm{P}\,dx = \int \frac{-1}{x-2}\,dx = -\ln(x-2)$$

$$\therefore\ \mathrm{IF} = e^{-\ln(x-2)} = e^{\ln\left\{(x-2)^{-1}\right\}} = (x-2)^{-1}$$

$$= \frac{1}{x-2}$$

$$\therefore\ y.\frac{1}{x-2} = \int (x-2)^2\,.\,\frac{1}{(x-2)}\,dx$$

$$= \int (x-2)\,dx$$

$$= \frac{(x-2)^2}{2} + \mathrm{C}$$

$$\therefore\ y = \frac{(x-2)^3}{2} + \mathrm{C}(x-2)\ \text{... General solution.}$$

When $x = 4$, $y = 10$

$$10 = \frac{8}{2} + \mathrm{C}.2 \quad \therefore\ 2\mathrm{C} = 6 \quad \therefore\ \mathrm{C} = 3$$

$$\therefore\ 2y = (x-2)^3 + 6(x-2)$$

**62**

Finally, for this part of the programme, here is a short revision exercise.

**Revision Exercise**

Solve the following:

1. $\dfrac{dy}{dx} + 3y = e^{4x}$
2. $x\dfrac{dy}{dx} + y = x \sin x$
3. $\tan x\dfrac{dy}{dx} + y = \sec x$

*Work through them all: then check your results with those given in frame 63.*

**63**

*Results:*

1. $y = \dfrac{e^{4x}}{7} + C\,e^{-3x}$  (IF $= e^{3x}$)

2. $xy = \sin x - x\cos x + C$  (IF $= x$)

3. $y\sin x = x + C$  (IF $= \sin x$)

□□□□□□□□□□□□□□□□□□□□□□□□□□□□□□□□□□□□□□□□

There is just one other type of equation that we must consider. Here is an example: let us see how it differs from those we have already dealt with.

To solve $$\frac{dy}{dx} + \frac{1}{x}.y = xy^2$$

Note that if it were not for the factor $y^2$ on the right-hand side, this equation would be of the form $\dfrac{dy}{dx} + Py = Q$ that we know of old.

To see how we deal with this new kind of equation, we will consider the general form, so move on to frame 64.

**64**

*Bernoulli's equation.* Equations of the form

$$\frac{dy}{dx} + Py = Qy^n$$

where, as before, P and Q are functions of $x$ (or constants).

The trick is the same every time:

(i) Divide both sides by $y^n$. This gives

$$y^{-n}\frac{dy}{dx} + Py^{1-n} = Q$$

(ii) Now put $z = y^{1-n}$

so that, differentiating, $\dfrac{dz}{dx} = \dots\dots\dots\dots\dots\dots$

**65**

So we have

$$\boxed{\frac{dz}{dx} = (1-n)\,y^{-n}\frac{dy}{dx}}$$

$$\frac{dy}{dx} + \mathrm{P}\,y = \mathrm{Q}\,y^n \qquad \text{(i)}$$

$$\therefore\; y^{-n}\frac{dy}{dx} + \mathrm{P}\,y^{1-n} = \mathrm{Q} \qquad \text{(ii)}$$

Put $z = y^{1-n}$ so that $\dfrac{dz}{dx} = (1-n)\,y^{-n}\dfrac{dy}{dx}$

If we now multiply (ii) by $(1-n)$ we shall convert the first term into $\dfrac{dz}{dx}$.

$$(1-n)\,y^{-n}\frac{dy}{dx} + (1-n)\,\mathrm{P}\,y^{1-n} = (1-n)\,\mathrm{Q}$$

Remembering that $z = y^{1-n}$ and that $\dfrac{dz}{dx} = (1-n)\,y^{-n}\dfrac{dy}{dx}$, this last line can now be written

$$\frac{dz}{dx} + \mathrm{P}_1\,z = \mathrm{Q}_1$$

with $\mathrm{P}_1$ and $\mathrm{Q}_1$ functions of $x$.

This we can now solve by use of an integrating factor in the normal way.

Finally, having found $z$, we convert back to $y$ using $z = y^{1-n}$.

*Let us see this routine in operation – so on to frame 66.*

**66**

*Example 1.* Solve $\dfrac{dy}{dx} + \dfrac{1}{x}y = x\,y^2$.

(i) Divide through by $y^2$, giving ...............................

**67**

$$\boxed{y^{-2}\frac{dy}{dx} + \frac{1}{x}\,y^{-1} = x}$$

(ii) Now put $z = y^{1-n}$, i.e. in this case $z = y^{1-2} = y^{-1}$

$$z = y^{-1} \quad \therefore\; \frac{dz}{dx} = -y^{-2}\frac{dy}{dx}$$

(iii) Multiply through the equation by $(-1)$, to make the first term $\dfrac{dz}{dx}$.

$$-y^{-2}\frac{dy}{dx} - \frac{1}{x}y^{-1} = -x$$

so that $\dfrac{dz}{dx} - \dfrac{1}{x}z = -x$ which is of the form $\dfrac{dz}{dx} + \mathrm{P}\,z = \mathrm{Q}$ so that you can now solve the equation by the normal integrating factor method. What do you get?

*When you have done it, move on to the next frame.*

$$y = (Cx - x^2)^{-1}$$

**68**

Check the working:

$$\frac{dz}{dx} - \frac{1}{x}z = -x$$

$$\text{IF} = e^{\int P\,dx} \qquad \int P\,dx = \int -\frac{1}{x}dx = -\ln x$$

$$\therefore \text{IF} = e^{-\ln x} = e^{\ln(x^{-1})} = x^{-1} = \frac{1}{x}$$

$$z.\text{IF} = \int \text{Q.IF}\,dx \quad \therefore z\frac{1}{x} = \int -x.\frac{1}{x}\,dx$$

$$\therefore \frac{z}{x} = \int -1\,dx = -x + C$$

$$\therefore z = Cx - x^2$$

But $\quad z = y^{-1} \quad \therefore \frac{1}{y} = Cx - x^2 \quad \therefore \underline{y = (Cx - x^2)^{-1}}$

Right! Here is another.

*Example 2.* Solve $\quad x^2y - x^3\frac{dy}{dx} = y^4 \cos x$

First of all, we must re-write this in the form $\frac{dy}{dx} + Py = Qy^n$

So, what do we do?

---

Divide both sides by $(-x^3)$

**69**

giving $\quad \frac{dy}{dx} - \frac{1}{x}.y = -\frac{y^4 \cos x}{x^3}$

Now divide by the power of $y$ on the R.H.S., giving ......................

---

$$y^{-4}\frac{dy}{dx} - \frac{1}{x}y^{-3} = -\frac{\cos x}{x^3}$$

**70**

Next we make the substitution $z = y^{1-n}$ which, in this example, is $z = y^{1-4} = y^{-3}$

$$\therefore z = y^{-3} \text{ and } \therefore \frac{dz}{dx} = \text{................}$$

**71**

$$\frac{dz}{dx} = -3\,y^{-4}\,\frac{dy}{dx}$$

If we now multiply the equation by $(-3)$ to make the first term into $\frac{dz}{dx}$, we have

$$-3\,y^{-4}\,\frac{dy}{dx} + 3\,\frac{1}{x}.\,y^{-3} = \frac{3\cos x}{x^3}$$

$$\text{i.e.}\ \frac{dz}{dx} + \frac{3}{x}z = \frac{3\cos x}{x^3}$$

This you can now solve to find $z$ and so back to $y$.

*Finish it off and then check with the next frame.*

**72**

$$y^3 = \frac{x^3}{3\sin x + C}$$

For:

$$\frac{dz}{dx} + \frac{3}{x}.\,z = \frac{3\cos x}{x^3}$$

$$\text{IF} = e^{\int \text{P}\,dx} \qquad \int \text{P}\,dx = \int \frac{3}{x}dx = 3\ln x$$

$$\therefore\ \text{IF} = e^{3\ln x} = e^{\ln(x^3)} = x^3$$

$$z.\text{IF} = \int \text{Q.IF}\,dx$$

$$\therefore\ z\,x^3 = \int \frac{3\cos x}{x^3}\,x^3\,dx$$

$$= \int 3\cos x\,dx$$

$$\therefore\ z\,x^3 = 3\sin x + C$$

But, in this example, $z = y^{-3}$

$$\therefore \frac{x^3}{y^3} = 3\sin x + C$$

$$\therefore\ y^3 = \frac{x^3}{3\sin x + C}$$

*Let us look at the complete solution as a whole, so on to frame 73.*

**73**

Here it is:

To solve $$x^2 y - x^3 \frac{dy}{dx} = y^4 \cos x$$

$$\therefore \frac{dy}{dx} - \frac{1}{x}y = -\frac{y^4 \cos x}{x^3}$$

$$\therefore y^{-4}\frac{dy}{dx} - \frac{1}{x}y^{-3} = -\frac{\cos x}{x^3}$$

Put $z = y^{1-n} = y^{1-4} = y^{-3} \quad \therefore \frac{dz}{dx} = -3\,y^{-4}\frac{dy}{dx}$

Equation becomes

$$-3\,y^{-4}\frac{dy}{dx} + \frac{3}{x}.\,y^{-3} = \frac{3\cos x}{x^3}$$

$$\text{i.e.} \quad \frac{dz}{dx} + \frac{3}{x}.z = \frac{3\cos x}{x^3}$$

IF $= e^{\int P\,dx}$ $$\int P\,dx = \int \frac{3}{x}dx = 3\ln x$$

$$\therefore \text{IF} = e^{3\ln x} = e^{\ln(x^3)} = x^3$$

$$\therefore z\,x^3 = \int \frac{3\cos x}{x^3}x^3\,dx$$

$$= \int 3\cos x\,dx$$

$$\therefore z\,x^3 = 3\sin x + C$$

But $z = y^{-3}$

$$\therefore \frac{x^3}{y^3} = 3\sin x + C$$

$$\therefore y^3 = \frac{x^3}{3\sin x + C}$$

They are all done in the same way. Once you know the trick, the rest is very straightforward.

*On to the next frame.*

---

**74**

Here is one for you to do entirely on your own.

*Example 3.* Solve $$2y - 3\frac{dy}{dx} = y^4\,e^{3x}$$

Work through the same steps as before. When you have finished, check your working with the solution in frame 75.

**75**

$$y^3 = \frac{5\,e^{2x}}{e^{5x} + A}$$

Solution in detail:

$$2y - 3\frac{dy}{dx} = y^4\,e^{3x}$$

$$\therefore \frac{dy}{dx} - \frac{2}{3}y = -\frac{y^4\,e^{3x}}{3}$$

$$\therefore y^{-4}\frac{dy}{dx} - \frac{2}{3}y^{-3} = -\frac{e^{3x}}{3}$$

Put $z = y^{1-4} = y^{-3} \quad \therefore \frac{dz}{dx} = -3\,y^{-4}\frac{dy}{dx}$

Multiplying through by $(-3)$, the equation becomes

$$-3\,y^{-4}\frac{dy}{dx} + 2\,y^{-3} = e^{3x}$$

$$\text{i.e. } \frac{dz}{dx} + 2\,z = e^{3x}$$

$\text{IF} = e^{\int P\,dx} \qquad \int P\,dx = \int 2\,dx = 2x \quad \therefore \quad \text{IF} = e^{2x}$

$$\therefore z\,e^{2x} = \int e^{3x}\,e^{2x}\,dx = \int e^{5x}\,dx$$

$$= \frac{e^{5x}}{5} + C$$

But $z = y^{-3} \quad \therefore \frac{e^{2x}}{y^3} = \frac{e^{5x} + A}{5}$

$$\therefore \; y^3 = \frac{5\,e^{2x}}{e^{5x} + A}$$

*On to frame 76.*

**76**

Finally, one further example for you, just to be sure.

*Example 4.* Solve $y - 2x\frac{dy}{dx} = x(x+1)\,y^3$

First re-write the equation in standard form $\frac{dy}{dx} + P\,y = Q\,y^n$

This gives ..........................................

**77**

$$\frac{dy}{dx} - \frac{1}{2x}.y = -\frac{(x+1)\,y^3}{2}$$

*Now off you go and complete the solution. When you have finished, check with the working in frame 78.*

**78**

$$\boxed{y^2 = \frac{6x}{2x^3 + 3x^2 + A}}$$

Solution:

$$\frac{dy}{dx} - \frac{1}{2x}.y = -\frac{(x+1)y^3}{2}$$

$$\therefore y^{-3}\frac{dy}{dx} - \frac{1}{2x}.y^{-2} = -\frac{(x+1)}{2}$$

Put $\quad z = y^{1-3} = \underline{y^{-2}} \quad \therefore \frac{dz}{dx} = -2y^{-3}\frac{dy}{dx}$

Equation becomes

$$-2y^{-3}\frac{dy}{dx} + \frac{1}{x}y^{-2} = (x+1)$$

$$\text{i.e. } \frac{dz}{dx} + \frac{1}{x}.z = x + 1$$

$$\text{IF} = e^{\int P\,dx} \qquad \int P\,dx = \int \frac{1}{x}dx = \ln x$$

$$\therefore \text{IF} = e^{\ln x} = x$$

$$z.\text{IF} = \int Q.\text{IF}\,dx \quad \therefore zx = \int (x+1)x\,dx$$

$$= \int (x^2 + x)\,dx$$

$$\therefore zx = \frac{x^3}{3} + \frac{x^2}{2} + C$$

But $z = y^{-2}$ $\qquad \therefore \dfrac{x}{y^2} = \dfrac{2x^3 + 3x^2 + A}{6}$

$$\therefore y^2 = \frac{6x}{2x^3 + 3x^2 + A}$$

□□□□□□□□□□□□□□□□□□□□□□□□□□□□□□□□□□□□□□□□

There we are. You have now reached the end of this programme, except for the Test Exercise that follows. Before you tackle it, however, read down the Revision Sheet presented in the next frame. It will remind you of the main points that we have covered in this programme on first order differential equations.

*Turn on then to frame 79.*

## 79 Revision Sheet

1. The *order* of a differential equation is given by the highest derivative present.
   An equation of *order n* is derived from a function containing *n arbitrary constants.*

2. *Solution of first order differential equations.*

   (a) By direct integration: $\frac{dy}{dx} = f(x)$

   $$\text{gives} \quad y = \int f(x)\, dx$$

   (b) By separating the variables: $F(y) . \frac{dy}{dx} = f(x)$

   $$\text{gives} \int F(y)\, dy = \int f(x)\, dx$$

   (c) Homogeneous equations: Substitute $y = vx$

   $$\text{gives} \quad v + x\frac{dv}{dx} = F(v)$$

   (d) Linear equations: $\frac{dy}{dx} + Py = Q$

   Integrating factor, $IF = e^{\int P\, dx}$

   and remember that $e^{\ln F} = F$

   $$\text{gives} \quad y\ IF = \int Q . IF\, dx$$

   (e) Bernoulli's equation: $\frac{dy}{dx} + Py = Qy^n$

   Divide by $y^n$: then put $z = y^{1-n}$

   Reduces to type (d) above.

□□□□□□□□□□□□□□□□□□□□□□□□□□□□□□□□□□□□□□□□□

If there is any section of the work about which you are not perfectly clear, turn back to that part of the programme and go through it again. Otherwise, turn on now to the Test Exercise in frame 80.

**80**

The questions in the test exercise are similar to the equations you have been solving in the programme. They cover all the methods, but are quite straightforward.

Do not hurry: take your time and work carefully and you will find no difficulty with them.

## Test Exercise – XXIV

Solve the following differential equations:

1. $x\dfrac{dy}{dx} = x^2 + 2x - 3$

2. $(1 + x)^2\dfrac{dy}{dx} = 1 + y^2$

3. $\dfrac{dy}{dx} + 2y = e^{3x}$

4. $x\dfrac{dy}{dx} - y = x^2$

5. $x^2\dfrac{dy}{dx} = x^3 \sin 3x + 4$

6. $x \cos y\dfrac{dy}{dx} - \sin y = 0$

7. $(x^3 + xy^2)\dfrac{dy}{dx} = 2y^3$

8. $(x^2 - 1)\dfrac{dy}{dx} + 2xy = x$

9. $\dfrac{dy}{dx} + y \tanh x = 2 \sinh x$

10. $x\dfrac{dy}{dx} - 2y = x^3 \cos x$

11. $\dfrac{dy}{dx} + \dfrac{y}{x} = y^3$

12. $x\dfrac{dy}{dx} + 3y = x^2y^2$

## Further Problems – XXIV

Solve the following equations.

I. *Separating the variables*

1. $x(y-3)\dfrac{dy}{dx}=4y$

2. $(1+x^3)\dfrac{dy}{dx}=x^2y$ given that $x=1$ when $y=2$.

3. $x^3+(y+1)^2\dfrac{dy}{dx}=0$

4. $\cos y+(1+e^{-x})\sin y\dfrac{dy}{dx}=0$, given that $y=\pi/4$ when $x=0$.

5. $x^2(y+1)+y^2(x-1)\dfrac{dy}{dx}=0$

II. *Homogeneous equations*

6. $(2y-x)\dfrac{dy}{dx}=2x+y$, given that $y=3$ when $x=2$.

7. $(xy+y^2)+(x^2-xy)\dfrac{dy}{dx}=0$

8. $(x^3+y^3)=3xy^2\dfrac{dy}{dx}$

9. $y-3x+(4y+3x)\dfrac{dy}{dx}=0$

10. $(x^3+3xy^2)\dfrac{dy}{dx}=y^3+3x^2y$

III. *Integrating factor*

11. $x\dfrac{dy}{dx}-y=x^3+3x^2-2x$

12. $\dfrac{dy}{dx}+y\tan x=\sin x$

13. $x\dfrac{dy}{dx}-y=x^3\cos x$, given that $y=0$ when $x=\pi$.

14. $(1+x^2)\dfrac{dy}{dx}+3xy=5x$, given that $y=2$ when $x=1$.

15. $\dfrac{dy}{dx}+y\cot x=5\,e^{\cos x}$, given that $y=-4$ when $x=\pi/2$.

IV *Transformations.* Make the given substitutions and work in much the same way as for first order homogeneous equations.

16. $(3x + 3y - 4)\dfrac{dy}{dx} = -(x + y)$ Put $x + y = v$

17. $(y - xy^2) = (x + x^2y)\dfrac{dy}{dx}$ Put $y = \dfrac{v}{x}$

18. $(x - y - 1) + (4y + x - 1)\dfrac{dy}{dx} = 0$ Put $v = x - 1$

19. $(3y - 7x + 7) + (7y - 3x + 3)\dfrac{dy}{dx} = 0$ Put $v = x - 1$

20. $y(xy + 1) + x(1 + xy + x^2y^2)\dfrac{dy}{dx} = 0$ Put $y = \dfrac{v}{x}$

V. *Bernoulli's equation*

21. $\dfrac{dy}{dx} + y = xy^3$

22. $\dfrac{dy}{dx} + y = y^4 e^x$

23. $2\dfrac{dy}{dx} + y = y^3(x - 1)$

24. $\dfrac{dy}{dx} - 2y \tan x = y^2 \tan^2 x$

25. $\dfrac{dy}{dx} + y \tan x = y^3 \sec^4 x$

VI. *Miscellaneous.* Choose the appropriate method in each case.

26. $(1 - x^2)\dfrac{dy}{dx} = 1 + xy$

27. $xy\dfrac{dy}{dx} - (1 + x)\sqrt{(y^2 - 1)} = 0$

28. $(x^2 - 2xy + 5y^2) = (x^2 + 2xy + y^2)\dfrac{dy}{dx}$

29. $\dfrac{dy}{dx} - y \cot x = y^2 \sec^2 x$, given $y = -1$ when $x = \pi/4$.

30. $y + (x^2 - 4x)\dfrac{dy}{dx} = 0$

## VII. *Further examples*

31. Solve the equation $\frac{dy}{dx} - y \tan x = \cos x - 2x \sin x$, given that $y = 0$ when $x = \pi/6$.

32. Find the general solution of the equation
$$\frac{dy}{dx} = \frac{2xy + y^2}{x^2 + 2xy}$$

33. Find the general solution of $(1 + x^2)\frac{dy}{dx} = x(1 + y^2)$.

34. Solve the equation $x\frac{dy}{dx} + 2y = 3x - 1$, given that $y = 1$ when $x = 2$.

35. Solve $x^2 \frac{dy}{dx} = y^2 - xy\frac{ay}{dx}$, given that $y = 1$ when $x = 1$.

36. Solve $\frac{dy}{dx} = e^{3x - 2y}$, given that $y = 0$ when $x = 0$.

37. Find the particular solution of $\frac{dy}{dx} + \frac{1}{x} \cdot y = \sin 2x$, such that $y = 2$ when $x = \pi/4$.

38. Find the general solution of $y^2 + x^2 \frac{dy}{dx} = xy \frac{dy}{dx}$.

39. Obtain the general solution of the equation
$$2xy\frac{dy}{dx} = x^2 - y^2$$

40. By substituting $z = x - 2y$, solve the equation
$$\frac{dy}{dx} = \frac{x - 2y + 1}{2x - 4y}$$
given that $y = 1$ when $x = 1$.

41. Find the general solution of $(1 - x^3)\frac{dy}{dx} + x^2 y = x^2(1 - x^3)$.

42. Solve $\frac{dy}{dx} + \frac{y}{x} = \sin x$, given that $y = 0$ at $x = \pi/2$.

43. Solve $\frac{dy}{dx} + x + xy^2 = 0$, given $y = 0$ when $x = 1$.

44. Determine the general solution of the equation

$$\frac{dy}{dx} + \left\{\frac{1}{x} - \frac{2x}{1-x^2}\right\} y = \frac{1}{1-x^2}$$

45. Solve $(1+x^2)\dfrac{dy}{dx} + xy = (1+x^2)^{3/2}$

46. Solve $x(1+y^2) - y(1+x^2)\dfrac{dy}{dx} = 0$, given $y = 2$ at $x = 0$.

47. Solve $\dfrac{r \tan\theta}{a^2 - r^2} \cdot \dfrac{dr}{d\theta} = 1$, given $r = 0$ when $\theta = \pi/4$.

48. Solve $\dfrac{dy}{dx} + y \cot x = \cos x$, given that $y = 0$ when $x = 0$.

49. Use the substitution $y = \dfrac{v}{x}$, where $v$ is a function of $x$ only, to transform the equation

$$\frac{dy}{dx} + \frac{y}{x} = xy^2$$

into a differential equation in $v$ and $x$. Hence find $y$ in terms of $x$.

50. The rate of decay of a radio-active substance is proportional to the amount A remaining at any instant. If $A = A_0$ at $t = 0$, prove that, if the time taken for the amount of the substance to become $\frac{1}{2}A_0$ is T, then $A = A_0\, e^{-(t \ln 2)/T}$. Prove also that the time taken for the amount remaining to be reduced to $\frac{1}{20} A_0$ is 4·32 T.

# Programme 25

# SECOND ORDER DIFFERENTIAL EQUATIONS

**1**

Many practical problems in engineering give rise to second order differential equations of the form

$$a\frac{d^2y}{dx^2} + b\frac{dy}{dx} + cy = f(x)$$

where $a, b, c$ are constant coefficients and $f(x)$ is a given function of $x$. By the end of this programme you will have no difficulty with equations of this type.

Let us first take the case where $f(x) = 0$, so that the equation becomes

$$a\frac{d^2y}{dx^2} + b\frac{dy}{dx} + cy = 0$$

Let $y = u$ and $y = v$ (where $u$ and $v$ are functions of $x$) be two solutions of the equation.

$$\therefore\ a\frac{d^2u}{dx^2} + b\frac{du}{dx} + cu = 0$$

and $$a\frac{d^2v}{dx^2} + b\frac{dv}{dx} + cv = 0$$

Adding these two lines together, we get

$$a\left(\frac{d^2u}{dx^2} + \frac{d^2v}{dx^2}\right) + b\left(\frac{du}{dx} + \frac{dv}{dx}\right) + c(u + v) = 0$$

Now $\frac{d}{dx}(u + v) = \frac{du}{dx} + \frac{dv}{dx}$ and $\frac{d^2}{dx^2}(u + v) = \frac{d^2u}{dx^2} + \frac{d^2v}{dx^2}$, therefore the equation can be written

$$a\frac{d^2}{dx^2}(u + v) + b\frac{d}{dx}(u + v) + c(u + v) = 0$$

which is our original equation with $y$ replaced by $(u + v)$.

i.e. <u>If $y = u$ and $y = v$ are solutions of the equation $a\frac{d^2y}{dx^2} + b\frac{dy}{dx} + cy = 0$, so also is $y = u + v$.</u>

This is an important result and we shall be referring to it later, so make a note of it in your record book.

*Turn on to frame 2.*

**2**

Our equation was $a\dfrac{d^2y}{dx^2} + b\dfrac{dy}{dx} + cy = 0$. If $a = 0$, we get the first order equation of the same family

$$b\frac{dy}{dx} + cy = 0 \quad \text{i.e.} \quad \frac{dy}{dx} + ky = 0 \quad \text{where} \quad k = \frac{c}{b}$$

Solving this by the method of separating the variables, we have

$$\frac{dy}{dx} = -ky \quad \therefore \int \frac{dy}{y} = -\int k\, dx$$

which gives ......................

**3**

$$\boxed{\ln y = -kx + c}$$

$$\therefore\ y = e^{-kx+c} = e^{-kx}.\, e^{c} = \mathrm{A}\, e^{-kx} \quad \text{(since } e^{c} \text{ is a constant)}$$

$$\text{i.e. } y = \mathrm{A}\, e^{-kx}$$

If we write the symbol $m$ for $-k$, the solution is $\underline{y = \mathrm{A}\, e^{mx}}$

In the same way, $y = \mathrm{A}\, e^{mx}$ will be a solution of the second order equation $a\dfrac{d^2y}{dx^2} + b\dfrac{dy}{dx} + cy = 0$, if it satisfies this equation.

Now, if $\qquad y = \mathrm{A}\, e^{mx}$

$$\frac{dy}{dx} = \mathrm{A}m\, e^{mx}$$

$$\frac{d^2y}{dx^2} = \mathrm{A}m^2\, e^{mx}$$

and substituting these expressions for the differential coefficients in the left-hand side of the equation, we get ............................

*On to frame 4.*

**4**

$$a\,\mathrm{A}m^2 e^{mx} + b\,\mathrm{A}m\, e^{mx} + c\,\mathrm{A}\, e^{mx} = 0$$

Right. So dividing both sides by $\mathrm{A}\,e^{mx}$, we obtain

$$am^2 + bm + c = 0$$

which is a quadratic equation giving two values for $m$. Let us call these

$$m = m_1 \quad \text{and} \quad m = m_2$$

i.e. $y = \mathrm{A}\,e^{m_1 x}$ and $y = \mathrm{B}\,e^{m_2 x}$ are two solutions of the given equation.

Now we have already seen that if $y = u$ and $y = v$ are two solutions so also is $y = u + v$.

$\therefore$ If $y = \mathrm{A}\,e^{m_1 x}$ and $y = \mathrm{B}\,e^{m_2 x}$ are solutions, so also is

$$\underline{y = \mathrm{A}\,e^{m_1 x} + \mathrm{B}\,e^{m_2 x}}$$

*Note* that this contains the necessary two arbitrary constants for a second order differential equation, so there can be no further solution.

*Move to frame 5.*

**5** The solution, then, of $a\dfrac{d^2y}{dx^2} + b\dfrac{dy}{dx} + cy = 0$ is seen to be

$$\underline{y = \mathrm{A}\,e^{m_1 x} + \mathrm{B}\,e^{m_2 x}}$$

where A and B are two arbitrary constants and $m_1$ and $m_2$ are the roots of the quadratic equation $am^2 + bm + c = 0$.

This quadratic equation is called the *auxiliary equation* and is obtained directly from the equation $a\dfrac{d^2y}{dx^2} + b\dfrac{dy}{dx} + cy = 0$, by writing $m^2$ for $\dfrac{d^2y}{dx^2}$, $m$ for $\dfrac{dy}{dx}$, 1 for $y$.

*Example:* For the equation $2\dfrac{d^2y}{dx^2} + 5\dfrac{dy}{dx} + 6y = 0$, the auxiliary equation is $2m^2 + 5m + 6 = 0$.

In the same way, for the equation $\dfrac{d^2y}{dx^2} + 3\dfrac{dy}{dx} + 2y = 0$, the auxiliary equation is ..............................

*Then on to frame 6.*

**6**

$$\boxed{m^2 + 3m + 2 = 0}$$

Since the auxiliary equation is always a quadratic equation, the values of $m$ can be determined in the usual way.

e.g. if $m^2 + 3m + 2 = 0$

$$(m + 1)(m + 2) = 0 \quad \therefore\ m = -1 \text{ and } m = -2$$

$\therefore$ the solution of $\dfrac{d^2y}{dx^2} + 3\dfrac{dy}{dx} + 2y = 0$ is

$$\underline{y = \mathrm{A}\,e^{-x} + \mathrm{B}\,e^{-2x}}$$

In the same way, if the auxiliary equation were $m^2 + 4m - 5 = 0$, this factorizes into $(m + 5)(m - 1) = 0$ giving $m = 1$ or $-5$, and in this case the solution would be ......................

**7**

$$\boxed{y = \mathrm{A}\,e^{x} + \mathrm{B}\,e^{-5x}}$$

The type of solution we get, depends on the roots of the auxiliary equation.

(i) *Real and different roots*

*Example 1.* $\dfrac{d^2y}{dx^2} + 5\dfrac{dy}{dx} + 6y = 0$

Auxiliary equation: $m^2 + 5m + 6 = 0$

$$\therefore\ (m + 2)(m + 3) = 0 \quad \therefore\ m = -2 \text{ or } m = -3$$

$\therefore$ Solution is $\quad \underline{y = \mathrm{A}\,e^{-2x} + \mathrm{B}\,e^{-3x}}$

*Example 2.* $\dfrac{d^2y}{dx^2} - 7\dfrac{dy}{dx} + 12y = 0$

Auxiliary equation: $m^2 - 7m + 12 = 0$

$$(m - 3)(m - 4) = 0 \quad \therefore\ m = 3 \text{ or } m = 4$$

So the solution is ......................

*Turn to frame 8.*

**8**

$$y = A e^{3x} + B e^{4x}$$

Here you are. Do this one.

Solve the equation $\frac{d^2y}{dx^2} + 3\frac{dy}{dx} - 10y = 0$

*When you have finished, move on to frame 9.*

**9**

$$y = A e^{2x} + B e^{-5x}$$

Now consider the next case.

(ii) *Real and equal roots* to the auxiliary equation.

Let us take $\frac{d^2y}{dx^2} + 6\frac{dy}{dx} + 9y = 0$.

The auxiliary equation is: $m^2 + 6m + 9 = 0$

$$\therefore\ (m + 3)(m + 3) = 0 \quad \therefore\ m = -3 \text{ (twice)}$$

If $m_1 = -3$ and $m_2 = -3$ then these would give the solution $y = A e^{-3x} + B e^{-3x}$ and their two terms would combine to give $y = Ce^{-3x}$. But every second order differential equation has two arbitrary constants, so there must be another term containing a second constant. In fact, it can be shown that $y = Kxe^{-3x}$ also satisfies the equation, so that the complete general solution is of the form $y = A e^{-3x} + Bxe^{-3x}$

$$\text{i.e. } \underline{y = e^{-3x}(A + Bx)}$$

*In general, if the auxiliary equation has real and equal roots, giving* $m = m_1$ *(twice), the solution of the differential equation is*

$$\underline{y = e^{m_1 x}(A + Bx)}$$

*Make a note of this general statement and then turn on to frame 10.*

**10**

Here is an example:

*Example 1.* Solve $\dfrac{d^2y}{dx^2} + 4\dfrac{dy}{dx} + 4y = 0$

Auxiliary equation: $m^2 + 4m + 4 = 0$

$$(m + 2)(m + 2) = 0 \quad \therefore\ m = -2 \text{ (twice)}$$

The solution is: $\underline{y = e^{-2x}(A + Bx)}$

Here is another:

*Example 2.* Solve $\dfrac{d^2y}{dx^2} + 10\dfrac{dy}{dx} + 25\,y = 0$

Auxiliary equation: $m^2 + 10m + 25 = 0$

$$(m + 5)^2 = 0 \quad \therefore\ m = -5 \text{ (twice)}$$

$$\underline{y = e^{-5x}(A + Bx)}$$

Now here is one for you to do:

Solve $\dfrac{d^2y}{dx^2} + 8\dfrac{dy}{dx} + 16y = 0$

*When you have done it, move on to frame 11.*

---

**11**

$$\boxed{y = e^{-4x}(A + Bx)}$$

Since if $\dfrac{d^2y}{dx^2} + 8\dfrac{dy}{dx} + 16y = 0$

the auxiliary equation is

$$m^2 + 8m + 16 = 0$$

$$\therefore\ (m + 4)^2 = 0 \quad \therefore\ m = -4 \text{ (twice)}$$

$$\therefore\ \underline{y = e^{-4x}(A + Bx)}$$

So, for *real and different roots* $m = m_1$ and $m = m_2$ the solution is

$$y = A\,e^{m_1x} + B\,e^{m_2x}$$

and for *real and equal roots* $m = m_1$ (twice) the solution is

$$y = e^{m_1x}(A + Bx)$$

Just find the values of $m$ from the auxiliary equation and then substitute these values in the appropriate form of the result.

*Move to frame 12.*

**12** (iii) *Complex roots* to the auxiliary equation.

Now let us see what we get when the roots of the auxiliary equation are complex.

Suppose $m = a \pm j\beta$, i.e. $m_1 = a + j\beta$ and $m_2 = a - j\beta$. Then the solution would be of the form

$$y = Ce^{(a+j\beta)x} + De^{(a-j\beta)x}$$
$$= Ce^{ax}.e^{j\beta x} + De^{ax}.e^{-j\beta x}$$
$$= e^{ax}\{Ce^{j\beta x} + De^{-j\beta x}\}$$

Now from our previous work on complex numbers, we know that

$$e^{jx} = \cos x + j \sin x$$
$$e^{-jx} = \cos x - j \sin x$$

and that
$$\begin{cases} e^{j\beta x} = \cos \beta x + j \sin \beta x \\ e^{-j\beta x} = \cos \beta x - j \sin \beta x \end{cases}$$

Our solution above can therefore be written

$$y = e^{ax}\{C(\cos \beta x + j \sin \beta x) + D(\cos \beta x - j \sin \beta x)\}$$
$$= e^{ax}\{(C + D)\cos \beta x + j(C - D)\sin \beta x\}$$
$$y = e^{ax}\{A \cos \beta x + B \sin \beta x\}$$

where $A = C + D$
$B = j(C - D)$

$\therefore$ If $m = a \pm j\beta$, the solution can be written in the form

$$\underline{y = e^{ax}\{A \cos \beta x + B \sin \beta x\}}$$

*Example:* If $m = -2 \pm j3$,

then $y = e^{-2x}\{A \cos 3x + B \sin 3x\}$

Similarly, if $m = 5 \pm j2$,

then $y = \ldots\ldots\ldots\ldots$

**13**

$$y = e^{5x}\ [\mathrm{A}\cos 2x + \mathrm{B}\sin 2x]$$

Here is one of the same kind:

Solve $$\frac{d^2y}{dx^2} + 4\frac{dy}{dx} + 9y = 0$$

Auxiliary equation: $m^2 + 4m + 9 = 0$

$$\therefore\ m = \frac{-4 \pm \sqrt{(16-36)}}{2} = \frac{-4 \pm \sqrt{-20}}{2}$$

$$= \frac{-4 \pm 2\mathrm{j}\sqrt{5}}{2} = -2 \pm \mathrm{j}\sqrt{5}$$

In this case $\alpha = -2$ and $\beta = \sqrt{5}$

Solution is: $y = e^{-2x}\ (\mathrm{A}\cos\sqrt{5}x + \mathrm{B}\sin\sqrt{5}x)$

Now you can solve this one:

$$\frac{d^2y}{dx^2} - 2\frac{dy}{dx} + 10y = 0$$

*When you have finished it, move on to frame 14.*

**14**

$$y = e^{x}\ (\mathrm{A}\cos 3x + \mathrm{B}\sin 3x)$$

Just check your working:

$$\frac{d^2y}{dx^2} - 2\frac{dy}{dx} + 10y = 0$$

Auxiliary equation: $m^2 - 2m + 10 = 0$

$$m = \frac{2 \pm \sqrt{(4-40)}}{2}$$

$$= \frac{2 \pm \sqrt{-36}}{2} = 1 \pm \mathrm{j}3$$

$$y = e^{x}\ (\mathrm{A}\cos 3x + \mathrm{B}\sin 3x)$$

*Turn to frame 15.*

**15** Here is a *summary* of the work so far.

Equations of the form $a\dfrac{d^2y}{dx^2} + b\dfrac{dy}{dx} + cy = 0$

Auxiliary equation: $am^2 + bm + c = 0$

(i) *Roots real and different* $m = m_1$ and $m = m_2$

Solution is $\underline{y = Ae^{m_1x} + Be^{m_2x}}$

(ii) *Real and equal roots* $m = m_1$ (twice)

Solution is $\underline{y = e^{m_1x}(A + Bx)}$

(iii) *Complex roots* $m = \alpha \pm j\beta$

Solution is $\underline{y = e^{\alpha x}(A\cos\beta x + B\sin\beta x)}$

In each case, we simply solve the auxiliary equation to establish the values of $m$ and substitute in the appropriate form of the result.

*On to frame 16.*

**16** Equations of the form $\dfrac{d^2y}{dx^2} \pm n^2y = 0$

Let us now consider the special case of the equation $a\dfrac{d^2y}{dx^2} + b\dfrac{dy}{dx} + cy = 0$ when $b = 0$.

i.e. $a\dfrac{d^2y}{dx^2} + cy = 0$ i.e. $\dfrac{d^2y}{dx^2} + \dfrac{c}{a}y = 0$

and this can be written as $\dfrac{d^2y}{dx^2} \pm n^2y = 0$ to cover the two cases when the coefficient of $y$ is positive or negative.

(i) If $\dfrac{d^2y}{dx^2} + n^2y = 0$, $m^2 + n^2 = 0$ $\therefore m^2 = -n^2$ $\therefore m = \pm jn$

(This is like $m = \alpha \pm j\beta$, when $\alpha = 0$ and $\beta = n$)

$\therefore \underline{y = A\cos nx + B\sin nx}$

(ii) If $\dfrac{d^2y}{dx^2} - n^2y = 0$, $m^2 - n^2 = 0$ $\therefore m^2 = n^2$ $\therefore m = \pm n$

$\therefore \underline{y = Ce^{nx} + De^{-nx}}$

This last result can be written in another form which is sometimes more convenient, so turn on to the next frame and we will see what it is.

**17**

You will remember from your work on hyperbolic functions that

$$\cosh nx = \frac{e^{nx} + e^{-nx}}{2} \quad \therefore\ e^{nx} + e^{-nx} = 2\cosh nx$$

$$\sinh nx = \frac{e^{nx} - e^{-nx}}{2} \quad \therefore\ e^{nx} - e^{-nx} = 2\sinh nx$$

Adding these two results: $2e^{nx} = 2\cosh nx + 2\sinh nx$

$$\therefore\ e^{nx} = \cosh nx + \sinh nx$$

Similarly, by subtracting: $e^{-nx} = \cosh nx - \sinh nx$

Therefore, the solution of our equation, $y = \text{C}e^{nx} + \text{D}e^{-nx}$, can be written

$$y = \text{C}(\cosh nx + \sinh nx) + \text{D}(\cosh nx - \sinh nx)$$
$$= (\text{C} + \text{D})\cosh nx + (\text{C} - \text{D})\sinh nx$$

i.e. $\underline{y = \text{A}\cosh nx + \text{B}\sinh nx}$

*Note.* In this form the two results are very much alike:

(i) $\dfrac{d^2y}{dx^2} + n^2y = 0 \qquad y = \text{A}\cos nx + \text{B}\sin nx$

(ii) $\dfrac{d^2y}{dx^2} - n^2y = 0 \qquad y = \text{A}\cosh nx + \text{B}\sinh nx$

Make a note of these results in your record book.

*Then, next frame.*

**18**

Here are some examples:

*Example 1.* $\dfrac{d^2y}{dx^2} + 16y = 0 \qquad \therefore\ m^2 = -16 \quad \therefore\ m = \pm \text{j}4$

$$\therefore\ y = \underline{\text{A}\cos 4x + \text{B}\sin 4x}$$

*Example 2.* $\dfrac{d^2y}{dx^2} - 3y = 0 \qquad \therefore\ m^2 = 3 \quad \therefore\ m = \pm\sqrt{3}$

$$y = \underline{\text{A}\cosh\sqrt{3}x + \text{B}\sinh\sqrt{3}x}$$

Similarly

*Example 3.* $\dfrac{d^2y}{dx^2} + 5y = 0$

$$y = \dots\dots\dots\dots$$

*Then turn on to frame 19.*

**19**

$$y = A \cos \sqrt{5}x + B \sin \sqrt{5}x$$

And now this one:

*Example 4.* $\dfrac{d^2y}{dx^2} - 4y = 0 \qquad \therefore\ m^2 = 4 \quad \therefore\ m = \pm 2$

$y = \ldots\ldots\ldots\ldots\ldots\ldots$

**20**

$$y = A \cosh 2x + B \sinh 2x$$

Now before we go on to the next section of the programme, here is a revision exercise on what we have covered so far. The questions are set out in the next frame. Work them all before checking your results.

*So on you go to frame 21.*

**21** **Revision Exercise**

Solve the following:

1. $\dfrac{d^2y}{dx^2} - 12\dfrac{dy}{dx} + 36y = 0$
2. $\dfrac{d^2y}{dx^2} + 7y = 0$
3. $\dfrac{d^2y}{dx^2} + 2\dfrac{dy}{dx} - 3y = 0$
4. $2\dfrac{d^2y}{dx^2} + 4\dfrac{dy}{dx} + 3y = 0$
5. $\dfrac{d^2y}{dx^2} - 9y = 0$

*For the answers, turn to frame 22.*

**22**

*Results*

1. $y = e^{6x}(A + Bx)$
2. $y = A\cos\sqrt{7}x + B\sin\sqrt{7}x$
3. $y = Ae^{x} + Be^{-3x}$
4. $y = e^{-x}\left(A\cos\dfrac{x}{\sqrt{2}} + B\sin\dfrac{x}{\sqrt{2}}\right)$
5. $y = A\cosh 3x + B\sinh 3x$

*By now, we are ready for the next section of the programme, so turn on to frame 23.*

**23**

So far we have considered equations of the form

$$a\frac{d^2y}{dx^2} + b\frac{dy}{dx} + cy = f(x) \text{ for the case where } f(x) = 0$$

If $f(x) = 0$, then $am^2 + bm + c = 0$ giving $m = m_1$ and $m = m_2$ and the solution is in general $y = Ae^{m_1x} + Be^{m_2x}$.

In the equation $a\dfrac{d^2y}{dx^2} + b\dfrac{dy}{dx} + cy = f(x)$, the substitution $y = Ae^{m_1x} + Be^{m_2x}$ would make the left-hand side zero. Therefore, there must be a further term in the solution which will make the L.H.S. equal to $f(x)$ and not zero. The complete solution will therefore be of the form $y = Ae^{m_1x} + Be^{m_2x} + X$, where X is the extra function yet to be found.

$y = Ae^{m_1x} + Be^{m_2x}$ is called the *complementary function* (C.F.)

$y = X$(a function of $x$)" " " *particular integral* (P.I.)

*Note* that the complete general solution is given by

<u>general solution = complementary function + particular integral</u>

Our main problem at this stage is how are we to find the particular integral for any given equation? This is what we are now going to deal with.

*So on then to frame 24.*

**24** To solve an equation $a\frac{d^2y}{dx^2} + b\frac{dy}{dx} + cy = f(x)$

(i) The *complementary function* is obtained by solving the equation with $f(x) = 0$, as in the previous part of this programme. This will give one of the following types of solution:

(i) $y = \text{A}\,e^{m_1x} + \text{B}\,e^{m_2x}$ (ii) $y = e^{m_1x}(\text{A} + \text{B}x)$

(iii) $y = e^{\alpha x}(\text{A}\cos\beta x + \text{B}\sin\beta x)$ (iv) $y = \text{A}\cos nx + \text{B}\sin nx$

(v) $y = \text{A}\cosh nx + \text{B}\sinh nx$

(ii) The *particular integral* is found by assuming the general form of the function on the right-hand side of the given equation, substituting this in the equation, and equating coefficients. An example will make this clear:

*Example:* Solve $\frac{d^2y}{dx^2} - 5\frac{dy}{dx} + 6y = x^2$

(i) *To find the C.F.* solve L.H.S. = 0, i.e. $m^2 - 5m + 6 = 0$

$\therefore (m-2)(m-3) = 0 \quad \therefore m = 2$ or $m = 3$

$\therefore$ Complementary function is $\underline{y = \text{A}\,e^{2x} + \text{B}\,e^{3x}}$ (i)

(ii) *To find the P.I.* we assume the general form of the R.H.S. which is a second degree function. Let $y = \text{C}x^2 + \text{D}x + \text{E}$.

Then $\frac{dy}{dx} = 2\text{C}x + \text{D}$ and $\frac{d^2y}{dx^2} = 2\text{C}$

Substituting these in the given equation, we get

$$2\text{C} - 5(2\text{C}x + \text{D}) + 6(\text{C}x^2 + \text{D}x + \text{E}) = x^2$$
$$2\text{C} - 10\text{C}x - 5\text{D} + 6\text{C}x^2 + 6\text{D}x + 6\text{E} = x^2$$
$$6\text{C}x^2 + (6\text{D} - 10\text{C})x + (2\text{C} - 5\text{D} + 6\text{E}) = x^2$$

Equating coefficients of powers of $x$, we have

$[x^2]$ $6\text{C} = 1$ $\therefore \text{C} = \frac{1}{6}$

$[x]$ $6\text{D} - 10\text{C} = 0$ $\therefore 6\text{D} = \frac{10}{6} = \frac{5}{3}$ $\therefore \text{D} = \frac{5}{18}$

[CT] $2\text{C} - 5\text{D} + 6\text{E} = 0$ $\therefore 6\text{E} = \frac{25}{18} - \frac{2}{6} = \frac{19}{18}$ $\therefore \text{E} = \frac{19}{108}$

$\therefore$ Particular integral is $\underline{y = \frac{x^2}{6} + \frac{5x}{18} + \frac{19}{108}}$ (ii)

Complete general solution = C.F. + P.I.

General solution is $\underline{y = \text{A}\,e^{2x} + \text{B}\,e^{3x} + \frac{x^2}{6} + \frac{5x}{18} + \frac{19}{108}}$

This frame is quite important, since all equations of this type are solved in this way. *On to frame 25.*

**25**

We have seen that to find the particular integral, we assume the general form of the function on the R.H.S. of the equation and determine the values of the constants by substitution in the whole equation and equating coefficients. These will be useful:

| If $f(x) = k$ | ... ... | Assume $y = C$ |
|---|---|---|
| $f(x) = kx$ | ... ... | ” $y = Cx + D$ |
| $f(x) = kx^2$ | ... ... | ” $y = Cx^2 + Dx + E$ |
| $f(x) = k \sin x$ *or* $k \cos x$ | | ” $y = C \cos x + D \sin x$ |
| $f(x) = k \sinh x$ *or* $k \cosh x$ | | ” $y = C \cosh x + D \sinh x$ |
| $f(x) = e^{kx}$ | ... ... | ” $y = Ce^{kx}$ |

This list will cover all the cases you are likely to meet at this stage.

So if the function on the R.H.S. of the equation is $f(x) = 2x^2 + 5$, you would take as the assumed P.I.,

$$y = \ldots\ldots\ldots\ldots$$

**26**

$$\boxed{y = Cx^2 + Dx + E}$$

Correct, since the assumed P.I. will be the general form of the second degree function.

What would you take as the assumed P.I. in each of the following cases:

1. $f(x) = 2x - 3$
2. $f(x) = e^{5x}$
3. $f(x) = \sin 4x$
4. $f(x) = 3 - 5x^2$
5. $f(x) = 27$
6. $f(x) = 5 \cosh 4x$

*When you have decided all six, check your answers with those in frame 27.*

**27**

*Answers*

| | | |
|---|---|---|
| 1. $f(x) = 2x - 3$ | P.I. is of the form | $y = Cx + D$ |
| 2. $f(x) = e^{5x}$ | " " " " " | $y = Ce^{5x}$ |
| 3. $f(x) = \sin 4x$ | " " " " " | $y = C\cos 4x + D\sin 4x$ |
| 4. $f(x) = 3 - 5x^2$ | " " " " " | $y = Cx^2 + Dx + E$ |
| 5. $f(x) = 27$ | " " " " " | $y = C$ |
| 6. $f(x) = 5\cosh 4x$ | " " " " " | $y = C\cosh 4x + D\sinh 4x$ |

All correct? If you have made a slip with any one of them, be sure that you understand where and why your result was incorrect before moving on.

*Next frame.*

**28**

Let us work through a few examples. Here is the first.

*Example 1.* Solve $\dfrac{d^2y}{dx^2} - 5\dfrac{dy}{dx} + 6y = 24$

(i) C.F. Solve L.H.S. = 0 $\therefore m^2 - 5m + 6 = 0$

$$\therefore (m-2)(m-3) = 0 \quad \therefore m = 2 \text{ and } m = 3$$

$$\therefore \underline{y = Ae^{2x} + Be^{3x}} \qquad \text{(i)}$$

(ii) P.I. $f(x) = 24$, i.e. a constant. Assume $y = C$

Then $\dfrac{dy}{dx} = 0$ and $\dfrac{d^2y}{dx^2} = 0$

Substituting in the given equation

$$0 - 5(0) + 6C = 24 \qquad C = 4$$

$$\therefore \text{P.I. is } \underline{y = 4} \qquad \text{(ii)}$$

General solution is $y = \text{C.F.} + \text{P.I.}$

$$\text{i.e. } y = \underbrace{Ae^{2x} + Be^{3x}}_{\text{C.F.}} + \underbrace{4}_{\text{P.I.}}$$

Now another:

*Example 2.* Solve $\dfrac{d^2y}{dx^2} - 5\dfrac{dy}{dx} + 6y = 2\sin 4x$

(i) C.F. This will be the same as in the last example, since the L.H.S. of this equation is the same.

$$\text{i.e. } \underline{y = Ae^{2x} + Be^{3x}}$$

(ii) P.I. The general form of the P.I. in this case will be ......................

**29**

$$y = C\cos 4x + D\sin 4x$$

*Note:* Although the R.H.S. is $f(x) = 2\sin 4x$, it is necessary to include the full general function $y = C\cos 4x + D\sin 4x$ since in finding the differential coefficients the cosine term will also give rise to $\sin 4x$.

So we have

$$y = C\cos 4x + D\sin 4x$$

$$\frac{dy}{dx} = -4C\sin 4x + 4D\cos 4x$$

$$\frac{d^2y}{dx^2} = -16C\cos 4x - 16D\sin 4x$$

We now substitute these expressions in the L.H.S. of the equation and by equating coefficients, find the values of C and D.

Away you go then.

*Complete the job and then move on to frame 30.*

**30**

$$C = \frac{2}{25};\quad D = -\frac{1}{25};\quad y = \frac{1}{25}(2\cos 4x - \sin 4x)$$

Here is the working:

$$-16C\cos 4x - 16D\sin 4x + 20C\sin 4x - 20D\cos 4x$$
$$+ 6C\cos 4x + 6D\sin 4x = 2\sin 4x$$

$$(20C - 10D)\sin 4x - (10C + 20D)\cos 4x = 2\sin 4x$$

$$\left.\begin{array}{ll} 20C - 10D = 2 & 40C - 20D = 4 \\ 10C + 20D = 0 & 10C + 20D = 0 \end{array}\right\} 50C = 4 \quad \therefore C = \frac{2}{25}$$

$$\therefore D = -\frac{1}{25}$$

In each case the P.I. is $y = \frac{1}{25}(2\cos 4x - \sin 4x)$

The C.F. was $y = Ae^{2x} + Be^{3x}$

The general solution is

$$y = Ae^{2x} + Be^{3x} + \frac{1}{25}(2\cos 4x - \sin 4x)$$

**31** Here is an example we can work through together.

Solve $$\frac{d^2y}{dx^2} + 14\frac{dy}{dx} + 49y = 4e^{5x}$$

First we have to find the C.F. To do this we solve the equation ................

**32**

$$\boxed{\frac{d^2y}{dx^2} + 14\frac{dy}{dx} + 49y = 0}$$

Correct. So start off by writing down the auxiliary equation, which is ...................

**33**

$$\boxed{m^2 + 14m + 49 = 0}$$

This gives $(m + 7)(m + 7) = 0$, i.e. $m = -7$ (twice).

$$\therefore \text{ The C.F. is } y = e^{-7x}(\text{A} + \text{B}x) \qquad \text{(i)}$$

Now for the P.I. To find this we take the general form of the R.H.S. of the given equation, i.e. we assume $y =$ ......................

**34**

$$\boxed{y = \text{C}e^{5x}}$$

Right. So we now differentiate twice, which gives us

$$\frac{dy}{dx} = \text{....................} \quad \text{and} \quad \frac{d^2y}{dx^2} = \text{....................}$$

## 35

$$\frac{dy}{dx} = 5Ce^{5x}; \quad \frac{d^2y}{dx^2} = 25Ce^{5x}$$

The equation now becomes

$$25Ce^{5x} + 14.5Ce^{5x} + 49Ce^{5x} = 4e^{5x}$$

Dividing through by $e^{5x}$: $25C + 70C + 49C = 4$

$$144C = 4 \quad \therefore \; C = \frac{1}{36}$$

$$\text{The P.I. is } \underline{y = \frac{e^{5x}}{36}} \qquad \text{(ii)}$$

So there we are. The C.F. is $y = e^{-7x}(A + Bx)$

and the P.I. is $y = \dfrac{e^{5x}}{36}$

and the complete general solution is therefore ........................

## 36

$$y = e^{-7x}(A + Bx) + \frac{e^{5x}}{36}$$

Correct, for in every case, the general solution is the sum of the complementary function and the particular integral.
Here is another.

Solve $$\frac{d^2y}{dx^2} + 6\frac{dy}{dx} + 10y = 2\sin 2x$$

(i) *To find C.F.* solve L.H.S. = 0 $\therefore \; m^2 + 6m + 10 = 0$

$$\therefore \; m = \frac{-6 \pm \sqrt{(36-40)}}{2} = \frac{-6 \pm \sqrt{-4}}{2} = -3 \pm j$$

$$\underline{y = e^{-3x}(A\cos x + B\sin x)} \qquad \text{(i)}$$

(ii) *To find P.I.* assume the general form of the R.H.S.

i.e. $y =$ ....................

*On to frame 37.*

**37**

$$\boxed{y = C\cos 2x + D\sin 2x}$$

Do not forget that we have to include the cosine term as well as the sine term, since that will also give $\sin 2x$ when the differential coefficients are found.

As usual, we now differentiate twice and substitute in the given equation $\frac{d^2y}{dx^2} + 6\frac{dy}{dx} + 10y = 2\sin 2x$ and equate coefficients of $\sin 2x$ and of $\cos 2x$.

Off you go then. Find the P.I. on your own.

*When you have finished, check your result with that in frame 38*

**38**

$$\boxed{y = \frac{1}{15}(\sin 2x - 2\cos 2x)}$$

For if $\quad y = C\cos 2x + D\sin 2x$

$$\therefore \quad \frac{dy}{dx} = -2C\sin 2x + 2D\cos 2x$$

$$\therefore \quad \frac{d^2y}{dx^2} = -4C\cos 2x - 4D\sin 2x$$

Substituting in the equation gives

$$-4C\cos 2x - 4D\sin 2x - 12C\sin 2x + 12D\cos 2x$$
$$+ 10C\cos 2x + 10D\sin 2x = 2\sin 2x$$

$$(6C + 12D)\cos 2x + (6D - 12C)\sin 2x = 2\sin 2x$$

$$6C + 12D = 0 \quad \therefore C = -2D$$

$$6D - 12C = 2 \quad \therefore 6D + 24D = 2 \quad \therefore 30D = 2 \quad \therefore D = \frac{1}{15}$$

$$\therefore C = -\frac{2}{15}$$

$$\text{P.I. is } y = \frac{1}{15}(\sin 2x - 2\cos 2x) \qquad \text{(ii)}$$

So the C.F. is $y = e^{-3x}(A\cos x + B\sin x)$

and the P.I. is $y = \frac{1}{15}(\sin 2x - 2\cos 2x)$

The complete general solution is therefore

$y = \ldots\ldots\ldots\ldots\ldots\ldots$

**39**

$$y = e^{-3x}(\text{A} \cos x + \text{B} \sin x) + \frac{1}{15}(\sin 2x - 2\cos 2x)$$

Before we do another example, list what you would assume for the P.I. in an equation when the R.H.S. function was

(1) $f(x) = 3 \cos 4x$

(2) $f(x) = 2e^{7x}$

(3) $f(x) = 3 \sinh x$

(4) $f(x) = 2x^2 - 7$

(5) $f(x) = x + 2e^x$

*Jot down all five results before turning to frame 40 to check your answers.*

**40**

(1) $y = \text{C} \cos 4x + \text{D} \sin 4x$

(2) $y = \text{C}e^{7x}$

(3) $y = \text{C} \cosh x + \text{D} \sinh x$

(4) $y = \text{C}x^2 + \text{D}x + \text{E}$

(5) $y = \text{C}x + \text{D} + \text{E}e^x$

Note that in (5) we use the general form of both the terms.

General form for $x$ is $\text{C}x + \text{D}$

" " " $e^x$ is $\text{E}e^x$

$\therefore$ The general form of $x + e^x$ is $y = \text{C}x + \text{D} + \text{E}e^x$

Now do this one all on your own.

Solve $$\frac{d^2y}{dx^2} - 3\frac{dy}{dx} + 2y = x^2$$

Do not forget: find (i) the C.F. and (ii) the P.I. Then the general solution is $y$ = C.F. + P.I.

Off you go.

*When you have finished completely, turn to frame 41.*

**41**

$$y = Ae^x + Be^{2x} + \frac{1}{4}(2x^2 + 6x + 7)$$

Here is the solution in detail.

$$\frac{d^2y}{dx^2} - 3\frac{dy}{dx} + 2y = x^2$$

(i) C.F. $m^2 - 3m + 2 = 0 \quad \therefore (m-1)(m-2) = 0 \quad \therefore m = 1$ or $2$

$$\therefore y = Ae^x + Be^{2x} \qquad \text{(i)}$$

(ii) P.I.

$$y = Cx^2 + Dx + E$$

$$\therefore \frac{dy}{dx} = 2Cx + D$$

$$\therefore \frac{d^2y}{dx^2} = 2C$$

$$2C - 3(2Cx + D) + 2(Cx^2 + Dx + E) = x^2$$

$$2Cx^2 + (2D - 6C)x + (2C - 3D + 2E) = x^2$$

$$2C = 1 \quad \therefore C = \frac{1}{2}$$

$$2D - 6C = 0 \quad \therefore D = 3C \quad \therefore D = \frac{3}{2}$$

$$2C - 3D + 2E = 0 \quad \therefore 2E = 3D - 2C = \frac{9}{2} - 1 = \frac{7}{2} \quad \therefore E = \frac{7}{4}$$

$$\therefore \text{P.I. is } y = \frac{x^2}{2} + \frac{3x}{2} + \frac{7}{4} = \frac{1}{4}(2x^2 + 6x + 7) \qquad \text{(ii)}$$

General solution:

$$y = Ae^x + Be^{2x} + \frac{1}{4}(2x^2 + 6x + 7)$$

*Next frame.*

**42**

***Particular solutions.*** The last result was $y = Ae^x + Be^{2x} + \frac{1}{4}(2x^2 + 6x + 7)$ and as with all second order differential equations, this contains two arbitrary constants A and B. These can be evaluated when the appropriate extra information is provided.

e.g. In this example, we might have been told that at $x = 0$, $y = \frac{3}{4}$ and $\frac{dy}{dx} = \frac{5}{2}$.

*It is important* to note that the values of A and B can be found only from the complete general solution and not from the C.F. as soon as you obtain it. This is a common error so do not be caught by it. Get the complete general solution before substituting to find A and B.

In this case, we are told that when $x = 0$, $y = \frac{3}{4}$, so inserting these values gives ............................

*Turn on to frame 43.*

**43**

$$\boxed{A + B = -1}$$

For: $\frac{3}{4} = A + B + \frac{7}{4} \quad \therefore \; A + B = -1$

We are also told that when $x = 0, \frac{dy}{dx} = \frac{5}{2}$, so we must first differentiate the general solution,

$$y = Ae^x + Be^{2x} + \frac{1}{4}(2x^2 + 6x + 7)$$

to obtain an expression for $\frac{dy}{dx}$.

So, $\frac{dy}{dx} = \dots\dots\dots\dots\dots\dots$

**44**

$$\boxed{\frac{dy}{dx} = Ae^x + 2Be^{2x} + \frac{1}{2}(2x + 3)}$$

Now we are given that when $x = 0, \; \frac{dy}{dx} = \frac{5}{2}$

$$\therefore \; \frac{5}{2} = A + 2B + \frac{3}{2} \quad \therefore \; A + 2B = 1$$

So we have
and

$$\left.\begin{aligned} A + B &= -1 \\ A + 2B &= 1 \end{aligned}\right\}$$

and these simultaneous equations give:

A = ................... ; B = ...................

*Then on to frame 45.*

**45**

$$\boxed{A = -3;\ B = 2}$$

Substituting these values in the general solution

$$y = Ae^{x} + Be^{2x} + \frac{1}{4}(2x^2 + 6x + 7)$$

gives the *particular solution*

$$\underline{y = 2e^{2x} - 3e^{x} + \frac{1}{4}(2x^2 + 6x + 7)}$$

And here is one for you, all on your own.

Solve the equation $\frac{d^2y}{dx^2} + 4\frac{dy}{dx} + 5y = 13e^{3x}$ given that when $x = 0$, $y = \frac{5}{2}$ and $\frac{dy}{dx} = \frac{1}{2}$. Remember:

(i) Find the C.F.; (ii) Find the P.I.;
(iii) The general solution is $y$ = C.F. + P.I.;
(iv) Finally insert the given conditions to obtain the particular solution.

*When you have finished, check with the solution in frame 46.*

**46**

$$\boxed{y = e^{-2x}(2\cos x + 3\sin x) + \frac{e^{3x}}{2}}$$

For: $$\frac{d^2y}{dx^2} + 4\frac{dy}{dx} + 5y = 13e^{3x}$$

(i) C.F. $m^2 + 4m + 5 = 0 \quad \therefore m = \frac{-4 \pm \sqrt{(16-20)}}{2} = \frac{-4 \pm j2}{2}$

$$\therefore m = -2 \pm j \quad \therefore \underline{y = e^{-2x}(A\cos x + B\sin x)} \qquad \text{(i)}$$

(ii) P.I. $y = Ce^{3x} \quad \therefore \frac{dy}{dx} = 3Ce^{3x},\ \frac{d^2y}{dx^2} = 9Ce^{3x}$

$$\therefore 9Ce^{3x} + 12Ce^{3x} + 5Ce^{3x} = 13e^{3x}$$

$$26C = 13 \quad \underline{\therefore C = \frac{1}{2} \quad \therefore \text{P.I. is } y = \frac{e^{3x}}{2}} \qquad \text{(ii)}$$

General solution $\underline{y = e^{-2x}(A\cos x + B\sin x) + \frac{e^{3x}}{2};\ x = 0,\ y = \frac{5}{2}}$

$$\therefore \frac{5}{2} = A + \frac{1}{2} \quad \therefore A = 2 \qquad y = e^{-2x}(2\cos x + B\sin x) + \frac{e^{3x}}{2}$$

$$\frac{dy}{dx} = e^{-2x}(-2\sin x + B\cos x) - 2e^{-2x}(2\cos x + B\sin x) + \frac{3e^{3x}}{2}$$

$$x = 0,\ \frac{dy}{dx} = \frac{1}{2} \quad \therefore \frac{1}{2} = B - 4 + \frac{3}{2} \quad \therefore B = 3$$

$\therefore$ Particular solution is $\underline{y = e^{-2x}(2\cos x + 3\sin x) + \frac{e^{3x}}{2}}$

**47**

Since the C.F. makes the L.H.S. = 0, it is pointless to use as a P.I. a term already contained in the C.F. If this occurs, multiply the assumed P.I. by $x$ and proceed as before. If this too is already included in the C.F., multiply by a further $x$ and proceed as usual.

*Example:* Solve $\dfrac{d^2y}{dx^2} - 2\dfrac{dy}{dx} - 8y = 3e^{-2x}$

(i) C.F. $m^2 - 2m - 8 = 0 \quad \therefore (m+2)(m-4) = 0 \quad \therefore m = -2$ or $4$

$$y = Ae^{4x} + Be^{-2x} \qquad \text{(i)}$$

(ii) P.I. The general form of the R.H.S. is $Ce^{-2x}$, but this term in $e^{-2x}$ is already contained in the C.F. Assume $y = Cxe^{-2x}$, and continue as usual.

$$y = Cxe^{-2x}$$

$$\frac{dy}{dx} = Cx(-2e^{-2x}) + Ce^{-2x} = Ce^{-2x}(1-2x)$$

$$\frac{d^2y}{dx^2} = Ce^{-2x}(-2) - 2Ce^{-2x}(1-2x) = Ce^{-2x}(4x-4)$$

Substituting in the given equation, we get

$$Ce^{-2x}(4x-4) - 2.Ce^{-2x}(1-2x) - 8Cxe^{-2x} = 3e^{-2x}$$

$$(4C + 4C - 8C)x - 4C - 2C = 3$$

$$-6C = 3 \quad \therefore C = -\frac{1}{2}$$

$$\text{P.I. is } y = -\frac{1}{2}xe^{-2x} \qquad \text{(ii)}$$

General solution $\qquad y = Ae^{4x} + Be^{-2x} - \dfrac{xe^{-2x}}{2}$

So remember, if the general form of the R.H.S. is already included in the C.F., multiply the assumed general form of the P.I. by $x$ and continue as before.

Here is one final example for you to work.

Solve $\qquad \dfrac{d^2y}{dx^2} + \dfrac{dy}{dx} - 2y = e^x$

*Finish it off and then turn to frame 48.*

**48**

$$\boxed{y = Ae^x + Be^{-2x} + \frac{xe^x}{3}}$$

Here is the working:

To solve $$\frac{d^2y}{dx^2} + \frac{dy}{dx} - 2y = e^x$$

(i) C.F. $m^2 + m - 2 = 0$

$$(m-1)(m+2) = 0 \quad \therefore m = 1 \text{ or } -2$$

$$\therefore \underline{y = Ae^x + Be^{-2x}} \qquad \text{(i)}$$

(ii) P.I. Take $y = Ce^x$. But this is already included in the C.F. Therefore, assume $y = Cxe^x$.

Then $$\frac{dy}{dx} = Cxe^x + Ce^x = Ce^x(x+1)$$

$$\frac{d^2y}{dx^2} = Ce^x + Cxe^x + Ce^x = Ce^x(x+2)$$

$$\therefore Ce^x(x+2) + Ce^x(x+1) - 2Cxe^x = e^x$$

$$C(x+2) + C(x+1) - 2Cx = 1$$

$$3C = 1 \quad \therefore C = \frac{1}{3}$$

$$\text{P.I. is } \underline{y = \frac{xe^x}{3}} \qquad \text{(ii)}$$

and so the general solution is

$$\underline{y = Ae^x + Be^{-2x} + \frac{xe^x}{3}}$$

You are now almost at the end of this programme. Before you work through the Test Exercise, however, look down the revision sheet given in frame 49. It lists the main points that we have established during this programme, and you may find it very useful.

*So on now to frame 49.*

## Revision Sheet

**49**

1. Solution of equations of the form $a\dfrac{d^2y}{dx^2} + b\dfrac{dy}{dx} + cy = f(x)$

2. Auxiliary equation: $am^2 + bm + c = 0$

3. Types of solutions:

   (a) Real and different roots $\quad m = m_1$ and $m = m_2$

   $y = \text{A}e^{m_1x} + \text{B}e^{m_2x}$

   (b) Real and equal roots $\quad m = m_1$ (twice)

   $y = e^{m_1x}(\text{A} + \text{B}x)$

   (c) Complex roots $\quad m = \alpha \pm \text{j}\beta$

   $y = e^{\alpha x}(\text{A}\cos\beta x + \text{B}\sin\beta x)$

4. Equations of the form $\dfrac{d^2y}{dx^2} + n^2y = 0$

   $y = \text{A}\cos nx + \text{B}\sin nx$

5. Equations of the form $\dfrac{d^2y}{dx^2} - n^2y = 0$

   $y = \text{A}\cosh nx + \text{B}\sinh nx$

6. General solution

   $y$ = complementary function + particular integral

7. (i) To find C.F. solve $a\dfrac{d^2y}{dx^2} + b\dfrac{dy}{dx} + cy = 0$

   (ii) To find P.I. assume the general form of the R.H.S.
   *Note:* If the general form of the R.H.S. is already included in the C.F., multiply by $x$ and proceed as before, etc. Determine the complete general solution before substituting to find the values of the arbitrary constants A and B.

*Now all that remains is the Test Exercise, so on to frame 50.*

**50**

The Test Exercise contains eight differential equations for you to solve, similar to those we have dealt with in the programme. They are quite straightforward, so you should have no difficulty with them.

Set your work out neatly and take your time: this will help you to avoid making unnecessary slips.

**Test Exercise – XXV**

Solve the following:

1. $\dfrac{d^2y}{dx^2} - \dfrac{dy}{dx} - 2y = 8$

2. $\dfrac{d^2y}{dx^2} - 4y = 10e^{3x}$

3. $\dfrac{d^2y}{dx^2} + 2\dfrac{dy}{dx} + y = e^{-2x}$

4. $\dfrac{d^2y}{dx^2} + 25y = 5x^2 + x$

5. $\dfrac{d^2y}{dx^2} - 2\dfrac{dy}{dx} + y = 4 \sin x$

6. $\dfrac{d^2y}{dx^2} + 4\dfrac{dy}{dx} + 5y = 2e^{-2x}$, given that at $x = 0, y = 1$ and $\dfrac{dy}{dx} = -2$.

7. $3\dfrac{d^2y}{dx^2} - 2\dfrac{dy}{dx} - y = 2x - 3$

8. $\dfrac{d^2y}{dx^2} - 6\dfrac{dy}{dx} + 8y = 8e^{4x}$

## Further Problems – XXV

Solve the following equations:

1. $2\dfrac{d^2y}{dx^2} - 7\dfrac{dy}{dx} - 4y = e^{3x}$

2. $\dfrac{d^2y}{dx^2} - 6\dfrac{dy}{dx} + 9y = 54x + 18$

3. $\dfrac{d^2y}{dx^2} - 5\dfrac{dy}{dx} + 6y = 100 \sin 4x$

4. $\dfrac{d^2y}{dx^2} + 2\dfrac{dy}{dx} + y = 4 \sinh x$

5. $\dfrac{d^2y}{dx^2} + \dfrac{dy}{dx} - 2y = 2 \cosh 2x$

6. $\dfrac{d^2y}{dx^2} - 6\dfrac{dy}{dx} + 10y = 20 - e^{2x}$

7. $\dfrac{d^2y}{dx^2} + 4\dfrac{dy}{dx} + 4y = 2 \cos^2 x$

8. $\dfrac{d^2y}{dx^2} - 4\dfrac{dy}{dx} + 3y = x + e^{2x}$

9. $\dfrac{d^2y}{dx^2} - 2\dfrac{dy}{dx} + 3y = x^2 - 1$

10. $\dfrac{d^2y}{dx^2} - 9y = e^{3x} + \sin 3x$

11. For a horizontal cantilever of length $l$, with load $w$ per unit length, the equation of bending is

$$\mathrm{EI}\frac{d^2y}{d^2x} = \frac{w}{2}(l - x)^2$$

where E, I, $w$ and $l$ are constants. If $y = 0$ and $\dfrac{dy}{dx} = 0$ at $x = 0$, find $y$ in terms of $x$. Hence find the value of $y$ when $x = l$.

12. Solve the equation

$$\frac{d^2x}{dt^2} + 4\frac{dx}{dt} + 3x = e^{-3t}$$

given that at $t = 0$, $x = \frac{1}{2}$ and $\frac{dx}{dt} = -2$.

13. Obtain the general solution of the equation

$$\frac{d^2y}{dt^2} + 4\frac{dy}{dt} + 5y = 6 \sin t$$

and determine the amplitude and frequency of the steady-state function.

14. Solve the equation

$$\frac{d^2x}{dt^2} - 3\frac{dx}{dt} + 2x = \sin t$$

given that at $t = 0$, $x = 0$ and $\frac{dx}{dt} = 1$.

15. Solve $\frac{d^2y}{dx^2} + 3\frac{dy}{dx} + 2y = 3 \sin x$, given that when $x = 0$, $y = -0{\cdot}9$ and $\frac{dy}{dx} = -0{\cdot}7$.

16. Obtain the general solution of the equation

$$\frac{d^2y}{dx^2} + 6\frac{dy}{dx} + 10y = 50x$$

17. Solve the equation

$$\frac{d^2x}{dt^2} + 2\frac{dx}{dt} + 2x = 85 \sin 3t$$

given that when $t = 0$, $x = 0$ and $\frac{dx}{dt} = -20$. Show that the values of $t$ for stationary values of the steady-state solution are the roots of $6 \tan 3t = 7$.

18. Solve the equation $\frac{d^2y}{dx^2} = 3 \sin x - 4y$, given that $y = 0$ at $x = 0$ and that $\frac{dy}{dx} = 1$ at $x = \pi/2$. Find the maximum value of $y$ in the interval $0 < x < \pi$.

19. A mass suspended from a spring performs vertical oscillations and the displacement $x$ (cm) of the mass at time $t$ (s) is given by

$$\tfrac{1}{2}\frac{d^2x}{dt^2} = -48x$$

If $x = \dfrac{1}{6}$ and $\dfrac{dx}{dt} = 0$ when $t = 0$, determine the period and amplitude of the oscillations.

20. The equation of motion of a body performing damped forced vibrations is $\dfrac{d^2x}{dt^2} + 5\dfrac{dx}{dt} + 6x = \cos t$. Solve this equation, given that $x = 0{\cdot}1$ and $\dfrac{dx}{dt} = 0$ when $t = 0$. Write the steady-state solution in the form $K \sin(t + \alpha)$.

# Programme 26

## OPERATOR D METHODS

**1** **Operator D**

$$\frac{d}{dx}(x^n) = n.x^{n-1}$$

$$\frac{d}{dx}(\sin x) = \cos x$$

$$\frac{d}{dx}(u+v) = \frac{du}{dx} + \frac{dv}{dx}$$

These results, and others like them, you have seen and used many times in the past in your work on differentiation.

The symbol $\frac{d}{dx}$, of course, can have no numerical value of its own, nor can it exist alone. It merely indicates the process or operation of finding the differential coefficient of the function to which it is attached, and as such it is called an *operator*.

For example, $\frac{d}{dx}(e^{5x})$ denotes that we are carrying out the operation of finding the differential coefficient of $e^{5x}$ with respect to $x$, which in fact gives us $\frac{d}{dx}(e^{5x}) =$ ..................

**2**

$$\boxed{\frac{d}{dx}(e^{5x}) = 5e^{5x}}$$

Also, $\left\{\frac{d}{dx}\right\}^2$, or $\frac{d^2}{dx^2}$ as it is written, denotes that the same operation is to be carried out twice – so obtaining the second differential coefficient of the function that follows.

Of course, there is nothing magic about the symbol $\frac{d}{dx}$. We could use any symbol to denote the same process and, for convenience, we do, in fact, often use the letter D to indicate the same operation.

$$\text{i.e. } D \equiv \frac{d}{dx}$$

So that $\frac{dy}{dx}$ can be written D$y$.

and

$$D(\sin x) = \cos x$$

$$D(e^{kx}) = ke^{kx}$$

$$D(x^2 + 6x - 5) = 2x + 6 \quad \text{etc., etc.}$$

So that $\quad D(\sinh x) =$ ..................

*Turn to frame 3.*

**3**

$$\boxed{\text{D}(\sinh x) = \cosh x}$$

Similarly, $\text{D}(\tan x) = \sec^2 x,\quad \text{D}(\ln x) = \frac{1}{x},$

$$\text{D}(\cosh 5x) = 5 \sinh 5x.$$

Naturally, all the rules of differentiation still hold good.

e.g. $\text{D}(x^2 \sin x) = x^2 \cos x + 2x \sin x$ (product rule)

and similarly, by the quotient rule,

$$\text{D}\left\{\frac{\sin 5x}{x+1}\right\} = \ldots\ldots\ldots\ldots$$

---

**4**

$$\boxed{\text{D}\left\{\frac{\sin 5x}{x+1}\right\} = \frac{(x+1)5\cos 5x - \sin 5x}{(x+1)^2}}$$

In the same way, $\text{D}^2\{x^3\} = \text{D}\{\text{D}(x^3)\} = \text{D}\{3x^2\} = 6x.$

So: The symbol D denotes the first differential coefficient,
$\text{D}^2$ " " second " "
$\text{D}^3$ " " third " "

and, if $n$ is a positive integer, $\text{D}^n$ denotes ..................

---

**5**

$$\boxed{\text{the } n^{\text{th}} \text{ differential coefficient}}$$

Correct.

(i) $\text{D}^2(3\sin x + \cos 4x) = \text{D}(3\cos x - 4\sin 4x)$
$= \underline{-3\sin x - 16\cos 4x}$

(ii) $\text{D}^2(5x^4 - 7x^2 + 3) = \text{D}(20x^3 - 14x)$
$= \underline{60x^2 - 14}$

All very easy: it just means that we are using a different symbol to represent the same operators of old.

$$\text{D}\ (e^{2x} + 5\sin 3x) = 2e^{2x} + 15\cos 3x$$
$$\text{D}^2(e^{2x} + 5\sin 3x) = 4e^{2x} - 45\sin 3x$$
$$\text{D}^3(e^{2x} + 5\sin 3x) = 8e^{2x} - 135\cos 3x \quad \text{etc.}$$

Here are some for you to do.

Find (i) $\text{D}\ (4e^{5x} - 2\cos 3x) = \ldots\ldots\ldots\ldots$
(ii) $\text{D}^2(\sinh 5x + \cosh 3x) = \ldots\ldots\ldots\ldots$
(iii) $\text{D}^3(5x^4 - 3x^3 + 7x^2 + 2x - 1) = \ldots\ldots\ldots\ldots$

*When you have finished, turn to frame 6.*

**6**

(i) $20e^{5x} + 6 \sin 3x$
(ii) $25 \sinh 5x + 9 \cosh 3x$
(iii) $120x - 18$

The special advantage of using a single letter as an operator is that it can be manipulated algebraically.

*Example 1.* $(D + 4)\{\sin x\} = D\{\sin x\} + 4 \sin x$

$= \underline{\cos x + 4 \sin x}$

i.e. we just multiply out in the usual way.

*Example 2.* $(D + 3)^2\{\sin x\} = (D^2 + 6D + 9)\{\sin x\}$

$= D^2\{\sin x\} + 6D\{\sin x\} + 9 \sin x$

$D\ (\sin x) = \cos x$

$D^2(\sin x) = -\sin x$

$= \underline{-\sin x + 6 \cos x + 9 \sin x}$

Similarly $(D - 3)\{\cos 2x\} = \ldots\ldots\ldots\ldots$

**7**

$-2 \sin 2x - 3 \cos 2x$

For $(D - 3)\{\cos 2x\} = D\{\cos 2x\} - 3 \cos 2x$

$= \underline{-2 \sin 2x - 3 \cos 2x}$

Similarly,

(i) $(D + 4)\{e^{3x}\} = D\{e^{3x}\} + 4e^{3x}$

$= 3e^{3x} + 4e^{3x} = \underline{7e^{3x}}$

(ii) $(D^2 - 5D + 4)\{x^2 + 4x - 1\}$

$= 2 - 5(2x + 4) + 4(x^2 + 4x - 1)$

$= 2 - 10x - 20 + 4x^2 + 16x - 4$

$= 4x^2 + 6x - 22$

$D\ (x^2 + 4x - 1) = 2x + 4$

$D^2(x^2 + 4x - 1) = 2$

Now you determine this one:

$(D^2 - 7D + 3)\{\sin 3x + 2 \cos 3x\} = \ldots\ldots\ldots\ldots$

*When you are satisfied with your result, turn on to frame 8.*

**8**

$$\boxed{36 \sin 3x - 33 \cos 3x}$$

Since $\quad D(\sin 3x + 2\cos 3x) = 3\cos 3x - 6\sin 3x$

and $\quad D^2(\sin 3x + 2\cos 3x) = -9\sin 3x - 18\cos 3x$

$$\therefore (D^2 - 7D + 3)\{\sin 3x + 2\cos 3x\}$$
$$= -9\sin 3x - 18\cos 3x - 21\cos 3x + 42\sin 3x + 3\sin 3x + 6\cos 3x$$
$$= \underline{36\sin 3x - 33\cos 3x}$$

*Remember* that the operator can be manipulated algebraically if required.

Here is one more:

$$(D^2 + 5D + 4)\{5e^{2x}\} = \dots\dots\dots\dots$$

**9**

$$\boxed{90e^{2x}}$$

Since

$$(D^2 + 5D + 4)\{5e^{2x}\} = D^2\{5e^{2x}\} + 5D\{5e^{2x}\} + 4\{5e^{2x}\}$$

Now $D\{5e^{2x}\} = 10e^{2x}$ and $D^2\{5e^{2x}\} = 20e^{2x}$

$$(D^2 + 5D + 4)\{5e^{2x}\} = 20e^{2x} + 50e^{2x} + 20e^{2x}$$
$$= \underline{90e^{2x}}$$

*or* we could have said:

$$(D^2 + 5D + 4)\{5e^{2x}\} = (D+4)(D+1)\{5e^{2x}\}$$
$$= (D+4)\{10e^{2x} + 5e^{2x}\}$$
$$= (D+4)\{15e^{2x}\}$$
$$= 30e^{2x} + 60e^{2x}$$
$$= \underline{90e^{2x}}$$

*On now to the next frame.*

**10** **The inverse operator $\frac{1}{D}$**

We define the inverse operator $\frac{1}{D}$ as being one, the effect of which is cancelled out when operated upon by the operator D. That is, the inverse operator $\frac{1}{D}$ is the reverse of the operator D, and since D indicates the process of differentiation, then $\frac{1}{D}$ indicates the process of ....................

**11**

integration

Right, though our definition of $\frac{1}{D}$ is a little more precise than that. Here it is:

*Definition:* The inverse operator $\frac{1}{D}$ denotes integration with respect to $x$, *omitting the arbitrary constant of integration.*

$$\text{e.g.}\quad \frac{1}{D}\{\sin x\} = -\cos x$$

$$\frac{1}{D}\{e^{3x}\} = \frac{e^{3x}}{3}$$

$$\frac{1}{D}\{x^4\} = \text{.......................}$$

**12**

$$\frac{1}{D}\{x^4\} = \frac{x^5}{5}$$

Similarly,

$$\frac{1}{D}\{\sinh 3x + \cosh 2x\} = \frac{\cosh 3x}{3} + \frac{\sinh 2x}{2}$$

and

$$\frac{1}{D}\left\{x + \frac{1}{x}\right\} = \frac{x^2}{2} + \ln x$$

Therefore, we have that

(i) the operator D indicates the operation of ....................

(ii) " " $\frac{1}{D}$ " " " " ....................

*Turn on to frame 13.*

**13**

| D denotes differentiation | |
|---|---|
| $\frac{1}{D}$ " | integration |

Of course, $\frac{1}{D^2} = \left(\frac{1}{D}\right)^2$ and $\frac{1}{D^2}\{f(x)\}$ therefore indicates the result of integrating the function $f(x)$ twice with respect to $x$, the arbitrary constants of integration being omitted.

$$\text{e.g.}\quad \frac{1}{D^2}\{x^2 + 5x - 4\} = \frac{1}{D}\left\{\frac{x^3}{3} + \frac{5x^2}{2} - 4x\right\}$$

$$= \frac{x^4}{12} + \frac{5x^3}{6} - \frac{4x^2}{2}$$

$$= \underline{\frac{x^4}{12} + \frac{5x^3}{6} - 2x^2}$$

*Note* that the constant of integration is omitted at each stage of integration.

So $\quad \frac{1}{D^2}\{\sin 3x - 2\cos x\} = \ldots\ldots\ldots\ldots\ldots$

**14**

$$\boxed{2\cos x - \frac{\sin 3x}{9}}$$

Since $\quad \frac{1}{D^2}\{\sin 3x - 2\cos x\} = \frac{1}{D}\left\{-\frac{\cos 3x}{3} - 2\sin x\right\}$

$$= -\frac{\sin 3x}{9} + 2\cos x$$

$$\therefore\ \frac{1}{D^2}\{\sin 3x - 2\cos x\} = \underline{2\cos x - \frac{\sin 3x}{9}}$$

Here is a short exercise. Work all the following and then check your results with those in frame 15.

(i) $D(\sin 5x + \cos 2x) = \ldots\ldots\ldots\ldots\ldots$

(ii) $D(x^2 e^{3x}) = \ldots\ldots\ldots\ldots\ldots$

(iii) $\frac{1}{D}\left(2x^2 + 5 + \frac{2}{x}\right) = \ldots\ldots\ldots\ldots\ldots$

(iv) $\frac{1}{D}(\cosh 3x) = \ldots\ldots\ldots\ldots\ldots$

(v) $\frac{1}{D^2}(3x^2 + \sin 2x) = \ldots\ldots\ldots\ldots\ldots$

*When you have completed all five, move on to frame 15.*

**15** Here are the results in detail.

(i) $D(\sin 5x + \cos 2x) = \underline{5\cos 5x - 2\sin 2x}$

(ii) $D(x^2 e^{3x}) = x^2\, 3e^{3x} + 2x\, e^{3x}$

$= \underline{e^{3x}(3x^2 + 2x)}$

(iii) $\frac{1}{D}\left(2x^2 + 5 + \frac{2}{x}\right) = \underline{\frac{2x^3}{3} + 5x + 2\ln x}$

(iv) $\frac{1}{D}(\cosh 3x) = \underline{\frac{\sinh 3x}{3}}$

(v) $\frac{1}{D^2}(3x^2 + \sin 2x) = \frac{1}{D}\left(x^3 - \frac{\cos 2x}{2}\right)$

$= \underline{\frac{x^4}{4} - \frac{\sin 2x}{4}}$

*You must have got those right, so on now to frame 16.*

**16** Before we can really enjoy the benefits of using the operator D, we have to note three very important theorems, which we shall find most useful a little later when we come to solve differential equations by operator D methods. Let us look at the first.

*Theorem I*

$$\underline{F(D)\{e^{ax}\} = e^{ax}\, F(a)} \quad \text{.............} \quad (I)$$

where $a$ is a constant, real or complex

$$D\{e^{ax}\} = a\, e^{ax}$$

$$D^2\{e^{ax}\} = a^2\, e^{ax}$$

$$\therefore\ \underline{(D^2 + D)\{e^{ax}\}} = a^2 e^{ax} + a\, e^{ax} = \underline{e^{ax}(a^2 + a)}$$

Note that the result is the original expression with D replaced by $a$. This applies to any function of D operating on $e^{ax}$.

*Example 1.* $(D^2 + 2D - 3)\{e^{ax}\} = e^{ax}(a^2 + 2a - 3)$

This sort of thing works every time: the $e^{ax}$ comes through to the front and the function of D becomes the same function of $a$, i.e. D is replaced by $a$.

So $\quad (D^2 - 5)\{e^{2x}\} = \text{..................}$

*Turn to frame 17.*

**17**

$$\boxed{(D^2 - 5)\{e^{2x}\} = -e^{2x}}$$

Similarly, $(2D^2 + 5D - 2)\{e^{3x}\} = e^{3x}(2.9 + 5.3 - 2) = e^{3x}(18 + 15 - 2)$

$$= \underline{31\,e^{3x}}$$

The rule applies whatever function of D is operating on $e^{ax}$.

e.g. $\dfrac{1}{D-2}\{e^{5x}\} = e^{5x}.\dfrac{1}{5-2} = \dfrac{e^{5x}}{3}$

e.g. $\dfrac{2}{D^2+3}\{e^{3x}\} = e^{3x}.\dfrac{2}{9+3} = \dfrac{e^{3x}}{12}$

e.g. $\dfrac{1}{D^2-4D-1}\{e^{-2x}\} = e^{-2x}\dfrac{1}{(-2)^2-4(-2)-1}$

$$= e^{-2x}\frac{1}{4+8-1} = \frac{e^{-2x}}{11}$$

So $(D^2 - 5D + 4)\{e^{4x}\} = \ldots\ldots\ldots\ldots\ldots\ldots$

**18**

$$\boxed{0}$$

for $(D^2 - 5D + 4)\{e^{4x}\} = e^{4x}(4^2 - 5.4 + 4)$

$$= e^{4x}(16 - 20 + 4) = \underline{0}$$

Right, and in the same way,

$$\frac{1}{D^2+6D-2}\{e^{3x}\} = \ldots\ldots\ldots\ldots\ldots\ldots$$

**19**

$$\boxed{\frac{e^{3x}}{25}}$$

for $\dfrac{1}{D^2+6D-2}\{e^{3x}\} = e^{3x}.\dfrac{1}{9+18-2}$

$$= \underline{\frac{e^{3x}}{25}}$$

*Fine. Turn on now to frame 20.*

**20** Just for practice, work the following:

(i) $(D^2 + 4D - 3)\{e^{2x}\} = \ldots\ldots\ldots$

(ii) $\dfrac{1}{D^2 + 4}\{e^{-3x}\} = \ldots\ldots\ldots$

(iii) $(D^2 - 7D + 2)\{e^{x/2}\} = \ldots\ldots\ldots$

(iv) $\dfrac{1}{D^2 - 3D - 2}\{e^{5x}\} = \ldots\ldots\ldots$

(v) $\dfrac{1}{(D-3)(D+4)}\{e^{-x}\} = \ldots\ldots\ldots$

*When you have finished, check your results with those in the next frame.*

**21** *Results*

(i) $(D^2 + 4D - 3)\{e^{2x}\} = e^{2x}(4 + 8 - 3) = \underline{9e^{2x}}$

(ii) $\dfrac{1}{D^2 + 4}\{e^{-3x}\} = e^{-3x} \cdot \dfrac{1}{9 + 4} = \underline{\dfrac{e^{-3x}}{13}}$

(iii) $(D^2 - 7D + 2)\{e^{x/2}\} = e^{x/2}\left(\frac{1}{4} - \frac{7}{2} + 2\right)$

$$= e^{x/2}\left(\frac{9}{4} - \frac{7}{2}\right) = \underline{-\frac{5}{4}e^{x/2}}$$

(iv) $\dfrac{1}{D^2 - 3D - 2}\{e^{5x}\} = e^{5x} \cdot \dfrac{1}{25 - 15 - 2}$

$$= \underline{\frac{e^{5x}}{8}}$$

(v) $\dfrac{1}{(D-3)(D+4)}\{e^{-x}\} = e^{-x} \cdot \dfrac{1}{(-1-3)(-1+4)}$

$$= e^{-x}\,\frac{1}{(-4)(3)}$$

$$= \underline{-\frac{e^{-x}}{12}}$$

All correct?

*Turn on now then to the next part of the programme that starts in frame 22.*

*Theorem II*

$$\underline{F(D)\{e^{ax}V\} = e^{ax}F(D+a)\{V\}} \quad \ldots\ldots\ldots\ldots \text{(II)}$$

where $a$ is a constant, real or complex,
and V is a function of $x$.

Consider $(D^2 + D + 5)\{e^{ax}V\}$

$$D\{e^{ax}V\} = e^{ax}D\{V\} + a\,e^{ax}V$$
$$= e^{ax}[D\{V\} + aV]$$
$$D^2\{e^{ax}V\} = e^{ax}[D^2\{V\} + aD\{V\}] + a\,e^{ax}[D\{V\} + aV]$$
$$= e^{ax}[D^2\{V\} + 2aD\{V\} + a^2V]$$

Therefore

$$(D^2 + D + 5)\{e^{ax}V\} = e^{ax}[D^2\{V\} + 2aD\{V\} + a^2V] + e^{ax}[D\{V\} + aV] + 5e^{ax}V$$
$$= e^{ax}[(D^2 + 2Da + a^2)\{V\} + (D+a)\{V\} + 5V]$$
$$= e^{ax}[(D+a)^2 + (D+a) + 5]\{V\}$$

which is the original function of D with D replaced by $(D + a)$.

So, for a function of D operating on $\{e^{ax}V\}$, where V is a function of $x$, the $e^{ax}$ comes through to the front and the function of D becomes the same function of $(D + a)$ operating on V.

$$\underline{F(D)\{e^{ax}V\} = e^{ax}F(D+a)\{V\}}$$

An example or two will make this clear.

(1) $(D + 4)\{e^{3x}x^2\}$ In this case, $a = 3$ and $V = x^2$

$$= e^{3x}\{(D+3) + 4\}\{x^2\}$$
$$= e^{3x}(D+7)\{x^2\} = e^{3x}(2x + 7x^2)$$
$$= \underline{(7x^2 + 2x)e^{3x}}$$

(2) $(D^2 + 2D - 3)\{e^{2x}\sin x\}$

$$= e^{2x}[(D+2)^2 + 2(D+2) - 3]\,.\,\{\sin x\}$$
$$= e^{2x}(D^2 + 4D + 4 + 2D + 4 - 3)\{\sin x\}$$
$$= e^{2x}(D^2 + 6D + 5)\{\sin x\} \qquad \begin{cases} D(\sin x) = \cos x \\ D^2(\sin x) = -\sin x \end{cases}$$
$$= \underline{e^{2x}[4\sin x + 6\cos x]}$$

And, in much the same way,

(3) $(D^2 - 5)\{e^{5x}\cos 2x\} = \ldots\ldots\ldots\ldots\ldots\ldots$

**23**

$$4e^{5x}(4\cos 2x - 5\sin 2x)$$

for: $(D^2 - 5)\{e^{5x}\cos 2x\}$

$= e^{5x}[(D+5)^2 - 5].\{\cos 2x\}$

$= e^{5x}[D^2 + 10D + 25 - 5]\{\cos 2x\}$

$= e^{5x}[D^2 + 10D + 20]\{\cos 2x\}$ $\quad D(\cos 2x) = -2\sin 2x$

$\quad D^2(\cos 2x) = -4\cos 2x$

$= e^{5x}(-4\cos 2x - 20\sin 2x + 20\cos 2x)$

$= \underline{4e^{5x}(4\cos 2x - 5\sin 2x)}$

Now here is another:

$$\frac{1}{D^2 - 8D + 16}\{e^{4x}x^2\}$$

$$= e^{4x}\frac{1}{(D+4)^2 - 8(D+4) + 16}\{x^2\}$$

$$= e^{4x}\frac{1}{D^2 + 8D + 16 - 8D - 32 + 16}\{x^2\}$$

$$= e^{4x}\frac{1}{D^2}\{x^2\} \qquad \frac{1}{D}(x^2) = \frac{x^3}{3}$$

$$= \underline{\frac{e^{4x}x^4}{12}} \qquad \frac{1}{D^2}(x^2) = \frac{x^4}{12}$$

Now this one: they are all done the same way.

$$(D^2 - 3D + 4)\{e^{-x}\cos 3x\}$$

The first step is to ....................

**24**

(i) bring the $e^{-x}$ through to the front
(ii) replace D by $(D-1)$

Right, so we get

$(D^2 - 3D + 4)\{e^{-x}\cos 3x\}$

$= e^{-x}[(D-1)^2 - 3(D-1) + 4].\{\cos 3x\}$

$= e^{-x}(D^2 - 2D + 1 - 3D + 3 + 4).\{\cos 3x\}$

$= e^{-x}(D^2 - 5D + 8)\{\cos 3x\}$ $\quad \begin{cases} D(\cos 3x) = -3\sin 3x \\ D^2(\cos 3x) = -9\cos 3x \end{cases}$

$=$ ........................................

*When you have sorted that out, turn on to frame 25.*

**25**

$$\boxed{e^{-x}(15 \sin 3x - \cos 3x)}$$

Now let us look at this one.

$$\frac{1}{D^2 + 4D + 5}\{x^3 e^{-2x}\} \qquad \text{Here } a = -2 \text{ and } V = x^3$$

$$= e^{-2x}\frac{1}{(D-2)^2 + 4(D-2) + 5}\{x^3\}$$

$$= e^{-2x}\frac{1}{D^2 - 4D + 4 + 4D - 8 + 5}\{x^3\}$$

$$= e^{-2x}\frac{1}{D^2 + 1}\{x^3\}$$

and we are now faced with the problem of how to deal with $\frac{1}{D^2+1}\{x^3\}$

Remember that operators behave algebraically.

$$e^{-2x}\frac{1}{D^2+1}\{x^3\} = e^{-2x}(1 + D^2)^{-1}\{x^3\}$$

and $(1 + D^2)^{-1}$ can be expanded by the binomial theorem.

$$\therefore\ (1 + D^2)^{-1} = \ldots\ldots\ldots\ldots$$

**26**

$$\boxed{(1 + D^2)^{-1} = 1 - D^2 + D^4 - D^6 + \ldots}$$

$$e^{-2x}(1 + D^2)^{-1}\{x^3\}$$
$$= e^{-2x}(1 - D^2 + D^4 - D^6 + \ldots).\{x^3\}$$

$D(x^3) = 3x^2$
$D^2(x^3) = 6x$
$D^3(x^3) = 6$
$D^4(x^3) = 0$ etc.

$$= e^{-2x}(x^3 - 6x + 0 - 0 \ldots)$$
$$= \underline{e^{-2x}(x^3 - 6x)}$$

Here is another.

$$\frac{1}{D^2+3}\{x^2\} = \frac{1}{3}\,\frac{1}{1 + \frac{D^2}{3}}\{x^2\}$$

$$= \frac{1}{3}\left(1 + \frac{D^2}{3}\right)^{-1}\{x^2\}$$

$$= \frac{1}{3}\left(1 - \frac{D^2}{3} + \frac{D^4}{9} - \frac{D^6}{27}\ldots\right)\{x^2\}$$

$$= \ldots\ldots\ldots\ldots\ldots\ldots\ldots\ldots$$

Note we take out the factor 3 to reduce the denominator to the form $(1 + u)$

*On to frame 27.*

## 27

$$\frac{1}{3}\left(x^2 - \frac{2}{3}\right)$$

Similarly

$$\frac{1}{D^2-2}\{x^4\} = -\frac{1}{2}\,\frac{1}{1-\frac{D^2}{2}}\{x^4\}$$

$$= -\frac{1}{2}\left(1 - \frac{D^2}{2}\right)^{-1}\{x^4\}$$

$$= -\frac{1}{2}\left(1 + \frac{D^2}{2} + \frac{D^4}{4} + \frac{D^6}{8} + \ldots\right)\{x^4\}$$

$$= \ldots\ldots\ldots\ldots\ldots\ldots\ldots\ldots$$

*Finish it off. Then move on to frame 28.*

## 28

$$-\frac{1}{2}(x^4 + 6x^2 + 6)$$

Right. So far we have seen the use of the first two theorems.

*Theorem I* $\quad F(D)\{e^{ax}\} = \ldots\ldots\ldots\ldots$

*Theorem II* $\quad F(D)\{e^{ax}V\} = \ldots\ldots\ldots\ldots$

*Check your results with the next frame.*

## 29

$$F(D)\{e^{ax}\} = e^{ax}F(a)$$
$$F(D)\{e^{ax}V\} = e^{ax}F(D+a)\{V\}$$

Now for Theorem III

*Theorem III*

$$F(D^2)\begin{Bmatrix}\sin ax\\ \cos ax\end{Bmatrix} = F(-a^2)\begin{Bmatrix}\sin ax\\ \cos ax\end{Bmatrix} \quad \ldots\ldots\ldots\ldots \quad \text{(III)}$$

If a function of $D^2$ is operating on $\sin ax$ or on $\cos ax$ (or both) the $\sin ax$ or the $\cos ax$ is unchanged and $D^2$ is everywhere replaced by $(-a^2)$. Note that this applies only to $D^2$ and *not* to D.

*Example 1.* $(D^2+5)\{\sin 4x\} = (-16+5)\sin 4x = -11\sin 4x$

Just as easy as that!

*Example 2.* $\dfrac{1}{D^2-3}\{\cos 2x\} = \dfrac{1}{-4-3}\cos 2x = -\dfrac{1}{7}\cos 2x$

*Example 3.* $\dfrac{1}{D^2+4}\{\sin 3x + \cos 3x\} = \dfrac{1}{-9+4}(\sin 3x + \cos 3x)$

$$= -\frac{1}{5}(\sin 3x + \cos 3x)$$

*Example 4.* $(2D^2-1)\{\sin x\} = \ldots\ldots\ldots\ldots\ldots$

**30**

$$\boxed{-3 \sin x}$$

for $(2D^2 - 1)\{\sin x\} = [2(-1) - 1]\{\sin x\} = \underline{-3 \sin x}$

If the value of $a$ differs in two terms, each term is operated on separately.

e.g. $\dfrac{1}{D^2 + 2}\{\sin 2x + \cos 3x\}$

$$= \frac{1}{D^2 + 2}\{\sin 2x\} + \frac{1}{D^2 + 2}\{\cos 3x\}$$

$$= \frac{1}{-4 + 2}\{\sin 2x\} + \frac{1}{-9 + 2}\{\cos 3x\}$$

$$= \underline{-\frac{\sin 2x}{2} - \frac{\cos 3x}{7}}$$

So therefore $\dfrac{1}{D^2 - 5}\{\sin x + \cos 4x\}$

$= \dots\dots\dots\dots\dots\dots$

**31**

$$\boxed{-\frac{\sin x}{6} - \frac{\cos 4x}{21}}$$

Here it is:

$$\frac{1}{D^2 - 5}\{\sin x + \cos 4x\}$$

$$= \frac{1}{D^2 - 5}\{\sin x\} + \frac{1}{D^2 - 5}\{\cos 4x\}$$

$$= \frac{1}{-1 - 5}\{\sin x\} + \frac{1}{-16 - 5}\{\cos 4x\}$$

$$= \underline{-\frac{\sin x}{6} - \frac{\cos 4x}{21}}$$

Here are those three theorems again:

*Theorem I* $\quad F(D)\{e^{ax}\} = e^{ax} F(a)$ ............ (I)

*Theorem II* $\quad F(D)\{e^{ax} V\} = e^{ax} F(D + a)\{V\}$ ............ (II)

*Theorem III* $\quad F(D^2)\begin{Bmatrix}\sin ax \\ \cos ax\end{Bmatrix} = F(-a^2)\begin{Bmatrix}\sin ax \\ \cos ax\end{Bmatrix}$ ............ (III)

Be sure to copy these down into your record book. You will certainly be using them quite a lot from now on.

*We have now reached the stage where we can use this operator* D *to our advantage, so* *turn now to frame 32.*

**32**

**Solution of differential equations by operator D methods**

The reason why we have studied the operator D is mainly that we can now use these methods to help us solve differential equations.

You will remember from your previous programme that the general solution of a second order differential equation with constant coefficients, consists of two distinct parts.

general solution = complementary function + particular integral.

(i) The C.F. was easily found by solving the auxiliary equation, obtained from the given equation by writing $m^2$ for $\frac{d^2y}{dx^2}$, $m$ for $\frac{dy}{dx}$, and 1 for $y$.

This gave a quadratic equation, the type of roots determining the shape of the C.F.

(a) Roots real and different $\quad y = Ae^{m_1x} + Be^{m_2x}$

(b) Roots real and equal $\quad y = e^{m_1x}(A + Bx)$

(c) Roots complex $\quad y = e^{\alpha x}(A\cos\beta x + B\sin\beta x)$

(ii) The P.I. has up to now been found by ..........................

**33**

. . . assuming the general form of the function $f(x)$ on the R.H.S., substituting in the given equation and determining the constants involved by equating coefficients.

In using operator D methods, the C.F. is found from the auxiliary equation as before, but we now have a useful way of finding the P.I. A few examples will show how we go about it.

*Example 1.* $\quad \frac{d^2y}{dx^2} + 4\frac{dy}{dx} + 3y = e^{2x}$

(i) C.F. $m^2 + 4m + 3 = 0 \quad \therefore (m+1)(m+3) = 0 \quad \therefore m = -1$ or $-3$.

$$y = Ae^{-x} + Be^{-3x}$$

(ii) P.I. First write the equation in terms of the operator D

$$D^2y + 4Dy + 3y = e^{2x}$$

$$(D^2 + 4D + 3)y = e^{2x}$$

$$y = \frac{1}{D^2 + 4D + 3}\{e^{2x}\}$$

and, applying theorem I, we get

$$y = \text{...................}$$

**34**

$$\boxed{y = \frac{e^{2x}}{15}}$$

for $\quad y = e^{2x} \dfrac{1}{4 + 8 + 3} = \dfrac{e^{2x}}{15}$

So C.F. is $\quad y = Ae^{-x} + Be^{-3x}$

and P.I. is $\quad y = \dfrac{e^{2x}}{15}$

So the complete general solution is

$$y = \ldots\ldots\ldots\ldots\ldots\ldots$$

**35**

$$\boxed{y = Ae^{-x} + Be^{-3x} + \frac{e^{2x}}{15}}$$

Correct. Notice how automatic it all is when using the operator D. Here is another.

Solve $\quad \dfrac{d^2y}{dx^2} + 6\dfrac{dy}{dx} + 9y = e^{5x}$

(i) First find the C.F. which is

$$y = \ldots\ldots\ldots\ldots\ldots\ldots$$

**36**

$$\boxed{y = e^{-3x}(A + Bx)}$$

since $m^2 + 6m + 9 = 0 \quad \therefore (m + 3)^2 = 0 \quad \therefore m = -3$ (twice)

$$y = e^{-3x}(A + Bx)$$

(ii) To find the P.I., write the equation in operator D form

$$D^2y + 6Dy + 9y = e^{5x}$$

$$(D^2 + 6D + 9)y = e^{5x}$$

$$y = \frac{1}{D^2 + 6D + 9}\{e^{5x}\}$$

and by theorem I $\quad y = e^{5x} \dfrac{1}{25 + 30 + 9} = \dfrac{e^{5x}}{64}$

C.F. is $y = e^{-3x}(A + Bx)$

P.I. is $y = \dfrac{e^{5x}}{64}$

$\therefore$ General solution is

$$y = \ldots\ldots\ldots\ldots\ldots\ldots$$

*On to frame 37.*

**37**

$$y = e^{-3x}(\mathrm{A} + \mathrm{B}x) + \frac{e^{5x}}{64}$$

Now that you see how it works, solve this one in the same way.

Solve $$\frac{d^2y}{dx^2} + 4\frac{dy}{dx} + 5y = e^{-x}$$

(i) C.F. $m^2 + 4m + 5 = 0 \quad \therefore m = \dfrac{-4 \pm \sqrt{(16-20)}}{2}$

$$= \frac{-4 \pm \sqrt{-4}}{2} = -2 \pm \mathrm{j}$$

$$y = \ldots\ldots\ldots\ldots\ldots\ldots$$

**38**

$$y = e^{-2x}(\mathrm{A} \cos x + \mathrm{B} \sin x)$$

(ii) Now for the P.I.

$$\mathrm{D}^2y + 4\mathrm{D}y + 5y = e^{-x}$$

$$\therefore (\mathrm{D}^2 + 4\mathrm{D} + 5)y = e^{-x}$$

$$y = \ldots\ldots\ldots\ldots\ldots\ldots$$

Now finish it off and obtain the complete general solution.

*When you have it, move on to frame 39.*

**39**

$$y = e^{-2x}(\mathrm{A} \cos x + \mathrm{B} \sin x) + \frac{e^{-x}}{2}$$

for the P.I. is

$$y = \frac{1}{\mathrm{D}^2 + 4\mathrm{D} + 5}\{e^{-x}\} \qquad a = -1$$

$$= e^{-x}\frac{1}{1 - 4 + 5} = \frac{e^{-x}}{2} \qquad \text{i.e. } y = \frac{e^{-x}}{2}$$

$\therefore$ General solution is $\underline{y = e^{-2x}(\mathrm{A} \cos x + \mathrm{B} \sin x) + \dfrac{e^{-x}}{2}}$

Now here is one for you to do all on your own.

Solve $$\frac{d^2y}{dx^2} + 7\frac{dy}{dx} + 12y = 5e^{2x}$$

*When you have finished it, turn on to frame 40 and check your result.*

**40**

$$y = \mathrm{A}\,e^{-3x} + \mathrm{B}\,e^{-4x} + \frac{e^{2x}}{6}$$

Since (i) C.F. $m^2 + 7m + 12 = 0 \ \therefore (m+3)(m+4) = 0 \ \therefore m = -3$ or $-4$

$$y = \mathrm{A}\,e^{-3x} + \mathrm{B}\,e^{-4x}$$

(ii) P.I. $\mathrm{D}^2y + 7\mathrm{D}y + 12y = 5\,e^{2x}$

$$(\mathrm{D}^2 + 7\mathrm{D} + 12)y = 5\,e^{2x}$$

$$y = \frac{1}{\mathrm{D}^2 + 7\mathrm{D} + 12}\{5\,e^{2x}\}$$

$$y = 5\,e^{2x}\,\frac{1}{4 + 14 + 12} = \frac{5\,e^{2x}}{30} = \frac{e^{2x}}{6}$$

General solution: $\underline{y = \mathrm{A}\,e^{-3x} + \mathrm{B}\,e^{-4x} + \frac{e^{2x}}{6}}$

Now if we were told that at $x = 0, y = \frac{7}{6}$ and $\frac{dy}{dx} = -\frac{5}{3}$, we could differentiate and substitute, and find the values of A and B. So off you go and find the *particular solution* for these given conditions.

*Then on to frame 41.*

---

**41**

$$y = 2\,e^{-3x} - e^{-4x} + \frac{e^{2x}}{6}$$

for $x = 0, y = \frac{7}{6} \quad \therefore \frac{7}{6} = \mathrm{A} + \mathrm{B} + \frac{1}{6} \quad \therefore \mathrm{A} + \mathrm{B} = 1$

$$\frac{dy}{dx} = -3\mathrm{A}\,e^{-3x} - 4\mathrm{B}\,e^{-4x} + \frac{2\,e^{2x}}{6}$$

$x = 0, \frac{dy}{dx} = -\frac{5}{3} \quad \therefore -\frac{5}{3} = -3\mathrm{A} - 4\mathrm{B} + \frac{1}{3} \quad \therefore 3\mathrm{A} + 4\mathrm{B} = 2$

$$\left.\begin{aligned} 3\mathrm{A} + 4\mathrm{B} &= 2 \\ 3\mathrm{A} + 3\mathrm{B} &= 3 \end{aligned}\right\} \therefore \mathrm{B} = -1, \mathrm{A} = 2$$

$\therefore$ Particular solution is

$$\underline{y = 2\,e^{-3x} - e^{-4x} + \frac{e^{2x}}{6}}$$

So (i) the C.F. is found from the auxiliary equation as before,

(ii) the P.I. is found by applying operator D methods to the original equation.

*Now turn on to frame 42.*

**42** Now what about this one?

Solve $$\frac{d^2y}{dx^2} + 3\frac{dy}{dx} + 2y = \sin 2x$$

(i) C.F. $m^2 + 3m + 2 = 0 \quad \therefore (m+1)(m+2) = 0 \quad \therefore m = -1 \text{ or } -2$

$$y = \text{A}\,e^{-x} + \text{B}\,e^{-2x}$$

(ii) P.I. $(\text{D}^2 + 3\text{D} + 2)y = \sin 2x$

$$y = \frac{1}{\text{D}^2 + 3\text{D} + 2}\{\sin 2x\}$$

By theorem III we can replace $\text{D}^2$ by $-a^2$, i.e. in this case by $-4$, but the rule says nothing about replacing D by anything.

$$y = \frac{1}{-4 + 3\text{D} + 2}\{\sin 2x\}$$

$$y = \frac{1}{3\text{D} - 2}\{\sin 2x\}$$

Now comes the trick! If we multiply top and bottom of the function of D by $(3\text{D} + 2)$ we get $y = \ldots\ldots\ldots\ldots\ldots\ldots$

---

**43**

$$y = \frac{3\text{D} + 2}{9\text{D}^2 - 4}\{\sin 2x\}$$

Correct, and we can now apply theorem III again to the $\text{D}^2$ in the denominator, giving:

$$y = \frac{3\text{D} + 2}{-36 - 4}\{\sin 2x\} = \frac{3\text{D} + 2}{-40}\{\sin 2x\}$$

Now the rest is easy, for $\text{D}(\sin 2x) = 2\cos 2x$

$$\therefore \quad y = -\frac{1}{40}(6\cos 2x + 2\sin 2x)$$

$$\text{i.e.} \quad y = -\frac{1}{20}(3\cos 2x + \sin 2x)$$

So C.F. is $y = \text{A}\,e^{-x} + \text{B}\,e^{-2x}$

P.I. is $y = -\frac{1}{20}(3\cos 2x + \sin 2x)$

$\therefore$ General solution is

$$y = \text{A}\,e^{-x} + \text{B}\,e^{-2x} - \frac{1}{20}(3\cos 2x + \sin 2x)$$

*Note* that when we were faced with $\frac{1}{3\text{D} - 2}\{\sin 2x\}$, we multiplied top and bottom by $(3\text{D}+2)$ to give the difference of two squares on the bottom. so that we could then apply theorem III again. Remember that move: it is very useful.

*Now on to frame 44.*

**44**

Here is another example.

Solve $$\frac{d^2y}{dx^2} + 10\frac{dy}{dx} + 25y = 3\cos 4x$$

(i) Find the C.F. You do that.

$$y = \dots\dots\dots\dots$$

**45**

$$\boxed{y = e^{-5x}(A + Bx)}$$

Since $m^2 + 10m + 25 = 0 \quad \therefore (m+5)^2 = 0 \quad \therefore m = -5$ (twice)

$$y = e^{-5x}(A + Bx)$$

(ii) Now for the P.I.

$$(D^2 + 10D + 25)y = 3\cos 4x$$

$$y = \frac{1}{D^2 + 10D + 25}\{3\cos 4x\}$$

Now apply theorem III, which gives us on the next line

$$y = \dots\dots\dots\dots$$

**46**

$$\boxed{y = \frac{1}{-16 + 10D + 25}\{3\cos 4x\}}$$

since, in this case, $a = 4 \quad \therefore -a^2 = -16 \quad \therefore D^2$ is replaced by $-16$.
Simplifying the result gives

$$y = \frac{1}{10D + 9}\{3\cos 4x\}$$

Now then, what do we do next?

*When you have decided, turn on to frame 47.*

## 47

**We multiply top and bottom by (10D − 9)**

Correct – in order to give $D^2$ in the denominator.

So we have

$$y = \frac{10D - 9}{(10D + 9)(10D - 9)}\{3\cos 4x\}$$

$$= \frac{10D - 9}{100D^2 - 81}\{3\cos 4x\}$$

We can now apply theorem III, giving

$$y = \ldots\ldots\ldots\ldots\ldots$$

## 48

$$\boxed{y = \frac{1}{1681}(120\sin 4x + 27\cos 4x)}$$

Here it is:

$$y = \frac{10D - 9}{100D^2 - 81}\{3\cos 4x\}$$

$$= \frac{10D - 9}{-1600 - 81}\{3\cos 4x\}$$

$$= -\frac{1}{1681}(10D - 9)\{3\cos 4x\}$$

$D(3\cos 4x) = -12\sin 4x$

$$= -\frac{1}{1681}(-120\sin 4x - 27\cos 4x)$$

$$y = \frac{1}{1681}(120\sin 4x + 27\cos 4x)$$

So C.F.: $y = e^{-5x}(A + Bx)$

P.I.: $y = \dfrac{1}{1681}(120\sin 4x + 27\cos 4x)$

Therefore, the general solution is

$$y = \ldots\ldots\ldots\ldots\ldots$$

*Now turn on to frame 49.*

$$y = e^{-5x}(A + Bx) + \frac{1}{1681}(120 \sin 4x + 27 \cos 4x)$$

Let us look at the complete solution. Here it is:

To solve $\quad \dfrac{d^2y}{dx^2} + 10\dfrac{dy}{dx} + 25y = 3 \cos 4x$

(i) C.F. $\quad m^2 + 10m + 25 = 0 \quad \therefore (m + 5)^2 = 0 \quad \therefore m = -5$ (twice)

$$\therefore y = e^{-5x}(A + Bx)$$

(ii) P.I. $\quad (D^2 + 10D + 25)y = 3 \cos 4x$

$$y = \frac{1}{D^2 + 10D + 25}\{3 \cos 4x\}$$

$$y = \frac{1}{-16 + 10D + 25}\{3 \cos 4x\}$$

$$= \frac{1}{10D + 9}\{3 \cos 4x\}$$

$$= \frac{10D - 9}{100D^2 - 81}\{3 \cos 4x\}$$

$$= \frac{10D - 9}{-1600 - 81}\{3 \cos 4x\}$$

$$= -\frac{1}{1681}(-120 \sin 4x - 27 \cos 4x)$$

$$y = \frac{1}{1681}(120 \sin 4x + 27 \cos 4x)$$

Therefore, the general solution is

$$\underline{y = e^{-5x}(A + Bx) + \frac{1}{1681}(120 \sin 4x + 27 \cos 4x)}$$

That is it. Now you can do this one in very much the same way.

Solve $\quad \dfrac{d^2y}{dx^2} - 4\dfrac{dy}{dx} + 13y = 2 \sin 3x$

*Find the complete general solution and then check your solution with that given in the next frame.*

**50** Here is the solution in detail.

$$\frac{d^2y}{dx^2} - 4\frac{dy}{dx} + 13y = 2 \sin 3x$$

(i) C.F. $m^2 - 4m + 13 = 0 \quad \therefore m = \dfrac{4 \pm \sqrt{(16-52)}}{2}$

$$= \frac{4 \pm \sqrt{-36}}{2} = 2 \pm j3$$

$$\therefore y = e^{2x}(\text{A} \cos 3x + \text{B} \sin 3x)$$

(ii) P.I. $(\text{D}^2 - 4\text{D} + 13)y = 2 \sin 3x$

$$y = \frac{1}{\text{D}^2 - 4\text{D} + 13}\{2 \sin 3x\}$$

$$= \frac{1}{-9 - 4\text{D} + 13}\{2 \sin 3x\}$$

$$= \frac{1}{4(1 - \text{D})}\{2 \sin 3x\}$$

$$= \frac{2}{4} \cdot \frac{1}{1 - \text{D}}\{\sin 3x\}$$

$$= \frac{1}{2} \cdot \frac{1 + \text{D}}{1 - \text{D}^2}\{\sin 3x\}$$

$$= \frac{1}{2} \cdot \frac{1 + \text{D}}{1 - (-9)}\{\sin 3x\}$$

$$= \frac{1}{20}(1 + \text{D})\{\sin 3x\}$$

$$y = \frac{1}{20}(\sin 3x + 3 \cos 3x)$$

General solution is

$$y = e^{2x}(\text{A} \cos 3x + \text{B} \sin 3x) + \frac{1}{20}(\sin 3x + 3 \cos 3x)$$

Now let us consider the following example.

Solve $\quad \dfrac{d^2y}{dx^2} - 6\dfrac{dy}{dx} + 5y = e^{2x} \sin 3x$

(i) First find the C.F. in the usual way. This comes to

$y = \ldots\ldots\ldots\ldots\ldots\ldots$

*On to frame 51.*

**51**

$$\boxed{y = \mathrm{A}\,e^{x} + \mathrm{B}\,e^{5x}}$$

Since $m^2 - 6m + 5 = 0 \quad \therefore (m-1)(m-5) = 0 \quad \therefore m = 1$ or $5$

$$\therefore y = \mathrm{A}\,e^{x} + \mathrm{B}\,e^{5x}$$

Now for the P.I.

$$(\mathrm{D}^2 - 6\mathrm{D} + 5)y = e^{2x}\sin 3x$$

$$y = \frac{1}{\mathrm{D}^2 - 6\mathrm{D} + 5}\{e^{2x}\sin 3x\}$$

This requires an application of theorem II

$$\mathrm{F}(\mathrm{D})\{e^{ax}\mathrm{V}\} = e^{ax}\,\mathrm{F}(\mathrm{D}+a)\{\mathrm{V}\} \qquad \text{Here } a = 2, \quad \mathrm{V} = \sin 3x$$

So the $e^{2x}$ comes through to the front and the function of D becomes the same function of $(\mathrm{D}+a)$, i.e. $(\mathrm{D}+2)$, and operates on V, i.e. $\sin 3x$

$$y = e^{2x}\frac{1}{(\mathrm{D}+2)^2 - 6(\mathrm{D}+2) + 5}\{\sin 3x\}$$

$$= e^{2x}\frac{1}{\mathrm{D}^2 + 4\mathrm{D} + 4 - 6\mathrm{D} - 12 + 5}\{\sin 3x\}$$

$$= e^{2x}\frac{1}{\mathrm{D}^2 - 2\mathrm{D} - 3}\{\sin 3x\}$$

Now, applying theorem III, gives

$$y = \dots\dots\dots\dots$$

---

**52**

$$\boxed{y = e^{2x}\frac{1}{-9 - 2\mathrm{D} - 3}\{\sin 3x\}}$$

$$\therefore y = e^{2x}\frac{1}{-2\mathrm{D} - 12}\{\sin 3x\} = \frac{-e^{2x}}{2}\cdot\frac{1}{\mathrm{D}+6}\{\sin 3x\}$$

$$y = -\frac{e^{2x}}{2}\cdot\frac{1}{\mathrm{D}+6}\{\sin 3x\}$$

Now what? Multiply top and bottom by ..................

---

**53**

$$\boxed{\mathrm{D} - 6}$$

Right.

$$\therefore y = -\frac{e^{2x}}{2}\cdot\frac{\mathrm{D}-6}{\mathrm{D}^2 - 36}\{\sin 3x\}$$

$$= -\frac{e^{2x}}{2}\cdot\frac{\mathrm{D}-6}{-9 - 36}\{\sin 3x\}$$

$$= \frac{e^{2x}}{90}(\mathrm{D}-6)\{\sin 3x\}$$

So the P.I. is finally

$$y = \dots\dots\dots\dots$$

**54**

$$y = \frac{e^{2x}}{30}(\cos 3x - 2\sin 3x)$$

So C.F.: $y = Ae^x + Be^{5x}$

P.I.: $y = \frac{e^{2x}}{30}(\cos 3x - 2\sin 3x)$

$\therefore$ General solution:

$$\underline{y = Ae^x + Be^{5x} + \frac{e^{2x}}{30}(\cos 3x - 2\sin 3x)}$$

This is an example of the use of theorem II. Usually, we hope to be able to solve the given equation by using theorems I or III, but where this is not possible, we have to make use of theorem II.

Let us work through another example.

Solve $\frac{d^2y}{dx^2} - y = x^2 e^x$

(i) Find the C.F. What do you make it?

$y = \ldots\ldots\ldots\ldots\ldots\ldots$

**55**

$$y = Ae^x + Be^{-x}$$

since $m^2 - 1 = 0 \quad \therefore m^2 = 1 \quad \therefore m = 1 \text{ or } -1$

Now for the P.I.

$$(D^2 - 1)y = x^2 e^x$$

$$y = \frac{1}{D^2 - 1}\{x^2 e^x\}$$

Applying theorem II, the $e^x$ comes through to the front, giving

$$y = e^x \frac{1}{(D+1)^2 - 1}\{x^2\}$$

$$= e^x \frac{1}{D^2 + 2D + 1 - 1}\{x^2\}$$

$$= e^x . \frac{1}{D} . \frac{1}{D+2}\{x^2\}$$

$$= \frac{e^x}{2} . \frac{1}{D} . \frac{1}{1 + D/2}\{x^2\}$$

$$= \frac{e^x}{2} . \frac{1}{D} . (1 + D/2)^{-1}\{x^2\}$$

Now expand $(1 + D/2)^{-1}$ as a binomial series, and we get

$$y = \frac{e^x}{2} . \frac{1}{D} . \left(\ldots\ldots\ldots\ldots\ldots\ldots\right)\{x^2\}$$

*On to frame 56.*

**56**

$$y = \frac{e^x}{2} \cdot \frac{1}{D} \cdot \left(1 - \frac{D}{2} + \frac{D^2}{4} \ldots\right)\{x^2\}$$

But $\quad D\{x^2\} = 2x; \; D^2\{x^2\} = 2; \; D^3\{x^3\} = 0 \quad$ etc.

$$\therefore \; y = \frac{e^x}{2} \cdot \frac{1}{D}\left\{x^2 - x + \frac{1}{2}\right\}$$

and since $\frac{1}{D}$ denotes integration, omitting the constant of integration,

then $\quad y = \frac{e^x}{2}\left(\text{....................}\right)$

**57**

$$y = \frac{e^x}{2}\left(\frac{x^3}{3} - \frac{x^2}{2} + \frac{x}{2}\right)$$

So the general solution is

$$\underline{y = A e^x + B e^{-x} + \frac{e^x}{2.}\left(\frac{x^3}{3} - \frac{x^2}{2} + \frac{x}{2}\right)}$$

Now here is one for you to do on your own. Tackle it in the same way.

Solve $\quad \frac{d^2y}{dx^2} - 6\frac{dy}{dx} + 9y = x^3 e^{3x}$

*Find the complete general solution and then check with the next frame.*

**58**

$$y = e^{3x}\left(A + Bx + \frac{x^5}{20}\right)$$

(i) C.F. $\quad y = e^{3x}(A + Bx)$

(ii) P.I. $\quad y = \frac{1}{D^2 - 6D + 9}\{x^3 e^{3x}\}$

$$= e^{3x}\frac{1}{(D+3)^2 - 6(D+3) + 9}\{x^3\}$$

$$= e^{3x}\frac{1}{D^2 + 6D + 9 - 6D - 18 + 9}\{x^3\}$$

$$= e^{3x}\frac{1}{D^2}\{x^3\}$$

$$= e^{3x}\frac{1}{D}\left\{\frac{x^4}{4}\right\}$$

$$= e^{3x} \cdot \frac{x^5}{20} \qquad \therefore \; y = \frac{x^5 e^{3x}}{20}$$

$\therefore$ General solution is $\quad y = e^{3x}(A + Bx) + \frac{x^5 e^{3x}}{20}$

$$\underline{y = e^{3x}\left(A + Bx + \frac{x^5}{20}\right)}$$

*Now move on to frame 59.*

**59**

**Special cases**

By now, we have covered the general methods that enable us to solve the vast majority of second order differential equations with constant coefficients. There are still, however, a few tricks that are useful when the normal methods break down. Let us see one or two in the following examples.

*Example 1.* $\frac{d^2y}{dx^2} + 4\frac{dy}{dx} + 3y = 5$

(i) C.F. $m^2 + 4m + 3 = 0 \quad \therefore (m+1)(m+3) = 0 \quad \therefore m = -1 \text{ or } -3$

$$\therefore y = A e^{-x} + B e^{-3x}$$

(ii) P.I. $(D^2 + 4D + 3)y = 5$

$$y = \frac{1}{D^2 + 4D + 3}\{5\}$$

This poses a problem, for none of the three theorems specifically applies to the case when $f(x)$ is a constant.

Have you any ideas as to how we can make progress?

*When you have thought about it, turn on to frame 60.*

**60**

We have $\quad y = \frac{1}{D^2 + 4D + 3}\{5\}$

The trick is to introduce a factor $e^{0x}$ with the constant 5 and since $e^{0x} = e^0 = 1$, this will not alter its value. So we have:

$$y = \frac{1}{D^2 + 4D + 3}\{5e^{0x}\}$$

We can now apply theorem I to the function. The $e^{0x}$ comes through to the front, the function of D becoming the same function of $a$ which, in this case, is 0.

$$y = e^{0x}\frac{1}{0 + 0 + 3}\{5\}$$

$$= e^{0x} \cdot \frac{5}{3} \qquad \text{and since } e^{0x} = 1,$$

$$y = \frac{5}{3}$$

So the general solution is:

$$\underline{y = A e^{-x} + B e^{-3x} + \frac{5}{3}}$$

*Now for another. Turn on to frame 61.*

**61**

Here is another example.

*Example 2.* $\dfrac{d^2y}{dx^2} + 2\dfrac{dy}{dx} = 5$

(i) C.F. $m^2 + 2m = 0 \quad \therefore m(m+2) = 0 \quad \therefore m = 0$ or $-2$

$$\therefore y = A e^{0x} + B e^{-2x} \qquad \therefore y = A + B e^{-2x}$$

(ii) P.I. $(D^2 + 2D)y = 5$

$$y = \frac{1}{D^2 + 2D}\{5\}$$

If we try the same trick again, i.e. introduce a factor $e^{0x}$ and apply theorem I, we get

$$y = \ldots\ldots\ldots\ldots\ldots\ldots$$

**62**

$$\boxed{y = ?}$$

for $y = \dfrac{1}{D^2 + 2D}\{5\}$ becomes $y = \dfrac{1}{D^2 + 2D}\{5\,e^{0x}\}$

$$y = e^{0x}\frac{1}{0+0}\{5\} \text{ which is infinite!}$$

So our first trick breaks down in this case.

However, let us try another approach.

$$y = \frac{1}{D^2 + 2D}\{5\}$$

$$= \frac{1}{D(D+2)}\{5\}$$

$$= \frac{1}{D}\cdot\frac{1}{(D+2)}\{5\}$$

Now introduce the $e^{0x}$ factor and apply only the operator $\dfrac{1}{D+2}$

$$y = \frac{1}{D}\cdot\frac{1}{(D+2)}\{5\,e^{0x}\}$$

$$= \frac{1}{D}\cdot e^{0x}\frac{1}{0+2}\{5\}$$

$$= \frac{1}{D}\,\frac{1}{2}(5) \quad \text{since } e^{0x} = 1$$

$$y = \frac{1}{D}\left\{\frac{5}{2}\right\}$$

which is $y = \ldots\ldots\ldots\ldots\ldots\ldots\ldots\ldots\ldots$

## 63

$$y = \frac{5x}{2}$$

since $\frac{1}{D}$ denotes integration (with the constant of integration omitted).

*Note* that we can apply the operators one at a time if we so wish.

The C.F. was $\quad y = A + B e^{-2x}$

The P.I. was found thus: look at it again.

$$(D^2 + 2D)y = 5$$
$$y = \frac{1}{D^2 + 2D}\{5\}$$
$$= \frac{1}{D} \cdot \frac{1}{D+2}\{5\}$$
$$= \frac{1}{D} \cdot \frac{1}{D+2}\{5\, e^{0x}\}$$
$$= \frac{1}{D} \cdot e^{0x} \frac{1}{0+2}\{5\} \qquad \text{by theorem I}$$
$$= \frac{1}{D}\left\{1\left(\frac{5}{2}\right)\right\} = \frac{1}{D}\left\{\frac{5}{2}\right\}$$
$$y = \frac{5x}{2}$$

General solution is

$$\underline{y = A + B e^{-2x} + \frac{5x}{2}}$$

Now here is another one. Let us work through it together.

*On to frame 64.*

## 64

*Example 3.* $\quad \dfrac{d^2y}{dx^2} - 16y = e^{4x}$

(i) C.F. $\quad m^2 - 16 = 0 \quad \therefore m^2 = 16 \quad \therefore m = \pm 4$

$$y = A e^{4x} + B e^{-4x}$$

(ii) P.I. $\quad (D^2 - 16)y = e^{4x}$

$$y = \frac{1}{D^2 - 16}\{e^{4x}\}$$

Theorem I applied to this breaks down, giving $\frac{1}{0}$ again.

$\therefore$ Introduce a factor 1 with the $e^{4x}$

$$y = \frac{1}{D^2 - 16}\{e^{4x}\ 1\}$$

We now apply theorem II and on the next line we get

$$y = \ldots\ldots\ldots\ldots\ldots$$

*Turn to frame 65.*

**65**

$$y = e^{4x} \frac{1}{(D+4)^2 - 16}\{1\}$$

i.e. the $e^{4x}$ comes throug.. to the front and the function of D becomes the same function of (D + 4).

Then

$$y = e^{4x} \frac{1}{D^2 + 8D + 16 - 16}\{1\}$$

$$= e^{4x} \frac{1}{D} \cdot \frac{1}{D+8}\{1\}$$

The function 1 can now be replaced by $e^{0x}$ and we can apply theorem I to the second operator $\frac{1}{D+8}$, which then gives us

$y = \ldots\ldots\ldots\ldots\ldots\ldots$

**66**

$$y = e^{4x} \frac{1}{D} \cdot \frac{1}{D+8}\{e^{0x}\}$$

$$= e^{4x} \frac{1}{D} e^{0x} \frac{1}{0+8}$$

$$= e^{4x} \frac{1}{D}\left\{\frac{1}{8}\right\} \qquad \text{(since } e^{0x} = 1\text{)}$$

$\therefore y = \ldots\ldots\ldots\ldots\ldots\ldots$

**67**

$$y = e^{4x} \frac{x}{8}$$

since $\frac{1}{D}$ denotes integration.

So we have:

C.F. $y = Ae^{4x} + Be^{-4x}$

P.I. $y = \frac{x e^{4x}}{8}$

$\therefore$ General solution

$$\underline{y = Ae^{4x} + Be^{-4x} + \frac{x e^{4x}}{8}}$$

Notice this trick then of introducing a factor 1 or $e^{0x}$ as required, so that we can use theorem I or II as appropriate.

There remains one further piece of work that can be very useful in the solution of differential equations, so turn on to frame 68 and we will see what it is all about.

## 68

Consider $$\frac{d^2y}{dx^2} + 4y = 3 \sin 2x$$

(i) C.F. $m^2 + 4 = 0 \quad \therefore m^2 = -4 \quad \therefore m = \pm j2$

$$y = A \cos 2x + B \sin 2x$$

(ii) P.I. $(D^2 + 4)y = 3 \sin 2x$

$$y = \frac{1}{D^2 + 4}\{3 \sin 2x\}$$

The constant factor 3 can be brought to the front to simplify the work.

$$y = 3 . \frac{1}{D^2 + 4}\{\sin 2x\}$$

If we now apply theorem III (since we are operating on a sine term) we get

$$y = \ldots\ldots\ldots\ldots$$

## 69

$$\boxed{y = 3 . \frac{1}{-4 + 4}\{\sin 2x\} = 3 . \frac{1}{0}\{\sin 2x\}}$$

and theorem III breaks down since it produces the factor $\frac{1}{0}$.

Our immediate problem therefore is what to do in a case like this. Let us think back to some previous work.

From an earlier programme on complex numbers, you will remember that

$$e^{j\theta} = \cos \theta + j \sin \theta$$

so that $\cos \theta$ = the real part of $e^{j\theta}$, written $\mathscr{R}\{e^{j\theta}\}$

and $\sin \theta$ = the imaginary part of $e^{j\theta}$, written $\mathscr{I}\{e^{j\theta}\}$.

In our example, we could write

$$\sin 2x = \mathscr{I}\{\ldots\ldots\}$$

**70**

$$\boxed{\sin 2x = \mathcal{I}\{e^{j2x}\}}$$

So we can work this way:

$$y = 3.\frac{1}{D^2+4}\{\sin 2x\} = 3.\frac{1}{D^2+4}\mathcal{I}\{e^{j2x}\} = 3.\mathcal{I}\frac{1}{D^2+4}\{e^{j2x}\}$$

Theorem I now gives

$$y = 3\mathcal{I}\, e^{j2x}.\frac{1}{(j2)^2+4} = 3\mathcal{I}\, e^{j2x}\frac{1}{-4+4}$$

$$= 3\,\mathcal{I}\, e^{j2x}\frac{1}{0} \text{ so this does not get us very far.}$$

Since this does not work, we now introduce a factor 1 and try theorem II.

$$y = 3\mathcal{I}\frac{1}{D^2+4}\{e^{j2x}.1\} = \text{........................}$$

**71**

$$\boxed{y = 3\mathcal{I}e^{j2x}\frac{1}{(D+j2)^2+4}\{1\}}$$

$$\therefore\ y = 3\mathcal{I}e^{j2x}\frac{1}{D^2+j4D-4+4}\{1\}$$

$$= 3\,\mathcal{I}e^{j2x}\frac{1}{D}.\frac{1}{D+j4}\{e^{0x}\} \qquad \text{putting } e^{0x} \text{ for 1.}$$

$$= 3\mathcal{I}e^{j2x}\frac{1}{D}.\,e^{0x}\frac{1}{0+j4} \qquad \text{theorem I on second operator.}$$

$$= 3\mathcal{I}e^{j2x}\frac{1}{D}\left\{\frac{1}{j4}\right\} = 3\mathcal{I}e^{j2x}.\frac{x}{j4}$$

$$= \frac{3}{4}\mathcal{I}\frac{x}{j}(\cos 2x + j\sin 2x) \qquad \text{writing } e^{j2x} \text{ back into its trig. form.}$$

$$= \frac{3}{4}\mathcal{I}\left(\frac{x\cos 2x}{j} + x\sin 2x\right)$$

$$= \frac{3}{4}\mathcal{I}\left(x\sin 2x - j\,x\cos 2x\right) \qquad \therefore\ y = -\frac{3x\cos 2x}{4}$$

That seems rather lengthy, but we have set it out in detail to show every step. It is really quite straightforward and a very useful method. So finally we have

C.F. $y = A\cos 2x + B\sin 2x$ P.I. $y = -\dfrac{3x\cos 2x}{4}$

General solution $\underline{y = A\cos 2x + B\sin 2x - \dfrac{3x\cos 2x}{4}}$

Look through the last example again and then solve this following equation in much the same way.

Solve $\dfrac{d^2y}{dx^2} + 9y = \cos 3x.$

*When you have finished, turn on to frame 72 and check your result.*

**72**

Solution: $y = \text{A}\cos 3x + \text{B}\sin 3x + \dfrac{x\sin 3x}{6}$

Here are the steps in detail:

(i) You will have had no trouble with the complementary function

$$y = \text{A}\cos 3x + \text{B}\sin 3x$$

(ii) Now for the particular integral:-

$$(\text{D}^2 + 9)y = \cos 3x \qquad \therefore\ y = \frac{1}{\text{D}^2 + 9}\{\cos 3x\}$$

Theorem III breaks down. Therefore use $\cos 3x + \text{j}\sin 3x = e^{\text{j}3x}$

i.e. $\cos 3x = \mathcal{R}\{e^{\text{j}3x}\}$

$$y = \mathcal{R}\ \frac{1}{\text{D}^2 + 9}\{e^{\text{j}3x}\}$$

Theorem I breaks down. Therefore introduce a factor 1 and use theorem II.

$$y = \mathcal{R}\frac{1}{\text{D}^2 + 9}\{e^{\text{j}3x}.1\}$$

$$= \mathcal{R}\ e^{\text{j}3x}\ \frac{1}{(\text{D} + \text{j}3)^2 + 9}\{1\}$$

$$= \mathcal{R}\ e^{\text{j}3x}\frac{1}{\text{D}^2 + \text{j}6\text{D} - 9 + 9}\{1\}$$

$$= \mathcal{R}\ e^{\text{j}3x}\ \frac{1}{\text{D}}.\ \frac{1.}{(\text{D} + \text{j}6)}\{e^{0x}\} \qquad e^{0x} = 1$$

Operate on $e^{0x}$ with the second operator $\dfrac{1}{(\text{D} + \text{j}6)}$ using theorem I.

$$y = \mathcal{R}\ e^{\text{j}3x}\ \frac{1}{\text{D}}\ e^{0x}\ \frac{1}{\text{j}6}$$

$$= \mathcal{R}\ e^{\text{j}3x}\ \frac{1}{\text{D}}\left\{\frac{1}{\text{j}6}\right\} \quad = \mathcal{R}\ e^{\text{j}3x}\ \frac{x}{\text{j}6}$$

$$= \mathcal{R}\ \frac{x}{\text{j}6}(\cos 3x + \text{j}\sin 3x)$$ writing $e^{\text{j}3x}$ back in trig. form.

$$= \mathcal{R}\left\{-\frac{\text{j}x}{6}(\cos 3x + \text{j}\sin 3x)\right\}$$

$$= \mathcal{R}\left\{\frac{-\text{j}x\cos 3x}{6} + \frac{x\sin 3x}{6}\right\}$$

$$y = \frac{x\sin 3x}{6}$$

Then, combining the C.F. and the P.I. we have the general solution

$$y = \text{A}\cos 3x + \text{B}\sin 3x + \frac{x\sin 3x}{6}$$

*Note*. These special methods come to your aid when the usual ones break down, so remember them for future reference.

*Turn to frame 73.*

**73**

You have now completed this programme on the use of operator D methods for solving second order differential equations. All that remains is the Test Exercise, but before you tackle that, here is a brief summary of the items we have covered.

**Summary Sheet**

1. *Operator D* $\quad \mathrm{D} \equiv \dfrac{d}{dx}; \quad \mathrm{D}^2 \equiv \dfrac{d^2}{dx^2}; \quad \mathrm{D}^n \equiv \dfrac{d^n}{dx^n}$

2. *Inverse operator* $\quad \dfrac{1}{\mathrm{D}} \equiv \int \ldots dx$, omitting the constant of integration.

3. *Theorem I* $\quad \mathrm{F(D)}\{e^{ax}\} = e^{ax}.\mathrm{F}(a)$

4. *Theorem II* $\quad \mathrm{F(D)}\{e^{ax}\mathrm{V}\} = e^{ax}\,\mathrm{F}(\mathrm{D}+a)\{\mathrm{V}\}$

5. *Theorem III* $\quad \mathrm{F}(\mathrm{D}^2)\begin{Bmatrix}\sin ax\\ \cos ax\end{Bmatrix} = \mathrm{F}(-a^2)\begin{Bmatrix}\sin ax\\ \cos ax\end{Bmatrix}$

6. *General solution*

   $y$ = complementary function + particular integral

7. *Other useful items* (where appropriate)

   (i) Introduction of a factor 1 or $e^{0x}$

   (ii) Use of $e^{\mathrm{j}\theta} = \cos\theta + \mathrm{j}\sin\theta$

   i.e. $\cos\theta = \mathcal{R}\{e^{\mathrm{j}\theta}\}$

   $\sin\theta = \mathcal{I}\{e^{\mathrm{j}\theta}\}$

Revise any part of the programme that you feel needs brushing up before working through the Test Exercise.

When you are ready, turn on to the next frame and solve the equations given in the exercise. They are all straightforward and similar to those you have been doing in the programme, so you will have no difficulty with them.

*On to frame 74.*

**74**

Work through the whole of the exercise below. Take your time and work carefully. The equations are just like those we have been dealing with in the programme: there are no tricks to catch you out.

So off you go.

**Test Exercise – XXVI**

Solve the following equations:

1. $\dfrac{d^2y}{dx^2} + 3\dfrac{dy}{dx} + 2y = e^{4x}$

2. $\dfrac{d^2y}{dx^2} + 4\dfrac{dy}{dx} + 4y = 5e^{-3x}$

3. $\dfrac{d^2y}{dx^2} + 4\dfrac{dy}{dx} + 3y = \cos 3x$

4. $\dfrac{d^2y}{dx^2} - 4\dfrac{dy}{dx} + 5y = \sin 4x$

5. $\dfrac{d^2y}{dx^2} + 2\dfrac{dy}{dx} + 2y = e^x \sin 2x$

6. $\dfrac{d^2y}{dx^2} + 4\dfrac{dy}{dx} + 4y = x^3 e^{2x}$

7. $\dfrac{d^2y}{dx^2} + y = 3e^x + 5e^{2x}$

8. $\dfrac{d^2y}{dx^2} + 6\dfrac{dy}{dx} + 8y = 2\sin x + \sin 3x$

9. $\dfrac{d^2y}{dx^2} + 25y = \sin 5x$

10. $\dfrac{d^2y}{dx^2} - 2\dfrac{dy}{dx} - 3y = 2e^{3x}$.

*Well done.*

**Further Problems – XXVI**

*Note:* Where hyperbolic functions occur, replace them by their corresponding exponential expressions.
Employ operator-D methods throughout.

Solve the following equations by the use of the operator D.

1. $D^2y + 2Dy - 3y = 4e^{-3x}$
2. $D^2y + 3Dy + 2y = xe^{-x}$
3. $D^2y + y = \sin x$
4. $D^2y - 2Dy + y = \sin x + x^2$
5. $D^2y - 3Dy + 2y = -4e^x \sinh x$ ----- $\left\{ \begin{array}{l} \text{given at } x = 0, \\ y = 2 \text{ and } Dy = 0. \end{array} \right.$
6. $D^2y - 5Dy + 6y = e^{3x}$
7. $D^2y - 5Dy + 6y = e^{4x} \sin 3x$
8. $D^2y + 4Dy + 5y = x + \cos 2x$
9. $D^2y + 2Dy + 5y = 17 \cos 2x$
10. $D^2y + 4Dy + 5y = 8 \cos x$
11. $D^2y + 2aDy + a^2y = x^2 e^{-ax}$
12. $D^2y + Dy + y = xe^x + e^x \sin x$
13. $D^2y - 6Dy + 9y = e^{3x} + e^{-3x}$
14. $D^2y + 4Dy + 4y = \cosh 2x$
15. $D^2y + 6Dy + 9y = e^{-3x} \cosh 3x$
16. $D^2y - Dy - 6y = xe^{3x}$
17. $D^2y + 4Dy + 5y = 8 \cos^2 x$
18. $D^2y + 2Dy + 5y = 34 \sin x \cos x$
19. $2D^2y + Dy - y = e^x \sin 2x$
20. $D^2y + 2Dy + 5y = x + e^{-x} \cos 3x$
21. $D^2y - 2Dy + 4y = e^x \sin 3x$
22. $D^2y - 4Dy + 4y = e^{2x}$
23. $D^2y - 9y = \cosh 3x + x^2$
24. $D^2y + 3Dy + 2y = e^{-x} \cos x$
25. $D^2y + 2Dy + 2y = x^2 e^{-x}$

# Programme 27

## STATISTICS

**1**

## Introduction

*Statistics* is concerned with the collection, ordering and analysis of data. *Data* consist of sets of recorded observations or values. Any quantity that can have a number of values is a *variable.* A variable may be of one of two kinds:

(a) **Discrete** – a variable that can be counted, or for which there is a fixed set of values,
(b) **Continuous** – a variable that can be measured on a continuous scale, the result depending on the precision of the measuring instrument, or the accuracy of the observer.

A statistical exercise normally consists of four stages:

(a) collection of data by counting or measuring,
(b) ordering and presentation of the data in a convenient form,
(c) analysis of the collected data,
(d) interpretation of the results and conclusions formulated.

State whether each of the following is a discrete or continuous variable:

(a) the number of components in a machine,
(b) the capacity of a container,
(c) the size of workforce in a factory,
(d) the speed of rotation of a shaft,
(e) the temperature of a coolant.

**2**

(a) and (c) discrete; (b), (d) and (e) continuous

## Arrangement of data

*Example*: The contents of each of 30 packets of washers are recorded.

| | | | | | | | | | |
|---|---|---|---|---|---|---|---|---|---|
| 28 | 31 | 29 | 27 | 30 | 29 | 29 | 26 | 30 | 28 |
| 28 | 29 | 27 | 26 | 32 | 28 | 32 | 31 | 25 | 30 |
| 27 | 30 | 29 | 30 | 28 | 29 | 31 | 27 | 28 | 28 |

We can appreciate this set of numbers better if we now arrange the values in ascending order, writing them still in 3 lines of 10. If we do this, we get ..........................................................

**3**

| 25 | 26 | 26 | 27 | 27 | 27 | 27 | 28 | 28 | 28 |
|---|---|---|---|---|---|---|---|---|---|
| 28 | 28 | 28 | 28 | 29 | 29 | 29 | 29 | 29 | 29 |
| 30 | 30 | 30 | 30 | 30 | 31 | 31 | 31 | 32 | 32 |

Some values occur more than once. Therefore, we can form a table showing how many times each value occurs.

| Value | Number of times |
|---|---|
| 25 | 1 |
| 26 | 2 |
| 27 | 4 |
| 28 | 7 |
| 29 | 6 |
| 30 | 5 |
| 31 | 3 |
| 32 | 2 |

The number of occasions on which any particular value occurs is called its *frequency*, denoted by the symbol $f$.

The total frequency is therefore ............

**4**

30 the total number of readings

**Tally diagram** When dealing with large numbers of readings, instead of writing all the values in ascending order, it is more convenient to compile a *tally diagram*, recording the range of values of the variable and adding a stroke for each occurrence of the reading, thus

| Variable ($x$) | Tally marks | Frequency ($f$) |
|---|---|---|
| 25 | / | 1 |
| 26 | // | 2 |
| 27 | //// | 4 |
| 28 | ~~////~~ // | 7 |
| 29 | ~~////~~ / | 6 |
| 30 | ~~////~~ | 5 |
| 31 | /// | 3 |
| 32 | // | 2 |

It is usual to denote the variable by $x$ and the frequency by $f$. The right-hand column gives the *frequency distribution* of the values of the variable.

*Exercise*: The number of components per hour turned out on a lathe was measured on 40 occasions

| | | | | | | | | | |
|---|---|---|---|---|---|---|---|---|---|
| 18 | 17 | 21 | 18 | 19 | 17 | 18 | 20 | 16 | 17 |
| 19 | 19 | 16 | 17 | 15 | 19 | 17 | 17 | 20 | 18 |
| 17 | 18 | 19 | 19 | 18 | 19 | 18 | 18 | 19 | 20 |
| 18 | 15 | 18 | 17 | 20 | 18 | 16 | 17 | 18 | 17 |

Compile a tally diagram and so determine the frequency distribution of the values. This gives ......................

**5**

| Variable ($x$) | Tally marks | Frequency ($f$) |
|---|---|---|
| 15 | // | 2 |
| 16 | /// | 3 |
| 17 | ~~////~~ ~~////~~ | 10 |
| 18 | ~~////~~ ~~////~~ // | 12 |
| 19 | ~~////~~ /// | 8 |
| 20 | //// | 4 |
| 21 | / | 1 |
| | | $n = \Sigma f = 40$ |

**Grouped data** If the range of values of the variable is large, it is often helpful to consider these values arranged in regular groups, or *classes*.

*Example*: The number of overtime hours per week worked by employees at a factory are as follows

| | | | | | | | | | |
|---|---|---|---|---|---|---|---|---|---|
| 45 | 31 | 46 | 25 | 57 | 39 | 42 | 55 | 20 | 37 |
| 40 | 59 | 11 | 38 | 34 | 22 | 62 | 33 | 48 | 43 |
| 57 | 37 | 43 | 51 | 29 | 41 | 35 | 66 | 45 | 32 |
| 44 | 47 | 42 | 46 | 54 | 65 | 17 | 35 | 53 | 27 |
| 38 | 22 | 33 | 39 | 45 | 32 | 43 | 41 | 57 | 45 |

Lowest value of the variable = 11
Highest value of the variable = 66 $\}$ $\therefore$ Arrange 7 classes of 10 h each.

To determine the frequency distribution, we set up a table as follows

| Overtime hours ($x$) | Tally marks | Frequency ($f$) |
|---|---|---|
| 10–19 | | |
| 20–29 | | |
| 30–39 | | |
| etc | | |

Complete the frequency distribution.

**6**

| Overtime hours ($x$) | Tally marks | Frequency ($f$) |
|---|---|---|
| 10–19 | // | 2 |
| 20–29 | ~~////~~ / | 6 |
| 30–39 | ~~////~~ ~~////~~ //// | 14 |
| 40–49 | ~~////~~ ~~////~~ ~~////~~ // | 17 |
| 50–59 | ~~////~~ /// | 8 |
| 60–70 | /// | 3 |
| | | $n = \Sigma f = 50$ |

**Grouping with continuous data** In this last example using discrete data, there is no difficulty in allocating any given value to its appropriate group, since, for example, there is no value between, say, 29 and 30. However, with continuous data, the variable is measured on ...................... and may well have values lying between 29 and 30, e.g. 29.7, 29.8, etc.

**7**

a continuous scale

In practice, where the values of the variable are all given to the same number of significant figures or decimal places, there is no trouble and we form the groups accordingly.

*Example:* The lengths (in mm) of 40 spindles were measured with the following results

| | | | | | | | | | |
|---|---|---|---|---|---|---|---|---|---|
| 20·90 | 20·57 | 20·86 | 20·74 | 20·82 | 20·63 | 20·53 | 20·89 | 20·75 | 20·65 |
| 20·71 | 21·03 | 20·72 | 20·41 | 20·94 | 20·75 | 20·79 | 20·65 | 21·08 | 20·89 |
| 20·50 | 20·88 | 20·97 | 20·78 | 20·61 | 20·92 | 21·07 | 21·16 | 20·80 | 20·77 |
| 20·82 | 20·72 | 20·60 | 20·90 | 20·86 | 20·68 | 20·75 | 20·88 | 20·56 | 20·94 |

Lowest value = 20·41
Highest value = 21.16
$\therefore$ Form classes from 20·40 to 21·20 at 0·10 intervals.

| Length (mm) ($x$) | Tally marks | Frequency ($f$) |
|---|---|---|
| 20·40–20·49 | | |
| 20·50–20·59 | | |
| 20·60–20·69 | | |
| etc. | | |

Complete the table and so determine the frequency distribution ................

**8**

| Length (mm) ($x$) | Tally marks | Frequency ($f$) |
|---|---|---|
| 20·40–20·49 | / | 1 |
| 20·50–20·59 | //// | 4 |
| 20·60–20·69 | ~~////~~ / | 6 |
| 20·70–20·79 | ~~////~~ ~~////~~ | 10 |
| 20·80–20·89 | ~~////~~ //// | 9 |
| 20·90–20·99 | ~~////~~ / | 6 |
| 21·00–21·09 | /// | 3 |
| 21·10–21·20 | / | 1 |
| | | $n = \Sigma f = 40$ |

Note that the last class is slightly larger than the others, but this has negligible effect, since there are very few entries in the end classes.

**Relative frequency** In the frequency distribution just determined, if the frequency of any one class is compared with the sum of the frequencies of all classes (i.e. the total frequency), the ratio is the *relative frequency* of that class. The result is generally expressed as a percentage.

Add a fourth column to the table above showing the relative frequency of each class expressed as a percentage.

*Check with the next frame.*

**9**

| Length (mm) ($x$) | Frequency ($f$) | Relative frequency (%) |
|---|---|---|
| 20·40–20·49 | 1 | 2·5 |
| 20·50–20·59 | 4 | 10·0 |
| 20·60–20·69 | 6 | 15·0 |
| 20·70–20·79 | 10 | 25·0 |
| 20·80–20·89 | 9 | 22·5 |
| 20·90–20·99 | 6 | 15·0 |
| 21·00–21·09 | 3 | 7·5 |
| 21·10–21·20 | 1 | 2·5 |
| $n = \Sigma f = 40$ | | 100·0% |

The sum of the relative frequencies of all classes must add up to the whole and therefore has a value 1 or 100%.

**Rounding off data** If the value 21·7 is expressed to two significant figures, the result is, of course, 22. Similarly, 21·4 is rounded off to 21. There are various ways of dealing with 21·5, which is exactly half-way between 21 and 22. With these 'middle values', the commonest rule is to round off to the next highest value, i.e. 21·5 is rounded off to 22 and 42·5 to 43.

Therefore, when a result is quoted to two significant figures as 37 on a continuous scale, this includes all possible values between 36·500000... and 37·499999... (37·5 itself would be rounded off to 38). This could be expressed by saying that 37 includes all values between 36·5 and $37{\cdot}5^{-}$, the small negative sign in the index position indicating that the value is just under 37·5 without actually reaching it.

So 42 includes all values between ....... and .......

31·4 includes all values between ....... and .......

17·63 includes all values between ....... and .......

| | | |
|---|---|---|
| 42 | : 41·5 | and $42{\cdot}5^-$ |
| 31·4 | : 31·35 | and $31{\cdot}45^-$ |
| 17·63 | : 17·625 | and $17{\cdot}635^-$ |

*Exercise* The thicknesses of 20 samples of steel plate are measured and the results (in millimetres) to two significant figures are as follows

| | | | | | | | | | |
|---|---|---|---|---|---|---|---|---|---|
| 7·3 | 7·1 | 6·6 | 7·0 | 7·8 | 7·3 | 7·5 | 6·2 | 6·9 | 6·7 |
| 6·5 | 6·8 | 7·2 | 7·4 | 6·5 | 6·9 | 7·2 | 7·6 | 7·0 | 6·8 |

Compile a table showing the frequency distribution and the relative frequency distribution for regular classes of 0·2 mm from 6·2 mm to 7·9 mm.

**11**

| Thickness (mm) ($x$) | Frequency ($f$) | Relative frequency (%) |
|---|---|---|
| 6·2–6·4 | 1 | 5 |
| 6·5–6·7 | 4 | 20 |
| 6·8–7·0 | 6 | 30 |
| 7·1–7·3 | 5 | 25 |
| 7·4–7·6 | 3 | 15 |
| 7·7–7·9 | 1 | 5 |
| $n = \Sigma f = 20$ | | 100% |

**Class boundaries** In the example above, the values of the variable are given to 2 significant figures. With the usual rounding off procedure, each class in effect extends from 0·05 below the first stated value of the class to just under 0·05 above the second stated value of the class. So, the class

7·1–7·3 includes all values between 7·05 and $7{\cdot}35^-$

7·4–7·6 includes all values between 7·35 and $7{\cdot}65^-$ etc.

We can use this example to define a number of terms we shall need. Let us consider in particular the class labelled 7·1–7·3.

(a) The class values stated in the table are the *lower* and *upper limits* of the class and their difference gives the *class width.*
(b) The *class boundaries* are 0·05 below the lower class limit and 0·05 above the upper class limit, that is

the lower class boundary is $7{\cdot}1 - 0{\cdot}05 = 7{\cdot}05$

the upper class boundary is $7{\cdot}3 + 0{\cdot}05 = 7{\cdot}35$.

(c) The *class interval* is the difference between the upper and lower class boundaries.

$$\text{class interval} = \text{upper class boundary} - \text{lower class boundary}$$
$$= 7{\cdot}35 - 7{\cdot}05 = 0{\cdot}30$$

Where the classes are regular, the class interval can also be found by substracting any lower class limit from the lower class limit of the following class.
(d) The *central value* (or mid-value) of the class is the average of the upper and lower class boundaries. So, in the particular class we are considering, the central value is ............

7·20

For the central value = $\frac{1}{2}$ (7·05 + 7·35) = 7·20.

We can summarise these terms in the following diagram, using the class 7·1–7·3 (inclusive) as our example.

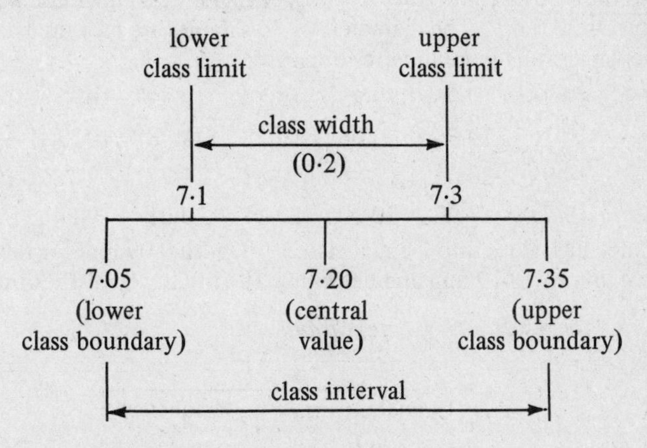


So now you can complete the following table:

| Thickness (mm) ($x$) | Central value | Lower class boundary | Upper class boundary |
|---|---|---|---|
| 6·2–6·4 | | | |
| 6·5–6·7 | | | |
| 6·8–7·0 | | | |
| 7·1–7·3 | | | |
| 7·4–7·6 | | | |
| 7·7–7·9 | | | |

**13**

| Thickness (mm) ($x$) | Central value | Lower class boundary | Upper class boundary |
|---|---|---|---|
| 6·2–6·4 | 6·3 | 6·15 | 6·45 |
| 6·5–6·7 | 6·6 | 6·45 | 6·75 |
| 6·8–7·0 | 6·9 | 6·75 | 7·05 |
| 7·1–7·3 | 7·2 | 7·05 | 7·35 |
| 7·4–7·6 | 7·5 | 7·35 | 7·65 |
| 7·7–7·9 | 7·8 | 7·65 | 7·95 |

*Exercise:* A machine is set to produce metal washers of nominal diameter 20·0 mm. The diameters of 34 samples are measured and the following results in millimetres obtained.

| | | | | | | | | |
|---|---|---|---|---|---|---|---|---|
| 19·63 | 19·82 | 19·96 | 19·75 | 19·86 | 19·82 | 19·61 | 19·97 | 20·07 |
| 19·89 | 20·16 | 19·56 | 20·05 | 19·72 | 19·96 | 19·68 | 19·87 | 19·90 |
| 19·73 | 19·93 | 20·03 | 19·86 | 19·81 | 19·77 | 19·78 | 19·75 | 19·87 |
| 19·66 | 19·77 | 19·99 | 20·00 | 20·11 | 20·01 | 19·84 | | |

Arrange the values into 7 equal classes of width 0·09 mm for the range 19·50 mm to 20·19 mm and determine the frequency distribution.

*Check your result with the next frame.*

**14**

| Diameter (mm) ($x$) | Frequency ($f$) |
|---|---|
| 19·50–19·59 | 1 |
| 19·60–19·69 | 4 |
| 19·70–19·79 | 7 |
| 19·80–19·89 | 9 |
| 19·90–19·99 | 6 |
| 20·00–20·09 | 5 |
| 20·10–20·19 | 2 |
| $n = \Sigma f = 34$ | |

So
(a) the class having the highest frequency is ...............
(b) the lower class boundary of the third class is ...............
(c) the upper class boundary of the seventh class is ...............
(d) the central value of the fifth class is ...............

| (a) 19·80–19·89; | (b) 19·695; |
|---|---|
| (c) 20·195; | (d) 19·945 |

**Histograms**

*Frequency histogram* A histogram is a graphical representation of a frequency distribution, in which vertical rectangular blocks are drawn so that

(a) the centre of the base indicates the central value of the class and
(b) the area of the rectangle represents the class frequency.

If the class intervals are regular, the frequency is then denoted by the height of the rectangle.

*Example:* Measurement of the lengths of 50 brass rods gave the following frequency distribution

| Length (mm) ($x$) | Lower class boundary | Upper class boundary | Central value | Frequency ($f$) |
|---|---|---|---|---|
| 3·45–3·47 | 3·445 | 3·475 | 3·460 | 2 |
| 3·48–3·50 | 3·475 | 3·505 | 3·490 | 6 |
| 3·51–3·53 | 3·505 | 3·535 | 3·520 | 12 |
| 3·54–3·56 | 3·535 | 3·565 | 3·550 | 14 |
| 3·57–3·59 | 3·565 | 3·595 | 3·580 | 10 |
| 3·60–3·62 | 3·595 | 3·625 | 3·610 | 5 |
| 3·63–3·65 | 3·625 | 3·655 | 3·640 | 1 |

First, we draw a base line and on it mark a scale of $x$ on which we can indicate the central values of the classes. Do that for a start.

**16**

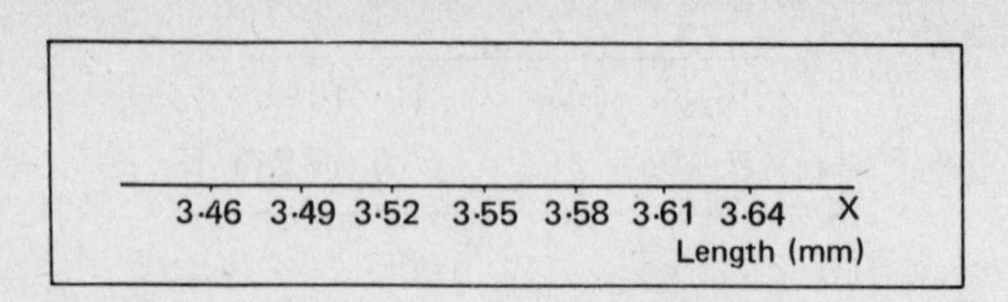


Since the classes are of regular class interval, the class boundaries will coincide with the points midway between the central values, thus:

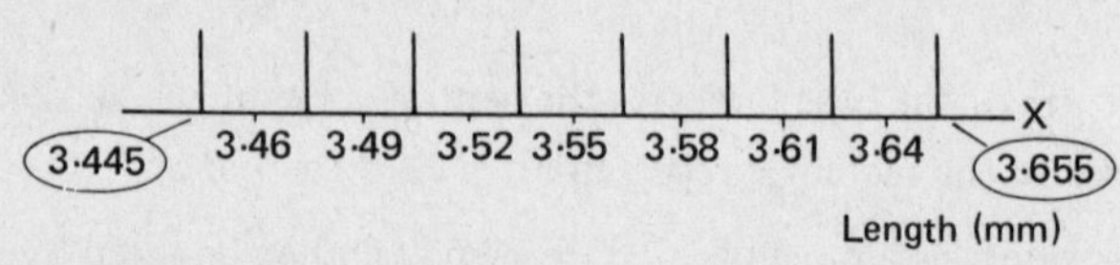


Note that the lower boundary of the first class extends to 3·445 and that the upper boundary of the seventh class extends to 3·655. Because the class intervals are regular, we can now erect a vertical scale to represent class frequencies and rectangles can be drawn to the appropriate height. Complete the work and give it a title.

**17**

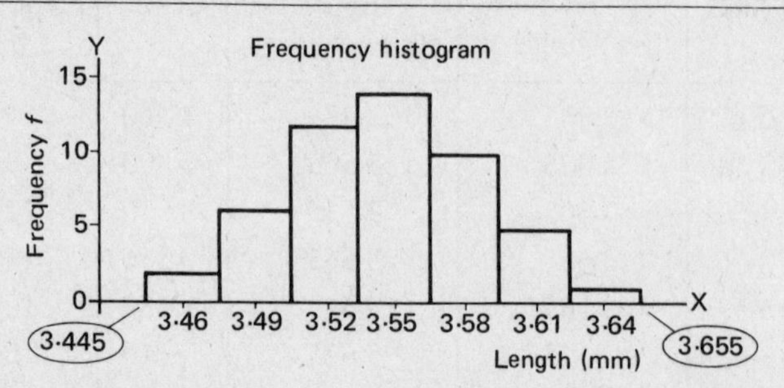


*Relative frequency histogram* The same diagram can be made to represent the relative frequency distribution by replacing the scale of frequency along the $y$-axis with a scale of relative frequencies (percentages).

Considerable information can be gleaned from a histogram on sight. For instance, we see that the class having the highest frequency is the ................class.

**18**

fourth

and the class with the lowest frequency is the ................ class.

19

seventh

Most of the whole range of values are clustered within the middle classes and knowledge of the centre region of the histogram is important. We can put a numerical value on this by determining a *measure of central tendency* which we shall now deal with.

*So, on to the next frame.*

20

**Measures of central tendency**

There are three common measures of central tendency, the *mean, mode* and *median*, of a set of observations and we shall discuss each of them in turn.

**Mean** The arithmetic mean $\bar{x}$ of a set of $n$ observations $x$ is simply their average, i.e. $\text{mean} = \dfrac{\text{sum of the observations}}{\text{number of observations}}$ $\quad \therefore \bar{x} = \dfrac{\Sigma x}{n}$

When calculating the mean from a frequency distribution, this becomes

$$\text{mean} = \bar{x} = \frac{\Sigma xf}{n} = \frac{\Sigma xf}{\Sigma f}$$

For example, for the following frequency distribution, we need to add a third column showing the values of the product $x \times f$, after which the mean can be found.

| Variable ($x$) | Frequency ($f$) | Product ($xf$) |
|---|---|---|
| 15 | 1 | |
| 16 | 4 | |
| 17 | 9 | |
| 18 | 10 | |
| 19 | 6 | |
| 20 | 2 | |

So, $\bar{x}$ = ................................

## 21

17·69

for $n = \Sigma f = 32$ and $\Sigma xf = 566$ $\quad \therefore \bar{x} = \dfrac{\Sigma xf}{n} = \dfrac{566}{32} = 17{\cdot}69.$

When calculating the mean from a frequency distribution with grouped data, the central value, $x_m$, of the class is taken as the $x$ value in forming the product $xf$. So, for the frequency distribution

| Variable ($x$) | 12–14 | 15–17 | 18–20 | 21–23 | 24–26 | 27–29 |
|---|---|---|---|---|---|---|
| Frequency ($f$) | 2 | 6 | 9 | 8 | 4 | 1 |

$\bar{x} =$ .................

## 22

$\bar{x} = 18$

since $n = \Sigma f = 30$ and $\Sigma x_m f = 540$ $\quad \therefore \bar{x} = \dfrac{\Sigma x_m f}{n} = \dfrac{540}{30} = 18.$

Here is one more.

Measurement in millimetres of 60 bolts gave the following frequency distribution.

| Length $x$ (mm) | 30·2 | 30·4 | 30·6 | 30·8 | 31·0 | 31·2 | 31·4 |
|---|---|---|---|---|---|---|---|
| Frequency $f$ | 3 | 7 | 12 | 17 | 11 | 8 | 2 |

The mean $\bar{x} =$ ...............

$\bar{x} = 30{\cdot}79$

| Length (mm) $(x)$ | Frequency $(f)$ | Product $(xf)$ |
|---|---|---|
| 30·2 | 3 | 90·6 |
| 30·4 | 7 | 212·8 |
| 30·6 | 12 | 367·2 |
| 30·8 | 17 | 523·6 |
| 31·0 | 11 | 341·0 |
| 31·2 | 8 | 249·6 |
| 31·4 | 2 | 62·8 |

$n = \Sigma f = 60$

$\Sigma xf = 1847{\cdot}6$

$$\therefore \bar{x} = \frac{\Sigma xf}{n} = \frac{1847{\cdot}6}{60}$$

$$= 30{\cdot}79$$

$$\therefore \bar{x} = 30{\cdot}79$$

**Coding for calculating the mean**

We can save ourselves some of the tedious work by using a system of *coding* which involves converting the values $x$ into simpler values for the calculation and then converting back again for the final result. An example will show the method in detail: we will use the same frequency distribution as above.

First, we choose a convenient value of $x$ (near the middle of the range) and substract this from each value of $x$ to give the second column

| Length (mm) $(x)$ | Deviation from chosen value $(x - 30{\cdot}8)$ | In units of 0·2 mm $\left(x_c = \dfrac{x - 30{\cdot}8}{0{\cdot}2}\right)$ |
|---|---|---|
| 30·2 | –0·6 | |
| 30·4 | –0.4 | |
| 30·6 | –0·2 | |
| 30·8 | 0 | |
| 31·0 | 0·2 | |
| 31·2 | 0·4 | |
| 31·4 | 0·6 | |

Then we change the values into an even simpler form by dividing the values in column 2 by 0·2 mm. These are entered in column 3 and give the coded values of $x$, i.e. $x_c$

All we have to do then is to add column 4 showing the class frequencies (from the table above) and then column 5 containing values of the product $x_c f$. Using the last two columns, the mean value of $x_c$ can be found as before. $\bar{x}_c =$ ...............

**24**

$\bar{x}_c = -0.0333$

| Length (mm) $(x)$ | Deviation from chosen value $(x - 30{\cdot}8)$ | In units of 0·2 mm $\left(x_c = \dfrac{x - 30{\cdot}8}{0{\cdot}2}\right)$ | Frequency $(f)$ | Product $(x_c f)$ |
|---|---|---|---|---|
| 30·2 | –0·6 | –3 | 3 | –9 |
| 30·4 | –0·4 | –2 | 7 | –14 |
| 30·6 | –0·2 | –1 | 12 | –12 |
| 30·8 | 0 | 0 | 17 | 0 |
| 31·0 | 0·2 | 1 | 11 | 11 |
| 31·2 | 0·4 | 2 | 8 | 16 |
| 31·4 | 0·6 | 3 | 2 | 6 |

$n = \Sigma f = 60$; $\Sigma x_c f = -2{\cdot}0$

$$\therefore\ \bar{x}_c = \frac{-2}{60} = -0{\cdot}0333$$

**25**

So we have $\bar{x}_c = -0{\cdot}0333$. Now we have to retrace our steps back to the original units of $x$.

*De-coding* In the coding procedure, our last step was to divide by 0·2. We therefore now multiply by 0·2 to return to the correct units of $(x - 30{\cdot}8)$

$$\therefore\ \bar{x} - 30{\cdot}8 = -0{\cdot}0333 \times 0{\cdot}2 = -0{\cdot}00667$$

We now add the 30·8 to both sides

$$\therefore\ \bar{x} = 30{\cdot}8 - 0{\cdot}00667 = 30{\cdot}79333 \quad \therefore\ \underline{\bar{x} = 30{\cdot}79}$$

Note that in de-coding, we reverse the operations used in the original coding process. The value (30·8) which was subtracted from all values of $x$ is near the centre of the range of $x$ values and is therefore sometimes referred to as a *false mean.*

**Coding with a grouped frequency distribution**

The method is precisely the same, except that we work with the centre values of the classes of $x$, i.e. $x_m$, for the calculation purposes.

*Exercise:* The thicknesses of 50 spacing pieces were measured.

| Thickness $x$ (mm) | 2·20–2·22 | 2·23–2·25 | 2·26–2·28 | 2·29–2·31 | 2·32–2·34 |
|---|---|---|---|---|---|
| Frequency $f$ | 1 | 5 | 8 | 15 | 12 |

| Thickness $x$ (mm) | 2·35–2·37 | 2·38–2·40 |
|---|---|---|
| Frequency $f$ | 7 | 2 |

Using coding, determine the mean value of the thickness. $\bar{x}$ = ................

**26**

$\bar{x} = 2{\cdot}307$ mm

| Thickness (mm) $(x)$ | Central value $(x_m)$ | Deviation from chosen value $(x_m - 2{\cdot}30)$ | In units of 0·03 mm $\left(x_c = \dfrac{x_m - 2{\cdot}30}{0{\cdot}03}\right)$ | Frequency $(f)$ | Product $(x_c f)$ |
|---|---|---|---|---|---|
| 2·20–2·22 | 2·21 | –0·09 | –3 | 1 | –3 |
| 2·23–2·25 | 2·24 | –0·06 | –2 | 5 | –10 |
| 2·26–2·28 | 2·27 | –0·03 | –1 | 8 | –8 |
| 2·29–2·31 | 2·30 | 0 | 0 | 15 | 0 |
| 2·32–2·34 | 2·33 | 0·03 | 1 | 12 | 12 |
| 2·35–2·37 | 2·36 | 0·06 | 2 | 7 | 14 |
| 2·38–2·40 | 2·39 | 0·09 | 3 | 2 | 6 |

$n = \Sigma f = 50$

$\Sigma x_c f = 11$

$$\therefore \bar{x}_c = \frac{\Sigma x_c f}{n} = \frac{11}{50} = 0{\cdot}22$$

$$\therefore \bar{x}_m - 2{\cdot}30 = \bar{x}_c \times 0{\cdot}03 = 0{\cdot}22 \times 0{\cdot}03 = 0{\cdot}0066$$

$$\therefore \bar{x}_m = 2{\cdot}30 + 0{\cdot}0066 = 2{\cdot}3067 \quad \therefore \underline{\bar{x} = 2{\cdot}307}$$

*Now let us deal with the mode, so move on to the next frame.*

## 27 Mode of a set of data

The *mode* of a set of data is that value of the variable that occurs most often. For instance, in the set of values 2, 2, 6, 7, 7, 7, 10, 13, the mode is clearly 7. There could, of course, be more than one mode in a set of observations, e.g. 23, 25, 25, 25, 27, 27, 28, 28, 28, has two modes, 25 and 28, each of which appears three times.

### Mode of a grouped frequency distribution

*Example:* The masses of 50 castings gave the following frequency distribution.

| Mass $x$ (kg) | 10–12 | 13–15 | 16–18 | 19–21 | 22–24 | 25–27 | 28–30 |
|---|---|---|---|---|---|---|---|
| Frequency $f$ | 3 | 7 | 16 | 10 | 8 | 5 | 1 |

If we draw the histogram, using the central values as the mid-points of the bases of the rectangles, we obtain ..............................

## 28

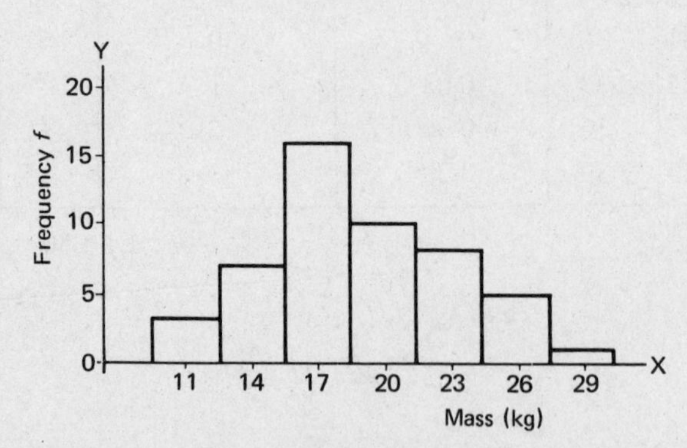


The class having the highest frequency is the ............... class

**29**

third

The modal class is therefore the third class, with boundaries 15·5 and 18·5 kg. The value of the mode itself lies somewhere within that range and its value can be found by a simple construction.

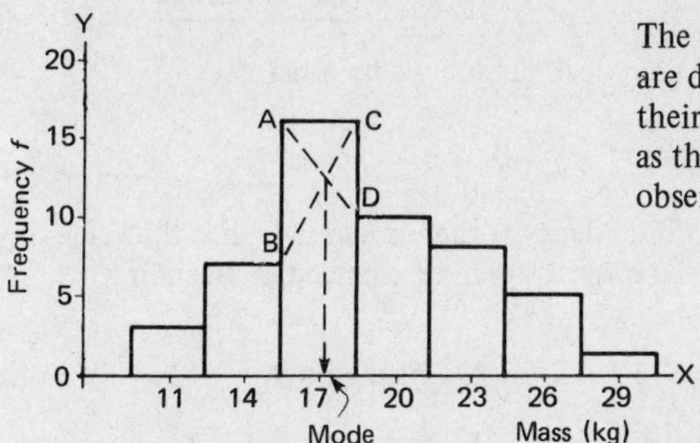


The two diagonal lines AD and BC are drawn as shown. The $x$ value of their point of intersection is taken as the mode of the set of observations.

Carry out the construction on your histogram and you will find that, for this set of observations, the mode = ................

**30**

mode = 17·3

The value of the mode can also be calculated.

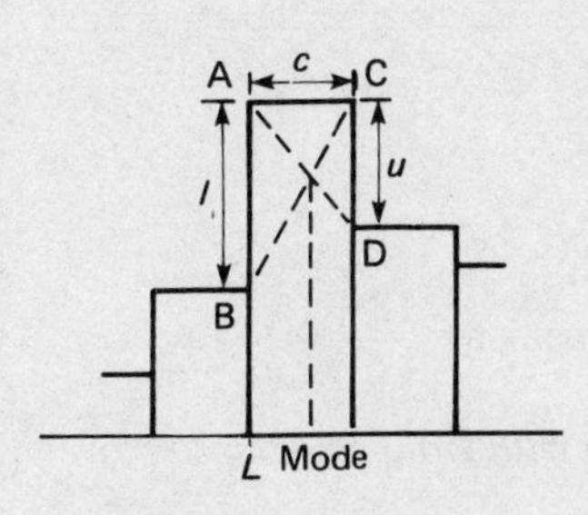


If $L$ = lower boundary value,

$l$ = AB = difference in frequency on the lower boundary,

$u$ = CD = difference in frequency on the upper boundary,

$c$ = class interval,

then the mode $= L + \left(\dfrac{l}{l+u}\right)c$.

In the example we have just considered, $L$ = ................ ; $l$ = ................ ;

$u$ = ................ ; $c$ = ................ .

## 31

$$\boxed{L = 15{\cdot}5;\; l = 16 - 7 = 9;\; u = 16 - 10 = 6;\; c = 3}$$

Then the mode $= 15{\cdot}5 + \left(\dfrac{9}{9+6}\right) \times 3$

$= 15{\cdot}5 + 1{\cdot}8 = 17{\cdot}3$ $\quad\therefore$ $\underline{\text{Mode} = 17{\cdot}3 \text{ kg}}$

This result agrees with the graphical method we used before.

**Median of a set of data**

The third measure of central tendency is the *median*, which is the value of the middle term when all the observations are arranged in ascending or descending order.

For example, with 4, 7, 8, 9, 12, 15, 26, the median = 9.

↑ median

Where there is an even number of values, the median is then the average of the two middle terms, e.g. 5, 6, 10, 12, 14, 17, 23, 30 has a median of

$$\frac{12 + 14}{7} = 13.$$

Similarly, for the set of values 12, 4, 14, 23, 8, 15, 17, 10, 15, 6, the median is ................

## 32

$$\boxed{14{\cdot}5}$$

for arranging the terms in order, we have

$$4, 6, 9, 10, 13, 16, 18, 18, 20, 23 \text{ and median} = \frac{13 + 16}{2} = \underline{14{\cdot}5}$$

*Now we must see how we can get the median with grouped data, so move on to frame 33.*

**33**

**Median with grouped data**

Since the median is the value of the middle term, it divides the frequency histogram into two equal areas. This fact gives us a method for determining the median.

*Example:* The temperature of a component was monitored at regular intervals on 80 occasions. The frequency distribution was as follows.

| Temperature $x$ (°C) | 30·0–30·2 | 30·3–30·5 | 30·6–30·8 | 30·9–31·1 |
|---|---|---|---|---|
| Frequency $f$ | 6 | 12 | 15 | 20 |

| Temperature $x$ (°C) | 31·2–31·4 | 31·5–31·7 | 31·8–32·0 |
|---|---|---|---|
| Frequency $f$ | 13 | 9 | 5 |

First we draw the frequency histogram.

**34**

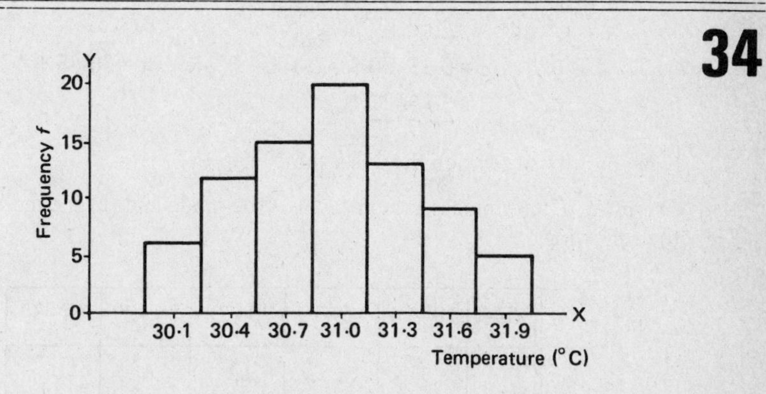


The median is the average of the 40th and 41st terms and, if we count up the frequencies of the rectangles, these terms are included in the ................ class.

**35**

fourth

If we insert a dotted line to represent the value of the median, it will divide the area of the histogram into two equal parts.

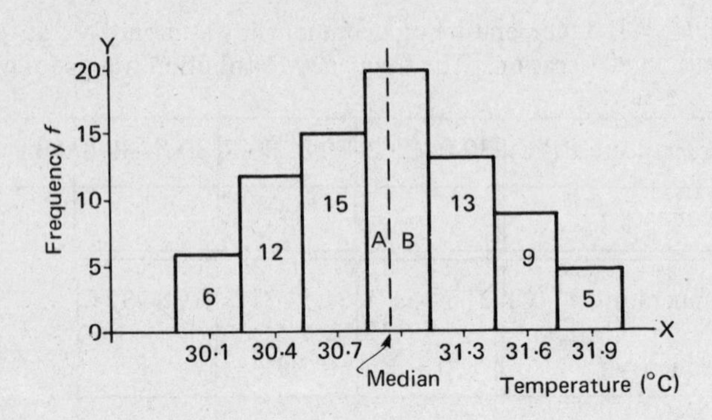


$$6 + 12 + 15 + A = B + 13 + 9 + 5$$

$\therefore\ 33 + A = B + 27$ But $A + B = 20\ \therefore\ B = 20 - A$

$\therefore\ 33 + A = 20 - A + 27$ $\therefore\ 2A = 14\ \therefore\ A = 7$

$$\therefore \text{ Width of } A = \frac{7}{20} \times \text{class interval}$$

$= 0{\cdot}35 \times 0{\cdot}3 = 0{\cdot}105$ $\therefore$ Median $= 30{\cdot}85 + 0{\cdot}105$

$= \underline{30{\cdot}96\,°\text{C}}$

Now, by way of revision, here is a short exercise.

*Exercise:* Determine (a) the mean, (b) the mode and (c) the median of the following

| $x$ | 5–9 | 10–14 | 15–19 | 20–24 | 25–29 | 30–34 |
|---|---|---|---|---|---|---|
| $f$ | 4 | 9 | 16 | 12 | 6 | 3 |

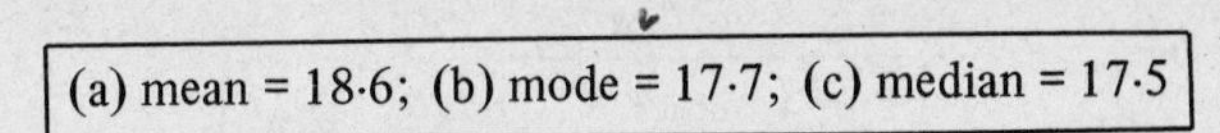
(a) mean = 18·6; (b) mode = 17·7; (c) median = 17·5

**Dispersion**

The mean, mode and median give important information regarding the general mass of the observations recorded. They do not, however, tell us anything about the dispersion of the values around the central value.

The set 26, 27, 28, 29, 30 has a mean of 28
and 5, 19, 20, 36, 60 also has a mean of 28

These two sets have the same mean, but clearly the first is more tightly arranged around the mean than is the second. We therefore need a measure to indicate the spread of the values about the mean.

**Range** The simplest value to indicate the spread or dispersion is the *range* which is merely the difference between the highest and lowest values in the set of observations. In the two cases quoted above, the range of set 1 is 30 – 26 = 4, while that of set 2 is 60–5 = 55. The disadvantage of the range is that it deals only with the extreme values: it does not take into account the behaviour of the intermediate values.

**Standard deviation** The *standard deviation from the mean* is used widely in statistics to indicate the degree of dispersion. It takes into account the deviation of every value from the mean and it is found as follows.

(a) The mean $\bar{x}$ of the set of $n$ values, is first calculated.

(b) The deviation of each of these $n$ values, $x_1, x_2, x_3 \ldots x_n$ from the mean is calculated and the results squared, i.e. $(x_1 - \bar{x})^2; (x_2 - \bar{x})^2; (x_3 - \bar{x})^2; \ldots (x_n - \bar{x})^2$.

(c) The average of these results is then found and the result is called the *variance* of the set of observations,

$$\text{i.e. variance} = \frac{(x_1 - \bar{x})^2 + (x_2 - \bar{x})^2 + \ldots + (x_n - \bar{x})^2}{n}$$

(d) The square root of the variance gives the standard deviation, denoted by the Greek letter 'sigma'

$$\text{standard deviation} = \sigma = \sqrt{\left[\frac{\Sigma(x - \bar{x})^2}{n}\right]}$$

So, for the set of values 3, 6, 7, 8, 11, 13, the mean $\bar{x}$ = ................ , the variance = ................ and the standard deviation = ................

**37**

$\bar{x} = 8{\cdot}0$; variance $= 10{\cdot}67$; $\sigma = 3.27$

for

| $x$ | $f$ |
|---|---|
| 3 | 1 |
| 6 | 1 |
| 7 | 1 |
| 8 | 1 |
| 11 | 1 |
| 13 | 1 |

$\Sigma x = 48$
$n = 6$
$\therefore \underline{\bar{x} = 8}$

| $x - 8$ | $(x - 8)^2$ |
|---|---|
| $-5$ | 25 |
| $-2$ | 4 |
| $-1$ | 1 |
| 0 | 0 |
| 3 | 9 |
| 5 | 25 |

$$\therefore \Sigma(x - \bar{x})^2 = 64$$

$$\text{variance} = \frac{\Sigma(x - \bar{x})^2}{n} = \frac{64}{6} = \underline{10{\cdot}67}$$

$$\therefore \text{standard deviation} = \sigma = \sqrt{10{\cdot}67} = 3{\cdot}266 \quad \therefore \underline{\sigma = 3{\cdot}27}$$

*Alternative formula for the standard deviation*

The formula $\sigma = \sqrt{\left[\dfrac{\Sigma(x - \bar{x})^2}{n}\right]}$ can also be written in another and more convenient form $\sigma = \sqrt{\left[\dfrac{\Sigma x^2}{n} - (\bar{x})^2\right]}$ which requires only the mean and the squares of the values of $x$. For grouped data, this then becomes $\sigma = \sqrt{\left[\dfrac{\Sigma x^2 f}{n} - (\bar{x})^2\right]}$ and the working is even more simplified if we use the coding procedure as we did earlier in the programme.

*Move on to the next frame for a typical example.*

**38**

*Example:* The lengths of 70 bars were measured and the following frequency distribution obtained.

| Length $x$ (mm) | 21·2–21·4 | 21·5–21·7 | 21·8–22·0 | 22·1–22·3 |
|---|---|---|---|---|
| Frequency $f$ | 3 | 5 | 10 | 16 |

| Length $x$ (mm) | 22·4–22·6 | 22·7–22·9 | 23·0–23·2 |
|---|---|---|---|
| Frequency $f$ | 18 | 12 | 6 |

First we prepare a table with the following headings:

| Length (mm) $(x)$ | Central value $(x_m)$ | Deviation from chosen value $(x_m - 22{\cdot}2)$ | Units of 0·3 $\left(x_c = \dfrac{x_m - 22{\cdot}2}{0{\cdot}3}\right)$ | Freq. $f$ | $x_c f$ | | |
|---|---|---|---|---|---|---|---|
| | | | | | | | |

Leave room on the right-hand side for two more columns yet to come. Now you can complete the six columns shown and from the values entered, determine the coded value of the mean and also the actual mean.

*Do that and then turn on.*

## 39

So far, the work looks like this:

| Length (mm) $(x)$ | Central value $(x_m)$ | Deviation from chosen value $(x_m - 22{\cdot}2)$ | Units of 0·3 mm $\left(x_c = \frac{x_m - 22{\cdot}2}{0{\cdot}3}\right)$ | Freq $f$ | $x_c f$ | | |
|---|---|---|---|---|---|---|---|
| 21·2–21·4 | 21·3 | –0·9 | –3 | 3 | –9 | | |
| 21·5–21·7 | 21·6 | –0·6 | –2 | 5 | –10 | | |
| 21·8–22·0 | 21·9 | –0·3 | –1 | 10 | –10 | | |
| 22·1–22·3 | 22·2 | 0 | 0 | 16 | 0 | | |
| 22·4–22·6 | 22·5 | 0·3 | 1 | 18 | 18 | | |
| 22·7–22·9 | 22·8 | 0·6 | 2 | 12 | 24 | | |
| 23·0–23·2 | 23·1 | 0·9 | 3 | 6 | 18 | | |

$n = \Sigma f = 70$

$\Sigma x_c f = 31$

$$\text{Coded mean} = \bar{x}_c = \frac{\Sigma x_c f}{n} = \frac{31}{70} = 0{\cdot}4429 \text{ (in units of } 0{\cdot}3)$$

$$\therefore \bar{x}_m - 22{\cdot}2 = 0{\cdot}4429 \times 0{\cdot}3 = 0{\cdot}1329$$

$$\bar{x}_m = 22{\cdot}2 + 0{\cdot}1329 = 22{\cdot}333 \qquad \therefore \underline{\bar{x} = 22{\cdot}33}$$

Now we can complete the remaining two columns. Head these $x_c^2$ and $x_c^2 f$. Fill in the appropriate values and using $\sigma_c = \sqrt{\left[\frac{\Sigma x_c^2 f}{n} - (\bar{x}_c)^2\right]}$ we can find the coded value of the standard deviation $(\sigma_c)$.

*Complete the table then and determine the coded standard deviation on your own.*

**40**

Finally the table now becomes:

| Length (mm) $(x)$ | Central value $(x_m)$ | Deviation from chosen value $(x_m - 22{\cdot}2)$ | Units of 0·3 mm $x_c = \dfrac{x_m - 22{\cdot}2}{0{\cdot}3}$ | Freq. $f$ | $x_c f$ | $x_c^2$ | $x_c^2 f$ |
|---|---|---|---|---|---|---|---|
| 21·2–21·4 | 23·3 | –0·9 | –3 | 3 | –9 | 9 | 27 |
| 21·5–21·7 | 21·6 | –0·6 | –2 | 5 | –10 | 4 | 20 |
| 21·8–22·0 | 21·9 | –0·3 | –1 | 10 | –10 | 1 | 10 |
| 22·1–22·3 | 22·2 | 0 | 0 | 16 | 0 | 0 | 0 |
| 22·4–22·6 | 22·5 | 0·3 | 1 | 18 | 18 | 1 | 18 |
| 22·7–22·9 | 22·8 | 0·6 | 2 | 12 | 24 | 4 | 48 |
| 23·0–23·2 | 23·1 | 0·9 | 3 | 6 | 18 | 9 | 54 |
| | | | | 70 | 31 | | 177 |

Now $\sigma_c = \sqrt{\left[\dfrac{\Sigma x_c^2 f}{n} - (\bar{x}_c)^2\right]}$ and from the table, $n = 70$ and $\Sigma x_c^2 f = 177$.

Also, from the previous work, $\bar{x}_c = 0{\cdot}443$. $\quad \therefore \sigma_c = \ldots\ldots\ldots\ldots$

**41**

$$\boxed{\sigma_c = 1{\cdot}527 \text{ (in units of 0·3)}}$$

$\therefore \sigma = 1{\cdot}527 \times 0{\cdot}3 = 0{\cdot}4581 \qquad \therefore \underline{\sigma = 0{\cdot}458}$

*Note* that, in calculating the standard deviation, we do not restore the 'false mean' substracted from the original values, since the standard deviation is relative to the mean and not to the zero or origin of the set of observations.

*Now on to something different.*

## 42 Frequency polygons

If the centre points of the tops of the rectangular blocks of a frequency histogram are joined, the resulting figure is a *frequency polygon.* If the polygon is extended to include the mid-points of the zero frequency classes at each end of the histogram, then the area of the complete polygon is equal to the area of the histogram and therefore represents the total frequency of the variable.

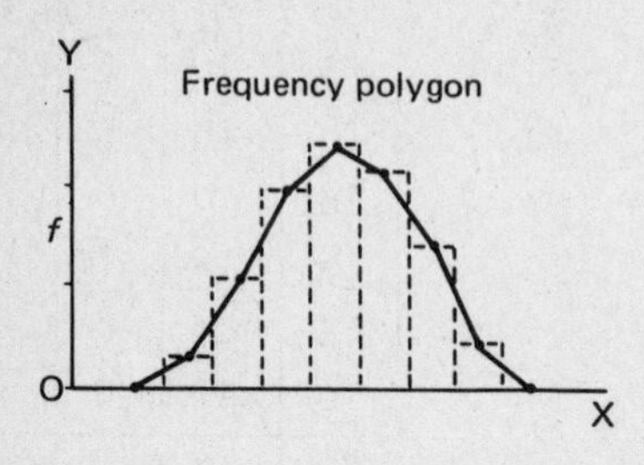


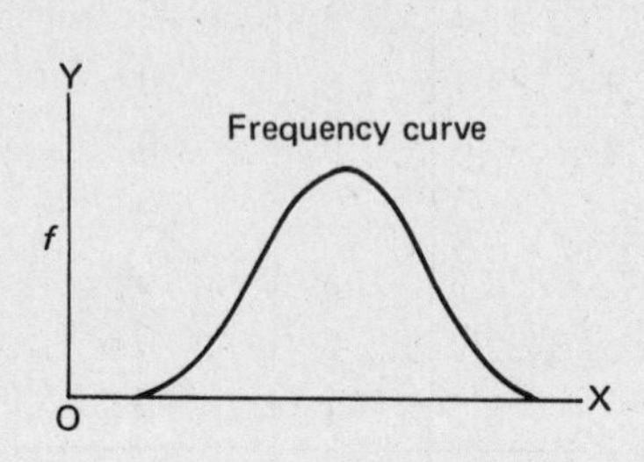


### Frequency curves

If the frequency polygon is 'smoothed out', or, if we plot the frequency against the central value of each class and draw a smooth curve, the result is a frequency curve.

### Normal distribution curve

When very large numbers of observations are made and the range is divided into a very large number of 'narrow' classes, the resulting frequency curve, in many cases, approximates closely to a standard curve known as the *normal distribution curve*, which has a characteristic bell-shaped formation.

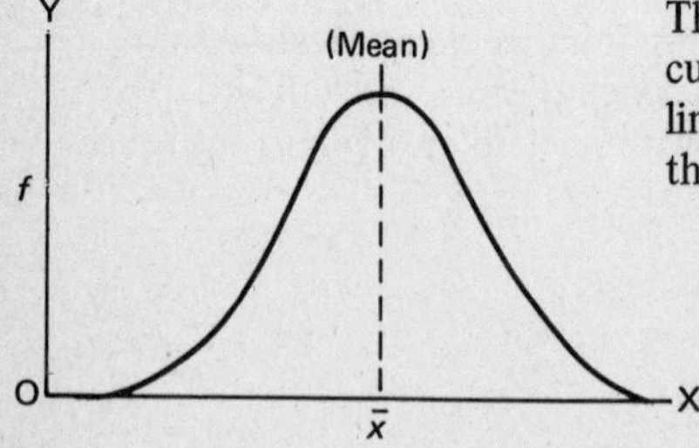


The normal distribution curve (or normal curve) is symmetrical about its centre line which coincides with the mean $\bar{x}$ of the observations.

There is, in fact, a close connection between the standard deviation (s.d.) from the mean of a set of values and the normal curve.

(a) *Values within 1 s.d. of the mean*

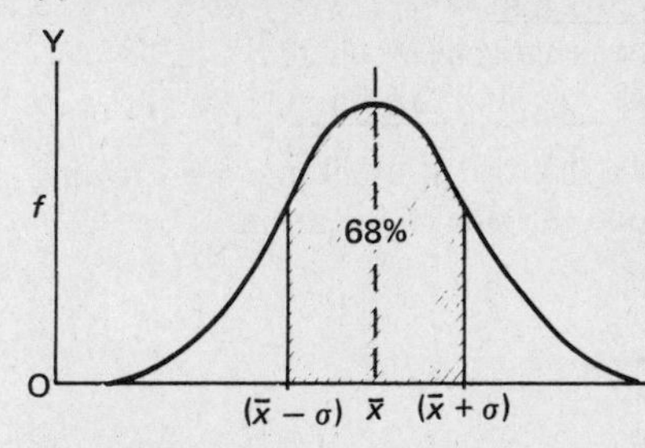


The normal curve has a complicated equation, but it can be shown that the shaded area is 68% of the total area under the normal curve, i.e. 68% of the observations occur within the range $(\bar{x}-\sigma)$ to $(\bar{x}+\sigma)$.

*Example:* On a manufacturing run to produce 1000 bolts of nominal length 32·5 mm. sampling gave a mean of 32·58 and a s.d. of 0·06 mm. From this information, $\bar{x} = 32{\cdot}58$ mm and $\sigma = 0{\cdot}06$ mm

$$\left.\begin{array}{l}\bar{x}-\sigma = 32{\cdot}58 - 0{\cdot}06 = 32{\cdot}52\\ \bar{x}+\sigma = 32{\cdot}58 + 0{\cdot}06 = 32{\cdot}52\end{array}\right\}$$

$\therefore$ 68% of the bolts, i.e. 680, are likely to have lengths between 32·52 mm and 32·64 mm

(b) *Values within 2 s.d. of the mean*

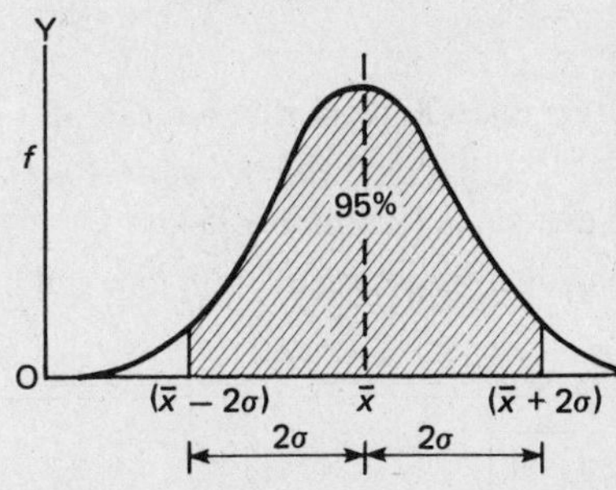


Between $(\bar{x}-2\sigma)$ and $(\bar{x}+2\sigma)$ the shaded area accounts for 95% of the area of the whole figure, i.e. 95% of the observations occur between these two values.

(c) *Values within 3 s.d. of the mean*

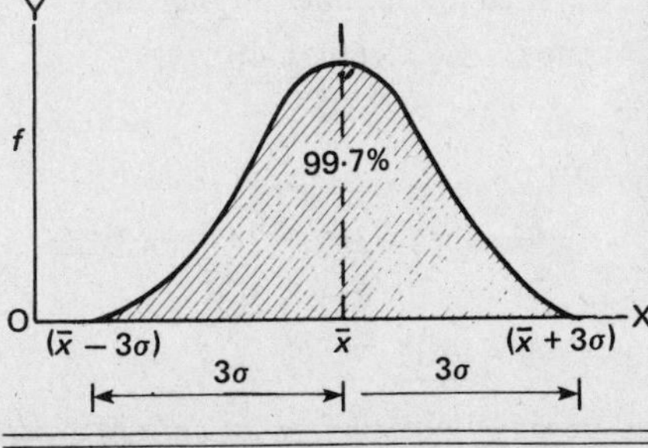


Between $(\bar{x}-3\sigma)$ and $(\bar{x}+3\sigma)$ the shaded area is 99·7% of the total area under the normal curve and therefore 99·7% of the observations occur within this range, i.e. almost all the values occur within $3\sigma$ or 3 s.d. of the mean.

Therefore, in our previous example where $\bar{x} = 32{\cdot}58$ and $\sigma = 0{\cdot}06$ mm,

68% of the bolts are likely to have lengths between 32·52 and 32·64 mm
95% of the bolts are likely to have lengths between 32·5 – 0·12 and 32·5 + 0·12 i.e. between 32·38 and 32·62 mm
99·7%, i.e. almost all, are likely to have lengths between 32·5 – 0·18 and 32·5 + 0·18, i.e. between 32·32 and 32·68 mm

We can enter the same information in a slightly different manner, dividing the figure into columns of $1\sigma$ width on each side of the mean.

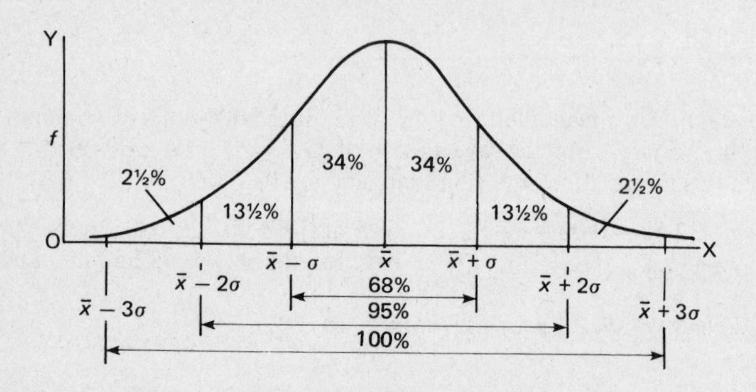


*Exercise:* Measurement of the diameters of 1600 plugs gave a mean of 74·82 mm and a standard deviation of 0·14 mm. Calculate

(a) the number of plugs likely to have diameters less than 74·54 mm,
(b) the number of plugs likely to have diameters between 74·68 mm and 75·10 mm.

## 43

(a) 40; (b) 1304

for (a) $74{\cdot}54 = 74{\cdot}82 - 2(0{\cdot}14) = \bar{x} - 2\sigma$ ∴ number of plugs with diameters less than 74·54 mm = 2½% x 1600 = 40

(b) $74{\cdot}68 = 74{\cdot}82 - 0{\cdot}14 = \bar{x} - \sigma$

$75{\cdot}10 = 74{\cdot}82 + 2(0{\cdot}14) = \bar{x} + 2\sigma$

∴ Number of plugs between these values = 34% + 34% + 13½%
= 81½%

81½% of 1600 = 1304

**Standardised normal curve**

This is the same shape as the normal curve, but the axis of symmetry now becomes the vertical axis with a scale of relative frequency. The horizontal axis now carries a scale of $z$-values indicated as multiples of the standard deviation $\sigma$. The area under the standardised normal curve has a total value of 1.

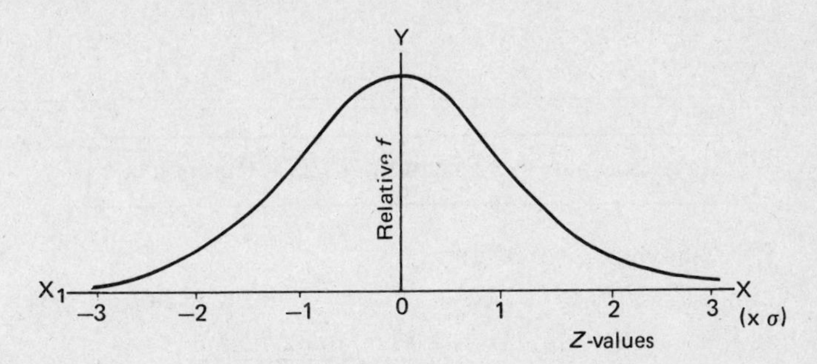


The standardised normal curve is obtained from the normal curve by the substitution $z = \dfrac{x - \bar{x}}{\sigma}$ and it converts the original distribution into one with zero mean and unit standard deviation. The equation of the standardised normal curve is somewhat complicated

$$y = \phi(z) = \frac{1}{\sqrt{(2\pi)}}\, e^{-\frac{1}{2}z^2}$$

but, if required, the curve can be drawn from the following table.

| $z$ ($\sigma$) | 0 | 0·2 | 0·4 | 0·6 | 0·8 | 1·0 | 1·2 | 1·4 |
|---|---|---|---|---|---|---|---|---|
| Relative $f$ | 0·399 | 0·391 | 0·368 | 0·333 | 0·290 | 0·242 | 0·194 | 0·150 |

| $z$ ($\sigma$) | 1·6 | 1·8 | 2·0 | 2·2 | 2·4 | 2·6 | 2·8 | 3·0 |
|---|---|---|---|---|---|---|---|---|
| Relative $f$ | 0·111 | 0·080 | 0·054 | 0·035 | 0·022 | 0·014 | 0·008 | 0·004 |

Since the curve is symmetrical about $z = 0$, the curve for negative values of $z$ is the mirror image of that for positive values of $z$.

*One further problem:* Components are machined to a nominal diameter of 32·65 mm. A sample of 400 components gave a mean diameter of 32·66 mm with a s.d. of 0·02 mm. If a total of 3500 components were produced, calculate

(a) the limits between which all the diameters are likely to lie,

(b) the total number of components with diameters between 32·64 and 32·72 mm.

**44**

(a) 32·60 mm to 32·72 mm; (b) 2940 components

(a) $\bar{x} = 32{\cdot}66$ mm; $\sigma = 0{\cdot}02$ mm

Limits $(\bar{x} - 3\sigma)$ to $(\bar{x} + 3\sigma) = (32{\cdot}66 - 0{\cdot}06)$ to $(32{\cdot}66 + 0{\cdot}06)$

$= \underline{32{\cdot}60 \text{ to } 32{\cdot}72 \text{ mm}}$

(b) $32{\cdot}64 = 32{\cdot}66 - 0{\cdot}02 = (\bar{x} - \sigma)$

$32{\cdot}72 = 32{\cdot}66 + 0{\cdot}06 = (\bar{x} + 3\sigma)$

Between $(\bar{x} - \sigma)$ and $(\bar{x} + 3\sigma)$ there are $(34 + 34 + 13\frac{1}{2} + 2\frac{1}{2})\%$

i.e. 84% of the values. 84% of 3500 = $\underline{2940 \text{ components}}$

As usual, check down the Revision Summary that follows and refer back to any points that may need clarification. All that remains after that is to work through the Test Exercise. You should have no trouble.

## Revision Summary

**45**

1. *Data* (a) Discrete data – values that can be precisely counted.
   (b) Continuous data – values measured on a continuous scale.

2. *Grouped data*

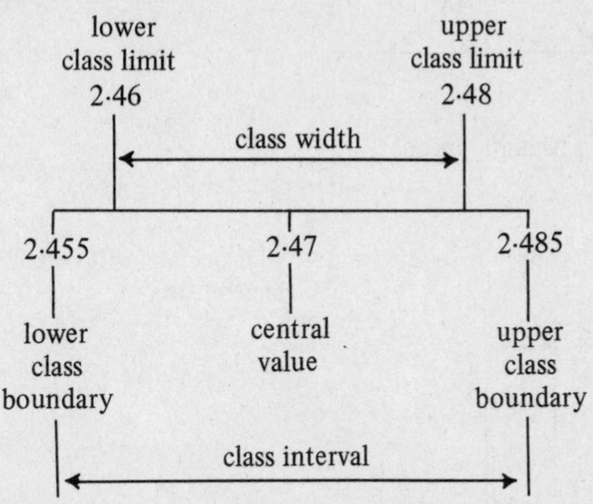


3. *Frequency* ($f$) – the number of occasions on which each value, or class, occurs.

   *Relative frequency* – each frequency expressed as a percentage or fraction of the total frequency.

4. *Histogram* – graphical representation of a frequency or relative frequency distribution.

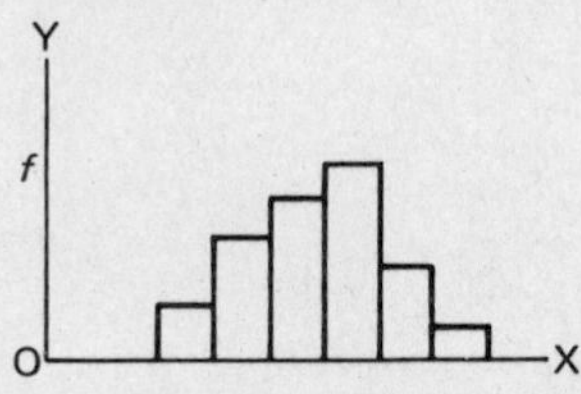


The frequency of any one class is given by the *area* of its column.
If the class intervals are constant, the height of the rectangle indicates the frequency on the vertical scale.

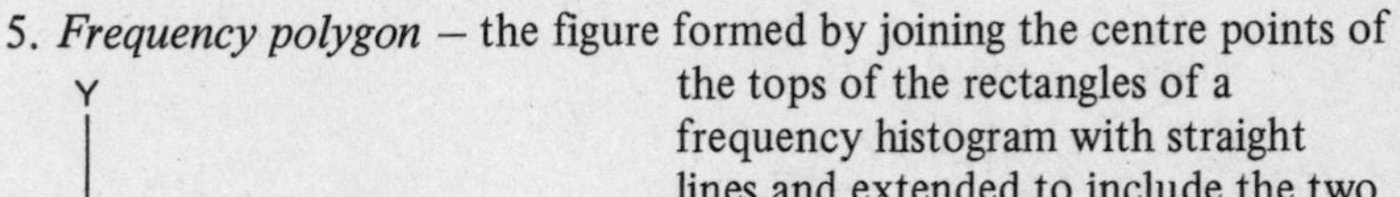

5. *Frequency polygon* – the figure formed by joining the centre points of the tops of the rectangles of a frequency histogram with straight lines and extended to include the two zero frequency columns on the sides.

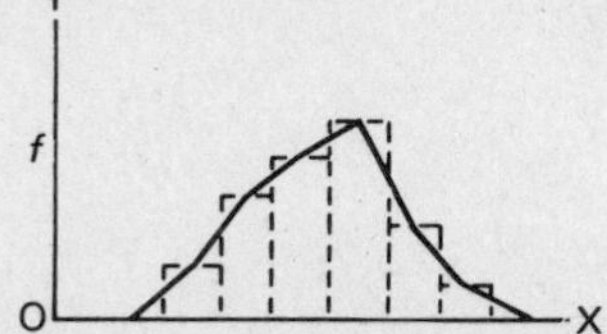

6. *Frequency curve* – obtained by 'smoothing' the boundary of the frequency polygon, or by plotting centre values and joining with a smooth curve.

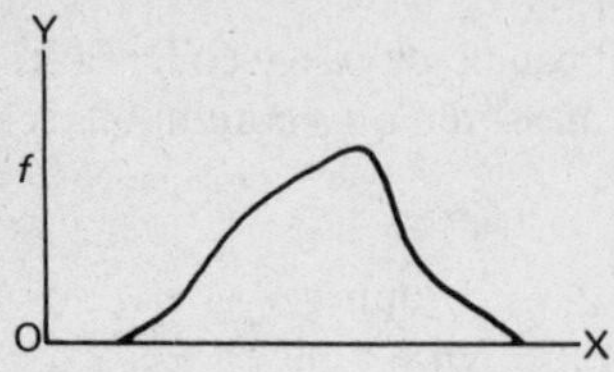


7. *Mean* (arithmetic mean) $\bar{x} = \dfrac{\Sigma xf}{n} = \dfrac{\Sigma xf}{\Sigma f}$

8. *Mode* – the value of the variable that occurs most often. For gouped distribution,

$$\text{mode} = L + \left(\frac{l}{l+u}\right)c.$$

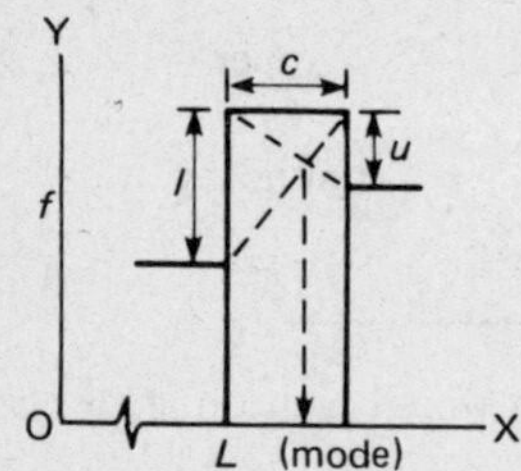


9. *Median* – the value of the middle term when all values are put in ascending or descending order. With an even number of terms, the median is the average of the two middle terms.

10. *Standard deviation*

$$\sigma = \sqrt{\left[\frac{\Sigma x^2 f}{n} - (\bar{x})^2\right]}$$

$n$ = number of observations
$\bar{x}$ = mean
$\sigma$ = standard deviation from the mean.

With coding, $\bar{x}_c$ = coded mean

$$\sigma_c = \sqrt{\left[\frac{\Sigma x_c{}^2 f}{n} - (\bar{x}_c)^2\right]}$$

11. *Normal distribution curve* – large numbers of observations.

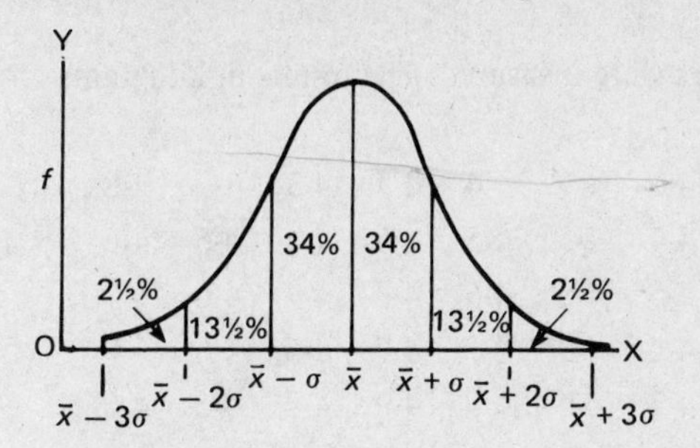


Symmetrical about the mean.
68% of observations lie within ± 1 s.d. of the mean.
95% of observations lie within ± 2 s.d. of the mean.
99·7% of observations lie within ± 3 s.d. of the mean.

12. *Standardised normal curve* – The axis of symmetry of the normal curve becomes the vertical axis with a scale of relative frequency. The horizontal axis carries a scale of $z$-values indicated as multiples of the standard deviation. The curve therefore represents a distribution with zero mean and unit standard deviation.

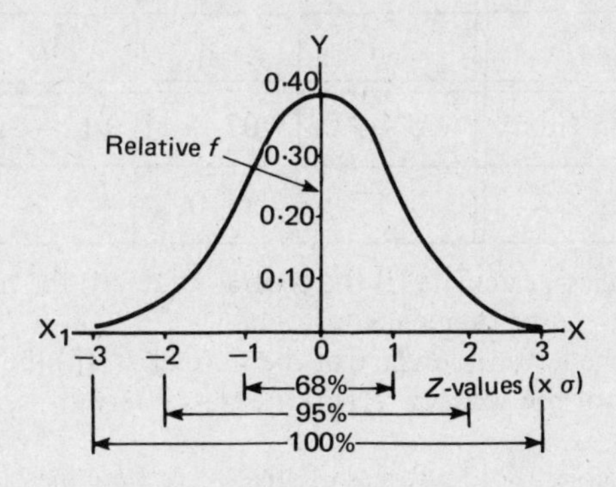

## 46 Test Exercise – XXVII

1. The masses of 50 castings were measured. The results in kilograms were as follows:

| | | | | | | | | | |
|---|---|---|---|---|---|---|---|---|---|
| 4·6 | 4·7 | 4·5 | 4·6 | 4·7 | 4·4 | 4·8 | 4·3 | 4·2 | 4·8 |
| 4·7 | 4·5 | 4·7 | 4·4 | 4·5 | 4·5 | 4·6 | 4·4 | 4·6 | 4·6 |
| 4·8 | 4·3 | 4·8 | 4·5 | 4·5 | 4·6 | 4·6 | 4·7 | 4·6 | 4·7 |
| 4·4 | 4·6 | 4·5 | 4·4 | 4·3 | 4·7 | 4·7 | 4·6 | 4·6 | 4·8 |
| 4·9 | 4·4 | 4·5 | 4·7 | 4·4 | 4·5 | 4·9 | 4·7 | 4·5 | 4·6 |

(a) Arrange the data in 8 equal classes between 4·2 and 4·9 mm.
(b) Determine the frequency distribution.
(c) Draw the frequency histogram.

2. The diameters of 75 rollers gave the following frequency distribution:

| Diameter $x$ (mm) | 8·82–8·86 | 8·87–8·91 | 8·92–8·96 | 8·97–9·01 |
|---|---|---|---|---|
| Frequency $f$ | 1 | 8 | 16 | 18 |
| Diameter $x$ (mm) | 9·02–9·06 | 9·07–9·11 | 9·12–9·16 | 9·17–9·21 |
| Frequency $f$ | 15 | 10 | 5 | 2 |

(a) For each class, calculate (i) the central value, (ii) the relative frequency.
(b) Draw the relative frequency histogram.
(c) State (i) the lower boundary of the third class, (ii) the upper boundary of the sixth class, (iii) the class interval.

3. The thicknesses of 40 samples of steel plate were measured.

| Thickness $x$ (mm) | 9·60–9·80 | 9·90–10·1 | 10·2–10·4 | 10·5–10·7 |
|---|---|---|---|---|
| Frequency $f$ | 1 | 4 | 10 | 11 |
| Thickness $x$ (mm) | 10·8–11·0 | 11·1–11·3 | 11·4–11·6 | |
| Frequency $f$ | 7 | 4 | 3 | |

Using coding procedure, calculate

(a) the mean, (b) the standard deviation, (c) the mode, (d) the median of the set of values given.

4. The lengths of 50 copper plugs gave the following frequency distribution.

| Length $x$ (mm) | 14·0–14·2 | 14·3–14·5 | 14·6–14·8 | 14·9–15·1 |
|---|---|---|---|---|
| Frequency $f$ | 2 | 4 | 9 | 15 |

| Length $x$ (mm) | 15·2–15·4 | 15·5–15·7 | 15·8–16·0 |
|---|---|---|---|
| Frequency $f$ | 11 | 6 | 3 |

(a) Calculate the mean and the standard deviation.
(b) For a full batch of 2400 plugs, calculate (i) the limits between which all the lengths are likely to occur, (ii) the number of plugs with lengths greater than 15·09 mm.

**Further Problems – XXVII**

1. The number of components processed in one hour on a new machine was recorded on 40 occasions.

| | | | | | | | | | |
|---|---|---|---|---|---|---|---|---|---|
| 66 | 86 | 79 | 74 | 84 | 72 | 81 | 78 | 68 | 74 |
| 80 | 71 | 91 | 62 | 77 | 86 | 87 | 72 | 80 | 77 |
| 76 | 83 | 75 | 71 | 83 | 67 | 94 | 64 | 82 | 78 |
| 77 | 67 | 76 | 82 | 78 | 88 | 66 | 79 | 74 | 64 |

(a) Divide the set of values into seven equal width classes from 60 to 94.
(b) Calculate (i) the frequency distribution, (ii) the mean, (iii) the standard deviation.

2. The lengths, in millimetres, of 40 bearings were determined with the following results:

| | | | | | | | | | |
|---|---|---|---|---|---|---|---|---|---|
| 16·6 | 15·3 | 16·3 | 14·2 | 16·7 | 17·3 | 18·2 | 15·6 | 14·9 | 17·2 |
| 18·7 | 16·4 | 19·0 | 15·8 | 18·4 | 15·1 | 17·0 | 18·9 | 18·3 | 15·9 |
| 13·6 | 18·3 | 17·2 | 18·0 | 15·8 | 19·3 | 16·8 | 17·7 | 16·8 | 17·9 |
| 17·3 | 16·6 | 15·3 | 16·4 | 17·3 | 16·9 | 14·7 | 16·2 | 17·4 | 15·6 |

(a) Group the data into six equal width classes between 13·5 and 19·4 mm.
(b) Obtain the frequency distribution.
(c) Calculate (i) the mean, (ii) the standard deviation.

3. The masses of 80 brass junctions gave the following frequency distribution:

| Mass $x$ (kg) | 4·12–4·16 | 4·17–4·21 | 4·22–4·26 | 4·27–4·31 |
|---|---|---|---|---|
| Frequency $f$ | 5 | 12 | 16 | 20 |

| Mass $x$ (kg) | 4·32–4·36 | 4·37–4·41 | 4·42–4·46 |
|---|---|---|---|
| Frequency $f$ | 14 | 9 | 4 |

(a) Calculate (i) the mean, (ii) the standard deviation.
(b) For a batch of 1800 such components, calculate (i) the limits between which all the masses are likely to lie, (ii) the number of junctions with masses greater than 4·36 kg.

4. The values of the resistance of 90 carbon resistors were determined.

| Resistance $x$ (MΩ) | 2·35 | 2·36 | 2·37 | 2·38 | 2·39 | 2·40 | 2·41 |
|---|---|---|---|---|---|---|---|
| Frequency $f$ | 3 | 10 | 19 | 20 | 18 | 13 | 7 |

Calculate (a) the mean, (b) the standard deviation, (c) the mode and (d) the median of the set of values.

5. Forty concrete cubes were subjected to failure tests in a crushing machine.

   Failure loads were as follows:

| Load $x$ (kN) | 30·6 | 30·8 | 31·0 | 31·2 | 31·4 | 31·6 |
|---|---|---|---|---|---|---|
| Frequency $f$ | 2 | 8 | 14 | 10 | 4 | 2 |

Calculate (a) the mean, (b) the standard deviation, (c) the mode and (d) the median of the set of results.

6. The time taken by employees to complete an operation was recorded on 80 occasions.

| Time (min) | 10·0 | 10·5 | 11·0 | 11·5 | 12·0 | 12·5 | 13·0 |
|---|---|---|---|---|---|---|---|
| Frequency $f$ | 4 | 8 | 14 | 22 | 19 | 10 | 3 |

(a) Determine (i) the mean, (ii) the standard deviation, (iii) the mode and (iv) the median of the set of observations.

(b) State (i) the class interval, (ii) the lower boundary of the third class, (iii) the upper boundary of the seventh class.

7. Components are machined to a nominal diameter of 32·65 mm. A sample batch of 400 components gave a mean diameter of 32·66 mm with a standard deviation of 0·02 mm. For a production total of 2400 components, calculate
   (a) the limits between which all the diameters are likely to lie,
   (b) the number of acceptable components if those with diameters less than 32·62 mm or greater than 32·68 mm are rejected.

8. The masses of 80 castings were determined with the following results:

| Mass $x$ (kg) | 7·3 | 7·4 | 7·5 | 7·6 | 7·7 | 7·8 |
|---|---|---|---|---|---|---|
| Frequency $f$ | 4 | 13 | 21 | 23 | 14 | 5 |

(a) Calculate (i) the mean mass, (ii) the standard deviation from the mean.
(b) For a batch of 2000 such castings, determine (i) the likely limits of all the masses and (ii) the number of castings likely to have a mass greater than 7·43 kg.

9. The heights of 120 pivot blocks were measured.

| Height $x$ (mm) | 29·4 | 29·5 | 29·6 | 29·7 | 29·8 | 29·9 |
|---|---|---|---|---|---|---|
| Frequency $f$ | 6 | 25 | 34 | 32 | 18 | 5 |

(a) Calculate (i) the mean height and (ii) the standard deviation.
(b) For a batch of 2500 such blocks, calculate (i) the limits between which all the heights are likely to lie and (ii) the number of blocks with heights greater than 29·52 mm.

10. A machine is set to produce bolts of nominal diameter 25·0 mm. Measurement of the diameters of 60 bolts gave the following frequency distribution:

| Diameter $x$ (mm) | 23·3–23·7 | 23·8–24·2 | 24·3–24·7 | 24·8–25·2 |
|---|---|---|---|---|
| Frequency $f$ | 2 | 4 | 10 | 17 |

| Diameter $x$ (mm) | 25·3–25·7 | 25·8–26·2 | 26·3–26·7 |
|---|---|---|---|
| Frequency $f$ | 16 | 8 | 3 |

(a) Calculate (i) the mean diameter and (ii) the standard deviation from the mean.
(b) For a full run of 3000, calculate (i) the limits between which all the diameters are likely to lie, (ii) the number of bolts with diameters less than 24·45 mm.

# Programme 28

## PROBABILITY

**1**

## Introduction

In very general terms, *probability* is a measure of the likelihood that a particular *event* will occur in any one *trial*, or experiment, carried out in prescribed conditions. Each separate possible result from a trial is called an *outcome*.

The ability to predict likely occurrences has obvious applications, e.g. in insurance matters and in industrial quality control and the efficient use of resources.

**Notation** The probability that a certain event $A$ will occur is denoted by $P(A)$. For example, if $A$ represents the event that a component, picked at random from stock, is faulty, written $A = \{\text{faulty component}\}$, then

$P(A)$ denotes . . . . . . . . . . . . . . .

**2**

| the probability of picking a faulty component |
|---|

**Sampling** In a manufacturing run, it would be both time-consuming and uneconomical to subject every single component produced to full inspection. It is usual, therefore, to examine a sample batch of components, taken at random, as being representative of the whole output. The larger the *random sample*, the more nearly representative of the whole *population* (total output of components) is it likely to be.

**Types of probability** The determination of probability may be undertaken from two approaches: (a) *empirical* (or experimental) probability, and (b) *classical* (or theoretical) probability. Let us look at each of these in turn.

*So move on.*

**Empirical probability**

3

*Empirical probability* is based on previous known results. The relative frequency of the number of times the event has previously occurred is taken as an indication of likely occurrences in the future.

*Example* A random sample batch of 240 components is subjected to strict inspection and 20 items are found to be defective. Therefore, if we pick any one component at random from this sample, the chance of its being faulty is '20 in 240', i.e. '1 in 12'.

So, if $$A = \{\text{faulty component}\}$$

then $$P(A) = 1 \text{ in } 12 = \frac{1}{12} = 0.0833 = 8.33\%$$

The most usual forms are $P(A) = \dfrac{1}{12}$ or $P(A) = 0.0833$

Therefore, a run of 600 components from the same machine would be likely to contain .......................... defectives.

4

$$\boxed{50}$$

for $$P\,(\text{faulty component}) = P(A) = \frac{1}{12}$$

Therefore, in 600 components, the likely number of defectives $x$ is

simply $$x = 600 \times \frac{1}{12} = \underline{50}$$

**Expectation** The result does not assert that there will be exactly 50 defectives in any run of 600 components, but, having found the probability of the event occurring in any one trial $\left(\dfrac{1}{12}\right)$, we can use it to predict the likely number of times the event will occur in $N$ similar trials. This *expectation $E$* is defined by the product of the number of trials $N$ and the probability $P(A)$ that the event $A$ will occur in any one trial,

i.e. $$\underline{E = N \times P(A)} \qquad (1)$$

*On to the next.*

**5**

**Success or failure** Throughout, we are concerned with the probability of the occurrence of a particular event. When it does occur in any one trial, we record a *success*: when it fails to occur, we record a *failure* – whatever the defined event may be.

If, in $N$ trials, there are $x$ successes, there will also be $(N - x)$ failures, so that

$$x + (N - x) = N$$

$$\therefore \frac{x}{N} + \frac{N - x}{N} = 1$$

But, $\frac{x}{N} = P(\text{success}) = P(A)$ and $\frac{N - x}{N} = P(\text{failure}) = P(\text{not } A)$

$$\therefore \underline{P(A) + P(\text{not } A) = 1}$$

The event $\{\text{not } A\}$ is called the complement of event $\{A\}$ and is often written $\{\bar{A}\}$, i.e. $\underline{P(A) + P(\bar{A}) = 1}$ (2)

$\therefore$ If the probability of picking a defective in any trial is $\frac{1}{5}$, the probability of not picking a defective is ....................

**6**

$$\boxed{\frac{4}{5}}$$

**Sample size** The size of the original sample from which the probability figure was established affects the reliability of the result.

Probabilities derived from small samples seldom reflect the probabilities associated with the whole population. The larger the sample, the more reliable are the results obtained.

*Example* Fifteen per cent of castings are found to be outside prescribed tolerances. Determine the number of acceptable items likely to be present in a batch of 120 such castings.

That is easy enough. $E =$ .............

**7**

102

for 15% are rejects. $\therefore$ 85% are acceptable.

$\therefore$ If event $A = \{\text{acceptable casting}\}$, $P(A) = 85\% = \dfrac{17}{20}$

$N = 120$ and $E = N \times P(A) = 120 \times \dfrac{17}{20} = 102$

$\therefore$ Expected number acceptable = 102

**Multiple samples** A single random sample of $n$ components taken from a whole population of $N$ components is not necessarily representative of the parent population. Another random sample of $n$ components from the same population could well include a different number of defectives from that in the first sample.

We have not got a production line on hand, but we can simulate the same kind of problem with the aid of a pack of playing cards (jokers removed). Then, the total population $N$ is ....................

**8**

the number of cards in the whole pack, i.e. 52

If we shuffle well and then deal out a random sample of 12 cards ($n$), we can count how many cards in the sample are (say) spades ($x$). These take the place of the defective components.

Replace the 12-card sample to the pack; shuffle well and then take a second sample of 12 cards. Counting the spades in this second sample will most likely give a different total from the result of the first trial.

If we average the two results, dividing the combined number of spades in the two trials by the number of trials, i.e. 2, we are effectively considering a larger sample.

*Let us extend this into a useful experiment, which we can now carry out – so move on to the next frame.*

## 9 *Experiment*

To determine the number of spades in each of 40 trials of a random 12-card sample from a full deck of playing cards and to compile a cumulative proportion (running average) of the results.

*Procedure* Start with a full deck of playing cards, excluding the jokers.

(a) Shuffle the cards thoroughly.
(b) Deal a random sample of 12 cards.
(c) From the sample, count and record the number of cards that are spades.
(d) Return the sample cards to the pack and shuffle thoroughly.
(e) Repeat the process for a total of 40 such trials.
(f) Compile a table showing
  (i) the number of the trial, $r$ (1 to 40)
  (ii) the number of spade cards in each trial, $x$
  (iii) the cumulative total of spade cards, cum.$x$
  (iv) the cumulative proportion (running average) of the results.

$$\text{Cum. prop.} = \frac{\text{cumulative total of spade cards in } r \text{ trials}}{\text{number of trials to date, i.e. } r}$$

(g) Display graphically the distribution of the number of spades at each trial (1 to 40) and the running average for the 40 trials.

Your table of results will look like this, but with different entries

| Trial $r$ | Spades $x$ | Cum. $x$ | Running average |
|---|---|---|---|
| 1 | 2 | 2 | 2.00 |
| 2 | 5 | 7 | 3.50 |
| 3 | 0 | 7 | 2.33 |
| • | • | • | • |
| • | • | • | • |
| • | • | • | • |

*Carry on and see what you get.*

**10**

The results from one such experiment gave the following distribution of spades. Yours will be different, but will eventually lead to the same conclusion.

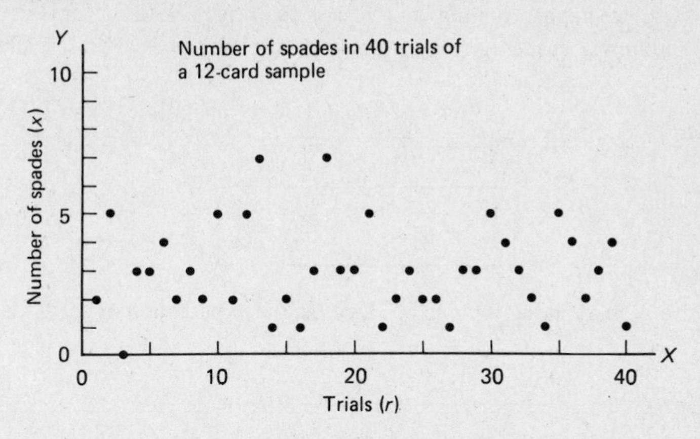


Plotting the running average $\frac{\text{cum. } x}{r}$ against $r$, we have the result shown in the next frame. Complete yours likewise.

**11**

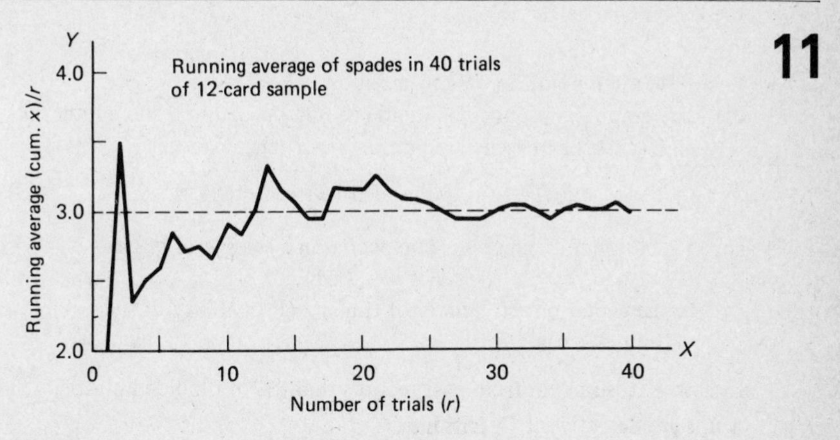


We see that

(a) the running average fluctuates in the early stages

(b) as the number of trials is increased, the up-dated running average settles much more closely to ......................

## 12

the value 3

Of course, we happen to know in this case that the pack of 52 cards in fact contains 13 spades, giving a probability that any one card is a spade as ..........................................

## 13

$\frac{13}{52}$ i.e $\frac{1}{4}$; $P(\text{spade}) = \frac{1}{4}$

Therefore, in a random sample of 12 cards, the expectation of spades is

$E = $ ..............................

## 14

$$E = n \times P(\text{spades}) = 12 \times \frac{1}{4} = 3$$

which is, in fact, the value that the graph settles down to as the number of trials increases.

So the main points so far are as follows;

(a) The empirical probability of an event $A$ occurring is the number, $x$, of successes experienced in $n$ previous trials, divided by $n$,

i.e. $$P(A) = \frac{x}{n} \quad \text{(the relative frequency)} \qquad (3)$$

(b) The number of successes $E$ expected in a sample of $m$ trials is

$$E = m \times P(A)$$

i.e. Expectation = (Number of trials) × (Probability of success in any one trial)

*Example* It is known from past records that 8% of moulded plastic items are defective. Determine

(a) the probability that any one item is: (i) defective, (ii) acceptable.

(b) the number of acceptable items likely to be found in a sample batch of 4500.

*Complete it. No snags.*

**15**

$$\text{(a) (i) } \frac{2}{25}, \quad \text{(ii) } \frac{23}{25}; \qquad \text{(b) } 4140$$

for, let $A = \{\text{defective}\}$ and $B = \{\text{acceptable}\}$.

Then (a) (i) $P(A) = \frac{8}{100} = \underline{\frac{2}{25}}$; (ii) $P(B) = \frac{92}{100} = \underline{\frac{23}{25}}$

(b) $E = m \times P(B) = 4500 \times \frac{23}{25} = \underline{4140}$

Note that, as always, $P(A) + P(B) = P(A) + P(\bar{A}) = 1$

Now let us turn our attention to *classical probability*

**16**

**Classical probability**

The classical approach to probability is based on a consideration of the theoretical number of ways in which it is possible for an event $A$ to occur.

As before,

(a) In any trial, each separate possible result is called an .....................

(b) The particular occurrence being looked for in a trial is the .............

(c) When it occurs, we have a ..................; when it does not occur, we have a ......................................

**17**

(a) outcome; (b) event;
(c) success; failure

To this list, we now add

(d) The classical probability $P$ of an event $A$ occurring is defined by

$$P(A) = \frac{\text{number of ways in which event } A \text{ can occur}}{\text{total number of all possible outcomes}} \qquad (4)$$

*Example* If we consider the chance result of rolling a normal unbiased die, the total number of possible outcomes is ..............................

**18**

six, i.e. 1, 2, 3, 4, 5, 6

If the event we are considering is the throwing of a 6, this can occur in only one way out of the six possible outcomes.

$\therefore\ P(\text{six}) =$ .................... and $P(\text{not six}) =$ .........................

**19**

$P(\text{six}) = \frac{1}{6}$; $P(\text{not six}) = \frac{5}{6}$

If, out of $n$ possible outcomes of a trial, it is possible for an event $A$ to occur in $x$ ways and for the event $A$ not to occur in $y$ ways, then

$$n = x + y \quad \text{and} \quad P(A) = \frac{x}{n} = \frac{x}{x+y} \qquad (5)$$

**Certain and impossible events**

(a) If an event $A$ is certain to occur every time, then

$$x = n,\ \ y = 0\ \ \therefore\ P(A) = \frac{n}{n} = 1$$

(b) If an event $A$ cannot possibly occur at any time, then

$$x = 0,\ \ y = n\ \ \therefore P(A) = \frac{0}{n} = 0$$

$$\therefore\ P(\text{certainty}) = 1; \qquad P(\text{impossibility}) = 0 \qquad (6)$$

In most cases, probability values lie between these extreme values.

So –

If one card is drawn at random from a full pack of playing cards, the probability of obtaining
(a) a heart is .................; (b) a king is ...................;
(c) a card other than a king is .................; (d) a black card is ...........

**20**

(a) $\frac{13}{52} = \frac{1}{4}$; (b) $\frac{4}{52} = \frac{1}{13}$

(c) $\frac{48}{52} = \frac{12}{13}$; (d) $\frac{26}{52} = \frac{1}{2}$

**Mutually exclusive and mutually non-exclusive events**

(a) *Mutually exclusive events* are events which cannot occur together. For example, in rolling a die, the event of throwing a six and that of throwing a five cannot occur at the same time. Similarly, when drawing a card from a pack, the event of drawing an ace and also that of drawing a king cannot occur in the same single trial.

(b) *Mutually non-exclusive events* are events that can occur simultaneously. For example, in rolling a die, the event of obtaining a multiple of 3 and the event of obtaining a multiple of 2 can occur together if a 6 is thrown. Similarly, with cards, the event of drawing a black suit and that of drawing a queen can occur together if the ......................... or the ....................... is drawn

**21**

queen of clubs; queen of spades

*Exercise* State whether the following pairs of events are mutually exclusive or non-exclusive.

(a) $A = \{\text{ace}\}$; $B = \{\text{black card}\}$
(b) $A = \{\text{heart}\}$; $B = \{\text{ace of spades}\}$
(c) $A = \{\text{red card}\}$; $B = \{\text{jack}\}$
(d) $A = \{\text{5 of clubs}\}$; $B = \{\text{10 of clubs}\}$
(e) $A = \{\text{card} < 10\}$; $B = \{\text{king of diamonds}\}$

**22**

(a) and (c) non-exclusive; (b), (d) (e) exclusive

**Addition law of probability**

If there are $n$ possible outcomes to a trial, of which $x$ give an event $A$ and $y$ give an event $B$, then, provided the events $A$ and $B$ are *mutually exclusive*, the probability $P$ of either event $A$ or event $B$ occurring – but clearly not both – is

$$P(A \text{ or } B) = \frac{x+y}{n}$$

$$= \frac{x}{n} + \frac{y}{n} = P(A) + P(B)$$

$$\therefore P(A \text{ or } B) = P(A) + P(B) \qquad (7)$$

If the events $A$ and $B$ are *non-exclusive*, so that events $A$ and $B$ can occur together, then the probability of $A$ or $B$ occurring is given by

$$P(A \text{ or } B) = P(A) + P(B) - P(A \text{ and } B) \qquad (8)$$

For example, in rolling a die,

the probability of scoring a multiple of 3 (i.e. 3 or 6) = $\frac{2}{6} = P(A)$

the probability of scoring a multiple of 2 (i.e. 2, 4, 6) = $\frac{3}{6} = P(B)$

So the probability of scoring a multiple of 3 or a multiple of 2 would seem to be ..................................

**23**

$$P(A) + P(B) = \frac{2}{6} + \frac{3}{6} = \frac{5}{6}$$

*But wait.* One outcome (6) appears in both sets of outcomes, i.e. the events are not mutually exclusive, and this outcome has been counted twice. The probability of scoring one of the sixes must therefore be removed.

$$P(\text{six}) = \frac{1}{6} = P(A \text{ and } B)$$

Therefore, in this example $P(A \text{ or } B) =$ ..............................

**24**

$$\boxed{P(A \text{ or } B) = \frac{2}{3}}$$

for $P(A) = \frac{2}{6}$; $P(B) = \frac{3}{6}$; $P(A \text{ and } B) = \frac{1}{6}$

$$\therefore P(A \text{ or } B) = P(A) + P(B) - P(A \text{ and } B) = \frac{2}{6} + \frac{3}{6} - \frac{1}{6} = \underline{\frac{2}{3}}$$

*Example* A single card is drawn from a pack of 52 playing cards.

(a) If event $A$ = {drawing an ace} and event $B$ = {drawing a seven} the probability of drawing either an ace or a seven, i.e.

$P(A \text{ or } B) =$ ....................

**25**

$$\boxed{P(A \text{ or } B) = P(\text{ace or seven}) = \frac{2}{13}}$$

for these are mutually exclusive events in any one trial

$$\therefore P(A \text{ or } B) = P(A) + P(B)$$

4 aces in the pack $\therefore P(A) = P(\text{ace}) = \frac{4}{52} = \frac{1}{13}$

4 sevens in the pack $\therefore P(B) = P(\text{seven}) = \frac{4}{52} = \frac{1}{13}$

$$\therefore P(A \text{ or } B) = P(\text{ace or seven}) = \frac{1}{13} + \frac{1}{13} = \underline{\frac{2}{13}}$$

(b) If event $A$ = {drawing a king} and event $B$ = {drawing a red card}

then $P(A \text{ or } B) =$ ....................

**26**

$$P(A \text{ or } B) = P(\text{king or red card}) = \frac{7}{13}$$

These are non-exclusive events since we could draw the king of hearts or the king of diamonds.

$$\therefore P(A \text{ or } B) = P(A) + P(B) - P(A \text{ and } B)$$

4 kings in the pack $\quad \therefore P(A) = P(\text{king}) = \frac{4}{52} = \frac{1}{13}$

26 red cards in the pack $\therefore P(B) = P(\text{red card}) = \frac{26}{52} = \frac{1}{2}$

2 red kings in the pack $\therefore P(A \text{ and } B) = P(\text{red king}) = \frac{2}{52} = \frac{1}{26}$

$$\therefore P(A \text{ or } B) = \frac{1}{13} + \frac{1}{2} - \frac{1}{26} = \frac{2 + 13 - 1}{26} = \frac{14}{26} = \underline{\frac{7}{13}}$$

Take it in steps, then all is well. *Next frame.*

**27**

So, for *mutually exclusive* events $A$ and $B$, i.e. $A$ or $B$ can occur, but not both at the same time,

$$P(A \text{ or } B) = \text{..........................}$$

**28**

$$P(A \text{ or } B) = P(A) + P(B)$$

and for *mutually non-exclusive* events $A$ and $B$, i.e. $A$ or $B$ or both can occur in any one trial,

$$P(A \text{ or } B) = \text{..........................}$$

**29**

$$\boxed{P(A \text{ or } B) = P(A) + P(B) - P(A \text{ and } B)}$$

Make a note of these results, if you have not already done so.

**Independent events and dependent events**

(a) Events are *independent* when the occurrence of one event does not affect the probability of the occurrence of the second event. For example, in rolling a die on two occasions, the outcome of the first throw will not affect the probability of throwing a six on the second throw.

(b) Events are *dependent* when one event does affect the probability of the occurrence of the second.

Thus, the probability of drawing an ace from a pack of cards is $\dfrac{4}{52} = \dfrac{1}{13}$.

If the card is replaced so that the pack is complete and shuffled, the probability of drawing an ace on the second occasion is similarly $\dfrac{1}{13}$ (Independent events).

However, if the ace is drawn on the first cutting and *not* replaced, the probability of drawing an ace on the second occasion is now

. . . . . . . . . . . . . . . . .

**30**

$$\boxed{\frac{3}{51}}$$

since there are now only 3 remaining aces in the incomplete pack of 51 cards (Dependent events).

**Multiplication law of probabilities**

Let us consider the probability of the occurrence of both events $A$ and $B$ where event $A = \{\text{throwing a six}\}$ when rolling a die and
event $B = \{\text{drawing an ace}\}$ from a pack of cards.

These are clearly independent events

$$P(A) = P(\text{six}) = \frac{1}{6}; \qquad P(B) = P(\text{ace}) = \frac{4}{52} = \frac{1}{13}$$

If we now both roll the die and draw a card as one trial, then there are 6 possible outcomes from the die and for each one of these there are 52 possible outcomes from the cards, giving $(6 \times 52)$ outcomes altogether.

There are 4 possibilities of obtaining a six and an ace together, so the probability of event $A$ and event $B$ occurring is

$$P(A \text{ and } B) = \frac{4}{6 \times 52} = \frac{1 \times 4}{6 \times 52} = \frac{1}{6} \times \frac{4}{52} = P(A) \times P(B)$$

So, when $A$ and $B$ are *independent events*

$$P(A \text{ and } B) = P(A) \times P(B) \qquad (9)$$

*Note this result.*

**31** **Conditional probability**

We are concerned here with the probability of an event $B$ occurring, given that an event $A$ has already taken place. This is denoted by the symbol $P(B|A)$

If $A$ and $B$ are *independent* events, the fact that event $A$ has already occurred will not affect the probability of event $B$. In that case,

$$P(B|A) = \ldots\ldots\ldots\ldots$$

$$\boxed{P(B|A) = P(B)} \qquad (10)$$

**32**

If $A$ and $B$ are *dependent* events, then event $A$ having occurred will affect the probability of the occurrence of event $B$. Let us see an example.

*Example* A box contains five 10 Ω resistors and twelve 30 Ω resistors. The resistors are all unmarked and of the same physical size.

(a) If one resistor is picked out at random, determine the probability of its resistance being 10 Ω.
(b) If this first resistor is found to be 10 Ω and it is retained on one side, find the probability that a second selected resistor will be of 30 Ω resistance.

Let $A = \{10\ \Omega \text{ resistor}\}$ and $B = \{30\ \Omega \text{ resistor}\}$

(a) $n = 5 + 12 = 17$ Then $P(A) =$ ..................

---

**33**

$$\boxed{P(A) = \frac{5}{17}}$$

(b) The box now contains four 10 Ω resistors and twelve 30 Ω resistors. Then the probability of $B$, $A$ having occurred,

$= P(B|A) =$ .........................

---

**34**

$$\boxed{P(B|A) = \frac{12}{16} = \frac{3}{4}}$$

So the probability of getting a 10 Ω resistor at the first selection, retaining it, and getting a 30 Ω resistor at the second selection is

$P(A \text{ and } B\ |A) =$ ......................

---

**35**

$$\boxed{\frac{15}{68}}$$

for $P(A \text{ and } B|A) = P(A) \times P(B|A) = \frac{5}{17} \times \frac{3}{4} = \underline{\frac{15}{68}}$

So, if $A$ and $B$ are independent events

$$P(A \text{ and } B) = P(A) \times P(B)$$

and if $A$ and $B$ are dependent events

$$P(A \text{ and } B) = P(A) \times P(B|A) \qquad (11)$$

*Make a note of these results. Then on to the next frame.*

**36**

*Example* A box contains 100 copper plugs, 27 of which are oversize and 16 undersize. A plug is taken from the box, tested and replaced: a second plug is then similarly treated. Determine the probability that (i) both plugs are acceptable, (ii) the first is oversize and the second undersize, (iii) one is oversize and the other undersize.

Let $A = \{\text{oversize plug}\}$; $B = \{\text{undersize plug}\}$

$N = 100$; 27 oversize; 16 undersize; $\therefore$ 57 acceptable

(i) $P_1$ (first plug acceptable) $= \frac{57}{100}$

$P_2$ (second acceptable) $= \frac{57}{100}$

$\therefore P_{12}$ (first acceptable and second acceptable) = .................

**37**

$$\boxed{P_1 \times P_2 = \frac{57}{100} \times \frac{57}{100} = \frac{3249}{10000} = 0.3249}$$

(ii) $P_1$ (first oversize) $= \frac{27}{100}$

$P_2$ (second undersize) $= \frac{16}{100}$

$\therefore P_{12}$ (first oversize and second undersize) = ....................

38

$$P_1 \times P_2 = \frac{27}{100} \times \frac{16}{100} = \frac{432}{10000} = 0.0432$$

(iii) This section, of course, includes part (ii) of the problem, but also covers the case when the first is undersize and the second oversize.

$$P_3 \text{ (first undersize)} = \frac{16}{100}; \quad P_4 \text{ (second oversize)} = \frac{27}{100}$$

$\therefore P_{34}$ (first undersize and second oversize)

$$= \frac{16}{100} \times \frac{27}{100} = \frac{432}{10000} = 0.0432$$

$\therefore P$ (one oversize and one undersize)

$= P$ {(first oversize and second undersize) or (first undersize and second oversize)} = ...............................

39

$$P_{12} + P_{34} = \frac{432}{10000} + \frac{432}{10000} = \frac{864}{10000} = 0.0864$$

One must be careful to read the precise requirements of the problem. Then the solution is straightforward, the main tools being:

Independent events

$P(A \text{ or } B) = P(A) + P(B)$ i.e. 'or' is associated with +

$P(A \text{ and } B) = P(A) \times P(B)$ i.e. 'and' is associated with ×

Dependent events

$P(A \text{ or } B) = P(A) + P(B)$; $P(A \text{ and } B) = P(A) \times P(B \mid A)$

Now one more on your own. It is the last example repeated with one important variation.

*Example* A box contains 100 copper plugs, 27 oversize and 16 undersize. A plug is taken, tested but *not* replaced: a second plug is then treated similarly. Determine the probability that (i) both plugs are acceptable, (ii) the first is oversize and the second undersize, (iii) one is oversize and the other undersize.

*Complete all three sections and then check with the next frame.*

**40**

See if you agree.

Let $A = \{\text{oversize plug}\}$; $B = \{\text{undersize plug}\}$

$N = 100$; 27 oversize; 16 undersize; $\therefore$ 57 acceptable

(i) $P_1$ (first plug acceptable) $= \dfrac{57}{100}$ Plug *not* replaced

$P_2$ (second acceptable) $= \dfrac{56}{99}$

$\therefore P_{12}$ (first acceptable and second acceptable)

$$= \frac{57}{100} \times \frac{56}{99} = \frac{3192}{9900} = \underline{0.2739}$$

(ii) $P_1$ (first oversize) $= \dfrac{27}{100}$ Plug *not* replaced

$P_2$ (second undersize) $= \dfrac{16}{99}$

$\therefore P_{12}$ (first oversize and second undersize)

$$= \frac{27}{100} \times \frac{16}{99} = \frac{432}{9900} = \underline{0.0436}$$

(iii) Here again, we must include the two cases of either (first oversize and second undersize) or (first undersize and second oversize). The first of these we have already covered above.

$$P_{12} = \frac{432}{9900} = 0.0436.$$

Now we have

$P_3$ (first undersize) $= \dfrac{16}{100}$ Plug *not* replaced

$P_4$ (second oversize) $= \dfrac{27}{99}$

$\therefore P_{34}$ (first undersize and second oversize)

$$= \frac{16}{100} \times \frac{27}{99} = \frac{432}{9900} = 0.0436$$

$\therefore P$ (one oversize and one undersize) =

$P\{$(first oversize and second undersize) or (first undersize and second oversize)$\}$ $= P_{12} + P_{34}$

$= 0.0436 + 0.0436 = \underline{0.0872}$

*Now let us move on to something rather different.*

**Discrete probability distribution**

**41**

If we consider a single trial of tossing a coin, there are two equally likely possible outcomes, a 'head' or a 'tail' (i.e. a not-head). Listing the possible outcomes in successive trials, we have:

1 trial: T H
2 trials: TT TH HT HH
3 trials: TTT TTH THT THH HTT HTH HHT HHH

Similarly with 4 trials, the possible outcomes would be those detailed for 3 trials with an extra T or H attached in each case. The possible outcomes with 4 trials are therefore ..........................

**42**

| TTTT | TTHT | THTT | THHT | HTTT | HTHT | HHTT | HHHT |
|---|---|---|---|---|---|---|---|
| TTTH | TTHH | THTH | THHH | HTTH | HTHH | HHTH | HHHH |

If we concentrate on $x$, the number of heads at the conclusion of each set of trials, and $f$, the frequency of $x$, we can tabulate the results, thus:

| Number of trials | Number of heads $x$ | 0 | 1 | 2 | 3 | 4 | Number of possible outcomes |
|---|---|---|---|---|---|---|---|
| 1 | $f$ | 1 | 1 | | | | 2 |
| | $P$ | $\frac{1}{2}$ | $\frac{1}{2}$ | | | | |
| 2 | $f$ | 1 | 2 | 1 | | | 4 |
| | $P$ | $\frac{1}{4}$ | $\frac{2}{4}$ | $\frac{1}{4}$ | | | |
| 3 | $f$ | | | | | | |
| | $P$ | | | | | | |
| 4 | $f$ | | | | | | |
| | $P$ | | | | | | |

Completing the table for 3 and 4 trials, we get .......................

**43**

| Trials | Heads $x$: | 0 | 1 | 2 | 3 | 4 | Outcomes |
|---|---|---|---|---|---|---|---|
| 3 | $f$ | 1 | 3 | 3 | 1 | | 8 |
| | $P$ | $\frac{1}{8}$ | $\frac{3}{8}$ | $\frac{3}{8}$ | $\frac{1}{8}$ | | |
| 4 | $f$ | 1 | 4 | 6 | 4 | 1 | 16 |
| | $P$ | $\frac{1}{16}$ | $\frac{4}{16}$ | $\frac{6}{16}$ | $\frac{4}{16}$ | $\frac{1}{16}$ | |

Summarising the probabilities, we have

| Trials | Heads $x$: | 0 | 1 | 2 | 3 | 4 | Outcomes |
|---|---|---|---|---|---|---|---|
| 1 | $P$ | $\frac{1}{2}$ | $\frac{1}{2}$ | | | | 2 |
| 2 | $P$ | $\frac{1}{4}$ | $\frac{2}{4}$ | $\frac{1}{4}$ | | | 4 |
| 3 | $P$ | $\frac{1}{8}$ | $\frac{3}{8}$ | $\frac{3}{8}$ | $\frac{1}{8}$ | | 8 |
| 4 | $P$ | $\frac{1}{16}$ | $\frac{4}{16}$ | $\frac{6}{16}$ | $\frac{4}{16}$ | $\frac{1}{16}$ | 16 |

There is clearly a pattern here and we soon recognise that the probability values are the separate terms of the expansion of $(\frac{1}{2} + \frac{1}{2})^n$ where $n$ is the number of trials in each case, i.e. the terms of the *binomial expansion* of $(a + b)^n$ where $a = b = \frac{1}{2}$.

You will remember that the binomial coefficients are given by Pascal's triangle

$n = 0$    1

1    1   1

2    1   2   1

3    1   3   3   1

4    1   4   6   4   1

So with 5 trials, the probabilities of obtaining $x$ heads are

..............................................

44

$n = 5$

| $x$ | 0 | 1 | 2 | 3 | 4 | 5 |
|---|---|---|---|---|---|---|
| $f$ | 1 | 5 | 10 | 10 | 5 | 1 |
| $P = \frac{f}{2^n}$ | $\frac{1}{32}$ | $\frac{5}{32}$ | $\frac{10}{32}$ | $\frac{10}{32}$ | $\frac{5}{32}$ | $\frac{1}{32}$ |

Note that, as ever, the total probability

$$\Sigma \frac{f}{2^n} = \frac{1+5+10+10+5+1}{32} = \frac{32}{32} = \underline{1}$$

We cannot get very far in studying probability without a brief revision of *permutations* and *combinations*. Therefore, let us make a fresh start in the next frame.

45

**Permutations and combinations**

(a) **Permutations** If we have $n$ different objects, we can arrange them in different orders of selection. Each different *ordered* arrangement is called a *permutation*.

For example, permutations of the three letters A, B, C taken together are ABC, ACB, BAC, BCA, CAB, CBA and we say that $^3P_3 = 6$. The upper 3 denotes the number of items from which the arrangements are made: the lower 3 indicates the number of items we are using in each arrangement.

Similarly, taking the same three letters A, B, C two at a time, we can form the permutations ..........................................
so that $\quad ^3P_2 =$ ..........................................

(You remember that the order of the letters *does* matter.
BA is a different permutation from AB, etc.)

**46**

$$\boxed{\text{AB, BA, AC, CA, BC, CB} \quad \therefore \quad {}^3P_2 = 6}$$

In general, to find ${}^nP_r$, i.e. the number of permutations of $n$ different items taken $r$ at a time, we can argue as follows.

Of the $r$ positions in each ordered arrangement,
the first place can be filled in $n$ different ways,
the second place can be filled in $(n-1)$ different ways,
the third place can be filled in $(n-2)$ different ways,
the $r^{\text{th}}$ place can be filled in $(n-r+1)$ different ways.
Therefore, the $r$ places can be filled in
$n(n-1)(n-2)\ldots(n-r+1)$ different ways

i.e. $$\underline{{}^nP_r = n(n-1)(n-2)\ldots(n-r+1)} \qquad (12)$$

So, ${}^5P_4 = \ldots\ldots\ldots\ldots\ldots\ldots$

**47**

$$\boxed{120}$$

For ${}^5P_4$, $n = 5$; $r = 4$ $\therefore {}^5P_4 = 5 \times 4 \times 3 \times 2 = \underline{120}$

If $r \neq n$, we can multiply and divide result (12) above by

$$(n-r)(n-r-1)\ldots 3 \times 2 \times 1$$

Then $${}^nP_r = \frac{n(n-1)(n-2)\ldots(n-r+1) \times (n-r)(n-r-1)\ldots 3 \times 2 \times 1}{(n-r)(n-r-1)\ldots 3 \times 2 \times 1}$$

$$\therefore \underline{{}^nP_r = \frac{n!}{(n-r)!}} = \frac{\text{factorial } n}{\text{factorial } (n-r)} \qquad (13)$$

Using result (13), $${}^3P_3 = \frac{3!}{0!} = \frac{3 \times 2 \times 1}{0!} = \frac{6}{0!}$$

But what do we mean by 0! ?

In frame 45, we found that ${}^3P_3 = 6 \therefore {}^3P_3 = \dfrac{6}{0!} = 6 \quad \therefore \underline{0! = 1}$ (14)

In that case ${}^nP_n = \dfrac{n!}{0!} = n!$ $\qquad \therefore \underline{{}^nP_n = n!}$ (15)

So ${}^{10}P_4 = \ldots\ldots\ldots\ldots\ldots\ldots$

**48**

$\boxed{5040}$

for $\quad {}^{10}P_4 = \dfrac{10!}{6!} = 10 \times 9 \times 8 \times 7 = \underline{5040}$

(b) **Combinations** Returning to our three letters A, B, C we now make selections without regard to the order of the letters in each group, i.e. AB is now the same as BA, etc. Each group is called a *combination*, and ${}^nC_r$, where $n$ is the total number of items and $r$ is the number in each selection, gives the number of possible combinations.

For example, ${}^3C_3 = \{ABC\}$. (Just that. Not BCA, ACB, . . . etc.)

$$\therefore {}^3C_3 = 1$$

Also $\quad {}^3C_2 = \ldots\ldots\ldots\ldots\ldots\ldots$

**49**

$\boxed{{}^3C_2 = \{AB, BC, CA\} \text{ i.e. } {}^3C_2 = 3}$

Note that AB and BA are different permutations, but are *not* different combinations.

In general, to find ${}^nC_r$, i.e. the number of combinations of $n$ different items taken $r$ at a time, we can proceed as follows.

Any one combination of $r$ items can be re-arranged within itself to give ${}^rP_r = r!$ different permutations. If all the ${}^nC_r$ groups are so re-arranged, we should have a total of $({}^nC_r \times r!)$ different permutations, i.e. the total number of permutations from $n$ items taken $r$ at a time.

i.e. $\quad {}^nC_r \times r! = {}^nP_r = \dfrac{n!}{(n-r)!}$

$$\therefore {}^nC_r = \frac{n!}{(n-r)!\,r!} \qquad (16)$$

This is sometimes written as $\dbinom{n}{r}$

So ${}^{10}C_6 = \ldots\ldots\ldots\ldots\ldots\ldots\ldots\ldots\ldots\ldots$

**50**

$\boxed{210}$

$$\text{since } {}^{10}C_6 = \frac{10!}{4!\,6!} = \frac{10 \times 9 \times 8 \times 7}{4 \times 3 \times 2 \times 1} = \underline{210}$$

*Exercise* Evaluate the following

(a) $0!$ (b) $1!$ (c) ${}^6P_3$ (d) ${}^8P_5$

(e) ${}^4P_4$ (f) ${}^5P_1$ (g) ${}^{10}P_3$ (h) ${}^7C_4$

(i) ${}^9C_6$ (j) ${}^3C_1$ (k) ${}^3C_3$ (l) ${}^6C_2$

No troubles: good practice. Check with the next frame.

**51**

| | | | |
|---|---|---|---|
| (a) 1 | (b) 1 | (c) 120 | (d) 6720 |
| (e) 24 | (f) 5 | (g) 720 | (h) 35 |
| (i) 84 | (j) 3 | (k) 1 | (l) 15 |

**General binomial distribution**

Tossing a coin has but two possible outcomes: 'heads' or 'tails'.
Rolling a die has six possible outcomes, each equally likely to occur.

If we regard throwing a six as a success and throwing any value other than a six as a failure, then

$$P(\text{six}) = P(\text{success}) = \frac{1}{6}$$

$$P(\text{not six}) = P(\text{failure}) = \frac{5}{6}$$

Let us denote the probability of a success by $p$
and the probability of a failure by $q$

Then $p + q = 1$ and in this example $p = \frac{1}{6}$ and $q = \frac{5}{6}$.

Let us consider the probabilities of success in rolling a die in $n$ trials.

(i) *No six* Probability of no success in any one trial $= q$.
There are $n$ trials $\therefore P(\text{no success}) =$ ..................................

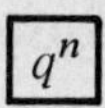

**52**

(ii) *One six* If the first $(n - 1)$ trials are failures and the last trial a success, the probability of this occurring is $q^{n-1}p$. But, of course, this one success may occur in any one of the $n$ trials. $\therefore$ Probability of one success $= n\,q^{n-1}p$.

(iii) *Two sixes* If the first $(n - 2)$ trials are failures and the last two trials are successes, the probability of this occurring is $q^{n-2}p^2$. But the two successes may occur in the $n$ trials in ${}^nC_2$, i.e. $\dfrac{n!}{(n-2)!\,2!}$ ways. Therefore the probability of two successes in $n$ trials $= \dfrac{n!}{(n-2)!\,2!}\,q^{n-2}p^2$.

– and so on –

(iv) *r sixes* By the same argument, $r$ successes may occur in $n$ trials in ${}^nC_r$ i.e. $\dfrac{n!}{(n-r)!\,r!}$ ways, so that the probability of $r$ success in $n$ trials = ..............................................

---

$$\boxed{\frac{n!}{(n-r)!\,r!}\,q^{n-r}p^r}$$

**53**

So, in general, the probability of obtaining $r$ successes in $n$ trials (independent events) is

$$P(r \text{ successes}) = \frac{n!}{(n-r)!\,r!}\,q^{n-r}p^r \qquad (17)$$

*Example 1* A die is rolled five times. Determine the probability of obtaining three sixes.

Here $n = 5$, $r = 3$, $P(\text{six}) = p = \dfrac{1}{6}$. $P(\text{not six}) = q = \dfrac{5}{6}$

$\therefore P(3 \text{ sixes}) =$ ...........................

**54**

$$\boxed{\frac{125}{3888} = 0.0322}$$

for $P(\text{3 sixes}) = \dfrac{5!}{2!\,3!}q^2p^3 = \dfrac{5!}{2!\,3!}\left(\dfrac{5}{6}\right)^2\left(\dfrac{1}{6}\right)^3$

$$= \frac{5 \times 4}{2 \times 1}\left(\frac{5^2}{6^5}\right) = \frac{250}{7776} = \underline{0.0322}$$

*Example 2* Twenty per cent of items produced on a machine are outside stated tolerances. Determine the probability distribution of the number of defectives in a pack of five items.

The probability distribution is the set of probabilities for $x = 0, 1, 2, 3, 4, 5$ successes, i.e. defectives in this case. We have $p = 0.2$; $q = 0.8$ and we compile the table shown.

| $x$ | 0 | 1 | 2 | 3 | 4 | 5 |
|---|---|---|---|---|---|---|
| $P$ | $q^5$ | $5q^4p$ | $10q^3p^2$ | $10q^2p^3$ | $5qp^4$ | $p^5$ |
| | $\left(\frac{4}{5}\right)^5$ | $5\left(\frac{4}{5}\right)^4\left(\frac{1}{5}\right)$ | $10\left(\frac{4}{5}\right)^3\left(\frac{1}{5}\right)^2$ | ... | ... | ... |
| | $\frac{4^5}{5^5}$ | $\frac{5 \times 4^4}{5^5}$ | ... | ... | ... | ... |
| | 0.3277 | ... | ... | ... | ... | ... |

Complete the table and check with the next frame.

**55**

| $x$ | 0 | 1 | 2 | 3 | 4 | 5 |
|---|---|---|---|---|---|---|
| $P$ | $q^5$ | $5q^4p$ | $10q^3p^2$ | $10q^2p^3$ | $5qp^4$ | $p^5$ |
| | $\left(\frac{4}{5}\right)^5$ | $5\left(\frac{4}{5}\right)^4\left(\frac{1}{5}\right)$ | $10\left(\frac{4}{5}\right)^3\left(\frac{1}{5}\right)^2$ | $10\left(\frac{4}{5}\right)^2\left(\frac{1}{5}\right)^3$ | $5\left(\frac{4}{5}\right)\left(\frac{1}{5}\right)^4$ | $\left(\frac{1}{5}\right)^5$ |
| | $\frac{4^5}{5^5}$ | $\frac{5 \times 4^4}{5^5}$ | $\frac{10 \times 4^3}{5^5}$ | $\frac{10 \times 4^2}{5^5}$ | $\frac{5 \times 4}{5^5}$ | $\frac{1}{5^5}$ |
| | 0.3277 | 0.4096 | 0.2048 | 0.0512 | 0.0064 | 0.0003 |

*Move on to frame 56.*

56

In this case, then, we have the results:
Probability distribution $n = 5$; $p = 0.2$; $q = 0.8$

| Successes $x$ | 0 | 1 | 2 | 3 | 4 | 5 |
|---|---|---|---|---|---|---|
| Probability $P$ | 0.3277 | 0.4096 | 0.2048 | 0.0512 | 0.0064 | 0.0003 |

and these can be displayed as a probability histogram.

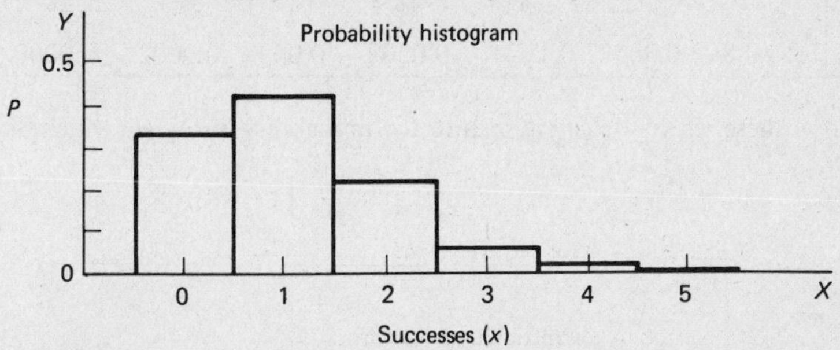


(a) The probability of any particular outcome is given by the height of each column, but since the columns are 1 unit wide, the probability is also represented by the area of each column.
(b) The total probability is 1, i.e. the total area of the probability histogram, is also 1.

Now for another example to work on your own.

*Example 3* A run of 600 components was found to contain 75 defectives. Determine the probability distribution of the number of defectives in a random sample set of 6 components.

First we find $p = \ldots\ldots\ldots\ldots$; $q = \ldots\ldots\ldots\ldots$

57

$$p = \frac{1}{8}; \quad q = \frac{7}{8}$$

So we have $p = \dfrac{75}{600} = \dfrac{1}{8}$; $q = \dfrac{7}{8}$; $n = 6$

and now we have to find the probabilities of 0, 1, 2, . . . 5, 6 successes (defectives) in any sample set of 6 components.
Proceed exactly as before. Complete the table on your own and then check the results with the next frame.

**58**

| $x$ | 0 | 1 | 2 | 3 | 4 | 5 | 6 |
|---|---|---|---|---|---|---|---|
| $P$ | $q^6$ | $6q^5p$ | $15q^4p^2$ | $20q^3p^3$ | $15q^2p^4$ | $6qp^5$ | $p^6$ |
| | Substituting $q = \frac{7}{8}$ and $p = \frac{1}{8}$ gives | | | | | | |
| | $\frac{7^6}{8^6}$ | $\frac{6 \times 7^5}{8^6}$ | $\frac{15 \times 7^4}{8^6}$ | $\frac{20 \times 7^3}{8^6}$ | $\frac{15 \times 7^2}{8^6}$ | $\frac{6 \times 7}{8^6}$ | $\frac{1}{8^6}$ |
| | 0.4488 | 0.3847 | 0.1374 | 0.0262 | 0.0028 | 0.0002 | 0.0000 |

From these results we can now draw the probability histogram which is

..........................................................................

**59**

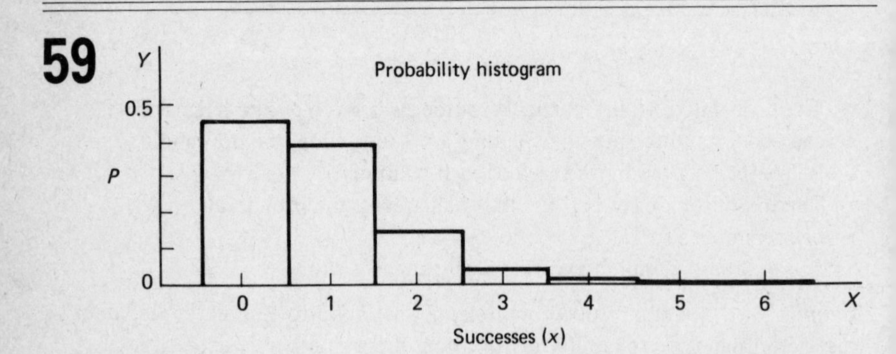


We can use the probability distribution obtained above to determine the theoretical frequencies of sets of 6 components having 0, 1, 2, . . . defectives per set in a total of $m$ sets.

Frequency $f = m \times P$ where $m$ = number of trials (sets)

$\therefore$ If $m = 5000$, the theoretical frequencies are

| $x$ | 0 | 1 | 2 | 3 | 4 | 5 | 6 |
|---|---|---|---|---|---|---|---|
| $P$ | 0.4488 | 0.3847 | 0.1374 | 0.0262 | 0.0028 | 0.0002 | 0.0000 |
| $f$ | | | | | | | |

The bottom line becomes ...............................................

**60**

| $x$ | 0 | 1 | 2 | 3 | 4 | 5 | 6 |
|---|---|---|---|---|---|---|---|
| $f$ | 2244 | 1924 | 687 | 131 | 14 | 1 | 0 |

The total value of $f$ in this case is, of course, 5000 (to 3 sig. figs) and the total value of $P$ is ..................................

**61**

1

**Mean and standard deviation of a probability distribution**

The mean and standard deviation of an empirical probability distribution are often denoted by $m$ (or $\overline{x}$) and $s$, while those obtained from a theoretical probability distribution are denoted by the Greek letters $\mu$ and $\sigma$ respectively.

*Mean of a probability distribution*

By definition $\mu = \dfrac{\Sigma(fx)}{m}$ and the results of the last example in which $m = 5000$ give

$\mu =$ ........................................

**62**

$$\boxed{\mu = 0.75}$$

$$\text{for } \mu = \frac{\Sigma(fx)}{m} = \frac{(2244 \times 0) + (1924 \times 1) + (687 \times 2) + \ldots + (1 \times 5) + (0 \times 6)}{5000}$$

$$= \frac{0 + 1924 + 1374 + 393 + 56 + 5 + 0}{5000}$$

$$= \frac{3752}{5000} = 0.750 \quad \therefore \underline{\mu = 0.750} \text{ to 3 sig. figs}$$

$$\text{Now } f = m \times P \quad \therefore P = \frac{f}{m}$$

$$\text{Then } \mu = \frac{\Sigma(fx)}{m} = \Sigma\left(\frac{f}{m} \cdot x\right) = \Sigma(Px) \text{ since } \frac{f}{m} = P$$

$$\therefore \underline{\mu = \Sigma(Px)} \qquad (18)$$

which is simpler still – and notice, does not directly depend on the value of $m$ or on the frequencies $f$.
We had

| $x$ | 0 | 1 | 2 | 3 | 4 | 5 | 6 |
|---|---|---|---|---|---|---|---|
| $P$ | 0.4488 | 0.3847 | 0.1374 | 0.0262 | 0.0028 | 0.0002 | 0.0000 |
| $Px$ | | | | | | | |

If we now form the products $Px$ and sum the results

$\mu =$ ................................

**63**

$\boxed{\mu = 0.750}$

Values of $Px$ are

| $Px$ | 0 | 0.3847 | 0.2748 | 0.0786 | 0.0112 | 0.0010 | 0.0000 |
|---|---|---|---|---|---|---|---|

$$\therefore \mu = \Sigma(Px) = 0.7503 = \underline{0.750} \text{ to 3 sig. figs}$$

which agrees with the result of our previous calculation.

There are also two very simple and useful formulae for the mean and standard deviation of a binomial probability distribution.

$$\text{Mean} \quad \underline{\mu = np} \qquad (19)$$

$$\text{Standard deviation } \underline{\sigma = \sqrt{(npq)}} \qquad (20)$$

where $n$ = number of possible outcomes in any single trial
$p$ = probability of a success in any single trial
$q$ = probability of a failure in any single trial.

*Make a note of these for future use.*

**64**

*Example 1* In the example we have just completed, $n = 6$ and $p = \frac{1}{8}$.

Then $\mu = np = 6 \times \frac{1}{8} = 0.7500 \quad \underline{\therefore \mu = 0.750}$

which again agrees with the two previous results for the mean of the probability distribution.

Using equation (20) for the standard deviation with $n = 6, p = \frac{1}{8}$, $q = \frac{7}{8}$

$$\sigma = \sqrt{(npq)} = \sqrt{\left(6 \times \frac{1}{8} \times \frac{7}{8}\right)} = \sqrt{\left(\frac{42}{64}\right)} = 0.810 \quad \underline{\therefore \sigma = 0.810}$$

It is just as easy as that!

Now another example by way of revision.

*Example 2* Twelve per cent of a batch of transistors are defective. Determine the binomial probability distribution that a packet of five transistors will contain up to 5 defectives. Calculate the mean and standard deviation of the distribution.
When you have completed it, check with the next frame.

**65** Here is the working.

$P(\text{defective}) = p = 0.12;\quad q = 0.88;\quad n = 5;$
$x$ = number of defectives

| $x$ | 0 | 1 | 2 | 3 | 4 | 5 |
|---|---|---|---|---|---|---|
| $P$ | $q^5$<br>0.5277 | $5q^4p$<br>0.3598 | $10q^3p^2$<br>0.0981 | $10q^2p^3$<br>0.0134 | $5qp^4$<br>0.0009 | $p^5$<br>0.0000 |

(a) Mean $\mu = np = 5 \times 0.12 = 0.6$ $\qquad \underline{\mu = 0.60}$

(b) Standard deviation $\sigma = \sqrt{(npq)} = \sqrt{(5 \times 0.12 \times 0.88)}$
$= 0.528 \qquad \underline{\sigma = 0.528}$

To wind up this section of the work, complete the following formulae – without reference to your notes!

(a) ${}^nP_r$ = ......................................... = ............................
(b) ${}^nP_n$ = ............................
(c) ${}^nC_r$ = ....................................
(d) $P(r$ successes in $n$ trials$)$ = ......................................
(e) $\mu$ = ........................... = ................................
(f) $\sigma$ = ............................

**66** Here they are

(a) ${}^nP_r = n(n-1)(n-2)\ldots(n-r+1) = \dfrac{n!}{(n-r)!}$

(b) ${}^nP_n = n!$

(c) ${}^nC_r = \dfrac{n!}{(n-r)!\,r!}$

(d) $P(r \text{ successes in } n \text{ trials}) = \dfrac{n!}{(n-r)!\,r!}\,q^{n-r}p^r$

(e) $\mu = \Sigma(Px) = np$

(f) $\sigma = \sqrt{(npq)}$

*Now on to the next stage.*

## The Poisson probability distribution

**67**

Another theoretical probability distribution is given by the *Poisson probability function*

$$P(x=r)=\frac{e^{-\mu}\mu^r}{r!} \qquad (21)$$

where $\mu$ = the mean of the theoretical distribution
and $r$ = the number of successes of event $A$.

The total probability, putting $r = 0, 1, 2, \ldots$ is therefore

$$\text{total } P = \frac{e^{-\mu}\mu^0}{0!} + \frac{e^{-\mu}\mu^1}{1!} + \frac{e^{-\mu}\mu^2}{2!} + \ldots$$

$$= e^{-\mu}\left\{1 + \mu + \frac{\mu^2}{2!} + \frac{\mu^3}{3!} + \ldots\right\}$$

$$= e^{-\mu} \times e^{\mu} = e^0 = 1 \qquad \therefore \underline{\text{total } P = 1}$$

The distribution is particularly useful when $p$, the probability of success in any one trial, is small and $n$, the number of trials, is large, for then $\frac{e^{-\mu}\mu^r}{r!}$ approaches very closely to the value of ${}^nC_r q^{n-r} p^r$.
That is, the Poisson distribution approximates closely to the binomial distribution when, in practice, $n \geqslant 50$ and $p \leqslant \frac{1}{10}$.

*Example 1* A machine produces on average 2% defectives. In a random sample of 60 items, determine the probability of there being three defectives.

$$n = 60; \quad p = \frac{2}{100} = 0.02 \qquad \mu = \ldots\ldots\ldots\ldots$$

---

**68**

$$\boxed{\mu = np = 60 \times 0.02 = 1.2}$$

Then $\quad P(x=3) = \dfrac{e^{-\mu}\mu^3}{3!} = \ldots\ldots\ldots\ldots\ldots\ldots$

**69**

$$\boxed{P = 0.0867}$$

for $P = \dfrac{e^{-1.2}1.2^3}{3!} = 0.0867$

If we use the binomial expression ${}^nC_r\, q^{n-r}p^r$,

$${}^{60}C_3 q^{57}p^3 = \frac{60 \times 59 \times 58}{1 \times 2 \times 3} \times 0.98^{57} \times 0.02^3$$

and we get 0.0865, so the agreement is close.

*Move on for another example.*

---

**70** *Example 2* Items processed on a certain machine are found to be 1% defective. Determine the probabilities of obtaining 0, 1, 2, 3, 4 defectives in a random sample batch of 80 such items.

$$P(x = r) = \frac{e^{-\mu}\mu^r}{r!}$$

$\mu = np$; $n = 80$; $p = 0.01$; $\therefore \mu = 80 \times 0.01 = \underline{0.8}$

Now we can calculate the Poisson probability distribution for $r = 0$ to 4, which is ..........................................................

---

**71**

| $x$ | 0 | 1 | 2 | 3 | 4 |
|---|---|---|---|---|---|
| $P$ | 0.4493 | 0.3595 | 0.1438 | 0.0383 | 0.0077 |

For example, $P(x = 2) = \dfrac{e^{-0.8}0.8^2}{2!} = \dfrac{(0.4493)(0.64)}{2} = 0.1438$

and so for the others.

If we compare these results with the binomial distribution for the same data, we get

| $x = r$ | Probabilities | | Difference |
|---|---|---|---|
| | Binomial | Poisson | |
| 0 | 0.4475 | 0.4493 | 0.0018 |
| 1 | 0.3616 | 0.3595 | 0.0021 |
| 2 | 0.1443 | 0.1438 | 0.0005 |
| 3 | 0.0379 | 0.0383 | 0.0004 |
| 4 | 0.0074 | 0.0077 | 0.0003 |

As we see, the two sets of results agree closely. *Move on.*

**72**

*Example 3* Fifteen per cent of carbon resistors drawn from stock are, on average, outside acceptable tolerances. Determine the probabilities of obtaining 0, 1, 2, 3, 4 defectives in a random batch of 20 such resistors, expressed as (a) a binomial distribution, (b) a Poisson distribution.

There is nothing new here, so you will have no trouble.

Dealing with part (a), the binomial distribution is

..............................................................

**73**

*Binomial distribution:*

| $x$ | 0 | 1 | 2 | 3 | 4 |
|---|---|---|---|---|---|
| $P$ | 0.0388 | 0.1368 | 0.2293 | 0.2428 | 0.1821 |

for $p = 0.15$; $q = 0.85$; $n = 20$; $x$ = defectives

$$P(x = r) = \frac{n!}{(n-r)!\,r!}\,q^{n-r}p^r$$

Now for the second part (b), the Poisson distribution is

..............................................................

**74** *Poisson distribution:*

| $x$ | 0 | 1 | 2 | 3 | 4 |
|---|---|---|---|---|---|
| $P$ | 0.0498 | 0.1494 | 0.2240 | 0.2240 | 0.1680 |

for $n = 20$; $p = 0.15$; $\mu = np = 20 \times 0.15 = 3.0$

$$P(x = r) = \frac{e^{-\mu}\mu^r}{r!}$$

Comparing the two sets of results as before, we now have

| $x = r$ | Probabilities | | Difference |
|---|---|---|---|
| | Binomial | Poisson | |
| 0 | 0.0388 | 0.0498 | 0.0110 |
| 1 | 0.1368 | 0.1494 | 0.0126 |
| 2 | 0.2293 | 0.2240 | 0.0053 |
| 3 | 0.2428 | 0.2240 | 0.0188 |
| 4 | 0.1821 | 0.1680 | 0.0141 |

**75**

Finally, let us recall the results of the last two examples

| | $x$ | Binomial | Poisson | Difference | Percentage difference |
|---|---|---|---|---|---|
| Ex. 2 | 0 | 0.4475 | 0.4493 | 0.0018 | 0.4022 |
| $n = 80$ | 1 | 0.3616 | 0.3595 | 0.0021 | 0.5808 |
| $p = 0.01$ | 2 | 0.1443 | 0.1438 | 0.0005 | 0.3465 |
| | 3 | 0.0379 | 0.0383 | 0.0004 | 1.0554 |
| | 4 | 0.0074 | 0.0077 | 0.0003 | 4.0541 |
| Ex. 3 | 0 | 0.0388 | 0.0498 | 0.0110 | 28.3505 |
| $n = 20$ | 1 | 0.1368 | 0.1494 | 0.0126 | 9.2105 |
| $p = 0.15$ | 2 | 0.2293 | 0.2240 | 0.0053 | 2.3114 |
| | 3 | 0.2428 | 0.2240 | 0.0188 | 7.7430 |
| | 4 | 0.1821 | 0.1680 | 0.0141 | 7.7430 |

In example 3, the differences between the binomial and Poisson probabilities are considerably greater than in example 2 – and this is even more apparent if we compare percentage differences. The reason for this is ........................................................ (Suggestions?)

**76**

> For the Poisson distribution to represent closely the binomial distribution, the size of the sample ($n$) should be large and the probability ($p$) of obtaining a success in any one trial very small.

The basic data of these two problems differ in these respects. In practice, for the satisfactory use of the Poisson distribution, we should have $p$ ............................ and $n$ ............................................

**77**

$p \leqslant 0.1$ and $n \geqslant 50$

*Now on to the next topic.*

**78**

## Continuous probability distributions

The binomial and Poisson distributions refer to discrete events, e.g. the number of successes probable in a trial. Where continuous variables are involved, e.g. measurements of length, mass, time, etc., we are concerned with the probability that a particular dimension lies between certain limiting values of that variable and for this we refer to the *normal distribution curve*.

### Normal distribution curve (or normal curve)

We introduced the normal curve on page 833 in Programme 27 as the limiting curve to which a relative frequency polygon approaches as the number of classes is greatly increased.

From a more theoretical approach, the equation of the normal curve is, in fact,

$$y = \frac{1}{\sigma(2\pi)} e^{-\frac{1}{2}(x-\mu)^2/\sigma^2}$$

where $\mu$ = mean and $\sigma$ = standard deviation of the distribution, an equation not at all easy to deal with! In practice, it is convenient to convert a normal distribution into a standardised normal distribution having a mean of 0 and a standard deviation of 1.

**Standard normal curve** The conversion from normal distribution to standard normal distribution is achieved by the substitution $z = \dfrac{x-\mu}{\sigma}$ which effectively moves the distribution curve along the $x$-axis and reduces the scale of the horizontal units by dividing by $\sigma$. To keep the total area under the curve at unity, we multiply the $y$-values by $\sigma$. The equation of the standardised normal curve then becomes

$$y = \phi(z) = \frac{1}{\sqrt{(2\pi)}} e^{-z^2/2}$$

$z = \dfrac{x-\mu}{\sigma}$ is called the *standardised normal variate*,

$\phi(z)$ is the *probability density function*.

*Make a note of these.*

## Standard normal curve

**79**

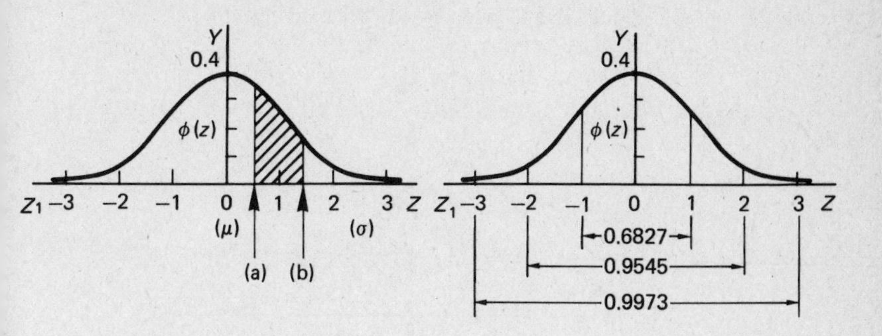


Note the following:

(i) Mean $\mu = 0$.

(ii) $z$-values are in standard deviation units.

(iii) Total area under the curve from $z = -\infty$ to $z = +\infty = 1$.

(iv) Area between $z = a$ and $z = b$ represents the probability that $z$ lies between the values $z = a$ and $z = b$, i.e.

$$P(a \leqslant z \leqslant b) = \text{area shaded.}$$

(v) The probability of a value of $z$ being
between $z = -1$ and $z = 1$ is 68.27% = 0.6827
between $z = -2$ and $z = 2$ is 95.45% = 0.9545
between $z = -3$ and $z = 3$ is 99.73% = 0.9973.

In each case the probability is given by ..........................................

**80**

the area under the curve between the stated limits.

Similarly, the probability of a randomly selected value of $z$ lying between $z = 0.5$ and $z = 1.5$ is given by the area shaded.

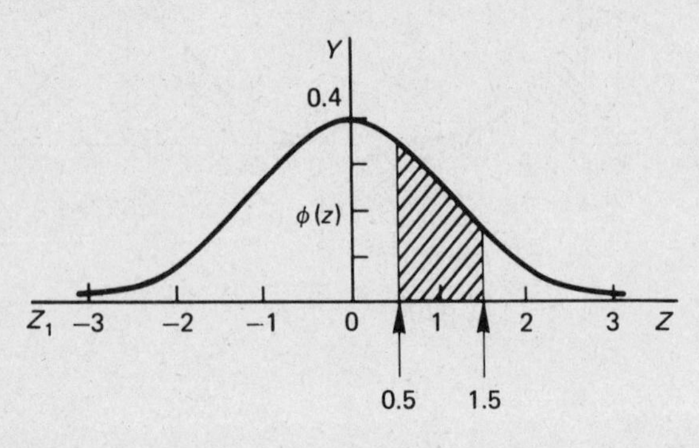


That is $$P(0.5 \leqslant z \leqslant 1.5) = \int_{0.5}^{1.5} \frac{1}{\sqrt{(2\pi)}} e^{-z^2/2} \, dz$$

This integral cannot be evaluated by ordinary means, so we use a table giving the area under the standard normal curve from $z = 0$ to $z = z_1$ as shown on the next page.

*Then move on to frame 82.*

# 81

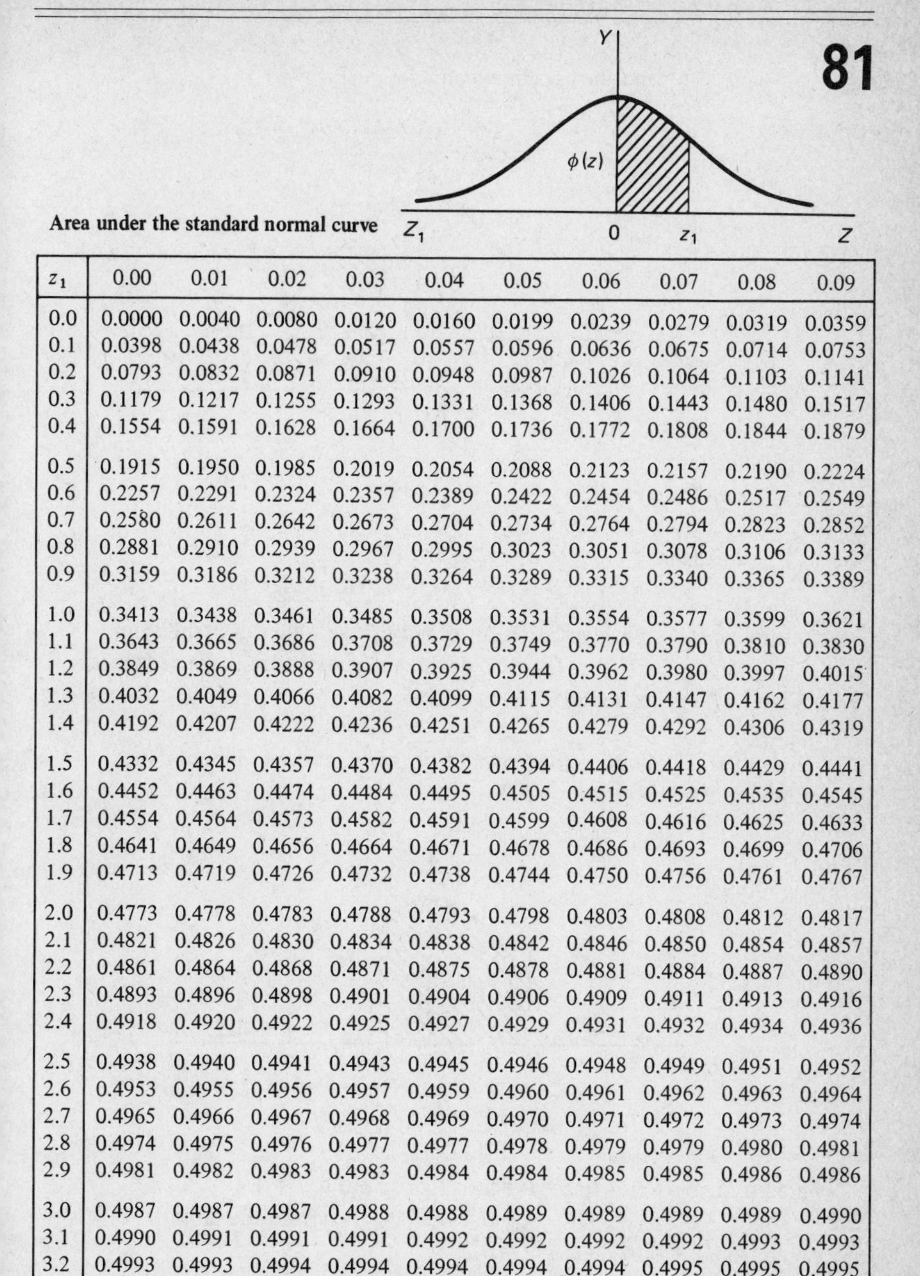


**Area under the standard normal curve**

| $z_1$ | 0.00 | 0.01 | 0.02 | 0.03 | 0.04 | 0.05 | 0.06 | 0.07 | 0.08 | 0.09 |
|---|---|---|---|---|---|---|---|---|---|---|
| 0.0 | 0.0000 | 0.0040 | 0.0080 | 0.0120 | 0.0160 | 0.0199 | 0.0239 | 0.0279 | 0.0319 | 0.0359 |
| 0.1 | 0.0398 | 0.0438 | 0.0478 | 0.0517 | 0.0557 | 0.0596 | 0.0636 | 0.0675 | 0.0714 | 0.0753 |
| 0.2 | 0.0793 | 0.0832 | 0.0871 | 0.0910 | 0.0948 | 0.0987 | 0.1026 | 0.1064 | 0.1103 | 0.1141 |
| 0.3 | 0.1179 | 0.1217 | 0.1255 | 0.1293 | 0.1331 | 0.1368 | 0.1406 | 0.1443 | 0.1480 | 0.1517 |
| 0.4 | 0.1554 | 0.1591 | 0.1628 | 0.1664 | 0.1700 | 0.1736 | 0.1772 | 0.1808 | 0.1844 | 0.1879 |
| 0.5 | 0.1915 | 0.1950 | 0.1985 | 0.2019 | 0.2054 | 0.2088 | 0.2123 | 0.2157 | 0.2190 | 0.2224 |
| 0.6 | 0.2257 | 0.2291 | 0.2324 | 0.2357 | 0.2389 | 0.2422 | 0.2454 | 0.2486 | 0.2517 | 0.2549 |
| 0.7 | 0.2580 | 0.2611 | 0.2642 | 0.2673 | 0.2704 | 0.2734 | 0.2764 | 0.2794 | 0.2823 | 0.2852 |
| 0.8 | 0.2881 | 0.2910 | 0.2939 | 0.2967 | 0.2995 | 0.3023 | 0.3051 | 0.3078 | 0.3106 | 0.3133 |
| 0.9 | 0.3159 | 0.3186 | 0.3212 | 0.3238 | 0.3264 | 0.3289 | 0.3315 | 0.3340 | 0.3365 | 0.3389 |
| 1.0 | 0.3413 | 0.3438 | 0.3461 | 0.3485 | 0.3508 | 0.3531 | 0.3554 | 0.3577 | 0.3599 | 0.3621 |
| 1.1 | 0.3643 | 0.3665 | 0.3686 | 0.3708 | 0.3729 | 0.3749 | 0.3770 | 0.3790 | 0.3810 | 0.3830 |
| 1.2 | 0.3849 | 0.3869 | 0.3888 | 0.3907 | 0.3925 | 0.3944 | 0.3962 | 0.3980 | 0.3997 | 0.4015 |
| 1.3 | 0.4032 | 0.4049 | 0.4066 | 0.4082 | 0.4099 | 0.4115 | 0.4131 | 0.4147 | 0.4162 | 0.4177 |
| 1.4 | 0.4192 | 0.4207 | 0.4222 | 0.4236 | 0.4251 | 0.4265 | 0.4279 | 0.4292 | 0.4306 | 0.4319 |
| 1.5 | 0.4332 | 0.4345 | 0.4357 | 0.4370 | 0.4382 | 0.4394 | 0.4406 | 0.4418 | 0.4429 | 0.4441 |
| 1.6 | 0.4452 | 0.4463 | 0.4474 | 0.4484 | 0.4495 | 0.4505 | 0.4515 | 0.4525 | 0.4535 | 0.4545 |
| 1.7 | 0.4554 | 0.4564 | 0.4573 | 0.4582 | 0.4591 | 0.4599 | 0.4608 | 0.4616 | 0.4625 | 0.4633 |
| 1.8 | 0.4641 | 0.4649 | 0.4656 | 0.4664 | 0.4671 | 0.4678 | 0.4686 | 0.4693 | 0.4699 | 0.4706 |
| 1.9 | 0.4713 | 0.4719 | 0.4726 | 0.4732 | 0.4738 | 0.4744 | 0.4750 | 0.4756 | 0.4761 | 0.4767 |
| 2.0 | 0.4773 | 0.4778 | 0.4783 | 0.4788 | 0.4793 | 0.4798 | 0.4803 | 0.4808 | 0.4812 | 0.4817 |
| 2.1 | 0.4821 | 0.4826 | 0.4830 | 0.4834 | 0.4838 | 0.4842 | 0.4846 | 0.4850 | 0.4854 | 0.4857 |
| 2.2 | 0.4861 | 0.4864 | 0.4868 | 0.4871 | 0.4875 | 0.4878 | 0.4881 | 0.4884 | 0.4887 | 0.4890 |
| 2.3 | 0.4893 | 0.4896 | 0.4898 | 0.4901 | 0.4904 | 0.4906 | 0.4909 | 0.4911 | 0.4913 | 0.4916 |
| 2.4 | 0.4918 | 0.4920 | 0.4922 | 0.4925 | 0.4927 | 0.4929 | 0.4931 | 0.4932 | 0.4934 | 0.4936 |
| 2.5 | 0.4938 | 0.4940 | 0.4941 | 0.4943 | 0.4945 | 0.4946 | 0.4948 | 0.4949 | 0.4951 | 0.4952 |
| 2.6 | 0.4953 | 0.4955 | 0.4956 | 0.4957 | 0.4959 | 0.4960 | 0.4961 | 0.4962 | 0.4963 | 0.4964 |
| 2.7 | 0.4965 | 0.4966 | 0.4967 | 0.4968 | 0.4969 | 0.4970 | 0.4971 | 0.4972 | 0.4973 | 0.4974 |
| 2.8 | 0.4974 | 0.4975 | 0.4976 | 0.4977 | 0.4977 | 0.4978 | 0.4979 | 0.4979 | 0.4980 | 0.4981 |
| 2.9 | 0.4981 | 0.4982 | 0.4983 | 0.4983 | 0.4984 | 0.4984 | 0.4985 | 0.4985 | 0.4986 | 0.4986 |
| 3.0 | 0.4987 | 0.4987 | 0.4987 | 0.4988 | 0.4988 | 0.4989 | 0.4989 | 0.4989 | 0.4989 | 0.4990 |
| 3.1 | 0.4990 | 0.4991 | 0.4991 | 0.4991 | 0.4992 | 0.4992 | 0.4992 | 0.4992 | 0.4993 | 0.4993 |
| 3.2 | 0.4993 | 0.4993 | 0.4994 | 0.4994 | 0.4994 | 0.4994 | 0.4994 | 0.4995 | 0.4995 | 0.4995 |

**82** We need to find the area between $z = 0.5$ and $z = 1.5$

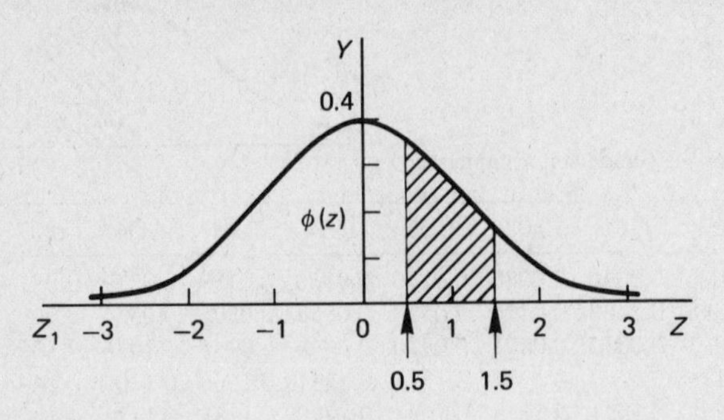


From the table

area from $z = 0$ to $z = 1.5 = 0.4332$

area from $z = 0$ to $z = 0.5 = 0.1915$

$\therefore$ area from $z = 0.5$ to $z = 1.5 = 0.2417$

$$\therefore P(0.5 \leqslant z \leqslant 1.5) = \underline{0.2417 = 24.17\%}$$

Although the table gives areas for only positive values of $z$, the symmetry of the curve enables us to deal equally well with negative values. Now for some examples.

*Example 1* Determine the probability that a random value of $z$ lies between $z = -1.4$ and $z = 0.7$.

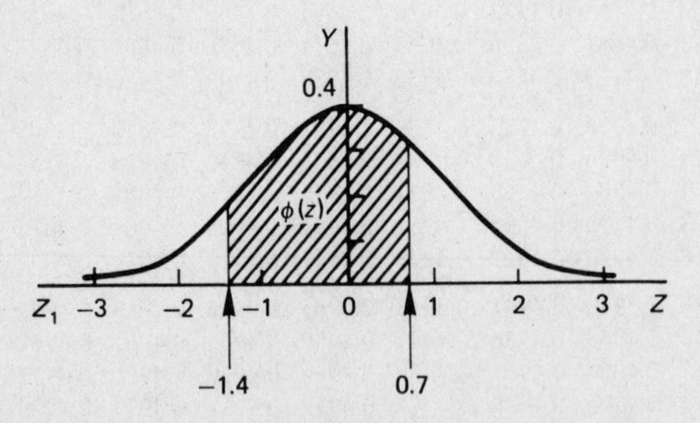


area from $z = -1.4$ to $z = 0$ = area from $z = 0$ to $z = 1.4$

$= 0.4192$ (from the table)

area from $z = 0$ to $z = 0.7 = 0.2580$

$\therefore$ area from $z = -1.4$ to $z = 0.7$ = ....................................

83

$$\boxed{0.4192 + 0.2580 = 0.6772}$$

$$\therefore P(-1.4 \leqslant z \leqslant 0.7) = \underline{0.6772 = 67.72\%}$$

*Example 2* Determine the probability that a value of $z$ is greater than 2.5

Draw a diagram: then it is easy enough. Finish it off.

84

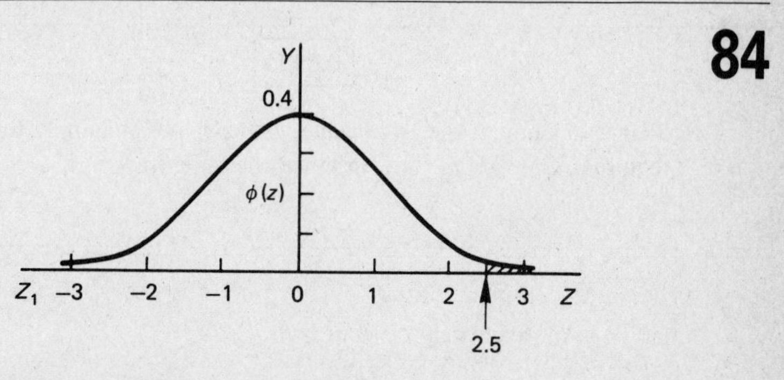


The total area from $z = 0$ to $z = \infty$ is 0.5000.
The area from $z = 0$ to $z = 2.5 = 0.4938$ (from the table).
$\therefore$ the area for $z \geqslant 2.5 = 0.5000 - 0.4938 = 0.0062$

$$\therefore P(z \geqslant 2.5) = \underline{0.0062 = 0.62\%}$$

*Example 3* The mean diameter of a sample of 400 rollers is 22.50 mm and the standard deviation 0.50 mm. Rollers are acceptable with diameters 22.36 ± 0.53 mm. Determine the probability of any one roller being within the acceptable limits.

We have $\mu = 22.50$ mm; $\sigma = 0.50$ mm

Limits of $x$ $x_1 = 22.36 - 0.53 = 21.83$ mm
$x_2 = 22.36 + 0.53 = 22.89$ mm

Using $z = \dfrac{x - \mu}{\sigma}$ we convert $x_1$ and $x_2$ into $z_1$ and $z_2$

$z_1 = $ ......................; $z_2 = $ .........................

## 85

$$\boxed{z_1 = -1.34; \; z_2 = 0.78}$$

Now, using the table, we can find the area under the normal curve between $z = -1.34$ and $z = 0.78$ which gives us the required result.

$$P(21.83 \leqslant x \leqslant 22.89) = P(-1.34 \leqslant z \leqslant 0.78) = \ldots\ldots\ldots\ldots$$

## 86

$$\boxed{0.6922}$$

So that is it. Convert the given values into $z$-values and apply the table as required. They are all done in much the same way.

## 87

Finally one more entirely on your own.

*Example 4* A thermostat set to switch at 20°C operates at a range of temperatures having a mean of 20.4°C and a standard deviation of 1.3°C. Determine the probability of its operating at temperatures between 19.5°C and 20.5°C.

Complete it and then check the working with the next frame.

**88**

$$P(19.5 \leqslant x \leqslant 20.5) = 0.2868$$

Here it is:

$$x_1 = 19.5 \quad \therefore z_1 = \frac{19.5 - 20.4}{1.3} = -\frac{0.9}{1.3} = -0.692$$

$$x_2 = 20.5 \quad \therefore z_2 = \frac{20.5 - 20.4}{1.3} = \frac{0.1}{1.3} = 0.077$$

Area ($z = -0.69$ to $z = 0$) = area ($z = 0$ to $z = 0.69$)
= 0.2549
area ($z = 0$ to $z = 0.08$) = 0.0319
$\therefore$ Area ($z = -0.69$ to $z = 0.08$) = 0.2868
$\therefore P(19.5 \leqslant x \leqslant 20.5) = P(-0.692 \leqslant z \leqslant 0.077) = \underline{0.2868}$

That brings us to the end of this introduction to probability.
Check down the Revision Summary before dealing with the Test Exercise.
There are no tricks so you will have no difficulty.

89

## Revision Summary

1. *Types of probability:* empirical (experimental); classical (theoretical).
2. *Expectation:* $E = m \times P(A)$ $\quad m$ = number of trials
3. *Complement* of $\{A\} = \{\text{not } A\} = \{\bar{A}\}$
4. $P(A) + P(\bar{A}) = 1$
5. *Classical:* $P(A) = \dfrac{\text{number of ways in which event } A \text{ can occur}}{\text{total number of all possible outcomes}}$
6. $P(\text{certainty}) = 1$; $P(\text{impossibility}) = 0$
7. *Addition law:*
   (a) mutually exclusive events, $P(A \text{ or } B) = P(A) + P(B)$
   (b) non-exclusive events, $P(A \text{ or } B) = P(A) + P(B) - P(A \text{ and } B)$
8. *Multiplication law:*
   (a) independent events: $P(A \text{ and } B) = P(A) \times P(B)$
   (b) dependent events: $P(A \text{ and } B) = P(A) \times P(B \mid A)$
9. *Permutations:* (a) ${}^nP_r = n(n-1)(n-2)\ldots(n-r+1 = \dfrac{n!}{(n-r)!}$
   (b) ${}^nP_n = n!$
   *Combinations:* ${}^nC_r = \dfrac{n!}{(n-r)!\,r!} = \dbinom{n}{r}$
10. *Binomial distribution*
    $P(r \text{ successes in } n \text{ trials}) = \dfrac{n!}{(n-r)!\,r!}\,q^{n-r}p^r$
11. *Probability distribution:* $\mu = \dfrac{\Sigma(fx)}{m} = \Sigma(Px) = np$
    $\sigma = \sqrt{(npq)}$
12. *Poisson distribution:* $P(x = r) = \dfrac{e^{-\mu}\mu^r}{r!}$
13. *Standard normal curve:* $y = \phi(z) = \dfrac{1}{\sqrt{(2\pi)}}\,e^{-z^2/2}$

    $z = \dfrac{x-\mu}{\sigma}$ = *standard normal variate*

    $\phi(z)$ = *probability density function.*

    $P(a \leqslant z \leqslant b)$ = area under the standard normal curve between $z = a$ and $z = b$.

## Text Exercise – XXVIII

**90**

1. Twelve per cent of a type of plastic bushes are rejects. Determine
   (a) the probability that any one item drawn at random is (i) defective, (ii) acceptable
   (b) the number of acceptable bushes likely to be found in a sample batch of 4000.

2. A box contains 12 transistors of type $A$ and 18 of type $B$, all identical in appearance. If one transistor is taken at random, tested and returned to the box, and a second transistor then treated in the same manner, determine the probability that
   (a) the first is type $A$ and the second type $B$,
   (b) both are of type $A$, (c) neither is of type $A$.

3. A packet contains 100 washers, 24 of which are brass, 36 copper and the remainder steel. One washer is taken at random, retained, and a second washer similarly drawn. Determine the probability that
   (a) both washers are steel,
   (b) the first is brass and the second copper,
   (c) one is brass and one is steel.

4. A large stock of resistors is known to have 20% defectives. If 5 resistors are drawn at random, determine (a) the probabilities that (i) none is defective, (ii) at least two are defective,
   (b) the mean and standard deviation of the distribution.

5. A firm, on average, receives 4 enquiries per week relating to its new product. Determine the probability that the number of enquiries in any one week will be (a) none, (b) two, (c) 3 or more.

6. A machine delivers rods having a mean length of 18.0 mm and a standard deviation of 1.33 mm. If the lengths are normally distributed, determine the number of rods between 16.0 mm and 19.0 mm long likely to occur in a run of 300.

## Further Problems – XXVIII

*Binomial distribution*

1. A box contains a large number of transistors, 30% of which are type $A$ and the rest type $B$. A random sample of 4 transistors is taken. Determine the probabilities that they are
   (a) all of type $A$,
   (b) all of type $B$,
   (c) two of type $A$ and two of type $B$,
   (d) three of type $A$ and one of type $B$.

2. A milling machine produces products with an average of 4% rejects. If a random sample of 5 components is taken, determine the probability that it contains
   (a) no reject,
   (b) less than 2 rejects.

3. If 12% of resistors produced in a run are defective, determine the probability distribution of defectives in a random sample of 5 resistors.

4. Production of steel rollers includes, on average, 8% defectives. Determine the probability that a random sample of 6 rollers contains
   (a) 2 defectives,
   (b) less than 3 defectives.

5. A machine produces, on average, 95% of mouldings within tolerance values. Determine the probability that a random sample of 5 mouldings shall contain
   (a) no defective,
   (b) more than one defective.

6. A large stock of resistors has 80% within tolerance values. If 7 resistors are drawn at random, determine the probability that
   (a) at least 5 are acceptable,
   (b) all 7 are acceptable.

7. Twenty per cent of the output from a production run are rejects. In a random sample of 5 items, determine the probability of there being
   (a) 0, 1, 2, 3, 4, 5 rejects,

(b) more than 1 reject,
(c) less than 4 rejects.

8. A production line produces 6% defectives. For a random sample of 10 components, determine the probability of obtaining
   (a) no defective,
   (b) 2 defectives,
   (c) more than 3 defectives.

*Poisson distribution*

9. Small metal springs are packed in boxes of 100 and 0.5% of the total output of springs are defective. Determine the probability that any one box chosen at random shall have
   (a) no defective,
   (b) 2 or more defectives.

10. In a long production run, 1% of the components are normally found to be defective. In a random sample of 10 components, determine the probability that there will be fewer than 2 defectives in the sample.

11. Three per cent of stampings are rejects. For a production run of 400, determine
    (a) the mean and standard deviation of the probability distribution,
    (b) the probability of obtaining 10 rejects.

12. If 2% of a certain brand of light bulbs are defective, find the probability that, in a random sample of 80 bulbs, a total of 0, 1, 2, 3, and more than 5 will be defective.

13. Brass terminals are packed in boxes of 200. If the production process is known to produce 1.5% defectives on average, determine the probability that a box chosen at random will contain
    (a) no defective,
    (b) 1 defective,
    (c) 2 defectives,
    (d) 3 defectives,
    (e) more than 3 defectives.

14. A product output is known to be 1% defective. In a random sample of 400 components, determine the probability of including
    (a) 2 or less defectives,
    (b) 7 or more defectives.

15. If 4% of the total output of panels are sub-standard, find the probability of obtaining 0, 1, 2, 3, 4-or-more defectives in a random sample batch of 60 panels.

16. Six per cent, on average, of manufactured spindles are outside stated tolerances. The spindles are randomly packed in boxes of 50 and dispatched in cartons of 100 boxes. For any one carton, determine the number of boxes likely to contain 0, 1, 2, 3, 4, 5-or-more defectives.

*Normal distribution*

17. Boxes of screws, nominally containing 250, have a mean content of 248 screws with a standard deviation of 7. If the contents are normally distributed, determine the probability that a randomly chosen box will contain fewer than 240 screws.

18. Samples of 10 A fuses have a mean fusing current of 9.9 A and a standard deviation of 1.2 A. Determine the probability of a fuse blowing with a current
    (a) less than 7.0 A,
    (b) between 8.0 A and 12.0 A.

19. Resistors of a certain type have a mean resistance of 420 Ω with a standard deviation of 12 Ω. Determine the percentage of resistors having resistance values
    (a) between 400 Ω and 430 Ω,
    (b) equal to 450 Ω.

20. Washers formed on a machine have a mean diameter of 12.60 mm with a standard deviation of 0.52 mm. Determine the number of washers in a random sample of 400 likely to have diameters between 12.00 mm and 13.50 mm.

21. The life of a drill bit has a mean of 16 hours and a standard deviation of 2.6 hours. Assuming a normal distribution, determine the probability of a sample bit lasting for
    (a) more than 20 hours,
    (b) less than 14 hours.

22. A type of bearing has an average life of 1500 hours and a standard deviation of 40 hours. Assuming a normal distribution, determine the number of bearings in a batch of 1200 likely
    (a) to fail before 1400 hours,
    (b) to last for more than 1550 hours.

23. Telephone calls from an office are monitored and found to have a mean duration of 452 s and a standard deviation of 123 s. Determine
    (a) the probability of the length of a call being between 300 s and 480 s,
    (b) the proportion of calls likely to last for more than 720 s.

24. Light bulbs, having a mean life of 2400 hours and standard deviation of 62 hours, are used for a consignment of 4000 bulbs. Determine
    (a) the number of bulbs likely to have a life in excess of 2500 hours,
    (b) the percentage of bulbs with a life length between 2300 hours and 2500 hours,
    (c) the probability of any one bulb having a life of 2500 hours (to the nearest hour).

# ANSWERS

# ANSWERS

## Test Exercise I (page 32)

1. (i) $-j$, (ii) $j$, (iii) $1$, (iv) $-1$
2. (i) $29 - j2$, (ii) $-j2$, (iii) $111 + j56$, (iv) $1 + j2$
3. (i) $5{\cdot}831\ \angle 59^\circ 3'$, (ii) $6{\cdot}708\ \angle 153^\circ 26'$, (iii) $6{\cdot}403\ \angle 231^\circ 24'$
4. (i) $-3{\cdot}5355(1 + j)$, (ii) $3{\cdot}464 - j2$
5. $x = 10{\cdot}5$, $y = 4{\cdot}3$
6. (i) $10\, e^{j0{\cdot}650}$, (ii) $10\, e^{-j0{\cdot}650}$; $2{\cdot}303 + j0{\cdot}650$, $2{\cdot}303 - j0{\cdot}650$
7. $j e$

## Further Problems I (page 33)

1. (i) $115 + j133$, (ii) $2{\cdot}52 + j0{\cdot}64$, (iii) $\cos 2x + j \sin 2x$
2. $(22 - j75)/41$
3. $0{\cdot}35 + j0{\cdot}17$
4. $0{\cdot}7, 0{\cdot}9$
5. $-24{\cdot}4 + j22{\cdot}8$
6. $1{\cdot}2 + j1{\cdot}6$
8. $x = 18, y = 1$
9. $a = 2, b = -20$
10. $x = \pm 2, y = \pm 3/2$
12. $a = 1{\cdot}5, b = -2{\cdot}5$
13. $\sqrt{2}\, e^{j2{\cdot}3562}$
14. $2{\cdot}6$
16. $R = (R_2 C_3 - R_1 C_4)/C_4$; $L = R_2 R_4 C_3$
18. $E = (1811 + j1124)/34$
20. $2 + j3, -2 + j3$

## Test Exercise II (page 67)

1. $5{\cdot}831\ \underline{|210^\circ 58'}$
2. (i) $-1{\cdot}827 + j0{\cdot}813$, (ii) $3{\cdot}993 - j3{\cdot}009$
3. (i) $36\ \underline{|197^\circ}$, (ii) $4\ \underline{|53^\circ}$
4. $8\ \underline{|75^\circ}$
5. $2\ \underline{|88^\circ}$, $2\ \underline{|208^\circ}$, $2\ \underline{|328^\circ}$; p.r. $= 2\ \underline{|328^\circ}$

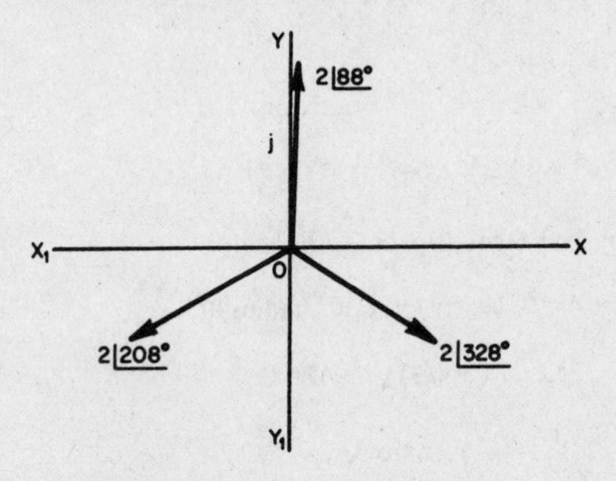


6. $\sin 4\theta = 4 \sin\theta \cos\theta - 8 \sin^3\theta \cos\theta$
7. $\cos^4\theta = \dfrac{1}{8}[\cos 4\theta + 4\cos 2\theta + 3]$
8. (i) $x^2 + y^2 - 8x + 7 = 0$
   (ii) $y = \dfrac{x+2}{\sqrt{3}}$

## Further Problems II (page 68)

1. $x = 0{\cdot}27$, $y = 0{\cdot}53$
2. $-3 + j\sqrt{3}$; $-j2\sqrt{3}$
3. $3{\cdot}606\ \underline{|56^\circ 19'}$, $2{\cdot}236\ \underline{|296^\circ 34'}$; $24{\cdot}2 - j71{\cdot}6$; $75{\cdot}6e^{-j1{\cdot}244}$
4. $1{\cdot}336\,(\underline{|27^\circ}, \underline{|99^\circ}, \underline{|171^\circ}, \underline{|243^\circ}, \underline{|315^\circ})$
   $1{\cdot}336\,(e^{j0{\cdot}4712}, e^{j1{\cdot}7279}, e^{j2{\cdot}9845}, e^{-j2{\cdot}0420}, e^{-j0{\cdot}7854})$

5. $2{\cdot}173 + j0{\cdot}899,\ 2{\cdot}351\ e^{j0{\cdot}392}$
6. $\sqrt{2}(1+j),\ \sqrt{2}(-1+j),\ \sqrt{2}(-1-j),\ \sqrt{2}(1-j)$
7. $1\underline{|36^\circ},\ 1\underline{|108^\circ},\ 1\underline{|180^\circ},\ 1\underline{|252^\circ},\ 1\underline{|324^\circ};\ e^{j0{\cdot}6283}$
8. $x = -4$ and $x = 2 \pm j3{\cdot}464$
9. $1\underline{|102^\circ 18'},\ 1\underline{|222^\circ 18'},\ 1\underline{|342^\circ 18'};\ 0{\cdot}953 - j0{\cdot}304$
11. $1{\cdot}401\ (\underline{|58^\circ 22'},\ \underline{|130^\circ 22'},\ \underline{|202^\circ 22'},\ \underline{|274^\circ 22'},\ \underline{|346^\circ 22'});$
    p.r. $= 1{\cdot}36 - j0{\cdot}33 = 1{\cdot}401\ e^{-j0{\cdot}2379}$
12. $-0{\cdot}36 + j0{\cdot}55,\ -1{\cdot}64 - j2{\cdot}55$
13. $-je$, i.e. $-j2{\cdot}718$
14. $\sin 7\theta = 7s - 56s^3 + 112s^5 - 64s^7\ (s \equiv \sin\theta)$
15. $\frac{1}{32}[10 - 15\cos 2x + 6\cos 4x - \cos 6x]$
16. $x^2 + y^2 + \frac{20}{3}x + 4 = 0$; centre $\left(-\frac{10}{3}, 0\right)$, radius $8/3$
17. $x^2 + y^2 - (1+\sqrt{3})x - (1+\sqrt{3})y + \sqrt{3} = 0,$
    centre $\left(\frac{1+\sqrt{3}}{2}, \frac{1+\sqrt{3}}{2}\right)$, radius $\sqrt{2}$
18. $x^2 + y^2 = 16$
19. (i) $2x^2 + 2y^2 - x - 1 = 0$, (ii) $x^2 + y^2 + 2x + 2y = 0$
20. (i) $x^2 + y^2 - 4x = 0$, (ii) $x^2 + y^2 + x - 2 = 0$
22. (i) $y = 3$, (ii) $x^2 + y^2 = 4k^2$

**Test Exercise III (page 97)**

1. 67·25
2. 19·40

3.

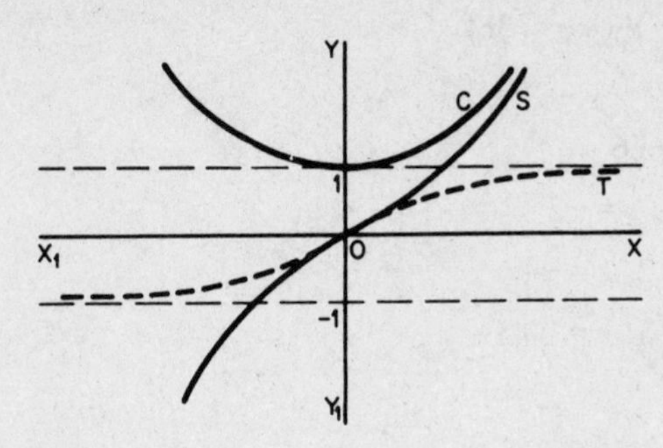


4. $-\coth A$

5. (i) 1·2125, (ii) $\pm$ 0·6931

6. $x = 0{\cdot}3466$

7. (i) $y = 224$, (ii) $x = \pm\, 48{\cdot}12$

8. $\sin x \cosh y - \mathrm{j} \cos x \sinh y$

**Further Problems III (page 98)**

2. $x = 0$, $x = 0{\cdot}549$

5. (i) 0·9731, (ii) 1·317

7. (i) $0{\cdot}9895 + \mathrm{j}0{\cdot}2498$, (ii) $0{\cdot}3210 + \mathrm{j}0{\cdot}3455$

10. $x = 0$, $x = \frac{1}{2} \ln 2$

12. $x = 0{\cdot}3677$ or $-1{\cdot}0986$

14. $1{\cdot}528 + \mathrm{j}0{\cdot}427$

18. 1·007

**Test Exercise IV (page 135)**

1. (a) 4, (b) 18

2. Equations not independent

3. $x = 3$, $y = -2$, $z = -1$

4. $k = 3$ or $-25$

5. $x = 3$, 1·654, $-6{\cdot}654$

**Further Problems IV (page 136)**

1. (i) 144, (ii) 0
2. (i) 0, (ii) 666
3. $x = 5,\ y = 4,\ z = -2$
4. $x = 2{\cdot}5,\ y = 3,\ z = -4$
5. $x = 2,\ y = 1{\cdot}5,\ z = -3{\cdot}5$
6. 4 or $-14$
7. 5 or $-2{\cdot}7$
8. (a) 0 or $\pm\sqrt{2}$, (b) $(a - b)(b - c)(c - a)(a + b + c)$
9. $x = 1$ or $x = -5 \pm \sqrt{34}$
10. $x = -1{\cdot}5$
11. $-2(a - b)(b - c)(c - a)(a + b + c)$
12. $i_2 = 5{\cdot}2$
13. $(a + b + c)^2\,(a - b)(b - c)(c - a)$
14. 2 or $-16/3$
15. $(x - y)(y - z)(z - x)(x + y + z)$
16. $x = -3$ or $\pm\sqrt{3}$
17. $x = \dfrac{(2M_1 + M_2)W}{M_1(M_1 + 2M_2)}$
18. $i_1 = 0,\ i_2 = 2,\ i_3 = 3$

20. $\theta = \dfrac{7\pi}{12}$, or $\dfrac{11\pi}{12}$

**Test Exercise – V (page 182)**

1. (a) $\begin{bmatrix} 5 & 9 & 8 & 10 \\ 10 & 8 & 6 & 7 \end{bmatrix}$ (b) $\begin{bmatrix} -1 & -1 & 4 & -4 \\ -8 & 6 & -6 & 1 \end{bmatrix}$
2. (a) $\begin{bmatrix} 18 & 0 & 12 \\ 3 & 15 & -9 \end{bmatrix}$ (b) $\begin{bmatrix} -4 & 82 \\ 54 & -12 \end{bmatrix}$ (c) $\begin{bmatrix} 21 & 43 & -19 \\ 48 & 0 & 32 \\ -17 & 35 & -37 \end{bmatrix}$

3. $\mathbf{A}^T = \begin{bmatrix} 2 & 1 & 8 \\ 3 & 7 & 0 \\ 5 & 4 & 6 \end{bmatrix}$ $\mathbf{A}^T \cdot \mathbf{I} = \begin{bmatrix} 2 & 1 & 8 \\ 3 & 7 & 0 \\ 5 & 4 & 6 \end{bmatrix}$ i.e. $\mathbf{A}^T$

4. $|a| = 0$

5. (a) 105; (b) $\begin{bmatrix} -35 & -21 & 14 \\ 5 & -3 & 13 \\ 40 & -3 & -22 \end{bmatrix}$

6. $\begin{bmatrix} 7.5 & -1.5 & -4.75 \\ -4 & 1 & 2.75 \\ -2.5 & 0.5 & 1.75 \end{bmatrix}$

7. $\begin{bmatrix} 2 & 4 & -5 \\ 1 & -3 & 1 \\ 3 & 5 & 3 \end{bmatrix} \cdot \begin{bmatrix} x_1 \\ x_2 \\ x_3 \end{bmatrix} = \begin{bmatrix} -7 \\ 10 \\ 2 \end{bmatrix}$

8. $x_1 = 2;\ x_2 = -3;\ x_3 = 5$

9. $x_1 = -1;\ x_2 = -3;\ x_3 = 4$

10. $\lambda_1 = 1 \quad \mathbf{x}_1 = \{-2 \quad 1 \quad 0\}$

    $\lambda_2 = 2 \quad \mathbf{x}_2 = \{-2 \quad 1 \quad 1\}$

    $\lambda_3 = 4 \quad \mathbf{x}_3 = \{\ 0 \quad 1 \quad 1\}$

**Further Problems – V (page 184)**

1. (a) $\begin{bmatrix} 11 & 8 \\ 8 & 9 \end{bmatrix}$ (b) $\begin{bmatrix} 3 & -4 \\ -1 & -7 \end{bmatrix}$ (c) $\begin{bmatrix} 38 & 58 \\ 17 & 26 \end{bmatrix}$ (d) $\begin{bmatrix} 46 & 14 \\ 59 & 18 \end{bmatrix}$

2. (a) **C**; (b) **A**; (c) **B**; (d) **–I**

3. (a) $\begin{bmatrix} -3 & 2 \\ 2 & -1 \end{bmatrix}$ (b) $\begin{bmatrix} 2 & 3.5 \\ 0.7 & 1.3 \end{bmatrix}$ (c) $\begin{bmatrix} -2 & -1.3 \\ 1.5 & 0.9 \end{bmatrix}$

4. $k = -2$

5\. (a) $\begin{bmatrix} 2 & -3 & -1 \\ 1 & 4 & 2 \\ 1 & -1 & 1 \end{bmatrix} \cdot \begin{bmatrix} x_1 \\ x_2 \\ x_3 \end{bmatrix} = \begin{bmatrix} 2 \\ 3 \\ 5 \end{bmatrix}$

(b) $\begin{bmatrix} 1 & -2 & -1 & 3 \\ 2 & 3 & 0 & 1 \\ 1 & 0 & -4 & -2 \\ 0 & -1 & 3 & 1 \end{bmatrix} \cdot \begin{bmatrix} x_1 \\ x_2 \\ x_3 \\ x_4 \end{bmatrix} = \begin{bmatrix} 10 \\ 8 \\ 3 \\ -7 \end{bmatrix}$

6\. $i_1 = 2;\ \ i_2 = 3;\ \ i_3 = 1$

7\. $x = 1;\ \ y = 2;\ \ z = -1$

8\. $i_1 = 2;\ \ i_2 = 1;\ \ i_3 = 0$

9\. No unique solution

10\. $x_1 = 2;\ \ x_2 = 1.5;\ \ x_3 = -3.5$

11\. $i_1 = 0.5;\ \ i_2 = 1.5;\ \ i_3 = 1.0$

12\. $i_1 = 3.0;\ \ i_2 = 2.5;\ \ i_3 = -4.0$

13\. $i_1 = 2.26;\ \ i_2 = 0.96;\ \ i_3 = 0.41$

14\. $i_1 = 12.5;\ \ i_2 = 7.5;\ \ i_3 = -20$

15\. $i_1 = \dfrac{(Z_2 + Z_3)V}{Z};\ \ i_2 = \dfrac{Z_3 V}{Z};\ \ i_3 = \dfrac{Z_2 V}{Z}$

where $Z = Z_1 Z_2 + Z_2 Z_3 + Z_3 Z_1$

16\. $\lambda_1 = 1 \quad \mathbf{x}_1 = \{0 \quad 1 \quad -1\}$

$\lambda_2 = 2 \quad \mathbf{x}_2 = \{1 \quad 1 \quad -1\}$

$\lambda_3 = 4 \quad \mathbf{x}_3 = \{3 \quad 5 \quad 1\}$

17\. $\lambda_1 = -1 \quad \mathbf{x}_1 = \{1 \quad 0 \quad -1\}$

$\lambda_2 = 1 \quad \mathbf{x}_2 = \{1 \quad -1 \quad 1\}$

$\lambda_3 = 5 \quad \mathbf{x}_3 = \{1 \quad 1 \quad 1\}$

18\. $\lambda_1 = 1 \quad \mathbf{x}_1 = \{1 \quad 0 \quad -1\}$

$\lambda_2 = 2 \quad \mathbf{x}_2 = \{2 \quad 1 \quad 0\}$

$\lambda_3 = 3 \quad \mathbf{x}_3 = \{1 \quad 2 \quad 1\}$

19. $\lambda_1 = 1 \quad x_1 = \{4 \quad 1 \quad -2\}$
 $\lambda_2 = 3 \quad x_2 = \{-2 \quad 1 \quad 0\}$
 $\lambda_3 = 4 \quad x_3 = \{-2 \quad 1 \quad 1\}$
20. $\lambda_1 = -2 \quad x_1 = \{3 \quad 3 \quad -5\}$
 $\lambda_2 = 3 \quad x_2 = \{3 \quad -2 \quad 0\}$
 $\lambda_3 = 6 \quad x_3 = \{1 \quad 1 \quad 1\}$

**Test Exercise VI (page 215)**

1. $2i - 5j, -4i + j, 2i + 4j$; $AB = \sqrt{29}$, $BC = \sqrt{17}$, $CA = \sqrt{20}$
2. (i) $-8$, (ii) $-2i - 7j - 18k$
3. $(0{\cdot}2308,\ 0{\cdot}3077,\ 0{\cdot}9230)$
4. (i) 6, $\theta = 82°44'$; (ii) $47{\cdot}05$, $\theta = 19°31'$

**Further Problems VI (page 216)**

1. $\overline{OG} = \frac{1}{3}(10i + 2j)$
2. $\frac{1}{\sqrt{50}}(3, 4, 5)$; $\frac{1}{\sqrt{14}}(1, 2, -3)$; $\theta = 80°5'$
3. Moduli: $\sqrt{74}$, $3\sqrt{10}$, $2\sqrt{46}$; D.C's: $\frac{1}{\sqrt{74}}(3, 7, -4)$,
 $\frac{1}{3\sqrt{10}}(1, -5, -8)$, $\frac{1}{\sqrt{46}}(6, -2, 12)$; Sum $= 10i$
4. 8, $17i - 7j + 2k$, $\theta = 66°36'$
5. (i) $-7$, (ii) $7(i - j - k)$, (iii) $\cos\theta = -0{\cdot}5$
6. $\cos\theta = -0{\cdot}4768$
7. (i) 7, $5i - 3j - k$; (ii) 8, $11i + 18j - 19k$
8. $-\frac{3}{\sqrt{155}}i + \frac{5}{\sqrt{155}}j + \frac{11}{\sqrt{155}}k$; $\sin\theta = 0{\cdot}997$
9. $\frac{2}{\sqrt{13}}, \frac{-3}{\sqrt{13}}, 0$; $\frac{5}{\sqrt{30}}, \frac{1}{\sqrt{30}}, \frac{-2}{\sqrt{30}}$

10. $6\sqrt{5}$; $\frac{-2}{3\sqrt{5}}, \frac{4}{3\sqrt{5}}, \frac{5}{3\sqrt{5}}$

11. (i) 0, $\theta = 90°$; (ii) 68·53, (−0·1459, −0·5982, −0·7879)

12. $4i - 5j + 11k$; $\frac{1}{9\sqrt{2}}(4, -5, 11)$

13. (i) $i + 3j - 7k$, (ii) $-4i + j + 2k$ (iii) $13(i + 2j + k)$,
(iv) $\frac{\sqrt{6}}{6}(i + 2j + k)$

**Test Exercise VII (page 239)**

1. (i) $2 \sec^2 2x$, (ii) $30(5x + 3)^5$, (iii) $\sinh 2x$,
(iv) $\frac{2x - 3}{(x^2 - 3x - 1) \ln 10}$, (v) $-3 \tan 3x$,
(vi) $12 \sin^2 4x \cos 4x$, (vii) $e^{2x}(3 \cos 3x + 2 \sin 3x)$,
(viii) $\frac{2x^3(x + 2)}{(x + 1)^3}$, (ix) $\frac{e^{4x} \sin x}{x \cos 2x}\left[4 + \cot x - \frac{1}{x} + 2 \tan 2x\right]$

2. $\frac{3}{4}, -\frac{25}{64}$

3. $-\frac{3x^2 + 4y^2}{3y^2 + 8xy}$

4. $\tan\frac{\theta}{2}$, $1\Big/\left(12 \sin\frac{\theta}{2} \cos^3\frac{\theta}{2}\right)$

**Further Problems VII (page 240)**

1. (i) $\frac{2}{\cos 2x}$, (ii) $\sec x$, (iii) $4 \cos^4 x \sin^3 x - 3 \cos^2 x \sin^5 x$

2. (i) $\frac{x \sin x}{1 + \cos x}\left[\frac{1}{x} + \cot x + \frac{\sin x}{1 + \cos x}\right]$, (ii) $\frac{-4x}{1 - x^4}$

4. $\frac{y^2 - x^2}{y^2 - 2xy}$

5. (i) $5 \sin 10x\, e^{\sin^2 5x}$, (ii) $\frac{2}{\sinh x}$, (iii) $\frac{x^2 - 1}{x^2 - 4}$

6. (i) $2x\cos^2 x - 2x^2 \sin x \cos x$, (ii) $\dfrac{2}{x} - \dfrac{x}{1-x^2}$,

   (iii) $\dfrac{e^{2x}\ln x}{(x-1)^3}\left[2 + \dfrac{1}{x\ln x} - \dfrac{3}{x-1}\right]$

8. $-4, -42$

12. $-\dfrac{1}{\sqrt{3}}, -\dfrac{8\sqrt{3}}{9}$; $x^2 + y^2 - 2y = 0$

14. $-\tan\theta$; $\dfrac{1}{3a\sin\theta\cos^4\theta}$

15. $-\cot^3\theta$; $-\cot^2\theta\,\text{cosec}^5\theta$

**Test Exercise VIII (page 265)**

1. $\theta = 37°46'$
2. $16y + 5x = 94$, $5y = 16x - 76$
3. $y = x$
4. $y = 2{\cdot}598x - 3{\cdot}849$
5. $R = 477$; C: $(-470, 50{\cdot}2)$
6. $R = 5{\cdot}59$; C: $(-3{\cdot}5, 2{\cdot}75)$

**Further Problems VIII (page 266)**

1. $20y = 125x - 363$; $y = 2x$
2. $y + 2x = 2$; $2y = x + 4$; $x = 1, y = 0$
3. $\dfrac{x\cos\theta}{13} + \dfrac{y\sin\theta}{5} = 1$; $5y = 13\tan\theta.x - 144\sin\theta$; ON.OT = 144
4. $\dfrac{3x+y}{x+3y}$; $3y + 5x = 14$
5. $R = {y_1}^2/c$
6. $5y + 8x = 43$
7. $a^2\cos^3 t\sin t$
8. $\dfrac{2a^2 - b^2}{a}$; $b$

9. (i) $y = x$; $y = -x$, (ii) $R = -\sqrt{2}$, (iii) $(1, -1)$
10. (i) $R = -6{\cdot}25$; C: $(0, -2{\cdot}25)$
    (ii) $R = 1$; C: $(2, 0)$
    (iii) $R = -11{\cdot}68$; C: $(12{\cdot}26, -6{\cdot}5)$
11. $R = -0{\cdot}177$
14. $R = 2{\cdot}744$
17. $\rho = t$; $(h, k) = (\cos t, \sin t)$
18. $R = -10{\cdot}54$, C: $(11, -3{\cdot}33)$
20. (i) $y = \pm \dfrac{x}{\sqrt{2}}$, (iii) $R = 0{\cdot}5$

**Test Exercise IX (page 294)**

1. (i) $130°$, (ii) $-37°$
2. (i) $\dfrac{3}{\sqrt{(-9x^2 - 12x - 3)}}$, (ii) $\dfrac{-1}{x\sqrt{(1-x^2)}} - \dfrac{\cos^{-1} x}{x^2}$,

   (iii) $\dfrac{2x^2}{4 + x^2} + 2x \tan^{-1}\left(\dfrac{x}{2}\right)$, (iv) $\dfrac{-3}{\sqrt{(9x^2 - 6x)}}$,

   (v) $\dfrac{-\sin x}{\sqrt{(\cos^2 x + 1)}}$, (vi) $\dfrac{5}{1 - 25x^2}$
3. (i) $y_{max} = 10$ at $x = 1$; $y_{min} = 6$ at $x = 3$; P of I at $(2, 8)$
   (ii) $y_{max} = -2$ at $x = -1$; $y_{min} = 2$ at $x = 1$
   (iii) $y_{max} = e^{-1} = 0{\cdot}3679$ at $x = 1$; P of I at $(2, 0{\cdot}271)$

**Further Problems IX (page 295)**

1. (i) 1, (ii) $2\sqrt{(1 - x^2)}$
3. (i) $\dfrac{2}{\sqrt{x}(1 + 4x)}$, (ii) $\dfrac{2}{1 - x^2}$
4. (i) $\left(\dfrac{11}{3}, -\dfrac{250}{27}\right)$; (ii) $(-0{\cdot}25, -4{\cdot}375)$
5. $y_{max} = 0$ at $x = \dfrac{1}{3}$; $y_{min} = 4$ at $x = 1$

6. $y_{max}$ at $x = 2$; $y_{min}$ at $x = 3$; P of I at $x = \sqrt{6}$

7. $y_{max} = \frac{16}{5}$ at $x = -\frac{11}{5}$; $y_{min} = 0$ at $x = 1$

8. $x = 1{\cdot}5$

10. $\frac{dy}{dx} = \sqrt{2}.e^{-x} \cos\left(x + \frac{\pi}{4}\right)$

11. (i) $y_{max}$ at $\left(\frac{2}{3}, \frac{1}{27}\right)$, $y_{min}$ at $(1, 0)$; P of I at $\left(\frac{5}{6}, \frac{1}{54}\right)$

(ii) $y_{max}$ at $(2 - \sqrt{2}, 3 - 2\sqrt{2})$; $y_{min}$ at $(2 + \sqrt{2}, 3 + 2\sqrt{2})$

(iii) P of I at $(n\pi, n\pi)$

12. (i) $\pm 0{\cdot}7071$, (ii) 0, (iii) $\pm 1{\cdot}29$

13. 0·606

14. $v = \sqrt{\frac{gT}{3w}}$

16. $y_{max} = 0{\cdot}514$

17. 17·46 cm

18. $\theta = 77°$

20. A = C, B = 0

**Test Exercise X (page 320)**

1. (i) $\frac{\partial z}{\partial x} = 12x^2 - 5y^2$ $\qquad$ $\frac{\partial z}{\partial y} = -10xy + 9y^2$

$\frac{\partial^2 z}{\partial x^2} = 24x$ $\qquad$ $\frac{\partial^2 z}{\partial y^2} = -10x + 18y$

$\frac{\partial^2 z}{\partial y.\partial x} = -10y$ $\qquad$ $\frac{\partial^2 z}{\partial x.\partial y} = -10y$

(ii) $\frac{\partial z}{\partial x} = -2\sin(2x + 3y)$ $\qquad$ $\frac{\partial z}{\partial y} = -3\sin(2x + 3y)$

$\frac{\partial^2 z}{\partial x^2} = -4\cos(2x + 3y)$ $\qquad$ $\frac{\partial^2 z}{\partial y^2} = -9\cos(2x + 3y)$

$\frac{\partial^2 z}{\partial y.\partial x} = -6\cos(2x + 3y)$ $\qquad$ $\frac{\partial^2 z}{\partial x.\partial y} = -6\cos(2x + 3y)$

(iii) $\dfrac{\partial z}{\partial x} = 2x\, e^{x^2-y^2}$ $\qquad \dfrac{\partial z}{\partial y} = -2y\, e^{x^2-y^2}$

$\dfrac{\partial^2 z}{\partial x^2} = 2\, e^{x^2-y^2}\,(2x^2+1)$ $\qquad \dfrac{\partial^2 z}{\partial y^2} = 2\, e^{x^2-y^2}\,(2y^2-1)$

$\dfrac{\partial^2 z}{\partial y.\partial x} = -4xy\, e^{x^2-y^2}$ $\qquad \dfrac{\partial^2 z}{\partial x.\partial y} = -4xy\, e^{x^2-y^2}$

(iv) $\dfrac{\partial z}{\partial x} = 2x^2 \cos(2x+3y) + 2x \sin(2x+3y)$

$\dfrac{\partial^2 z}{\partial x^2} = (2-4x^2)\sin(2x+3y) + 8x\cos(2x+3y)$

$\dfrac{\partial^2 z}{\partial y.\partial x} = -6x^2 \sin(2x+3y) + 6x\cos(2x+3y)$

$\dfrac{\partial z}{\partial y} = 3x^2 \cos(2x+3y)$

$\dfrac{\partial^2 z}{\partial y^2} = -9x^2 \sin(2x+3y)$

$\dfrac{\partial^2 z}{\partial x.\partial y} = -6x^2 \sin(2x+3y) + 6x\cos(2x+3y)$

2. (i) 2V
3. P decreases 375 W
4. ± 2·5 %

**Further Problems X (page 321)**

10. $\pm\, 0{\cdot}67\, E \times 10^{-5}$ approx.
12. $\pm\,(x+y+z)\,\%$
13. $y$ decreases by 19% approx.
14. ± 4·25%
16. 19%
18. $\delta y = y\,\{\delta x.\, p \cot(px+a) - \delta t.\, q \tan(qt+b)\}$

**Test Exercise XI (page 340)**

1. (i) $\dfrac{4xy-3x^2}{3y^2-2x^2}$, (ii) $\dfrac{e^x \cos y - e^y \cos x}{e^x \sin y + e^y \sin x}$,

(iii) $\dfrac{5\cos x\cos y - 2\sin x\cos x}{5\sin x\sin y + \sec^2 y}$

2. V decreases at 0·419 $cm^3/s$

3. $y$ decreases at 1·524 cm/s

4. $\dfrac{\partial z}{\partial r} = (4x^3 + 4xy)\cos\theta + (2x^2 + 3y^2)\sin\theta$

$\dfrac{\partial z}{\partial \theta} = r\{(2x^2 + 3y^2)\cos\theta - (4x^3 + 4xy)\sin\theta\}$

**Further Problems XI (page 341)**

2. $3x^2 - 3xy$

3. $\tan\theta = 17/6 = 2{\cdot}8333$

9. (i) $\dfrac{1-y}{x+2}$, (ii) $\dfrac{8y - 3y^2 + 4xy - 3x^2y^2}{2x^3y - 2x^2 + 6xy - 8x}$, (iii) $\dfrac{y}{x}$

14. $a = -\dfrac{5}{2}$, $b = -\dfrac{3}{2}$

16. $\dfrac{\cos x\,(5\cos y - 2\sin x)}{5\sin x\sin y + \sec^2 y}$

17. $\dfrac{y\cos x - \tan y}{x\sec^2 y - \sin x}$

20. (i) $-\left\{\dfrac{2xy + y\cos xy}{x^2 + x\cos xy}\right\}$, (ii) $-\left\{\dfrac{xy + \tan xy}{x^2}\right\}$

**Test Exercise XII (page 389)**

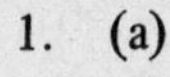

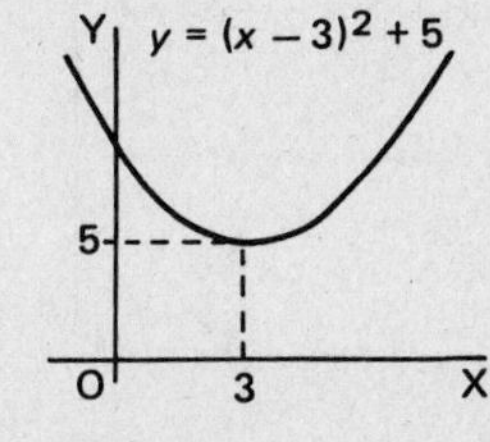

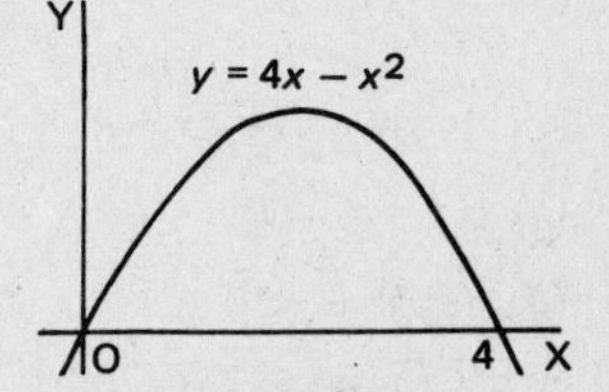

(c)

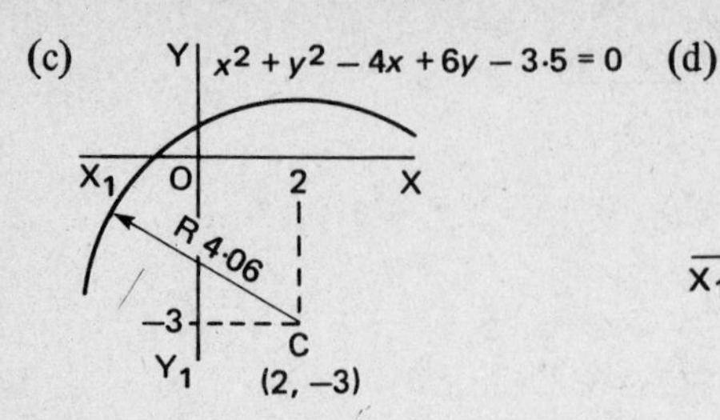


(d)

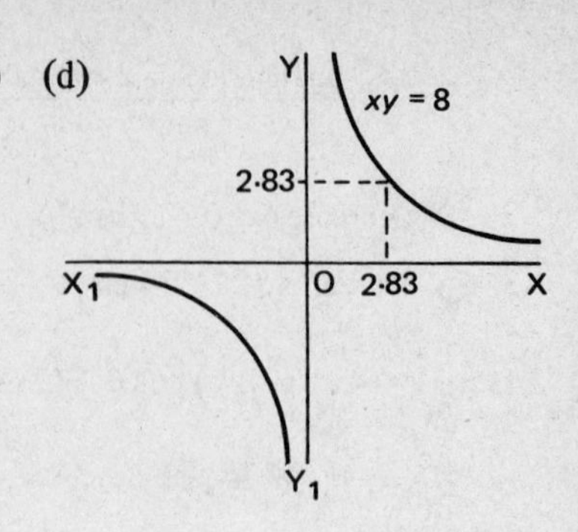


(e)

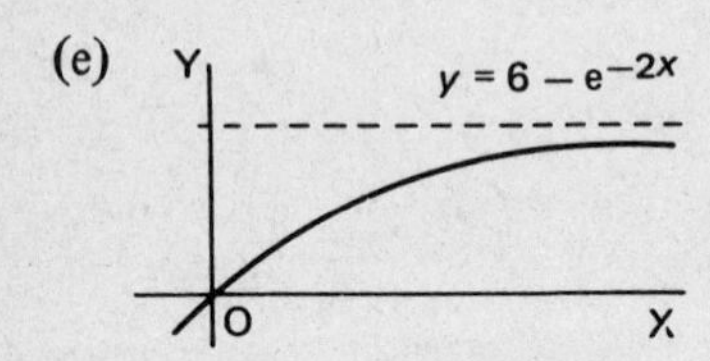


2. (a) $x = -3$; $x = 3$; $y = -x$
 (b) $x = 4$; $y = x + 3$; $y = -x - 3$

3. (i) no symmetry

 (ii) $\begin{cases} x = 0, y = -7.5 \\ y = 0, x = 3 \text{ or } -5 \end{cases}$

 (iii) asymptotes: $x = -2$ and $y = x$

 (iv) no turning points

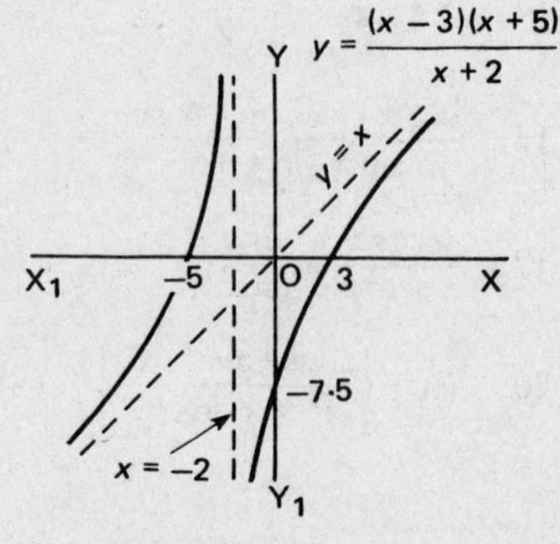


4.

| | x-axis | y-axis |
|---|---|---|
| (a) $\dfrac{y}{x} = A + Bx$ | $x$ | $\dfrac{y}{x}$ |
| (b) $\ln(y - x) = kx + \ln A$ | $x$ | $\ln(y - x)$ |
| (c) $xy = A - By$ | $y$ | $xy$ |
| (d) $(xy)^2 = x^2 + k$ | $x^2$ | $(xy)^2$ |

5. $V = 0.137P^{0.543}$

**Further Problems XII (page 390)**

1. (a) $x = 0$ (b) $x = \pm 2$; $y = \pm 2$

   (c) $x = 3$ (d) $x = -2$; $x = -3$; $y = 1$

   (e) $x = \pm 1$; $y = \pm 1$ (f) $x = 2$; $y = \pm 1$

2. (a) $x = 0$; $y = x$; $y = -x$

   (b) $x = 4$; $y = 0$

   (c) $y = 0$; $y = x - 1$; $y = -x - 1$

3. (a)

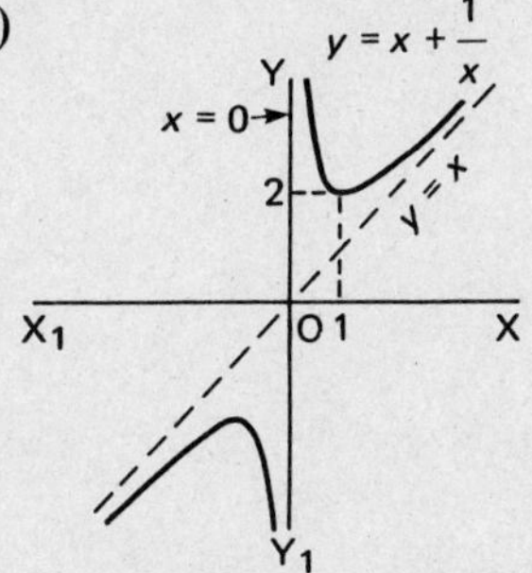


(b)

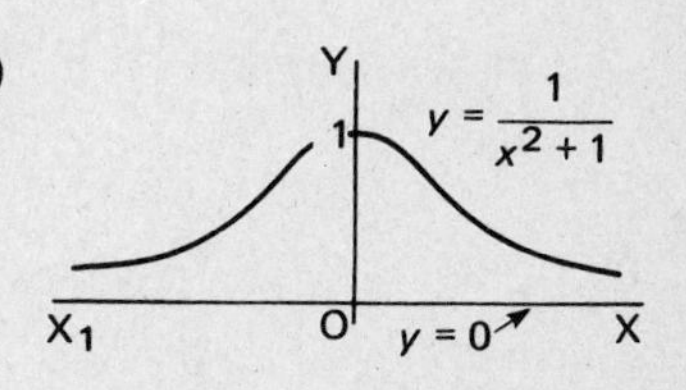


(c)

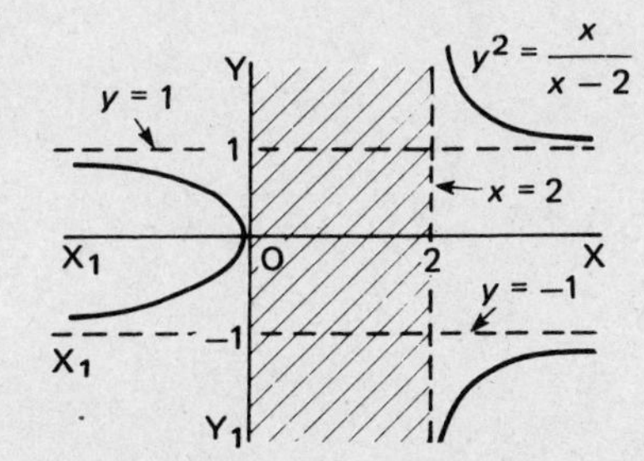


(d)

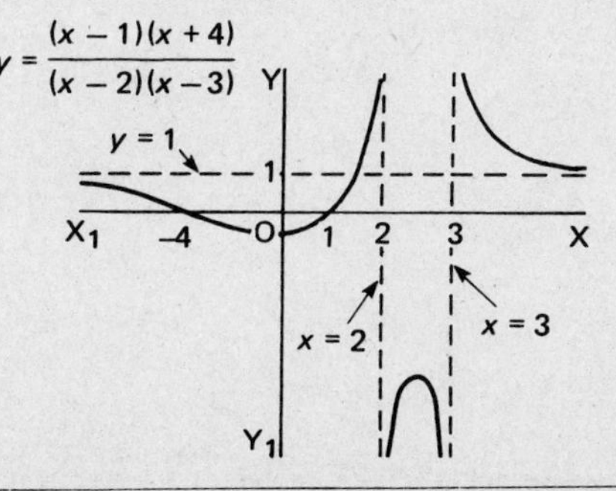

(e)

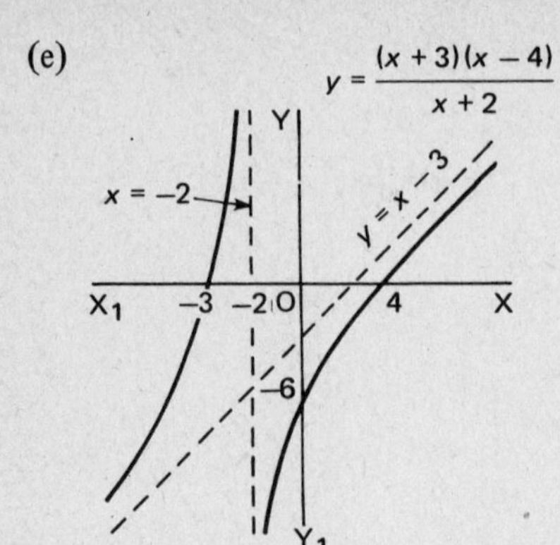


(f)

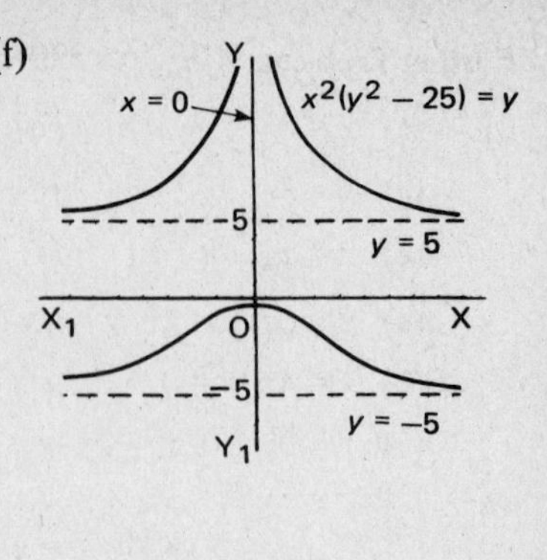


(g)

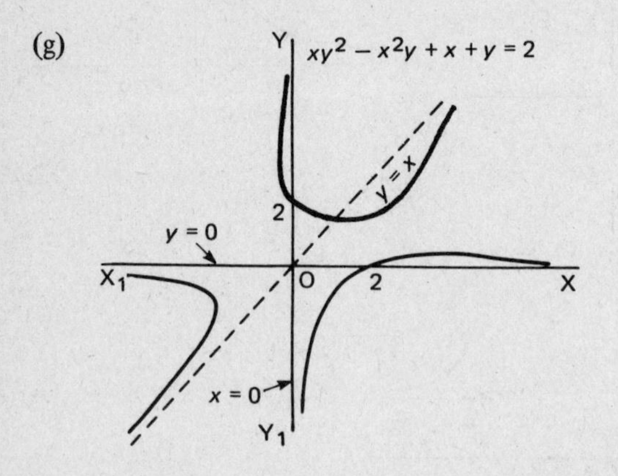


4. $y = 1.08 + 0.48x^2$
5. $P = 2.48\sqrt{W} + 3.20$
6. $R = 1.08 + \dfrac{0.047}{d^2}$
7. $y = \dfrac{4.5}{1 - 0.005x^2}$
8. $pv^{1.39} = 118$
9. $I = 8.63V^{0.75}$
10. $y = 0{\cdot}044x + 3.28 \ln x$

11. $y = 1.63x^2 + \dfrac{3.55}{x}$

12. $V = 96t^{0.67}$

13. $T = 0.045D^{1.58}$

**Test Exercise XIII (page 420)**

1. 230
2. 2·488, 25·945
3. 1812
4. (i) convergent, (ii) divergent, (iii) divergent, (iv) convergent
5. (i) convergent for all values of $x$.

   (ii) convergent for $-1 \leqslant x \leqslant 1$

   (iii) convergent for $-1 \leqslant x \leqslant 1$

**Further Problems XIII (page 421)**

1. $\dfrac{n}{3}(4n^2 - 1)$
2. $\dfrac{n(3n+1)}{4(n+1)(n+2)}$
3. $\dfrac{n}{4}(n+1)(n+4)(n+5)$
4. (i) $\dfrac{n}{3}(n+1)(n+5)$, (ii) $\dfrac{1}{4}(n^2+3n)(n^2+3n+4)$
5. 2
6. $S_n = \dfrac{10}{3}\left\{1 + \dfrac{(-1)^{n+1}}{2^n}\right\}$; $S_\infty = \dfrac{10}{3}$
7. (i) 0·6, (ii) 0·5
8. (i) diverges, (ii) diverges, (iii) converges, (iv) converges
9. $-1 \leqslant x \leqslant 1$

11. $-1 \leqslant x \leqslant 1$
12. All values of $x$
13. $-1 \leqslant x \leqslant 1$
16. (i) convergent, (ii) divergent, (iii) divergent, (iv) divergent
18. (i) convergent, (ii) convergent
19. $1 \leqslant x \leqslant 3$
20. $\frac{n}{6}(n+1)(4n+5)+2^{n+2}-4$

**Test Exercise XIV (page 450)**

1. $f(x)=f(0)+xf'(0)+\frac{x^2}{2!}f''(0)+\ldots$
2. $1-x^2+\frac{x^4}{3}-\frac{2x^6}{45}+\ldots$
3. $1+\frac{x^2}{2}+\frac{5x^4}{24}+\ldots$
5. $x+x^2+\frac{5x^3}{6}+\frac{x^4}{2}+\ldots$
6. 1·0247
7. (i) $-\frac{1}{10}$, (ii) $\frac{2}{9}$, (iii) $-\frac{1}{2}$
8. 0·85719

**Further Problems XIV (page 451)**

3. (i) $-\frac{1}{10}$, (ii) $\frac{1}{3}$, (iii) $\frac{1}{2}$, (iv) $-\frac{1}{6}$, (v) 2
6. $-\frac{1}{4}$
7. $-\frac{3}{2}-\frac{5x}{2}-\frac{11x^2}{4}-\frac{13x^3}{4}$

9. $\frac{2}{3}$

10. (i) $-\frac{1}{2}$, (ii) $\frac{1}{2}$, (iii) 2

11. $\frac{(n-r+2)x}{r-1}$; 1·426

13. $\ln\cos x = -\frac{x^2}{2} - \frac{x^4}{12} - \ldots$

16. (i) $-\frac{1}{6}$, (ii) $\frac{1}{2}$

17. $1 - \frac{7x}{2} + 8x^2$

19. $x^2 - x^3 + \frac{11x^4}{12}$; max. at $x = 0$

**Test Exercise XV (page 482)**

1. $-e^{\cos x} + C$
2. $2\sqrt{x}\,(\ln x - 2) + C$
3. $\tan x - x + C$
4. $\frac{x\sin 2x}{2} - \frac{x^2\cos 2x}{2} + \frac{\cos 2x}{4} + C$
5. $\frac{2e^{-3x}}{13}\left\{\sin 2x - \frac{3}{2}\cos 2x\right\} + C$
6. $-\cos x + \frac{2\cos^3 x}{3} - \frac{\cos^5 x}{5} + C$
7. $\frac{3x}{8} + \frac{\sin 2x}{4} + \frac{\sin 4x}{32} + C$
8. $2\ln(x^2 + x + 5) + C$
9. $\frac{1}{3}(1 + x^2)^{3/2} + C$
10. $\frac{9}{2}\ln(x-5) - \frac{5}{2}\ln(x-3) + C$

11. $2\ln(x-1)+\tan^{-1}x+C$

12. $-\left(\frac{\cos 8x}{16}+\frac{\cos 2x}{4}\right)+C$

**Further Problems XV (page 483)**

1. $\ln\{A(x-1)(x^2+x+1)\}+C$

2. $\frac{1}{2}$

3. $-\ln(1+\cos^2 x)+C$

4. $\frac{1}{\sqrt{3}}-\frac{\pi}{6}$

5. $\frac{\pi^2}{4}$

6. $C-\frac{2}{(x^2+x+1)^{1/2}}$

7. $\frac{2}{3}\ln(x-1)-\frac{1}{3}\ln(x^2+x+1)+C$

8. $\frac{x^2}{2}-x+\ln(x+1)+C$

9. $2\ln(x-1)+\tan^{-1}x+C$

10. $2$

11. $\frac{-2n^{p+3}}{(p+1)(p+2)(p+3)}$

12. $3\ln(x-2)+\frac{1}{2}\ln(x^2+1)-5\tan^{-1}x+C$

13. $\frac{1}{2}$

14. $\frac{(\sin^{-1}x)^2}{2}+C$

15. $\frac{1}{4}(2\ln 3-\pi)$

16. $\pi^2-4$

17. $\frac{\pi^3}{6} - \frac{\pi}{4}$

18. $\frac{\pi}{4} - \frac{1}{2}$

19. $-\frac{1}{x} - \tan^{-1} x + C$

20. $\frac{1}{3}(1 + x^2)^{3/2} + C$

21. $\ln(x + 1) - \ln(x - 2) - \frac{2}{x - 2} + C$

22. $\frac{1}{10}(e^{2\pi} - 1) = 53{\cdot}45$

23. $\frac{1}{24}$

24. $\frac{1}{13}\left\{3\, e^{\pi/3} - 2\right\}$

25. $-\frac{2}{3\omega}$

26. $\frac{\tan^3 x}{3} + C$

27. $\frac{1}{2}\ln(x - 4) - \frac{1}{10}\ln(5x + 2) + C$

28. $\ln(x + 2) + C$

29. $2\ln(x + 5) + \frac{3}{2}\ln(x^2 + 9) - \frac{4}{3}\tan^{-1}\left(\frac{x}{3}\right) + C$

30. $\ln(9x^2 - 18x + 17)^{1/18} + C$

31. $2x^2 + \ln\{(x^2 - 1)/(x^2 + 1)\} + C$

32. $\frac{1}{9}\left\{3x^2\ln(1 + x^2) - 2x^3 + 6x - 6\tan^{-1} x + C\right\}$

33. $\ln(\cos\theta + \sin\theta) + C$

34. $\tan\theta - \sec\theta + C$

35. $\frac{1}{4}\ln(x-1)+\frac{1}{5}\ln(x-2)-\frac{9}{20}\ln(x+3)+C$

36. $\frac{1}{6}$

37. $\frac{2}{3}\ln 2-\frac{5}{18}$

38. $3\ln x+\frac{1}{2}\ln(x^2+4)-\frac{1}{2}\tan^{-1}\left(\frac{x}{2}\right)+C$

39. $\ln x-\tan^{-1}x-\frac{1}{x}+C$

**Test Exercise XVI (page 514)**

1. $\sin^{-1}\left(\frac{x}{7}\right)+C$

2. $\frac{1}{\sqrt{29}}\ln\left\{\frac{2x+3-\sqrt{29}}{2x+3+\sqrt{29}}\right\}+C$

3. $\frac{1}{\sqrt{2}}\tan^{-1}\{(x+2)\sqrt{2}\}+C$

4. $\frac{1}{\sqrt{3}}\sinh^{-1}\left(\frac{x\sqrt{3}}{4}\right)+C$

5. $\frac{1}{10}\ln\left\{\frac{x+9}{1-x}\right\}+C$

6. $\frac{5}{8}\left\{\sin^{-1}\left(\frac{2x+1}{\sqrt{5}}\right)+\frac{2(2x+1)}{5}\sqrt{(1-x-x^2)}\right\}+C$

7. $\frac{1}{\sqrt{5}}\cosh^{-1}\left(\frac{x+1}{\sqrt{(21/5)}}\right)+C$

8. $\frac{1}{\sqrt{3}}\tan^{-1}(\sqrt{3}\tan x)+C$

9. $\frac{1}{\sqrt{13}}\ln\left\{\frac{\sqrt{13}-3+2\tan x/2}{\sqrt{13}+3-2\tan x/2}\right\}+C$

10. $\ln\left\{\frac{1+\tan x/2}{1-\tan x/2}\right\}+C$

**Further Problems XVI (page 515)**

1. $\frac{1}{2\sqrt{21}} \ln\left\{\frac{x+6-\sqrt{21}}{x+6+\sqrt{21}}\right\} + C$

2. $\frac{1}{4\sqrt{11}} \ln\left\{\frac{2\sqrt{11}+x+6}{2\sqrt{11}-x-6}\right\} + C$

3. $\frac{1}{\sqrt{11}} \tan^{-1}\left(\frac{x+7}{\sqrt{11}}\right) + C$

4. $\frac{1}{2}\ln(x^2+4x+16) - \frac{5}{\sqrt{3}}\tan^{-1}\left(\frac{x+2}{2\sqrt{3}}\right) + C$

5. $\sinh^{-1}\left(\frac{x+6}{2\sqrt{3}}\right) = \ln\left\{\frac{x+6+\sqrt{(x^2+12x+48)}}{2\sqrt{3}}\right\} + C$

6. $\sin^{-1}\left(\frac{x+7}{\sqrt{66}}\right) + C$

7. $\cosh^{-1}\left(\frac{x+8}{2\sqrt{7}}\right) + C$

8. $6\sqrt{(x^2-12x+52)} + 31\sinh^{-1}\left(\frac{x-6}{4}\right) + C$

9. $\frac{2\sqrt{3}}{3}\tan^{-1}\left\{\frac{\sqrt{3}}{3}\tan^{-1}\left(\frac{x}{2}\right)\right\} + C$

10. $\frac{\sqrt{5}.\pi}{20} = 0{\cdot}3511$

11. $\frac{1}{\sqrt{5}}\ln\left\{\frac{2x+5-\sqrt{5}}{2x+5+\sqrt{5}}\right\} + C$

12. $\frac{3x^2}{2} - 4x + 4\tan^{-1}x + C$

13. $\frac{x+1}{2}\sqrt{(3-2x-x^2)} + 2\sin^{-1}\left(\frac{x+1}{2}\right) + C$

14. $\pi$

15. $\cosh^{-1}\left(\frac{x-2}{5}\right) + C$

16. $\frac{1}{6}\tan^{-1}\left\{\frac{2}{3}\tan x\right\}+C$

17. $\frac{1}{5}\ln\left\{\frac{2\tan x-1}{\tan x+2}\right\}+C$

18. $\frac{\pi}{2}-1$

19. $3\sin^{-1}x-\sqrt{(1-x^2)}+C$

20. $\frac{4}{\sqrt{3}}\tan^{-1}\left\{\sqrt{3}\tan\frac{x}{2}\right\}-x+C$

21. $\frac{5}{2}\ln(x+2)-\frac{3}{4}\ln(x^2+4)+\tan^{-1}\left(\frac{x}{2}\right)+C$

22. $\frac{1}{3\sqrt{5}}\tan^{-1}\left(\frac{\sqrt{5}\tan x}{3}\right)+C$

23. $\sqrt{(x^2+9)}+2\ln\{x+\sqrt{(x^2+9)}\}+C$

24. $\frac{1}{\sqrt{2}}\cosh^{-1}\left(\frac{4x-7}{3}\right)+C$

25. $\frac{\pi}{2}$

26. $\frac{1}{2\sqrt{2}}\ln\left\{\frac{\sqrt{2}\tan\theta-1}{\sqrt{2}\tan\theta+1}\right\}+C$

27. $\sqrt{(x^2+2x+10)}+2\sinh^{-1}\left(\frac{x+1}{3}\right)+C$

28. $8\sin^{-1}\left(\frac{x+1}{4}\right)+\frac{x+1}{2}\sqrt{(15-2x-x^2)}+C$

29. $\frac{1}{8a^3}(\pi+2)$

30. $\frac{1}{3\sqrt{2}}\tan^{-1}\left(\frac{x}{a\sqrt{2}}\right)+\frac{1}{6}\ln\left\{\frac{(x+a)^2}{x^2+2a^2}\right\}+C$

**Test Exercise XVII (page 528)**

1. $e^{2x}\left\{\frac{x^3}{2}-\frac{3x^2}{4}+\frac{3x}{4}-\frac{3}{8}\right\}+C$

2. (i) $\dfrac{5\pi}{256}$, (ii) $\dfrac{8}{315}$

3. $\dfrac{2a^7}{35}$

4. $I_n = \dfrac{1}{n-1} \tan^{n-1} x - I_{n-2}$

5. $\dfrac{3\pi}{256}$

**Further Problems XVII (page 529)**

2. $-\dfrac{1}{7} s^6 c - \dfrac{6}{35} s^4 c - \dfrac{8}{35} s^2 c - \dfrac{16}{35} c + C_1$ where $\left\{ \begin{array}{l} s \equiv \sin x \\ c \equiv \cos x \end{array} \right\}$

3. $\dfrac{2835}{8}$

5. $I_3 = \dfrac{3\pi^2}{4} - 6$; $I_4 = \dfrac{\pi^3}{2} - 12\pi + 24$

6. $I_n = x^n e^x - n\, I_{n-1}$; $I_4 = e^x (x^4 - 4x^3 + 12x^2 - 24x + 24)$

7. $\dfrac{1328\sqrt{3}}{2835}$

10. $I_6 = -\dfrac{\cot^5 x}{5} + \dfrac{\cot^3 x}{3} - \cot x - x + C$

11. $I_3 = x\{(\ln x)^3 - 3(\ln x)^2 + 6 \ln x - 6\} + C$

12. $\dfrac{4}{3}$

**Test Exercise XVIII (page 550)**

1. 70·12

2. $\dfrac{80}{\pi} + 2\pi = 31{\cdot}75$

3. $\dfrac{3}{2} \ln 6 = 2{\cdot}688$

4. 73·485

5. $\frac{1}{2}\mathrm{R}\,\mathrm{I}^2$

6. 132·3

**Further Problems XVIII (page 551)**

1. 2·4

2. 1

3. $3\pi$

4. $\frac{1}{4}$

5. 0

7. 2

8. $\frac{1}{2}v_0\, i_0 \cos a$

9. $\sqrt{\left\{\frac{\mathrm{E}^2}{\mathrm{R}^2} + \frac{1}{2}\mathrm{I}^2\right\}}$

11. $\ln(2^{11}.3^{-6}) - 1$

12. $a^2\left(\ln 2 - \frac{2}{3}\right)$

15. $a(1 - 2e^{-1})$

16. 2·83

17. 39·01

18. $\sqrt{\left\{\frac{1}{2}(\mathrm{I}_1^2 + \mathrm{I}_2^2)\right\}}$

20. 1·361

**Test Exercise XIX (page 575)**

1. (0·75, 1·6)

2. (0·4, 0)

3. $5\pi^2 a^3$

4. $\dfrac{e^2 + 7}{8}$

5. $70{\cdot}35\,\pi$

6. $\dfrac{5.\pi^2}{8}$

7. $2\sqrt{2}.\pi\left(\dfrac{e^\pi - 2}{5}\right)$

**Further Problems XIX (page 576)**

1. $\dfrac{3}{16} + \dfrac{1}{2}\ln 2$

2. (i) 2·054, (i) 66·28

3. $\dfrac{64\pi a^3}{15}$

4. (i) (0·4, 1), (ii) (0·5, 0)

6. 24

7. $\dfrac{17}{12}$

8. $-\dfrac{19}{20}$

9. $\dfrac{11}{5}$

10. $A = 2{\cdot}457$, $V = 4\pi\sqrt{3}$, $\bar{y} = 1{\cdot}409$

12. (i) 8, (ii) $\dfrac{64\pi}{3}$, (iii) $\dfrac{4}{3}$

13. 1·175

16. $V = 25{\cdot}4\ \text{cm}^3$, $A = 46{\cdot}65\ \text{cm}^2$

17. $S = 15{\cdot}31\,a^2$, $y = 1{\cdot}062\,a$

**Test Exercise XX (page 611)**

1. (i) $I_z = \frac{ab\rho}{12}(b^2 + a^2)$,

   (ii) $I_{AB} = \frac{ab\rho}{3}(a^2 + b^2)$, $k = \sqrt{\frac{a^2 + b^2}{3}}$

2. $k = \frac{l}{\sqrt{2}}$

3. (i) $\frac{1}{\ln 4}$, (ii) $\frac{6}{\ln 2}$

5. $\frac{w\,a^3}{8}$, $0{\cdot}433\,a$

**Further Problems XX (page 612)**

2. $\frac{1}{2}\mathrm{M}\,a^2$

6. (i) $\sqrt{\frac{4ac}{5}}$, (ii) $\sqrt{\frac{3c^2}{7}}$

9. $\frac{a^4}{12}$

10. $I = M\left\{\frac{h^2}{10} + \frac{3r^2}{20}\right\}$; $k = \sqrt{\left\{\frac{h^2}{10} + \frac{3r^2}{20}\right\}}$

12. $\frac{\pi ab^3}{4}$

14. $\frac{2wa^3}{3}$, $\frac{3\pi a}{16}$

15. (i) $\frac{1}{3}\sqrt{(e^2 + e + 1)}$, (ii) $\sqrt{\frac{e-2}{e-1}}$

16. $51{\cdot}2\,w$

17. 9·46 cm

19. $\frac{(15\pi - 32)a}{4(3\pi - 4)}$

**Test Exercise XXI (page 632)**

1. 0·946
2. 0·926
3. 26·7
4. 1·188
5. 1·351

**Further Problems XXI (page 633)**

1. 0·478
2. 0·091
3. (i) 0·6, (ii) 6·682, (iii) 1·854
4. 560
5. 15·86
6. 0·747
7. 28·4
8. 28·92
9. 0·508
10. $\frac{\sqrt{2}}{4}\int_0^{\pi/2} \sqrt{(9 + \cos 2\theta)}.d\theta$; 4·99
11. $\tan^{-1}x = x - \frac{x^3}{3} + \frac{x^5}{5} - \frac{x^7}{7}$; 0·076
12. (i) 0·5314, (ii) 0·364
13. 2·422
14. 2·05

## Test Exercise XXII (page 658)

1. $\dfrac{56}{3\pi^3}$

2. (i) $r = 2 \sin \theta$

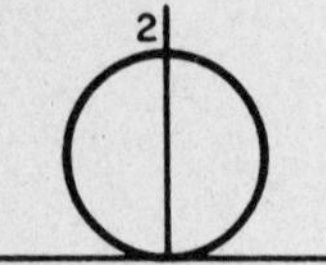
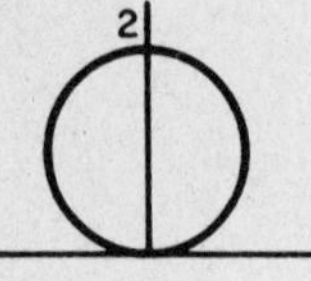


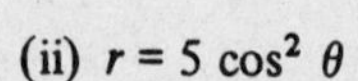
(ii) $r = 5 \cos^2 \theta$

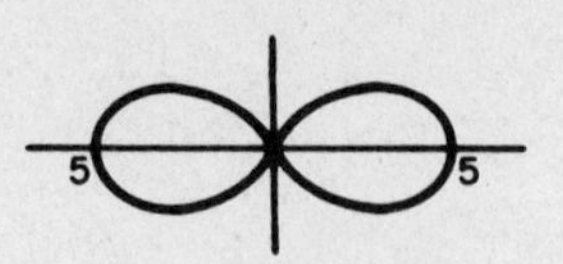


(iii) $r = \sin 2\theta$

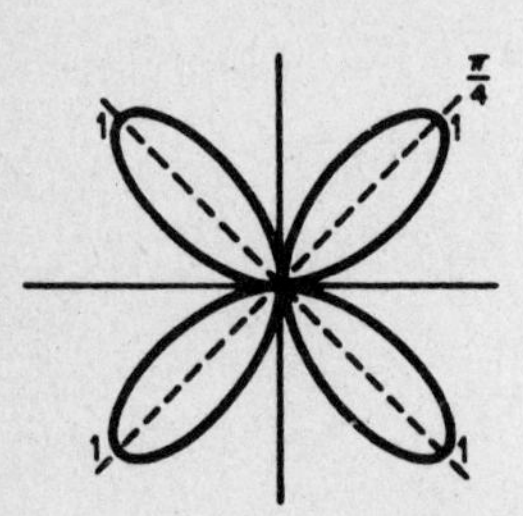


(iv) $r = 1 + \cos \theta$

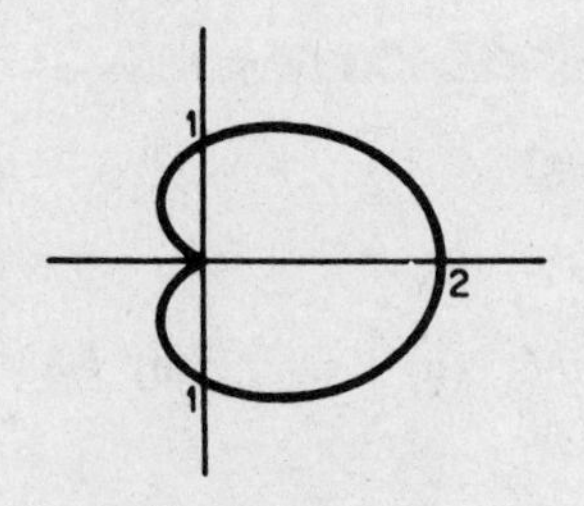


(v) $r = 1 + 3 \cos \theta$

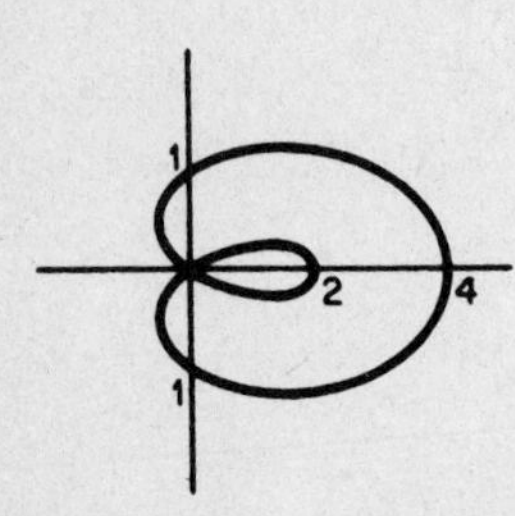


(vi) $r = 3 + \cos \theta$

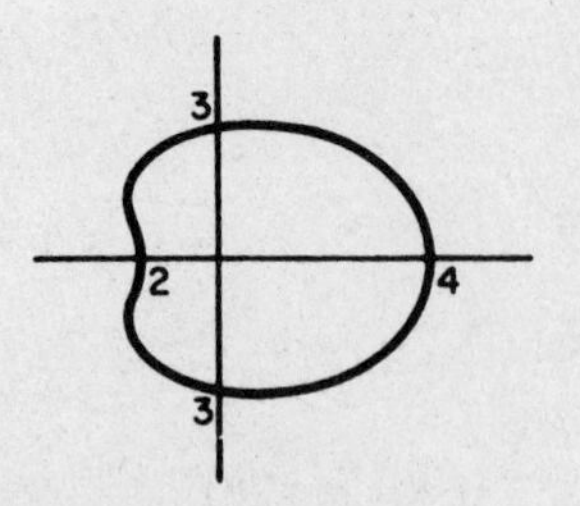


3. $\dfrac{40\pi}{3}$

4. 8

5. $\dfrac{32\pi a^2}{5}$

**Further Problems XXII (page 659)**

1. $A = \frac{3\pi}{16}$; $V = \frac{2\pi}{21}$
2. $3\pi$
3. $\frac{4a^2}{3}$
4. $\frac{13\pi}{8} + 3$
5. $\frac{2}{3}$
6. $\frac{20\pi}{3}$
7. $\frac{8\pi a^3}{3}$
9. $\frac{3\pi a}{2}$
10. $\frac{5\pi}{2}$
11. $21{\cdot}25\,a$
12. $\frac{3\pi a}{2}$
14. $\frac{a}{b}\left\{\sqrt{(b^2+1)}\right\}(e^{b\theta_1} - 1)$ ; $\frac{a^2}{4b}(e^{2b\theta_1} - 1)$
15. $\pi a^2 (2 - \sqrt{2})$

**Test Exercise XXIII (page 686)**

1. (i) 0, (ii) 0
2. (i) $-1$, (ii) 120, (iii) $\frac{17}{4}$

3. 3·67
4. 170·67
5. $\frac{11\pi}{4} + 6$
6. 36

**Further Problems XXIII (page 687)**

1. $\frac{1}{3}$
2. $\frac{243\pi}{2}$
3. 4·5
4. $\frac{abc}{3}(b^2 + c^2)$
5. $\frac{\pi r^3}{3}$
6. 4·5
7. $\pi + 8$
8. 26
9. $\frac{22}{3}$
10. $\frac{1}{8}\left(\frac{\pi}{2} + 1\right)$
11. $\frac{1}{3}$
12. $A = 2\int_0^{\pi/6}\int_0^{2\cos 3\theta} r\,dr\,d\theta = \frac{\pi}{3}$
13. $4\pi\left\{\frac{1}{\sqrt{2}} - \frac{1}{\sqrt{5}}\right\}$

14. $\frac{64}{9}(3\pi - 4)$

15. $M = \int_0^{\pi}\int_0^{a(1+\cos\theta)} r^2 \sin\theta \, dr \, d\theta = \frac{4a^3}{3}$; $h = \frac{16a}{9\pi}$

16. (i) $\frac{1}{2}\pi ab$, (ii) $\frac{1}{8}\pi ab^3$: centroid $\left(0, \frac{4b}{3\pi}\right)$

17. 19·56

18. $\frac{b^2}{4}(c^2 - a^2)$

19. $\frac{a^2}{6}(2\pi + 3\sqrt{3})$

20. 232

**Test Exercise XXIV (page 728)**

1. $y = \frac{x^2}{2} + 2x - 3\ln x + C$

2. $\tan^{-1} y = C - \frac{1}{1+x}$

3. $y = \frac{e^{3x}}{5} + C\,e^{-2x}$

4. $y = x^2 + Cx$

5. $y = -\frac{x\cos 3x}{3} + \frac{\sin 3x}{9} - \frac{4}{x} + C$

6. $\sin y = Ax$

7. $y^2 - x^2 = Ax^2 y$

8. $y(x^2 - 1) = \frac{x^2}{2} + C$

9. $y = \cosh x + \frac{C}{\cosh x}$

10. $y = x^2 (\sin x + C)$
11. $xy^2 (Cx + 2) = 1$
12. $y = 1/(Cx^3 + x^2)$

**Further Problems XXIV (page 729)**

1. $x^4 y^3 = A e^y$
2. $y^3 = 4(1 + x^3)$
3. $3x^4 + 4(y + 1)^3 = A$
4. $(1 + e^x) \sec y = 2\sqrt{2}$
5. $x^2 + y^2 + 2x - 2y + 2 \ln (x - 1) + 2 \ln (y + 1) = A$
6. $y^2 - xy - x^2 + 1 = 0$
7. $xy = A e^{y/x}$
8. $x^3 - 2y^3 = Ax$
9. $A(x - 2y)^5 (3x + 2y)^3 = 1$
10. $(x^2 - y^2)^2 = Axy$
11. $2y = x^3 + 6x^2 - 4x \ln x + Ax$
12. $y = \cos x (A + \ln \sec x)$
13. $y = x(1 + x \sin x + \cos x)$
14. $(3y - 5)(1 + x^2)^{3/2} = 2\sqrt{2}$
15. $y \sin x + 5 e^{\cos x} = 1$
16. $x + 3y + 2 \ln (x + y - 2) = A$
17. $x = Aye^{xy}$
18. $\ln \{4y^2 + (x - 1)^2\} + \tan^{-1} \left\{\frac{2y}{x - 1}\right\} = A$
19. $(y - x + 1)^2 (y + x - 1)^5 = A$
20. $2x^2 y^2 \ln y - 2xy - 1 = Ax^2 y^2$

21. $\dfrac{2}{y^2} = 2x + 1 + Ce^{2x}$

22. $\dfrac{1}{y^3} = \dfrac{3e^x}{2} + Ce^{3x}$

23. $y^2(x + Ce^x) = 1$

24. $\dfrac{\sec^2 x}{y} = C - \dfrac{\tan^3 x}{3}$

25. $\cos^2 x = y^2(C - 2\tan x)$

26. $y\sqrt{(1 - x^2)} = A + \sin^{-1} x$

27. $x + \ln Ax = \sqrt{(y^2 - 1)}$

28. $\ln(x - y) = A + \dfrac{2x^2}{(y - x)^2} - \dfrac{4x}{y - x}$

29. $y = \dfrac{\sqrt{2}\sin 2x}{2(\cos x - \sqrt{2})}$

30. $(x - 4)y^4 = Ax$

31. $y = x\cos x - \dfrac{\pi}{8}\sec x$

32. $(x - y)^3 - Axy = 0$

33. $2\tan^{-1} y = \ln(1 + x^2) + A$

34. $2x^2 y = 2x^3 - x^2 - 4$

35. $y = e^{\frac{x - y}{x}}$

36. $3e^{2y} = 2e^{3x} + 1$

37. $4xy = \sin 2x - 2x\cos 2x + 2\pi - 1$

38. $y = Ae^{y/x}$

39. $x^3 - 3xy^2 = A$

40. $x^2 - 4xy + 4y^2 + 2x - 3 = 0$

41. $y(1 - x^3)^{-1/3} = -\dfrac{1}{2}(1 - x^3)^{2/3} + C$

42. $xy + x\cos x - \sin x + 1 = 0$

43. $2\tan^{-1} y = 1 - x^2$

44. $y = \dfrac{x^2 + C}{2x(1 - x^2)}$

45. $y\sqrt{(1 + x^2)} = x + \dfrac{x^3}{3} + C$

46. $1 + y^2 = A(1 + x^2)$

47. $\sin^2\theta\,(a^2 - r^2) = \dfrac{a^2}{2}$

48. $y = \dfrac{1}{2}\sin x$

49. $y = \dfrac{1}{x(A - x)}$

**Test Exercise XXV (page 761)**

1. $y = A\,e^{-x} + B\,e^{2x} - 4$
2. $y = A\,e^{2x} + B\,e^{-2x} + 2\,e^{3x}$
3. $y = e^{-x}\,(A + Bx) + e^{-2x}$
4. $y = A\cos 5x + B\sin 5x + \dfrac{1}{125}(25x^2 + 5x - 2)$
5. $y = e^{x}\,(A + Bx) + 2\cos x$
6. $y = e^{-2x}\,(2 - \cos x)$
7. $y = Ae^{x} + B\,e^{-x/3} - 2x + 7$
8. $y = A\,e^{2x} + B\,e^{4x} + 4x\,e^{4x}$

**Further Problems XXV (page 762)**

1. $y = A\,e^{4x} + B\,e^{-x/2} - \dfrac{e^{3x}}{7}$
2. $y = e^{3x}\,(A + Bx) + 6x + 6$
3. $y = 4\cos 4x - 2\sin 4x + A\,e^{2x} + B\,e^{3x}$

4. $y = e^{-x}(Ax + B) + \frac{e^x}{2} - x^2 e^{-x}$

5. $y = A e^x + B e^{-2x} + \frac{e^{2x}}{4} - \frac{x e^{-2x}}{3}$

6. $y = e^{3x}(A \cos x + B \sin x) + 2 - \frac{e^{2x}}{2}$

7. $y = e^{-2x}(A + Bx) + \frac{1}{4} + \frac{1}{8}\sin 2x$

8. $y = A e^x + B e^{3x} + \frac{1}{9}(3x + 4) - e^{2x}$

9. $y = e^x(A \cos 2x + B \sin 2x) + \frac{x^2}{3} + \frac{4x}{9} - \frac{7}{27}$

10. $y = A e^{3x} + B e^{-3x} - \frac{1}{18}\sin 3x + \frac{1}{6} x e^{3x}$

11. $y = \frac{wx^2}{24\,EI}\{x^2 - 4lx + 6l^2\}$ ; $y = \frac{wl^4}{8\,EI}$.

12. $x = \frac{1}{2}(1 - t)\,e^{-3t}$

13. $y = e^{-2t}(A \cos t + B \sin t) - \frac{3}{4}(\cos t - \sin t)$;

    amplitude $\frac{3\sqrt{2}}{4}$, frequency $\frac{1}{2\pi}$

14. $x = -\frac{1}{2}e^t + \frac{1}{5}e^{2t} + \frac{1}{10}(\sin t + 3 \cos t)$

15. $y = e^{-2x} - e^{-x} + \frac{3}{10}(\sin x - 3 \cos x)$

16. $y = e^{-3x}(A \cos x + B \sin x) + 5x - 3$

17. $x = e^{-t}(6 \cos t + 7 \sin t) - 6 \cos 3t - 7 \sin 3t$

18. $y = \sin x - \frac{1}{2}\sin 2x$; $y_{max} = 1{\cdot}299$ at $x = \frac{2\pi}{3}$

19. $T = \frac{\pi}{2\sqrt{6}} = 0{\cdot}641$ s; $A = \frac{1}{6}$

20. $x = \frac{1}{10}\{e^{-3t} - e^{-2t} + \cos t + \sin t\}$;

    Steady state: $x = \frac{\sqrt{2}}{10}\sin\left(t + \frac{\pi}{4}\right)$

**Test Exercise XXVI (page 801)**

1. $y = \text{A}\,e^{-x} + \text{B}\,e^{-2x} + \dfrac{e^{4x}}{30}$

2. $y = e^{-2x}(\text{A} + \text{B}x) + 5\,e^{-3x}$

3. $y = \text{A}\,e^{-x} + \text{B}\,e^{-3x} - \dfrac{1}{30}(\cos 3x - 2\sin 3x)$

4. $y = e^{2x}(\text{A}\cos x + \text{B}\sin x) + \dfrac{1}{377}(16\cos 4x - 11\sin 4x)$

5. $y = e^{-x}(\text{A}\cos x + \text{B}\sin x) - \dfrac{e^{x}}{65}(8\cos 2x - \sin 2x)$

6. $y = e^{-2x}(\text{A} + \text{B}x) + \dfrac{e^{2x}}{16}\left(x^3 - \dfrac{3x^2}{2} + \dfrac{9x}{8} - \dfrac{3}{8}\right)$

7. $y = \text{A}\cos x + \text{B}\sin x + \dfrac{3e^{x}}{2} + e^{2x}$

8. $y = \text{A}\,e^{-2x} + \text{B}\,e^{-4x} - \dfrac{2}{85}\{6\cos x - 7\sin x\}$
$\qquad - \dfrac{1}{325}\{18\cos 3x + \sin 3x\}$

9. $y = \text{A}\cos 5x + \text{B}\sin 5x - \dfrac{x\cos 5x}{10}$

10. $y = \text{A}\,e^{-x} + \text{B}\,e^{3x} + \dfrac{x\,e^{3x}}{2}$

**Further Problems XXVI (page 802)**

1. $y = \text{A}\,e^{x} + \text{B}\,e^{-3x} - x\,e^{-3x}$

2. $y = \text{A}\,e^{-x} + \text{B}\,e^{-2x} + e^{-x}\left(\dfrac{x^2}{2} - x\right)$

3. $y = \text{A}\cos x + \text{B}\sin x - \dfrac{x}{2}\cos x$

4. $y = e^{x}(\text{A} + \text{B}x) + \dfrac{1}{2}\cos x + x^2 + 4x + 6$

5. $y = 1 + e^{2x}(1 - 2x)$

6. $y = \text{A}\,e^{2x} + \text{B}\,e^{3x} + x\,e^{3x}$

7. $y = \text{A}\,e^{2x} + \text{B}\,e^{3x} - e^{4x}(9\cos 3x + 7\sin 3x)/130$

8. $y = e^{-2x}(\mathrm{A}\cos x + \mathrm{B}\sin x) + \frac{x}{5} - \frac{4}{25} + (8\sin 2x + \cos 2x)/65$

9. $y = e^{-x}(\mathrm{A}\cos 2x + \mathrm{B}\sin 2x) + \cos 2x + 4\sin 2x$

10. $y = e^{-2x}(\mathrm{A}\cos x + \mathrm{B}\sin x) + \cos x + \sin x$

11. $y = e^{-ax}\left(\frac{x^4}{12} + \mathrm{A} + \mathrm{B}x\right)$

12. $y = e^{-x/2}\left(\mathrm{A}\cos\frac{\sqrt{3}}{2}x + \mathrm{B}\sin\frac{\sqrt{3}}{2}x\right) + \frac{e^x}{3}(x-1) - \frac{e^x}{13}(3\cos x - 2\sin x)$

13. $y = e^{3x}(\mathrm{A} + \mathrm{B}x) + \frac{x^2 e^{3x}}{2} + \frac{e^{-3x}}{36}$

14. $y = e^{-2x}(\mathrm{A} + \mathrm{B}x) + \frac{e^{2x}}{32} + \frac{x^2 e^{-2x}}{4}$

15. $y = e^{-3x}(\mathrm{A} + \mathrm{B}x) + \frac{1}{18}(1 + e^{-6x})$

16. $y = \mathrm{A}\,e^{3x} + \mathrm{B}\,e^{-2x} + e^{3x}(5x^2 - 2x)/50$

17. $y = e^{-2x}(\mathrm{A}\cos x + \mathrm{B}\sin x) + \frac{4}{5} + \frac{4}{65}(8\sin 2x + \cos 2x)$

18. $y = e^{-x}(\mathrm{A}\cos 2x + \mathrm{B}\sin 2x) + \sin 2x - 4\cos 2x$

19. $y = \mathrm{A}\,e^{x/2} + \mathrm{B}\,e^{-x} - e^x(3\sin 2x + 5\cos 2x)/68$

20. $y = e^{-x}(\mathrm{A}\cos 2x + \mathrm{B}\sin 2x) + \frac{x}{5} - \frac{2}{25} - \frac{e^{-x}\cos 3x}{5}$

21. $y = e^x(\mathrm{A}\cos\sqrt{3}x + \mathrm{B}\sin\sqrt{3}x) - \frac{e^x\sin 3x}{6}$

22. $y = e^{2x}\left(\mathrm{A} + \mathrm{B}x + \frac{x^2}{2}\right)$

23. $y = e^{3x}\left(\mathrm{A} + \frac{x}{12}\right) + e^{-3x}\left(\mathrm{B} - \frac{x}{12}\right) - \frac{1}{81}(9x^2 + 2)$

24. $y = e^{-x}\left(\mathrm{A} + \frac{1}{2}\sin x - \frac{1}{2}\cos x\right) + \mathrm{B}\,e^{-2x}$

25. $y = e^{-x}(\mathrm{A}\cos x + \mathrm{B}\sin x + x^2 - 2)$

## Test Exercise XXVII (page 841)

1. (a), (b)

| Mass (kg) $x$ | Frequency $f$ |
|---|---|
| 4.2 | 1 |
| 4.3 | 3 |
| 4.4 | 7 |
| 4.5 | 10 |
| 4.6 | 12 |
| 4.7 | 10 |
| 4.8 | 5 |
| 4.9 | 2 |
| $n = \Sigma f = 50$ | |

(c)

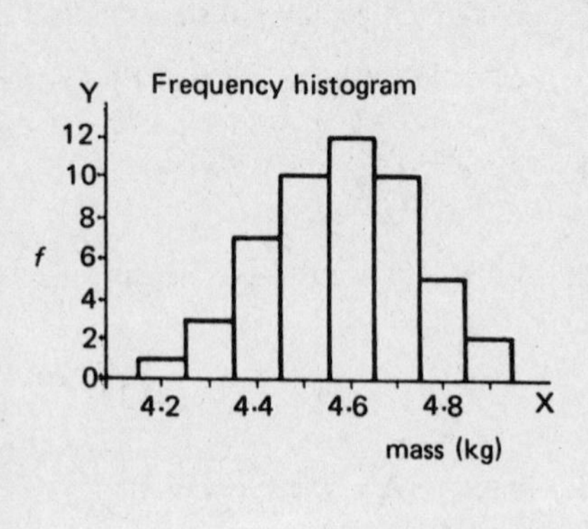


2. (a)

| Diam. (mm) $x$ | $x_m$ | Rel. $f$ (%) |
|---|---|---|
| 8.82–8.86 | 8.84 | 1.33 |
| 8.87–8.91 | 8.89 | 10.67 |
| 8.92–8.96 | 8.94 | 21.33 |
| 8.97–9.01 | 8.99 | 24.00 |
| 9.02–9.06 | 9.04 | 20.00 |
| 9.07–9.11 | 9.09 | 13.33 |
| 9.12–9.16 | 9.14 | 6.67 |
| 9.17–9.21 | 9.19 | 2.67 |
| | | 100.00% |

(b)

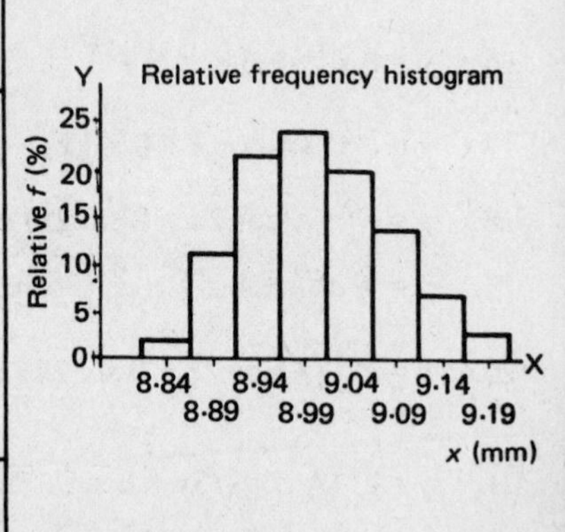


(c) (i) 8.917 mm. (ii) 9.115 mm. (iii) 0.05 mm

3. (a) $\bar{x} = 10.62$ mm; (b) $\sigma = 0.437$ mm; (c) mode = 10.51 mm; (d) median = 10.59 mm

4. (a) $\bar{x} = 15.05$ mm; $\sigma = 0.371$ mm
   (b) (i) 13.94 mm to 16.16 mm; (b) 384

**Further Problems XXVII (page 843)**

1. (a), (b)

| $x$ | $f$ |
|---|---|
| 60–64 | 3 |
| 65–69 | 5 |
| 70–74 | 7 |
| 75–79 | 11 |
| 80–84 | 8 |
| 85–89 | 4 |
| 90–94 | 2 |
| $n = \Sigma f = 40$ | |

(ii) $\bar{x} = 76.5$

(iii) $\sigma = 7.73$

2. (a), (b)

| Length (mm) $x$ | Frequency $f$ |
|---|---|
| 13.5–14.4 | 2 |
| 14.5–15.4 | 5 |
| 15.5–16.4 | 9 |
| 16.5–17.4 | 13 |
| 17.5–18.4 | 7 |
| 18.5–19.4 | 4 |
| $n = \Sigma f = 40$ | |

(c) (i) $\bar{x} = 16.75$ mm

(ii) $\sigma = 1.299$ mm

3. (a) (i) $\bar{x} = 4.283$ kg; (ii) $\sigma = 0.078$ kg
   (b) (i) 4.050 kg to 4.516 kg; (ii) 288
4. (a) $\bar{x} = 2.382$ MΩ; (b) $\sigma = 0.0155$ MΩ; (c) mode = 2.378 MΩ;
   (d) median = 2.382 MΩ
5. (a) $\bar{x} = 31.06$ kN (b) $\sigma = 0.237$ kN; (c) mode = 31.02 kN;
   (d) median = 31.04 kN
6. (a) (i) $\bar{x} = 12.04$ min; (ii) $\sigma = 0.724$ min; (iii) mode = 11.6 min;
   (iv) median = 11.57 min
   (b) (i) 0.5 min; (ii) 10.75 min; (iii) 13.25 min

7. (a) 32.60 to 32.72 mm, (b) 1956.
8. (a) (i) $\bar{x} = 7.556$ kg; (ii) $\sigma = 0.126$ kg
   (b) (i) 7.18 kg to 7.94 kg; (ii) 320
9. (a) (i) $\bar{x} = 29.64$ mm; (ii) $\sigma = 0.123$ mm
   (b) (i) 29.27 mm to 30.01 mm; (iii) 2100
10. (a) (i) $\bar{x} = 25.14$ mm; (ii) $\sigma = 0.690$ mm
    (b) (i) 23.07 mm to 27.21 mm; (ii) 480

**Test Exercise XXVIII (page 894)**

1. (a) (i) 0.12; (ii) 0.88; (b) 3520
2. (a) 0.2400; (b) 0.1600; (c) 0.3600
3. (a) 0.1576; (b) 0.0873; (c) 0.1939
4. (a) (i) 0.3277; (ii) 0.2627; (b) $\mu = 1.0$, $\sigma = 0.8944$
5. (a) 0.0183; (b) 0.1465; (c) 0.7619
6. 212

**Further Problems XXVIII (page 895)**

1. (a) 0.0081; (b) 0.2401; (c) 0.2646; (d) 0.0756
2. (a) 0.8154; (b) 0.9852
3. 0.5277, 0.3598, 0.0981, 0.0134, 0.0009, 0.0000
4. (a) 0.0688; (b) 0.9915
5. (a) 0.7738; (b) 0.0226
6. (a) 0.8520; (b) 0.2097
7. (a) 0.3277, 0.4096, 0.2408, 0.0512, 0.0064, 0.0003
   (b) 0.2627; (c) 0.9933
8. (a) 0.5386; (b) 0.0988; (c) 0.0020
9. (a) 0.6065; (b) 0.0902
10. 0.9953
11. (a) $\mu = 12$, $\sigma = 3.412$; (b) 0.1048
12. 0.202, 0.323, 0.258, 0.138, 0.006
13. (a) 0.0498; (b) 0.1494; (c) 0.2240; (d) 0.2240; (e) 0.3528
14. (a) 0.2381; (b) 0.1107
15. 0.0907, 0.2177, 0.2613, 0.2090, 0.2213
16. 5, 15, 22, 22, 17, 18
17. 0.1271

18. (a) 0.0078; (b) 0.9028
19. (a) 74.9%; (b) 0.14%
20. 333
21. (a) 0.0618; (b) 0.2206
22. (a) 7; (b) 127
23. (a) 0.4835; (b) 1.46%
24. (a) 215; (b) 89.3%; (c) 0.0022

# INDEX